5th Symposium on Prospects in the Physics of Discrete Symmetries 2016

Journal of Physics: Conference Series Volume 873

Warsaw, Poland
28 November - 3 December 2016

Editors:

Bohdan Grzadkowski
Jan Kalinowski
Maria Krawczyk

ISBN: 978-1-5108-4559-6
ISSN: 1742-6588

Printed from e-media with permission by:

Curran Associates, Inc.
57 Morehouse Lane
Red Hook, NY 12571

Some format issues inherent in the e-media version may also appear in this print version.

Printed by Curran Associates, Inc. (2017)

For permission requests, please contact the Institute of Physics
at the address below.

Institute of Physics
Dirac House, Temple Back
Bristol BS1 6BE UK

Phone: 44 1 17 929 7481
Fax: 44 1 17 920 0979

techtracking@iop.org

Additional copies of this publication are available from:

Curran Associates, Inc.
57 Morehouse Lane
Red Hook, NY 12571 USA
Phone: 845-758-0400
Fax: 845-758-2633
Email: curran@proceedings.com
Web: www.proceedings.com

Table of contents

Volume 873

Fifth Symposium on Prospects in the Physics of Discrete Symmetries
28 November to 3 December 2016, Warsaw, Poland

Accepted papers received: 3 July 2017
Published online: 27 July 2017

Preface

Plenary lectures

Special session

Parallel talks
T, C, P, CP and CPT symmetries

Emergence of symmetries from entangelment

Discrete symmetries in cosmology

Role of symmetries beyond the Standard Model

Strings, branes, extra dimensions and discrete symmetries

New results from LHC: new facilities

IOP Conf. Series: Journal of Physics: Conf. Series **873** (2017) 011001 doi:10.1088/1742-6596/873/1/011001

Fifth Symposium on Prospects in the Physics of Discrete Symmetries (DISCRETE2017)

Preface

The DISCRETE 2014: Fifth Symposium on Prospects in the Physics of Discrete Symmetries took place at the Faculty of Physics, University of Warsaw, Pasteura 5, 02-093 Warsaw, Poland, from Tuesday, November 28 2016 to Saturday, December 3 2016. It has been organised by the Faculty of Physics of the University of Warsaw and the Division for Fundamental Interactions of the Polish Physical Society. This is the fifth Edition of the DISCRETE conference series, which is a biannual event, having been held previously in Valencia (Discrete2008), Rome (Discrete2010), Lisbon (Discrete2012) and London (Discrete2014).

The topics covered at the DISCRETE series of conferences are: T, C, P, CP symmetries; accidental symmetries (B, L conservation); CPT symmetry, decoherence and entangled states, Lorentz symmetry breaking (phenomenology and current bounds); neutrino mass and mixing; implications for cosmology and astroparticle physics, dark matter searches; experimental prospects at LHC, new facilities.

The Symposium was opened by the Vice-Dean of the Faculty of Physics Professor Zygmunt Lalak. Following the tradition of DISCRETE series of conferences, the talks were organised in plenary sessions with a total of 24 invited talks, and parallel sessions with a total of 61 talks, including both invited and selected contributions from the submitted abstracts. On Thursday, December 1 2016, a special session as a part of the DISCRETE 2016 Symposium and the Leopold Infeld Colloquium took place during which Andrzej Trautman from University of Warsaw gave a personal account on *Gravitational waves: early theoretical work*, Michal Bejger from Nicolaus Copernicus Astronomical Centre presented *The first detections of gravitational waves from binary black holes* and Sergio Ferrara from CERN gave a talk *Supergravity and its Legacy: Personal Prelude and the Play.*

The speakers of the plenary sessions were: José Bernabéu, Tatsuya Nakada, José Ignatio Lattore, Beatrix Hiesmayr, Francesco Gonnella, Carl Bender, Sarben Sarkar, Alessandro Calandri, Silvio Donato, Antonio Di Domenico, Nikolaos Mavromatos, Nazilla mahmoudi, Margarida (Gui) Rebelo, Apostolos Pilaftsis, Jacob Noordmans, Anna Soter, Carlos Munoz, Graham Ross, Vasiliki MitsouEwa Rondio, Serguey Petcov, Eberhard Widmann, Gerald Gabrielse, Vincenzo Branchina and Gustavo Branco. Talks in parallel sessions have been grouped in 7 thematic streams:

T, C, P, CP and CPT symmetries (conveners: Antonio Di Domenico and Andrzej Kupsc),
Emergence of symmetries from entanglement (conveners: Antonio Di Domenico and Andrzej Kupsc),
Discrete symmetries in cosmology (convener: Roberto Ruiz de Austri),
Neutrino masses, mixing and discrete symmetries (convener: Magdalena Posiadala-Zezula)
Discrete symmetries and models of flavour mixing (convener: Margarida (Gui) Rebelo)
Role of symmetries beyond the Standard Model (convener: Zygmunt Lalak)
Strings, branes, extra dimensions and discrete symmetries (Tomasz Taylor and Apostolos Pilaftsis).

The Symposium was attended by about 120 participants. Among the social events, there was a welcome reception on November 28, a social dinner in the Old Town of Warsaw on December 1 and a Chopin Concert in the Staszic Palace on December 2. The next symposium of the series will be organised by Stefan Meyer Institute of the Austrian Academy of Science in Vienna (November 2018)

The organisers wish to thank the conveners, speakers and participants of DISCRETE 2016 for providing an intellectually challenging and stimulating atmosphere through presentations and discussions during this meeting. As editors, we are indebted to all the contributors of this volume for their prompt and diligent submission of very high quality papers, covering the state of the art of the respective fields.

Warsaw, June 2017
The Editors
Bohdan Grzadkowski, Jan Kalinowski and Maria Krawczyk (Faculty of Physics, University of Warsaw)

It is with our great sadness to inform that our best friend and colleague Maria Krawczyk passed away on May 24th, 2017.

Local Organising Committee

Drazen Glavan - *scientific secretary*
Bohdan Grządkowski - *vice-chair*
Jan Kalinowski - *vice-chair*
Maria Krawczyk - *chair*
Zygmunt Lalak
Marek Olechowski
Stefan Pokorski
Janusz Rosiek
Krzysztof Turzyński

Conference Secretariat

Luiza Wiślicz-Iwańczyk e-mail: luiza.wislicz@fuw.edu.pl

Session Conveners

Roberto Ruiz de Austri Bazan (Valencia)
Antonio Di Domenico (Rome)
Andrzej Kupsc (Uppsala)
Zygmunt Lalak (Warsaw)
Apostolos Pilaftsis (Manchester)
Magdalena Posiadala-Zezula (Warsaw)
Margarida (Gui) Rebelo (Lisbon)
Rui Santos (Lisbon)
Tomasz Taylor (Boston and Warsaw)

International Advisory Committee

Francisco del Aguila (Granada)
Luis Alvarez-Gaume (CERN)
Jose Bernabeu (Valencia)
Gustavo C. Branco (Lisbon)
Andrzej J. Buras (Munich)
Antonio Di Domenico (Rome)
Jonathan R. Ellis (CERN & KCL)
Domenec Espriu (Barcelona)
David G. Hitlin (Caltech)
Cecilia Jarlskog (Lund)
Antonio Masiero (Padua)
Nikolaos E. Mavromatos (KCL)
Vasiliki A. Mitsou (Valencia)
Hitoshi Murayama (Berkeley)
Keith Olive (Minnesota)
Antonio Pich (Valencia)
Apostolos Pilaftsis (Manchester)
Mariano Quiros (Barcelona)
Subir Sarkar (Oxford)
Fabio Zwirner (Padua)

DISCRETE Symposia Steering Committee

J. Bernabeu (Valencia) - *Chair*

IOP Conf. Series: Journal of Physics: Conf. Series **873** (2017) 011002 doi:10.1088/1742-6596/873/1/011002

G. Branco (IST Lisbon)
A. Di Domenico (U. Roma Sapienza)
G. Isidori (LN Frascati)
N.E. Mavromatos (King's College London)
V.A. Mitsou (IFIC Valencia)
M.N. Rebelo (IST Lisbon)

Plenary Speakers

C. Bender (Washington U.,St Louis)
J. Bernabeu (IFIC, Valencia)
V. Branchina (U. Catania)
G. Branco (IST, Lisbon)
M. Bejger (CAMK, Warsaw)
R. Demkowicz-Dobrzanski (UW, Warsaw)
A. Di Domenico (INFN, Rome)
A. Dragon (UW, Warsaw)
S. Ferrara (CERN; INFN, Frascati;UCLA)
G. Gabrielse (Harvard U., CERN)
B. Garbrecht (TUM, Munich)
B. Hiesmayr (U. Vienna)
J. Latorre (U. Barcelona)
F. Mahmoudi (Lyon U., CERN)
J. Marton (SMISP, Vienna)
N. Mavromatos (King's Coll., London)
V. Mitsou (IFIC, Valencia)
P. Moskal (UJ, Krakow)
C. Munnoz (Autonoma U., Madrid)
T. Nakada (EPFL-ISIC, Lausanne)
J. Noordmans (U. Algarve)
S. Petcov (INFN, Trieste & SISSA, Trieste & Tokyo U., IPMU)
A. Pilaftsis (Manchester U.)
M. Rebelo (IST, Lisbon)
E. Rondio (NCBJ, Warsaw)
G. Ross (U. Oxford)
S. Sarkar (King's Coll., London)
A. Soter (MPQ, Garching)
A. Trautman (UW, Warsaw)
E. Widmann (SMISP, Wien)

IOP Conf. Series: Journal of Physics: Conf. Series **873** (2017) 011002 doi:10.1088/1742-6596/873/1/011002

Peer review statement

All papers published in this volume of *Journal of Physics: Conference Series* have been peer reviewed through processes administered by the proceedings Editors. Reviews were conducted by expert referees to the professional and scientific standards expected of a proceedings journal published by IOP Publishing.

IOP Conf. Series: Journal of Physics: Conf. Series **873** (2017) 012001 doi:10.1088/1742-6596/873/1/012001

Separate T, CP, CPT Asymmetries in Neutral Meson Transitions

José Bernabéu

Departamento de Física Teórica, Universidad de Valencia,
e IFIC, Joint Centre CSIC-UV, E-46100 Burjassot,

Jose.bernabeu@uv.es

Abstract. Symmetries, and symmetry breakings, in the laws of physics play a crucial role in fundamental science. Parity and charge conjugation violations prompted the consideration of chiral fields in the construction of the Standard Model, whereas CP-violation needed at least three families of quarks leading to flavour physics. In this lecture I will discuss the conceptual basis and the present experimental results for a direct evidence of separate reversal-in-time T, CP and CPT genuine asymmetries in decaying particles like neutral meson transitions, using quantum entanglement and the decay as a filtering measurement. The eight transitions associated to the flavour-CP eigenstate decay products of entangled neutral mesons have demonstrated with impressive significance a separate evidence of TRV and CPV in B_d-physics, whereas a CPTV asymmetry shows a 2-σ effect interpreted as an upper limit. Novel CPTV observables are discussed for K and B_d transitions. Their observation would lead to a change of paradigm beyond quantum field theory, however there is nothing in quantum mechanics forbidding CPTV. A clean methodology to disentangle CPTV effects in the Hamiltonian dynamics and the ω-effect weakening entanglement in a given experiment is discussed.

1. The Fundamental Role of Symmetry Breaking

In everyday life, we observe symmetries of objects, as characteristic features of geometric forms, material objects or biological bodies. In all cases the symmetry is related to their invariance under definite transformations. We say that an object is symmetric if, after a transformation is applied, it remains "invariant". In physics we are interested in the symmetries of the physical laws, and their breakings, which have played a crucial role in the advance of knowledge. Conservation laws lead to regularity and order in the description of physical phenomena even when there is no input from dynamics. In the second half of the XXth century, a conceptual revolution appeared when the requirement of local gauge symmetries was understood as originating the interactions themselves with their charges. The emergence of the Standard Model(SM) $SU(3)_C$ x $SU(2)_L$ x $U(1)_Y$ prompted a spectacular development of particle physics with precise experimental results in agreement with theoretical predictions. The symmetry is exact for massless particles only, which is evidently wrong in Nature. This apparent contradistinction was solved by the Brout-Englert-Higgs mechanism [1] of spontaneous breaking of the electroweak gauge symmetry $SU(2)_L$ x $U(1)_Y$ into $U(1)_{e.m.}$. This methodology for explaining the origin of mass leaves as a remnant the Higgs boson, discovered [2] at CERN by the ATLAS and CMS Collaborations.

With respect to the discrete symmetries, P, C, CP and T have been observed to be separately broken by weak interactions. Their breakings have had a fundamental imprint in the construction of the SM:

- Parity violation first observed in nuclear beta decay [3] and later, together with the violation of charge conjugation, in charged pion decays [4], led to the V-A theory for charged current weak interactions [5] and the role of chirality as the essential ingredient of requiring different transformation properties under the gauge group for left- and right- handed fermionic fields. The chiral fields are the starting components of the theory in the SM. The electroweak unification [6, 7, 8] culminated with the discovery [9] of neutral current weak interactions by the Gargamelle experiment at CERN.

- CP Violation was unexpectedly discovered in 1964 [10] first in the decay of the long lived neutral kaon K_L into two pions, a decay forbidden by the conservation law. The proposal to incorporate this breaking into the SM by means of enlarging the particle content of the theory to at least three quark families and using the Cabibbo-Kobayashi-Maskawa Quark Mixing [11, 12] was clearly supported by the discovery and results of CPV in the B_d-system in the B-Factories [13, 14] in 2001. The physical phenomena associated to the existence of several fermion families of quarks and leptons are nowadays included in the so-called flavour physics.

- Time reversal violation is more difficult to be directly observed. Contrary to P, C and CP, reversal-in-time is a transformation which, in the space of physical quantum states, is implemented by an antiunitary operator. This has two important implications: i) there is no conservation law associated to this symmetry; ii) the test of its invariance in a given transition i → f requires its comparison with the f → i transition. We may expect that, according with the CPT theorem [15], T should be violated as well as CP in the systems in which the CP symmetry violation has been observed. However, kaons and B mesons are unstable particles and, as emphasized by Wolfenstein [16] and others, the decay is irreversible and "the T-transform of a decaying state is not a physical state". This NO-GO argument suggests that a direct evidence for TRV is impossible for unstable particles. The bypass to this argument came [17, 18] in 1999 from the strategy of not including explicitly the decay products in the TRV asymmetry and using: i) the decay as a filtering measurement of the projected meson state only; ii) quantum entanglement of the neutral meson system in order to transfer the information from the decaying meson to its (still alive) orthogonal partner; iii) the asymmetry is searched in the time evolution of this last partner Under a deep scrutiny of this conceptual basis by many authors, the first observation [19] of TRV in the time evolution of Bd's was announced in 2012 by the BABAR Collaboration, using the methodology that was previously defined in [20]. Its observation triggered [21, 22] a revisit to the conditions under which the asymmetry becomes a genuine observable of TRV. For the neutral kaon system there is a proposal in the DAPHNE-Factory using entanglement.

- Although P, C, CP and T symmetries have been observed to be separately broken, there is no experimental evidence of CPT Violation. Its observation would imply a change of paradigm to beyond QFT. There is, however, nothing at the level of quantum mechanics which forbids CPTV.

In Section 2 we discuss the implication of antiunitarity for the T and CPT symmetries and the wording, related to physics, for constructing true asymmetries valid for a symmetry test. The state of the art for genuine T, CP, CPT separate asymmetries in B_d transitions is presented in Section 3. Direct measurements of TRV in neutral kaon transitions at KLOE-2 are discussed in Section 4 and the direct tests of CPT in neutral kaons at KLOE-2 in Section 5. For entangled neutral mesons, the possibility of a ω-effect weakening entanglement, due to an ill-defined CPT operator, is analysed in Section 6. Conclusions and outlook are given in Section 7.

2. T and CPT Tests for unstable particles

In quantum mechanics, there is an operator U_{CP} implementing the CP-symmetry acting on the states of the physical system, such that

IOP Conf. Series: Journal of Physics: Conf. Series **873** (2017) 012001 doi:10.1088/1742-6596/873/1/012001

$$U_C\, Q\, U_C^+ = -Q,\ U_P\, \vec{r}\, U_P^+ = -\vec{r},\ U_P\, \vec{p}\, U_P^+ = -\vec{p},\ U_P\, \vec{s}\, U_P^+ = \vec{s} \tag{1}$$

The operator U_{CP} is both unitary and self adjoint and so observable with conservation laws: $K_L \rightarrow \pi\ \pi$ would be a forbidden process and its observation [10] was an undisputable proof of CP violation.

The operator U_T implementing reversal-in-time T symmetry is such that

$$U_T\, \vec{r}\, U_T^+ = \vec{r},\ U_T\, \vec{p} U_T^+ = -\vec{p},\ U_T \vec{s} U_T^+ = -\vec{s} \tag{2}$$

By considering the commutator $[r_j, p_k] = i\hbar\delta_{jk} I$ the operator U_T must be antiunitary: unitary for conserving probabilities, antilinear for complex conjugation. Antiunitarity introduces many interesting subtleties between the reference and T-transformed transition amplitudes for i → f

$$S_{i\rightarrow f} \xrightarrow{T} S_{U_T f \rightarrow U_T i} \tag{3}$$

so that T-violation (TRV) means an asymmetry under the interchange of in and out states.

In a similar way, the antiunitary U_{CPT} operator needs for its test as a symmetry not only the exchange of initial and final states, but also the transformation of particles into the corresponding antiparticles.

It is highly interesting to clarify the language used for asymmetries as a true proof of symmetry breaking:

i) DIRECT evidence of symmertry breaking

. For CP, violation of the conservation law

. For All T, CP, CPT → comparison between the probabilities for a reference process and its symmetry-transformed in a single experiment

. NOT by a fit of parameters describing symmetry breaking in a given theoretical framework.

ii) GENUINE asymmetries

. A set of observables, for each symmetry, in yes-no biunivocal correspondence with symmetry violation.

. NO Fake terms or controlled in the same experiment.

iii) SEPARATE T, CP, CPT asymmetries

. TRV independent of CPV or CPT invariance.

. Different information content from different transformed processes.

iv) TRANSITION processes

. Interest in going beyond "expectation values" of symmetry breaking.

. In 1st order of perturbation theory in SB, diagonal matrix elements vanish if perturbation is odd under non-perturbed quantum numbers: $< + \mid - \mid + > = < - \mid - \mid - > = 0$.

Well known examples of this result are the Stark effect in atomic physics or the TRV e. d. m. in particle physics.

We center the interest in neutral meson transitions of the $K^0 - \bar{K}^0$ and $B^0 - \bar{B}^0$ systems, for which a clear evidence of CPV exists.

But these particles are unstable and the decay is irreversible. As we have explained before, T and CPT are antiunitary needing however the exchange of initial and final states. As illustrated in a devastating statement by Wolfenstein [16]: "The T-reverse of a decaying state is not a physical state". The conceptual basis for bypassing this **no-go** argument came [17, 18] from not including the decay products in the asymmetry and write it in terms of initial and final meson states.

In addition, the decay should not be an essential ingredient for getting a non-vanishing value of the asymmetry. In precise terms,

a) Use the observed decay as a **quantum filtering measurement** of the meson state only. It is given by the orthogonal state to that which is not leading to the decay products.
b) Use **quantum entanglement** for **information transfer from the first decay to the (still alive) orthogonal partner. This tags the preparation of the initial meson state as the non-decay state.**
c) The test of the symmetries is made in the time evolution of the living partner FROM THE FIRST TO THE SECOND DECAY.

After a detailed scrutiny of the concept, Wolfenstein stated [23]: "it appears to be a true TRV effect".

3. Separate T, CP, CPT asymmetries in B_d transitions

For unstable particles, the decay products should not be included in the states building the asymmetries for T or CPT. However, these decay products are what we observe. Therefore, the problem is in the preparation and filtering of the appropriate initial and final meson states from the observed two decay channels. For (flavour, CP) eigenstate decay products, the Editors of Rev. Mod. Phys. decided to put "what is T-transformation experimentally" in the cover page of the volume in which the colloquium article [24] was published. This Figure 1 provides the explanation of what has to be measured in order to build a true genuine TRV asymmetry.

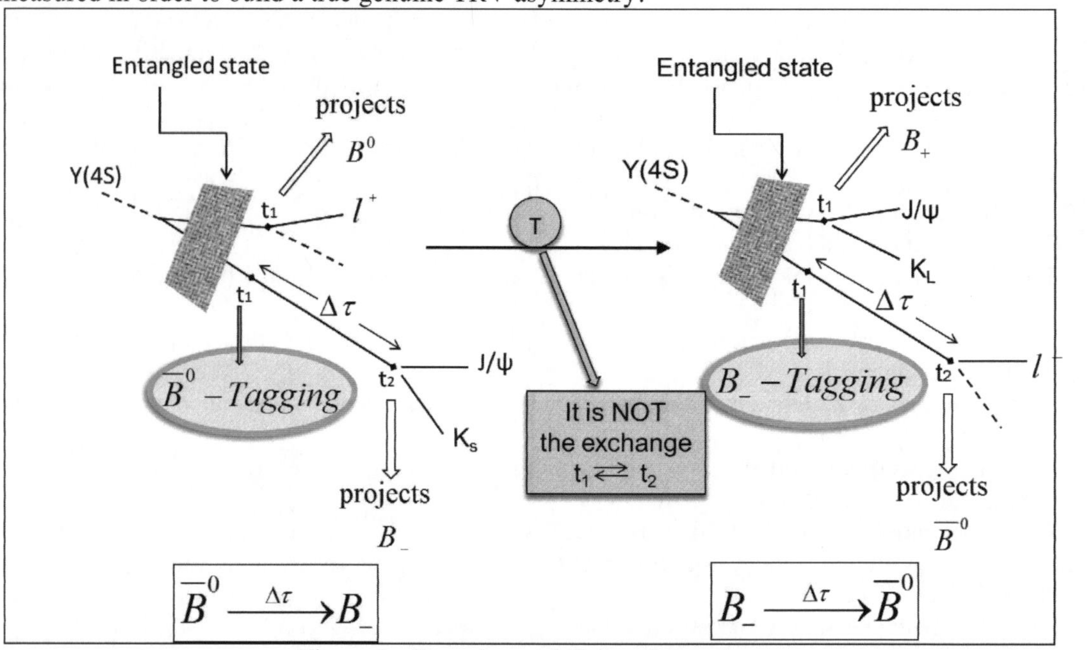

Figure 1.- Experimental T-transformation

The Figure has illustrated the Reference $\bar{B}^0 \to B_-$ transition, transformed by T into the transition $B_- \to \bar{B}^0$. The named neutral meson states B_+, B_- correspond to the filtered states by the CP = +, - eigenstate decay products J/ψ K_L, J/ψ K_S. In a B-Factory operating at the ϒ(4S) peak, our initial two-meson state is Einstein-Podolsky-Rosen entangled as

$$|\Psi_0\rangle = \frac{1}{\sqrt{2}}\Big(|B_d^0\rangle|\bar{B}_d^0\rangle - |\bar{B}_d^0\rangle|B_d^0\rangle\Big) = \frac{1}{\sqrt{2}(p_L q_H + p_H q_L)}\Big(|B_L\rangle|B_H\rangle - |B_H\rangle|B_L\rangle\Big), \tag{4}$$

which maintains its antisymmetric entangled character in the Hamiltonian eigenstate basis. This implies the antisymmetric character of the two meson state at all times and for any two independent linear combinations of entangled B_d^0 *and* $\bar{B}_d^0$. The corresponding evolution before the first decay is therefore trivial for perfect entanglement.

Given a decay "f", the partner meson is tagged by means of the decay amplitudes A_f from B^0 and $\bar{A}_f$ from $\bar{B}^0$ as

$$|B_{\nrightarrow f}\rangle = \frac{1}{\sqrt{|A_f|^2 + |\bar{A}_f|^2}} \left(\bar{A}_f |B_d^0\rangle - A_f |\bar{B}_d^0\rangle\right) \tag{5}$$

and the "filtered" state is its orthogonal. The **filtering identity** establishes the connection between the meson transition probability and the experimental "reduced" Intensity for the pair of decay channels (f,g)

$$\hat{I}(f,g;t) = \frac{|\langle g|T|B_{\nrightarrow f}(t)\rangle|^2}{|A_g|^2+|\bar{A}_g|^2} = |\langle B_{\nrightarrow g}^{\perp}|B_{\nrightarrow f}(t)\rangle|^2 \tag{6}$$

As seen in (6), there are **no fake terms** for a proof of symmetry breaking if the ratio of the two decay amplitudes $\bar{A}\,/\,A$ is a pure phase. This requirement may be controlled in the same experiment.

How many (Flavour, CP) decay channels we have at our disposal experimentally? 2 Flavours x 2 CP x 2 time-ordering = 8.

These 8 transitions may be grouped in two blocks with extremely important properties. Within each block, the 4 transitions correspond to one reference and the three different transitions transformed from the reference by the T, CP and CPT transformations separately. These amazing properties for (Flavour, CP) eigenstate decay channels of the entangled system of neutral mesons are represented in figure 2, where the time-dependent Intensities of the 8 transitions are organised appropriately in order to make the direct connection of the transitions by the three discrete symmetries apparent. The red (left) and blue (right) blocks connected by time exchange are **not** connected by the symmetries.

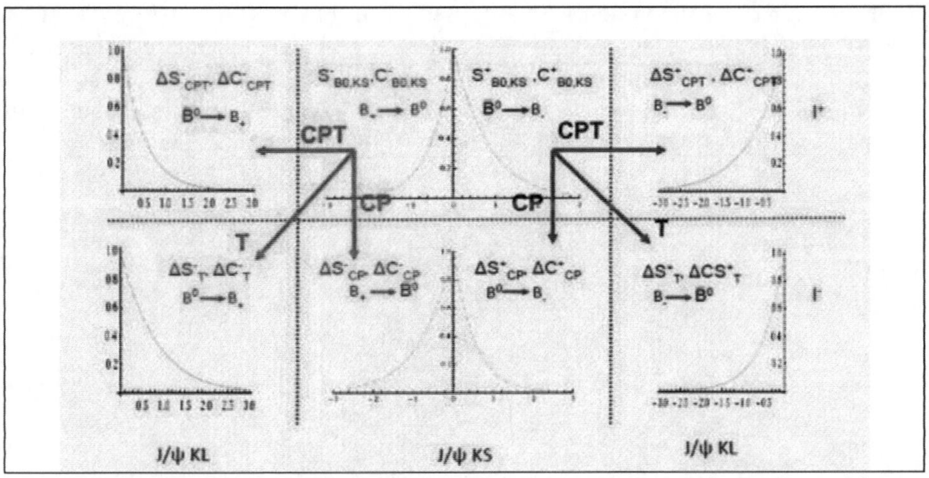

Figure 2 – Reference transition and Separate T, CP, CPT transformed transitions.

Using quantum mechanics for the time evolution of the two-state $B^0 - \bar{B}^0$ system and no particular Hamiltonian dynamics, each time dependent Intensity has the following general form

$$I\,(\Delta t) \sim e^{-\Gamma\Delta t}\left\{C_C\cos(\Delta m\Delta t) + S_c\sin(\Delta m\Delta t) + C_h\cosh(\Delta\Gamma\Delta t) + S_h\sinh(\Delta\Gamma\Delta t)\right\} \tag{7}$$

where the coefficients C_c, S_c, C_h and S_h depend of the selected transition. Without more dynamical theoretical framework, these coefficients are unknown. From the experiment we can construct asymmetry parameters ΔC_c, ΔS_c and ΔC_h (in the B_d-system $\Delta\mathsf{j}=0$) for each discrete symmetry T, CP,

CPT. In total, we have 9 asymmetry parameters with different information content within each block. Any non-vanishing value of the 3 asymmetry parameters for each symmetry is a bona-fide genuine proof of the violation of that symmetry.

In Ref. [25] the authors have worked out the explicit effective Hamiltonian dynamics associated to a Weisskopf-Wigner (WW) quantum description [26] of the meson-antimeson system. In this way the 9 asymmetry parameters can be written in terms of the parameters of the Hamiltonian matrix and the decay amplitudes. Two results are worth mentioning:

i) the asymmetries are genuine signals of symmetry breaking, as they should for any dynamics;
ii) the 9 asymmetry parameters appear with different information content.

Using data from the BABAR analysis [19], in which the intensities (7) were normalised by dividing by C_h with $\not{C}\Gamma = 0$, the authors of [25] have re-derived the 9 asymmetry parameters with some mild additional information. The resulting TRV and CPV asymmetries give the result

$$\Delta S_c^T = -0.687 \pm 0.020\,; \quad \Delta S_c^{CP} = -0.680 \pm 0.021 \tag{8}$$

These results constitute an impressive separate evidence of TRV and CPV. We emphasize that they are independent results and TRV is established without any assumption on either CPV or CPT invariance.

The analysis performed assuming perfect entanglement of the neutral meson system in which the two-body state is antisymmetric under the permutation exchange of the two mesons, has also been used to extract the CPTV asymmetry parameters. The unique antisymmetric state remains unaltered under the time evolution of the entangled system before the first decay, even with mixing under these conditions, given the time ordered (f, g) decays, the reversed time ordering (g, f) decays satisfy the exchange symmetry conditions

$$C_h\,(f,g) = C_h\,(g,f)\,;\; C_c\,(f,g) = C_c\,(g,f)\,;\; S_c\,(f,g) = -S_c\,(g,f) \tag{9}$$

Notice that the exchange symmetry has nothing to do with the discrete symmetries T, CP, CPT associated to the dynamics of the system. As a consequence, one has to consider the measurements associated with (f, g) and (g, f) decays as two independent experimental processes measuring the same properties, once equations (9) are imposed. The CPTV asymmetry is generated in the WW approach by the θ parameter in the diagonal Hamiltonian matrix elements distinguishing particle and antiparticle properties. The correlated fit to Re(θ) and Im(θ) is presented [25] in figure 3

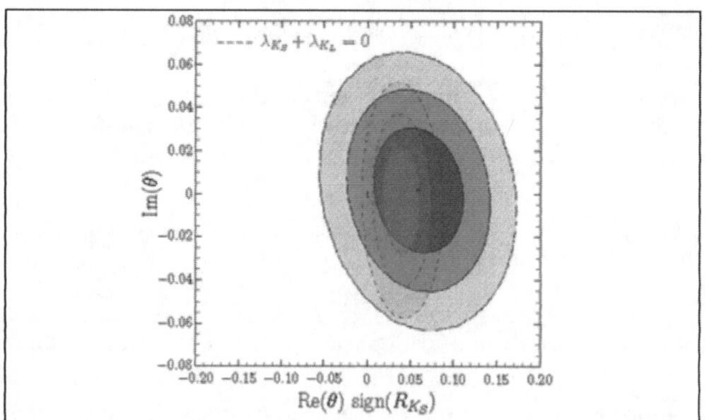

Figure 3 – Correlated fit to the CPTV Re(θ) and Im(θ)

and gives an intriguing 2σ effect for CPTV in Re(θ).

Under an ill-defined CPT operator expected in models of quantum gravity, the perfect entanglement can be weakened [27] by the presence of an ω-effect associated to the "wrong" exchange

IOP Conf. Series: Journal of Physics: Conf. Series **873** (2017) 012001 doi:10.1088/1742-6596/873/1/012001

symmetry of the neutral meson and antimeson. Equations (9) are no longer valid and their violation would then be a signal of this new form of CPT breaking.

4. Time reversal violation in neutral Kaon transitions

Similarly to the $B^0 - \bar{B}^0$ system, EPR correlations at a ϕ Factory can be exploited to study T-conjugated **transitions** between $K^0, \bar{K}^0$ states and the orthogonal K_+, K_- states filtered by CP eigenstate decay products

$$|i\rangle = \frac{1}{\sqrt{2}}\left[|K^0(\vec{p})\rangle|\bar{K}^0(-\vec{p})\rangle - |\bar{K}^0(\vec{p})\rangle|K^0(-\vec{p})\rangle\right]$$
$$= \frac{1}{\sqrt{2}}\left[|K_+(\vec{p})\rangle|K_-(-\vec{p})\rangle - |K_-(\vec{p})\rangle|K_+(-\vec{p})\rangle\right] \quad (10)$$

Using the decays as filtering measurements of the meson states and entanglement for the preparation of the living partner initial state, we identify the reference process, for example $K^0 \rightarrow K_-$

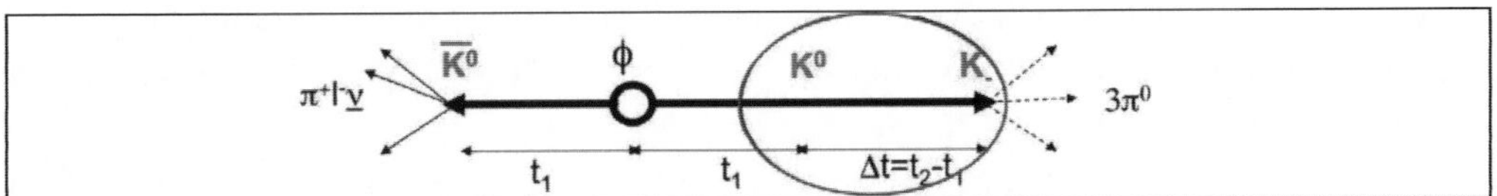

Figure 4 - Neutral Kaon transition $K^0 \rightarrow K_-$ filtered by the l^- decay first, $3\pi^0$ decay later

For this reference, the T-conjugated process is $K_- \rightarrow K^0$, separated from the CP conjugated $\bar{K}^0 \rightarrow K_-$ and the CPT-conjugated $K_- \rightarrow \bar{K}^0$. The T-conjugated process is associated to the time-ordered 2π decay first, l^+ decay later

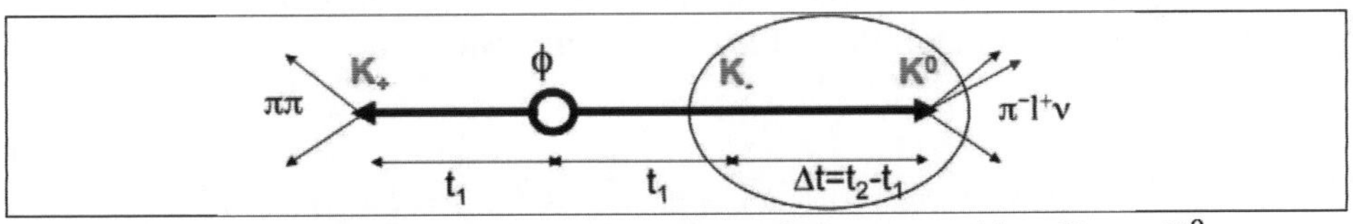

Figure 5 - The Reversal-in-Time Neutral Kaon transition $K_- \rightarrow K^0$

For a T-symmetry test one has a priori 4 possible comparisons of transitions as given in Table 1

Table 1- Neutral kaon transitions to be compared for a T symmetry test

T symmetry test			
Reference		T-conjugate	
Transition	Decay products	Transition	Decay products
$\bar{K}^0 \rightarrow K_-$	$(\ell^+, \pi^0\pi^0\pi^0)$	$K_- \rightarrow \bar{K}^0$	$(\pi^0\pi^0\pi^0, \ell^-)$
$K_+ \rightarrow K^0$	$(\pi^0\pi^0\pi^0, \ell^+)$	$K^0 \rightarrow K_+$	$(\ell^-, \pi\pi)$
$\bar{K}^0 \rightarrow K_+$	$(\ell^+, \pi\pi)$	$K_+ \rightarrow \bar{K}^0$	$(\pi^0\pi^0\pi^0, \ell^-)$
$K_- \rightarrow K^0$	$(\pi\pi, \ell^+)$	$K^0 \rightarrow K_-$	$(\ell^-, \pi\pi)$

One can then define [28] the following ratios of time dependent probabilities

$$\begin{aligned} R_1(\Delta t) &= P\left[\mathrm{K}^0(0) \to \mathrm{K}_+(\Delta t)\right]/P\left[\mathrm{K}_+(0) \to \mathrm{K}^0(\Delta t)\right] \\ R_2(\Delta t) &= P\left[\mathrm{K}^0(0) \to \mathrm{K}_-(\Delta t)\right]/P\left[\mathrm{K}_-(0) \to \mathrm{K}^0(\Delta t)\right] \\ R_3(\Delta t) &= P\left[\bar{\mathrm{K}}^0(0) \to \mathrm{K}_+(\Delta t)\right]/P\left[\mathrm{K}_+(0) \to \bar{\mathrm{K}}^0(\Delta t)\right] \\ R_4(\Delta t) &= P\left[\bar{\mathrm{K}}^0(0) \to \mathrm{K}_-(\Delta t)\right]/P\left[\mathrm{K}_-(0) \to \bar{\mathrm{K}}^0(\Delta t)\right] \end{aligned} \quad (11)$$

Any deviation from R = 1 constitutes a direct evidence of a violation of T-symmetry.

These theoretical ratios for meson state transitions can be connected to experimental ratios of intensities for pairs of decay channels as

$$\begin{aligned} R_1^{\mathrm{exp}}(\Delta t) &\equiv \frac{I(\ell^-,\pi\pi;\Delta t)}{I(3\pi^0,\ell^+;\Delta t)} = R_1(\Delta t) \times \frac{C(\ell^-,\pi\pi)}{C(3\pi^0,\ell^+)} \\ R_2^{\mathrm{exp}}(\Delta t) &\equiv \frac{I(\ell^-,3\pi^0;\Delta t)}{I(\pi\pi,\ell^+;\Delta t)} = R_2(\Delta t) \times \frac{C(\ell^-,3\pi^0)}{C(\pi\pi,\ell^+)} \\ R_3^{\mathrm{exp}}(\Delta t) &\equiv \frac{I(\ell^+,\pi\pi;\Delta t)}{I(3\pi^0,\ell^-;\Delta t)} = R_3(\Delta t) \times \frac{C(\ell^+,\pi\pi)}{C(3\pi^0,\ell^-)} \\ R_4^{\mathrm{exp}}(\Delta t) &\equiv \frac{I(\ell^+,3\pi^0;\Delta t)}{I(\pi\pi,\ell^-;\Delta t)} = R_4(\Delta t) \times \frac{C(\ell^+,3\pi^0)}{C(\pi\pi,\ell^-)} \end{aligned} \quad (12)$$

where the co-factors of equation (12) are related to the corresponding branching ratios. In practice, one has two measurable ratios with the two time-orderings either ¢t > 0 or ¢t < 0, due to the inverse connections

$$\begin{aligned} R_2^{\mathrm{exp}}(-\Delta t) &= \frac{1}{R_3^{\mathrm{exp}}(\Delta t)} = \frac{1}{R_3(\Delta t)} \times \frac{C(3\pi^0,\ell^-)}{C(\ell^+,\pi\pi)}, \\ R_4^{\mathrm{exp}}(-\Delta t) &= \frac{1}{R_1^{\mathrm{exp}}(\Delta t)} = \frac{1}{R_1(\Delta t)} \times \frac{C(3\pi^0,\ell^+)}{C(\ell^-,\pi\pi)}. \end{aligned} \quad (13)$$

In figures 6 we give [28] the Standard Model expectations for the two experimental ratios R_2^{exp} and R_4^{exp}, in arbitrary units, as function of ¢t for the two time orderings

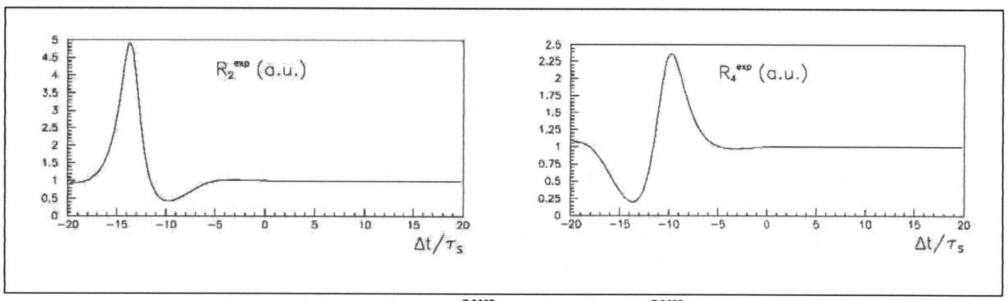

Figure 6 – The experimental ratios R_2^{exp} (left) and R_4^{exp}(right) for TRV expected in the Standard Model

One reaches the conclusion that this T-test could be feasible in KLOE-2 at DAPHNE with a luminosity of about 10 fb^{-1}.

5. Direct test of CPT in neutral kaon transitions

In theories beyond the quantum field theory paradigm, the CPT-Theorem may be not applicable and, in fact, CPT breaking appears in quantum gravity models. It would be of highest interest to construct explicit predictive theories incorporating CPT violation and phenomenological models, as a guide, to be tested by experiments.

The most immediate consequences of CPT invariance are the equality of masses, lifetimes and absolute values of charges and magnetic moments for a particle and its antiparticle. The neutral meson systems offer unique possibilities to test CPT invariance. Taking as figure of merit the fractional difference between the masses of a particle and its antiparticle, the best CPT Violation limits are

$$\begin{array}{ll} \left|m_{K^0} - m_{\overline{k}^0}\right| / m_K < 10^{-18} & [29] \\ \left|m_{B^0} - m_{\overline{B}^0}\right| / m_B < 10^{-14} & [25] \\ \left|m_p - m_{\overline{p}}\right| / m_p < 8 \,.\, 10^{-10} & [30] \end{array} \qquad (14)$$

In the following we discuss the CPT test at CPLEAR [29] from the time evolution of neutral kaons using the semileptonic decay charge asymmetry. The initial neutral kaon state is tagged at CPLEAR as the flavour state as indicated in figure 7

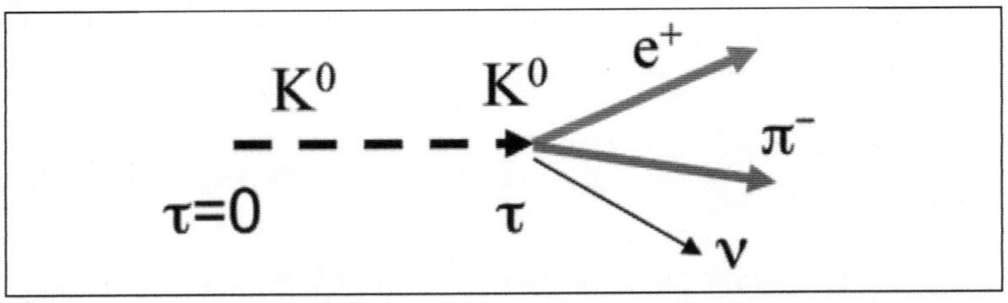

Figure 7.- Semiletonic decay from K^0 assuming the ¢S = ¢Q rule

From the four time-dependent semileptonic decay rates, including violations of the ¢S = ¢Q rule, one may construct the asymmetry defined in equation (15)

$$\begin{cases} A_\delta(\tau) = \dfrac{\overline{R}_+(\tau) - \alpha R_-(\tau)}{\overline{R}_+(\tau) + \alpha R_-(\tau)} + \dfrac{\overline{R}_-(\tau) - \alpha R_+(\tau)}{\overline{R}_-(\tau) + \alpha R_+(\tau)} \\ R_{+(-)}(\tau) = R\left(K^0{}_{t=0} \rightarrow (e^{+(-)}\pi^{-(+)}\nu)_{t=\tau}\right) \\ \overline{R}_{-(+)}(\tau) = R\left(\overline{K}^0{}_{t=0} \rightarrow (e^{-(+)}\pi^{+(-)}\nu)_{t=\tau}\right) \\ \alpha = 1 + 4\Re\varepsilon_L \end{cases} \qquad (15)$$

The experimental results for this time-dependent asymmetry are shown in figure 8

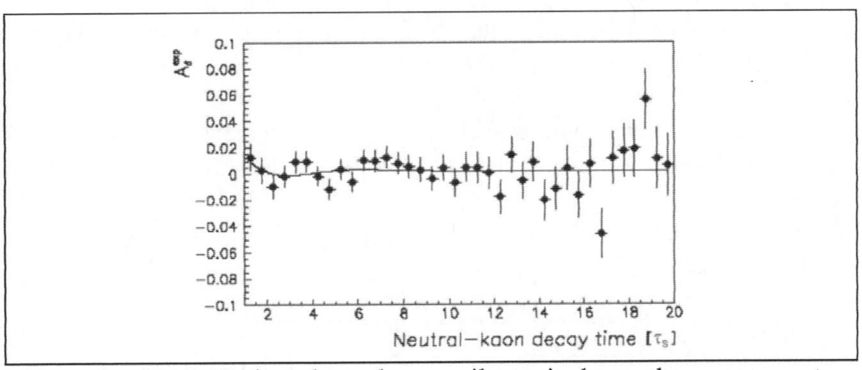

Figure 8.- CPLEAR time-dependent semileptonic decay charge asymmetry

In the time region above the interference effects $\tau >> \tau_S$, the asymmetry becomes a constant which disentangles the "δ" CPTV contribution to the diagonal matrix elements of the effective Hamiltonian in the Weisskopf-Wigner approach [26] and it is given by

$$A_\delta(\tau >> \tau_S) = 8\Re\delta \tag{16}$$

The fit [29] to the experimental data gives the value

$$R\delta = (0.30 \pm 0.33 \pm 0.06) \cdot 10^{-3} \tag{17}$$

At the Φ-Factory DAPHNE, the entanglement between the two neutral kaon states induced by their antisymmetric exchange allows a separate selection of observable asymmetries from either K_L or K_S states.
The charge asymmetry from these states is given by

$$A_{S,L} = \frac{\Gamma(K_{S,L}\to\pi^-e^+\nu)-\Gamma(K_{S,L}\to\pi^+e^-\bar{\nu})}{\Gamma(K_{S,L}\to\pi^-e^+\nu)+\Gamma(K_{S,L}\to\pi^+e^-\bar{\nu})} = 2\,[Re\,(\epsilon_K) \pm Re\,(\delta_K) - Re\,(y)] \tag{18}$$

where, in addition to the CPV and CPTV parameters in the Hamiltonian matrix, a possible CPTV in the flavour decay amplitude of Figure 7, parameterised by "y", is included. As seen, the difference between these two charge asymmetries A_S - A_L is able to disentangle CPTV effects in the diagonal matrix elements of the mass matrix.
The value of A_L is well known from hadronic machines, with the best precise result from the KTeV Collaboration [31]

$$A_L = (3.332 \pm 0.058_{stat} \pm 0.047_{syst}) \cdot 10^{-3} \tag{19}$$

The measurement of A_S was undertaken in the first period of the KLOE experiment with the result [32]

$$A_S = (1.5 \pm 9.6_{stat} \pm 2.9_{syst}) \cdot 10^{-3} \tag{20}$$

An important improvement of this result is expected at KLOE-2.

EPR correlations at a Φ-Factory can also be exploited to study CPT-conjugated **transitions** involving flavour $K^0, \bar{K}^0$ states and the orthogonal K_+, K_- states filtered by CP eigenstate decay products.

Using the decays as filtering measurements of the meson states and entanglement for the preparation of the living partner initial state, we identify the reference process, for example $K^0 \to K_-$ as in figure 4.

For this reference, the CPT-conjugated process is $K_- \to \bar{K}^0$, separated out from the CP-conjugated $\bar{K}^0 \to K_-$ and the T-conjugated $K_- \to K^0$. The CPT-conjugated process is associated to the time-ordered decay channels

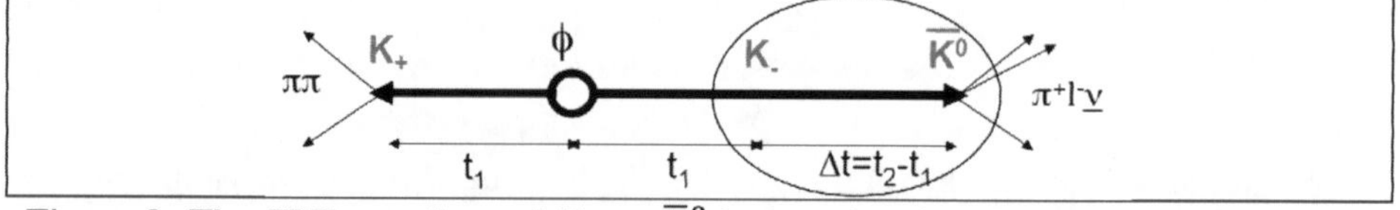

Figure 9- The CPT-transformed $K_- \to \bar{K}^0$ is associated to the 2π decay first, the l^- decay later.

For a CPT-symmetry test one has a priori 4 possible comparisons of transitions as given in table 2

Table 2.- Reference transitions and their CPT-transformed transitions

CPT symmetry test

Reference		$\mathcal{CPT}$-conjugate	
Transition	Decay products	Transition	Decay products
$K^0 \to K_+$	$(\ell^-, \pi\pi)$	$K_+ \to \bar{K}^0$	$(3\pi^0, \ell^-)$
$K^0 \to K_-$	$(\ell^-, 3\pi^0)$	$K_- \to \bar{K}^0$	$(\pi\pi, \ell^-)$
$\bar{K}^0 \to K_+$	$(\ell^+, \pi\pi)$	$K_+ \to K^0$	$(3\pi^0, \ell^+)$
$\bar{K}^0 \to K_-$	$(\ell^+, 3\pi^0)$	$K_- \to K^0$	$(\pi\pi, \ell^+)$

One can then define [33] the following ratios of time dependent probabilities

$$\begin{aligned}
R_{1,\mathcal{CPT}}(\Delta t) &= P\left[K_+(0) \to \bar{K}^0(\Delta t)\right] / P\left[K^0(0) \to K_+(\Delta t)\right] \\
R_{2,\mathcal{CPT}}(\Delta t) &= P\left[K^0(0) \to K_-(\Delta t)\right] / P\left[K_-(0) \to \bar{K}^0(\Delta t)\right] \\
R_{3,\mathcal{CPT}}(\Delta t) &= P\left[K_+(0) \to K^0(\Delta t)\right] / P\left[\bar{K}^0(0) \to K_+(\Delta t)\right] \\
R_{4,\mathcal{CPT}}(\Delta t) &= P\left[\bar{K}^0(0) \to K_-(\Delta t)\right] / P\left[K_-(0) \to K^0(\Delta t)\right]
\end{aligned} \quad (21)$$

Any deviation from $R_{i,CPT} = 1$ constitutes a direct evidence of a violation of CPT-symmetry.

These theoretical ratios for meson state transitions can be connected to experimental ratios of Intensities for pairs of decay channels as indicated in Table 2.

For visualisation purposes, we plot [33] in figure 10 these four ratios for CPT-Violation using a "δ" parameter in the Hamiltonian matrix with values Re(δ)=3.3 . 10^{-4} and Im(δ) =1.6 .10^{-5} (continous line -) and Im(δ) = 0 (broken line - -).

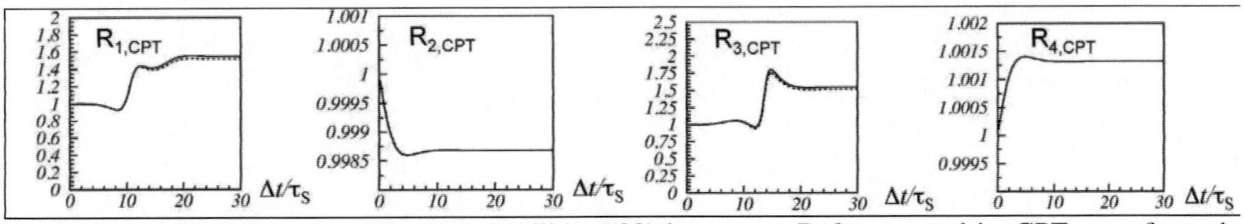

Figure 10.- The ratios of transition probabilities (20) between a Reference and its CPT- transformed for values of the δ parameter as explained in the text.

For the most interesting ¢t region larger than the short lifetime of K_S τ_S, one has

$$R_{2,CPT}(\Delta t \gg \tau_S) = 1 - 4\mathrm{Re}(\delta)$$

$$R_{4,CPT}(\Delta t \gg \tau_S) = 1 + 4\mathrm{Re}(\delta) \quad (22)$$

One cannot imagine a better clearcut test of CPT symmetry in transitions as the one shown in equations (22). This test is feasible at KLOE-2.

6. The ω-effect

In presence of decoherence and CPT breaking induced by quantum gravity effects, the CPT operator is "ill-defined" and the definition of the particle-antiparticle states could be modified. This in turn could induce a weakening of the entanglement [27] imposed by Bose statistics as an EPR correlation

$$\begin{aligned} |i\rangle &\propto \left(|K^0\rangle|\bar{K}^0\rangle - |\bar{K}^0\rangle|K^0\rangle\right) + \omega\left(|K^0\rangle|\bar{K}^0\rangle + |\bar{K}^0\rangle|K^0\rangle\right) \\ &\propto \left(|K_S\rangle|K_L\rangle - |K_L\rangle|K_S\rangle\right) + \omega\left(|K_S\rangle|K_S\rangle - |K_L\rangle|K_L\rangle\right) \end{aligned} \qquad (23)$$

In some microscopic models of space-time foam [34] with stochastic fluctuations of defect recoils in the propagation of particles, the magnitude of the ω-parameter in equation (23) can reach values $\omega \sim 10^{-4}$ - 10^{-5}.

The maximum sensitivity to ω is expected for a symmetric decay pair of the entangled state f1= f2= $\pi^+ \pi^-$, which is CP-violating from the dominant antisymmetric state of equation (23). As a consequence, one gets a fantastic enhancement of the ω-effect from the "wrong" component in equation (23) as ω/ε, where ε is the small CPV parameter in K-physics. In figure 11 we plot the expected double decay rate time-dependent Intensities for vanishing and non-vanishing values of ω accessible at KLOE-2.

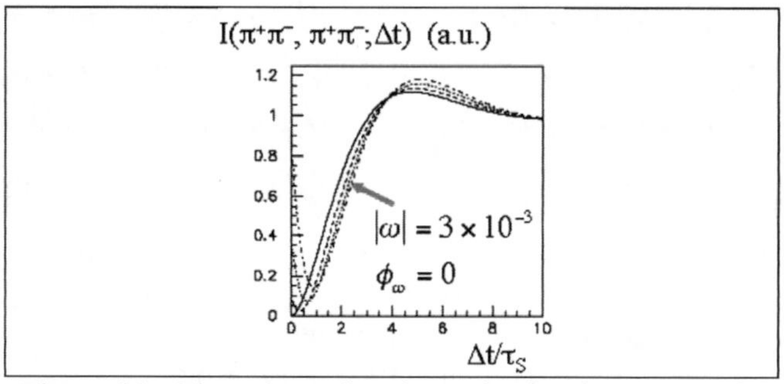

Figure 11.- Time-dependent intensity for the symmetric pair of decay channels ($\pi^+\pi^-$, $\pi^+\pi^-$) for vanishing and non-vanishing ω-parameter.

Previous measurement of the ω-effect by KLOE [35] for analysed data with 1.5 fb-1 led to the limits represented in figure 12

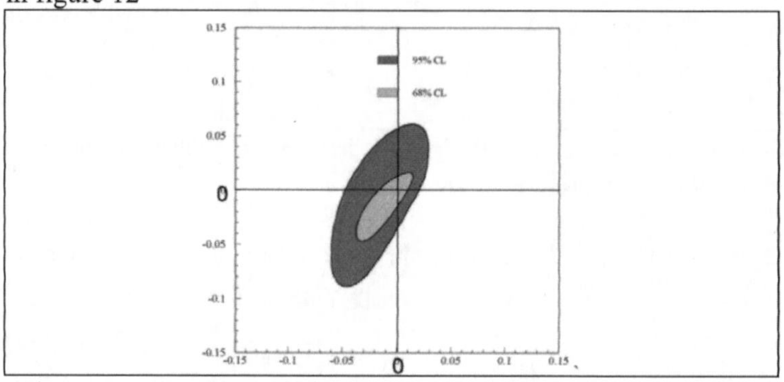

Figure 12.- Corraleted limits for Re (ω) and Im (ω) from KLOE

Explicitly the values for these limits are

$$\begin{aligned} \Re\omega &= \left(-1.6^{+3.0}_{-2.1\,STAT} \pm 0.4_{SYST}\right) \times 10^{-4} \\ \Im\omega &= \left(-1.7^{+3.3}_{-3.0\,STAT} \pm 1.2_{SYST}\right) \times 10^{-4} \\ |\omega| &< 1.0 \times 10^{-3} \quad \text{at } 95\% \text{ C.L.} \end{aligned} \tag{24}$$

where one notices the ample room for statistical improvement.

In the B-system the best limits for Re (ω) come from the equal-sign dilepton intensity, which again should vanish from the allowed component (ω= 0) of the entangled state at equal decay times. The analysis of this intensity in terms of ω led [36] to the limit

$$-0.0084 \leq \mathrm{Re}\,(\omega) \leq 0.0100 \tag{25}$$

at 95 % CL.

Recently a novel signal for separating both Re and Im (ω) in B-physics was found [37] when (flavour, CP) eigenstate decay channels are used from the entangled state (23). The presence of ω =/ 0 violates the (f, g) ↔ (g, f) Connection between the two time-orderings imposed by antisymmetry, as given in equation (9) of Section 3. As a consequence, the authors of [37] have proposed the measurements of

$$C(f, g) - C(g, f);\; S(f, g) + S(g, f) \tag{26}$$

to look for non-vanishing values of these quantities as a signal of ω≠0. The three figures 13 demonstrate the non-correlation of the values of the CPTV θ term due to the dynamical Violation in the Hamiltonian [CPT,H] ≠ 0 and the ω-parameter weakening entanglement of the neutral meson pair, with the results obtained by means of a careful analysis of experimental data from BABAR [19]

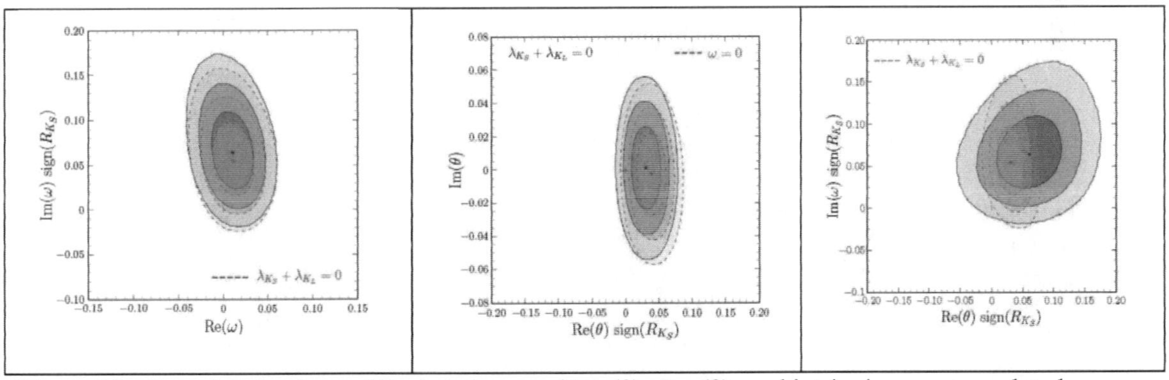

Figure 13.- Correlated values of Re (ω) -Imω and Re (θ) –Im (θ) and intriguing non-correlated Re (θ)-Im (ω) 2σ values

The present result Im (ω) = 0.06 ± 0.03 is the first measurement of this parameter in B-physics and an important improvement is expected with the sensitivity reachable at BELLE II.

7. Conclusions and Outlook

The breaking of the discrete symmetries has played a fundamental role in the understanding of the laws of physics. It emphasizes the importance of looking for genuine asymmetries of separate T, CP, CPT in transitions of neutral meson states. In dealing with the antiunitary T and CPT, entanglement is needed for bypassing the decay irreversibility.

IOP Conf. Series: Journal of Physics: Conf. Series **873** (2017) 012001 doi:10.1088/1742-6596/873/1/012001

The main concept involves the connection between the theoretical transition probabilities for meson states and the experimental intensities for the double decay rate of entangled states.

The flavour-CP transitions in entangled $B^0 - \bar{B}^0$ have demonstrated genuine separate asymmetries for T and CP with high statistical significance and, with a 2σ tension, compatibility with CPT invariance. This effect in the diagonal matrix elements of the mass matrix induces a CPTV asymmetry that can be further explored at BELLE II and, being CPV too, in the unconventional "cosine" term of the intensities at LHCb. KLOE-2 at the upgraded DAPHNE is in the best condition to accomplish a complete program of genuine separate asymmetries for T, CP and CPT in flavour-CP transitions for $K^0 - \bar{K}^0$. Possible fake effects can be controlled within the same experiment.

The measurement of the semileptonic charge asymmetry from K_S by KLOE-2 allows, by comparison with the known semileptonic charge asymmetry from K_L, the separation of CPTV in the mass matrix of $K^0 - \bar{K}^0$.

The best way to study the ω-effect weakening entanglement due to an ill-defined CPT operator is the CP-violating correlated decay (π+ π-, π+ π-) at KLOE-2. In the B-system, BELLE II can address the studies of the equal-sign lepton charge Intensity and the violation of the connection between the Intensity coefficients for the double decay rates of the time-ordered (flavour, CP) and (CP, flavour) channels as a bona-fide signal of the ω-effect. All these measurement have a distinctive signature from the dynamical CPTV in the effective Hamiltonian for the neutral meson system.

Acknowledgements

This research has been supported by MINECO Project FPA 2014-54459-P, Generalitat Valenciana Project GV PROMETEO II 2013-017 and Severo Ochoa Excellence Centre Project SEV 2014-0398.

References

[1] Englert F, Brout R, Phys.Rev.Lett. 13(1964) 321;
Higgs P W, Phys.Rev.Lett. 13(1964)
[2] ATLAS Collaboration, Phys. Lett. B 716(2012)1;
CMS Collaboration, Phys.Lett. B 716(2012)30
[3] Lee T D, Yang C -N, Phys.Rev. **104**(1956)254;
Wu C S et al., Phys.Rev. **105**(1957)141
[4] Garwin R L et al., Phys. Rev. 105(1957)1415
[5] Feynman R P, Gell-Mann M, Phys.Rev. 109(1958)193
[6] Glashow S L, Nucl. Phys. 22(1961)579
[7] Weinberg S, Phys. Rev. Lett. 19(1967)1264
[8] Salam A, Conf. Proc. C 680519(1968)367
[9] Gargamelle Neutrino Collaboration, Phys. Lett. B 46 (1973)138
[10] Christenson J H et al., Phys. Rev. Lett. 13(1964)138
[11] Cabibbo N, Phys.Rev.Lett. 10(1963)531.
[12] Kobayashi M, Maskawa T, Progr.Theor.Phys. 49(1973)652
[13] BABAR Collaboration, Phys.Rev.Lett. 87(2001)091801.
[14] Belle Collaboration, Phys.Rev.Lett. 87(2001)091802.
[15] Lüders G, Annals Phys. 2(1957)1;
Pauli W et al., "Niels Bohr and the development of Physics", Pergamon, NY(1995)
[16] Wolfenstein L, Phys.Rev.Lett. 83(1999)911.
[17] Banuls M C, Bernabeu J, Phys.Lett.B 464(1999)117
[18] Banuls M C, Bernabeu J, Nucl. Phys. B 590(2000)19
[19] BABAR Collaboration, Phys.Rev.Lett. 109(2012)211801

IOP Conf. Series: Journal of Physics: Conf. Series **873** (2017) 012001 doi:10.1088/1742-6596/873/1/012001

[20] Bernabeu J et al., JHEP 1208(2012)064
[21] Applebaum E et al., Phys.Rev. D 89(2014)076011
[22] Bernabeu J et al., Phys.Lett. B 728(2014)95
[23] Wolfenstein L, Int. J. Mod. Phys. E8(1999)501
[24] Bernabeu J, Martinez-Vidal F, Rev. Mod. Phys. 87(2015)165
[25] Bernabeu J et al., JHEP 1510(2015)139
[26] Weisskopf V, Wigner E P, Z. Phys. 63(1930)54;
Z. Phys. 65(1930)18
[27] Bernabeu J et al., Phys. Rev. Lett. 92(2004)131601
[28] Bernabeu J et al., Nucl. Phys. B 868(2013)102
[29] CPLEAR Collaboration, Phys. Lett. B 444(1998)52
[30] Soler A, ASACUSA Collaboration, these Proceedings
[31] KTeV Collaboration, Phys. Rev. Lett. 88(2002)181601
[32] KLOE Collaboration, Phys. Lett. B 636(2006)173
[33] Bernabeu J et al., JHEP 1510(2015)139
[34] Bernabeu J et al., Phys. Rev. D74(2006)045014
[35] KLOE Collaboration, Phys. Lett. B642 (2006)315; Found. Phys. 40(2010)852
[36] Alvarez E et al., JHEP0611 (2006)087
[37] Bernabeu J et al., arXiv: 1612.0565[hep-ph]

IOP Conf. Series: Journal of Physics: Conf. Series **873** (2017) 012002 doi:10.1088/1742-6596/873/1/012002

Nonlinear eigenvalue problems and $\mathcal{PT}$-symmetric quantum mechanics

Carl M. Bender

Department of Physics, Washington University, St. Louis, MO 63130, USA

E-mail: cmb@wustl.edu

Abstract. Semiclassical (WKB) techniques are commonly used to find the large-energy behavior of the eigenvalues of linear time-independent Schrödinger equations. In this talk we generalize the concept of an eigenvalue problem to nonlinear differential equations. The role of an eigenfunction is now played by a separatrix curve, and the special initial condition that gives rise to the separatrix curve is the eigenvalue. The Painlevé transcendents are examples of nonlinear eigenvalue problems, and semiclassical techniques are devised to calculate the behavior of the large eigenvalues. This behavior is found by reducing the Painlevé equation to the linear Schrödinger equation associated with a non-Hermitian $\mathcal{PT}$-symmetric Hamiltonian. The concept of a nonlinear eigenvalue problem extends far beyond the Painlevé equations to huge classes of nonlinear differential equations.

1. Introduction

This talk is focused on the notion of *stability*. We begin by reviewing the concept of $\mathcal{PT}$-symmetric quantum theory, and we show that $\mathcal{PT}$-symmetric systems that appear unstable may actually be stable because they are defined in a complex domain rather than a real domain. Next, we explain that stability (and instability) plays a crucial role in a conventional Schrödinger eigenvalue problem. (A small perturbation of an eigenvalue prevents the boundary conditions from being satisfied.) We then introduce the idea of a *nonlinear* eigenvalue problem in the context of an elementary example and show that the concepts of stability and eigenvalues are intimately linked. Finally, we consider some sophisticated nonlinear eigenvalue problems for the Painlevé transcendents and show that there is a connection between $\mathcal{PT}$-symmetric Hamiltonians and the Painlevé transcendents. Finally, we discuss eigenvalue problems that are posed for nonlinear equations beyond the Painlevé class of differential equations.

2. Brief review of $\mathcal{PT}$-symmetric quantum theory

In the DISCRETE2014 conference in London I presented a talk introducing the ideas of $\mathcal{PT}$-symmetric quantum theory [1]. In that talk I explained that in $\mathcal{PT}$-symmetric quantum theory the mathematical condition of Hermiticity, where the adjoint † represents combined complex conjugation and matrix transposition, is replaced by the weaker and more physical condition of $\mathcal{PT}$ symmetry (invariance under space-time reflection). Here, $\mathcal{P}$ is the space reflection operator, $\mathcal{P}: x \to -x,\, p \to -p$, and $\mathcal{T}$ is the time reversal operator $\mathcal{T}: x \to x,\, p \to -p,\, i \to -i$. Invariance under $\mathcal{PT}$ reflection is a physical condition because $\mathcal{P}$ and $\mathcal{T}$ are elements of the Lorentz group [2, 3]. Many laboratory studies of $\mathcal{PT}$-symmetric models

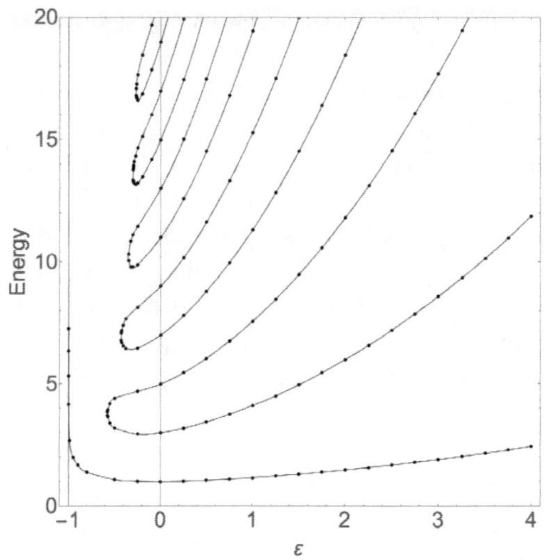

Figure 1. Real eigenvalues of the Hamiltonian $H = p^2 + x^2(ix)^\varepsilon$ plotted as functions of the parameter ε. In the region of *unbroken $\mathcal{PT}$ symmetry* $\varepsilon \geq 0$ the spectrum is real, positive, and discrete. However, in the region of *broken $\mathcal{PT}$ symmetry* $\varepsilon < 0$, the real eigenvalues merge pairwise and form complex-conjugate pairs. When $-1 < \varepsilon < 0$, there are a finite number of real positive eigenvalues and an infinite number of complex-conjugate pairs of eigenvalues. When $\varepsilon \leq -0.57793$, there is just one real eigenvalue, and as $\varepsilon \to -1^+$ this real eigenvalue becomes infinite.

have been published. These experiments have been performed in such diverse areas as optics [4, 5, 6, 7, 8, 9, 10, 11, 12, 13, 14, 15, 16], nuclear magnetic resonance [17], superconductivity [18, 19], microwave cavities [20, 21], laser physics [22, 23], atomic diffusion [24], electronic and mechanical systems [26, 27, 28, 29], optical graphene [30, 31], optical resonators [32, 33, 34], Bose-Einstein condensates [35, 36, 37], and metamaterials [38, 39, 40, 41].

A heavily studied one-parameter family of $\mathcal{PT}$-symmetric Hamiltonians is

$$H = p^2 + x^2(ix)^\varepsilon, \tag{1}$$

and the eigenvalues of this Hamiltonian are shown in Fig. 1. A proof that the eigenvalues of H are real was first given by Dorey, Dunning, and Tateo [42, 43]. Two special cases of H in (1) are

$$H = p^2 + ix^3 \quad \text{and} \quad H = p^2 - x^4, \tag{2}$$

which are obtained when $\varepsilon = 1$ and $\varepsilon = 2$. The latter Hamiltonian has an upside-down potential, and thus appears to be unstable. Nevertheless, it has bound states at *positive* discrete energies (see Fig. 1). The stability of this Hamiltonian has been established rigorously [44, 45].

A heuristic explanation for why an upside-down potential, such as that in (2), may be stable is that a $\mathcal{PT}$-symmetric quantum theory is an extension (an analytic continuation) of a conventional quantum theory into the complex plane. [The Hamiltonian H in (1) is an analytic continuation in ε of the conventional quantum-harmonic-oscillator Hamiltonian $p^2 + x^2$.] When we use the term *unbounded*, we mean that the energy is unbounded below. However, the complex numbers are not ordered, so the notions of $>$ (greater than) and $<$ (less than) do not apply. Thus, it is not obvious which complex potentials have bound states and which do not. One must distinguish between these two possibilities to determine if a system is stable.

The stability of a negative quartic potential has implications in quantum field theory. For example, the double-scaling limit of a O(N)-symmetric quartic scalar field theory appears flawed because an upside-down potential arises in that limit. However, this limit is perfectly acceptable as a $\mathcal{PT}$-symmetric theory [46, 47]. In quantum field theory the process of renormalization can introduce instabilities by making the Hamiltonian non-Hermitian. For example, for the Lee Model [48], Källén and Pauli showed that renormalization introduces non-Hermiticity, which appears to make the scattering matrix nonunitary and leads to ghost states [49]. The techniques of $\mathcal{PT}$-symmetric quantum theory fully resolve these problems [50]. Similarly, renormalization introduces non-Hermiticity in the Standard Model, which appears to make the Higgs vacuum state unstable. Again, $\mathcal{PT}$-symmetry resolves this problem [51]. This stabilization effect works at the classical level as well. An example of a classical instability is the runaway mode in the electromagnetic self-force problem. $\mathcal{PT}$-symmetric analysis resolves this problem [52].

The purpose of this paper is to examine the instabilities that arise in eigenvalue problems associated with classical nonlinear differential equations, and in particular the Painlevé transcendents. We will demonstrate that $\mathcal{PT}$ symmetry resolves these instabilities. Indeed, we will show that there is a deep connection between the Painlevé transcendents and the class of Hamiltonians in (1). To do so we will generalize the notion of an eigenvalue problem for a linear differential equation to an eigenvalue problem for a nonlinear differential equation. In the nonlinear context a separatrix plays the role of an eigenfunction and the initial conditions that give rise to the separatrix play the role of the eigenvalues.

3. Instabilities associated with eigenvalue problems

In quantum mechanics the eigenvalue problems for the linear second-order time-independent Schrödinger equation are well understood. This equation has the form

$$-y''(x) + V(x)y(x) = Ey(x), \tag{3}$$

where E is the eigenvalue. For a rising potential $V(x)$ the eigenfunction $y(x)$ is required to satisfy homogeneous boundary conditions at $\pm\infty$: $y(-\infty) = 0$, $y(\infty) = 0$. Semiclassical (WKB) techniques give accurate approximations to the large eigenvalues. The leading WKB approximation to the nth eigenvalue E_n is given by the phase-integral condition

$$\int_{x_{\rm L}}^{x_{\rm R}} dx\sqrt{E_n - V(x)} \sim \left(n + \tfrac{1}{2}\right)\pi \quad (n \gg 1), \tag{4}$$

where $x = x_{\rm L}$ and $x = x_{\rm R}$ are turning points that satisfy the equation $V(x) = E_n$. Typically, the semiclassical (high-energy) approximation to E_n has the form $E_n \sim ab^n \quad (n \gg 1)$, where a and b are determined by the condition (4). For example, the semiclassical approximation to the eigenvalues of the harmonic oscillator $V(x) = x^2$ is $E_n \sim 2n$ $(n \gg 1)$, and for the anharmonic oscillator $V(x) = x^4$ we get $E_n \sim [3\Gamma(3/4)\sqrt{\pi}/\Gamma(1/4)]^{4/3} n^{4/3}$ $(n \gg 1)$.

The solutions to the linear eigenvalue problem (3) have characteristic qualitative features. If $V(x)$ has a single minimum and rises as $x \to \pm\infty$ (like the potentials x^2 and x^4), the eigenfunctions exhibit distinct behaviors in each of five regions of x: When $x > x_{\rm R}$ and when $x < x_{\rm L}$, the eigenfunctions $y_n(x)$ decay exponentially as $|x| \to \infty$; these are the *classically forbidden* regions. However, when $x_{\rm L} < x < x_{\rm R}$, the eigenfunctions are oscillatory and $y_n(x)$ has exactly n nodes; this is the *classically allowed* region. Near $x_{\rm L}$ and $x_{\rm R}$, there is a transition between exponentially decreasing and oscillatory behavior, which is universally described by the Airy function $\mathrm{Ai}(x)$. Furthermore, the eigenfunctions exhibit unstable behavior: There is an abrupt change in the character of the solution to (3) as E moves away from an eigenvalue; an infinitesimal change in E away from an eigenvalue causes the eigenfunction to lose the property of square integrability. This is the instability referred to earlier.

Nonlinear differential equations can have eigenfunction-like solutions whose behaviors are strongly analogous to the behaviors of eigenfunctions for linear eigenvalue problems. This idea was originally proposed and investigated in Ref. [53]. It was shown that *a nonlinear differential equation may have a discrete set of critical initial conditions that give rise to unstable separatrix solutions.* We treat these initial conditions as *eigenvalues* and the unstable separatrices that come from these special initial conditions as the corresponding *eigenfunctions.* Our objective here will be to find the large-n (semiclassical) asymptotic behavior of the nth eigenvalue by using both numerical and analytic techniques.

A simple-looking first-order nonlinear differential equation that exhibits eigenfunction and eigenvalue behavior is

$$y'(x) = \cos[\pi x y(x)] \tag{5}$$

with the initial condition $y(0) = E$. The initial value E plays the role of an eigenvalue. Solutions to this equation for various initial conditions E are plotted in Fig. 2.

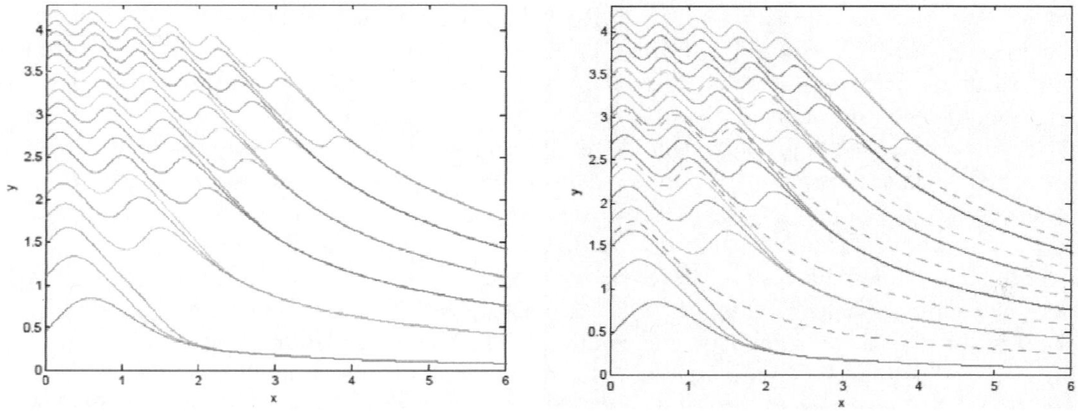

Figure 2. Left panel: Numerical solutions $y(x)$ to (5) for $0 \leq x \leq 6$ with initial conditions $y(0)$ ranging from 0 to 4.2. The solutions initially oscillate but abruptly change their character and decay smoothly and monotonically. In the decaying regime solutions merge into quantized bundles and the bundles decay like $1/x$ as $x \to \infty$. Right panel: Same as left panel except that the separatrix (eigenfunction) solutions (dashed lines) to (5) are shown. The separatrix solutions begin at the eigenvalues $E_1 = 1.6026$, $E_2 = 2.3884$, $E_3 = 2.9767$, $E_4 = 3.4675$, and $E_5 = 3.8975$. The separatrices are unstable; if $y(0)$ lies infinitesimally below (above) the eigenvalue, the solution rapidly diverges away from the separatrix and merges with the bundle of solutions below (above) the separatrix.

Figure 2 shows that solutions to the initial-value problem (5) have n maxima before vanishing like $1/t$ as $t \to \infty$. As the initial condition $y(0)$ increases past the special critical value E_n, the number of maxima changes from n to $n+1$. At these critical values the solution $y(t)$ to (5) is an unstable separatrix curve. We understand this instability as follows: If $y(0)$ lies infinitesimally below E_n, the solution merges with a bundle of stable solutions all having n maxima, and when $y(0)$ is infinitesimally above E_n, the solution merges with a bundle of stable solutions all having $n+1$ maxima. The separatrix curves are displayed in Fig. 2 (right panel) as dashed curves.

The nonlinear eigenvalue problem illustrated in Fig. 2 is qualitatively similar to the linear eigenvalue problem (3) for the time-independent Schrödinger equation. First, the separatrices are unstable with respect to a small change in the eigenvalue E_n; if $y(0) = E_n$ is increased or decreased slightly, $y(x)$ abruptly jumps from the asymptotic bundle of solutions on one side of the separatrix to the asymptotic bundle on the other side of the separatrix. Also, the eigenfunctions (separatrix curves) corresponding to the nth eigenvalue (the initial condition E_n) exhibit n oscillations in the analog of a classically allowed region before decreasing monotonically to 0 in the analog of a classically forbidden region. We can see in Fig 2 that this change from oscillatory to decaying behavior occurs over a narrow region that is the analog of a turning-point region.

An exact formula for the nth eigenvalue E_n is unknown, and without such a formula our objective is to determine the asymptotic behavior of the critical values E_n for large n. It was shown in Ref. [53] that for large n the nonlinear-differential-equation problem (5) reduces to a *linear* one-dimensional random-walk problem. This random-walk problem was solved exactly and it was thereby established analytically that the nth eigenvalue grows like an^b for large n:

$$E_n \sim 2^{5/6}\sqrt{n} \quad (n \to \infty). \tag{6}$$

Kerr found an alternative derivation of this remarkable semiclassical (high-energy) solution to this asymptotics problem and verified (6) [54]. [The numerical constant $a = 2^{5/6}$ and the exponent $b = 1/2$ are remarkable because there is no hint of such numbers in (5)!]

There is a huge class of *second-order* nonlinear differential-equation eigenvalue problems having discrete eigenvalues. In Refs. [53, 55] it was shown that the first two Painlevé transcendents lead to eigenvalue problems. (These are discussed in Sec. 4.) The Painlevé equations have the special property that their solutions have *movable* (*spontaneous*) singularities that are poles. There are no other kinds of movable singularities, such as branch points. Thus, the solutions to these differential equations are *meromorphic* (defined on a one-sheeted Riemann surface) and such solutions are readily studied by using numerical and asymptotic analysis. The large-eigenvalue behaviors of these equations can be found asymtotically by linearizing the eigenvalue problems. Specifically, for large n the nonlinear eigenvalues are approximated by the eigenvalues of a *linear* Schrödinger eigenvalue problem and the Hamiltonian for this problem belongs to the class of $\mathcal{PT}$-symmetric non-Hermitian Hamiltonians (1) [2].

In Ref. [55] it was proposed that one could extend the study of nonlinear-differential-equation eigenvalue problems beyond the Painlevé transcendents to more complicated differential equations such as the Thomas-Fermi equation $y''(x) = [y(x)]^{3/2}x^{1/2}$, and clearly the physical solution to this equation, which satisfies the boundary conditions $y(0) = 1$ and $y(+\infty) = 0$, is an eigenfunction solution. Like the solutions to (5), $y(x)$ vanishes algebraically (like $144x^{-3}$) as $x \to \infty$. This is an unstable separatrix solution and the specific value of $y'(0)$ that gives rise to this solution is an eigenvalue. If $y'(0)$ is larger than the critical value of $y'(0)$, the function $y(x)$ becomes singular at some point $x = a$ and blows up like $400a(x-a)^{-4}$ as $x \to a$; if $y'(0)$ is less than the critical value, the function $y(x)$ crosses 0 and becomes complex. Unfortunately, the movable singularity at $x = a$ is a *logarithmic branch-point* singularity [to verify this one must expand $y(x)$ to sixteenth order in powers of $(x-a)$] and therefore the solutions to this equation live on a Riemann surface having infinitely many sheets. Thus, for this equation it is not easy to find additional separatrix solutions.

However, we have found infinite numbers of nonlinear differential equations whose movable singularities are *algebraic* branch points, and these equations, which are discussed in Sec. 5, are analytically tractable. The features of these new kinds of nonlinear differential-eigenvalue problems are qualitatively similar to those of the Painlevé equations [56].

4. Instabilities associated with the Painlevé transcendents

The Painlevé transcendents are six second-order nonlinear differential equations whose movable (spontaneous) singularities are poles (and not branch points, essential singularities, or other kinds of singularities). Many papers and books have been written on these beautiful differential equations [57, 58, 59, 60, 61, 62, 63, 64] and these equations arise often in mathematical physics [65, 66, 67, 68, 69, 70, 71].

This section is focused mostly on the first and second Painlevé transcendents, referred to here as PI and PII. The initial-value problem (IVP) for the PI differential equation is

$$y''(t) = 6[y(t)]^2 + t, \qquad y(0) = c,\ y'(0) = b \tag{7}$$

and the IVP for PII (in which an arbitrary additive constant is set to 0) is

$$y''(t) = 2[y(t)]^3 + ty(t), \qquad y(0) = c,\ y'(0) = b. \tag{8}$$

For the fixed initial condition $y(0) = 0$ there is a discrete set of initial slopes $y'(0) = b_n$ that give rise to unstable separatrix solutions. Similarly, for fixed initial slope $y'(0) = 0$, there is a discrete set of initial values $y(0) = c_n$ that give rise to separatrix solutions. For Painlevé I the large-n asymptotic behavior of the eigenvalues b_n and c_n is

$$b_n \sim B_{\rm I} n^{3/5} \quad (n \to \infty) \quad \text{and} \quad c_n \sim C_{\rm I} n^{2/5} \quad (n \to \infty). \tag{9}$$

For Painlevé II the large-n asymptotic behavior of the eigenvalues b_n and c_n are

$$b_n \sim B_{\rm II} n^{2/3} \quad (n \to \infty) \quad \text{and} \quad c_n \sim C_{\rm II} n^{1/3} \quad (n \to \infty). \tag{10}$$

We have determined the coefficients $B_{\rm I}$, $C_{\rm I}$, $B_{\rm II}$, and $C_{\rm II}$ in these asymptotic behaviors analytically and numerically. The analytical calculation of these constants for PI and PII is done by reducing the nonlinear equations to the linear eigenvalue problems for the cubic and quartic $\mathcal{PT}$-symmetric Hamiltonians $H = \frac{1}{2}p^2 + 2ix^3$ and $H = \frac{1}{2}p^2 - \frac{1}{2}x^4$ [44] [see (2)].

Let us look at the PI equation. In Ref. [72] there is a brief asymptotic study of the first Painlevé transcendent (7). One can easily show that there are two possible asymptotic behaviors of the solution to this differential equation as $t \to -\infty$; the solutions can approach either $+\sqrt{-t/6}$ or $-\sqrt{-t/6}$. If the solution $y(t)$ approaches $-\sqrt{-t/6}$, the solution oscillates *stably* about this curve with gradually decreasing amplitude. However, while the curve $+\sqrt{-t/6}$ is another possible asymptotic behavior, this behavior is *unstable* and *nearby solutions tend to veer away from it.* The eigenfunctions of PI are those solutions that *do* approach the curve $+\sqrt{-t/6}$ as $t \to -\infty$. These separatrix solutions resemble the eigenfunctions of conventional quantum mechanics in that they exhibit n oscillations before settling down to this asymptotic behavior. However, because the PI equation is nonlinear, these oscillations are unbounded; the nth eigenfunction passes through $[n/2]$ double poles where it blows up, and only then does it smoothly approach the curve $+\sqrt{-t/6}$. (The symbol $[n/2]$ means greatest integer in $n/2$.)

To find the numerical solutions to the initial-value problem for the PI equation (7) for $t < 0$ we use Runge-Kutta to integrate down the negative-real axis. When we approach a double pole and the solution becomes large and positive, we estimate the location of the pole and integrate along a semicircle in the complex-t plane around the pole. We then continue integrating down the negative-real axis. We begin by choosing the fixed initial value $y(0) = 0$ and allow the initial slope $y'(0) = b$ to have increasingly positive values. (We only present results for positive initial slope; the behavior for negative initial slope is analogous.) We find that the particular choice of $y(0)$ is not crucial and that for *any* fixed $y(0)$ the large-n leading asymptotic behavior of the initial-slope eigenvalues b_n is the same.

Above the critical value $b_1 = 1.851854034$ (first eigenvalue) there is a continuous interval of b for which $y(t)$ first has a minimum and then has an infinite sequence of double poles (Fig. 3, left panel). However, if b increases past the next critical value $b_2 = 3.004031103$ (second eigenvalue), the character of the solutions changes abruptly and $y(t)$ oscillates stably about $-\sqrt{-t/6}$ (Fig. 3, right panel). When b exceeds the critical value $b_3 = 3.905175320$ (third eigenvalue), the solutions again exhibit an infinite sequence of poles (Fig. 4, left panel). When b increases past the fourth critical value $b_4 = 4.683412410$ (fourth eigenvalue), the solutions once again oscillate stably about $-\sqrt{-t/6}$ (Fig. 4, right panel). There is an infinite sequence of critical points (eigenvalues) at which the PI solutions alternate between infinite sequences of double poles and stable oscillation about $-\sqrt{-t/6}$.

The separatrix (eigenfunction) solutions that arise when $y'(0)$ is an eigenvalue have a completely different (and unstable) character from those in Figs. 3 and 4. These special solutions exhibit a *finite* number $[n/2]$ of double poles (analogous to the oscillatory behavior of quantum-mechanical bound-state eigenfunctions in the classically allowed region of a potential well) and then exhibit a turning-point-like transition in which the poles abruptly cease and $y(t)$

IOP Conf. Series: Journal of Physics: Conf. Series **873** (2017) 012002 doi:10.1088/1742-6596/873/1/012002

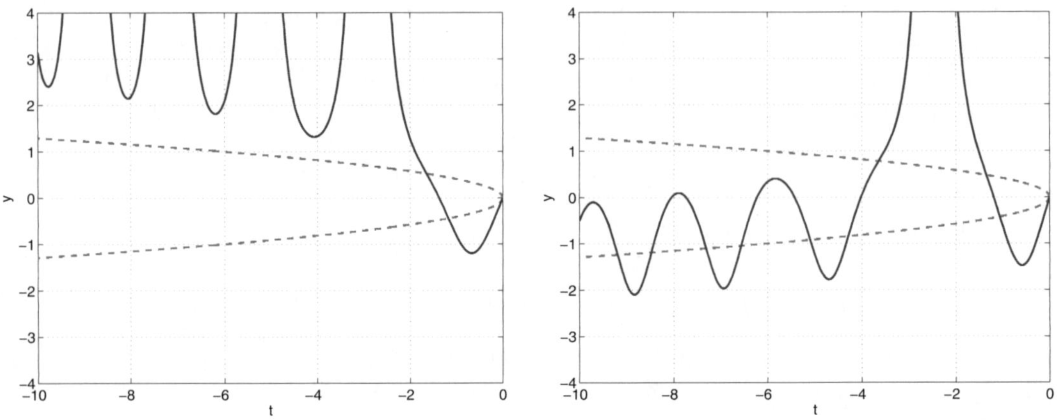

Figure 3. Typical behavior of solutions to the first Painlevé transcendent $y(t)$ for the initial conditions $y(0) = 0$ and $b = y'(0)$. Left panel:$b = 2.504031103$, which lies between the eigenvalues $b_1 = 1.851854034$ and $b_2 = 3.004031103$. Right panel: $b = 3.504031103$, which lies between the eigenvalues $b_2 = 3.004031103$ and $b_3 = 3.905175320$. The dashed curves are $y = \pm\sqrt{-t/6}$. In the left panel the solution $y(t)$ has an infinite sequence of double poles and in the right panel the solution oscillates stably about $-\sqrt{t/6}$.

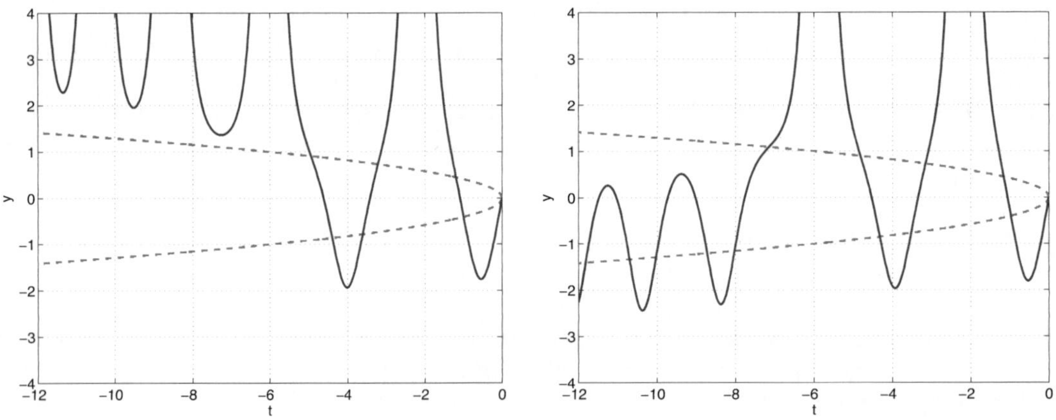

Figure 4. Solutions to the PI equation (7) for $y(0) = 0$ and $b = y'(0)$. Left panel: $b = 4.583412410$ (between the eigenvalues $b_3 = 3.905175320$ and $b_4 = 4.6834124103$). Right panel: $b = 4.783412410$ (between the eigenvalues $b_4 = 4.683412410$ and $b_5 = 5.383086722$).

exponentially decays towards the limiting curve $+\sqrt{-t/6}$. The solutions arising from the first and second critical points b_1 and b_2 are shown in Fig. 5, those arising from the third and fourth critical points b_3 and b_4 are shown in Fig. 6, and those arising from the tenth and eleventh critical points b_{10} and b_{11} are shown in Fig. 7. The critical points are analogous to eigenvalues because they give rise to *unstable* separatrix solutions; if $y'(0)$ changes by an infinitesimal amount above or below a critical value, the character of the solutions changes abruptly and the solutions exhibit the two possible kinds of generic behaviors shown in Figs. 3 and 4.

In Ref. [55] the constant $B_{\rm I}$ was determined numerically to great accuracy by applying fifth-order Richardson extrapolation to the first eleven eigenvalues. The value of $B_{\rm I}$ in (9) was found to an accuracy of one part in nine decimal places: $B_{\rm I} = 2.0921467\underline{4}$. The underlined digit lies in the range from 3 to 5 and $B_{\rm I}$ is accurate to one part in 2×10^8.

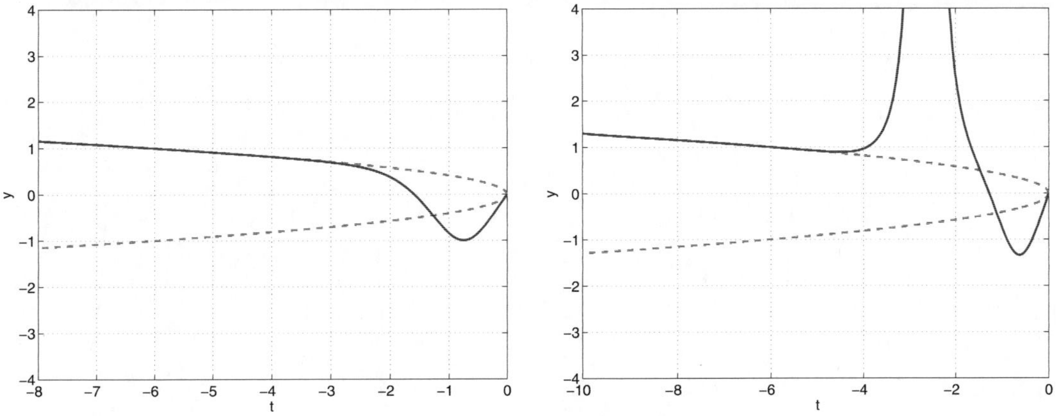

Figure 5. First two separatrix solutions (eigenfunctions) of Painlevé I with initial condition $y(0) = 0$. Left panel: $y'(0) = b_1 = 1.851854034$. Right panel: $y'(0) = b_2 = 3.004031103$. The dashed curves are $y = \pm\sqrt{-t/6}$.

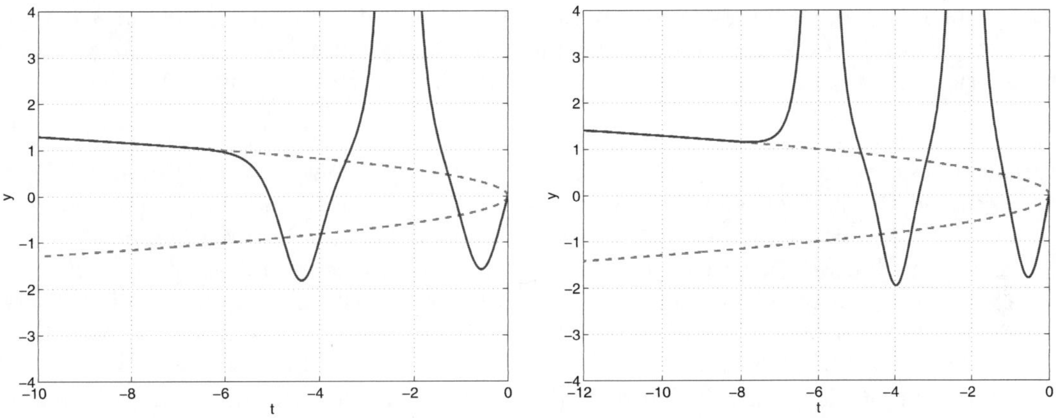

Figure 6. Third and fourth eigenfunctions of Painlevé I with initial condition $y(0) = 0$. Left panel: $y'(0) = b_3 = 3.905175320$. Right panel: $y'(0) = b_4 = 4.683412410$.

If the initial slope is held fixed at $y'(0) = 0$ and the initial value $y(0) = c$ is allowed to become increasingly negative, a new sequence of negative eigenvalues c_n appears for which the solutions resemble the eigenfunction separatrix solutions in Figs. 5–7. Fourth-order Richardson extrapolation applied to the first 15 eigenvalues reveals that for large n the sequence of initial-value eigenvalues c_n is asymptotic to $C_{\rm I} n^{2/5}$, where the constant $C_{\rm I}$ in (9) is $C_{\rm I} = -1.030484\underline{4}$ (see Ref. [55]). The last digit is accurate to an error of ± 1 so $C_{\rm I}$ is determined to one part in 10^7. We have done similar numerical studies for PII and PIV, but to save space we do not repeat them here.

Asymptotic analysis may be used to derive analytic expressions for these numerical results. We multiply the PI equation in (7) by $y'(t)$ and integrate from $t = 0$ to $t = x$ and get

$$H \equiv \tfrac{1}{2}[y'(x)]^2 - 2[y(x)]^3 = \tfrac{1}{2}[y'(0)]^2 - 2[y(0)]^3 + I(x), \tag{11}$$

where $I(x) = \int_0^x dt\, t y'(t)$. The path of integration is the same as that used to calculate $y(t)$ numerically; the path follows the negative-real axis until it gets near a pole, at which point it makes a semicircular loop in the complex-t plane to avoid the pole.

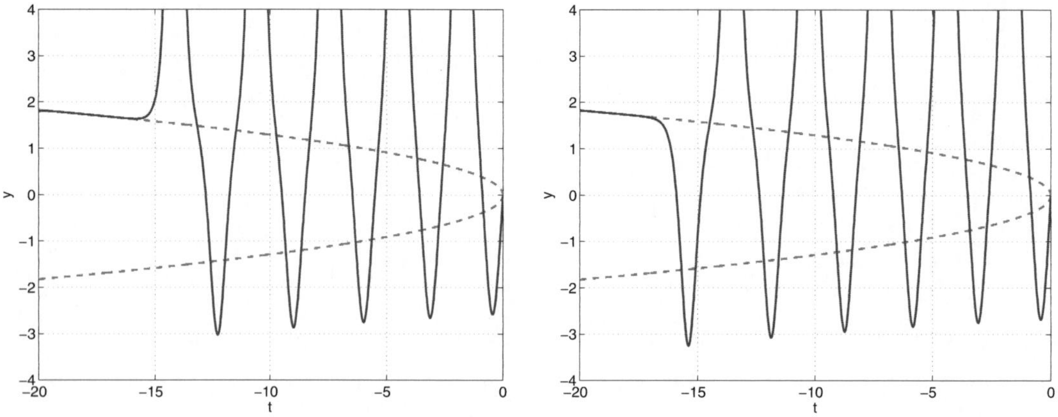

Figure 7. Tenth and eleventh eigenfunctions of Painlevé I with initial condition $y(0) = 0$. Left panel: $y'(0) = b_{10} = 8.244932302$. Right panel: $y'(0) = b_{11} = 8.738330156$. As n increases, the eigenfunctions pass through more double poles before exhibiting a turning-point-like transition and approaching the limiting curve $+\sqrt{-t/6}$ exponentially. This behavior is analogous to that of the eigenfunctions of a time-independent Schrödinger equation for a particle in a potential well; the higher-energy eigenfunctions exhibit many oscillations in the classically allowed region before entering the classically forbidden region, where they decay exponentially.

We calculate $I(x)$ for large-negative x in the classically allowed region (before the poles abruptly cease at the turning point) and find that on the real-x axis, as $n \to \infty$, $I(x)$ fluctuates and becomes small compared with H. This is not surprising because $I(x)$ receives many positive and negative contributions from the poles. We can see from the definition of $I(x)$ that $I'(x)$ vanishes when $y'(x)$ vanishes. Near the points where $y'(x)$ vanishes, we verify numerically that $I(x)$ is small compared with $-2[y(x)]^3$. Far from these points $y(x)$ becomes large, but so does $I(x)$. However, $-2[y(x)]^3$ blows up like a sixth-order pole and $I(x)$ blows up like a second-order pole. These asymptotic estimates are difficult to verify analytically, but numerical analysis confirms these results. These estimates are valid when x is large and negative but only in the classically allowed region and not as $x \to -\infty$.

By examining $I(x)$ as $x \to -\infty$, we can see a signal of an eigenvalue; as $y'(0) = b$ passes an eigenvalue, $I(x)$ goes from having positive to negative (or negative to positive) fluctuations, but at an eigenvalue $I(x)$ is smooth and does not fluctuate. For large n we treat the fluctuating quantity $I(x)$ as small and we interpret H as a time-independent quantum-mechanical Hamiltonian. [The isomonodromic properties of H when $I(x)$ is not neglected were studied in Ref. [62].]

To support these observations regarding the behavior of $I(x)$ on the real axis, we have done extremely detailed numerical studies of the distribution of poles and the behavior of PI in the complex-x plane. This gives a much clearer picture of the behavior of $I(x)$ for large x. We find that along the lines $x = re^{\pm i\pi/4}$, where r is real, the function $I(x)$ rapidly approaches 0 as $r \to \infty$. This provides strong evidence that the large-n (semiclassical) behavior of the eigenvalues is determined by solving the *linear* time-independent quantum-mechanical eigenvalue problem $\hat{H}\psi = E\psi$, where $\hat{H} = \frac{1}{2}\hat{p}^2 - 2\hat{x}^3$ along these lines in the complex-x plane. To find these eigenvalues we simply rotate $\hat{H}$ into the complex plane by 45^circ [73] and obtain the well-studied $\mathcal{PT}$-symmetric Hamiltonian [2, 3]

$$\hat{H} = \tfrac{1}{2}\hat{p}^2 + 2i\hat{x}^3. \tag{12}$$

The large eigenvalues of this Hamiltonian are found by using the complex WKB techniques discussed in Ref. [2]. For the class of $\mathcal{PT}$-symmetric Hamiltonians $\hat{H} = \frac{1}{2}\hat{p}^2 + g\hat{x}^2(i\hat{x})^\varepsilon$ $(\varepsilon \geq 0)$, the WKB approximation to the nth eigenvalue $(n \gg 1)$ is given by

$$E_n \sim \frac{1}{2}(2g)^{2/(4+\varepsilon)}\left[\frac{\Gamma\left(\frac{3}{2}+\frac{1}{\varepsilon+2}\right)\sqrt{\pi}\,n}{\sin\left(\frac{\pi}{\varepsilon+2}\right)\Gamma\left(1+\frac{1}{\varepsilon+2}\right)}\right]^{(2\varepsilon+4)/(\varepsilon+4)}. \tag{13}$$

For H in (12) we take $g = 2$ and $\varepsilon = 1$ and obtain the asymptotic behavior

$$E_n \sim 2\left[\sqrt{3\pi}\Gamma\left(\tfrac{11}{6}\right)n/\Gamma\left(\tfrac{1}{3}\right)\right]^{6/5} \quad (n \to \infty). \tag{14}$$

The Hamiltonian $\hat{H}$ in (12) is time independent, so we evaluate H in (11) for fixed $y(0)$ and large $y'(0) = b_n$ and obtain the result

$$b_n \sim \sqrt{2E_n} = B_{\rm I}n^{3/5} \quad (n \to \infty), \tag{15}$$

which verifies (9). We read off the analytic value of the constant $B_{\rm I}$:

$$B_{\rm I} = 2\left[\sqrt{3\pi}\Gamma\left(\tfrac{11}{6}\right)/\Gamma\left(\tfrac{1}{3}\right)\right]^{3/5}, \tag{16}$$

which agrees with the numerical result in (9). Also, if we take the initial slope $y'(0)$ to vanish and take the initial condition $y(0) = c_n$ to be large, we obtain an analytic expression for $C_{\rm I}$,

$$C_{\rm I} = -\left[\sqrt{3\pi}\Gamma\left(\tfrac{11}{6}\right)/\Gamma\left(\tfrac{1}{3}\right)\right]^{2/5}, \tag{17}$$

which verifies (10). These analytic results are in precise agreement with our numerical work on the large-n behavior of the eigenvalues.

To obtain analytic expressions for $B_{\rm II}$ in (10) and $C_{\rm II}$ in (10), we follow the same procedure as for PI. We multiply the PII equation by $y'(t)$ and integrate from $t = 0$ to $t = x$, where x is in the turning-point region (where the simple poles stop). The result is

$$H \equiv \tfrac{1}{2}[y'(x)]^2 - \tfrac{1}{2}[y(x)]^4 = \tfrac{1}{2}[y'(0)]^2 - \tfrac{1}{2}[y(0)]^4 + I(x), \tag{18}$$

where $I(x) = \int_0^x dt\, ty(t)y'(t)$. The path of integration follows the negative-real axis until it gets near a simple pole, at which point it makes a loop in the complex-t plane to avoid the pole. As before, along this path the integrand of $I(x)$ oscillates and because of cancellations we neglect $I(x)$ when n is large.

We treat H as the $\mathcal{PT}$-symmetric quantum-mechanical Hamiltonian

$$\hat{H} = \tfrac{1}{2}\hat{p}^2 - \tfrac{1}{2}\hat{x}^4 \tag{19}$$

and use (13) with $g = 1/2$ and $\epsilon = 2$ to obtain

$$E_n \sim \tfrac{1}{2}\left[3n\sqrt{2\pi}\Gamma\left(\tfrac{3}{4}\right)/\Gamma\left(\tfrac{1}{4}\right)\right]^{4/3} \tag{20}$$

for the large eigenvalues of $\hat{H}$. For the eigenvalues b_n we use

$$\sqrt{2E_n} \sim \left[3n\sqrt{2\pi}\Gamma\left(\tfrac{3}{4}\right)/\Gamma\left(\tfrac{1}{4}\right)\right]^{2/3} \quad (n \to \infty) \tag{21}$$

from which we identify the value of $B_{\rm II}$:

$$B_{\rm II}=\left[3\sqrt{2\pi}\Gamma\left(\tfrac{3}{4}\right)/\Gamma\left(\tfrac{1}{4}\right)\right]^{2/3}. \tag{22}$$

To calculate C_{II} we note that the initial value $y(0)$ is positive. If we neglect $I(x)$ and assume a vanishing initial slope, we see that the right side of (18) is negative. Thus, as we did for the cubic Hamiltonian $\frac{1}{2}\hat{p}^2-2\hat{x}^3$, we perform a complex rotation of the coupling constant to convert the quartic Hamiltonian to $\hat{H}=\frac{1}{2}\hat{p}^2+\frac{1}{2}\hat{x}^4$. Here, we obtain the conventional Hermitian quartic-anharmonic-oscillator Hamiltonian, which does not belong to the class of $\mathcal{PT}$-symmetric Hamiltonians $\hat{H}=\frac{1}{2}\hat{p}^2+g\hat{x}^2(i\hat{x})^\epsilon$. The WKB approximation gives the large-eigenvalue approximation

$$E_n\sim\left[3n\sqrt{\pi}\Gamma\left(\tfrac{3}{4}\right)/\Gamma\left(\tfrac{1}{4}\right)\right]^{4/3}\quad(n\to\infty) \tag{23}$$

from which we read off the value of C_{II}:

$$C_{II}=\left[3\sqrt{\pi}\Gamma\left(\tfrac{3}{4}\right)/\Gamma\left(\tfrac{1}{4}\right)\right]^{1/3}. \tag{24}$$

To summarize, we have found the remarkable result that the large-eigenvalue behavior of PI, PII, and PIV (not discussed here) are obtained by WKB analysis of (1) for $\varepsilon=1$, 2, 4. (We are planning further studies of PIII, PV, and PVI.) Finally, we note that the asymptotic analysis performed in this section has recently been verified at a more rigorous level [74].

5. Instabilities beyond the Painlevé transcendents

Let us consider a new class of nonlinear-differential-equation eigenvalue problems of the form

$$y''(x)=\frac{2M+2}{(M-1)^2}[y(x)]^M+x[y(x)]^N, \tag{25}$$

where M and N are integers. We refer to the members of this class as SP(M,N) (SP stands for super-Painlevé). These equations are a natural generalization of the first two Painlevé equations; PI in (4) is SP$(2,0)$ and PII in (5) is SP$(3,1)$. These equations give particularly interesting nonlinear eigenvalue problems when $N<M-1$.

This special class of equations appear to have algebraic but not logarithmic movable singularities. To see why this is so recall that the solution to PI can become singular at an arbitrary point $x=A$ and the leading asymptotic approximation to such a solution is $y(x)\sim(x-A)^{-2}$ $(x\to A)$. However, to show that the singularity at $x=A$ is a pole it is necessary to show that an expansion around $x=A$ is a Laurent series. To do this we substitute

$$y(x)=\frac{1}{(x-A)^2}\left[1+\sum_{n=1}^{\infty}a_n(x-A)^n\right] \tag{26}$$

into (4) and collect powers of $(x-A)$. If we solve recursively for the coefficients a_n, we find that the first five coefficients are $a_1=0$, $a_2=0$, $a_3=0$, $a_4=-A/10$, and $a_5=-1/6$. However, the key result is that a_6 is *undetermined* and thus is *arbitrary*. Since the series expansion for $y(x)$ contains *two arbitrary constants*, A and a_6, it follows that this series is the most general solution to the PI equation. To complete the argument one must show that the series expansion for $y(x)$ converges when $|x-A|$ is sufficiently small. This establishes that the expansion (26) is a Laurent series and verifies that the solutions to PI are meromorphic.

This analysis extends to the SP(M,N) equation, whose solutions can become singular at a point $x=A$. These singular solutions have the leading asymptotic behavior

$$y(x)\sim(x-A)^{-2/(M-1)}\quad(x\to A). \tag{27}$$

For simplicity, we study solutions that remain real on both sides of this singularity, so we assume that M is an *even* integer or $(M-1)/2$ is an odd integer. We seek an asymptotic expansion of $y(x)$ about the point $x=A$ of the form $y(x)=(x-A)^{-2/(M-1)}\left[1+\sum_{n=1}^{\infty}a_n(x-A)^{1/(M-1)}\right]$. (This expansion is actually valid for both even and odd M; for odd M half of the terms vanish.) We substitute this expansion into the SP(M,N) equation and compare like powers of $x-A$ to find the coefficients, we see that the a_n are uniquely determined up to $n=2(M+1)$ for even M and $n=M+1$ for odd M; at this value of n the coefficient a_n is *arbitrary*. Note that if $N\geq M-1$, the a_n term contains logarithmic terms.

As an example, we consider SP$(4,0)$. Here, the first seven coefficients vanish $a_1=a_2=a_3=a_4=a_5=a_6=a_7=0$ and the next two coefficients are $a_8=-9A/22$ and $a_9=0$. Next, we find that a_{10} is *not determined and thus is arbitrary.* The next few coefficients are $a_{11}=9/14$, $a_{12}=a_{13}=a_{14}=a_{15}=0$, $a_{16}=405A^2/4598$, $a_{17}=0$. From here on the coefficients depend on the choice of a_{10}: $a_{18}=-45Aa_{10}/154$, $a_{19}=-135A/847$, $a_{20}=6a_{10}^2/23$, $a_{21}=45a_{10}/154$, and so on. We have found the general solution because the series contains two arbitrary constants, namely A and a_{10}. We do not present a proof here, but numerical analysis shows that the coefficients in the series have only geometric growth for large n, so the radius of convergence of the series is nonzero. While the series (27) is not a Laurent series because it contains fractional powers of $(x-A)$, our analysis shows that the solutions to the class of equations SP(M,N) have only algebraic singularities, and because M is chosen to be even, we may seek solutions that are entirely real. In particular, we can seek eigenfunction (separatrix) solutions that are real.

Solutions to the SP(M,N) equation have one or two possible real asymptotic behaviors as $x\to-\infty$: $y(x)\sim\pm\left[-(M-1)^2x/(2M+2)\right]^{1/(M-N)}$. (There are two possible asymptotic curves when $M-N$ is even but only one when $M-N$ is odd.) As in the PI case, the upper curve is unstable and the lower curve is stable. Thus, the discrete eigenfunction (separatrix) solutions approach the upper curve. Here, we only consider here eigenfunction solutions for which $y(0)=0$. The eigenvalues E_n are the initial values of the slope $y'(0)$ that give rise to solutions that approach the upper curve. For example, for $M=4$ and $N=2$, the first eight eigenvalues are 2.4240, 4.5364, 6.2471, 7.7792, 9.1960, 10.5292, 11.7973, 13.0127.

The analysis used for the Painlevé equations shows that if we neglect the function $I(x)$, we can replace the SP(M,N) equation by a simpler equation generated by the *linear* Hamiltonian

$$\hat{H}=\tfrac{1}{2}\hat{p}^2+\tfrac{2}{(M-1)^2}\hat{x}^{M+1}.$$

In this approximation the term containing the parameter N has dropped out. The WKB formula (4) implies that for large n, the nth eigenvalue of $\hat{H}$ grows like $n^{(2M+2)/(M+3)}$ as $n\to\infty$. Thus, the nth separatrix eigenvalue $y'(0)$ grows like

$$y'(0)\sim n^{(M+1)/(M+3)}\quad(n\to\infty). \tag{28}$$

Our preliminary numerical studies show that (28) holds in some but not all cases. For $M=4$ we expect the large-n behavior $n^{5/7}$ and for $M=6$ we expect the behavior $n^{7/9}$. For $M=4$ and $N=2$ and for $M=6$ and $N=4$ this is what we find. Specifically, when $M=4$, there are no eigenfunctions for $N=0$, a full set of eigenfunctons qualitatively identical to those shown in Figs. 5-7 for $N=1$, and a half-set of eigenfunctions qualitatively identical to those shown in the right panels of Figs. 5 and 6 and the left panel of Fig. 7 when $N=2$. For the case $N=2$

the eigenvalues grow like $n^{5/7}$. However, when $N = 1$, the eigenvalues grow slightly less rapidly; the nth eigenvalue grows like an^b, where $a = 2.04$ and $b = 0.56$.

When, $M = 6$, there is a half-set of eigenfunctions when $N = 0$ and no eigenfunctions when $N = 1$. However, when $N = 2$, $N = 3$, and $N = 4$, there are full sets of eigenfunctions. For the case $N = 4$ the eigenvalues grow like $n^{7/9}$, but when $N < 4$, the eigenvalues grow slightly less rapidly. Interestingly, it is the *largest* value of N that gives an eigenspectrum whose asymptotic behavior is determined by a linear approximation in which the y^N term in the nonlinear equation can be neglected! Evidently, for smaller values of N, the y^N term cannot be neglected. Specifically, when $N = 0$, the nth eigenvalue grows like $2.43n^{0.41}$; when $N = 2$, the nth eigenvalue grows like $2.55n^{0.54}$; when $N = 3$, the nth eigenvalue grows like $1.69n^{0.65}$. However, when $N = 4$, the nth eigenvalue grows like $3.06n^{7/9}$.

We have demonstrated that there is a huge, rich, and remarkable class of nonlinear-differential-equation eigenvalue problems for which there exists an infinite discrete set of eigenvalues. These differential equations are generalizations of the Painlevé transcendents. In full generality, the differential equations resulting from extending PI and PII have the form

$$y''(x) = \frac{2M+2}{(M-1)^2}[y(x)]^M + xP(y) + Q(y),$$

where $P(y)$ and $Q(y)$ are polynomials in $y(x)$ of degree less than $M - 1$. Clearly, there are additional general classes of nonlinear eigenvalue problems beyond the Painlevé transcendents. These new kinds of problems warrant much intensive further analysis.

Acknowledgments
I thank my colleagues A. Fring, J. Komijani, and Q. Wang for their contributions to our research on nonlinear eigenvalue problems.

References
[1] Bender C M 2015 $\mathcal{PT}$-symmetric quantum theory *J. Phys. Conf. Series* **631** 012002 [Talk given at the 4th Symposium on Prospects in the Physics of Discrete Symmetries (DISCRETE2014).]
[2] Bender C M and Boettcher S 1998 Real spectra in non-Hermitian Hamiltonians having $\mathcal{PT}$ symmetry *Physical Review Letters* **80** 5243-5246
[3] Bender C M (2007) Making Sense of Non-Hermitian Hamiltonians *Rep. Prog. Phys.* **70** 947-1018
[4] Guo A, Salamo G J, Duchesne D, Morandotti F, Volatier-Ravat M, Aimez V, Siviloglou G A and Christodoulides D N 2009 Observation of $\mathcal{PT}$-symmetry breaking in complex optical potentials *Phys. Rev. Lett.* **103** 093902
[5] Longhi S 2009 Bloch oscillations in complex crystals with $\mathcal{PT}$ symmetry *Phys. Rev. Lett.* **103** 123601
[6] Rüter C E, Makris, El-Ganainy R, Christodoulides D N, Segev M and Kip D 2010 Observation of $\mathcal{PT}$ symmetry in optics *Nat. Phys.* **6** 192
[7] Longhi S 2010 Optical realization of relativistic non-Hermitian quantum mechanics *Phys. Rev. Lett.* **105** 013903
[8] Lin Z, Ramezani H, Eichelkraut T, Kottos T, Cao H and Christodoulides D N 2011 Unidirectional invisibility induced by $\mathcal{PT}$-symmetric periodic structures *Phys. Rev. Lett.* **106** 213901
[9] Feng L, Ayache M, Huang J, Xu Y-L, Lu M-H, Chen Y-F, Fainman Y and Scherer A 2011 Nonreciprocal light propagation in a silicon photonic circuit *Science* **333** 729
[10] Ramezani H, Christodoulides D N, Kovanis V, Vitebskiy I and Kottos T 2012 $\mathcal{PT}$-symmetric Talbot effects *Phys. Rev. Lett.* **109** 033902
[11] Regensberger A, Bersch C, Miri M-A, Onishchukov G and Christodoulides D N 2012 Parity-time synthetic photonic lattices *Nature* **488** 167
[12] Razzari L and Morandotti R 2012 Gain and loss mixed in the same cauldron *Nature* **488** 163
[13] Liang G Q and Chong Y D 2013 Optical resonator analog of a two-dimensional topological insulator *Phys. Rev. Lett.* **110** 203904
[14] Luo X, Huang J, Zhong H, Qin X, Xie Q, Kivshar Y S and C Lee C 2013 Pseudo-parity-time symmetry in optical systems *Phys. Rev. Lett.* **110** 243902

[15] Regensberger A, Miri M-A, Bersch C, Nager J, Onishchukov G, Christodoulides D N and Peschel U 2013 Observation of defect states in $\mathcal{PT}$-symmetric optical lattices *Phys. Rev. Lett.* **110** 223902
[16] Hang C, Huang G and Konotop V V 2013 $\mathcal{PT}$ symmetry with a system of three-level atoms *Phys. Rev. Lett.* **110**, 083604
[17] Zheng C, Hao L and Long G L 2013 Observation of a fast evolution in a parity-time-symmetric system *Phil. Trans. R. Soc.* A **371** 20120053
[18] Rubinstein J, Sternberg P and Ma Q 2007 Bifurcation diagram and pattern formation of phase slip centers in superconducting wires driven by electric currents *Phys. Rev. Lett.* **99** 167003
[19] Chtchelkatchev N, Golubov A, Baturina T and Vinokur V 2012 Stimulation of the fluctuation superconductivity by $\mathcal{PT}$ symmetry *Phys. Rev. Lett.* **109** 150405
[20] Dietz B, Harney H L, Kirillov O N, Miski-Oglu M, Richter A and Schäfer F 2011 Exceptional points in a microwave billiard with time-reversal invariance violation *Phys. Rev. Lett.* **106** 150403
[21] Bittner S, Dietz B, Günther U, Harney H L, Miski-Oglu M, Richter A and Schäfer F 2012 $\mathcal{PT}$ symmetry and spontaneous symmetry breaking in microwave billiards *Phys. Rev. Lett.* **108** 024101
[22] Chong Y D, Ge L and Stone A D 2011 $\mathcal{PT}$-symmetry breaking and laser-absorber modes in optical scattering systems *Phys. Rev. Lett.* **106** 093902
[23] Liertzer M, Ge L, Cerjan A, Stone A D, Tureci H and Rotter S 2012 Pump induced exceptional points in lasers *Phys. Rev. Lett.* **108** 173901
[24] Zhao K F, Schaden M and Wu Z 2010 Enhanced magnetic resonance signal of spin-polarized Rb atoms near surfaces of coated cells *Phys. Rev.* A **81** 042903
[25] Schindler J, Li A, Zheng M C, Ellis F M and Kottos T 2011 Experimental study of active LRC circuits with $\mathcal{PT}$ symmetries *Phys. Rev.* A **84** 040101(R)
[26] Bender N, Factor S, Bodyfelt J D, Ramezani H, Christodoulides D N, Ellis F M and Kottos T 2013 Observation of asymmetric transport in structures with active nonlinearities *Phys. Rev. Lett.* **110** 234101
[27] Schomerus H 2010 Quantum noise and self-sustained radiation of $\mathcal{PT}$-symmetric systems *Phys. Rev. Lett.* **104** 233601
[28] West C T, Kottos T and Prosen T 2010 $\mathcal{PT}$-symmetric wave chaos *Phys. Rev. Lett.* **104** 054102
[29] Bender C M, Berntson B, Parker D and Samuel E 2013 Observation of $\mathcal{PT}$ phase transition in a simple mechanical system *Am. J. Phys.* **81** 173
[30] Fagotti M, Bonatti C, Logoteta D, Marconcini P and Macucci M 2011 Armchair graphene nanoribbons: $\mathcal{PT}$-symmetry breaking and exceptional points without dissipation *Phys. Rev.* B **83** 241406(R)
[31] Szameit A, Rechtsman M C, Bahat-Treidel O and Segev M 2012 $\mathcal{PT}$-symmetry in honeycomb photonic lattices *Phys. Rev.* A **84** 021806
[32] Bender C M, Gianfreda M, Peng B, Özdemir S K and Yang L 2013 Twofold transition in $\mathcal{PT}$-symmetric coupled oscillators *Phys. Rev.* A **88** 062111
[33] Peng B, Özdemir S K, Lei F, Monifi F, Gianfreda M, Long G L, Fan S, Nori F, Bender C M and Yang L 2014 Parity-time-symmetric whispering-gallery microcavities *Nature Physics* **10** 394
[34] Peng B, Özdemir S K, Rotter S, Yilmaz H, Liertzer M, Monifi F, Bender C M, Nori F and Yang L 2014 Loss-induced suppression and revival of lasing *Science* **346**, 328
[35] Cartarius H and Wunner G 2012 Model of a $\mathcal{PT}$-symmetric Bose-Einstein condensate in a δ-function double-well potential *Phys. Rev.* A **86** 013612
[36] Kreibich M, Main J, Cartarius H and Wunner G (2013) Hermitian four-well potential as a realization of a $\mathcal{PT}$-symmetric system *Phys. Rev.* A **87** 051601
[37] Kartashov Y V, Konotop V V and Abdullaev F Kh 2013 Gap solitons in a spin-orbit-coupled Bose-Einstein condensate *Phys. Rev. Lett.* **111** 060402
[38] Castaldi G, Savoia S, Galdi V, Alu A and Engheta N 2013 $\mathcal{PT}$ metamaterials via complex coordinate transformation optics *Phys. Rev. Lett.* **110** 173901
[39] Lazarides N and Tsironis G P 2013 Gain-driven discrete breathers in $\mathcal{PT}$-symmetric nonlinear metamaterials *Phys. Rev. Lett.* **110** 053901
[40] Yin X and Zhang X 2013 Unidirectional light propagation at exceptional points *Nature Materials* **12** 175
[41] Feng L, Xu Y-L, Fegadolli W S, Lu M-H, Oliveira J E B, Almeida V R, Chen Y-F and Scherer A 2013 Experimental demonstration of a unidirectional reflectionless parity-time metamaterial at optical frequencies *Nature Materials* **12** 108
[42] Dorey P E, Dunning C, and Tateo R (2001) Spectral equivalences, Bethe ansatz equations, and reality properties in PT-symmetric quantum mechanics *J. Phys. A: Math. Gen.* **34**, 5679
[43] Dorey P E, Dunning C, and Tateo R (2007) The ODE/IM Correspondence *J. Phys. A: Math. Theor.* **40**, R205
[44] Ahmed Z, Bender C M, Berry M V (2005) Reflectionless Potentials and $\mathcal{PT}$ Symmetry *J. Phys. A: Math. Gen.* **38**, L627-L630

[45] Bender C M, Brody D C, Chen J-H, Jones H F, Milton K A, and Ogilvie M C (2006) Equivalence of a Complex $\mathcal{PT}$-Symmetric Quartic Hamiltonian and a Hermitian Quartic Hamiltonian with an Anomaly *Phys. Rev. D* **74**, 025016
[46] Bender C M, Moshe M, and Sarkar S (2013) $\mathcal{PT}$-symmetric interpretation of double-scaling *J. Phys. A: Math. Theor.* **46**, 102002
[47] Bender C M and Sarkar S (2013) Double-scaling limit of the O(N)-symmetric anharmonic oscillator *J. Phys. A: Math. Theor.* **46** 442001 (2013)
[48] Lee T D (1954) *Phys. Rev.* **95** 1329
[49] Källén G and Pauli W (1955) *Mat.-Fys. Medd.* **30**, No. 7
[50] Bender C M, Brandt S F, Chen J-H, and Wang Q (2005) Ghost Busting: $\mathcal{PT}$-symmetric interpretation of the Lee model *Physical Review D* **71** 025014
[51] Bender C M, Hook D W, Mavromatos N E, and Sarkar S (2016) $\mathcal{PT}$-symmetric interpretation of unstable effective potentials *J. Phys. A: Math. Theor.* **49** 45LT01
[52] Bender C M and Gianfreda M (2015) $\mathcal{PT}$-symmetric interpretation of the electromagnetic self-force *J. Phys. A: Math. Theor.* **48** 34FT01
[53] Bender C M, Fring A, and Komijani J (2014) Nonlinear eigenvalue problems *J. Phys. A: Math. Theor.* **47** 235204
[54] Kerr O S (2014) Comment on "Nonlinear eigenvalue problems" *J. Phys. A: Math. Theor.* **47** 368001 (2014)
[55] Bender C M and Komijani J (2015) Painlevé Transcendents and $\mathcal{PT}$-Symmetric Hamiltonians *J. Phys. A: Math. Theor.* **48** 475202
[56] Bender C M, Komijani J, and Wang Q (2017) Nonlinear eigenvalue problems *CRM Conference Series* (in press)
[57] Ince P L (1956) *Ordinary Differential Equations* (Dover, New York); Miles J W (1978) On the second Painlevé transcendent *Proc. Roy. Soc. London A* **361** 277-291; Holmes P and Spence D (1984) On a Painlevé-type boundary-value problem *Quart. J. Mech. Appl. Math.* **37** 525-538; Hastings S P and McLeod J B (2011) Classical methods in ordinary differential equations: With applications to boundary value problems *Graduate Studies in Math.* **129** (American Mathematical Society)
[58] A detailed study of the asymptotic behavior of the Painlevé transcendents may be found in Jimbo M and Miwa T (1981) *Physica D* **2** 407-448
[59] Separatrix behavior of the first Painlevé transcendent is mentioned briefly in Kapaev A A (1989) *Differential Equations* **24** 1107-1115; see also Kapaev A A (2002) Monodromy deformation approach to the scaling limit of the Painlevé first equation *CRM Proc. Lect. Notes* **32** 157-179
[60] Clarkson P A (2003) Painlevé Equations — Nonlinear special functions *J. Comp. Appl. Math.* **153** 127-140
[61] Maseoro P (2010) Essays on the Painlevé first equation and the cubic oscillator (PhD Thesis, SISSA)
[62] Kawai T and Takei Y (2005) Algebraic analysis of singular perturbation theory (American Mathematical Society, New York)
[63] Costin O, Costin R D, and Huang M (2013) Tronquée solutions of the Painlevé equation P_1 (unpublished)
[64] Fokas A S, Its A R, Kapaev A A, and Novokshonov V Y (2006) Painlevé transcendents: The Riemann-Hilbert approach (American Mathematical Society, New York)
[65] The spin-spin correlation function for the two-dimensional Ising model for temperatures near T_c is described by PIII. See Wu T T, McCoy B M, Tracy C A, and Barouch E (1976) Spin-spin correlation functions for the two-dimensional Ising model: exact theory in the scaling region *Phys. Rev. B* **13** 316-374
[66] For all temperatures the diagonal correlation function for the Ising model in two dimensions $\langle\sigma_{0,0}\sigma_{N,N}\rangle$ is given in terms of PVI. See Jimbo M and Miwa T (1980) Studies on holonomic quantum fields XVII *Proc. Jap. Acad.* **56A** 405 and (1981) **57A** 347
[67] Brézin E and Kazakov V A (1990) Exactly solvable field theories of closed strings *Phys. Lett. B* **236** 144-150
[68] Douglas M and Shenker S (1990) Strings in less than one dimension *Nucl. Phys. B* **335** 635-654
[69] Gross P and Migdal A (1990) A nonperturbative treatment of two-dimensional quantum gravity *Nucl. Phys. B* **340** 333-365
[70] Moore G (1990) Geometry of the string equations *Comm. Math. Phys.* **133** 261-304; Matrix models of 2D gravity and isomonodromic deformations *Prog. Theor. Phys. Suppl.* **102** 255-285
[71] Fokas A S, Its A R, and Kitaev A V (1992) The isomonodromy approach to matrix problems in 2D quantum gravity *Comm. Math. Phys.* **147** 395-430
[72] Bender C M and Orszag S A (1978) *Advanced Mathematical Methods for Scientists and Engineers* (McGraw Hill, New York), chap. 4
[73] See Reinhardt W P (1982)*Ann. Rev. Phys. Chem.* **33** 223 for a review of complex rotation of coordinates.
[74] Long W-G, Li Y-T, Liu S-Y, and Zhao Y-Q (2016) Real solutions of the first Painlevé equation with large initial data (arXiv:1612.01350 [math.CA])

IOP Conf. Series: Journal of Physics: Conf. Series **873** (2017) 012003 doi:10.1088/1742-6596/873/1/012003

Review of Standard Model Higgs results at the ATLAS experiment

Alessandro Calandri on behalf of the ATLAS Collaboration

Université Aix-Marseille - Centre de Physique des Particules de Marseille (CPPM)

E-mail: `alessandro.calandri@cern.ch`

Abstract. The investigation of the mechanism of electroweak (EW) symmetry breaking has been one of the main goals of the ATLAS experiment [1] research program at the CERN Large Hadron Collider (LHC). In the Standard Model (SM) of particle physics, the breaking of the EW symmetry is realised by introducing a complex doublet scalar field which is related to the existence of a neutral particle, the Higgs boson. The Higgs scalar field is responsible of the mass of the other particles; the mass of its mediator, the Higgs boson, is the only free parameter in the theory. In 2012, the ATLAS and CMS collaborations discovered a new particle consistent with the SM Higgs boson and two years later measured its mass, in the H→ $\gamma\gamma$ and H→ZZ* →4l channels, to be m_H=125.09±0.21 (stat) ±0.11 (sys) GeV with data collected in 2011 and 2012 (LHC Run 1). This document will cover the SM Higgs analyses performed with approximately 15 fb^{-1} of *pp* collision data collected until summer 2016 during the so-called Run 2 LHC data-taking at a center-of-mass energy ($\sqrt{s}$=13 TeV).

The documents will be structured as follows. Section 1 will describe the measurement of the Higgs boson properties in the bosonic decay channels (H→ $\gamma\gamma$ and H→ZZ* →4 leptons), while Section 2 will focus on the searches in the fermionic final states (H→$b\bar{b}$, $t\bar{t}$H and $H \to \mu\mu$). Finally, conclusions will be discussed in Section 3. There is also a vast number of di-Higgs searches, reported for example in References [9] and [10]. However, the signal sensitivity is still small and they will not be detailed in this document. The dominant Higgs boson production modes are gluon-gluon fusion (ggF), vector boson fusion (VBF), production in association with a vector boson V (VH, V=W or Z) and in association with top quarks ($t\bar{t}$H). The most sensitive Higgs boson decay channels at the LHC are the modes $\gamma\gamma$, ZZ* →4l, WW*, $\tau\tau$ and $b\bar{b}$. In addition, the production channel with the association of top quarks is of particulat interest in order to study the top-Yukawa coupling.

1. Higgs decay to bosons

1.1. $H \to \gamma\gamma$

Measurements of the Higgs boson properties have been performed in the H→ $\gamma\gamma$ channel using 13.3 fb^{-1} of proton-proton collisions data at $\sqrt{s}$=13 TeV [2]. The Higgs boson decay into photons, through a loop of heavy particles, is an important channel in order to study the properties of the Higgs boson. Although this is featured by a small branching ratio, a very large yield can be obtained with a narrow peak on top of a smoothly falling background.

1.1.1. Object definition and analysis selection Photon reconstruction is based on energy clusters in the electromagnetic calorimeter with transverse energy larger than 2.5 GeV in

$\Delta\eta \times \Delta\phi$=0.075×0.125. The reconstruction is optimized to separate electrons, unconverted photons and converted photons. The photon energy is then corrected using calibration factors that are extracted both from simulation and data. The photon identification relies on the information of the lateral and longitudinal shape of the electromagnetic shower in the calorimeter. Photon candidates are required to exhibit a lateral shower shape consistent with that from a single electromagnetic shower and to deposit only a small fraction of their energy in the hadronic calorimeter. The selection criteria applied to such photons require them to be isolated with transverse momentum greater than $0.35{\cdot}m_{\gamma\gamma}$ and $0.25{\cdot}m_{\gamma\gamma}$ and each with pseudorapidity smaller than 2.37 excluding the region 1.37<—η—<1.52. The diphoton candidate pair is defined by seeding the two photon candidates with the highest transverse energy. Starting from this pair, a neural-network algorithm based on track and primary vertex as well as on the pointing information from the electromagnetic calorimeter identifies the diphoton primary vertex among all reconstructed vertices.

1.1.2. Signal and background modelling The Higgs boson signal is extracted by employing a maximum-likelihood fit to the diphoton invariant mass spectrum in the range $105< m_{\gamma\gamma} <160$ GeV. The signal contribution to the fit is modelled as a double-sided Crystal Ball function centered on the previously measured Higgs boson mass of m_H=125.09 GeV. Due to the narrow width of the Higgs boson, the shape of this distribution is entirely driven by the energy and angular resolutions of the photons. The parameters of the model are extracted using a fit to a simulated H→ $\gamma\gamma$ sample. The background is dominated by irreducible non-resonant $\gamma\gamma$ production and reducible γ+jets and di-jet production.

1.1.3. Measurements of the fiducial, differential production cross sections and signal strengths The Higgs boson fiducial cross section has been measured for different phase spaces which are sensitive to the inclusive Higgs boson production, to the vector boson fusion (VBF) production and to the Higgs boson production in association with a vector boson (VH). An unbinned maximum likelihood fit is performed on the di-photon invariant mass spectrum in each of the fiducial regions defined above. The extracted cross section is also corrected to account for detector effects. The inclusive fiducial cross section is σ_{fid}=43.2 ±14.9 (stat)±4.9 (syst) fb for a Higgs boson of mass 125.09 GeV decaying to two photons. The Standard Model predicts σ_{SM}=62.8 ±3.4 (stat)±4.4 (syst) fb when employing the same fiducial region.

The differential cross sections are measured as a function of the diphoton transverse momentum, the rapidity of the diphoton system, the cosine of the angle between the beam axis and the photons in the Collins-Soper frame of the Higgs boson, the jet multiplicity, the transverse momentum of the leading jet, the azimuthal angle between the leading two jets and the invariant mass of the leading two jets. The differential cross section as a function of the jet multiplicity exhibits an overall good agreement between data and the Standard Model predictions constructed from the highest perturbative QCD order available (NNLO + Parton Shower).

The event selected in the diphoton analysis are divided into exclusive categories that are optimised to enhance the separation of the various Higgs boson production processes. Each category is enriched with events of a given production mode. Two $\mathrm{t\bar{t}H}$ categories are defined for events in which at least one top quark decays leptonically or in which both top quarks decay hadronically. Five categories are enriched in Higgs boson production in association with a vector boson: the VH dilepton category targets ZH production with Z decaying in leptons, the VH one-lepton production targets WH production with W→ $l\nu$, the VH missing transverse momentum category targets ZH production with Z→ $\nu\nu$ and W→ $l\nu$. Finally, one hadronic category targets VH production with a hadronically decaying vector boson. Two categories are created to enhance

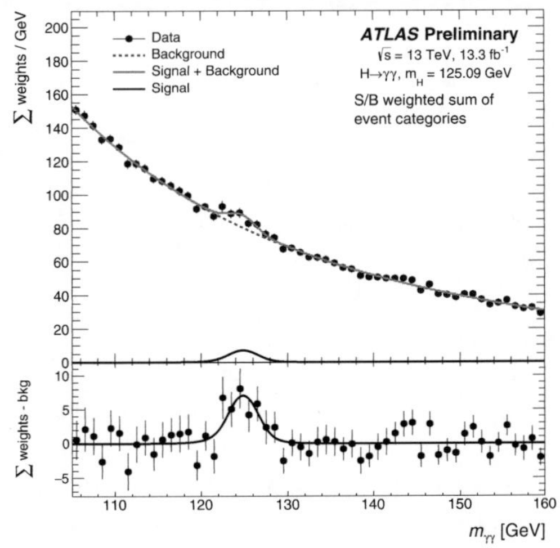

Figure 1. Invariant mass spectrum combining all production mode categories. The black data points show the measured distribution, where each event is weighted by the signal-to-background ratio of the event category it belongs to. The blue dashed curve shows the result of a background-only fit to the data, the red curve shows the signal+background distribution based on the fitted signal yields, and the black curve shows the signal component [2].

the contribution of Higgs boson production through vector boson fusion. The remaining events populate the untagged category and they are split into four additional sub-categories which contain events produced through gluon fusion. Figure 1 reports the invariant mass distributions for all categories combined. Each of them is weighted by the signal-to-background ratio of the event category to which it belongs. The statistical procedure to extract the signal yield and the cross section in each category relies on a maximum likelihood fit to the $m_{\gamma\gamma}$ spectrum. The signal strength μ, defined as the ratio of the measured cross section for a given category and the one predicted by the Standard Model for a given Higgs boson mass are shown in Figure 2 in addition to the global signal strength measured in Run 1. No significant deviations from the Standard Model expectations are observed.

1.2. $H \to ZZ^* \to 4l$

The Higgs boson decay to four leptons, $H \to ZZ^* \to 4l$ where l=e, μ, provides a good sensitivity for the measurement of its properties due to the very clean final state with electrons and muons. The largest background of the analysis throughout the full four-lepton invariant mass (m_{4l}) spectrum is represented by the continuum ZZ production. For m_{4l} >160 GeV there are also contributions from Z+jets and $t\bar{t}$ with two prompt leptons while the additional charged lepton candidates faking the signal arise from the decay of $b-$ or $c-$ quarks or from mis-identification of jets. The inclusive total and fiducial cross section measurements have been extracted with 11.6 fb^{-1} collected in the 2015 and 2016 data taking periods [3].

1.2.1. Object definition and analysis selection Electrons are reconstructed using information from the inner detector (ID) and the electromagnetic calorimeter. Electron candidates are built from clusters of energy associated with ID tracks. Background discrimination is based on shower shape information available from the highly segmented calorimeter, high-threshold transition radiation tracker (TRT) hits as well as compatibility of the tracking and calorimeter

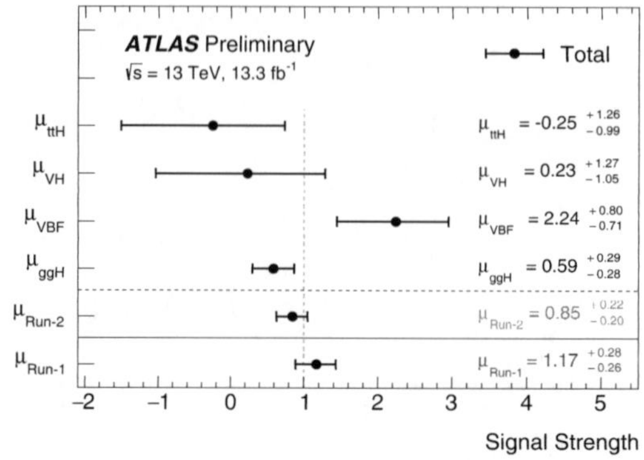

Figure 2. Production signal strength measured for the different Higgs boson production processes (gluon fusion, VBF, VH, ttH) and globally (Run 2) compared to the signal strength measured at 7 and 8 TeV (Run 1). The error bar shown is the total uncertainty [2].

information. This information is combined into a likelihood discriminant. Muons are reconstructed as tracks in the ID and the muon spectrometer (MS); their identification is based on the presence of matching tracks in the ID and in the MS. If the ID and the MS have track information, a combined muon track is formed by a global refit using the hit information from both the ID and MS detectors (combined muons) otherwise the muon momentum is measured either using the MS information only (standalone muons) in the very forward pseudorapidity region (2.5< $|\eta|$ <2.7) or with the partial MS track and the ID information (segment-tagged muons). Such segment-tagged muons are limited to the central barrel region ($|\eta|$ <0.1) which lacks MS geometrical coverage. In addition, in this central region a track measured in the ID is identified as a muon if its calorimetric energy deposit is compatible with a minimum ionising particle (calorimeter-tagged muons). Final State Radiation (FSR) photons from the QED process of the radiative photons from Z boson decays can be identified and incorporated into the four-lepton measurement.

The four-lepton events are selected and classified according to their final state: 4e, 2e2μ, 2μ2e, 4μ. Higgs boson candidates are creating by selecting two same-flavour opposite-sign lepton pairs in an event. Selection criteria are applied to the leptons in the event. A Z-mass-constrain kinematic fit is applied to the final state after the FSR correction to improve the m_{4l} resolution. All the events satisfying the above criteria are regarded as candidate signal events.

1.2.2. Fiducial cross section measurement Studies on the fiducial cross section measurements are based on a slightly different selection. For instance, the selected events in the mass region 115 < m_{4l} <130 GeV are used to extract the fiducial and total cross sections and the m_{4l} distribution is employed as a discriminant to increase the sensitivity of the signal. The detector correction factors are calculated from Monte Carlo simulations and are given by the ratio of the number of selected reconstructed events relative to the number of particle-level events for a given fiducial region. Probability density functions are produced separately for signal and background components and fiducial acceptance factors correspond to the theory-dependent fraction of H→ZZ* →4l decays that pass the fiducial selection. The fiducial cross sections can be extracted with a likelihood fit to the observed m_{4l} distribution in the signal mass window under different model assumptions.

In addition to the fiducial cross section measurement, in order to have larger sensitivity to different Higgs production modes, the events selected in the mass region 118 $< m_{4l} <$129 GeV are split into exclusive categories. Such categories are based on the presence of additional leptons and the number of jets used to discriminate among the different production modes. Events are first categorised in a VH-leptonic category by requiring at least one additional isolated lepton with a transverse momentum greater than 8 GeV. Events not passing this requirement are categorised in exclusive categories according to the number of jets, 0, 1, 2 or more jets. In the 2-jet category, the events are further split into low (m_{jj} <120 GeV) and high (m_{jj} >120 GeV) dijet invariant mass. This division enables to further optimise the separation between the VH-hadronic (where the vector boson decays to a quark pair) and the VBF production Higgs boson productions. Dedicated Boosted Decision Tree (BDT) algorithms are used for the 0-jet and the 1-jet categories. For the 0-jet category, the BDT is trained on kinematic variables based on the 4-lepton final state to separate the gluon fusion process from the SM background. For the 1-jet category, the BDT is trained to disentangle gluon fusion production from VBF. For the 2-jet category, two BDTs are trained to separate the gluon fusion production from the VH hadronic and from the VBF mode and are used in the low and high mass categories.

1.2.3. Background modelling The main background component in the H$\rightarrow$ZZ* $\rightarrow$4l final state is the non-resonant ZZ production. Additional background sources are the Z+jets and $t\bar{t}$ processes which are estimated in the signal region using different data-driven techniques according to the flavour of the sub-leading lepton pairs ($\mu\mu$ or ee). The ll+$\mu\mu$ component is estimated with the shape of the leading dilepton invariant mass in three control regions, defined by inverted selection requirements of the analysis, enriched in a different background component ($t\bar{t}$, Z+heavy flavour, Z+light flavour). The yield in the control region is extrapolated in the signal region using transfer factors. The ll+ee background is mostly due to the mis-identification of light-flavour jets as electrons, photon conversions, and the semileptonic decays of heavy flavour hadrons. Similarly to the ll+$\mu\mu$ component, ll+ee background yields are extracted in control regions and extrapolated in the signal region. A kinematic discriminant based on a BDT algorithm trained on input variables of the ZZ final state to disentangle the Higgs boson signal from the SM irreducible background is also employed. Figure 3 shows the mass spectrum for the leading dilepton pair, m_{12}, the sub-leading pair, m_{34}, the 4-lepton final state and the distributions of m_{12} versus m_{34} for data and signal and background Monte Carlo.

1.2.4. Results The measured total fiducial cross-section is σ^{4l}_{fid}=4.48$^{+1.01}_{-0.89}$ fb, in agreement with the SM expectations. Results on cross sections for different final states are also provided showing no tension with the SM predictions. The total cross section is obtained by extrapolating the fiducial cross section to the full phase-space and by taking into account the SM branching ratio $\mathcal{B}(H \rightarrow$4l), σ_{tot}=81$^{+18}_{-0.16}$ fb. The total measured cross section and the SM expectations are compatible within 1.6 standard deviations. The expected and observed yields in the various Higgs boson production categories (0-jet, 1-jet, 2-jet with m_{jj} > 120 GeV, 2-jet with m_{jj} < 120 GeV) are also extracted. The compatibility between the measured cross section for each production category and the SM predictions is at the level of 1.1 standard deviation. Figure 4 shows the observed and expected negative log-likelihood ratio scans for $\sigma_{ggF+ttH} \times \mathcal{B}(H \rightarrow ZZ^*)$, $\sigma_{VBF} \times \mathcal{B}(H \rightarrow ZZ^*)$ and $\sigma_{VH} \times \mathcal{B}(H \rightarrow ZZ^*)$ as well as the 2-dimensional compatibility checks among all the various production modes at 68% and 95% confidence level.

IOP Conf. Series: Journal of Physics: Conf. Series **873** (2017) 012003 doi:10.1088/1742-6596/873/1/012003

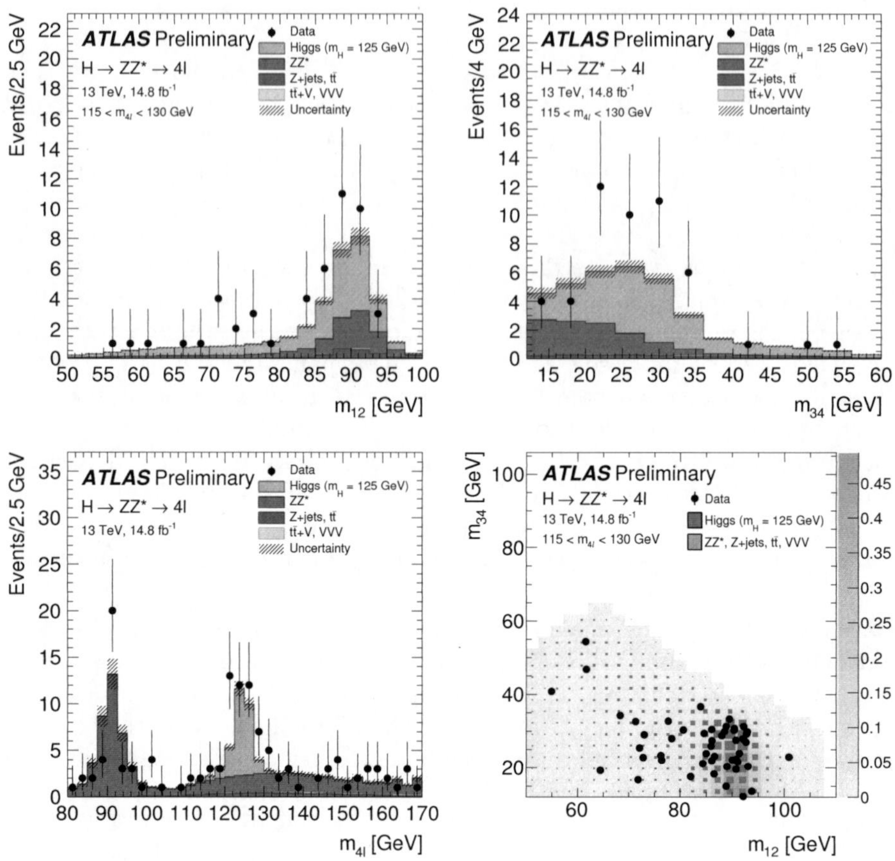

Figure 3. m_{12}, m_{34}, m_{4l}, m_{Z_1} *vs* m_{Z_2} distributions of the selected candidates compared to the background expectations in the low mass region. The signal contribution is shown for m_H=125 GeV as the blue histogram while the expected background contributions ZZ, Z+jets, $t\bar{t}$+V are also included.

2. Higgs decay to fermions

2.1. VH(H→$b\bar{b}$)

The process with the largest branching ratio (58%) for a SM Higgs boson of mass 125 GeV is H→$b\bar{b}$. The overwhelming background produced from multi-jet at the LHC makes a fully inclusive search unfeasible. That is the reason why the production modes under study for this final state are the ones with W and Z bosons (VH) where the presence of leptons in the final state provides a clean signature and enables an efficient triggering of the event. Results have been presented with an integrated luminosity of 13.2 fb^{-1} [4] from *pp* collision data collected in 2015 and 2016.

2.1.1. Object reconstruction, event categorisation and selection Events are selected where two high transverse momentum *b*-tagged jets are reconstructed together with 0, 1 or 2 charged leptons (electrons or muons). A set of kinematic observables is used as input to a multivariate algorithm for each analysis category. This discriminant is then used for the extraction of the signal yield and the signal strength in the fit. Electrons and muons are identified and reconstructed as described in Section 1.2.1. Jets are reconstructed from energy clusters in the hadronic calorimeter

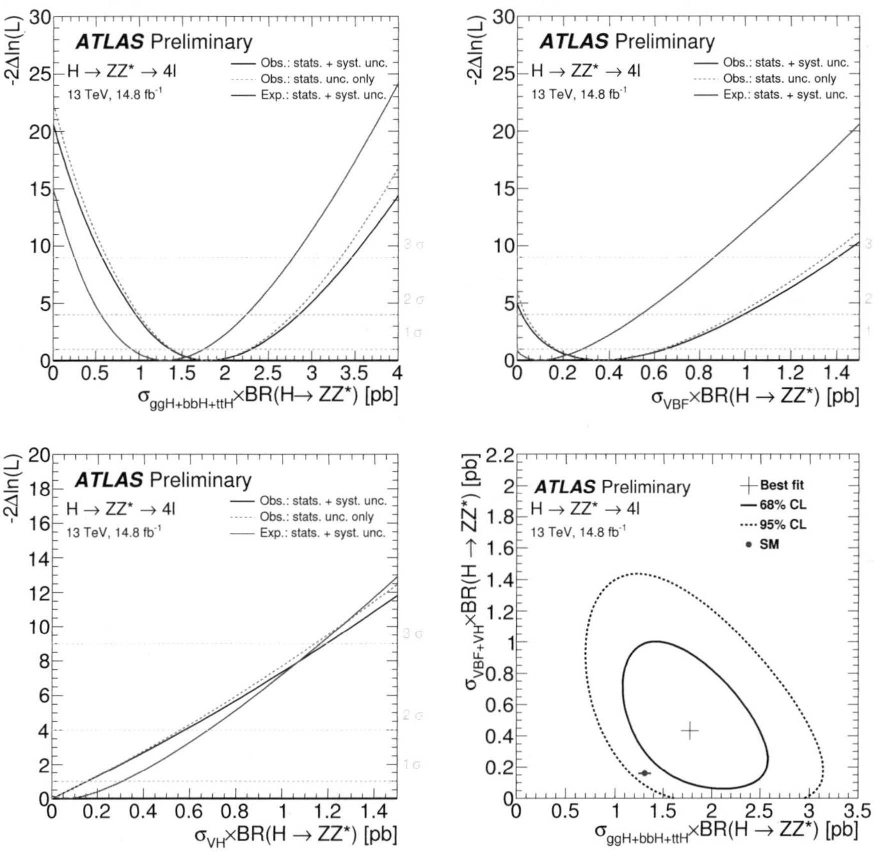

Figure 4. Observed and expected negative log-likelihood scans for $\sigma_{ggF+ttH} \times \mathcal{B}(H \to ZZ^*)$, $\sigma_{VBF} \times \mathcal{B}(H \to ZZ^*)$ and $\sigma_{VH} \times \mathcal{B}(H \to ZZ^*)$ with (black line) and without systematic uncertainties (red line). The expected SM negative log-likelihood scans are also shown. The green dashed lines indicate the value of the profile likelihood ratio corresponding to 1, 2 and 3 standard deviations for the parameter of interest. The plot on the bottom right side displays the negative log-likelihood contours at 68% and 95% confidence level in the $\sigma_{ggF+ttH} \times \mathcal{B}(H \to ZZ^*)$-$\sigma_{VBF+VH} \times \mathcal{B}(H \to ZZ^*)$. The SM prediction is represented by a dot in the 2-dimensional plane [3].

with the anti-k_t algorithm with a radius parameter of 0.4. The raw energy of the reconstructed jets are calibrated using a jet energy scale correction extracted from a combination of data and Monte Carlo templates. The b-tagging algorithm is based on information of the jet kinematics, the properties of the tracks within a jet and the presence of displaced vertices in the event. Lepton-jet overlap removals are also applied to avoid double-counting of reconstructed particles in the final state.

Events passing detector quality requirements are split into categories according to their lepton, jet multiplicity and vector boson transverse momentum. In the 0-lepton channel, events are selected using the missing transverse energy trigger and additional selection requirements are applied on the angular separation between jets to further suppress multi-jet background. On the 1-lepton channel, events are selected with single electron trigger and are required to contain at

IOP Conf. Series: Journal of Physics: Conf. Series **873** (2017) 012003 doi:10.1088/1742-6596/873/1/012003

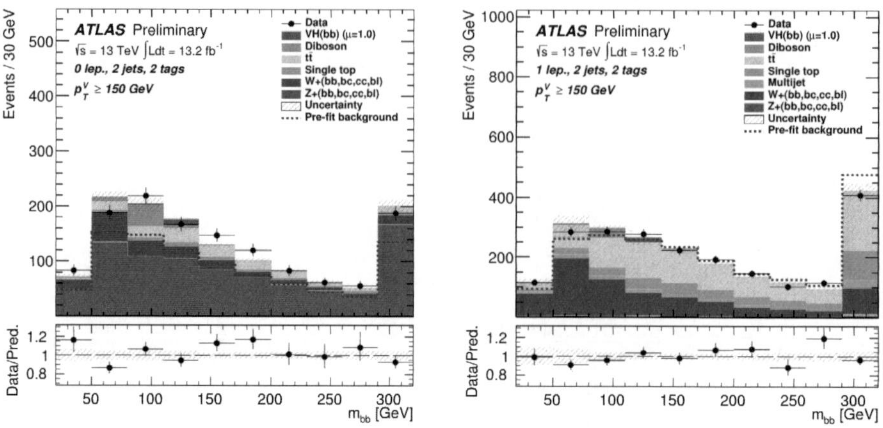

Figure 5. Invariant mass of the b-tagged jet pairs (m_{bb}) for the 0- and 1-lepton categories. The Higgs boson signal generated at m_H=125 GeV is shown as a filled histogram on top of the background components. The dashed histogram shows the total background as expected from the pre-fit Monte Carlo simulation. The size of the statistical and systematic uncertainty is indicated by hatched band around each bin. The ratio of the data to the sum of the signal and fitted background is also displayed [4].

least one lepton that is identified with tight criteria. In the 2-lepton selection, the same trigger requirement is applied as the one used in the 1-lepton category. Exactly two leptons identified with the loose criteria are required in the final state. Given that such leptons originate from the decay of the Z boson, the invariant mass of the dilepton system must be between 71 and 112 GeV, around the Z peak at 91 GeV.

The analysis strategy is based on a multivariate discriminant (BDT) trained in these three lepton categories for the 2- and 3-jet categories. For the 2-lepton channel, two BDTs are used depending on the transverse momentum of the vector boson. The discriminant is trained to separate the VH→b$\bar{b}$ signal against the sum of all background process. An additional BDT is also employed to separate the SM VZ→b$\bar{b}$ signal production against the sum of all background processes in the event. The input variables of the BDT have been chosen to improve the signal-to-background separation; the most powerful being the mass of the di-jet system, m_{bb} as shown in Figure 5.

The main background is the multi-jet production arising from QCD strong interactions at the LHC. This component contributes with very large cross section and is therefore the dominant challenge of the analysis. Background components are studied in the 0-, 1- and 2-lepton channels and specific strategies are carried out in the various categories. For instance, in the 1-lepton case, the background yields are extracted from control regions constructed separately for each of the 2- and 3-jet categories. Correction factors are then applied to extrapolate the yields and the shapes from the control region to the relevant signal region. In the 2-lepton region, requiring two isolated leptons with a dilepton invariant mass compatible with that of a Z boson suppresses the contribution from multi-jet events. The residual contribution is found to be negligible (0.3%) and does not have impact on the signal extraction in the fit.

2.1.2. Results The results are extracted using a profile likelihood fit to measure the signal strength in the 8 BDT regions. For all channels combined, the expected value of the signal

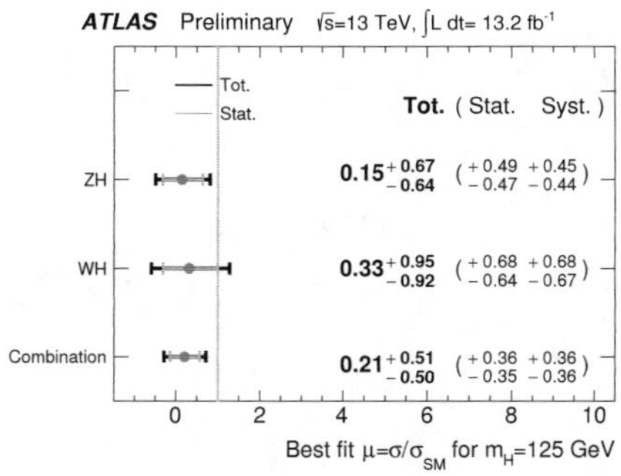

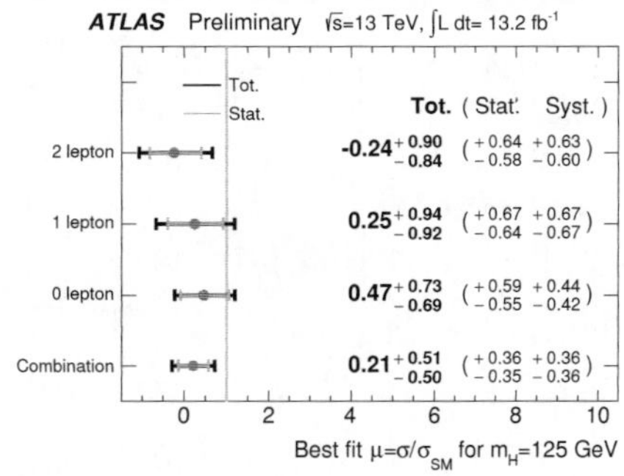

Figure 6. Fitted values of the Higgs bosons signal strengths for a Higgs boson mass m_H=125 GeV for the production and decay categories, 0-, 1- and 2-lepton channels and for their combinations [4].

strength is $\mu = 0.21 \pm 0.36$ (stat) ± 0.36 (sys). Figure 6 shows the signal strength μ (m_H=125 GeV) for the various Higgs production modes in the VH final state (ZH, WH and the combination), and for the 0-, 1- and 2-lepton channels as well as the combination.

2.2. $t\bar{t}H(H\to b\bar{b})$

The Higgs boson production associated with a pair of top quarks, $t\bar{t}$H, is very important for observing the top-Yukawa coupling of the Higgs boson to top quarks. Even though the $t\bar{t}$H production is largely sub-dominant compared to the gluon fusion at the LHC, the presence of top quarks in the final state offers a very distinctive signature. One of the Higgs boson decay channels contributing with the largest branching ratio (58% for m_H=125 GeV) is that with two b-quarks in the final state. The analysis described in this Section is performed with 13.3 fb^{-1} collected by the ATLAS experiment at $\sqrt{s}$=13 TeV [5] and focuses on the final state where the W bosons from the top quarks decay leptonically to electrons or muons. This enables to trigger the event and strongly suppresses the overwhelming multi-jet background arising in the case of hadronic W decays.

2.2.1. Event selection and analysis strategy Leptons and jets are identified and reconstructed as reported in Sections 1.2.1 and 2.1.1. Specific triggers are applied in the single lepton and dilepton channels. The events are classified in exclusive regions of jet and b-jet multiplicity in order to disentangle the signal component that is characterised by a large jet multiplicity with respect to the $t\bar{t}$ background. In the dilepton channel, the regions are five; the regions with (3 jets, 3 b-jets), (4 jets, 3 b-jets), (4 jets, 4 b-jets) are regarded as signal regions. In the single lepton case, the categorisation is exploited in nine categories, being the signal regions the ones with (5 jets, 4 b-jets), (6 jets, 3 b-jets) and (6 jets, 4 b-jets). The dominant background is $t\bar{t}$ with additional heavy flavours jets ($t\bar{t}b\bar{b}$, $t\bar{t}c\bar{c}$) as shown in Figure 7.

In the signal regions two boosted decision trees (BDT) are employed. The first BDT (*reconstruction BDT*) is trained to match the reconstructed jets to the partons from the top and Higgs decays. In each signal region, the information from the output of the reconstruction BDT is used with other kinematic variables into another multivariate algorithm (*classification BDT*) which separates events as being more signal- or background-like. The shapes of the classification BDT in the signal regions are used as inputs to the final fit to data. Similarly in the dilepton (3j, 3b) region, a neural network (NN) is applied to disentangle signal from background. In the control regions of the single lepton channel, the scalar sum of the transverse momentum of all

IOP Conf. Series: Journal of Physics: Conf. Series **873** (2017) 012003 doi:10.1088/1742-6596/873/1/012003

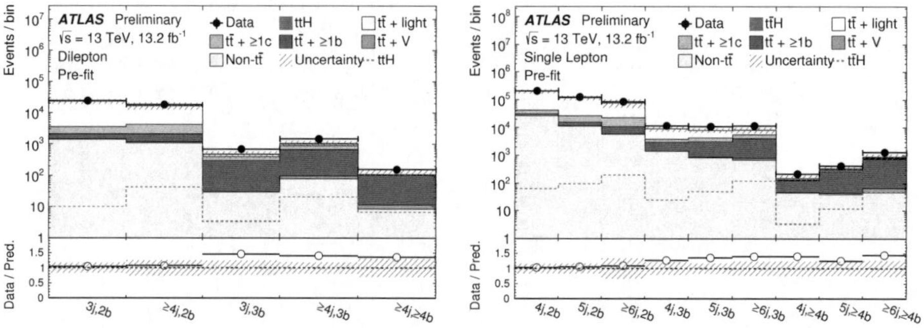

Figure 7. Expected and observed event yields in the dilepton channel and the single-lepton channel. The $t\bar{t}$ background is divided into four components, $t\bar{t}$+b-jets, $t\bar{t}$+c-jets, $t\bar{t}$+light jets, $t\bar{t}$+V. The non $t\bar{t}$ background includes contribution from Wt, W/Z+jets as discussed in Section 2.2.2. The signal contribution is shown as a dashed red line while the hatched area is the total uncertainty on the Monte Carlo prediction [5].

jets ($H_T^{had}{=}\sum_i p_T^i$) is used as discriminating variable while in the dilepton case, such scalar sum is also extended to leptons.

2.2.2. Signal and background modelling The $t\bar{t}$H production is treated with NLO simulation for the matrix element calculation interfaced to the Pythia 8 program for what concerns parton shower and hadronisation. The dominant $t\bar{t}$ background is categorised according to the flavour of the additional jets in the event. To model such component with the best possible precision and to ensure consistent data-Monte Carlo comparisons, detailed studies on the parton shower schemes applied and sequencial reweightings to higher-order Monte Carlo generations are performed. Other backgrounds are represented by Wt, W/Z + jets and are sub-dominant. In the single-lepton channel, an additional source of background is due to events with a misidentified or non-prompt leptons. This component is estimated in data.

2.2.3. Results The main source of systematic uncertainty affecting the measurement of the parameter of interest μ is the modelling of the $t\bar{t}$ background. An uncertainty of 6% is assumed for the inclusive $t\bar{t}$ cross section computed at NNLO+NNLL including the effects from the variation of the factorisation, resummation scales and the parton density functions. An additional uncertainty on the choice of the Monte Carlo generator, the showering and the hadronisation models is also applied. The statistical analysis is based on a likelihood function formed as a product of Poisson probability for each bin of the input distributions in the various signal and control regions. The parameter of interest of the fit is the Higgs boson signal strength, μ, being θ the set of nuisance parameters carrying the effects of the systematic uncertainties. These nuisance parameters are treated with Gaussian or log-normal priors and are correlated across all the regions and channels. The $t\bar{t}b\bar{b}$ normalisation is left free floating and its value is extracted by the fit. As shown in Figure 8, the distribution of the BDT discriminant in the signal region (6j, 4b) after the fit exhibits a good data-Monte Carlo agreement given that the $t\bar{t}b\bar{b}$ and $t\bar{t}c\bar{c}$ components are scaled by approximately 1.30 times the pre-fit value. As reported in Figure 8, the observed signal strength is $4.6^{+2.9}_{-2.3}$ in the dilepton data, $1.6^{+1.1}_{-1.1}$ in the single-lepton final state. The combined value is $2.1^{+1.0}_{-0.9}$. The error on the measurement is mostly dominated by the systematic uncertainty on the background modelling.

IOP Conf. Series: Journal of Physics: Conf. Series **873** (2017) 012003 doi:10.1088/1742-6596/873/1/012003

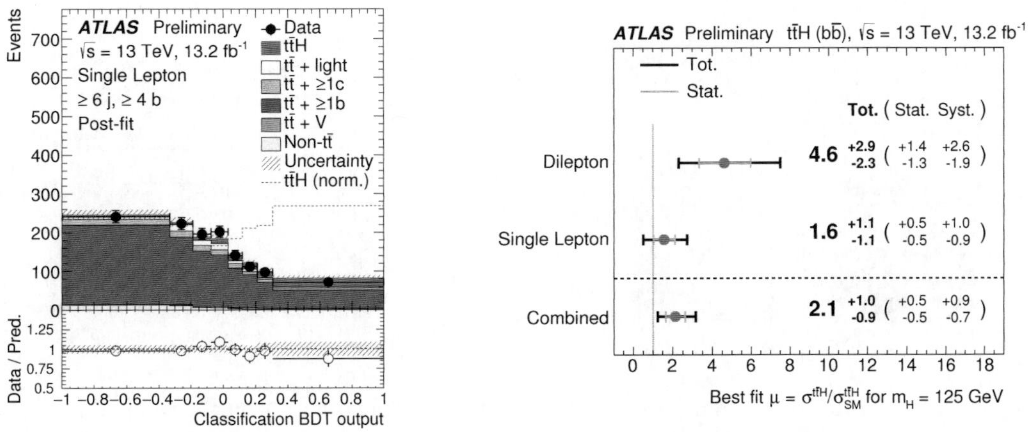

Figure 8. Post-fit comparison between data and prediction for the BDT discriminant in the single-lepton channel signal region (6j, 4b). The signal $t\bar{t}$H signal is normalized to the fitted signal strength μ extracted by the fit. The dashed line show the $t\bar{t}$H distribution normalized to the total yield. The hatched line correspond to the total uncertainty on the normalisation of the Monte Carlo predictions. Signal strength measurements on the right-side distribution in the individual channels and for the combination [5].

2.3. $t\bar{t}H$ in multilepton final states

This Section focuses on the search of the $t\bar{t}$H Higgs production using 13.2 fb^{-1} of data collected with the ATLAS detector at $\sqrt{s}$=13 TeV during 2015 and 2016 [6]. The search uses four final states distinguished by the number and flavour of leptons: two same-charge light leptons (electrons or muons) and no hadronically decaying τ leptons, two same-charge light leptons and one hadronically decay τ lepton, three light leptons and four light leptons. Based on the number of lepton candidates after the preselection requirements, events are placed into at most one of these categories. Additional requirements beyond the preselection classification are applied depending on the final state. Such categorisation is mostly sensitive to H$\rightarrow$WW* (mainly in the 0-τ channel) and H$\rightarrow \tau\tau$ (in the 1-τ channel).

Backgrounds are categorised into those in which all the selected leptons are produced in decays of electroweak bosons or τ leptons (prompt leptons) and those in which at least one lepton arises from another source. In such cases, the leptons arise from hadronic decay channels or photon conversions (non-prompt leptons), interactions in detector material (charge mis-identification) or wrong reconstruction of other particles (fake leptons). Signal-free control regions are defined with inverted selection requirements for each channel of the analysis. For each control region, the background yield is extracted and extrapolated in the signal region with transfer factors.

The expected backgrounds, the $t\bar{t}$H signal and the observed data yields in each category of the analysis are shown in Figure9. The best fit value of the signal strength is extracted with a maximum likelihood approach on the six categories which are treated with Poissonian terms in the likelihood parametrisation. Systematic uncertainties are implemented as nuisance parameters with priors which the fit is able to constrain. The best fit value of the $t\bar{t}$H signal strength combining all channels is μ=2.5$\pm$0.7 (stat) $\pm$ 1.0 (sys) as shown in Figure 9. The observed p-value associated with the background-only hypothesis is 2.2 standard deviations for m_H=125 GeV.

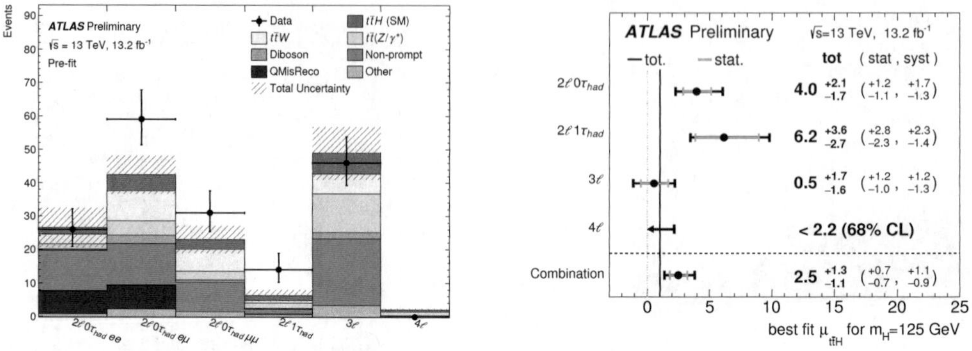

Figure 9. Background and signal predictions (pre-fit) and observed data for each signal region. The signal prediction is calculated for SM expectations ($\mu_{\mathrm{t\bar{t}H}}$=1). Best fit value on the right-side distribution of the t$\bar{\mathrm{t}}$H signal strength, $\mu_{\mathrm{t\bar{t}H}}$, split in analysis categories and for the combination. Since no events are observed in the 4 lepton case, upper limits on $\mu_{\mathrm{t\bar{t}H}}$ are shown [6].

2.4. Combination of the $t\bar{t}H$ analyses

The following decay modes are considered for the t$\bar{\mathrm{t}}$H combination [7]:

- H→ $\gamma\gamma$ in both the hadronic and leptonic t$\bar{\mathrm{t}}$ decay channels (Section 1.1);
- H→ WW*, $\tau\tau$, ZZ (Section 2.3);
- H→b$\bar{\mathrm{b}}$ in the single lepton and opposite sign dilepton t$\bar{\mathrm{t}}$ decay channels (Section 2.2).

The orthogonality among the various analyses is ensured by using a common lepton definition for the categorisation of the channels in H→ WW*, $\tau\tau$, ZZ* → leptons and H→b$\bar{\mathrm{b}}$ analyses. The combination is performed with the full dataset of *pp* collisions at $\sqrt{s}$=13 TeV corresponding to an integrated luminosity of 13.3 fb^{-1}. The distributions of the discriminant from each of the channels considered are combined to test for the presence of a signal assuming a Higgs boson mass of m_H=125 GeV. The statistical analysis is based on a binned likelihood function defined as the product of Poisson probability components over all bins considered in the analysis. The parameter of interest is the signal-strength μ and the nuisance parameters denote the set of continuous values that describe the effect of each systematic uncertainty on the signal and background expectations in each region. Such nuisance parameters adjust the expectations for signal and background according to the corresponding systematic uncertainties. The test statistics employed is the profile likelihood ratio and a correlation scheme is defined to treat coherently the nuisance parameters coming from different final states

Figure 10 summarises the observed fitted signal strength of the individual channels and the combination. The best fit value is found to be: $\mu_{\mathrm{t\bar{t}H}}$=1.8±0.7 and the compatibility of the individual signal strengths per channels with the combination is 7%. The likelihood profile for individual channels and for the combination is reported in Figure 10. The largest systematic uncertainty contributing to the measurement is the non-resonant t$\bar{\mathrm{t}}$ background modelling affecting the H→b$\bar{\mathrm{b}}$ final state. The combined expected (observed) significance of the Higgs boson signal in the t$\bar{\mathrm{t}}$H production mode is 2.8 (1.8) standard deviations.

2.5. H→ $\mu\mu$

The H→ $\mu\mu$ decay is a sensitive channel in which the Higgs coupling to second-generation fermions can be measured at the LHC with a clean signature in the final state. Given its

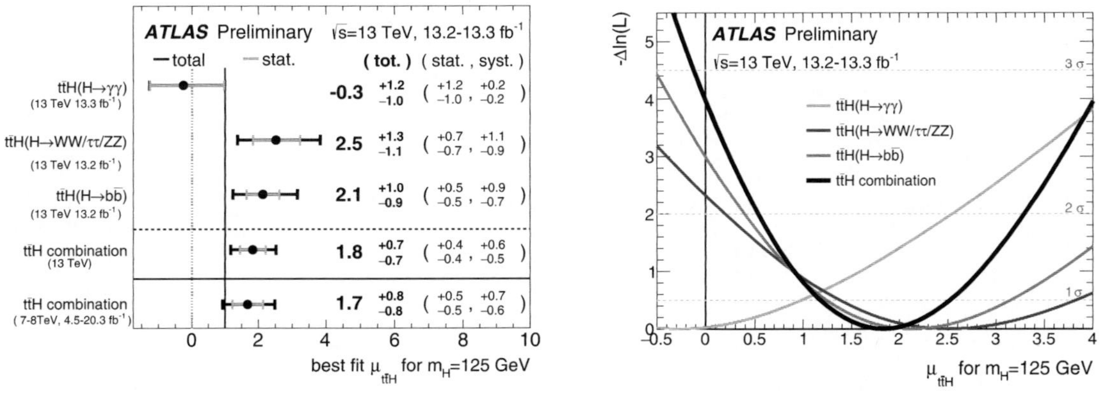

Figure 10. Observed signal strength, $\mu_{t\bar{t}H}$, measurements from the individual analyses and for their combination assuming m_H=125 GeV. The total, statistical and systematic uncertainties on $\mu_{t\bar{t}H}$ are shown. The SM expectation for $\mu_{t\bar{t}H}$ ($\mu_{t\bar{t}H}$=1) is shown as the black vertical line. The signal strength extracted for the Run 1 combination is also reported for comparison. Negative logarithmic likelihood ratio as a function of the signal strength for the various channels and for the combination, assuming m_H=125 GeV. The horizontal dashed lines correspond to 1, 2 and 3 standard deviations [7].

very small branching ratio, $\mathcal{B}(\mathrm{H}\rightarrow \mu\mu)=2.18\cdot 10^{-4}$, this process is very hard to observe at the LHC though. The search described in this Section is performed using data corresponding to an integrated luminosity of 13.2 fb^{-1} in *pp* collisions collected with the ATLAS detector [8].

2.5.1. Object definition and event selection Muons and electrons are identified according to the criteria described in Section 1.2.1. Light jets and jets containing *b*-hadrons are identified and reconstructed as reported in Section 2.1.1.

Events are required to have at least one primary vertex that has two associated tracks. If there is more than one vertex reconstructed in the event, the one with the largest squared track transverse momentum is chosen to be the primary vertex. Di-muon events are selected by requiring two opposite-charge muons with additional cuts on the transverse momentum on the leading and sub-leading muons. To improve data-Monte Carlo agreement for the muon transverse momentum, a reweighting method is adopted to correct the distribution of the $p_T^{\mu\mu}$ from Z+jets MC to match the data.

In order to increase the sensitivity of the search, the selected events are split into seven categories with different signal-to-background ratios. These categories are created on the muon pseudorapidity, $p_T^{\mu\mu}$ and VBF di-jet signatures. The latter is characterised by two forward jets with very large rapidity gap and little hadronic activity between them. The usage of a BDT discriminant is employed to define this VBF signal region.

2.5.2. Signal and background modelling The signal models for the gluon fusion and the VBF production mechanisms are obtained from simulated Higgs boson samples. This model is defined as the sum of a Crystal Ball function and a Gaussian core. The background is treated with a fit using the convolution of a Breit-Wigner function with a gaussian and an exponential distributions.

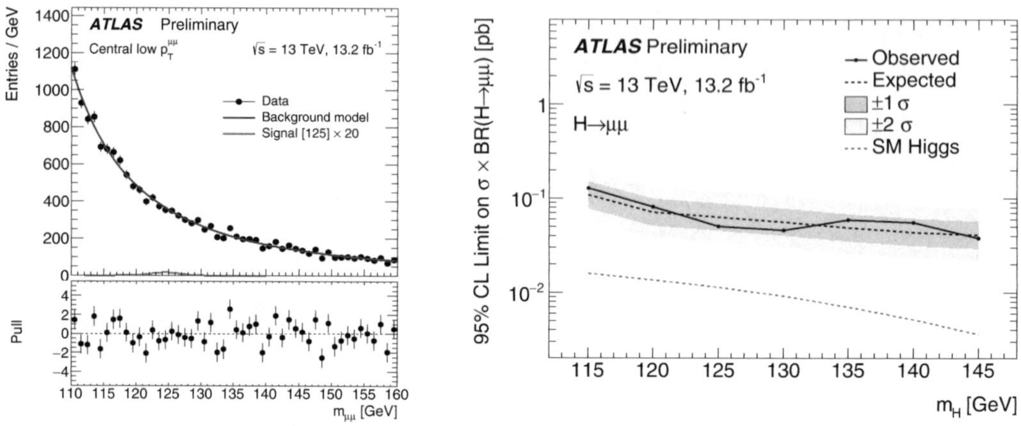

Figure 11. Background model fit to the $m_{\gamma\gamma}$ distribution in data for the low $p_T^{\mu\mu}$ category for the central pseudorapidity region. Statistical uncertainties are indicated on the data points. The expected signal is shown for m_H=125 GeV and is scaled by a factor 20. Observed and expected 95% upper limit on the $\sigma \times \mathcal{B}(H \to \mu\mu)$ as a function of m_H over the range 115-145 GeV. The dark- and light-shaded regions denote the $\pm 1\sigma$ and $\pm 2\sigma$ uncertainty bands on the expected limit [8].

In order to derive results, a binned maximum likelihood fit to the observed $m_{\mu\mu}$ distributions in the range 110-160 GeV is performed using the sum of the signal and background models. The fit is employed simultaneously in all seven categories. The $m_{\mu\mu}$ spectra for data and for the signal/background parametrisations is shown in Figure 11 for the low $p_T^{\mu\mu}$ in the central pseudorapidity region. Upper limits on $\sigma \times \mathcal{B}(H \to \mu\mu)$ are displayed in Figure 11. For the Higgs mass of 125.09 GeV, the observed (expected) upper limit on the signal strength at 95% confidence level is 4.4 (5.5) times the SM prediction.

3. Conclusions

Following the Higgs boson discovery in 2012 in the H→ $\gamma\gamma$, H→ZZ* →4l and H→WW* and the measurement of its properties in the H→ $\gamma\gamma$ and H→ZZ* →4l final states with the final LHC Run 1 dataset, recent years (2015 and 2016) have seen a focus on re-discovery of the Standard Model Higgs boson and the extraction of its properties in the bosonic channels (Section 1) together with the search of this particle in the fermionic final states (Section 2). Measurements collected so far with approximately 15 fb^{-1} of *pp* collisions at the ATLAS experiment are consistent with the SM predictions within uncertainties.

References

[1] ATLAS Collaboration, The ATLAS Experiment at the CERN Large Hadron Collider *JINST 3 (2008) 003.*
[2] ATLAS Collaboration, Measurement of fiducial, differential and production cross section in the H→ $\gamma\gamma$ channel with 13.3 fb^{-1} of 13 TeV proton-proton collision data with the ATLAS detector *ATLAS-CONF-2016-067.*
[3] ATLAS Collaboration, Study of the Higgs boson properties and search for high-mass scalar resonances in the H→ZZ* →4*l* decay channel at $\sqrt{s}$=13 TeV with the ATLAS detector *ATLAS-CONF-2016-079.*
[4] ATLAS Collaboration, Search for the Standard Model Higgs boson produced in association with a vector boson and decaying to a $b\bar{b}$ pair in *pp* collisions at 13 TeV using the ATLAS detector *ATLAS-CONF-2016-091.*
[5] ATLAS Collaboration, Search for the Standard Model Higgs boson produced in association with top quarks

and decaying into a $b\bar{b}$ pair in *pp* collisions at $\sqrt{s}$=13 TeV with the ATLAS detector *ATLAS-CONF-2016-080.*

[6] ATLAS Collaboration, Search for the associated production of a Higgs boson and a top quark pair in multilepton final states with the ATLAS detector *ATLAS-CONF-2016-058.*

[7] ATLAS Collaboration, Combination of the searches for Higgs boson production in association with top quarks in $\gamma\gamma$, multilepton, and $b\bar{b}$ decay channels at $\sqrt{s}$=13 TeV with the ATLAS detector *ATLAS-CONF-2016-068.*

[8] ATLAS Collaboration, Search for Higgs boson decaying into di-muon in *pp* collisions at $\sqrt{s}$=13 TeV with the ATLAS detector *ATLAS-CONF-2016-041.*

[9] ATLAS Collaboration, Search for pair production of Higgs boson in the *bbbb* final state using proton-proton collisions at $\sqrt{s}$=13 TeV with the ATLAS detector *ATLAS-CONF-2016-049.*

[10] ATLAS Collaboration, Search for Higgs boson pair production in the final state of $\gamma\gamma WW^*$ ($\rightarrow l\nu jj$) using 13.3 fb^{-1} of *pp* collision data recorded at $\sqrt{s}$=13 TeV with the ATLAS detector *ATLAS-CONF-2016-071.*

IOP Conf. Series: Journal of Physics: Conf. Series **873** (2017) 012004 doi:10.1088/1742-6596/873/1/012004

Higgs Physics at CMS

Silvio Donato on behalf of the CMS collaboration
Universität Zürich, Physik-Institut, Winterthurerstrasse 190, 8057 Zürich, CH
E-mail: silvio.donato@cern.ch

Abstract. The discovery of the Standard Model Higgs boson performed by the CMS and ATLAS collaborations during the LHC Run 1 has been an important success. This document is a short review of the search for the Higgs boson performed by the CMS collaboration during the LHC Run 1 and Run 2. In the first part, after a brief description of the Higgs boson production and decay channels, the Run-1 results are presented emphasizing the possible hints of New Physics. The main part of this document is devoted to the search for the Higgs boson with the 13-TeV data collected by the CMS experiment in 2015 and 2016, including the Standard Model searches as well as the Beyond Standard Model searches, such as the search for additional Higgs bosons and for resonant and non-resonant double Higgs boson production.

1. The Standard Model Higgs boson

1.1. Production and decay channels

In proton-proton colliders, the Standard Model (SM) Higgs boson can be produced through several production channels. The cross section of each channel depends on the center-of-mass energy $\sqrt{s}$ and on the Higgs boson mass. In the Run 2, since the LHC increased $\sqrt{s}$ from 8 TeV to 13 TeV, the 125-GeV SM Higgs boson cross section has increased in all the main production channels: gluon-gluon fusion (ggF, 19.3 pb $\rightarrow$ 43.9 pb), vector-boson fusion (VBF, 1.57 pb $\rightarrow$ 3.75 pb), vector-boson associated production (VH, 1.12 pb $\rightarrow$ 2.25 pb), and top-quark pair associated production (ttH, 0.13 pb $\rightarrow$ 0.51 pb) [1].

The branching ratios predicted by the SM for the main Higgs boson decay channels are $H \rightarrow bb\,(58\%)$, $H \rightarrow WW\,(21\%)$, $H \rightarrow \tau\tau\,(6.3\%)$, $H \rightarrow ZZ\,(2.6\%)$, and $H \rightarrow \gamma\gamma\,(0.23\%)$ [1].

2. Run 1 results

2.1. Higgs production and decay

In the Run 1, the Higgs boson was searched in many different final states with the CMS detector [2]. Figure 1 shows the signal strength ($\mu = \sigma/\sigma_{SM}$) measured in each analysis.

The analyses were combined according to the Higgs boson production channels, and the fitted values are shown in Fig. 2. All the results are compatible with the SM prediction within one standard deviation except for the ttH production channel ($\mu = 2.90^{+1.08}_{-0.94}$).

As shown in Fig. 3, the slight ttH excess was mainly driven by the same-sign two-leptons final-state analysis ($\mu = +5.3^{+2.1}_{-1.8}$), especially in the same-sign two-muons final states ($\mu = +8.5^{+3.3}_{-2.7}$) [4]. All the other measurements are in agreement with the SM expectation in 1.5 standard deviations.

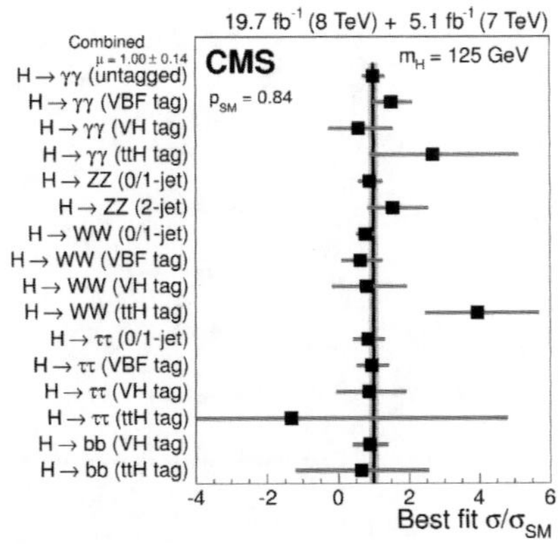

Figure 1. Signal strengths of SM Higgs boson searches in the Run-1 [3].

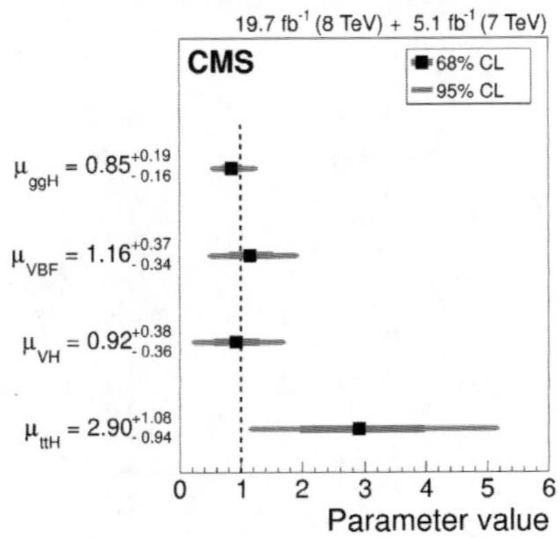

Figure 2. Run-1 signal strengths by the production modes [3].

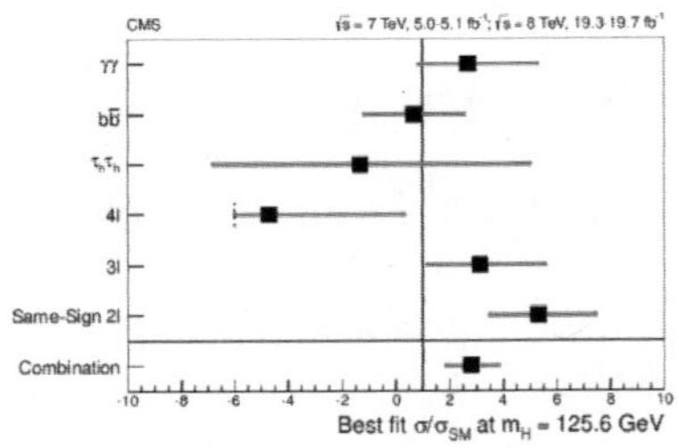

Figure 3. Run-1 signal strengths of ttH sub-channels [4].

The Higgs boson searches were also combined to measure the signal strength of the different decay channels. The results are shown in Fig. 4: the signal strengths are compatible with SM for all decay channels. The corresponding observed (expected) significances are

- $H \to ZZ : 6.5\,(6.3)$;
- $H \to \gamma\gamma : 5.6\,(5.3)$;
- $H \to WW : 4.7\,(5.4)$;
- $H \to \tau\tau : 3.8\,(3.9)$;
- $H \to bb : 2.0\,(2.6)$;
- $H \to \mu\mu :< 0.1\,(0.4)$ [3].

2.2. Higgs couplings

The Higgs boson searches were eventually combined to search for deviations from the SM in the Higgs boson couplings. Figure 5 shows the measured Higgs boson couplings to the fundamental particles with respect to the SM expectations. All the results are in agreement with the SM expectation.

2.3. Higgs mass

The Higgs mass was measured using the $H \to \gamma\gamma$ and $H \to ZZ$ channels and the combination of the two channels is shown in Figure 6. A slight tension of 1.6σ was reported between the two

IOP Conf. Series: Journal of Physics: Conf. Series **873** (2017) 012004 doi:10.1088/1742-6596/873/1/012004

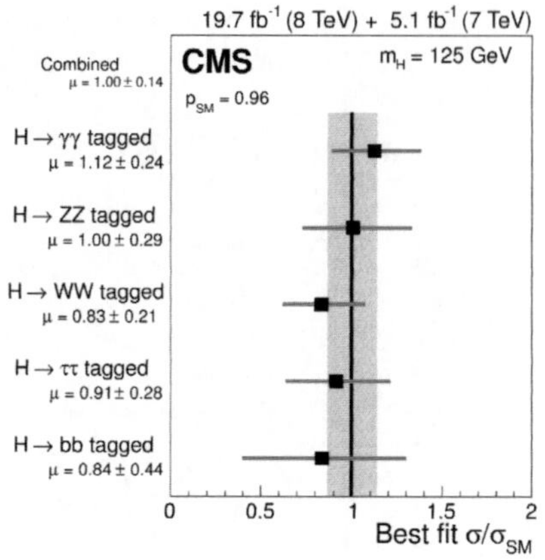

Figure 4. Run-1 signal strengths by decay channels [3].

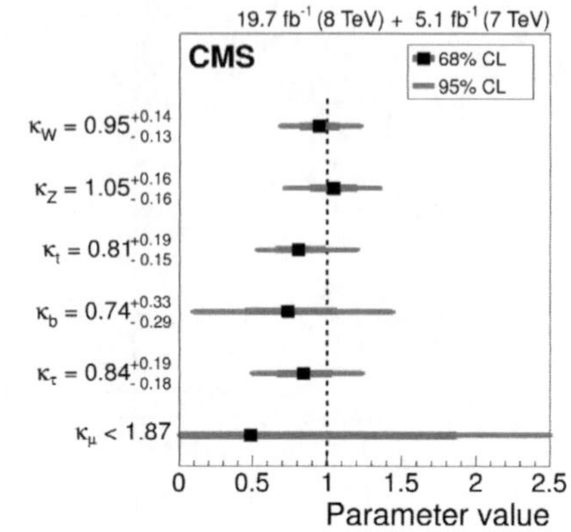

Figure 5. Higgs boson coupling modifiers fitted in the Run 1 [3].

measurements, as shown by the log-likelihood scan of $m_{\gamma\gamma} - m_{4\ell}$ in Fig. 7. The best-fit mass is

$$m_H = 125.03^{+0.26}_{-0.27}(\text{stat.})^{+0.14}_{-0.15}(\text{syst.})\,\text{GeV}.$$

The measurement is dominated by the statistical uncertainty, and hence we expect to largely improve the measurement in the LHC Run 2.

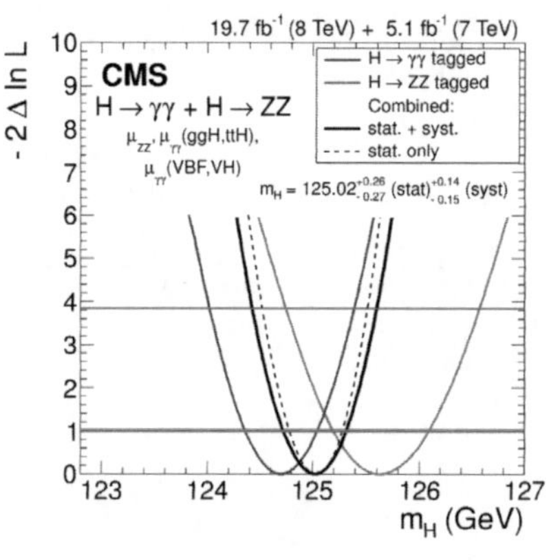

Figure 6. Log-likelihood scan of the Higgs boson mass for the Run-1 $H \to \gamma\gamma$, $H \to ZZ$, and their combination [3].

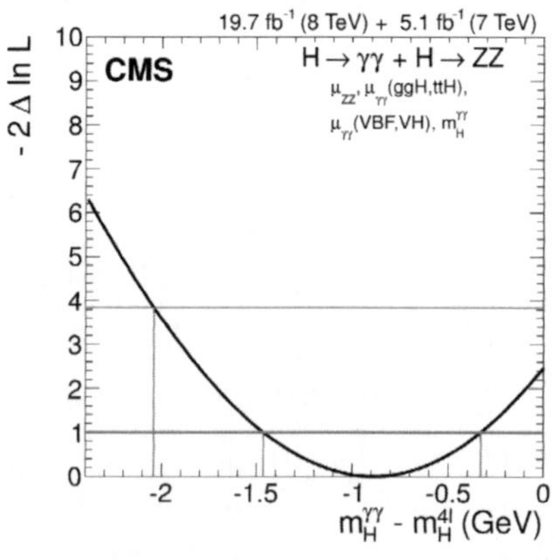

Figure 7. Log-likelihood scan for the difference between the Higgs boson mass measured in the Run-1 $H \to \gamma\gamma$ and $H \to ZZ$ [3].

2.4. Higgs spin and parity

The high purity of the $H \to ZZ \to 4\ell$ channel allows to test the validity of many spin and parity models. Figure 8 shows the probability distributions of the likelihood ratio between the 0^+ and 0^- hypotheses, for the two hypotheses: the experimental result excludes the 0^- parity hypothesis with 3.3 standard deviations. Likewise, many other spin/parity models were tested and the results are collected in Fig. 9. All results found are compatible with the SM within 2 standard deviations.

IOP Conf. Series: Journal of Physics: Conf. Series **873** (2017) 012004 doi:10.1088/1742-6596/873/1/012004

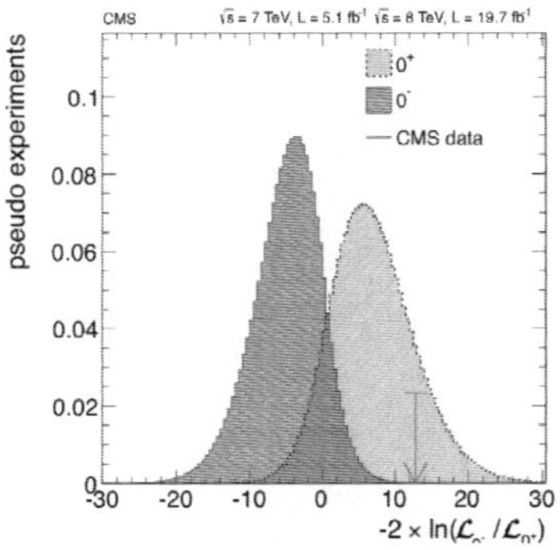

Figure 8. Likelihood ratio distribution for spin-0 Higgs boson with positive and negative parity, along with the Run-1 result [5].

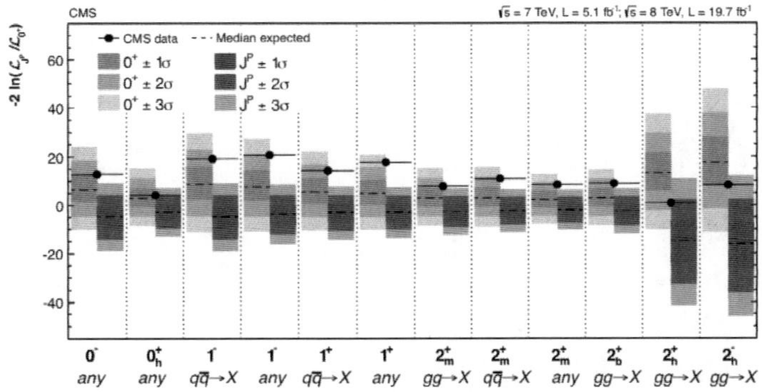

Figure 9. Likelihood ratio distribution for several spin-parity Higgs boson models, along with the Run-1 results [5].

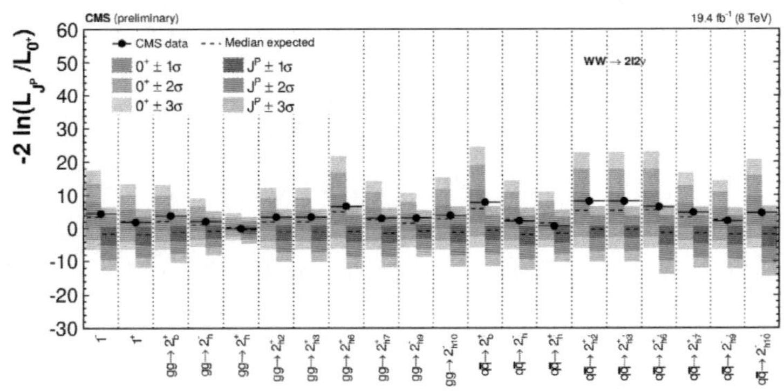

Figure 10. Likelihood ratio distribution for several spin-parity Higgs boson models, along with the Run-1 results [6].

Alternative spin-parity hypotheses were probed exploiting the $H \to WW \to 2\ell 2\nu$ channel. The results, shown in Fig. 10, are compatible with the SM expectations within one standard deviation.

2.5. Higgs width

At $m_H = 125\,\text{GeV}$, the width of Higgs boson predicted by the SM is of few MeV, but the experimental Higgs mass resolution is of few GeV. However, it can be measured indirectly exploiting the off-shell production of the Higgs boson, assuming that all properties of the Higgs boson are those predicted by the SM except for the Higgs boson width (Γ^H). Figure 11 compares the four-leptons invariant mass distribution expected for $\Gamma^H = \Gamma^H_{SM}$ and $\Gamma^H = 25 \times \Gamma^H_{SM}$. The likelihood scan of Γ^H measured by the CMS collaboration after the LHC Run 1 is shown in Fig. 12. It sets an observed (expected) upper limit of 22 MeV (33 MeV) at a 95% CL corresponding to 5.4 (8.0) times the SM value.

2.6. Higgs differential cross section

The large statistic of the Run-2 will allow to measure precisely the differential cross section of the production of the Higgs boson. The first preliminary results were obtained in Run 1 for the $H \to \gamma\gamma$ and $H \to ZZ \to 4\ell$. Figure 13 and 14 show, respectively, the $p_T(H)$ and $y(H)$

IOP Conf. Series: Journal of Physics: Conf. Series **873** (2017) 012004 doi:10.1088/1742-6596/873/1/012004

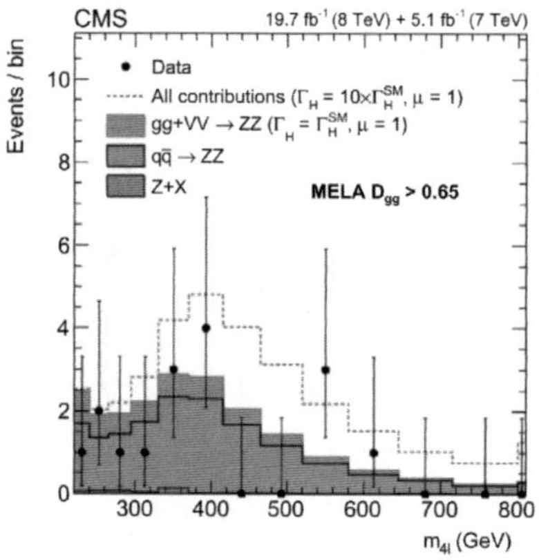

Figure 11. Four lepton mass distribution for $\Gamma = \Gamma_{SM}$ and $\Gamma = 25 \times \Gamma_{SM}$, in the $H \to ZZ$ channel [7].

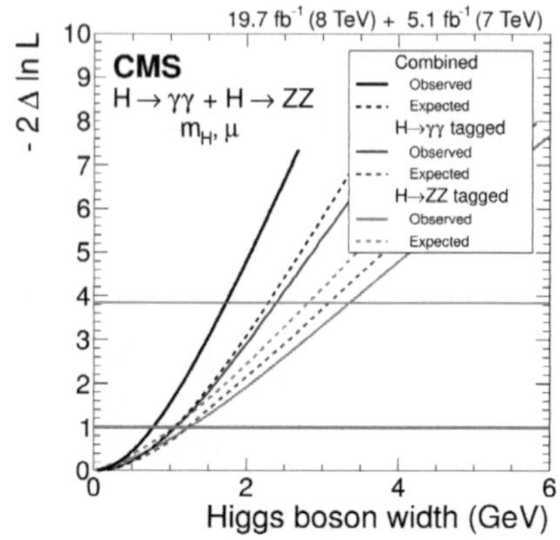

Figure 12. Log-likelihood scan for the Higgs boson width as measured in the Run 1 using the ZZ and $\gamma\gamma$ final states, and their combination [3].

differential cross section measured for the $H \to \gamma\gamma$ channel. Similarly, Figure 15 and 16 are for the $H \to ZZ \to 4\ell$ channel.

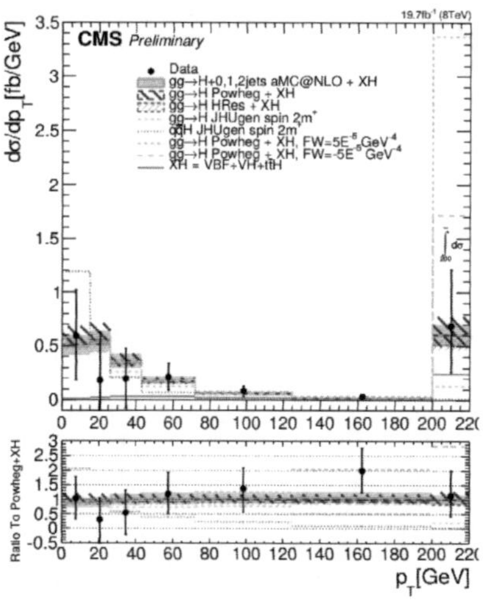

Figure 13. Differential cross section as a function of the Higgs boson momentum measured using the $H \to \gamma\gamma$ with 2012 data [8].

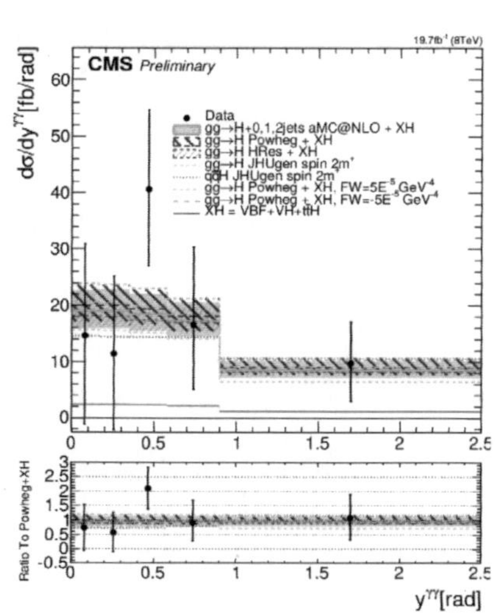

Figure 14. Differential cross section as a function of the Higgs boson rapidity measured using the $H \to \gamma\gamma$ with 2012 data [8].

2.7. Beyond Standard Model searches

Many Beyond Standard Model (BSM) Higgs boson searches were performed using the Run-1 dataset. Among them, a slight excess (2.5σ) was reported in the lepton flavour violation search $H \to \mu\tau$. The excess was equivalent to a $\mathrm{BR}(H \to \mu\tau) = 0.89^{+0.40}_{-0.37}\%$, as shown in Fig. 17. Figure 18 shows the $\tau\,\mu$ invariant mass distribution for the category with the decay $\tau \to e\nu_e\nu_\tau$ and one jet in the final state, assuming that the neutrinos have the same direction of the electron.

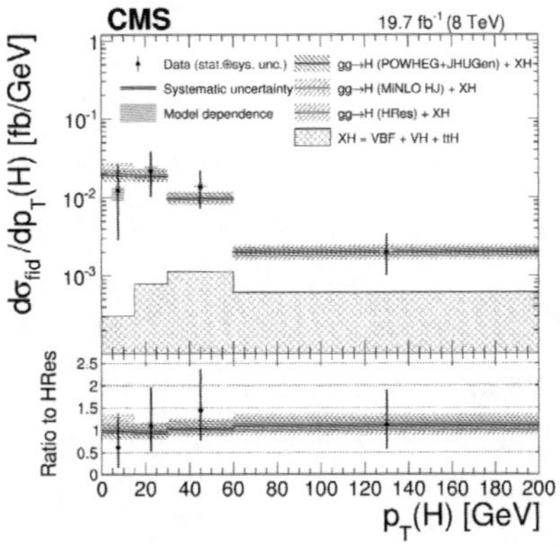

Figure 15. Differential cross section as a function of the Higgs boson momentum measured using the $H \to ZZ \to 4\ell$ with 2012 data [9].

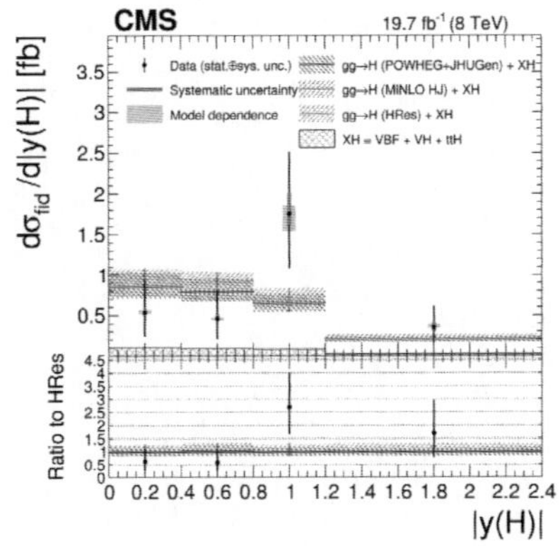

Figure 16. Differential cross section as a function of the Higgs boson rapidity measured using the $H \to ZZ \to 4\ell$ with 2012 data [9].

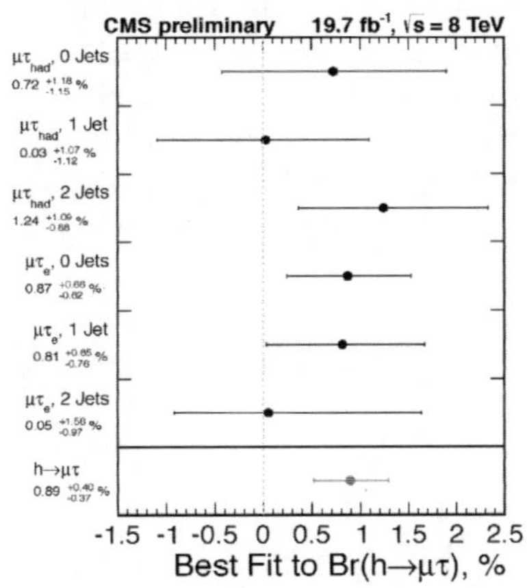

Figure 17. Branching ratio of $H \to \mu\tau$, obtained using the 2012 dataset [10].

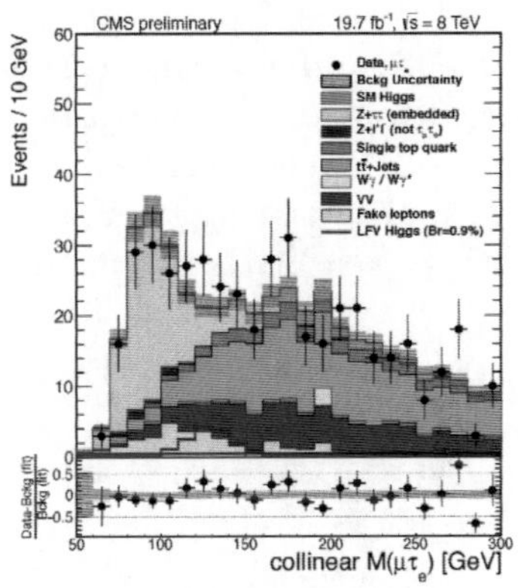

Figure 18. Invariant mass between μ and $\tau(e\nu_e\nu_\tau)$, assuming that the neutrinos have the same direction of the electron, using the 2012 dataset [10].

Another important way to search for new physics is the measurement of the effective couplings of the Higgs boson with gluons (κ_g) and photons (κ_γ), since BSM particles might modify these loop interactions. The effective couplings were measured combining all the SM Higgs boson searches and assuming the SM prediction for the couplings with the other fundamental particles. Figure 19 shows the 68%, 95%, and 99.7% CL limits measured for the coupling modifiers κ_g and κ_γ. The result is compatible with the SM expectation within one standard deviation and therefore no hint of New Physics was observed in the loops of the Higgs-gluon and Higgs-photon interactions.

3. Run 2 - SM Higgs

3.1. $H \to \gamma\gamma$

The di-photon Higgs decay was one of the main discovery channels in the Run 1. In the Run 2, the analysis has been performed splitting it into eight categories with different event topology

IOP Conf. Series: Journal of Physics: Conf. Series **873** (2017) 012004 doi:10.1088/1742-6596/873/1/012004

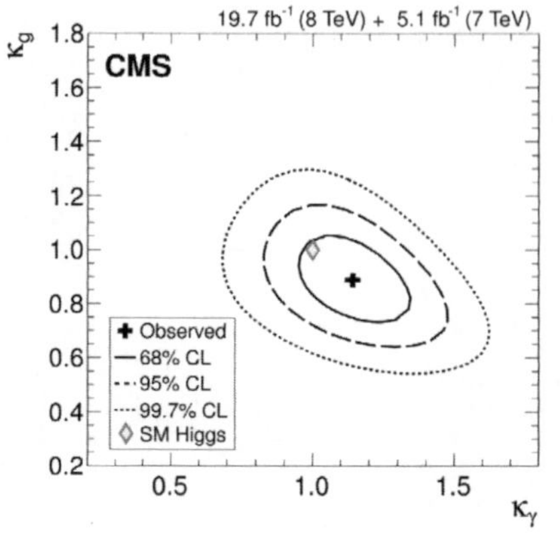

Figure 19. Confidence regions for modifier Higgs couplings with photons (κ_γ) and gluons (κ_g), obtained using the Run-1 dataset [3].

(VBF, ttH, untagged) and signal-to-background ratio. The latter separation has been obtained using a Boosted Decision Tree (BDT) that exploits the most discriminating variables, with the exception of $m_{\gamma\gamma}$ and correlated variables. The definition of the four categories with the untagged topology is shown in Fig. 20. The signal has been extracted fitting simultaneously the $m_{\gamma\gamma}$ distribution in each category, looking for the Higgs peak on top of a smooth falling background distribution. Figure 21 shows the $m_{\gamma\gamma}$ distribution combining all categories and weighting each of them according to its purity ($S/(S+B)$). The result of each category is

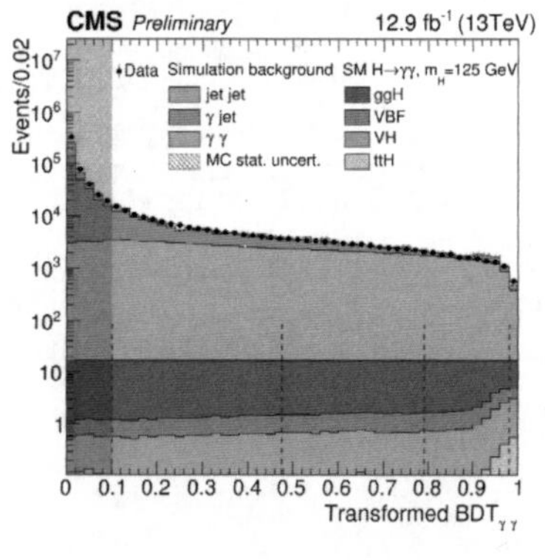

Figure 20. BDT distribution for the untagged event topology in the $H \to \gamma\gamma$ analysis and definition of the four untagged categories. The shaded region is not used in the data analysis [11].

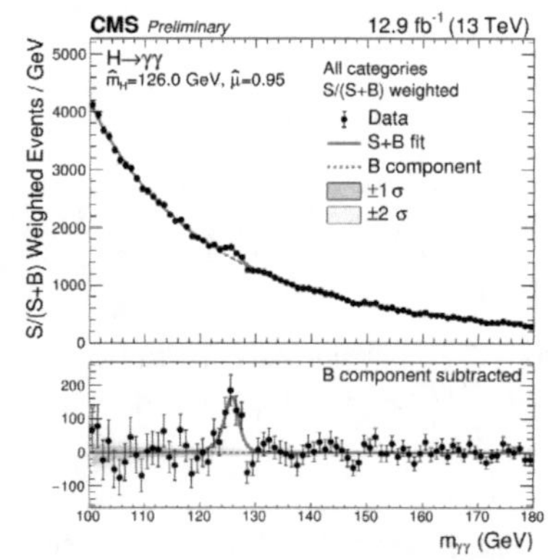

Figure 21. Di-photon invariant mass distribution combining all categories according to their purity [11].

shown in Fig. 22. The measured combined signal strength is $\mu = 0.95^{+0.21}_{-0.18}$. The background-only hypothesis is excluded with 6.2 standard deviations. Figure 23 shows the Higgs boson cross section measured as a function of the center-of-mass energy, along with the SM prediction.

3.2. $H \to ZZ \to 4\ell$

In the Run-1, the other main channel of the Higgs boson discovery was the $H \to ZZ \to 4\ell$. The Run-2 analysis has been divided into three final states (4μ, $4e$, $2e2\mu$) and in six event categories,

IOP Conf. Series: Journal of Physics: Conf. Series **873** (2017) 012004 doi:10.1088/1742-6596/873/1/012004

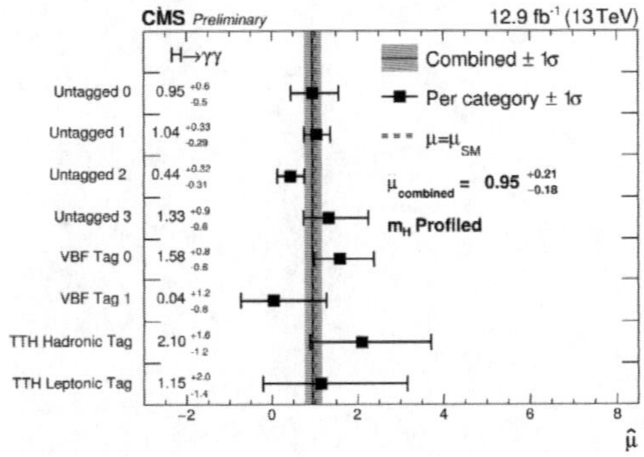

Figure 22. Signal strengths of the $H \to \gamma\gamma$ categories [11].

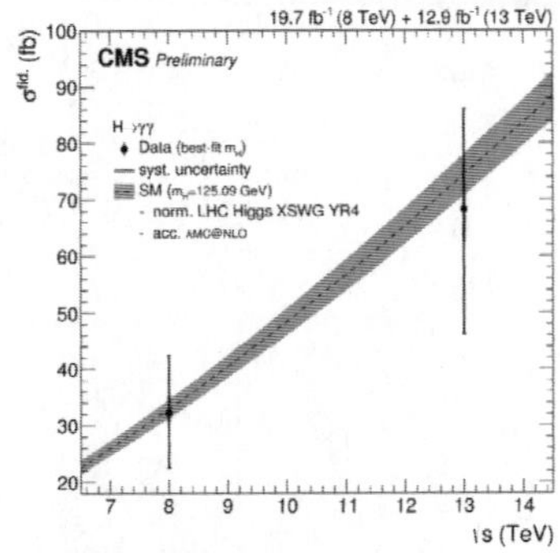

Figure 23. $H \to \gamma\gamma$ cross section as a function of $\sqrt{s}$ [11].

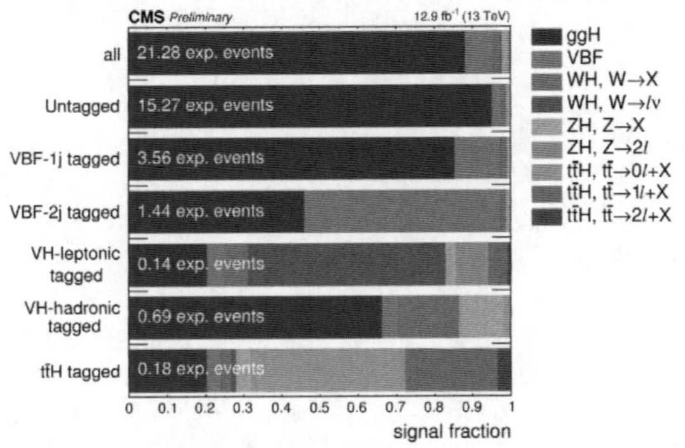

Figure 24. Signal composition of the $H \to 4\ell$ categories [12].

aiming to separate the different production channels (VBF-1jet, VBF-2jet, VH-leptonic, VH-hadronic, ttH, untagged). The signal composition of each event category is shown in Fig. 24. The overall four-leptons invariant mass distribution is shown Fig. 25. A matrix element discriminant (D_{kin}) has been exploited in order to increase the signal to background separation. The signal is extracted fitting the 2D distribution of ($m_{4\ell}$,D_{kin}) in 18 categories (3 final states $\times$ 6 event categories). Figure 26 is a superimposition of the 2D distributions of the 18 categories used in the final fit. The fitted global signal strength is $\mu = 0.99^{+0.33}_{-0.26}$ at $m_H = 125.09\,\mathrm{GeV}$, corresponding to an observation of the $H \to ZZ \to 4\ell$ with 6.2σ, as shown in Fig. 27. Figure 28 is the measurement of the Higgs boson cross section as a function of the center-of-mass energy.

Likewise the Run 1, the $H \to ZZ \to 4\ell$ channel has been used to measure also the differential cross-section, the mass, and the width. Figure 29 shows the Higgs boson cross section as a function of $p_T(H)$. The best fit of the Higgs boson mass using the 13-TeV dataset is $m_H = 124.50^{+0.48}_{-0.46}$. The observed (expected) 95% CL upper limit on the Higgs boson width is 41 MeV (32 MeV), as shown in Fig. 30.

3.3. ttH (multilepton)

In the multilepton ttH analysis, the Higgs boson has been searched in the ttH topology with at least two same-sign leptons in the final state. The signal includes the $H \to WW, ZZ, \tau\tau$ decays. The main backgrounds of this search are the ttW and ttZ productions: two BDTs have been trained in order to reject these backgrounds and are used in the final signal extraction. Other

IOP Conf. Series: Journal of Physics: Conf. Series **873** (2017) 012004 doi:10.1088/1742-6596/873/1/012004

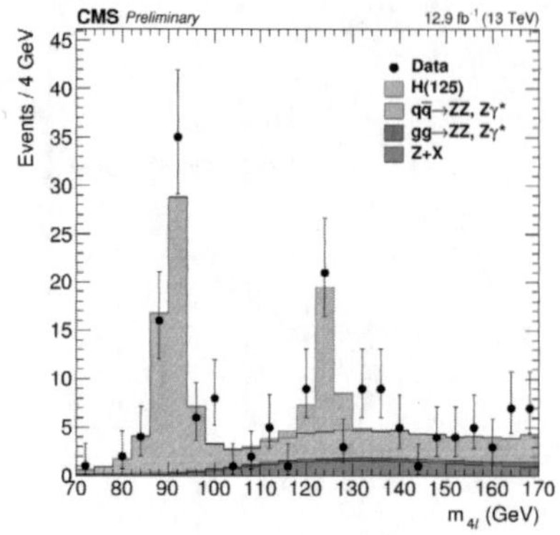

Figure 25. Four lepton invariant mass, combining all $H \to 4\ell$ categories [12].

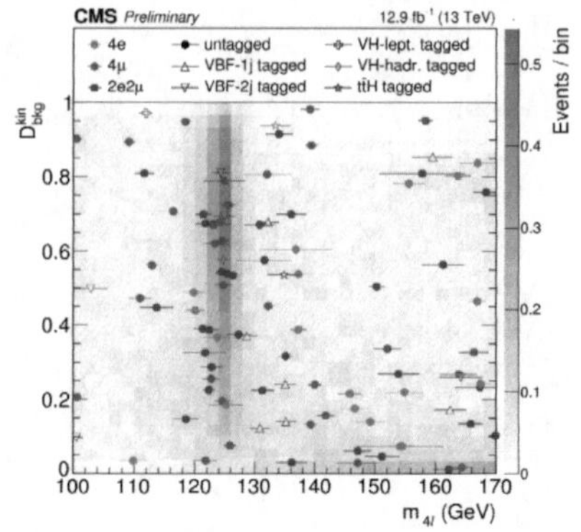

Figure 26. Four lepton invariant mass vs matrix element discriminant, for all $H \to 4\ell$ categories [12].

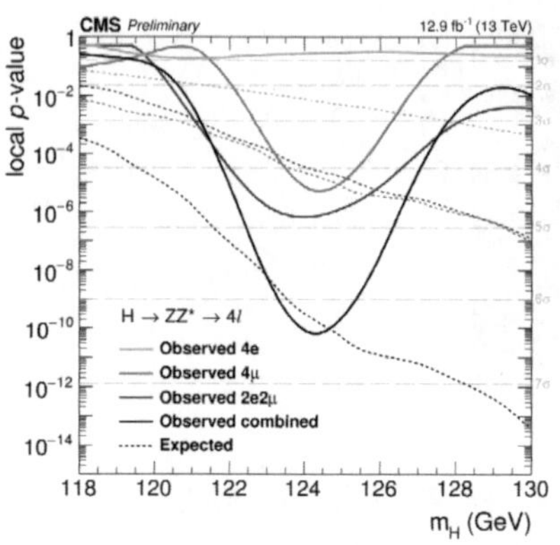

Figure 27. Local p-value for the $4e$, 4μ, $2e2\mu$ final states, and their combination [12].

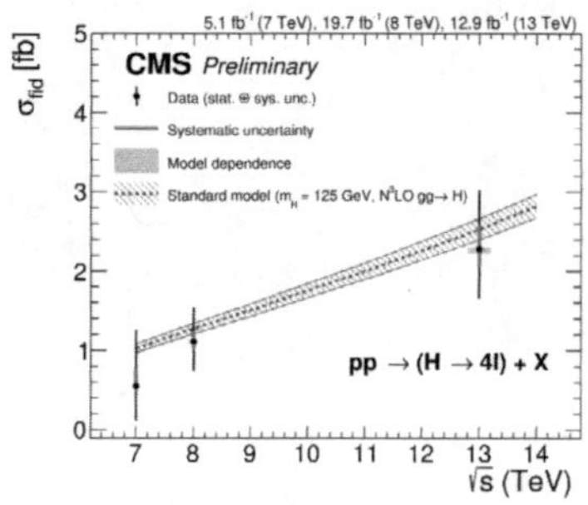

Figure 28. $H \to 4\ell$ cross section as a function of $\sqrt{s}$ [12].

backgrounds are the fake lepton identification from multijet events and the charge misidentified electrons. Figure 31 and 32 show the BDT distribution in the two and three leptons categories. The fitted combined signal strength is $\mu = 2.0^{+0.8}_{-0.7}$. The observed (expected) 95% CL upper limit is 3.4 (1.3) time the standard model cross section.

3.4. ttH(bb)

The $ttH(bb)$ analysis exploits the large branching ratio of $H \to bb$ (58%). The main background is top pairs production with additional jets. The analysis has been split into single and double lepton channels, and in categories depending on the number of jets and b-jets. A BDT has been trained to further split the categories into two parts (high and low S/B). Figure 33 shows the distribution of the BDT used to split one of the single-lepton categories. A Matrix Element Method (MEM) is used for the signal extraction, using the likelihood ratio of the event to be signal (tt+H) or background (tt+bb) as a discriminant. Figure 34 shows the MEM distribution for one of the single categories before and after the final fit. A background under-fluctuation has been reported: the fitted signal strength is $\mu = -0.19^{+0.80}_{-0.81}(\text{stat.})^{+0.66}_{-0.68}(\text{syst.})$. The fitted upper limits obtained in the fully leptonic and semi-leptonic categories are shown, respectively, in Fig. 35 and 36. The combination of all categories gives a 95% CL observed (expected) upper limit of

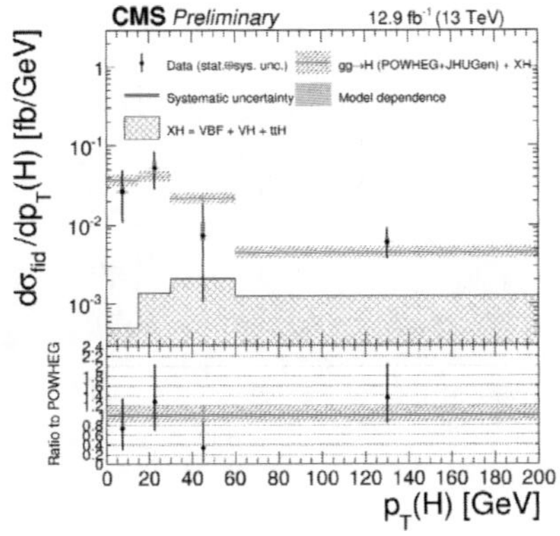

Figure 29. Differential cross section as a function of the Higgs boson momentum measured using the $H \to ZZ \to 4\ell$ [12].

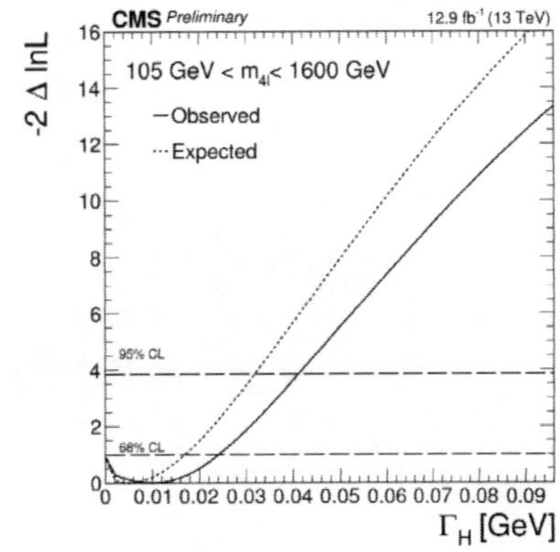

Figure 30. Observed and expected log-likelihood scan for the Higgs boson width using the $H \to 4\ell$ channel [12].

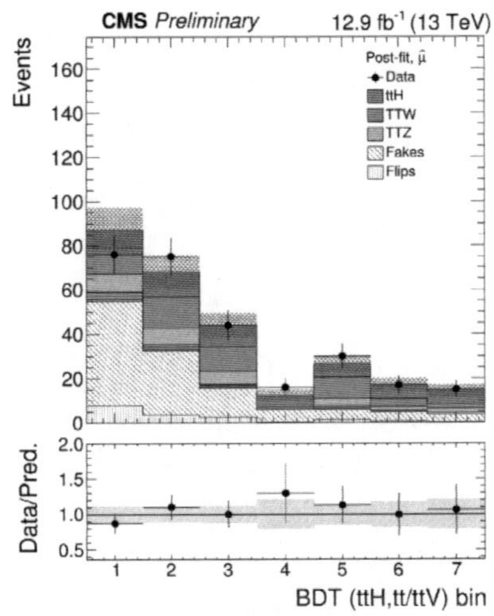

Figure 31. BDT distribution in the two same-sign lepton category [13].

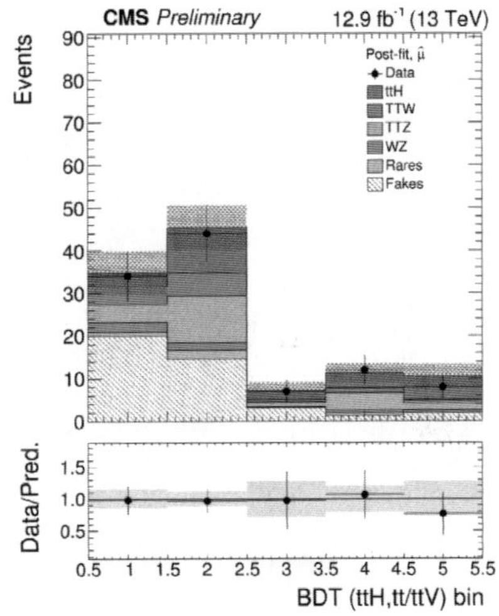

Figure 32. BDT distribution in the three lepton category [13].

1.5 (1.7) times the standard model cross section. Figure 38 shows is a combination of the final BDT distributions from all categories, as a function of the signal-to-background ratio.

3.5. VBF H(bb)

The $H \to bb$ has been searched also in VBF production. This signal topology is made of two b-jets from the Higgs boson decay and two forward/backward jets from the vector-boson fusion. As shown in Fig. 39, the di-jet invariant mass resolution has been improved using a b-jet energy regression. A BDT has been trained in order to split the analysis into several categories with a different signal-to-background ratio. The BDT distribution is shown in Fig. 40. The signal is extracted fitting the m_{bb} distribution in each category. Figure 41 shows the fit for the most sensitive category. A background under-fluctuation has been reported, as shown in Fig. 42 where the likelihood scan of the signal strength of the 13 TeV, 8 TeV, and the combination are reported. The 95% CL observed (expected) upper limit is 3.0 (5.0) times the SM cross-section.

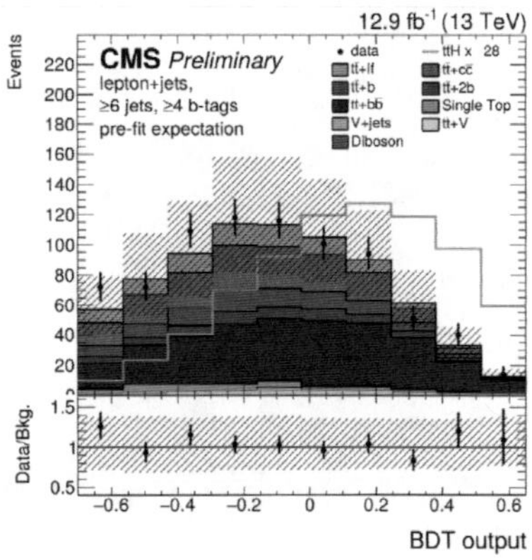

Figure 33. BDT distribution in the $\geq$ 6 jets and $\geq$ 4 b jets region. The signal median is used to split the region into two categories [14].

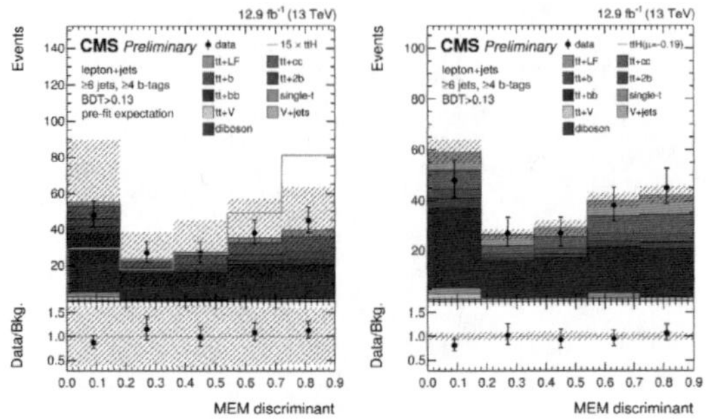

Figure 34. MEM discriminant before and after the final fit, in the $\geq$ 6 jets and $\geq$ 4 b jets region [14].

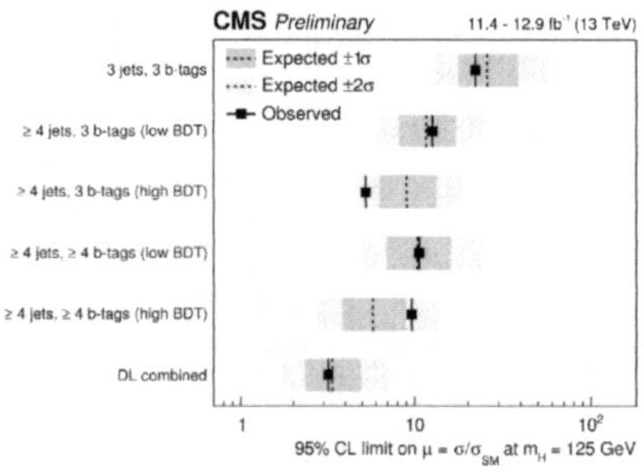

Figure 35. Upper limit for the fully-leptonic $ttH(bb)$ categories and their combination [14].

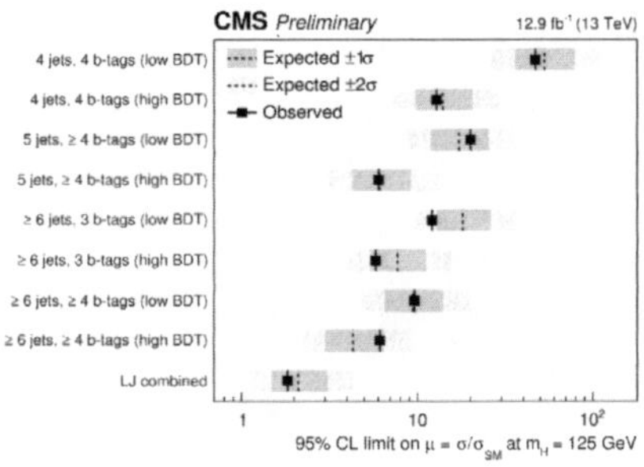

Figure 36. Upper limit for the semi-leptonic $ttH(bb)$ categories and their combination [14].

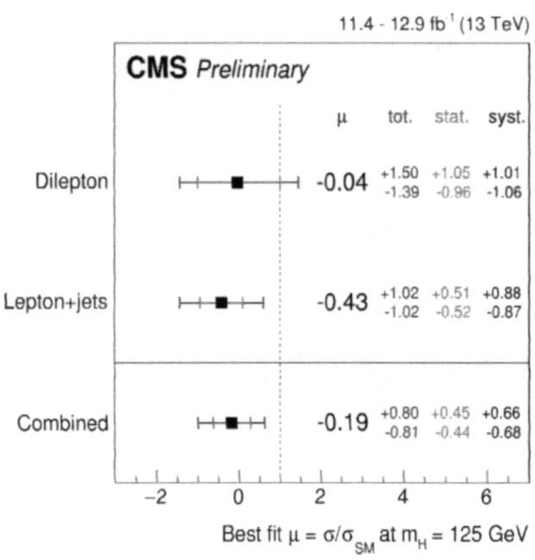

Figure 37. Signal strength of the fully-leptonic, semi-leptonic $ttH(bb)$, and their combination [14].

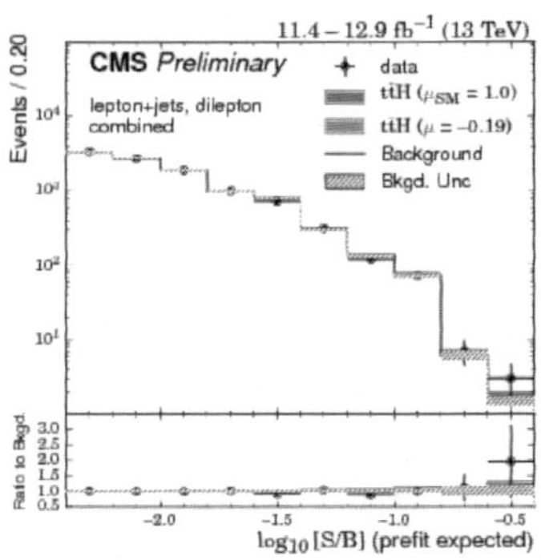

Figure 38. Combination of the $ttH(bb)$ MEM distributions in S/B bins [14].

4. Run 2 - Higgs pair production

The search for the double Higgs boson production - and hence the measurement of the self Higgs boson couplings - is one important target of the High Luminosity LHC. However, new Physics

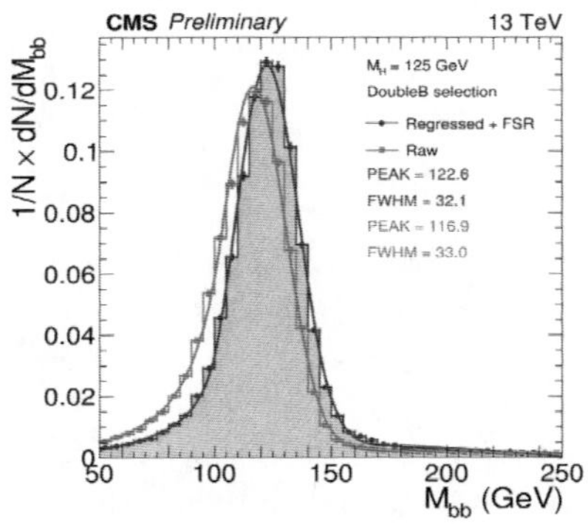

Figure 39. Di-jet invariant mass resolution for VBF $H(bb)$ before and after the jet-energy regression [15].

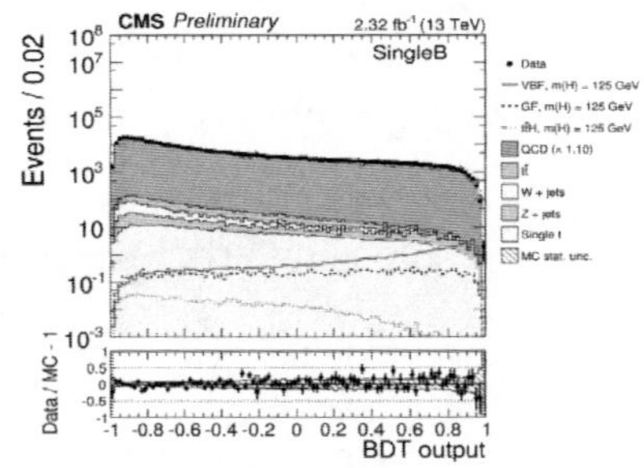

Figure 40. BDT distribution for VBF $H(bb)$ [15].

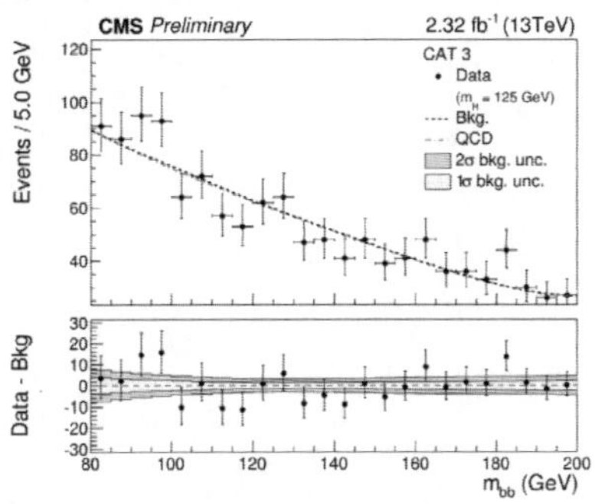

Figure 41. Di-jet invariant mass distribution for the VBF $H(bb)$ category with the highest sensitivity [15].

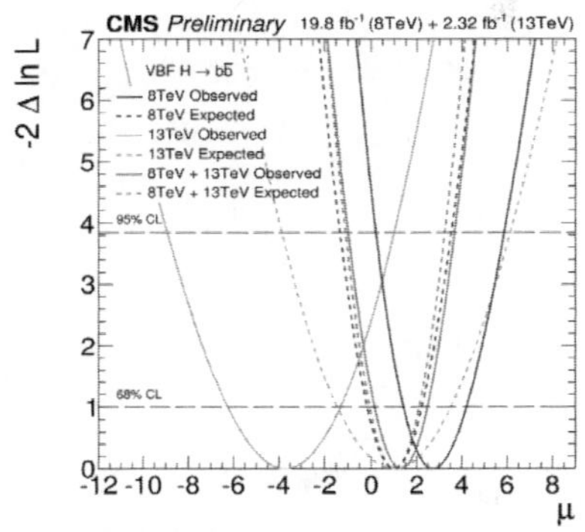

Figure 42. VBF $H(bb)$ signal strength using 8 TeV and 13 TeV datasets, and their combination [15].

might enhance this process or produce new particles decaying into a Higgs boson pair, making this process visible even at a lower luminosity. Therefore, the double Higgs boson production has been searched in the Run 2 in both the resonant and non-resonant final state.

4.1. $HH \to \gamma\gamma bb$

One possible way to search for the double Higgs boson production is the resonant $pp \to X \to HH \to \gamma\gamma bb$, exploiting both the $H \to \gamma\gamma$ to reduce QCD background and the $H \to bb$ in the attempt to have the largest possible signal yield. This channel is characterized by a very high purity but a low statistic. The 95% CL upper limit is shown in Fig. 43 as a function of the resonance mass.

The $pp \to HH \to \gamma\gamma bb$ has been also searched also in the non-resonant final states. The upper limit is shown in Fig. 44 and it corresponds roughly to 90 times the SM cross section.

4.2. $HH \to bbbb$

The search for the double Higgs in the fully hadronic final state exploits the high $H \to bb$ branching ratio (58%), but it is characterized by a large multijet background. In the resonant search, the signal has been extracted looking for a peak in the smooth di-Higgs invariant mass

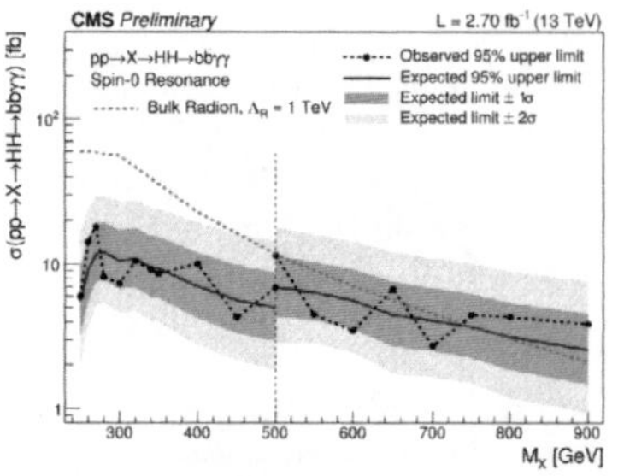

Figure 43. Upper limit for the search for non-resonant $pp \to X \to HH \to bb\gamma\gamma$ [16].

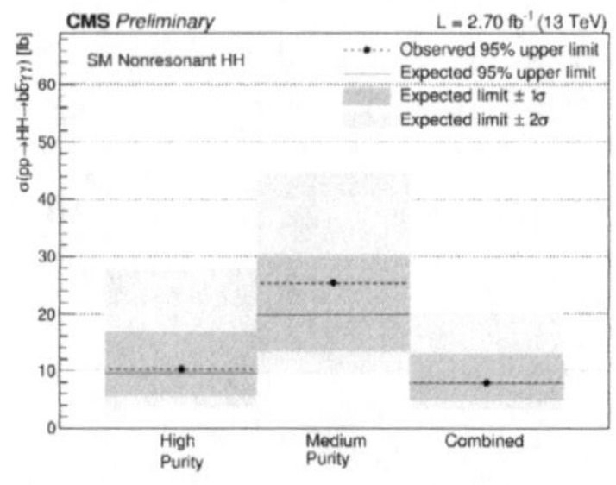

Figure 44. Upper limit for the search for non-resonant $pp \to HH \to bb\gamma\gamma$ [16].

distribution, as shown in Fig. 45 and 46. As shown if Fig. 47, a kinematic fit has been used in order to improve the resolution on the di-Higgs mass, exploiting the Higgs mass constraint. The results are shown in Fig. 48: no excess has been found. The $pp \to X \to HH \to bbbb$ is the analysis with the highest sensitivity at high resonance mass.

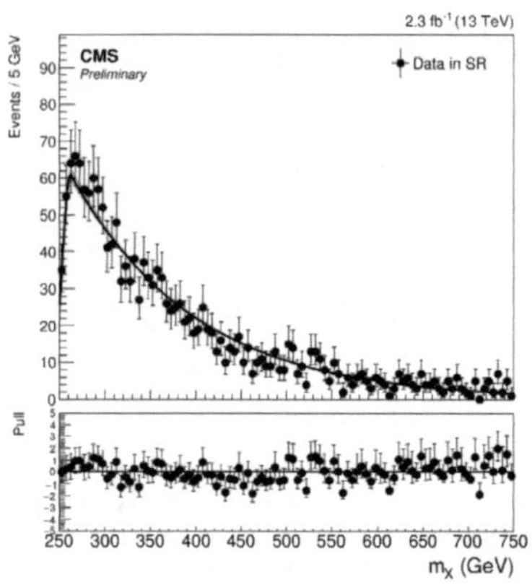

Figure 45. Four b jet invariant mass distribution in the low mass $pp \to X \to HH \to bbbb$ signal region [17].

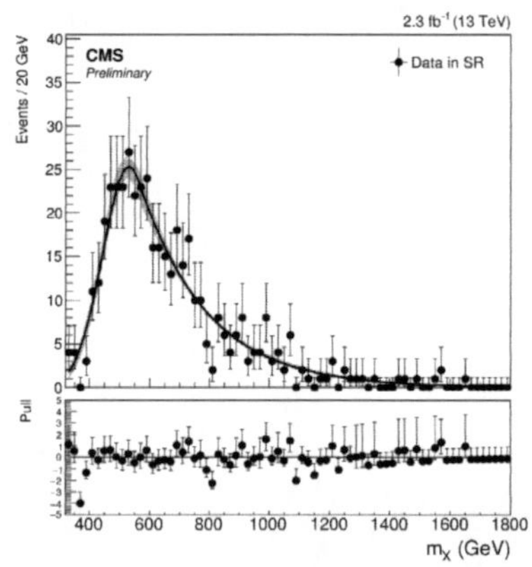

Figure 46. Four b jet invariant mass distribution in the high mass $pp \to X \to HH \to bbbb$ signal region [17].

The search of non-resonant $pp \to HH \to bbbb$ has been performed using a BDT, to select a signal-like phase-space, and a 2D fit of the mass of the two Higgs boson candidates, to perform the final signal extraction. Figure 49 shows the BDT distribution and Figure 50 show the unrolled 2D-mass fit. The background has been estimated in a fully data-driven way. The observed (expected) upper limit is $3.88\,\mathrm{pb}^{-1}$ ($3.49\,\mathrm{pb}^{-1}$), corresponding to roughly 340 times the standard model cross section.

4.3. $HH \to bb\ell\ell$

The search for $HH \to bb\ell\ell$ includes both the $HH \to bbW(\ell\nu)W(\ell\nu)$ and $HH \to bbZ(\ell\ell)Z(\nu\nu)$ signals: in both cases, the presence of two neutrinos in the final state degrades the invariant mass resolution on the Higgs boson decaying to leptons. The main background is the top-pair production decaying fully leptonically.

In the resonant search, two BDTs have been trained to separate the signals ($m_X = 400\,\mathrm{GeV}, 650\,\mathrm{GeV}$) from the background, as shown in Fig. 51. The phase space has been then

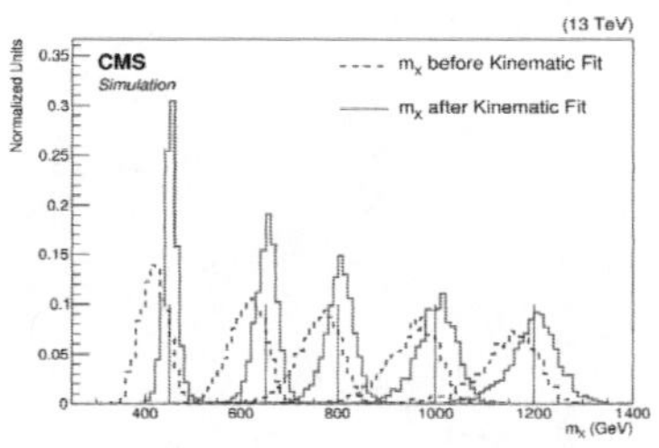

Figure 47. Four b-jet invariant mass resolution before and after the kinematic fit [17].

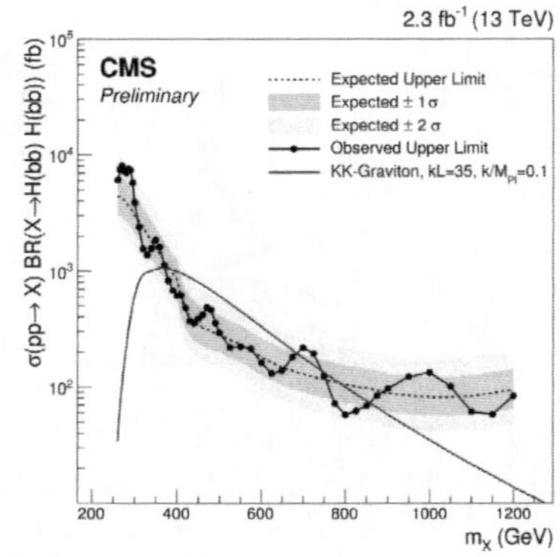

Figure 48. Upper limit of the search for resonant $pp \to X \to HH \to bbbb$ [17].

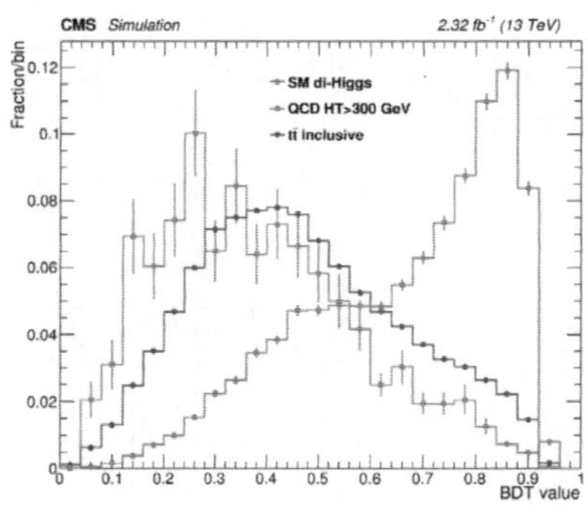

Figure 49. Signal and background BDT distribution in the search for non-resonant $pp \to HH \to bbbb$ [18].

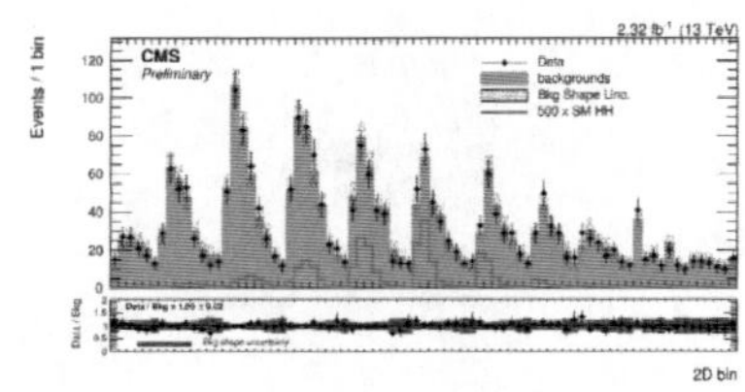

Figure 50. Signal and background BDT distribution in the search for non-resonant $pp \to HH \to bbbb$ [18].

divided in low/high BDT and inside/outside the mjj Higgs peak. Eventually, the signal has been extracted using a cut-and-count method applied to the region having with the highest signal-to-background ratio. The other regions have been used to estimate the background normalization. The 95% upper limits are shown in Fig. 52 as a function of the resonance mass.

The nonresonant search has been performed splitting the di-jet m_{bb} invariant mass, shown in Fig.53, in three parts and then fitting the BDT distribution in each of these regions, as shown in Fig. 54. The 95% CL upper limit is about $\sigma < 400\sigma_{SM}$.

4.4. $HH \to bb\tau\tau$

The $pp \to HH \to bb\tau\tau$ search has been divided into three sub-channels $\tau(\mu\nu\nu)\tau(\text{hadr.}\nu)$, $\tau(e\nu\nu)\tau(\text{hadr.}\nu)$, $\tau(\text{hadr.}\nu)\tau(\text{hadr.}\nu)$. A kinematic fit has been used in order to improve the $m_{\tau\tau bb}$ resolution - degraded by the presence of neutrinos in the final state. The main backgrounds are $t\bar{t}$, Drell-Yan, and the multijet events.

The resonant search is performed looking for a peak in the $m_{\tau\tau bb}$ distribution, as shown in Fig. 55. No significant excess have been found. The measured upper limits are shown in Fig. 56.

In the non-resonant analysis, a BDT has been trained in order to increase the signal-to-

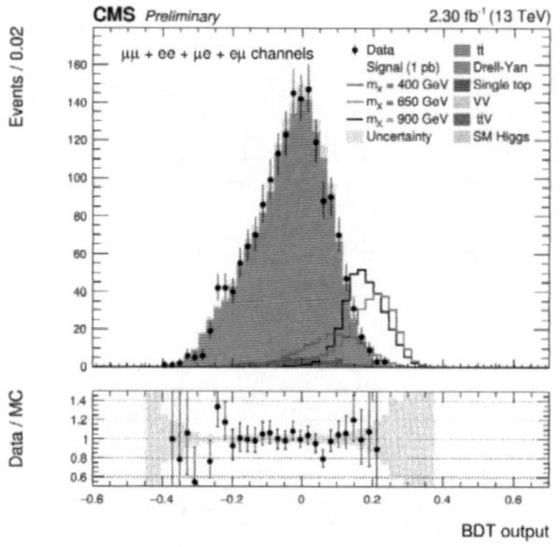

Figure 51. BDT distribution in the search for resonant $pp \to X \to HH \to bb\ell\ell$ [19].

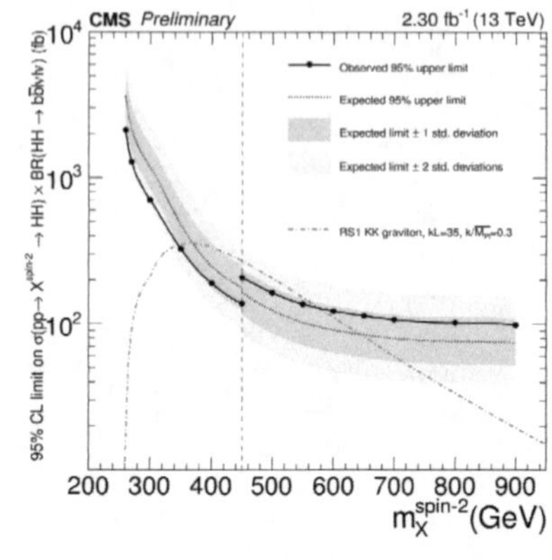

Figure 52. Upper limits as a function of the resonant $pp \to X \to HH \to bb\ell\ell$ mass [19].

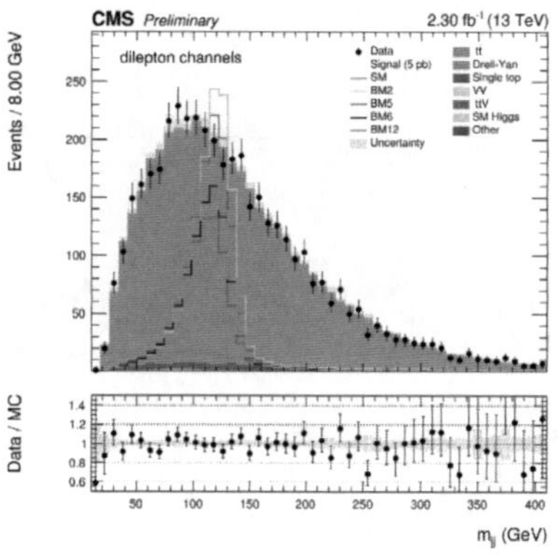

Figure 53. Di-jet invariant mass distribution in the search for non-resonant $pp \to HH \to bb\ell\ell$ [20].

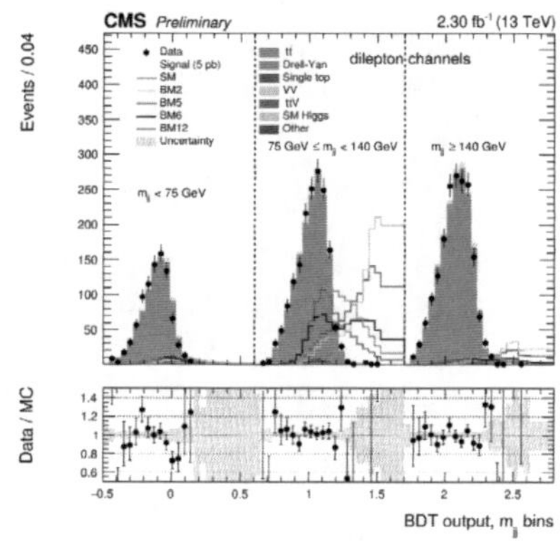

Figure 54. Di-jet invariant mass distribution in the search for non-resonant $pp \to HH \to bb\ell\ell$ [20].

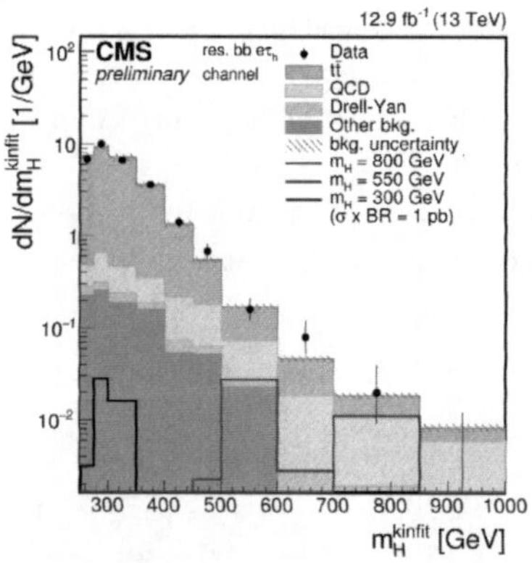

Figure 55. Invariant mass distribution, after the kinematic fit, of $X \to HH \to bb\tau\tau$ [21].

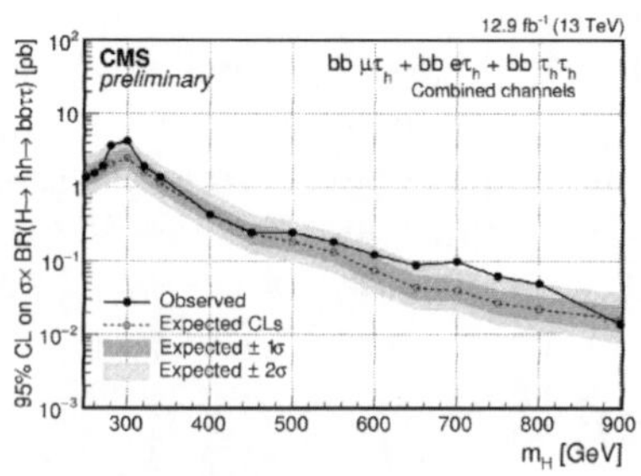

Figure 56. Upper limit for the search of $X \to HH \to bb\tau\tau$, as a function of the resonance mass [21].

background ratio. Figure 57 shows the BDT output distribution in one of the $\tau(\mu\nu\nu)\tau(\text{had.}\nu)$ subchannel. The signal has been extracted fitting the di-Higgs mass distribution, as shown in Fig. 58. The results have been interpreted as 95% CL limits on the cross-section, as a function of the coupling modifier $\kappa_\lambda = \lambda_{HHH}/\lambda^{SM}_{HHH}$. They are shown in Fig. 59. For $\kappa_\lambda = 1$, the observed (expected) 95% CL upper limit is about 200 (170) σ_{SM}.

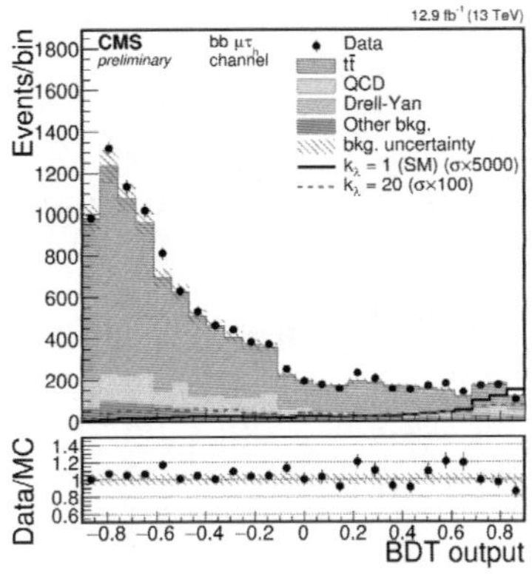

Figure 57. BDT distribution for the search of non-resonant $pp \to HH \to bb\tau\tau$ [22].

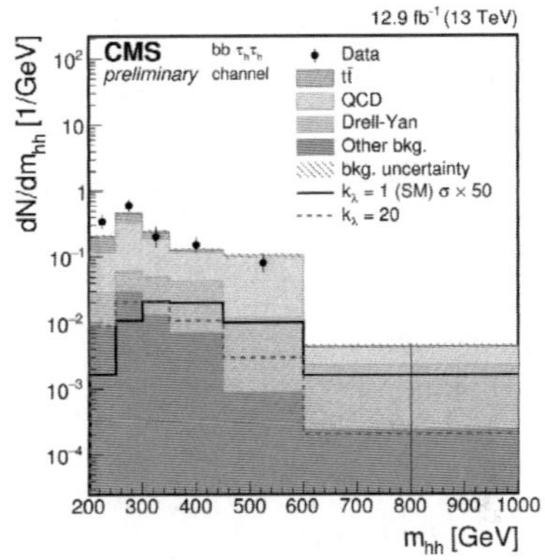

Figure 58. Higgs boson pair invariant mass distribution in the search of non-resonant $pp \to HH \to bb\tau\tau$ [22].

5. Run 2 - BSM Higgs

5.1. Higgs to invisible

If a massive weakly interacting particle couple with the Higgs boson, it would give the decay $H \to$ invisible, if $m_H > 2m_X$. This decay has been searched in different Higgs production channels: VBF H(inv.), $Z(\ell\ell)H$(inv.), $Z(bb)H$(inv.), $V(jj)H$(inv.), ggF H(inv.) + jets.

Figure 60 shows the upper limit on the H(inv.) branching ratio, for all channels and their combination. Figure 61 shows the likelihood scan of the branching ratio of $H \to$ inv. for the Run 1 and Run 2 analyses, and their combination. The observed (expected) upper limit for the combination is 24% (23%).

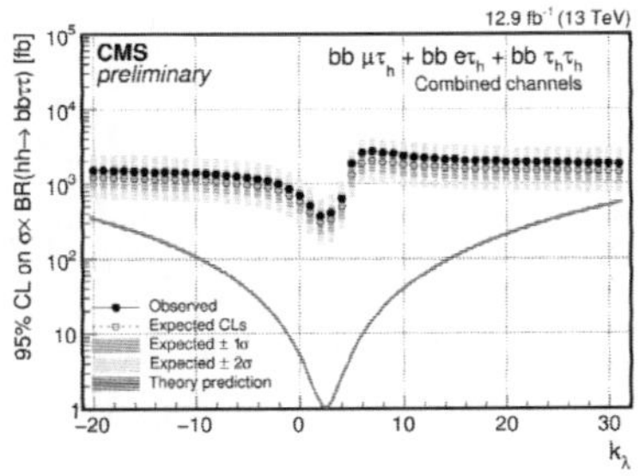

Figure 59. Cross-section 95% upper limit as a function of κ_λ obtained with the search for non-resonant $pp \to HH \to bb\tau\tau$ [22].

IOP Conf. Series: Journal of Physics: Conf. Series **873** (2017) 012004 doi:10.1088/1742-6596/873/1/012004

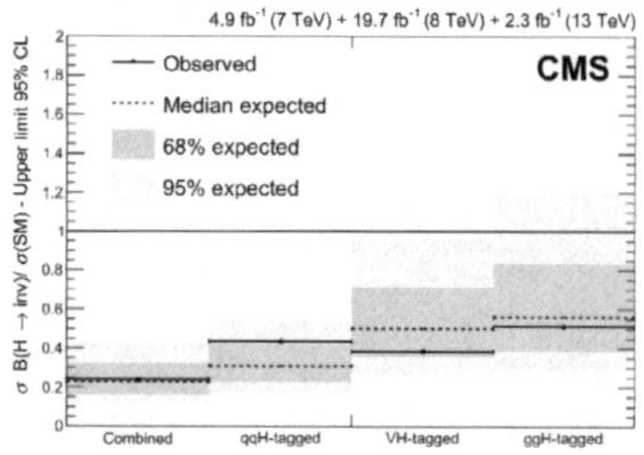

Figure 60. Upper limit on the invisible Higgs boson branching ratio, for each category and the combination [23].

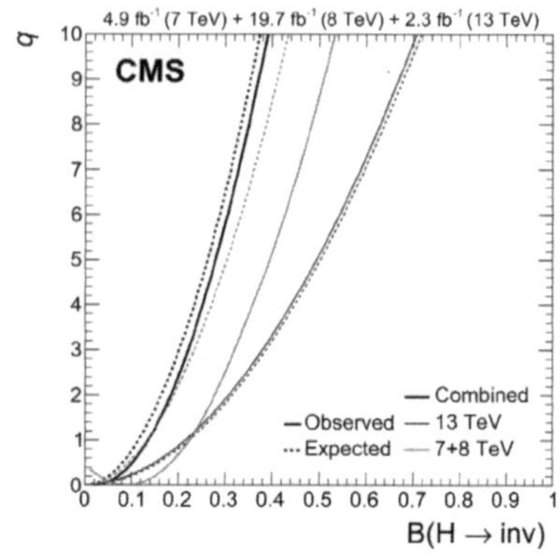

Figure 61. Log-likelihood scan as a function of invisible Higgs boson branching ratio, for the Run 1 and Run 2 analyses, and their combination [23].

5.2. $tH(bb)$

The search for the Higgs boson produced in association with a single top quark is a powerful way to probe the sign of the top quark Yukawa coupling (y_t). Figure 62 shows the two leading diagrams that produce the tH signal. In the SM, the interference of these two processes is destructive. Any alteration of y_t or y_W would increase the cross section: it increases of about a factor ten for $y_t = -1$.

The analysis is performed using two BDTs, where jets are assigned according to the tH and tt hypothesis. They are inputs of a final BDT, shown in Fig. 63, that is used for the signal extraction. The upper limit, as a function of y_t, is shown in Fig. 64. For $y_t = -1$, the observed (expected) upper limit is $\sigma < 6.0\,(6.4)\sigma_{y_t=-1}$.

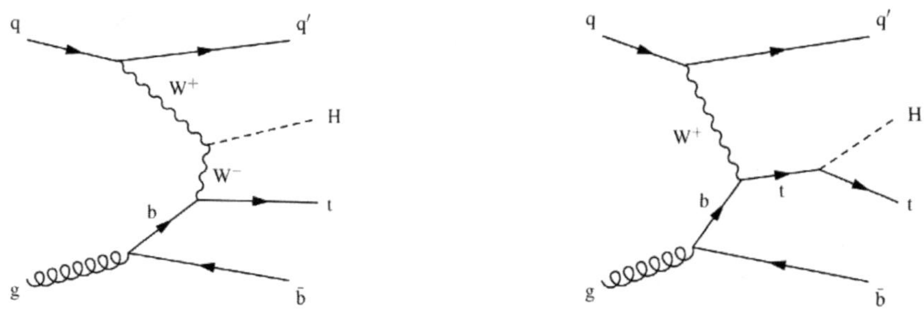

Figure 62. Two leading diagram of the tH production.

5.3. $H \to \tau\mu$ (lepton flavour violation)

In Run 1, a slight excess was reported in the lepton flavour violating channel $H \to \tau\mu$. The search has been repeated with the 2015 data and, oppositely to the Run 1 result, a background underfluctuation has been reported. Figure 65 shows the distribution of the invariant mass $m_{\mu\tau}$ for the analysis performed in Run 1. The upper limit obtained in each category is shown in Fig. 66. The observed (expected) combined upper limit is $\mu < 1.20\%(1.62\%)$, and it does not exclude the Run 1 excess ($\mu = 0.80 \pm 0.37\%$).

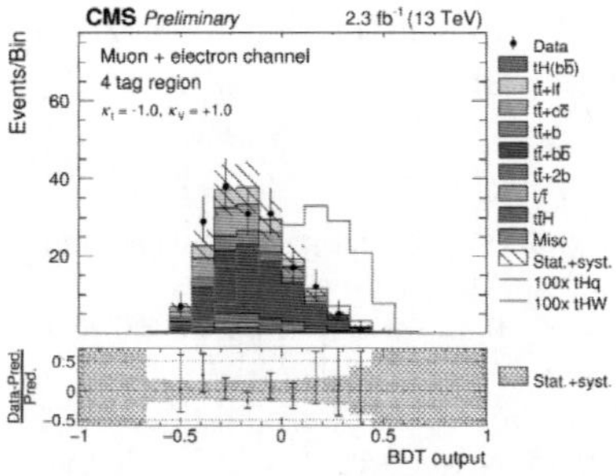

Figure 63. BDT distribution for the search for $tH(bb)$ [24].

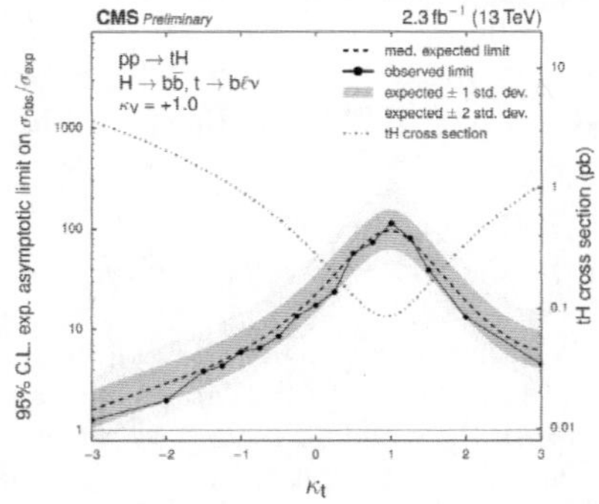

Figure 64. Cross-section 95% upper limit, normalized to the expectation, as a function of κ_t [24].

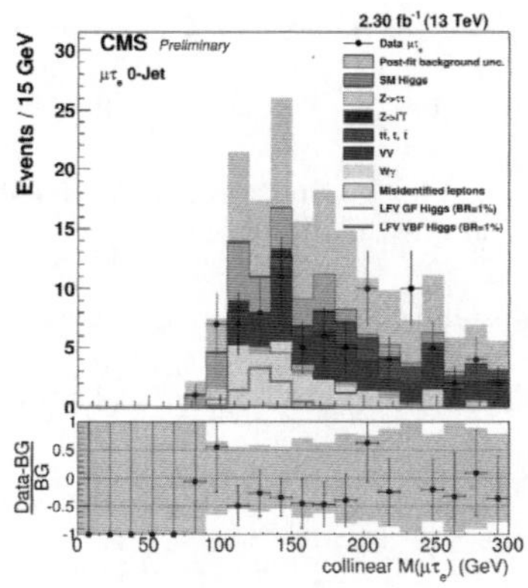

Figure 65. Invariant mass between μ and $\tau(e\nu_e\nu_\tau)$, assuming that the neutrinos have the same direction of the electron [25].

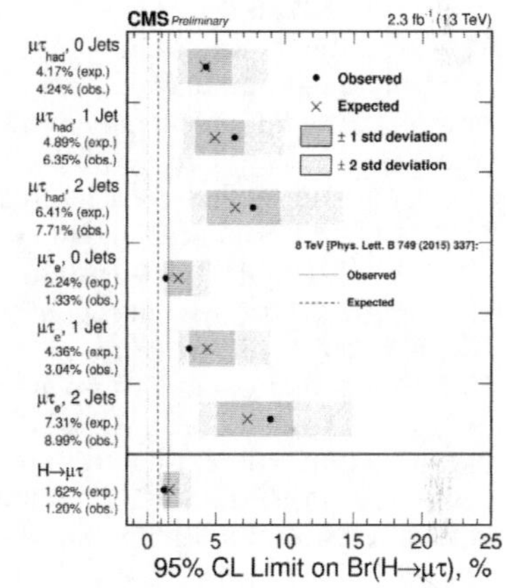

Figure 66. Branching ratio 95% upper limit of the $H \to \mu\tau$ decay [25].

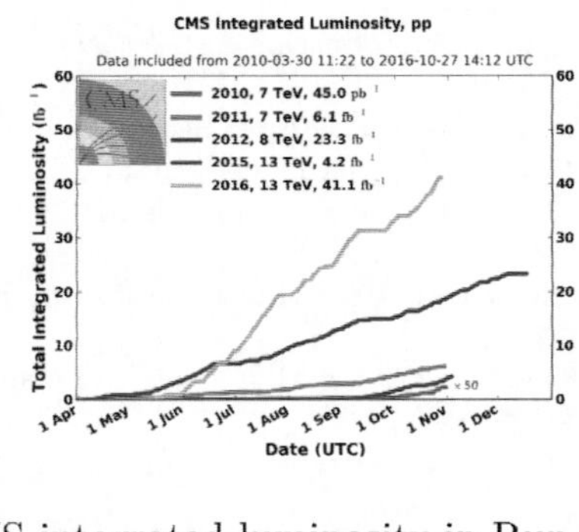

Figure 67. CMS integrated luminosity in Run 1 and Run 2 [26].

6. Conclusions and outlook

In Run 2, the Higgs boson has been observed at 13 TeV and the sensitivity of many analysis is already close to the Run 1. All properties confirm the SM expectation and no hint of new

Physics have been found. In 2016, LHC delivered more than 40 fb^{-1} of proton-proton collisions, as shown in Fig. 67, and LHC is expected to deliver more than 100 fb^{-1} by the end of Run 2. Many important CMS Higgs physics results are expected to come soon.

7. Bibliography

[1] LHC Higgs Cross Section Working Group Collaboration, "Handbook of LHC Higgs Cross Sections: 3. Higgs Properties".

[2] CMS Coll 2008 "The CMS Experiment at the CERN LHC". CMS Collaboration, *JINST* 3 **S08004**

[3] CMS Coll 2015 "Precise determination of the mass of the Higgs boson and tests of compatibility of its couplings with the standard model predictions using proton collisions at 7 and 8 TeV" *Eur. Phys. J.* C **75** 212

[4] CMS Coll 2015 "Search for the associated production of the Higgs boson with a top-quark pair" *J. High Energy Phys.* 09 **087** (2014)

[5] CMS Coll 2014 "Properties of the Higgs-like boson in the decay H to ZZ to 4l in pp collisions at $\sqrt{s}$ =7 and 8 TeV" *CMS PAS* **HIG-13-002**

[6] CMS Coll 2015 "Constraints on the spin-parity and anomalous HVV couplings of the Higgs boson in proton collisions at 7 and 8 TeV" *Phys. Rev.* D **92**

[7] CMS Coll 2014 "Constraints on the Higgs boson width from off-shell production and decay to Z-boson pairs", *Phys. Lett.* B**736** 64

[8] CMS Coll 2016 "Measurement of differential and integrated fiducial cross sections for Higgs boson production in the four-lepton decay channel in pp collisions " *JHEP* 04 **005**

[9] CMS Coll 2016 "Measurement of differential cross sections for Higgs boson production in the diphoton decay channel in pp collisions at $\sqrt{s}$ = 8 TeV" *Eur. Phys. J.* C **76** 13

[10] CMS Coll 2015 "Search for lepton-flavour-violating decays of the Higgs boson" *Phys. Lett.* B **749** 337

[11] CMS Coll 2016 "Updated measurements of Higgs boson production in the diphoton decay channel at $\sqrt{s}$ = 13 TeV" *CMS PAS* **HIG-16-020**

[12] CMS Coll 2016 "Measurements of properties of the Higgs boson and search for an additional resonance in the four-lepton final state at $\sqrt{s}$ = 13 TeV" *CMS PAS* **HIG-16-033**

[13] CMS Coll 2016 "Search for associated production of Higgs bosons and top quarks in multilepton final states at $\sqrt{s}$ = 13 TeV" *CMS PAS* **HIG-16-022**

[14] CMS Coll 2016 "Search for $t\bar{t}H$ production in the $H \to b\bar{b}$ decay channel with 2016 pp collision data at $\sqrt{s}$ = 13 TeV" *CMS PAS* **HIG-16-038**

[15] CMS Coll 2016 "Search for the standard model Higgs boson produced through vector boson fusion and decaying to $b\bar{b}$ with proton-proton collisions at $\sqrt{s}$ = 13 TeV" *CMS PAS* **HIG-16-003**

[16] CMS Coll 2016 "Search for H($b\bar{b}$)H($\gamma\gamma$) decays at $\sqrt{s}$ = 13 TeV " *CMS PAS* **HIG-16-032**

[17] CMS Coll 2016 "Search for resonant pair production of Higgs bosons decaying to two bottom quark-antiquark pairs in proton-proton collisions at 13 TeV" *CMS PAS* **HIG-16-002**

[18] CMS Coll 2016 "Search for non-resonant pair production of Higgs bosons in the

[19] CMS Coll 2016 "Search for resonant Higgs boson pair production in the $b\bar{b}\ell\nu\ell\nu$ final state at $\sqrt{s}$=13 TeV" *CMS PAS* **HIG-16-011**

[20] CMS Coll 2016 "Search for Higgs boson pair production in the $b\bar{b}\ell\nu\ell\nu$ final state at $\sqrt{s}$=13 TeV" *CMS PAS* **HIG-16-024**

[21] CMS Coll 2016 "Search for resonant Higgs boson pair production in the $b\bar{b}\tau^+\tau^-$ final state using 2016 data" *CMS PAS* **HIG-16-029** $b\bar{b}b\bar{b}$ final state with 13 TeV CMS data" *CMS PAS* **HIG-16-026**

[22] CMS Coll 2016 "Search for non-resonant Higgs boson pair production in the $bb\tau^+\tau^-$ final state using 2016 data" *CMS PAS* **HIG-16-028**

[23] CMS Coll 2016 "Searches for invisible decays of the Higgs boson in pp collisions at $\sqrt{s}$= 7, 8, and 13 TeV " *J. High Energy Phys.* 02 **135** (2017)

[24] CMS Coll 2016 "Search for $H \to b\bar{b}$ in association with a single top quark as a test of Higgs boson couplings at $\sqrt{s}$ = 13 TeV" *CMS PAS* **HIG-16-019**

[25] CMS Coll 2016 "Search for lepton flavour violating decays of the Higgs boson in the $\mu\tau$ final state at 13 TeV" *CMS PAS* **HIG-16-005**

[26] `https://twiki.cern.ch/twiki/bin/view/CMSPublic/LumiPublicResults`

Tests of discrete symmetries in the kaon system

Antonio Di Domenico

Dipartimento di Fisica, Sapienza Università di Roma,
and I.N.F.N. Sezione di Roma, P.le A. Moro, 2, I-00185 Rome, Italy

E-mail: antonio.didomenico@roma1.infn.it

Abstract. The status of ongoing experiments with kaons is reported, with a focus on the prospects for discrete symmetries tests.

1. Introduction

The first observation of a charged kaon dates back to an experiment performed with cosmic rays in the French Alps by Leprince-Ringuet and L'héritier [1] during the II World War, while the first evidence of a neutral kaon, i.e. the decay of a neutral particle into two charged particles, came few years later, in 1947, with the observation by Rochester and Butler of the decay of a neutral particle into two charged particles, the famous V-particles [2].

The discovery of K mesons and their properties had an important influence on the development of particle physics and its theoretical understanding culminated with the formulation of the Standard Model, and also provided beautiful tests of the basic principles of Quantum Mechanics. For instance the study of the properties of kaon decays led to the introduction of the strangeness quantum number by Gell-Mann and Nishijima, and to the explanation of the regeneration phenomena as one of the most beautiful manifestation of the superposition principle of Quantum Mechanics by Pais and Piccioni.

The kaon system is also one of the most suited physical system for the study of the fundamental discrete symmetries of Nature, exibhiting parity non-conservation, the $\tau - \theta$ anomaly phenomenon [3], a CP violation effect in the $K^0 - \bar{K}^0$ mixing, and also time reversal violation, and direct CP violation.

After more than 70 years from their first observation, kaons are still the subject of an intense experimental scrutiny, especially concerning discrete symmetries. The present paper reports the status of the ongoing experiments in the field, updating previous reports [4, 5]; for general and historical reviews the reader is referred to several excellent papers and books in the literature [6].

2. Rare kaon decays

2.1. The golden mode: $K \to \pi\nu\bar{\nu}$

Precise measurements in the Flavor sector can probe extremely high energy scales, in a complementary way with respect to the LHC high energy frontier. In fact, the present LHC searches at high transverse momenta exclude wide regions of the parameter space for New Physics (NP), thus increasing the importance of the study of rare processes sensitive to very high energy. In this respect the improved measurements of the branching ratios of the two golden modes $K^+ \to \pi^+\nu\bar{\nu}$ and $K_L \to \pi^0\nu\bar{\nu}$ will constitute one of the top highlights of flavour physics in the next years. $K^+ \to \pi^+\nu\bar{\nu}$ is CP conserving while $K_L \to \pi^0\nu\bar{\nu}$ is governed by CP violation. Both decays are very rare being FCNC processes and with the highest CKM suppression. These decays are theoretically very clean in SM, and the calculation of their branching ratios within the SM includes next-to-leading order (NLO) QCD corrections to the top quark contributions, NNLO QCD corrections to the charm contribution, and NLO electroweak corrections to both top and charm contributions. Moreover, extensive calculations of isospin breaking effects and non-perturbative effects have been performed. The most recent evaluations are

$$BR(K_L \to \pi^0\nu\bar{\nu}) = (3.4 \pm 0.6) \times 10^{-11} \quad (1)$$
$$BR(K^+ \to \pi^+\nu\bar{\nu}) = (8.4 \pm 1.0) \times 10^{-11} \quad (2)$$

obtained using the parametric expressions for the branching ratios of these two decays in terms of the CKM parameter inputs [7].

In some models of NP large enhancement with respect to the above SM predictions are possible. In this sense the $K \to \pi\nu\bar{\nu}$ modes play a key role in seeking NP beyond the SM. This is clear from Fig.1, which reproduces the BR values predicted by the SM and by some models of NP [8]. Apart from establishing a direct signal of NP, the correlation of the BR of the two modes can be exploited to probe the flavour structure of NP theories and, therefore, to distinguish among different classes of NP scenarios. The exclusion region given by the Grossman-Nir consistency condition [9], and limiting the BR ratio of the two decay modes in the correlation plane is also reported in Fig.1.

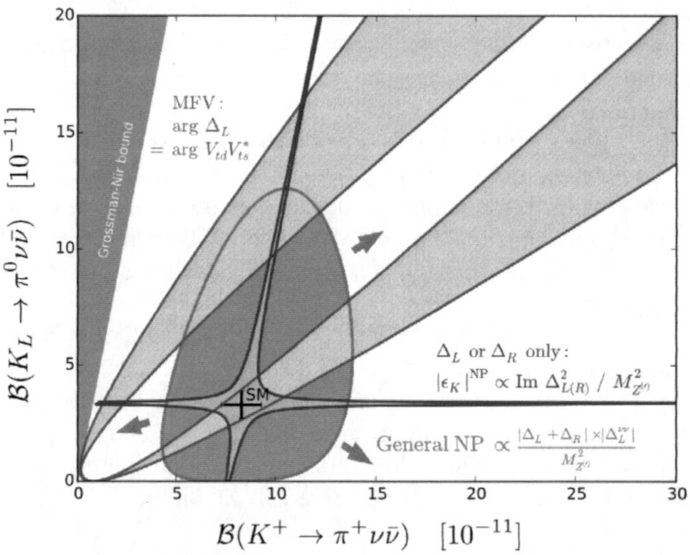

Figure 1. Correlation between the branching ratios of the charged and neutral $K \to \pi\nu\bar{\nu}$ decays showing the predictions of SM and some NP models [8].

Experimentally the charged decay mode has been observed by E959 and E787 experiments at BNL. The result $BR(K^+ \to \pi^+\nu\bar{\nu}) = (1.73 \pm 1.15) \times 10^{-10}$ is based on 7 events [10] and is compatible with the SM prediction. The CP-violating neutral $K_L \to \pi^0\nu\bar{\nu}$ decay has not been observed yet. An upper limit on the BR was set by the E391a collaboration at KEK [11]: $BR(K_L \to \pi^0\nu\bar{\nu}) < 260 \times 10^{-10}$ (90% C.L.).

IOP Conf. Series: Journal of Physics: Conf. Series **873** (2017) 012005 doi:10.1088/1742-6596/873/1/012005

2.2. NA62 experiment at CERN

The main goal of the NA62 experiment at CERN [12] is to collect about 100 $K^+ \to \pi^+\nu\bar{\nu}$ events with a signal-over-background ratio $S/B \sim 5$ by the end of 2018.

Its scheme is typical of a fixed-target experiment, with a proton beam from SPS hitting a target and producing an intense unseparated beam of charged hadrons selected in momentum at 75 GeV/c. The experimental method relies on exploiting a decay-in-flight technique, a long decay region with an extended detector and the event-by-event measurement of kaon momenta. Positive charged kaons are used in order to get, at the same proton energy and flux, higher kaon fluxes and lower pion background.

The NA62 analysis strategy is based on the accurate kinematic reconstruction of all the particles detected in the event to disentangle the signal from the huge amount of background processes, a precise timing to associate correctly the π^+ with the parent K^+, a system of efficient vetoes to reject events with γ's and μ's in the final state, effective particle identification systems to identify K^+ among other particles of the intense hadron beam and to distinguish π^+ from μ^+ and e^+ in the final states (see Fig.2).

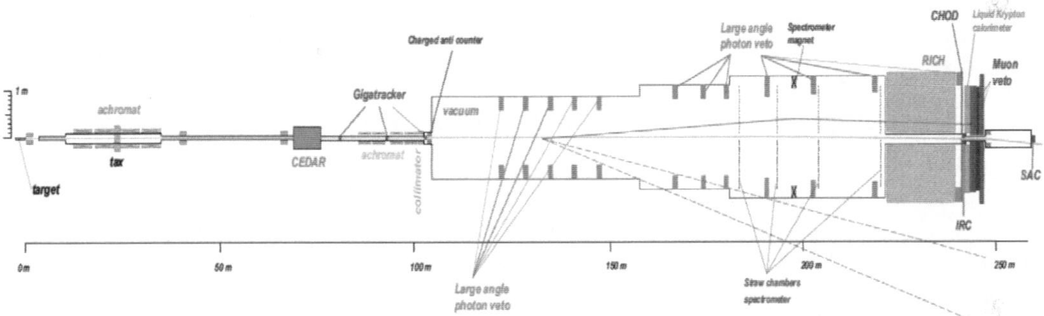

Figure 2. Drawing of the NA62 apparatus at SPS, CERN.

The trigger system and the beam line commissioning up to nominal intensity have been completed in 2015, together with a low intensity run with a minimum bias trigger for detector quality studies. The latter show good performance of the apparatus, in line with the expectations. After the completion of the detector commissioning, the Physics run started in 2016 and is currently ongoing.

Possible extension of the run starting from year 2021 with the aim of a broader Physics program is under discussion and will partly depend on the outcome of the present run.

2.3. KOTO experiment at J-PARC

The KOTO experiment [13] at the high-intensity proton accelerator facility J-PARC (Japan Proton Accelerator Research Complex) aims to observe for the first time the rare decay $K_L \to \pi^0\nu\bar{\nu}$, which is experimentally very challenging because both initial state and final decay products are completely neutral.

A decay volume for K_L is surrounded by particle detectors. The signature of a $K_L \to \pi^0\nu\bar{\nu}$ decay is two photons from a π^0 decay and no other visible particles in the final state. An electromagnetic calorimeter is placed downstream of the decay volume to detect the two photons. All the K_L decay modes except $K_L \to \pi^0\nu\bar{\nu}$ and $K_L \to \gamma\gamma$ have at least two charged particles, or two or more extra photons in the final state. These decays can be rejected by detecting additional particles with the surrounding detectors. The $K_L \to \gamma\gamma$ decays can be rejected by requiring a finite transverse momentum for the two photon system. In the case of $K_L \to \pi^0\nu\bar{\nu}$

decay, the two photon system has a finite transverse momentum due to the missing momentum of the two undetected neutrinos. The decay vertex is calculated and the $K_L \to \pi^0\nu\bar{\nu}$ decay reconstructed from two photons in the calorimeter, with the assumption that the two photons come from a π^0 decay on the line of flight.

The KOTO detector consists mainly of the CsI crystal calorimeter, the charged-particle veto counters (CV) in front of it, and the main-barrel photon veto counters (MB) surrounding the decay volume in the vacuum vessel (see Fig.3).

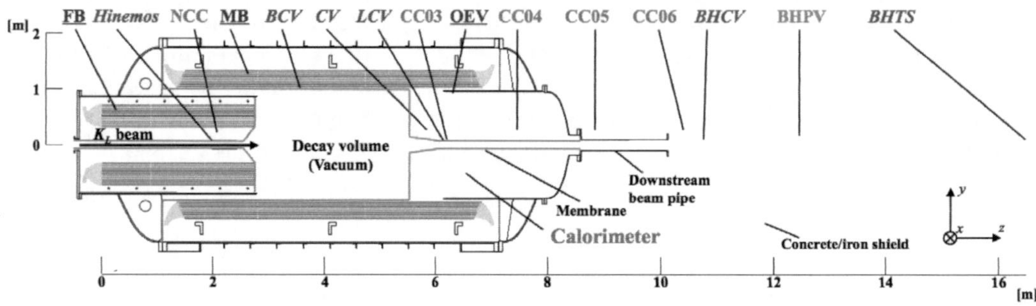

Figure 3. Sketch of the KOTO detector at J-PARC [13].

The analysis of the first 100 hours of the physics run in 2013 [13], after determining all the selection criteria and estimating background levels, showed one observed event in the signal region, as shown in Fig.4, while 0.34 ± 0.16 background events were expected. This result corresponds to an upper limit of 5.1×10^{-8} for the branching fraction of the $K_L \to \pi^0\nu\bar{\nu}$ decay at the 90% confidence level (C.L.), still higher than the previous limit set by the E391a experiment.

After several detector upgrades, e.g. the replacement of the vacuum pipe with a thinner one and the installation of new scintillator counters to improve the rejection of the $K_L \to \pi^+\pi^-\pi^0$ background; the installation of a beam profile monitor for improving the rejection of neutrons, and of an In-beam charged veto chamber to cope with higher beam intensities, in years 2015 and 2016 the KOTO experiment continued to collect data, and improved results are expected in the near future.

2.4. Searches for rare kaon decays at LHCb

The LHCb experiment recently proved to be competitive in the searches for rare kaon decays at LHC at least for the cleanest decay channels with a closed kinematics, a controllable background, and in the proximity of the interaction region like the $K_S \to \mu^+\mu^-$ decay. This is a Flavour Changing Neutral Current (FCNC) transition that has not yet been observed, and is suppressed in the SM, with an expected branching fraction [14] $BR(K_S \to \mu^+\mu^-) = (5.0 \pm 1.5) \times 10^{-12}$. NP effects might enhance this value.

The LHCb detector is a single-arm forward spectrometer covering the pseudorapidity range $2 < \eta < 5$. A search for the decay $K_L \to \mu^+\mu^-$ has been performed using a data sample of 1.0 fb^{-1} of pp collisions at $\sqrt{s} = 7$ TeV collected by the LHCb experiment at LHC. The observed number of candidates is consistent with no signal hypothesis, corresponding to an upper limit of $BR(K_S \to \mu^+\mu^-) < 11(9) \times 10^{-9}$ at 95 (90)% confidence level [15]. An analysis using 2 fb^{-1} of data further improves this limit and provides the following preliminary result: $BR(K_S \to \mu^+\mu^-) < 6.9(5.8) \times 10^{-9}$ at 95 (90)% CL [16].

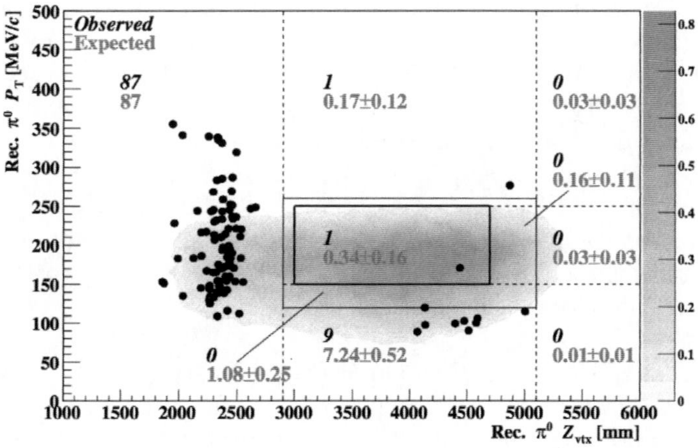

Figure 4. Reconstructed π^0 transverse momentum (P_T) versus decay vertex position (Z_{vtx}) of the events with all the analysis cuts imposed. The region surrounded with a thick solid line is the signal region. The black dots represent the data, and the contour indicates the distribution of the $K_L \to \pi^0 \nu\bar{\nu}$ decay from MC [13].

The LHCb trigger system has been partially modified to overcome one of the main limitation for these studies, and it will undergo a major upgrade in the next future, facilitating not only the study of $K_S \to \mu^+\mu^-$ but also of other rare kaon decays as $K_S \to e^+e^-\mu^+\mu^-$, $K_S \to e^+e^-e^+e^-$, and $K_S \to \pi^0\mu^+\mu^-$.

3. Testing CPT symmetry with entangled neutral kaons

3.1. Direct test of CPT symmetry in transitions

CPT symmetry at present appears to be the only discrete symmetry of Quantum Mechanics respected in Nature. This result has a very solid theoretical foundation in the well known *CPT* theorem [17, 18, 19, 20], ensuring exact *CPT* invariance for any quantum field theory formulated on flat space-time assuming (1) Lorentz invariance, (2) Locality, and (3) Unitarity (i.e. conservation of probability).

A violation of the *CPT* symmetry would have a dramatic impact on our present theoretical picture and would definitely constitute an unambiguous signal of a New Physics framework, thus strongly motivating both experimental searches and theoretical studies on this subject.

CPT invariance has been confirmed by all present experimental tests, particularly in the neutral kaon system where strong limits have been set to a variety of possible *CPT* violation effects which might arose in a quantum gravity scenario [21, 22, 23, 24, 25, 26, 27]. The best limits on the δ parameter expressing *CPT* violation in the K^0-$\bar{K}^0$ mixing matrix, i.e. in the *standard* Weisskopf-Wigner approach [28] are obtained in the CPLEAR experiment for $\Re\delta$ [29], and using the Bell-Steinberger relation for $\Im\delta$ [30, 31], yielding a stringent limit on the difference of mass terms for K^0 and $\bar{K}^0$: $|m_{K^0} - m_{\bar{K}^0}| < 4 \times 10^{-19}$ GeV at 95% c.l. [21].

The *CPT* violating probe has been, however, often limited to a difference of masses (and other intrinsic properties) for a particle and its anti-particle, i.e. to diagonal mass terms. In many physical phenomena the perturbing effect does not appear at first order in perturbation theory: it would be sufficient that the perturbation breaks a symmetry of the non-perturbed states. This vanishing effect at first order for the diagonal elements, like e.g. the case of the electric dipole moment for T violation, is not present for transitions (non-diagonal elements) [32].

In the following a recently proposed *CPT* test for transitions in the neutral kaon system will be discussed in some detail [33], where the exchange of *in* and *out* states (and *CP* conjugation), required for a direct and genuine *CPT* test, is performed exploiting the entanglement of the kaon pair produced at a ϕ-factory. This methodology has been proposed for a direct test of the T symmetry in the same context [34], similarly to the one adopted for the performed test in the B meson system at B-factories [35, 36, 37, 38, 32, 39]. The decay is not an essential ingredient for a non-vanishing effect and it is only used for filtering the appropriate initial and final states of the neutral kaon transition [40]. Explicitly, in the standard Weisskopf-Wigner approach to this system, the *CPT*-violating effects can be connected to the $\Re\delta$ parameter, a genuine *CPT*-violating effect independent of $\Delta\Gamma = \Gamma_S - \Gamma_L$, with Γ_S and Γ_L the widths of the physical states.

In order to implement this test, the Einstein-Podolsky-Rosen (EPR) entanglement of neutral mesons produced at a ϕ-factory must be exploited. In fact in this case the initial state of the neutral kaon pair produced in $\phi \to \mathrm{K}^0\bar{\mathrm{K}}^0$ decay can be rewritten in terms of any pair of orthogonal states $|\mathrm{K}_+\rangle$ and $|\mathrm{K}_-\rangle$:

$$|i\rangle = \frac{1}{\sqrt{2}}\{|\mathrm{K}^0\rangle|\bar{\mathrm{K}}^0\rangle - |\bar{\mathrm{K}}^0\rangle|\mathrm{K}^0\rangle\} = \frac{1}{\sqrt{2}}\{|\mathrm{K}_+\rangle|\mathrm{K}_-\rangle - |\mathrm{K}_-\rangle|\mathrm{K}_+\rangle\} . \tag{3}$$

Here one can consider the states $|\mathrm{K}_+\rangle$, $|\mathrm{K}_-\rangle$ defined as follows: $|\mathrm{K}_+\rangle$ is the state filtered by the decay into $\pi\pi$ ($\pi^+\pi^-$ or $\pi^0\pi^0$), a pure $CP = +1$ state; analogously $|\mathrm{K}_-\rangle$ is the state filtered by the decay into $3\pi^0$, a pure $CP = -1$ state. Their orthogonal states correspond to the states which cannot decay into $\pi\pi$ or $3\pi^0$, defined, respectively, as

$$\begin{aligned} |\widetilde{\mathrm{K}}_-\rangle &\propto [|\mathrm{K}_\mathrm{L}\rangle - \eta_{\pi\pi}|\mathrm{K}_\mathrm{S}\rangle] \\ |\widetilde{\mathrm{K}}_+\rangle &\propto [|\mathrm{K}_\mathrm{S}\rangle - \eta_{3\pi^0}|\mathrm{K}_\mathrm{L}\rangle] , \end{aligned} \tag{4}$$

with $\eta_{\pi\pi} = \langle\pi\pi|T|\mathrm{K}_\mathrm{L}\rangle/\langle\pi\pi|T|\mathrm{K}_\mathrm{S}\rangle$ and $\eta_{3\pi^0} = \langle 3\pi^0|T|\mathrm{K}_\mathrm{S}\rangle/\langle 3\pi^0|T|\mathrm{K}_\mathrm{L}\rangle$. With these definitions of states, it can be shown that the condition of orthogonality $\langle\mathrm{K}_-|\mathrm{K}_+\rangle = 0$, (i.e. $|\mathrm{K}_+\rangle \equiv |\widetilde{\mathrm{K}}_+\rangle$ and $|\mathrm{K}_-\rangle \equiv |\widetilde{\mathrm{K}}_-\rangle$) corresponds to assume negligible direct *CP* (or *CPT*) violation contributions, assumption quite well satisfied for neutral kaons [33]. The validity of the $\Delta S = \Delta Q$ rule is also assumed, so that the two flavor orthogonal eigenstates $|\mathrm{K}^0\rangle$ and $|\bar{\mathrm{K}}^0\rangle$ are identified by the charge of the lepton in semileptonic decays, i.e. a $|\mathrm{K}^0\rangle$ can decay into $\pi^-\ell^+\nu$ and not into $\pi^+\ell^-\bar{\nu}$, and vice-versa for a $|\bar{\mathrm{K}}^0\rangle$[1].

Thus, exploiting the perfect anticorrelation of the states implied by Eq. (3), it is possible to have a "flavor-tag"or a "CP-tag", i.e. to infer the flavor (K^0 or $\bar{\mathrm{K}}^0$) or the *CP* (K_+ or K_-) state of the still alive kaon by observing a specific flavor decay ($\pi^+\ell^-\nu$ or $\pi^-\ell^+\bar{\nu}$) or CP decay ($\pi\pi$ or $\pi^0\pi^0\pi^0$) of the other (and first decaying) kaon in the pair.

In this way one can experimentally access – for instance – the transition $\mathrm{K}^0 \to \mathrm{K}_+$, taken as reference, and the $\mathrm{K}_+ \to \mathrm{K}^0$, $\bar{\mathrm{K}}^0 \to \mathrm{K}_+$ and $\mathrm{K}_+ \to \bar{\mathrm{K}}^0$ transitions, i.e. the T, *CP* and *CPT* conjugated transitions, respectively. Specifically for the *CPT* symmetry test, one can *directly* compare the probabilities for the reference transition and the *CPT* conjugated one through the following ratios of probabilities:

$$\begin{aligned} R_{1,CPT}(\Delta t) &= P\left[\mathrm{K}_+(0) \to \bar{\mathrm{K}}^0(\Delta t)\right]/P\left[\mathrm{K}^0(0) \to \mathrm{K}_+(\Delta t)\right] \\ R_{2,CPT}(\Delta t) &= P\left[\mathrm{K}^0(0) \to \mathrm{K}_-(\Delta t)\right]/P\left[\mathrm{K}_-(0) \to \bar{\mathrm{K}}^0(\Delta t)\right] \\ R_{3,CPT}(\Delta t) &= P\left[\mathrm{K}_+(0) \to \mathrm{K}^0(\Delta t)\right]/P\left[\bar{\mathrm{K}}^0(0) \to \mathrm{K}_+(\Delta t)\right] \\ R_{4,CPT}(\Delta t) &= P\left[\bar{\mathrm{K}}^0(0) \to \mathrm{K}_-(\Delta t)\right]/P\left[\mathrm{K}_-(0) \to \mathrm{K}^0(\Delta t)\right] . \end{aligned} \tag{5}$$

[1] It is important to underline that both these assumptions can be relaxed for some specific observables, as discussed below

The measurement of any deviation from the prediction $R_{i,CPT}(\Delta t) = 1$ imposed by *CPT* invariance is a signal of *CPT* violation.
It is worth noting that for $\Delta t = 0$:

$$R_{1,CPT}(0) = R_{2,CPT}(0) = R_{3,CPT}(0) = R_{4,CPT}(0) = 1 \,, \tag{6}$$

i.e. the *CPT*-violating effect is built in the time evolution of the system, and it is absent at $\Delta t = 0$, within our approximations.
For $\Delta t \gg \tau_S$, assuming the presence of *CPT* violation only in the mass matrix ($\delta \neq 0$) and nothing else, one gets:

$$R_{2,CPT}(\Delta t \gg \tau_S) \simeq 1 - 4\Re\delta \tag{7}$$

$$R_{4,CPT}(\Delta t \gg \tau_S) \simeq 1 + 4\Re\delta \,, \tag{8}$$

i.e. the *CPT*-violating effect built in the time evolution reaches a "plateau" regime and dominates in this limit. It is a *genuine* effect because $\Re\delta$ does not depend on $\Delta\Gamma$ as an essential ingredient [32, 40].

At a ϕ-factory one can define two observable ratios:

$$R^{\rm exp}_{2,CPT}(\Delta t) \equiv \frac{I(\ell^-, 3\pi^0; \Delta t)}{I(\pi\pi, \ell^-; \Delta t)} \tag{9}$$

$$R^{\rm exp}_{4,CPT}(\Delta t) \equiv \frac{I(\ell^+, 3\pi^0; \Delta t)}{I(\pi\pi, \ell^+; \Delta t)} \,, \tag{10}$$

where $I(f_1, f_2; \Delta t)$ are the double decay rates into decay products f_1 and f_2 as a function of the difference of kaon decay times Δt [34, 42], with f_1 occurring before f_2 decay for $\Delta t > 0$, and viceversa for $\Delta t < 0$.
They are related to the $R_{i,CPT}(\Delta t)$ ratios defined in eqs (5) as follows, for $\Delta t \geq 0$:

$$\begin{aligned} R^{\rm exp}_{2,CPT}(\Delta t) &= R_{2,CPT}(\Delta t) \times D_{CPT} \\ R^{\rm exp}_{4,CPT}(\Delta t) &= R_{4,CPT}(\Delta t) \times D_{CPT} \end{aligned} \tag{11}$$

whereas for $\Delta t < 0$ one has:

$$\begin{aligned} R^{\rm exp}_{2,CPT}(\Delta t) &= R_{1,CPT}(|\Delta t|) \times D_{CPT} \\ R^{\rm exp}_{4,CPT}(\Delta t) &= R_{3,CPT}(|\Delta t|) \times D_{CPT} \,, \end{aligned} \tag{12}$$

with

$$D_{CPT} = \frac{|\langle 3\pi^0|T|{\rm K}_-\rangle|^2}{|\langle \pi\pi|T|{\rm K}_+\rangle|^2} = \frac{{\rm BR}\,({\rm K_L} \to 3\pi^0)}{{\rm BR}\,({\rm K_S} \to \pi\pi)} \frac{\Gamma_L}{\Gamma_S} \,, \tag{13}$$

where the last r.h.s. equality holds with a high degree of accuracy, at least $\mathcal{O}(10^{-7})$. The value of D_{CPT} can be therefore directly evaluated from branching ratios and lifetimes.

The explicit expressions of ratios (9) and (10) (neglecting higher order terms in small parameters and for not too large negative Δt) are:

$$\begin{aligned} R^{\rm exp}_{2,CPT}(\Delta t) &= \frac{P[{\rm K}^0(0) \to {\rm K}_-(\Delta t)]}{P[{\rm K}_-(0) \to \bar{\rm K}^0(\Delta t)]} \times D_{CPT} \\ &\simeq |1 - 2\delta|^2 \left|1 + 2\delta e^{-i(\lambda_S - \lambda_L)\Delta t}\right|^2 \times D_{CPT} \,, \end{aligned} \tag{14}$$

$$R^{\text{exp}}_{4,CPT}(\Delta t) = \frac{P[\bar{\text{K}}^0(0) \to \text{K}_-(\Delta t)]}{P[\text{K}_-(0) \to \text{K}^0(\Delta t)]} \times D_{CPT}$$
$$\simeq |1+2\delta|^2 \left|1 - 2\delta e^{-i(\lambda_S - \lambda_L)\Delta t}\right|^2 \times D_{CPT} \,. \qquad (15)$$

The expected behavior of the observables $R^{\text{exp}}_{2,CPT}(\Delta t)$ and $R^{\text{exp}}_{4,CPT}(\Delta t)$ as a function of Δt, and without the approximations of eqs.(14) and (15), is shown in figure 5, where – for visualization purposes – the probabilities involved have been evaluated fixing the *CPT* violating parameters $\Re\delta$ and $\Im\delta$ to a value different from zero, and equal to their present uncertainties [21], i.e. $\Re\delta = 3.3 \times 10^{-4}$ and $\Im\delta = 1.6 \times 10^{-5}$. In figure 6 a zoom of the $\Delta t > 0$ region, where the "plateau" regimes (7) and (8) dominate, is shown. Experimentally, this is the most interesting and statistically most populated region, where the best sensitivity to *CPT* violation effects can be reached by the KLOE-2 experiment [33, 34, 41].

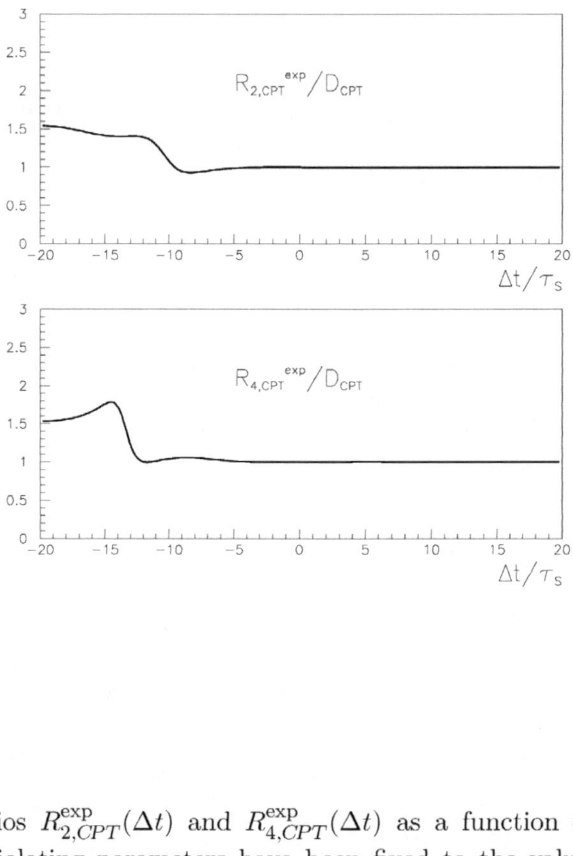

Figure 5. The ratios $R^{\text{exp}}_{2,CPT}(\Delta t)$ and $R^{\text{exp}}_{4,CPT}(\Delta t)$ as a function of Δt. For visualization purposes the *CPT* violating parameters have been fixed to the values $\Re\delta = 3.3 \times 10^{-4}$ and $\Im\delta = 1.6 \times 10^{-5}$.

The presence of direct *CP* violation contributions in the decay amplitudes, even though in principle could mimic *CPT* violation effects, turns out to be totally irrelevant for the plateau

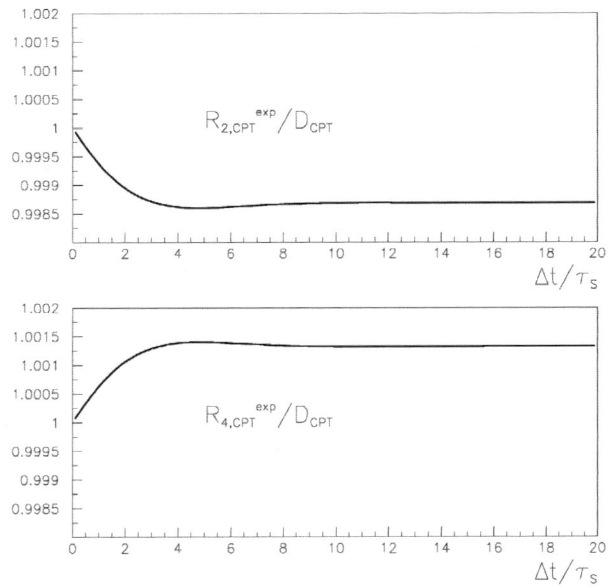

Figure 6. A zoom of the plots shown in Fig.1 in the region $0 \leq \Delta t \leq 20\tau_S$.

region $\Delta t \gg \tau_S$ (see detailed description in Ref. [33]). The effect of a possible violation of the $\Delta S = \Delta Q$ rule is also not affecting the *CPT* test in the same region with the double ratio defined as:

$$\frac{R^{\rm exp}_{2,CPT}(\Delta t \gg \tau_S)}{R^{\rm exp}_{4,CPT}(\Delta t \gg \tau_S)} = 1 - 8\Re\delta - 8\Re x_- \; . \tag{16}$$

with x_- describing *CPT* violation in the $\Delta S \neq \Delta Q$ semileptonic decay amplitudes. Therefore the double ratio (16) constitutes one of the most robust observables for the proposed *CPT* test. It is independent of D_{CPT}, and in the limit $\Delta t \gg \tau_S$ it exhibits a pure and genuine *CPT* violating effect, even without the assumptions of the validity of the $\Delta S = \Delta Q$ rule and of negligible contaminations from direct *CP* violation.

3.2. The KLOE-2 experiment at DAΦNE

KLOE-2 represents the continuation of the KLOE experiment with a new physics program mainly focused on the study of K short and η decays as well as on kaon interferometry and tests of discrete symmetries [41, 42]. The new data taking campaign aims to collect more than 5 fb^{-1}

of integrated luminosity at the upgraded DAΦNE, which is implementing an innovative collision scheme based on a *crab-waist* configuration [43, 44]. An improved KLOE detector [45] will allow in particular to perform CPT symmetry and quantum coherence tests using entangled neutral kaons with an unprecedented precision. Improvements of about one order of magnitude in almost all present limits on CPT violation and decoherence parameters are expected [41, 42, 23, 24]. The KLOE-2 data taking campaign started in November 2014, collected 3 fb^{-1} of integrated luminosity at the end of 2016, and the goal of 5 fb^{-1} is expected to be reached in the next 1-1.5 years.

KLOE-2 data are being analysed [46] to get a precise measurement of the two observable ratios $R^{\mathrm{exp}}_{2,CPT}(\Delta t)$ and $R^{\mathrm{exp}}_{4,CPT}(\Delta t)$. By considering a large Δt interval in the statistically most populated region, e.g. $0 \leq \Delta t \leq 300\ \tau_S$, a statistical sensitivity on the double ratio (16) of (3.0×10^{-3}) can be obtained for $L = 5\ \ \mathrm{fb}^{-1}$ [33].

4. Conclusions

Rare K decays constitute a unique probe of possible Physics beyond the SM, complementary to LHC. The two golden modes $\mathrm{K}^+ \to \pi^+\nu\bar{\nu}$ and $\mathrm{K_L} \to \pi^0\nu\bar{\nu}$ are addressed by the new generation of kaon experiments at hadron machines (NA62 and KOTO), both operational and presently collecting data. The first phase of the NA62 data taking campaign will be completed by the end of 2018.

The LHCb experiment turned out to be suited for the search of some rare $\mathrm{K_S}$ decays, and new results and improvements are expected after the detector upgrade.

A novel *CPT* test has been studied in the neutral kaon system based on the direct comparison of a transition probability with its *CPT* reverse transition. The proposed measurement is fully robust and might shed light on possible *CPT* violation mechanisms.

CPT symmetry and quantum mechanics tests with entangled neutral kaon pairs will be one of the key issues at KLOE-2. The newly proposed CPT test in transitions is being implemented, and new results are expected soon.

Acknowledgements

The author would like to warmly thank Prof. Maria Krawczyk, and all the organizing committee for the invitation to the DISCRETE 2016 symposium, and the very pleasant stay in Warsaw.

References

[1] L. Leprince-Ringuet and M. L'héritier, Comptes Rendus Acad. Sciences de Paris, séance du 13 Dec. 1944, p. 618.
[2] G. D. Rochester and C. C. Butler, Nature, **160**, 855 (1947).
[3] M. Baldo-Ceolin et al., Nuovo Cim. **6** (1957) 84.
[4] A. Di Domenico, J.Phys.Conf.Ser. 631 (2015) 012004.
[5] A. Di Domenico, J.Phys. Conf. Ser. 447 (2013) 012008.
[6] see e.g. G. C. Branco, L. Lavoura, and J. P. Silva, *CP violation*, Oxford Univ. Press, 1999;
M. Sozzi, *Discrete symmetries and CP violation*, Oxford Univ. Press, 2008;
T. K. Komatsubara, Progr. Part. Nucl. Phys. **67** (2012) 995;
M. Sozzi, J. Phys. Conf. Ser. 335 (2011) 012002.
[7] A. J. Buras, D. Buttazzo, J. Girrbach-Noe and R. Knegjens, JHEP 11 (2015) 033.
[8] A. J. Buras, D. Buttazzo, R. Knegjens, JHEP 11 (2015) 166.
[9] Y. Grossman and Y. Nir, Phys. Lett. B **398** (1997) 163
[10] A.V. Artamonov et al., Phys. Rev. D **79** (2009) 092004.
[11] J. K. Ahn et al., Phys. Rev. D **81** (2010) 072004.
[12] F. Gonnella, these proceedings.
[13] J. K. Ahn et al., KOTO collaboration, Prog. Theor. Exp. Phys. 2015, 00000; arXiv:1609.03637 [hep-ex] 12 Sep 2016.
[14] G. Ecker and A. Pich, Nucl. Phys. B **366** (1991) 189;
G. Isidori and R. Unterdorfer, JHEP 01 (2004) 009.

IOP Conf. Series: Journal of Physics: Conf. Series **873** (2017) 012005 doi:10.1088/1742-6596/873/1/012005

[15] R. Aaij et al., LHCb Collaboration, JHEP 01 (2013) 090.
[16] Miguel Ramos Pernas, presented at the KAON2016 conference.
[17] G. Lüeders, Ann. Phys. (NY) **2**, 1 (1957), reprinted in Ann. Phys. (NY) **281**, 1004 (2000)
[18] W. Pauli, in *Niels Bohr and the development of physics*, edited by W. Pauli, Pergamon, London, 1955, p.30
[19] J. S. Bell, Proc. R. Soc. London A **231**, 479 (1955)
[20] R. Jost, Helv. Phys. Acta **30**, 409 (1957)
[21] C. Patrignani et al. (Particle Data Group), Chin. Phys. **C 40**, 100001 (2016).
[22] F. Ambrosino et al., KLOE collaboration, Phys. Lett. B **642**, 315 (2006)
[23] A. Di Domenico and the KLOE collaboration, J. Phys. Conf. Ser. **171**, 012008 (2009)
[24] A. Di Domenico and the KLOE collaboration, Found. Phys. **40** , 852 (2010)
[25] D. Babusci et al., KLOE-2 Collaboration, Phys. Lett. B **730**, 89 (2014)
[26] R. Adler et al., CPLEAR collaboration, Phys. Lett. B **364**, 239 (1995).
[27] N. Mavromatos, these proceedings.
[28] V. Weisskopf and E. P. Wigner, Z. Phys. **63**, 54 (1930)
[29] A. Angelopoulos et al., CPLEAR collaboration, Phys. Lett. B **444**, 52 (1998)
[30] J. S. Bell and J. Steinberger, Proc. of the Oxford Int. Conf. on Elementary Particles (1965)
[31] F. Ambrosino et al., KLOE collaboration, JHEP **12**, 011 (2006)
[32] J. Bernabeu and F. Martinez-Vidal, Rev. Mod. Phys. **87**, 165 (2015)
[33] J. Bernabeu, A. Di Domenico, P. Villanueva-Perez, JHEP **10**, 139 (2015)
[34] J. Bernabeu, A. Di Domenico, P. Villanueva-Perez, Nucl. Phys. B. **868**, 102 (2013)
[35] M. C. Banuls and J. Bernabeu, Phys. Lett. B **464**, 117 (1999)
[36] M. C. Banuls and J. Bernabeu, Nucl. Phys. B **590**, 19 (2000)
[37] J. Bernabeu, F. Martínez-Vidal, P. Villanueva-Pérez, JHEP **08**, 064 (2012)
[38] J. P. Lees *et al.*, Babar Collaboration, Phys. Rev. Lett. **109**, 211801 (2012)
[39] J. Bernabeu, these proceedings.
[40] L. Wolfenstein, Int. J. Mod. Phys. E **8**, 501 (1999)
[41] G. Amelino-Camelia et al., Eur. Phys. J. C **68**, 619 (2010)
[42] A. Di Domenico (Editor), *Handbook on Neutral Kaon Interferometry at a ϕ-factory*, Frascati Phys. Ser. **43** (2007)
[43] M. Zobov et al., Phys. Rev. Lett. **104**, 174801 (2010).
[44] C. Milardi et al., JINST, **7**, 2012,T03002.
[45] E. De Lucia, these proceedings.
[46] A. Gajos, these proceedings.

IOP Conf. Series: Journal of Physics: Conf. Series **873** (2017) 012006 doi:10.1088/1742-6596/873/1/012006

Models and (some) Searches for CPT Violation: From Early Universe to the Present Era

Nick E. Mavromatos

King's College London, Department of Physics, Theoretical Particle Physics & Cosmology Group, Strand, London WC2R 2LS, UK

E-mail: Nikolaos.Mavromatos@kcl.ac.uk

Abstract. In the talk, I review theoretical models, inspired by quantum gravity, that may violate CPT symmetry. The amount of violation today (which is constrained severely by a plethora of experiments that I will not describe due to lack of space) need not be the same with the one that occurred in the Early Universe,. In certain models, one can obtain a precise temperature dependence of CPT violating effects, which is such that these effects are significant during the radiation era of the Universe, but are damped quickly so that they do not to affect nucleosynthesis and are negligible in the present epoch (that is, beyond experimental detection with the current experimental sensitivity). The CPT Violation (CPTV) in these models may arise from special properties of the background over which the fields of the model are propagating upon and be responsible for the generation of a matter-antimatter asymmetry, where any CP violation effects could only assist in the creation of the asymmetry, the dominant effect being CPTV. However, there are cases, where the CPTV arises as a consequence of an ill-defined CPT operator due to decoherence as a result of quantum gravity environmental degrees of freedom, inaccessible to a low-energy observer. I also discuss briefly the current-era phenomenology of some of the above models; in particular, for the ones involving decoherence-induced CPT violation, I argue that entangled states of neutral mesons (Kaons or B-systems) can provide smoking-gun sensitive tests or even falsify some of these models. If CPT is ill-defined one may also encounter violations of the spin-statistics theorem, with possible consequences for the Pauli Exclusion Principle.

1. Introduction and Summary

Invariance of a relativistic (*i.e.* Lorentz Invariant), local and unitary field theory Lagrangian under the combined transformations of Charge Conjugation (C), Parity (spatial reflexions, (P)) and reversal in Time (T), at any order, is guaranteed by the corresponding celebrated theorem [1]. This has important implications for the equality of masses and the absolute values of the various quantum numbers that characterise particles and antiparticles, and the equal amounts of matter and antimatter when they were created in the beginning of the Universe's evolution. On the other hand, today, there is an overwhelming dominance of matter over antimatter in the Cosmos, which calls for an explanation. Assuming the validity of the CPT theorem, A. Sakharov [2] has suggested that the observed matter-antimatter asymmetry in the Universe today is the result of *out of thermal equilibrium* processes in the early universe that violate C, CP and Baryon symmetries. The out of equilibrium assumption is a crucial one so that any asymmetry generated by the violation of C , B and CP symmetries in the expanding Universe is not washed out but remains to the current epoch, thereby explaining the observed dominance of matter over antimatter today.

Although all of Sakharov conditions are met qualitatively by the Standard Model (SM) of Particle Physics, unfortunately they are not valid quantitatively, meaning that the amount of CP violation within the Standard Model is some ten orders of magnitude smaller than the required one to produce the observable matter-antimatter asymmetry [3]. According to observations, the abundance of baryons over that of antibaryons is of order [4]

$$Y_{\Delta B} = \frac{n_B - n_{\bar{B}}}{n_\gamma} = (6.1 \pm 0.3) \times 10^{-10} \tag{1}$$

for temperatures $T \gtrsim 1$ GeV, where n_B is the number density of baryons, $n_{\bar{B}}$ is the number density of antibaryons and n_γ is the density of photons (proportional to the entropy density s of the Universe). This number was determined with accurate measurements of the Cosmic Microwave Background (CMB) radiation [5]. Out of equilibrium processes that generate a baryon asymmetry in the Universe are called collectively *Baryogenesis* [6]. Similarly, the generation of an asymmetry between leptons. and antileptons is known as *Leptogenesis* [7], and is expected to be of the same order of magnitude as $Y_{\Delta B}$.

Since the amount of CP violation in the SM is not sufficient to produce (1), one should look for models beyond the standard model that contain new sources of CP violation, that can be tuned so as to produce sufficient Baryogenesis and Leptogenesis. There is a plethora of such models in the current literature, ranging from supersymmetric theories, to theories with extra dimensions of space, including strings. There is no experimental evidence as yet for the realisation of such extended models in Nature.

Interesting models [8, 9] of Baryon asymmetry in the Universe involve a two stage process, during which one generates first a lepton asymmetry in the early Universe, by appropriate CP-violating non-equilibrium processes in the early Universe, which is then communicated to the baryon sector, at temperatures lower than those of the electroweak phase transition ($T \lesssim 100$ GeV), by means of Baryon-minus-Lepton-(B-L)-number-conserving sphaleron processes within the SM sector of the theory. In this important scenario of Leptogenesis as the path to Baryogenesis, pioneered by Fukugita and Yanagida [8, 9], the lepton abundance is produced by the decay of heavy right-handed Majorana neutrinos (and so represents physics beyond the Standard Model (BSM)). The difference in the branching ratios of the channels of production of leptons and antileptons is equal to the imaginary part of the interference term of tree-level and one-loop diagrams for the decay processes. For the interference to generate a non-zero CP violating phase, at least two generations of right-handed neutrinos are needed [8, 7]. In fact, at least to right-handed neutrinos are *also* required in the see-saw mechanism [10] for the generation of light neutrino masses, as necessitated by the observed phenomena of neutrino flavour oscillations among the light neutrinos, which require at least two of the active neutrinos to have a mass. Measurements on solar, atmospheric and reactor neutrinos have established that there are oscillations with distance of neutrino flavours [11]. The model of Fukugita and Yanagida, therefore, connects an explanation of leptogenesis with the see-saw mechanism of light neutrino masses observed in Nature. The model thus represents an economical extension of SM, since, at least for the purpose of generating baryon asymmetry it does not require other particles apart from the massive right-handed neutrinos.

In the framework of ref. [8], the right handed neutrinos are very massive, and thus have decayed today, leaving no other trace apart from the see-saw type masses of the active neutrinos. Shaposhnikov and collaborators [12] have also made a proposal for a Minimal extension of the SM, termed νMSM, involving three generations of right-handed neutrinos, whose masses though are much lower than the corresponding ones in the model of [8]. In fact, in the νMSM the two heavier neutrinos are almost degenerate, with masses of order $\mathcal{O}(1)$ GeV, while the lightest of

IOP Conf. Series: Journal of Physics: Conf. Series **873** (2017) 012006 doi:10.1088/1742-6596/873/1/012006

the right-handed neutrinos may have masses of order of a few keV, and has a life time longer than the age of the Universe. In this sense, the model can provide a natural candidate for dark matter (DM), although there is still a long way to go before a satisfactory phenomenology of the dark sector of the Universe, including the dark energy sector, is provided within the framework of νMSM [1]. The relatively light masses of the right-handed neutrinos in the νMSM makes the Baryogenesis process in this model rather complicated [13], and certainly dissociated from the Leptoigenesis path of [8]. From our point of view in this work we shall focus our attention to the model of [8], but taking into account of the coupling of the fermions in the model to the gravitational background; this will have interesting consequences.

However, we shall not simply consider the Robertson-Walker Universe. Indeed, an alternative approach to baryon asymmetry, which avoids altogether the Sakharov condition on out of equilibrium processes, is the one in which from the beginning, the Universe has a matter-antimatter asymmetry, as a result of *violations* of the CPT symmetry. Keeping locality and unitarity intact, violation of Lorentz symmetry seems the most plausible possibility for evading the CPT theorem, and in fact this lead to a sort of anti-CPT theorem by Greenberg [15], claiming that the assumption of Lorentz symmetry plays somehow a more fundamental rôle than the other assumptions of the CPT theorem, with the conclusion that CPT violation *implies* necessarily a violation of Lorentz invariance. However, in his proof, Greenberg did assume the existence of a well-defined scattering transfer matrix, and in this sense his anti-CPT theorem is equivalent to the original one on CPT symmetry. Explicit counter examples to these claims, where violations of CPT occur in a Lorentz invariant way, *e.g.* in some non-local theories, have been provided [17]. Moreover there are Lorentz-invariant models with intrinsic decoherence [18], and it is well known that the presence of decoherence may lead [19] in an ill-defined nature of the CPT operator in the effective theory, obtained after tracing over the appropriate environmental degrees of freedom. Such environmentally-induced CPT violation, due to intrinsic decoherence, may lead to interesting and unique effects in entangled states of mesons (ω-effect [20]), which we shall review at the end of talk.

Nevertheless, for our purposes, as far as Baryon asymmetry is concerned, we shall restrict ourselves to *spontaneous violations* of Lorentz symmetry, which in turn induce CPT violation. Such violations can be provided by the background geometry, for instance within the context of string-inspired models [21, 22], by certain flux fields which are constant in a given frame (to be identified with the co-moving frame of the observer in an expanding Universe framework).

For early pioneering works on matter-antimatter asymmetry generated by CPT Violating backgrounds within the SME framework we refer the reader to [23]. In that work, it was argued that, under certain circumstances, certain CPT violating terms within an SME effective Lagrangian can produce large baryon asymmetry, at Grand Unified temperatures, which is eventually diluted to the current value by sphaleron processes within the Standard Model sector.

An alternative source of CPT Violating interactions that could lead to matter-antimatter asymmetry is the coupling of the baryon (or B-L) number (anomalous) current to scalar curvature $\mathcal{R}$ of space-time through a CP violating interaction Lagrangian $\mathcal{L}$, that could occur, *e.g.* within some Supergravity theories [24, 25]:

$$\mathcal{L} = \frac{1}{M_*^2} \int d^4x \sqrt{-g} \, (\partial_\mu \mathcal{R}) \, J^\mu \tag{2}$$

[1] In this respect, we mention that an O(50) keV right-handed neutrino DM can play an important rôle in explaining galactic structures and resolving some of the tensions between ΛCDM model and observations at small (galactic) scales, especially if appropriate self-interactions among the keV-right-handed neutrinos are introduced [14]. The coincidence in the range of the allowed right-handed neutrino DM mass obtained in this approach, based on purely astrophysical reasons at galactic scales, with the one of the νMSM, induced by particle physics and DM cosmology reasons is intriguing.

where M_* is a cut-off in the effective field theory and J^μ could be the current associated with $B-L$ (L being the lepton number). There is an implicit choice of sign in front of the interaction (2), which has been fixed so as to ensure matter dominance. In this context, it has been shown that [24]

$$\frac{n_{B-L}}{s} = \frac{\dot{R}}{M_*^2 T_d}, \tag{3}$$

T_d being the freeze-out temperature for $B-L$ interactions. To leading order in M_*^{-2} we have $R = 8\pi G(1-3w)\rho$ where ρ is the energy density of matter and the equation of state is $p = w\rho$ where p is pressure. For radiation $w = 1/3$ and so in the radiation dominated era of the Friedmann-Robertson-Walker cosmology $R = 0$. However w is precisely 1/3 when $T^\mu_\mu = 0$. In general $T^\mu_\mu \propto \beta(g) F^{\mu\nu} F_{\mu\nu}$ where $\beta(g)$ is the beta function of the running gauge coupling g in a $SU(N_c$ gauge theory with N_c colours. This allows $w \neq 1/3$. Further issues and progress in gravitationally induced baryogenesis can be found in [25]. Recently some potential problems with the coupling (2) have been pointed out in [26], regarding instabilities of theories with such interactions.

The structure of the talk is as follows: in the next section 2, we discuss a model for matter-antimatter asymmetry in a string-inspired Universe, which is due to a CPT violating space-time geometry with torsion at early epochs. In this model, the torsion decays to negligible values today, in agreement with current phenomenology, but a lepton asymmetry has been frozen at a given temperature where the Universe undergoes a phase transition. The baryon asymmetry is then generated in such scenarios by means of B-L preserving processes in the SM sector of the model. In section 3, we discuss a different scenario for CPT-induced baryon asymmetry, which is based on string Universe models with bulk D-brane defects. The interesting feature of this second class of models is that there is an intrinsic CPT violation in such cases, due to unobserved (by a low-energy observer) degrees of freedom associated with the recoil of the D-brane defects on the brane Universe during their interaction with string matter. In these models, CPT violation is primarily associated with an ω-type effect [20], and there are different dispersion relations between particles and antiparticles as they propagate in the "medium" of D-brane defects [27], which can lead to matter-antimatter asymmetry under certain conditions to be discussed. The current Phenomenology of both types of CPT violation is then discussed briefly in section 4, where we also make a few remarks on possible (tiny) violations of the spin-statistics theorem in some models, especially the ones of section 3 entailing ω-type CPT-violating effects. Indeed, as we explain, one of the essential underlying assumptions of the spin-statistics theorem is CPT and Lorentz invariance, which if violated or not well-defined may lead to violations, which though may not be describable within a local effective field theory framework. A brief description of current experimental searches for (and bounds of) such spin-statistics violations is also given. Finally, in section 5, we present our conclusions and outlook.

2. A string-inspired Model with CPT Violating Geometry in the Early Universe

The basic Lagrangian of this model is based on the order α' (=Regge slope = string length unit) expansion of a low-energy string effective action, in the Einstein frame, compactified to four space-time dimensions [28]

$$\begin{aligned} L &= (-g)^{1/2} \Big\{ \frac{1}{16\pi G} \Big\{ -R + \frac{1}{2}\partial_\mu \Phi \partial^\mu \Phi - V(\Phi) - \frac{1}{6} e^{-2\Phi} H_{\rho\mu\nu} H^{\varrho\mu\nu} \\ &+ \sum_{J=1}^{f} \overline{\psi}_J \Big(i\frac{1}{2}\gamma^\mu \overleftrightarrow{\nabla}_\mu - \frac{1}{4} e^{-\Phi} \gamma_\sigma \gamma^5 \epsilon^{\mu\nu\rho\sigma} H_{\mu\nu\rho} \Big) \psi_J + \dots \Big\} \end{aligned} \tag{4}$$

IOP Conf. Series: Journal of Physics: Conf. Series **873** (2017) 012006 doi:10.1088/1742-6596/873/1/012006

where R is the Ricci scalar for the metric $g_{\mu\nu}$, $H_{\rho\mu\nu} = \partial_{[\rho} B_{\mu\nu]}$, with the brackets $[\ldots]$ denoting total antisymmetrization of the respective indices, is the field strength of the spin-one antisymmetric tensor (Kalb-Ramond) field $B_{\mu\nu} = -B_{\nu\mu}$ which, together with the spin two traceless symmetric metric tensor $g_{\mu\nu}$, and the spin-zero (scalar) dilaton Φ field, constitutes the massless gravitational multiplet of the (bosonic) string spectrum. Newton's gravitational constant G is related to the string scale $M_s = 1/\sqrt{\alpha'}$ and the compactification volume $V^{(c)}$ (in units of $\sqrt{\alpha'}$) via $8\pi\mathrm{G} = \alpha' V^{(c)}$. The fermions ψ_J indicate fermions that appear in extensions of the Standard Model of Particle Physics, with (massive) right handed Majorana neutrinos, and the $\ldots$ denotes all relevant interactions, whose explicit form is of no concern here. The ∇_μ denotes gravitational covariant derivative, as appropriate for the curved-space-time formulation. The coupling of the Kalb-Ramond field strength to the axial currents $J^5_\sigma \equiv \overline{\psi}_J \gamma_\sigma \gamma^5 \psi_J$, $J = 1, \ldots f$, is dictated by the fact that the quantity $e^{-\phi} H_{\mu\nu\rho}$ plays the rôle of (totally antisymmetric) torsion, given that one can absorb the H^2 terms in (4) into a generalised curvature scalar of a space-time with a torsionful spin connection [28]: $\overline{\Gamma}_{\mu\nu}^{\;\rho} = \Gamma_{\mu\nu}^{\rho} + e^{-\Phi} H_{\mu\nu}^{\;\rho} \neq \overline{\Gamma}_{\nu\mu}^{\;\rho}$, with $\Gamma_{\mu\nu}^{\rho}$ the ordinary (torsion-free) Christoffel symbol. In four space-time dimensions, one also has the following duality relation:

$$H_{\mu\nu\lambda} = e^{2\Phi}\, \epsilon_{\mu\nu\lambda}^{\quad\sigma}\, \partial_\sigma b\ , \qquad (5)$$

with $b(x)$ a pseudiscalar (Kalb-Ramond "axion") field.

As discussed in [22], among the solutions of the equations of motion of (4), there are those corresponding to a metric tensor $g_{\mu\nu}$ of a Robertson-Walker expanding universe, a constant dilaton field Φ, which without loss of generality is taken to be zero, and some fermion condensates, such that the temporal component of the sum over fermion species of the axial fermion current term is a non-zero constant vacuum expectation value, $\rangle J^{5\,0}\rangle \neq 0$, and satisfies the following equation of motion for the dual Kalb-Ramond field $b(x)$:

$$\partial^\mu \Big[\sqrt{-g}\, \epsilon_{\mu\nu\rho\sigma} \Big(\partial^\sigma b - \frac{1}{4} \langle J^{5\,\sigma} \rangle + \ldots \Big) \Big] = 0\ , \quad \langle J^{5\,\sigma} \rangle = c\, \delta^{0\sigma} \neq 0,\ c = \text{constant}\ . \qquad (6)$$

The cosmological solution of (6) is such that $b(t)$ is a linear function of cosmic time: $\dot{b} \equiv db/dt = c/4$. This in turn implies a constant torsion background $H_{ijk} =$ constant, $i, j, k = 1, 2, 3$ in the frame of a cosmic observer. Such solutions, violate Lorentz invariance, and by the "anti-CPT theorem" of [15] CPT would also be violated. Thus, it is natural to expect matter-antimatter asymmetry generation in such Universes [21, 22].

A few remarks are in order at this point. First, in extensions of the Standard Model with right-handed Majorana neutrinos we are considering here, the axial condensates $\langle J^{5\,0} \rangle \neq 0$ refer to all other fermion species except the right-handed Majorana neutrinos. Indeed, the condensate of the latter vanishes identically due to the properties of the Majorana spinors [22]. With this in mind, we concentrate from now on to the Heavy neutrino sector of the model (4), which will be responsible for leptogenesis as we shall explain.

The constant torsion background coupling with the axial fermions implies that we should study the following model of the right-handed heavy neutrinos [22]:

$$\mathcal{L} = i\overline{N}\partial\!\!\!/ N - \frac{m}{2}(\overline{N^c}N + \overline{N}N^c) - \overline{N}B\!\!\!/\gamma^5 N - Y_k \overline{L}_k \tilde{\phi} N + h.c.\ , \qquad (7)$$

where N is the heavy right-handed Majorana field, satisfying the Majorana condition $N^c = N$, with the superscript c denoting Dirac's charge conjugation, and L_k is the lepton SU(2) left-handed doublet field of the Standard Model, with k a generation index. The adjoint of the Higgs field is defined by the relation $\tilde{\phi}_i = \varepsilon_{ij}\phi_j$. Due to our constant H-torsion situation, the axial vector background is $B^\mu = \dot{b}\delta^{0\,\mu}$. This is understood in what follows. For our purposes,

IOP Conf. Series: Journal of Physics: Conf. Series **873** (2017) 012006 doi:10.1088/1742-6596/873/1/012006

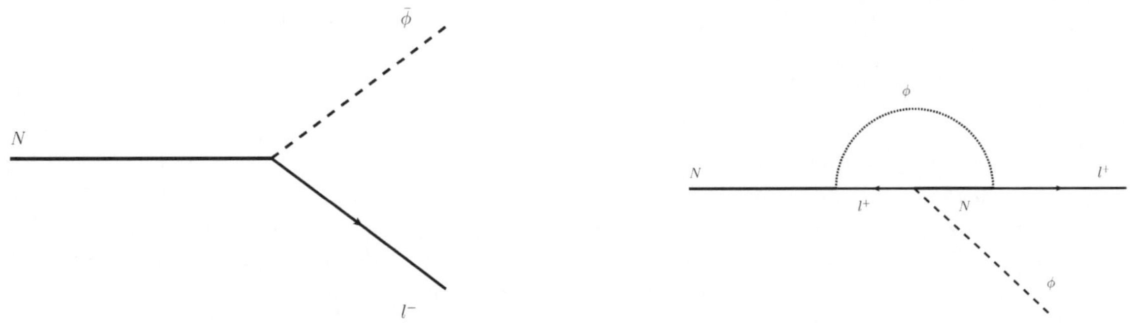

Figure 1. Tree- (left) and one-loop (right) decay amplitudes, corresponding to the Yukawa term in (7) that couples a right-handed neutrino to the Standard Model lepton sector. Analogous diagrams describe the decay in anti-leptons. Continuous undirected lines represent right-handed neutrinos, lines with an arrow are used to represent SM leptons, whilst dashed lines correspond to the Standard Model Higgs. The left diagrams are understood to be evaluated in the presence of an axial background field. The right diagram is the standard result of [8], leading to leptogenesis in conventional scenarios, due to CP violation in the lepton sector, in the case of more than one flavour of righti-handed neutrinos.

we shall restrict ourselves [22] to *only one generation* of right-handed neutrinos [2]. Notice that the model has a portal of communication of the right-handed neutrino sector with the Standard Model sector via the Yukawa interactions with couplings Y_k. In this work we do not specify the mechanism for generating a mass m for the heavy right-handed neutrino. The mass M will be determined self consistently by the requirement of the model generating sufficient leprogenesis [22].

The important point to notice is that, in the presence of a non-trivial constant (in the Robertson-Walker frame) Lorentz and CPT breaking axial background, B^0 = constant $\neq 0$ in (7), which in our case, as explained above, is geometrical in origin, there is a difference in the decay rates [22]

$$\begin{aligned}
&\Gamma_1(N \to \ell^- \phi^+) \neq \Gamma_2(N \to \ell^+ \phi^-) \,, \\
&\Gamma_1 = \sum_k \frac{|Y_k|^2}{32\pi^2} \frac{m^2}{\Omega} \frac{\Omega + B_0}{\Omega - B_0}, \quad \Gamma_2 = \sum_k \frac{|Y_k|^2}{32\pi^2} \frac{m^2}{\Omega} \frac{\Omega - B_0}{\Omega + B_0}\,, \\
&\Omega \equiv \sqrt{(B^0)^2 + m^2}\,,
\end{aligned} \tag{8}$$

where, since we anticipate the leptogenesis to occur at sufficiently high temperatures, above the electroweak symmetry breaking, the (massless) charged Higgs fields are present in the physical spectrum. It should be stressed that this difference already occurs for a single flavour of right-handed neutrinos, *at tree level* in a quantum field theory framework and will lead to a lepton asymmetry. The situation has to be contrasted with the conventional leptgenesis scenarios [8, 9, 7], based on the seesaw mechanism for generation of light neutrino masses in the Standard Model sector, where the presence of more than two flavours of heavy right handed

[2] This will be sufficient for leptogenesis, but of course in such a case one should consider alternative ways [29] to seesaw mechanism [10] in order to give masses to the active light neutrinos of the Standard Model sector (that are parts of the doublet L_k). In case one requires a seesaw type mass generation for neutrinos, then at least two right-handed neutrino generations are needed. In such a case, there are extra sources of CP violation, which lead to a generation of lepton asymmetry [8]. As we shall demonstrate in our model (7), due to the CPT and Lorentz violating background B^0, lepton asymmetry is generated even with a single generation of right-handed neutrinos.

IOP Conf. Series: Journal of Physics: Conf. Series **873** (2017) 012006 doi:10.1088/1742-6596/873/1/012006

neutrinos, necessitated by the seesaw mechanism, imply a CP violation, leading to a one-loop difference in the aforementioned decay rates (*cf.* fig. 1).

For weak backgrounds $B^0 \ll T$, where T is the temperature, the difference (8) has been estimated in [22] at the freezeout (decoupling) point, $T = T_D$, at which $H \simeq \Gamma_1 + \Gamma_2 = \sum_k \frac{|Y_k|^2}{16\pi^2}\frac{\Omega^2+B_0^2}{\Omega}$. We assumed that standard radiation-era cosmology [30] is not affected much by the presence of the background B^0, thereby setting $H = 1,66 T_D^2 \mathcal{N}^{1/2} m_P^{-1}$, where $\mathcal{N}$ is the effective number of degrees of freedom of all elementary particles and m_P the Planck mass. From the last equation one can estimate the decoupling temperature T_D in terms Ω, $|Y|$ and B_0.

$$T_D \simeq 6.2 \cdot 10^{-2} \frac{|Y|}{\mathcal{N}^{1/4}} \sqrt{\frac{m_P(\Omega^2 + B_0^2)}{\Omega}} \tag{9}$$

In order for the inverse decay to be suppressed by the Boltzmann factor, we have to impose [8] the further requirement that $T_D \leq \Omega$ when $\Gamma \simeq H$, leading to $z(\Omega^2 + B_0^2) \leq \Omega^3$, where $z = 3.8 \cdot 10^{-3} \frac{m_P |Y|^2}{\mathcal{N}^{1/2}}$. Upon the requirement that the bound is satisfied for all values of B_0, one obtains

$$m^2 \geq 1.09\, z^2. \tag{10}$$

In general, in our scenarios the Yukawa coupling Y is a free parameter. If we assume $|Y| \approx 10^{-5}$, $\mathcal{N} \approx 10^2$, as in standard leptogenesis models, we get an order of magnitude estimate for the lower bound of the right-handed Majorana neutrino mass:

$$\overline{m} \approx 100 \text{ TeV} . \tag{11}$$

The estimate of [22] for the total lepton number difference between particles and antiparticles ΔL, induced by (8), to leading order in the small quantity $B_0 \ll T_D \simeq m$, is [3]:

$$\Delta L^{TOT} = (2r-1) n_N = \frac{2\Omega B^0}{\Omega^2 + (B^0)^2} n_N \simeq \frac{2B^0}{m}\, n_N, \tag{12}$$

The high-temperature expansion of n_N yields for small $B_0 \ll T_D \simeq m$ [22]

$$n_N(T_D) = e^{-\beta \overline{m}} \left(\frac{\overline{m}}{2\pi\beta}\right)^{\frac{3}{2}} + \mathcal{O}(B_0^2)\Big|_{T \simeq T_D} \simeq 0.023\, \overline{m}^3. \tag{13}$$

The lepton asymmetry $\frac{\Delta L^{TOT}}{n_\gamma}$, where n_γ is the photon density, is expected to be of the same order of magnitude of the baryon asymmetry (1), given that in scenarios of baryogenesis via leptogenesis we are adopting here, the asymmetry in the lepton sector is communicated to the baryon sector via Baryon (B) and Lepton (L) number violating, B-L conserving, sphaleron processes in the Standard Model sector [3]. An order of magnitude estimate of the ratio $\frac{B_0}{\overline{m}}$

[3] The estimate was based on several simplifying assumptions, for instance the decaying right-handed neutrino is assumed initially at rest, with branching ratios given by $r = \frac{\Gamma_1}{\Gamma}$ and $1-r$. The decay of a single neutrino produces the lepton number asymmetry $\Delta L = r - (1-r) = 2r - 1 = \frac{2\Omega B_0}{\Omega^2 + B_0^2}$. Multiplying this quantity by the initial abundance of right-handed Majorana neutrinos at the temperature T_D one gets a rough estimate of the lepton number density. The density of the Majorana neutrinos $n_N = \sum_\lambda \frac{1}{(2\pi)^3} \int d^3p\, f(p,\lambda)$ can be estimated by using where λ is the helicity and $f(p,\lambda)$ Fermi-Dirac distribution function, which at high temperatures, relevant fr our case here, is well approximated by the Maxwell-Boltzmann function. Therefore we set $f(p,\lambda) = e^{-\beta\sqrt{m^2 + (p+\lambda B_0)^2}}$, with $\beta = 1/T$ the inverse temperature. The estimate (12) is then obtained upon making an appropriate high temperature expansion, assuming that the right-handed neutrino density distribution follows closely the equilibrium distribution for $T \geq T_D$ and drops rapidly to zero at lower temperatures $T \leq T_D$; furthermore the density of the sterile neutrino (normalised to the entropy density) is well approximated by a step-function.

can be found making use of the approximation $T_D \simeq m$ and retaining only first order terms in $\frac{B_0}{m}$. Recalling that the photon number density is [30] $n_\gamma \simeq \frac{2\zeta(3)}{\pi^2} T^3 \simeq 0.24 T^3$, we observe that, in this scenario, phenomenologically relevant baryogenesis (1) induced by a similar-order of magnitude lepton asymmetry, $\frac{\Delta L}{n_\gamma} \simeq 10^{-10}$, is achieved provided the ratio of the background field to the mass of the sterile neutrino is of order:

$$\frac{B_0}{\overline{m}} \simeq 5 \cdot 10^{-8} \,, \quad \text{at} \quad T = T_D \simeq \overline{m} \sim 100 \text{ TeV} \,. \tag{14}$$

The small value of this ratio also allows us to justify *a posteriori* neglecting higher powers of B_0 in the formulae above. From the lower bound (10) of 100 TeV that has been previously found for the mass, for the case where $Y = \mathcal{O}(10^{-5})$, we get an approximation for the smallest possible magnitude of the background field required in order for this mechanism to be effective $B_0 \simeq 1$ MeV. If other mechanisms contributed to the lepton asymmetry in the universe, or the Yukawa couplings assume smaller values, the minimum value of B_0 would be smaller than the one given here.

It goes without saying, that the proper estimation of the right-handed neutrino abundance, and thus the lepton asymmetry , should of course be obtained by solving the appropriate coupled system of Boltzmann equations for the right-handed neutrino and lepton abundances [7, 8, 9, 30], which get modifications in the presence of B_0 [22]. This is currently in progress. Nevertheless, the above qualitative estimate suffices for demonstrating the non-trivial role of the CPT violating background geometry in inducing a lepton asymmetry. We stress once more that the occurrence of leptogenesis here is due to decay processes at tree level, since the required CP violation is introduced by the Lorentz and CPT violating background field that enters in the external lines of the Feynman diagrams of fig. 1.

Having established a "torsionful"-geometry-induced matter-antimatter asymmetry, the next step is to discuss mechanisms for a significant reduction of the torsion background field B^0 today, where there are very stringent upper bounds on the magnitude of such Lorentz and CPT violating axial backgrounds. We should also avoid any significant modification of the Big-Bang Nucleosynthesis (BBN) conditions [30], which requires the background B_0 to be sufficiently weak already at BBN temperatures of order $\mathcal{O}(1$ MeV). Indeed, since the H-field couples universally to all fermionic species in the Standard model sector, it will do so to electrons. In such a case, the form of the corresponding terms in the action (4) will resemble the Standard Model Extension lagrangian of [16], with the b_μ coefficient corresponding to $b_0 = B_0$, for which there are very stringent upper bounds that we shall discuss later on in section 4. Such stringent limits, imply that one should have a proper mechanism in place, by means of which B^0 today should be very suppressed.

In [22] we remarked that such a situation can be encountered if the (string) Universe undergoes, after decoupling, at temperatures $T \lesssim T_D$, a phase transition such that the axial condensate $\langle J^{5\,0} \rangle \to 0$. In such a case, one observes from (6) that the Kalb-Ramond axion field $b(x)$ no longer varies linearly with the cosmic time but diminishes with the scale factor as a^{-3}. If one assumes a cooling law for the Universe, of the form $a \sim T^{-1}$, then the B^0 torsion field would scale with the temperature as T^3 in this scenario:

$$B_0 = c_0 \, T^3 \,, \quad c_0 > 0 \,, \quad \text{for} \quad T \lesssim T_D \,. \tag{15}$$

The parameter c_0 is a phenomenological, and can be constrained by requiring that B^0 today must be at most equal to the experimental upper bounds of the b_0 (temporal) axial Lorentz and CPT violating coefficient of the Standard Model Extension [16].

We remark at this point that models, in which the critical temperature for the destruction of the axial fermion condensate (6) is of the same order of magnitude as the right-handed neutrino

IOP Conf. Series: Journal of Physics: Conf. Series **873** (2017) 012006 doi:10.1088/1742-6596/873/1/012006

decoupling temperature $T_D = 100$ TeV, are not constrained at all by the current constraints on the upper bound of the b_μ coefficients of the Standard Model extension [16]. Indeed, taking into account the temperature of the Universe today (from the CMB measurements) is $T_{\rm CMB} = 2.725$ K $= 0.2348$ meV, we obtain from (15) and (14) [22] $c_0 = 1$ MeV(100 TeV)$^{-3}$ = 10^{-42} meV^{-2}. implying a current value of B_0 of order

$$B_{0\,\rm today} = \mathcal{O}\Big(10^{-44}\Big) \ \rm meV \ , \tag{16}$$

way too small for any experimental detection. Moreover, in this case, the value of B^0 at BBN temperature of a few MeV, $B^0(T = T_{\rm BBN}) \simeq 10^{-18}$ MeV, is also very small, so there are negligible effects of the background on the formation of the material elements. However, as we have already mentioned, in general the temperature at which destruction of the condensate occurs, could be significantly lower than the decoupling temperature for the right handed neutrinos, T_D; applying the experimental upper bounds on the magnitude of the background H-torsion field B^0 today, then, we can constrain the parameter c_0 in the cooling law (15). We shall come back to such an analysis in section 4.

In the next section 3, we discuss an alternative scenario for induction of a matter-antimatter asymmetry in the Universe [31], which may also provide a detailed string-theoretic mechanism for the destruction of the condensate (6), although the latter feature may not be required. The scenario pertains to brane universes, in which our world is viewed as a D(irichlet)-three-brane, with three large spatial dimensions, propagating in a bulk, which is populated by lower-dimensional D(irichlet) branes appropriately compactified, so that from the point of view of an observer on the three-brane they look effectively point-like defects.

3. "D-particle foam" Universe, Matter-Antimatter Asymmetry and Intrinsic (ω-effect-type) CPT Violation

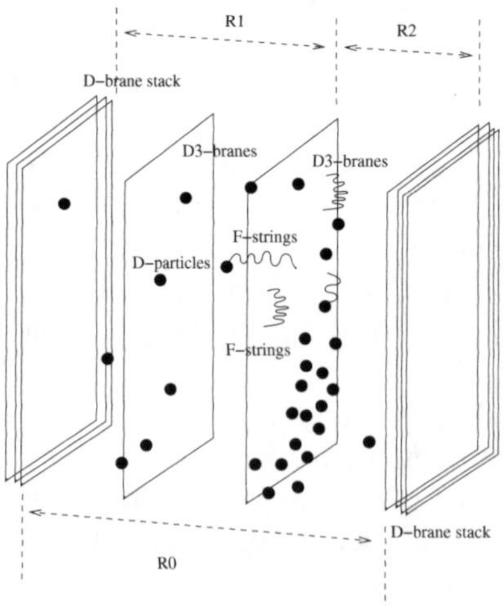

Figure 2. Schematic representation of a generic D-foam space-time model. In this picture, our brane world is represented as one of the D3-branes (Dirichlet branes with three large spatial dimensions) propagating in a higher-dimensional bulk space, which is punctured by inhomogeneous populations of D-particles. Open fundamental (F-)strings on the D3 branes interact dominantly with the D-particles only if they are electrically neutral.

D-foam models [32, 33] are stringy models of space-time foamy geometries, which involve a number of parallel brane universes, with three large spatial dimensions, propagating in higher-dimensional bulk geometries (see fig. 2). The required number of parallel brane worlds is determined by the initial target-space supersymmetry [32], which eventually is broken (the motion of the brane worlds in the bulk space is a factor in target-space supersymemtry breaking). One of the three branes represents our observable Universe. The model of ref. [32], acts as a prototype of a D-foam Universe, involving two stacks of D8-branes (which can be eventually compactified to four dimensions), each stack being attached to an orientifold plane. Owing to their special reflective properties, the latter provide a natural compactification of the bulk dimension. The bulk is punctured by effectively point-like D-branes (D-particles), which are either truly point-like (D0-branes), allowed in the type IA string theory, or 3-branes compactified on appropriate 3-cycles, with small radii of order of the string length, which are allowed in (the phenomenologically more relevant) type IIB string theories. The presence of a D-brane is essential due to gauge flux conservation, since an isolated D-particle cannot exist.

For an observer living on our brane world the D-particles will appear as space-time defects. Some of these D-particles will be bounded on the moving brane, others will pass through, and from the point of view of a brane observer they will appear as "flashing on and off" defects (which justifies the name "D-foam"). Open (F-)strings live on the brane world, representing Standard Model (SM) matter and they can interact in a topologically non-trivial way with the D-particle defects in the foam, only if they do not carry electric flux (electrically neutral excitations). Indeed, open strings that interact with D-particles can satisfy Dirichlet boundary conditions on the world-sheet when attached to them. Closed and open strings may be "cut" by D-particles, a process that involves capture of the incident open string and creation of stretched strings between the (recoiling) D-particle and the brane world (string "splitting"), and subsequent re-emission of the open string. In many type IIA string theories the D-particles are stable zero-space-dimensional defects. However for our purposes we will consider them to be present in string theories of phenomenological interest since, even when elementary D-particles cannot exist consistently, as is the case of type IIB string models. In the latter case, there can be "effectively point-like D-particles" formed by the compactification of higher dimensional D-branes [36] (*e.g.* three-branes wrapped around three-cycles, with relatively small radii). D-particles are electrically neutral and thus electric charge would not have been conserved if such processes had taken place. Hence, the D-particle foam is transparent to charged excitations of the Standard Model (or its effects are strongly suppressed on charged particles compared to the neutral excitations in type IIB theories [36]), leaving neutral particles, in particular neutrinos, susceptible to the foam effects. Recoil of the D-particle during such interactions creates appropriate distortion in the space-time geometry, which depend on the momenta of the incident string states.

The D-particles bulk population density is constrained at various epochs of the universe in a different way, given that it is in general *inhomogeneous*. For instance, today, such D-foam Universe cosmologies [34] may be restricted by requiring consistency with observations and in particular with the dark-sector energy budget dictated by ΛCDM fiducial cosmologies [35]. At earlier epochs, the density of D-particles can be different, *e.g.* much denser during the inflationary era [35]. In our context we shall be concerned with the induced asymmetry between matter and antimatter during matter-D-foam interactions, and thus we shall also constrain the density of such D-particles by the amount of CPTV that we observe.

As discussed in detail in [34] the density of D-particles on the brane world is permitted to be relatively large, even at late eras of the universe, given the fact that bulk D-particles exert forces on the brane universe with mixed sign contributions to the brane vacuum energy, depending on the distance of the bulk D-particles from the brane [32]. Such forces are due to stretched strings between the defect and the brane. These energy contributions depend only on the transverse components of the relative velocities of the defect with respect to the brane

IOP Conf. Series: Journal of Physics: Conf. Series **873** (2017) 012006 doi:10.1088/1742-6596/873/1/012006

worlds. For our purposes in this work we may therefore consider that statistically significant populations of D-particles existed in the early eras of the brane universe. However, as the time elapses, the brane universe, which propagates in the higher-dimensional bulk (*cf.* fig. 2), enters regions characterised by D-particle depletion, in such as way that the late eras cosmology of the universe is not affected. Nevertheless, as we shall discuss below, the early D-particle populations may still have important effects in generating neutrino-antineutrino populations differences (asymmetries), which are then communicated to the baryon sector via the standard sphaleron processes [3] or more generally B-L conserving processes in Grand Unified Theories.

To this end, we need to consider the *effective dispersion relation* of a (anti)neutrino field in a brane space-time punctured with statistically significant populations of D-particles [4]. The number density of (anti) neutrinos on the brane world is limited by the requirement that they do not overclose the universe. if neutrinos are assumed to have a chemical potential μ, then standard cosmological neutrino models predict that the number densities of a single flavour of relativistic neutrinos and antineutrinos in thermal equilibrium at temperature T_ν is estimated by [37]

$$n_{\nu\overline{\nu}} = T_\nu^3 \frac{3\zeta_3}{2\pi^2}\Big(1 + \frac{2\ln 2\, \mu_\nu^2}{3T_\nu^2\zeta_3} + \frac{\mu_\nu^4}{72T_\nu^4\zeta_3} + \mathcal{O}(\frac{\mu_\nu^6}{T_\nu^6})\Big) \tag{17}$$

upon making the standard assumption that $\mu_\nu \ll T_\nu$ for all neutrino flavours. The quantity $\xi_\nu \equiv \frac{\mu_\nu}{T_\nu}$ is called the degeneracy parameter and is invariant under cosmic expansion. if we assume that the electron-neutrino chemical potential is the only one with significant presence in the early universe, then BBN constraints imply $-0.04 < \xi_{\nu_e} < 0.07$. Thus, the order of magnitude of the neutrino plus antineutrino number density is agrees with naive standard estimate $n_{\nu\overline{\nu}} \sim \frac{3}{11}\, n_\gamma$, where n_γ is the photon density. Today, where the temperature of the universe is the CMB temperature, corresponding to an energy of $k_B T_0 \sim 2.35 \times 10^{-13}$ GeV, the density of neutrinos is found to be of order $n_{\nu\overline{\nu}}^{(0)} \sim 112$ cm^{-3} and scales roughly with the cubic power of the temperature: $n_{\nu\overline{\nu}} \sim n_{\nu\overline{\nu}}^{(0)} \left(\frac{T_\nu}{T_0}\right)^3$. So, for the decoupling temperatures of neutrinos, $k_B T_d \sim 10^{15}$ GeV, where we are interested in this work, in order to compute the frozen CPT Violating neutrino-antineutrino population differences, one obtains a number density of neutrino plus antineutrino populations of order

$$n_{\nu\overline{\nu}}(T = T_d \sim 10^{15}\ \text{GeV}) \sim 10^{85}\, cm^{-3}\,. \tag{18}$$

On the other hand, as already mentioned, there are no similar restrictions on the population of the D-particle defects on the brane, in view of the negative contributions on the potential energy of the brane universe by bulk D-particle populations [34]. Thus, at the early universe, at the above neutrino-decoupling temperatures, we may even assume D-particle densities of *one* defect per *string* volume on the three brane world, without overclosing the universe. The assumption that the string length can take on values in the phenomenologically acceptable (post LHC era) range $10^{-27} - 10^{-32}$ cm, corresponding to string mass scales from 10 TeV to 10^{18} GeV, yields a D-particle number density in the range

$$n_{\rm D}(T = T_d \sim 10^{15}\ \text{GeV}) \sim 10^{54} - 10^{96}\ \text{cm}^{-3} \tag{19}$$

respectively. Thus we observe that in order to be able to treat the D-particle populations as providing a more-or-less uniform "medium" over which neutrinos propagate, with non-trivial

[4] Here we are dealing with active neutrino species of the standard model. Heavy right handed Majorana neutrinos, as was the case of the previous section, are their own antiparticles, so their propagation in the D-foam, with which they interact as neutral under the standard model group, although non trivial does lead to this effect [22]. However they may have other effects, for instance can play a rôle in the destruction of the condensate (6) as we shall see below.

effective dispersion relations, we need to have at the decoupling temperature much higher densities of D-particles than those of neutrinos plus antineutrinos. Comparing (18) with (19), we observe that, if one assumes *one* D-particle per *three-dimensional string volume* on the brane, then this latter requirement excludes the low values of the string mass scale, implying an allowed range

$$10^{-5}M_P \leq M_s \leq 10^{-1}M_P \ , \tag{20}$$

with $M_P \sim 10^{19}$ GeV the four-dimensional Planck mass. One of course could have much more dense D-particle gases in the early universe, which would allow for lower string scales.

We will now estimate the *modification of the dispersion relations* of neutrinos in such a "media" of D-particles in the early universe [31]. The interaction of a string with a D-particle implies that at least one of the ends of the string is attached to the D-particle defect. Furthermore, the simultaneous creation of virtual strings stretched between the defect and the brane, describes the recoil of the D-particle. During the interaction time, the D-particle undergoes motion characterized by non-trivial velocities, $u_{\parallel} = \frac{g_s}{M_s}\Delta p_i = \frac{g_s}{M_s} r_i\, p_i$ along the brane longitudinal dimensions, where r_i denotes the proportion of the incident neutrino momentum that corresponds to the momentum transfer Δp_i during the scattering, and $v_\perp$ in directions transverse to the brane world [32].

As discussed in [39, 38, 33] the non-trivial capture and splitting of the open string excitation corresponding to neutrino during its interaction with the D-particle, and the recoil of the latter, result in a *local* effective metric distortion of the form:

$$ds^2 = g_{\mu\nu}dx^\mu dx^\nu = (\eta_{\mu\nu} + h_{\mu\nu})dx^\mu dx^\nu \ , \qquad h_{0i} = (u^a_{i\,\parallel}\sigma_a) \ , \tag{21}$$

where $u_{i\,\parallel}$ is the recoil velocity of the D-particle *on* the D-brane world, with $i = 1, 2, 3$ a spatial space-time index, σ_a are the 2×2 Pauli flavour matrices with $a = 1, 2, 3$ (assuming two-flavour oscillations for simplicity). On average over a population of stochastically fluctuating D-particles including flavour changes, one may have the conditions (26), the second of which in the case of flavour oscillations can be generalised to

$$\ll u^{\parallel}_{a,i} u^{\parallel}_{b,j} \gg = \sigma^2 \delta_{ij}\delta_{ab} \ . \tag{22}$$

(We still assume that $\ll u^{\parallel}_{a,i} \gg = 0$.) As a result of (22), on average, the flavour change during the interactions of neutrinos with the D-foam can be ignored. In such a case, any flavour structure in the metric (21) is ignored [5].

On considering string theory scattering amplitudes we find that the four momentum is conserved in the scattering of D-particles and strings. D-particles in the bulk exert forces on the vacuum energy of the brane world of mixed sign, depending on their relative distance. Thus, during the scattering process of a neutrino field with a D-particle, the vacuum energy of

[5] Ignoring the flavour structure, the metric (21) can be written as

$$ds^2 = dt^2 + 2u_i dx^i dt - \delta_{ij}dx^i dx^j \ , \tag{23}$$

This metric was determined from world-sheet conformal field theory considerations [39] and represents a dragging of the frame by the Gallilean (slowly moving) D-particle, which moves on a flat space-time background. As discussed in [31], upon a time coordinate change, this metric becomes the the so-called Gullstrand-Painlevé metric representing the geometry in the exterior of a Schwarzschild black hole, where the falling space into the black hole is represented as a Gallilean "river" on a flat space-time in which "relativistic fishes" swim. The river represents the frame of the recoiling D-particle, while the fishes are the relativistic matter strings:

$$ds^2 = dt_{\rm ff}^2 + 2u_i dx^i dt_{\rm ff} - \delta_{ij}(dx^i - u^i dt_{\rm ff})(dx^j - u^j dt_{\rm ff}) + \mathcal{O}(u^3). \tag{24}$$

Here $t_{\rm ff}$ is the time of a free-floating observer who is at rest at infinity (compared to the centre of the black hole).

IOP Conf. Series: Journal of Physics: Conf. Series **873** (2017) 012006 doi:10.1088/1742-6596/873/1/012006

the brane fluctuates by an amount $\Delta\mathcal{V}$ which depending on the process can be of either sign. From energy-momentum conservation, at each individual scattering event between a neutrino field and a recoiling D-particle, one could thus write:

$$p_{\rm before}^{\vec{}} + \vec{p}_{\rm after} + \frac{M_s}{g_s}\,\vec{u}_\parallel = 0\ , \quad E_{\rm before} = E_{\rm after} + \frac{1}{2}\frac{M_s}{g_s}\,\vec{u}_\parallel^2 + \Delta\mathcal{V} \tag{25}$$

where $(\vec{p},\,E)_{\rm before\,(after)}$ denote the incident (outgoing) neutrino momenta, energies repectively and we used the fact that the recoiling heavy D-particle of mass M_s/g_s (with M_s the string scale and $g_s < 1$ the string coupling, assumed weak, so that string perturbation theory applies) has a non-relativistic kinetic energy $\frac{1}{2}\frac{M_s}{g_s}\,\vec{u}_\parallel^2$. We have also assumed that the fraction of the neutrino momentum transfer in the direction perpendicular to the brane world is negligible. The importance of the term $\Delta\mathcal{V}$ not having a fixed sign in each individual scattering process is associated with the possibility of D-particle induced neutrino flavour oscillations [33].

Indeed, upon averaging $\langle\langle\ldots\rangle\rangle$ over a statistically significant number of events, due to multiple scatterings in a D-foam background, we may use the following stochastic hypothesis [33, 31]

$$\ll u_{i\,\parallel} \gg = 0\ , \qquad \ll u_{i\,\parallel} u_{j\,\parallel} \gg = \sigma^2\delta_{ij}\ . \tag{26}$$

implying that Lorentz invariance holds only as an average symmetry over large populations of D-particles in the foam. At a microscopic level, (26) translates to momentum conservation on average in (25), since $\ll \vec{u}_\parallel \gg = 0$. At an individual scattering process, if one represents the energy of the incident neutrino on-shell as $\sqrt{\overline{p}^2 + m_1^2}$, where $\overline{p}$ is the amplitude of the conserved spatial momentum of the neutrino, and the outgoing one as $\sqrt{\overline{p}^2 + m_2^2}$, we observe that the energy-conservation equation (25) implies in general $m_1 \neq m_2$. Which one is larger depends on the signature of the term $\frac{1}{2}\frac{M_s}{g_s}\,\vec{u}_\parallel^2 + \Delta\mathcal{V}$, which as mentioned is not of fixed sign, thereby allowing for neutrino oscillations to take place. The situation is somewhat analogous to the standard Mossbauer effect, where the emitted or absorbed photon from a nucleus of an atom bound in a solid may sometimes be free of nuclear recoil, in contrast to the case of gases, thereby attributing the phenomena of nuclear resonances to such recoil-free fraction of nuclear events. In our case the rle of the "nuclei bound in a lattice" is played by the D-particle lattice. In addition to the D-particle recoil energy during scattering with stringy matter, which would lead to energy losses for the neutrinos, there are vacuum energy fluctuations, as a consequence of the motion of bulk particles in the foam, thus the neutrino experiences losses and gains from the vacuum, which results in the induced flavour oscillations. The analogue of resonances in this case would correspond to the loss-and-gain-free fraction of events, in which the neutrino does not oscillate.

However, the effects of the D-foam go beyond the above-mentioned kinematical ones. On assuming isotropic momentum transfer, $r_i = r$ for all $i = 1, 2, 3$. The dispersion relation of a neutrino of mass m propagating on such a deformed isotropic space-time, then, reads:

$$p^\mu p^\nu g_{\mu\nu} = p^\mu p^\nu(\eta_{\mu\nu} + h_{\mu\nu}) = -m^2 \;\Rightarrow\; E^2 - 2E\vec{p}\cdot u_\parallel - \vec{p}^2 - m^2 = 0\ . \tag{27}$$

This on-shell condition implies that

$$E = \vec{u_\parallel}\cdot\vec{p} \pm \sqrt{(\vec{u}\cdot\vec{p})^2 + \vec{p}^2 + m^2}\ . \tag{28}$$

We take the average $\ll\cdots\gg$ over D-particle populations with the stochastic processes (22), (26). Hence we arrive at the following expression for an average neutrino energy in the D-foam background:

$$\begin{aligned} \ll E \gg &= \ll \vec{p}\cdot\vec{u} \gg \pm \ll \sqrt{p^2 + m^2 + (\vec{p}\cdot\vec{u})^2} \gg \\ &\simeq \pm\sqrt{p^2+m^2}\left(1 + \frac{1}{2}\sigma^2\right), \qquad p \gg m\ , \end{aligned} \tag{29}$$

for the active light neutrino species. The last relation in eq. (29) expresses the corrections due to the space-time distortion of the stochastic foam to the free neutrino propagation. It is this expression for the neutrino energies that should be used in the averaged energy-momentum conservation equation (25) that characterises a scattering event between a neutrino and a D-particle. On further making the assumption for the brane vacuum energy that $\ll \Delta\mathcal{V} \gg = 0$, the total combined effect on the energy-momentum dispersion relations, from both capture/splitting and metric distortion, can then be represented as:

$$\ll E_2 \gg = \pm\sqrt{p^2+m^2}\left(1+\frac{1}{2}\sigma^2\right) - \frac{1}{2}\frac{M_s}{g_s}\sigma^2 \tag{30}$$

Since antiparticles of spin 1/2 fermions can be viewed as "holes" with negative energies, we obtain from (25) and (29) the following dispersion relations between particles and antiparticles in this geometry (for Majorana neutrinos, the rles of particles /antiparticles are replaced by left/right handed fermions):

$$\begin{aligned} \ll E_\nu \gg &= \sqrt{p^2+m_\nu^2}\left(1+\frac{1}{2}\sigma^2\right) - \frac{1}{2}\frac{M_s}{g_s}\sigma^2 \\ \ll E_{\overline{\nu}} \gg &= \sqrt{p^2+m_\nu^2}\left(1+\frac{1}{2}\sigma^2\right) + \frac{1}{2}\frac{M_s}{g_s}\sigma^2 \end{aligned} \tag{31}$$

where $\overline{E} > 0$ represents the positive energy of a physical antiparticle. In our analysis above we have made the symmetric assumption that the recoil-velocities fluctuation strengths are the same between particle and antiparticle sectors [6].

There can thus be *local* CPTV in the sense that the effective dispersion relation between neutrinos and antineutrinos are different. This is a consequence of the local violation of Lorentz symmetry (LV), as a result of the non-trivial recoil velocities of the D-particle, leading to the LV space-time distortions (21). This difference between neutrino and antineutrino phase-space distribution functions in D-foam backgrounds generates a matter-antimatter lepton asymmetry in the relevant densities

$$\ll n - \overline{n} \gg = g_{d.o.f.}\int \frac{d^3p}{(2\pi)^3} \ll [f(E) - f(\overline{E})] \gg \ , \tag{32}$$

where $g_{d.o.f.}$ denotes the number of degrees of freedom of relativistic neutrinos, and $\ll \cdots \gg$ denotes an average over suitable populations of stochastically fluctuating D-particles (26).

As discussed in detail in [33], one may parameterise the momentum transfer by the fraction parameter of the incident momentum r, which is in turn assumed stochastic, that is

$$u_i = \frac{g_s}{M_s}\Delta p_i \rightarrow g_s\, r_i \frac{p_i}{M_s} \text{ ,no sum over i }, \quad \ll r_i \gg = 0\ , \quad \ll r_i r_j \gg = \Delta^2 \delta_{ij}\ . \tag{33}$$

In this case, the dispersion relations (31) are modified by the replacement of

$$\sigma^2 \rightarrow \frac{g_s^2}{M_s^2}\Delta^2 p^2\ , \tag{34}$$

which is now momentum dependent:

$$\begin{aligned} \ll E_\nu \gg &= \sqrt{p^2+m_\nu^2}\left(1+\frac{g_s^2}{2\,M_s^2}\Delta^2\,p^2\right) - \frac{g_s}{2\,M_s}\Delta^2\,p^2 \\ \ll E_{\overline{\nu}} \gg &= \sqrt{p^2+m_\nu^2}\left(1+\frac{g_s^2}{2\,M_s^2}\Delta^2\,p^2\right) + \frac{g_s}{2\,M_s}\Delta^2\,p^2 \end{aligned} \tag{35}$$

[6] Scenarios for which this symmetry was not asssumed have also been considered in an early work [33], and will lead to important phenomena regarding *intrinsic* CPT violation, which we shall discuss later.

IOP Conf. Series: Journal of Physics: Conf. Series **873** (2017) 012006 doi:10.1088/1742-6596/873/1/012006

In such a case, the lepton asymmetry can be calculated from the integral (32), for $\Delta^2 < 1$, but, in contrast to conventional point-like field theory models, where the upper limit of momentum integration can be extended to ∞, in D-foam models, due to (21), this is extended up to the value for which the D-particle recoil velocity approaches the value of the speed of light *in vacuo*, c=1 in our units, *i.e.* $p_{\rm max} \equiv |\vec{p}|_{\rm max} = \frac{M_s}{g_s\sqrt{\Delta^2}}$, where r is the stochastic variable satisfying (33). The resulting integrals in (32) then become:

$$\Delta n_\nu = \frac{g_{d.o.f.}}{2\pi^2}T^3\int_0^{\frac{M_s}{T\, g_s\sqrt{\Delta^2}}} d\tilde{u}\left(\frac{1}{1+e^{\tilde{u}-\tilde{u}^2\frac{g_s\Delta^2 T}{2M_s}}} - \frac{1}{1+e^{\tilde{u}+\tilde{u}^2\frac{g_s\Delta^2 T}{2M_s}}}\right) \simeq \frac{g_{d.o.f.}}{\pi^2}T^4\frac{\Delta^2\, g_s}{M_s} \quad (36)$$

The lepton asymmetry resulting from (36) freezes out at temperature T_d and is:

$$\Delta L(T < T_d) = \frac{\Delta n_\nu}{s} = \frac{2\Delta^2 g_s T_d}{M_s}. \quad (37)$$

From (37), we observe that for a freeze-out temperature $T_d \sim 10^{15}$ GeV, the phenomenological value $\Delta L \sim 10^{-10}$ is attained for

$$\frac{M_s}{g_s} \sim 10^{25}\,\Delta^2\,\text{GeV}\;. \quad (38)$$

For $\Delta^2 \sim 10^{-6}$ a Planck size D-particle mass $M_s/g_s \sim 10^{19}$ GeV is required so that the D-foam provides the physically observed Lepton and, thus, Baryon Asymmetry. For the unnaturally small $\Delta^2 < 10^{-21}$ one arrives at $M_s/g_s \sim 10$ TeV. For $\Delta^2 \sim \mathcal{O}(1)$ transplanckian D-particle masses are required. We should stress that the above conclusions were based on the assumption that the freeze-out temperature was the temperature at decoupling of neutrinos in standard big-bang cosmology.

Our approach to leptogenesis is distinguished from others in that a local effective field theoretical description is not adopted. Because of D-particle recoil when scattering off matter strings, the background of D-particles can be modelled as a stochastic medium, which goes bayond local field theory frameworks. The underlying string theoretic description provides the rigorous description of the scattering of D-particles. The D-particles backreact (as seen from infra-red divergences in perturbation theory) and change the metric which influences the space in which matter is moving.

The above described model has another important consequence, regarding the induction of an intrinsic CPT Violation (ω-effect [20]) which affects entangled states of particles, as discussed in [33], which we review briefly below.

An example of an (unnormalised) initial entangled quantum state $|i\rangle$ is given by

$$|\psi\rangle = |k,\uparrow\rangle^{(1)}\,|{-k},\downarrow\rangle^{(2)} - |k,\downarrow\rangle^{(1)}\,|{-k},\uparrow\rangle^{(2)} + \xi\,|k,\uparrow\rangle^{(1)}\,|{-k},\uparrow\rangle^{(2)} + \xi'\,|k,\downarrow\rangle^{(1)}\,|{-k},\downarrow\rangle^{(2)} \quad (39)$$

where $\left|M_L\left(\vec{k}\right)\right\rangle = |k,\uparrow\rangle$ in an actual situation may represent a neutral meson (Kaon K^0 or B^0 meson) and we have taken the momentum $\vec{k}$ to have only a non-zero component k in the x-direction for brevity and concnreteness; superscripts label the two separated detectors of the collinear meson pair, ξ and ξ' are complex constants and we have left the state $|\psi\rangle$ unnormalised.

In the case of neutral mesons, which (ignoring other quantum numbers, such as strangeness or beuaty) are treated as identical bosons, if the CPT operator was well defined as a quantum mechanical operator, one should have $\xi = \xi' = 0$ [40]. On the other hand, in case of quantum gravity environments, entailing loss of information for a low-energy observer, the CPT operator may not be well-defined [19], which, in the case of neutral mesons, implies an initial entangled state (39) with a non-trivial but *small* contamination $\xi = \xi' = \omega \neq 0$, with $\omega = |\omega| e^{i\Omega}$ a complex parameter, so that the antiparticle state is *perturbatively defined*, but the Einstein-Podoskly-Rosen (EPR) correlations associated with the decay products of the initial entangled state (39) get modified. This is the ω-effect [20], whose phenomenology will be discussed in the next section.

In our case of matter propagation in D-foam space-times, as a consequence of the effective loss of information for a low-energy observer, due to the recoil degrees of freedom of the stochastically fluctuating D-particles during their interaction with neutral meson matter, we shall face an ω-like situation, with $\omega \neq 0$, which we shall estimate. To this end, we first remark that, as a consequence of the recoil metric (21), the evolution of this state is governed by a hamiltonian $\widehat{H}$: [31]

$$\widehat{H} = g^{01}\left(g^{00}\right)^{-1}\widehat{k} - \left(g^{00}\right)^{-1}\sqrt{\left(g^{01}\right)^{2}k^{2} - g^{00}\left(g^{11}k^{2} + m^{2}\right)} \tag{40}$$

which is the natural generalisation of the standard Klein Gordon hamiltonian in a one particle situation. Moreover $\widehat{k}\,|\pm k, \uparrow\rangle = \pm k\,|k, \uparrow\rangle$ together with the corresponding relation for $\downarrow$.

The effect of space-time foam on the initial entangled state of two neutral mesons is conceptually difficult to isolate, given that the meson state is itself entangled with the bath. Nevertheless, in the context of our specific model, which is written as a stochastic hamiltonian, one can estimate the order of the associated ω-effect by applying non-degenerate perturbation theory to the states $|k, \uparrow\rangle^{(i)}$, $|k, \downarrow\rangle^{(i)}$, $i = 1, 2$. Although it would be more rigorous to consider the corresponding density matrices, traced over the unobserved gravitational degrees of freedom, in order to obtain estimates it will suffice formally to work with single-meson state vectors). Owing to the form of the hamiltonian (40) the gravitationally perturbed states will still be momentum eigenstates. The dominant features of a possible ω-effect can be seen from a term $\widehat{H_I}$ in the single-particle interaction hamiltonian

$$\widehat{H_I} = -\left(r_1\sigma_1 + r_2\sigma_2\right)\widehat{k} \tag{41}$$

which is the leading order contribution in the small stochastic parameters $r_i \ll 1$ that satisfy

$$\langle r_i\rangle = 0, \quad \langle r_i r_j\rangle = \Delta_i\,\delta_{ij}\,, \quad i, j = 1, 2\;. \tag{42}$$

In first order in perturbation theory the gravitational dressing of $|k, \downarrow\rangle^{(i)}$ leads to a state:

$$\left|k^{(i)}, \downarrow\right\rangle_{QG}^{(i)} = \left|k^{(i)}, \downarrow\right\rangle^{(i)} + \left|k^{(i)}, \uparrow\right\rangle^{(i)}\alpha^{(i)} \tag{43}$$

where

$$\alpha^{(i)} = \frac{{}^{(i)}\left\langle\uparrow, k^{(i)}\right|\widehat{H_I}\left|k^{(i)}, \downarrow\right\rangle^{(i)}}{E_2 - E_1} \tag{44}$$

and correspondingly for $\left|k^{(i)}, \uparrow\right\rangle^{(i)}$ the dressed state is obtained from (44) by exchanging $|\downarrow\rangle \leftrightarrow |\uparrow\rangle$ and $\alpha \rightarrow \beta$ where

$$\beta^{(i)} = \frac{{}^{(i)}\left\langle\downarrow, k^{(i)}\right|\widehat{H_I}\left|k^{(i)}, \uparrow\right\rangle^{(i)}}{E_1 - E_2} \tag{45}$$

Here the quantities $E_i = (m_i^2 + k^2)^{1/2}$ denote the energy eigenvalues, and $i = 1$ is associated with the up state and $i = 2$ with the down state. With this in mind the totally antisymmetric "gravitationally-dressed" state can be expressed in terms of the unperturbed single-particle states as:

$$
\begin{aligned}
&|k,\uparrow\rangle_{QG}^{(1)}\,|-k,\downarrow\rangle_{QG}^{(2)} - |k,\downarrow\rangle_{QG}^{(1)}\,|-k,\uparrow\rangle_{QG}^{(2)} = \\
&|k,\uparrow\rangle^{(1)}\,|-k,\downarrow\rangle^{(2)} - |k,\downarrow\rangle^{(1)}\,|-k,\uparrow\rangle^{(2)} \\
&+|k,\downarrow\rangle^{(1)}\,|-k,\downarrow\rangle^{(2)}\left(\beta^{(1)}-\beta^{(2)}\right) + |k,\uparrow\rangle^{(1)}\,|-k,\uparrow\rangle^{(2)}\left(\alpha^{(2)}-\alpha^{(1)}\right) \\
&+\beta^{(1)}\alpha^{(2)}\,|k,\downarrow\rangle^{(1)}\,|-k,\uparrow\rangle^{(2)} - \alpha^{(1)}\beta^{(2)}\,|k,\uparrow\rangle^{(1)}\,|-k,\downarrow\rangle^{(2)}
\end{aligned}
$$

It should be noted that for $r_i \propto \delta_{i1}$ the generated ω-like effect corresponds to the case $\xi = \xi'$ in (39) since $\alpha^{(i)} = -\beta^{(i)}$, while the ω-effect of [20], specific to neutral mesons, corresponds to $r_i \propto \delta_{i2}$ (and the generation of $\xi = -\xi'$) since $\alpha^{(i)} = \beta^{(i)}$. In the density matrix these cases can be distinguished by the off-diagonal terms.

These two cases are physically very different. In the case of ϕ-factories, the former corresponds to non-definite strangeness in the initial state of the neutral Kaons (seen explicitly when written in terms of $K_0 - \overline{K}_0$), and hence strangeness nonconservation in the initial decay of the ϕ-meson, while the latter conserves this quantum number. We remind the reader that in a stochastic quantum-gravity situation, strangeness, or, in that matter, the appropriate quantum number in the case of other neutral mesons, is not necessarily conserved, and this is reflected in the above-described general parametrisation of the interaction Hamiltonian (41) in "flavour" space.

As we discussed in [31], where we refer the interested reader, the (decoherent) time evolution of these two cases causes the appearence of terms with the opposite effects, as far as the quantrum numbers in question are concerned. Namely, the strangeness-conserving initial state leads to the appearance of CPT violating terms with a strangeness violating form, while an initially strangeness-violating combination generates, under evolution in the foam, a strangeness-conserving ω-effect of the form proposed in [20].

We next remark that on averaging the density matrix over the random variables r_i, we observe that only terms of order $|\omega|^2$ will survive, with the order of $|\omega|^2$ being

$$
|\omega|^2 = \mathcal{O}\left(\frac{1}{(E_1 - E_2)}(\langle\downarrow, k|H_I|k,\uparrow\rangle)^2\right) = \mathcal{O}\left(\frac{\Delta_2 k^2}{(E_1 - E_2)^2}\right) \sim \frac{\Delta_2 k^2}{(m_1 - m_2)^2} \tag{46}
$$

for the physically interesting case in which the momenta are of order of the rest energies (*i.e.* masses of the constituent states). The variance Δ_2 (and also Δ_1) is of the order of the square of the momentum transfer during the scattering of the single particle state off a space-time-foam defect [39] (*cf.* (34)), *i.e.*

$$
\Delta_2 = g_s^2\,\frac{\Delta^2 k^2}{M_s^2}\ , \tag{47}
$$

where M_s/g_s the D-particle defect rest mass (with M_s the string mass scale, and $g_s < 1$ the (perturbative) string coupling). The parameter Δ^2 is at present a phenomenological parameter, which is proportional to the probability of interaction of the string matter with the D-particles, and therefore the total cross section of such processes. It cannot be further determined due to the lack (at present) of a complete theory of (string/brane) quantum gravity. Thus, we arrive at the following estimate of the order of ω in this model of foam [31]:

$$
|\omega|^2 \sim g_s^2\,\frac{\Delta^2 k^4}{M_s^2 (m_1 - m_2)^2} \tag{48}
$$

We shall perform a phenomenological analysis of this phenomenon in the next section.

Before doing so, we would like to make a final remark on the baryogenesis via leptogenesis if we embed the models of the previous section 2, employing Majorana right handed massive neutrinos to initiate lepton asymmetry, in the current (D-foam) framework. Since Majorana neutrinos are their own antiparticles, it is only the corresponding helicity states that would exhibit differences between matter and antimatter à la (35), which however are not associated with a lepton asymmetry, as they do not have a well-defined lepton number [22]. In such a case therefore, the thermal equilibrium matter-antimatter asymmetries (35) could only refer to light neutrino species. Such asymmetries could co-exist with the mechanism of leptogenesis of [22], described in the previous sector, which however we consider the dominant one. In such models, the D-particle bulk foam could provide a way for destruction of the condensate (6), at a given era of the Universe, which can be fine tuned so as the current epoch axial CPT- and Lorentz-violating background B^0 is consistent with the stringent phenomenological bounds, to be examined below, and also with the BBN constraints.

4. Current Phenomenology of CPT Violation, inclusive of Spin-Statistics Violation searches

In this section we shall discuss some aspects of the current era phenomenology of the types of CPT Violation (CPTV) motivated by early Universe matter-antimatter asymmetry generation in the previous two sections. Above we examined two basic types of CPTV, one (section 2) induced by Lorentz violating backgrounds, which can be studied within the framework of effective field theories, and in particular the Standard Model Extension (SME) [16] parametrization. The other (section 3) goes beyond local effective field theories, and is linked to situations of environmental decoherence in which the CPT operator is not well-defined, leading to ω-type modifications of the Einstein-Podolsky-Rosen (EPR) correlators of entangled boson (specifically, meson) states in appropriate meson factories (ϕ-factories [41] or B-factories [42]).

4.1. CPT Violation within Standard Model Extension Framework

Let us commence our discussion with the phenomenology of CPTV within the SME framework [16]. There is a phethora of precision tests which are tabularted and regularly updated in ref. [43], where we refer the interested reader for details. For our purposes here we shall only concentrate on bounds of the axial background Lorentz and CPT Violating coefficient b_μ in the notation of [16, 43], and in particular its temporal component, which coincides with our axial background vector B_0 in (7). The first few terms in the fermion sector of the SME read [16, 43]:

$$\begin{aligned}
\mathcal{L}_{SME} &\ni \frac{1}{2} i\overline{\psi}\,\Gamma^\nu\,\partial_\nu\psi - \overline{\psi}\,\mathcal{M}\,\psi\,, \\
\mathcal{M} &\equiv m + a_\mu\gamma^\mu + b_\mu\,\gamma^5\gamma^\mu + \frac{1}{2}H^{\mu\nu}\sigma_{\mu\nu}\,, \quad \sigma_{\mu\nu} = \frac{1}{4}[\gamma_\mu, \gamma_\nu]\,, \\
\Gamma^\nu &\equiv \gamma^\nu + c^{\mu\nu}\,\gamma_\mu + d^{\mu\nu}\,\gamma_5\,\gamma_\mu + e^\nu + if^\nu\,\gamma_5 + \frac{1}{2}g^{\lambda\mu\nu}\,\sigma_{\lambda\mu}, \;. \qquad (49)
\end{aligned}$$

For our purposes we note that the terms proportional to a_μ and b_μ violate both Lorentz and CPT symmetries, unlike the $c_{\mu\nu}, d_{\mu\nu}$ and $H_{\mu\nu}$ terms that violate only Lorentz symmetries. Analogous observations can be made for the various terms inside the Γ^μ structure of (49). The main assumption behind the form of such operators is that an unknown physics at high energy scales could lead to a spontaneous breaking of Lorentz invariance by giving an expectation value to certain tensorial fields, which are not in the Standard Model (SM) spectrum. The interaction of these fields with operators composed from the SM fields, which are fully Lorentz-symmetric before the spontaneous breaking, will manifest itself as effective LV terms, which below the scale of the LV condensation would have the schematic form:

$$O^{\rm SM}_{\mu\,\nu\,\dots}\,C^{\mu\,\nu\,\dots} \to O^{\rm SM}\,\langle C^{\mu\,\nu\,\dots}\rangle\,, \qquad (50)$$

IOP Conf. Series: Journal of Physics: Conf. Series **873** (2017) 012006 doi:10.1088/1742-6596/873/1/012006

where $C^{\mu\nu\cdots}$ is an external field that undergoes condensation and $O^{\rm SM}$ is a SM field operator that transforms properly under the Lorentz group. The classification of [44] requires that the independent dimension-5 operators must be gauge invariant, Lorentz invariant after contraction with the background tensors $\langle C^{\mu_1\mu_2\cdots}\rangle$, not reducible to total derivatives or to lower-dimension operators by the use of equations of motion, and they should couple to an irreducible background tensor. Several experiments, of diverse origin, can be used in order to impose stringent constraints on the relevant SME coefficients, that range from searches for forbidden atomic transitions in precision experiments and studies of low-energy antiprotonic atoms and antimatter factories, to high-energy cosmic rays, nuclear spin precession and atomic and nuclear Electric Dipole Moments (EDM) measurements, as well as data on neutrino oscillations. In the following we shall discuss some phenomenological consequences of some of the above coefficients, especially in the context of neutral mesons, forbidden transitions in (anti)hydrogen molecules or atomic dipole moment measurements. Then we shall scale the appropriate bounds on b_μ back in time, in the context of our CPTV stringy Universe of section 2, in order to see whether (some of) our leptogenesis scenarios discussed above can be falsified.

We commence with SME tests in antiprotonic atoms [45], in particular antihydrogen ($\overline{\rm H}$) of interest in this conference. Motivated by the theoretical microscopic models of section 2, I shall restrict myself to constraining the b_μ coefficients of the SME (49) using spectroscopy, in particular looking for forbidden transitions, *e.g.* $1s \to 2s$. Within H spectrocopic measurements, the presence of a b_μ coefficient in the SME (49) leads to the relevant transition of the electron in the H atom. The sensitivity of the tests depend crucially whether the atoms are free or trapped in an external magnetic field. In the case of free H (and $\overline{\rm H}$), the frequency shift of the 1s-2s transition is a higher-loop quantum effect in the SME/Quantum-Electrodynamics (QED) lagrangian, and thus the effect is suppressed by the square of the fine structure constant, α^2: $\delta_{1s-2s}\nu^{\rm H} \simeq -\alpha^2\, b_3^e/8\pi$, *i.e.* the pertinent sensitivity of such experiments would be about five orders of magnitude smaller compared to tests involving the corresponding transitions in trapped H and $\overline{\rm H}$. However, in the latter tests, the corresponding frequency shifts are proportional to the difference $b_3^e - b_3^p$ of the third spatial component of b_μ between electrons (e) and protons(p) (in a frame where the direction of the external magnetic field is along the z axis). In view of the universal character of B^μ vectors due to background space-time geometries discussed in section 2, for this model the above difference would vanish. To cover ourselves against such cases, it is therefore imperative to either measure the sum of the coefficients $b_\mu^{e,p}$, or isolate them experimentally. The former can be achieved by examining hyperfine structure transitions in atomic (anti)matter. Indeed, within 1s transitions of H or $\overline{\rm H}$, one can determine the relevant energy shifts induced by b_μ [45]:

$$\Delta^H_{a\to b} \simeq (b^e_\mu + b^p_\mu)/\pi + \dots \tag{51}$$

where the ... denote contributions from the rest of the SME coefficients (49), which are not written explicitly here. Hyperfine transitions within the 1S level of H can be measured with accuracies exceeding 1 mHz in masers. So transitions of this type in trapped H and $\overline{\rm H}$ are interesting candidates for performing tests of Lorentz or CPT symmetry, although to achieve resolutions of 1 mHz in trapped antihydrogen does not seem feasible in the foreseeable future.

Another possibility would be to measure [45] radio-frequency transitions between states within the triplet of hyperfine levels in H and $\overline{\rm H}$, in particular the so called $d\rangle_1 \to |c\rangle_1$ transition at external magnetic fields of order B $\simeq$ 0.65 Tesla. The corresponding frequency shifts depend solely on b_3^p:

$$\Delta^{\rm H}_{c\to d} \simeq -b_3^p/\pi, \qquad \Delta^{\overline{\rm H}}_{c\to d} \simeq +b_3^p/\pi \tag{52}$$

where we took into account that under the action of CPT operation, which exchanges H and $\overline{\rm H}$, the coefficient of the b_3^p changes sign. Thus, comparison of the above spectroscopic measurement between trapped H and $\overline{\text{H}}$ would yield immediately a bound (or a value !) on b_3^p. If a frequency resolution of 1 mHz could be attained (which at present is far from being plausible), then, one could obtained $|b_3^p| \leq 10^{-27}$ GeV. Still such bounds are about four orders of magnitude smaller that the ones coming from masers. We also note that, although, clock-comparison experiments are able to resolve spectral lines to about 1 μHz, nevertheless, isolating b_i^p is very complicated due to the complex structure of the nuclei involved.

The above experiments are sensitive only to spatial components of Lorentz-violating couplings. Sensitivity to timelike couplings, b^0, would require appropriate boosts. On the other hand, in the context of the model (7) of section 2, such experiments can bound the combinations $\gamma \vec{v}_3 B^0$, where $\vec{v}$ is the current-era relative velocity of us (as local observers) with respect to the CMB (or Friedmann-Robertson-Walker) frame. Currently, we can quote the following bounds on b_μ coefficients for electrons [43]:

$$b^0 \leq 0.02\,\text{eV}\ , \quad |\vec{b}| \leq 10^{-21}\ \text{eV}\ . \tag{53}$$

It should be mentioned that the stringent limits on $|\vec{b}|$ have been obtained in measurements using torsion pendulum containing macroscopic numbers of polarised electrons [46]. New interactions, such as the above-mentioned Lorentz- and CPT-violating ones are then searched for (and bounded) by looking at the corresponding effects on the electron spin. Such tests may also be performed in man-made antihydorgen [47] or other anti-atoms, with the aim of providing direct comparison of CPT properties and thus tests of CPT invariance.

Next we describe the situation governing the constraints on the relevant dimension-5 terms of the SME lagrangian coming from EDM. These are generically found to be of order [44] $\leq 10^{-25}\, e\,\text{cm}$. The overall expression for the total EDM, due to the CP Violating conventional QED terms and the CPT Violating terms due to the presence of an appropriate Lorentz-violating background vector n^μ, is obtained from the effective Lagrangian

$$\mathcal{L}_{\text{EDM}} = -i\frac{1}{2}d_{\text{CP}}\overline{\psi}\,\sigma^{\mu\nu}\,F_{\mu\nu}(A)\psi + d_{\text{CPT}}\overline{\psi}\,\gamma^\mu\,\gamma^5\,F_{\mu\nu}(A)\,n^\nu\,\psi\ , \tag{54}$$

where $F_{\mu\nu}$ is the Maxwell field strength. The currently null result on the neutron dipole moment imposes the constraint $d_{\text{CP}} + d_{\text{CPT}} = 0$. The lagrangian (54) should be completed with the a^μ and b^μ SME terms (49), as well as the appropriate dimension-5 operators from the QED sector of the SME [44]:

$$\mathcal{L}_5 = \sum_{\text{fermion species}} \left[c^\mu\,\overline{\psi}\gamma^\lambda F_{\lambda\mu}\psi + d^\mu\,\overline{\psi}\gamma^\lambda\,\gamma^5\,F_{\lambda\mu}\psi + g^\mu\overline{\psi}\gamma^\lambda\tilde{F}_{\lambda\mu}\psi + f^\mu\,\overline{\psi}\gamma^\lambda\,\gamma^5\tilde{F}_{\lambda\mu}\psi\right]\ , \tag{55}$$

where $\tilde{F}_{\mu\nu}$ is the dual of the Maxwell tensor. The various terms in (55) have different transformation properties under the action of the discrete symmetries C, P and T, which, together with the corresponding terms of (49), are indicated in figure fig. 3, on the assumption that the vector backgrounds are time-like and invariant under C,P and T reflections [44].

Experimentally [48], one can disentangle CP-odd from CPT-odd operators, because of different suppression scales. Specifically, the former require helicity flip and are thus represented by dimenion-six operators in the SME effective lagrangian, with suppression by the CP breaking scale of order $1/\Lambda_{\text{CP}}^2$. Such operators imply spin precession in a magnetic field relative to the direction of $\mathbf{B} \times \mathbf{v}$. On the other hand, the CPT-odd operators are of dimension 5, as they do not require helicity flip, *e.g.* in the quark sector such operators are of the form $\overline{q}_{R(L)}\gamma^\nu\,\gamma^5\,F_{\nu\,\mu}\,q_{R(L)}$, and $\overline{q}_L\gamma^\nu\,\gamma^5\,F^a_{\nu\,\mu}\,\tau^a\,q_L$, where τ^a, $a = 1,2,3$ are the SU(2) generators of the weak interaction standard model group, and $F_{\mu\nu}$ and $F^a_{\mu\nu}$ are the U(1) and SU(2) gauge field strengths respectively. These operators are suppressed linearly by the CPT-breaking scale, $1/\Lambda_{\text{CPT}}$.

Coefficient	Operator	C	P	T
a^0	$\overline{\psi}\gamma_0\psi$	$-$	$+$	$+$
b^0	$\overline{\psi}\gamma_0\gamma_5\psi$	$+$	$-$	$+$
c^0	$F_{\lambda 0}\overline{\psi}\gamma^\lambda\psi$	$+$	$+$	$-$
d^0	$F_{\lambda 0}\overline{\psi}\gamma^\lambda\gamma^5\psi$	$-$	$-$	$-$
f^0	$\widetilde{F}_{\lambda 0}\overline{\psi}\gamma^\lambda\gamma^5\psi$	$-$	$+$	$+$
g^0	$\widetilde{F}_{\lambda 0}\overline{\psi}\gamma^\lambda\psi$	$+$	$-$	$+$

Figure 3. Tranformation properties of the various terms in (55) under the action of the discrete symmetries C, P and T. A + (−) indicates an even (odd) function. From ref. [44].

EDMs have been bounded with high precision in several occasions [48]:
(i) neutrons, with the bound $d_n < 3 \times 10^{-26}\,\mathrm{e\,cm}$,
(ii) diamagnetic atoms (such as Hg, Xe, ...) : their EDMs are induced by the EDMs of the valence nucleons; for the case of mercury EDM, one has the (approximate) relation: $d_{\rm Hg} \simeq -5 \times 10^{-4}(d_n + 0.1\,d_p) \sim -5 \times 10^{-4}\,d_n$. The last approximate relation implies that a signal consistent with CPT violation would occur, if a non zero d_n, $d_{\rm Hg}$ were found.
(iii) paramagnetic atoms (such as Tl, Cs, ...): their EDM are extremely suppressed as a result of the absence of a CPT-odd electron EDM.

In general, theoretical estimates of dimension-three operators induced by multiloop CP violating corrections in the standard model, imply the following bounds of the SME coefficients in (49) [48]

$$a^\mu, b^\mu \sim d^\mu \left(10^{-20} - 10^{-18}\right) {\rm GeV}^2 \ , \tag{56}$$

providing sensitivity to $d^\mu \le 10^{-12}\ {\rm GeV}^{-1}$ and thus $\Lambda_{\rm CPT} \sim \left(10^{11} - 10^{12}\right)$ GeV, if one takes into account the current bounds on b_μ [43] (*cf.* (53) below).

Higher Lorentz-violating background tensors, *e.g.* terms in SME effective lagrangian of the form $\mathcal{D}_{\mu\nu\rho}\overline{e}\,\gamma^\rho\,\gamma^5\,e\,F_{\mu\nu}$ can also be bounded experimentally with high accuracy, by looking [48] for corrections to the spin precession frequency of the form $\left(\mathcal{D}^{i[0\,k]} + \mathcal{D}^{k\,[0\,i]}\right)\,E_i\,B_k$, which changes sign under the reversal of the electric field E_i. The relative signal changes during the day as a result of the change of the Laboratory orientation relative to the tensor background.

We close this section by mentioning the interesting suggestion of ref. [49] on further tests of CPT symmetry due to the CPT-odd axial vector background b^μ, which has been of interest to us in section 2. According to this work, within the framework of Lorentz-violating extended electrodynamics, the Dirac equation for a bound electron in an external electromagnetic field has been considered, assuming the interaction with the background field b_μ. A Foldy-Wouthysen quasi-relativistic (1/c)-series expansion (truncated to order $1/c^2$) has been applied to obtain an effective Hamiltonian for the hydrogen atom and through this the relativistic Dirac eigenstates in a spherically-symmetric potential to second order in b^0. The b^0-induced CPT-odd corrections

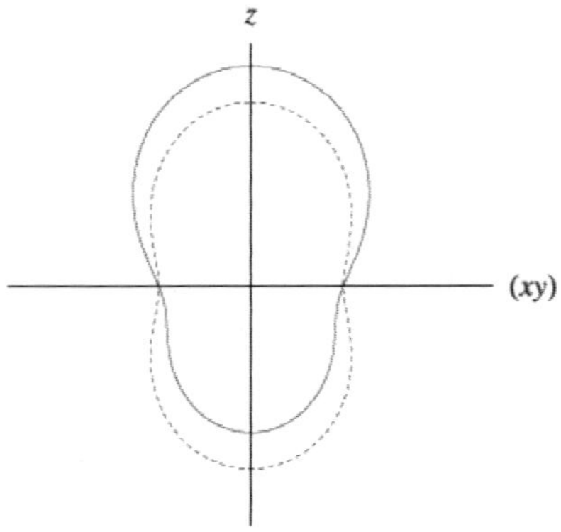

Figure 4. Anglular distribution for spontaneous radiation for the atomic transition $2p_{1/2,1/2} \to 1s_{1/2,-1/2}$ in the presence of a CPT-odd SME axial background vector b^μ. The dashed line indicates the standard electrodynamics $b^\mu = 0$ case. From ref. [49].

to the electromagnetic dipole moment operators of a bound electron have been calculated. Such corrections contribute to the anapole moment of the atomic orbital and may cause a specific asymmetry of the angular distribution of the radiation of a hydrogen atom, in particular the $2p_{1/2,1/2} \to 1s_{1/2,-1/2}$ (*cf.* fig. 4). The non-observation currently of such asymmetries leads to bounds of the magnitude of $|b^0|$: $|b^0| \le 2 \times 10^{-8}\, m_e\, c^2 \simeq 10^{-11}$ GeV, which are consistent with the general bounds (53) for the SME coefficient b_μ for electrons [43].

Finally we mention that, further tests of CPT invariance can be made by direct measurements of particle antiparticle mass and charge differences, which we are not going to discuss here. However, in the spirit of our cosmological model discussed in section 2, we do mention that, if the observed matter/antimatter asymmetry were due to a mass difference between particle and antiparticles, then, one may make the reasonable assumption that baryogenesis could be due to mass differences between quarks and antiquarks [50]. The latter nay depend linearly with temperature, $m_q(T) \sim gT$, as a consequence of known high-temperature properties of Quantum Chromodynamics (QCD). Furthermore, it is reasonable (although not strictly necessary) to assume that the quantk-antiquark differences today are bound by the current bound on proton-antiproton mass difference, which is of order 7×10^{-10} GeV, as provided in 2011 by the ASACUSA Collaboration [47]. Scaling back in temperature such differences, up to the respective decoupling temperature of the quarks, lead to baryon asymmetries that are much smaller than the observed one [50].

In this sense the model of [21] can still survive, given that, even if a $B^0 < 0.02$ eV is observed today, according to the current SME limits, the Universe may have undergone such a (or series of) phase transition at $T \sim 10^9$ GeV towards a smaller (or zero) H-torsion background. This is an (crude) example of how one can use current SME bounds to fit early universe cosmologies. In a similar spirit, the model of leptogenesis through CPT and CP violating decays of heavy right-handed neutrinos, discussed in section 2, based on the Lagrangian (7), with the assumption that the temperature of the destruction of the condensate (6) is of the same order as the decoupling temperature of right-handed neutrinos, $T_D \sim 100$ TeV (14), lies comfortably within the limits (53), given the scaling (15) of the background $B^0 = b^0$ and its current value (16). Reversing the logic, we can assume the upper bound of B^0 today (53), scale back in time with the scaling (15) up to BBN temperatures, $T_{\rm BBN} \sim \mathcal{O}(1)$ MeV, and then constrain the coefficient c_0 by the requirement that the BBN conditions are not disturb. The one can continue scaling back in the cosmic time, to check at which temperature range sufficient leptogenesis is produced, if at all.

IOP Conf. Series: Journal of Physics: Conf. Series **873** (2017) 012006 doi:10.1088/1742-6596/873/1/012006

Such procedures require of course detailed models of baryogenesis via leptogenesis, which we reserve for future studies.

4.2. Intrinsic CPT Violation in Quantum Decoherence Models - the ω-effect

We now come to examine the current phenomenology of the ω-effect [20], which is associated with a second type of CPTV, already discussed in the context of the D-foam model in section 3, in which the quantum CPT operator is not well-defined. We stress again that this latter type of CPTV goes beyond local effective field theories, and its most sensitive bounds can be placed in experimental facilities involving entangled states of neutral mesons, such as neutral Kaon(Φ) factories [41] or $B - \overline{B}$ meson factories [42].

In case of loss of information for a low energy observer, carried by degrees of freedom (d.o.f,) inaccessible to him/her due to quantum gravity environments (*e.g.* the gravitational reocoil d.o.f. in the D-foam example of section 3), the quantum operator that generates CPT symmetry may not be well defined [19]. The proof is obtained by *recuctio ad absurdum*, that is by first assuming the existence of a well-defined *unitary* and *invertible* CPT operator acting on density matrices (antiunitary if acting on state vectors): Θ, such that (a bar above an operator denotes a quantity pertaining to anti-matter states, obtained via the action of the CPT transformation):

$$\rho_{\rm out} = \$\, \overline{\rho}_{\rm in} \tag{57}$$

where the subscript "in" and "out" denotes asymptotic states (from now one we ignore the time arguments for brevity), and the density matrix is defined as $\rho = {\rm tr}|\psi\rangle\langle\psi|$, where the trace operation "tr" is over quantum states.inaccessible to a low energy observer.

. Using the "superscattering matrix" \$ (which is a linear operator acting on density matrices without an inverse, due to the existance of information loss in the problem), we may write:

$$\rho_{\rm out} = \$\, \rho_{\rm in} \;\Rightarrow\; \Theta\text{-}\rho_{\rm in} = \$\, \Theta^{-1}\, \overline{\rho}_{\rm out} \;\Rightarrow \overline{\rho}_{\rm in} = \Theta^{-1}\, \$\, \Theta^{-1}\, \overline{\Theta}_{\rm out}\ . \tag{58}$$

However, since $\overline{\rho}_{\rm out} = \$\, \overline{\rho}_{\rm in}$, the last relation on the right-hand-side of (58) implies

$$\overline{\rho}_{\rm in} = \Theta^{-1}\, \$\, \Theta^{-1}\, \$\, \overline{\rho}_{\rm in}\ . \tag{59}$$

But this is *impossible*, as it would imply that the superscattering operator \$ has an inverse $\Theta^{-1}\, \$\, \Theta^{-1}$, that contradicts the initial assumption of information loss [7]. This is the so called *strong form of (intrinsic) CPT violation*, which would imply a *microscopic time arrow.*

Nevertheless in nature there could be [19] a *weak form* of CPT invariance, according to which the microscopic time arrow does not show in any scattering experiments. Indeed, in such a case the experimentalist would be able to prepare initial pure quantum mechanical state vectors, and there should a well defined transition probability $\mathcal{P}$ from the initial pure state $|\psi\rangle$ to the final state $|\phi\rangle$, such that

$$\mathcal{P}\Big(\psi \to \phi\Big) = \mathcal{P}\Big(\theta^{-1}\, \phi \to \theta\, \psi\Big) \tag{60}$$

where the (antiunitary) CPT operator θ acts on "in" and "out" Hilbert spaces $\mathcal{H}$ vectors now, $\theta : \mathcal{H}_{\rm in} \to \mathcal{H}_{\rm out}$, and is such that

$$\Theta\, \rho = \theta\, \rho\, \theta^\dagger\ , \quad \theta^\dagger = -\theta^{-1}\ , \tag{61}$$

[7] In a theory without information loss, of course, \$ would factorise to the Heisenberg scattering matrix S as $\$ = S\, S^\dagger$, and would have a well-defined inverse, $\$^{-1} = \$^\dagger$, in which case the CPT operator would be well defined satisfying $\Theta = \$\, \Theta^{-1}\, \$$.

IOP Conf. Series: Journal of Physics: Conf. Series **873** (2017) 012006 doi:10.1088/1742-6596/873/1/012006

which in terms of the $ matrix can be written as

$$\$^\dagger = \Theta^{-1}\,\$\,\Theta^{-1} \; . \tag{62}$$

Whether there exists such a situation of weak form of CPT invariance is in general an experimental question. The ω-effect [20] is one way to answer this question experimentally, and, as we have seen in 3, it characterises the D-foam example. We next proceed to the phenomenology of this effect in entangled states of mesons, which, if observed, would constitute a "smoking-gun" evidence for such an intrinsic CPT violation.

We commence our discussion by briefly mentioning direct tests of Time reversal invariance within the Lorentz invariant standard model theory, using entangled neutral mesons, independently of CP and CPT violation. These have been initially proposed in [52], leading to the recent observation of *direct* T violation by the Ba-Bar collaboration [42], through the exchange of initial and final states in transitions that can only be connected by a T -symmetry transformation. For example, the transition $\overline{B}^0 \to B^-$ for the second B to decay, at time t_2, once the first B (entangled with the second) has been tagged at time t_1, is identified by reconstructing events in the time-ordered final states $(\ell^+ X, J/\psi\, K^0_s)$. The rate of this transition is then compared to that of the $B^- \to \overline{B}^0$ transition, that exchanges initial and final states, which is identified by the reconstruction of the final states $(J/\psi\, K^0_L, \ell^- X)$. Any observed difference between these two rates, would thus indicate direct observation of T violation, independent of CP properties [53]. This would also imply an independent test of CPT symmetry within the standard Model. Similar tests of T violation in entangled Kaon Φ factories have also been suggested [54], by identifying the appropriate reactions that exchange initial and final states.

However, if CPT is *intrinsically* violated, in the sense of being not well defined due to decoherence [19] induced by quantum gravity [56], the above-mentioned direct observation of T violation cannot constitute a test of decoherence-induced CPT breaking. This is because in such a case a distinct phenomenon, associated with the ill-defined nature of CPT operator, emerges, termed ω-effect [20]. If the ω effect were present, such direct T-violation tests using entangled states of B-mesons [55] would allow the experimenter to disentangle it from conventional CPT violating effects in the Hamiltonian, within the SME framework, and also to measure independently Imω and Reω. We shall comment briefly on this later in the section.

For the moment, let us concentrate first to the neutral Kaon system, where the effects as we shall see are dominant, although conceptually our analysis applies equally [57] to entangled B-meson factories as well, such as those of [42]. In a quantum-gravity induced decohered situation, the Neutral mesons K^0 and $\overline{K}^0$ should *no longer* be treated as *identical particles*. As a consequence [20], the initial entangled state in Φ factories $|i>$, after the Φ-meson decay, assumes the form:

$$|i\rangle = \mathcal{N}\Big[\Big(|K_S(\vec{k}), K_L(-\vec{k})\rangle - |K_L(\vec{k}), K_S(-\vec{k})\rangle\Big) + \omega\Big(|K_S(\vec{k}), K_S(-\vec{k})\rangle - |K_L(\vec{k}), K_L(-\vec{k})\rangle\Big)\Big] \; , \tag{63}$$

where $\omega = |\omega| e^{i\Omega}$ is a complex parameter, parametrizing the intrinsic CPTV modifications of the EPR correlations [20]. The ω-parameter controls the amount of contamination of the final C(odd) state by the "wrong" (C(even)) symmetry state. The appropriate observable (*c.f.* fig. 5) is the "intensity" $I(\Delta t) = \int_{\Delta t \equiv |t_1 - t_2|}^{\infty} |A(X,Y)|^2$, with $A(X,Y)$ the appropriate Φ decay amplitude [20], where one of the Kaon products decays to the final state X at t_1 and the other to the final state Y at time t_2 (with $t = 0$ the moment of the Φ decay).

It must be noticed that in Kaon factories there is a particularly good channel, the one with bi-pion states $\pi^+\pi^-$ as final decay products, which *enhances the sensitivity* to the ω-effect by three orders of magnitude. This is due to the fact that the relevant terms [20] in the intensity

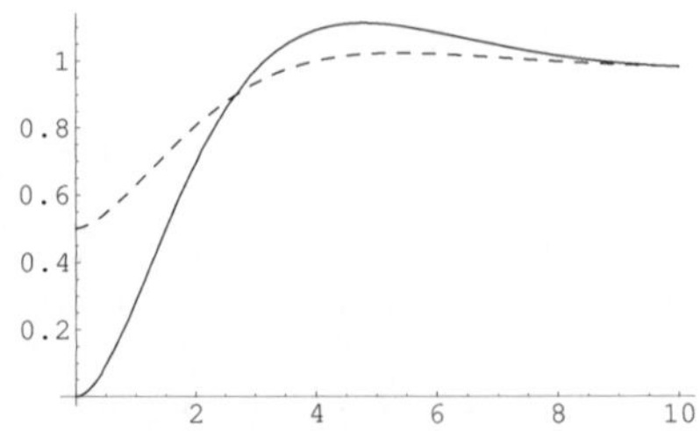

Figure 5. A characteristic case of the intensity $I(\Delta t)$ (vertical axis) as a function of Δt (horizontal axis), with $|\omega| = 0$ (solid line) vs $I(\Delta t)$ (dashed line) with $|\omega| = |\eta_{+-}|$, $\Omega = \phi_{+-} - 0.16\pi$, for definiteness [20].

$I(\Delta t)$ (c.f. fig. 5) contain the combination $\omega/|\eta_{+-}|$, where η_{+-} is the relevant CP-violating amplitude for the $\pi^+\pi^-$ states, which is of order 10^{-3}. The KLOE experiment bounds of the ω parameter are [51]:

$$\mathrm{Re}(\omega) = \left(-1.6^{+3.0}_{-2.1\,\mathrm{stat}} \pm 0.4_{\mathrm{syst}}\right) \times 10^{-4} , \quad \mathrm{Im}(\omega) = \left(-1.7^{+3.3}_{-3.0\,\mathrm{stat}} \pm 1.2_{\mathrm{syst}}\right) \times 10^{-4} . \tag{64}$$

At least an order of magnitude improvement is expected for upgraded facilities such as KLOE-2 at (the upgraded) DAΦNE-2 [51].

This sensitivity is not far from certain optimistic models of space time foam leading to ω-like effects [31]. Indeed, let us comparing these bounds to the D-foam case (48). First, we recall that successful leptogenesis from this class of models requires heavy D-particles masses (20). Assuming for definiteness D-particle masses of order of the Planck mass, $M_s/g_s \sim M_P$, we recall that the variance $\Delta^2 \sim O(10^{-6})$ (38) in order to have phenomenolgically acceptable leptogenesis. In the case of neutral kaons, with momenta of the order of the rest energies ($\sim$ 1 GeV) and mass differences $m_1 - m_2 \simeq 10^{-15}$ GeV, we observe from (48) that $|\omega| \sim 10^{-4}|\Delta|$ (whilst for B-mesons we have $|\omega| \sim 10^{-6}|\Delta|$). For $1 > \Delta \geq 10^{-3}$ these values for ω are not far below the sensitivity of current facilities, such as KLOE-2 at DAΦNE, and thus such leptogenesis CPTV models may be constrained experimentally in the foreseeable future.

In B-factories one can look for similar ω-like effects. Although in this case there is no particularly good channel to lead to enhancement of the sensitivity, as in the Φ-factories, nevertheless one gains in statistics, and hence interesting limits may also be obtained [57]. The presence of a quantum-gravity induced ω-effect in B systems is associated with a theoretical limitation on flavour tagging, namely the fact that in the absence of such effects the knowledge that one of the two-mesons in a meson factory decays at a given time through a flavour-specific channel determines unambiguously the flavour of the other meson at the same time. This is not true if intrinsic CPT Violation is present. One of the relevant observables [57] is given by the CP-violating semi-leptonic decay charge asymmetry (in equal-sign dilepton channel), with the first decay $B \to X\ell^{\pm}$ being time-separated from the second decay $B \to X'\ell^{\pm}$ by an interval Δt. In the absence of ω-effects, the intensity at equal decay times vanishes, $I_{\mathrm{sl}}(\ell^{\pm}, \ell^{\pm}, \Delta t = 0) = 0$, whilst in the presence of a complex $\omega = |\omega|e^{i\Omega}$, $I_{\mathrm{sl}}(\ell^{\pm}, \ell^{\pm}, \Delta t = 0) \sim |\omega|^2$. In such a case, the asymmetry observable exhibits a peak, whose position depends on $|\omega|$, while the shape of the curve itself depends on the phase Ω [57]. The analysis of [57], using the above charge asymmetry method and comparing with currently available experimental data, leads to the following bounds:

$$-0.0084 \leq \mathrm{Re}(\omega) \leq 0.0100 , \quad \text{at 95\% C.L.} . \tag{65}$$

Such tests for intrinsic CPT violation may be performed simultaneously with the above-mentioned observations of direct T violation, as they are completely independent. Quite recently,

IOP Conf. Series: Journal of Physics: Conf. Series **873** (2017) 012006 doi:10.1088/1742-6596/873/1/012006

we have embarked [55] on a detailed study of ω-effects in B_d-system, using the experimental procedure suggested in [52], and implemented in [42], for tests of T violation in etnangled meson systems independent of CP violation. We have identified how to probe the complex ω parameter in the entangled B_d-system using Flavour(f)-CP(g) eigenstate decay channels: the connection between the Intensities for the two time-ordered decays (f, g) and (g, f) is lost in the presence of a non-zero ω. Appropriate observables have been constructed allowing independent experimental determinations of Re(ω) and Im(ω), disentangled from CPT violation in the evolution Hamiltonian, Re(θ) and Im(θ), which parametrise CPTV within the SME local effective field theory frameworks. The general analysis of [55] has found

$$\begin{aligned} \mathrm{Im}\,(\omega) &= \pm(6.40 \pm 2.80) \times 10^{-2} \,, \quad \mathrm{Re}\,(\omega) = (1.09 \pm 1.60) \times 10^{-2} \,, \\ \mathrm{Im}\,(\theta) &= \pm(6.11 \pm 3.45) \times 10^{-2} \,, \quad \mathrm{Re}\,(\theta) = (0.99 \pm 1.98) \times 10^{-2} \,, \end{aligned} \tag{66}$$

where the 2.4 σ deviations from Im(ω) = 0 and Re(θ) = 0 are interpreted as upper bounds. These 2σ tensions have been shown to be uncorrelated [55].

At this juncture, I would like to point out that an observation of the ω-effect in both the Φ and B-factories could also provide an independent test of Lorentz symmetry properties of the intrinsic CPT Violation, namely whether the effect respects Lorentz symmetry. This is because, although the Φ particle in neutral Kaon factories is produced at rest, the corresponding Υ state in B-factories is boosted, and hence there is a frame change between the two experiments. If the quantum gravity ω-effect is Lorentz violating, as it may happen in certain models [31], then a difference in the value of ω between the two experiments should be expected.

Finally, since Lorentz Violation has been mentioned, I also point out that bounds of the LV SME coefficients a^μ (*cf.* eq. (49)) can be placed by measurements in the entangled Kaon Φ factories [51]. In particular by adopting the relevant SME terms to the quark sector, relevant for Kaon physics, one can bound differences $\Delta a^\mu = a^\mu_{q1} - a^\mu_{q2}$, where q_i, $i =, 2$ denote appropriate quark states. The current experimental limits for the coefficients Δ^μ_a are: from the KTeV Collaboration $\Delta_X, \Delta a_Z < 9.2 \times 10^{-22}$ GeV, while from the the KLOE Collaboration in the DaΦNE Φ factory [51] are less competitive but with the advantage that entangled meson factories have sensitivity to all four coefficients Δa^μ, in particular: $\Delta a^0 = (0.4 \pm 1.8) \times 10^{-17}$ GeV, from KLOE, with expected sensitivity at KLOE-2 in upgraded DAΦNE facilties for $\Delta a^{X,Y,Z} = \mathcal{O}(10^{-18})$ GeV. Unfortunately, entangled meson factories have only sensitivity to differences Δa^μ rather than absolute coefficients a^μ. Of course, if gravity acts universally for all quark species, such differences may be zero.

4.3. Spin Statistics & Ill-defined CPT Violation - Searches for Pauli Principle Violations

Before closing the talk I would like to make some remarks on another potential effect of quantum-gravity-decoherence-induced CPTV, namely tiny violations of the Spin-Statistics theorem [1] and in particular Pauli exclusion principle and related searches [58]. As with the CPT theorem, Spin-Statistics theorem is based on several underlying assumptions, which if relaxed, could lead to violations. One of the most dramatic consequences of such violations would be the evasion of Pauli principle, and recently there are dedicated experimental searches for it by the VIP,VIP2 collaborations [58], looking for spontaneous x-ray emissions of atoms (*e.g.* forbidden transitions $2p \to 1s$ in Copper atoms, which test the PEP for electrons).

The *Spin-Statistics Theorem* states that the wave function of a system of identical integer-spin particles has the same value when the positions of any two particles are swapped. Particles with wave functions symmetric under exchange are called bosons. The wave function of a system of identical half-integer spin particles changes sign when two particles are swapped. Particles with wave functions antisymmetric under exchange are called fermions. The theorem was first

IOP Conf. Series: Journal of Physics: Conf. Series **873** (2017) 012006 doi:10.1088/1742-6596/873/1/012006

proposed by Fierz in 1939, by Pauli in a more systematic formulation in 1940, and in a rigorous mathematical formalism, using quantum field theory path integrals, by Schwinger in 1950, where the underlying mathematical assumptions were made clear. It is this latter proof we shall follow in this talk, in order to discuss possible violations.

An important consequence of the spin-statistics theorem is that the wavefunction of two identical fermions is zero, hence two identical fermions (i.e. with all quantum numbers the same) cannot occupy the same state, which is the celebrated *Pauli exclusion Principle* (PEP), which was postulated by Pauli in 1925, without knowledge of the spin-statistics theorem at the time.

Another important consequence of the theorem is that in quantum field theory, Bosons obey commutation relations, whilst fermions obey anticommutation ones. Schwinger's proof of the Spin-Statistics theorem, using quantum field theory machinary, requires the following assumptions:

- (1) The theory has a Lorentz and CPT invariant Lagrangian and relativistic causality.
- (2) The vacuum is Lorentz-invariant (can be weakened).
- (3) The particle is a localized excitation. Microscopically, it is not attached to a string or domain wall.
- (4) The particle is propagating (has a non-infinite mass).
- (5) The particle is a real excitation, meaning that states containing this particle have a positive-definite norm and has positive energy.

Proof:

Consider generic quantum relativistic fields $\phi(x)$, where x denotes a generic space-time point. The object of interest is the two-point correlator

$$\mathcal{G}(x) = \langle 0|\phi(-x)\,\phi(0)|0\rangle \tag{67}$$

Let us denote the matrix corresponding to the rotation by π of the spin polarization of the field by $\mathcal{R}(\pi)$. Following Schwinger, we take the following steps towards the proof of the Spin-Statistics theorem:

STEP I : Formulate the quantum field theory at hand in Euclidean space time where the path integral makes rigorous sense. In this case spatial Lorentz transformations are ordinary rotations, but Boosts become also rotations in imaginary time, and hence a rotation by π in the plane (x(space) - t(time)) plane in Euclidean space-time is a CPT transformation in the language of Minkowski spacetime.

CPT transformation, if well defined, takes states in a path integral into their conjugates so, CPT invariance of the theory implies that the two-point correlator:

$$\langle 0|\mathcal{R}\phi(x)\,\phi(-x)|0\rangle \tag{68}$$

must be positive-definite at x=0 according to the positive-norm-state assumption (5) of the spin-statistics theorem. Propagating states, i.e. finite mass states, implies that this correlator is non-zero at space-like separations. Special relativity is needed to define space-like intervals of course, hence the Lorentz invariance assumptions (1) + (2) of the theorem.

STEP II: Lorentz Invariance allows fields to be transformed according to their spin, such that:

$$\langle 0|\mathcal{R}\mathcal{R}\,\phi(x)\,\mathcal{R}\,\phi(-x)|0\rangle = \pm\langle 0|\phi(-x)\,\mathcal{R}\phi(x)|0\rangle\ , \tag{69}$$

where $+(-)$ is for bosons/integer spin (fermions/half-integer spin).

STEP III : Using CPT invariance (which is equivalent to also assuming a well-defined CPT operator and which, as we have already mentioned, in Euclidean space-time is equivalent to rotational invariance) we may equate the rotated correlation function (69) to $\mathcal{G}(\mathrm{x})$ (67), thus obtaining:

$$\langle 0| \Big(\mathcal{R}\, \phi(x)\, \phi(y) - \phi(y)\, \mathcal{R}\phi(x) \Big) |0\rangle = 0, \tag{70}$$

for integer spins, and

$$\langle 0| \Big(\mathcal{R}\, \phi(x)\, \phi(y) + \phi(y)\, \mathcal{R}\phi(x) \Big) |0\rangle = 0, \tag{71}$$

for half-integer spins.

This completes the proof of the Spin-Statistics theorem. The theorem essentially implies that, since the operators are space-like separated, a different order can only create states that differ by a phase. The argument fixes the phase to be ?1 or 1 according to the spin. Since it is possible to rotate the space-like separated polarizations independently by local perturbations; the phase should not depend on the polarization in appropriately chosen field coordinates.

We should remark at this point that when a violation of the spin-statistics theorem appears, it is one or more of the above assumptions that they are violated. For instance, spinless anticommuting fields, which could exist in condensed matter models are not relativistic invariant; ghost fields in gauge theories are spinless fermions but they have negative norm. In 2+1 dimensional Chern-Simons theory has anyons (fractional spin) but in such a case the wave function of the planar system splits between the bulk and the boundary, and hence is somehow delocalised. One remark concerns quarks: Despite being attached to a confining string, QCD quarks can have a spin-statistics relation proven at short distances (ultraviolet limit) due to asymptotic freedom.

We have seen above tat the rôle of CPT and Lorentz invariance is crucial for the validity of the spin-statistics theorem. Although spontaneous violation of Lorentz and CPT symmetry ay be tolerated (see assumption (2), which may be relaxed), nevertheless if the CPT is ill defined, and there is a strong (intrinsic) form of CPT violation, as is the case of quantum-gravity induced decoherence (or D-foam situations, as in section 3), then there may be tiny violations of the spin-statistics theorem. In this respect, we recall that it was the Bose-Statistics requirement of the neutral mesons that resulted in the antisymmetric initial state (63) when $\omega = 0$. In D-foam situation, for instance, the matter excitation is dressed by open strings stretched between the defect and the brane world, and in this sense, assumption (3) of the theorem is violated, together with assumption (1), due to the ill-defined nature of the CPT operator. In such cases, there are hidden d.o.f. in a particle state, and thus an evasion, in case of fermions, of the PEP may be understood by the fact that the otherwise looking identical quantum states entering the PEP formulation, actually differ by hidden quantum numbers. Thus we believe that searches for intrinsic CPT violation could also be complemented by searches of PEP. The current searches of PEP [58] are mainly for charged particles (electrons), for which, as we have discussed previously, the effects of the D-particle foam are expected to be strongly suppressed for type IIB theories. Nevertheless, this is only one model and experiment should be independent of any theoretical assumptions, hence searches of PEP are equally encouraged as searches of intrinsic CPT violation.

5. Conclusions and Outlook

.

In the talk I discussed theoretical motivations for CPT Violation, both within the SME local effective field theory framework and outside of it, as a result of ill-defined CPT generators in theories of quantum gravity entailing decoherence of quantum matter. We have been concerned with generation mechanisms of matter-antimatter asymmetry in the Universe by means of either early geometries with CPT Violation, e.g. the string-inspired Kalb-Ramond torsion models discussed in section 2, or space-time foamy models of brane worlds propagating in bulk geometries punctured with D-particle (effectively) point-like defects, leading to a decoherence-induced ω-effect in entangled states of mesons (section 3).

By using detailed models of the early Universe with CPTV, and knowing the scaling law of the relevant CPTV quantities entering the induced matter-antimatter asymmetry with the temperature of the Universe, I have used the current stringy bounds of such parameters to constrain the models. Specifically, I argued that the presence of Lorentz and CPT Violating geometries in the early universe, rather than quantum gravity, may be responsible for the emergence of b^μ-like axial vector SME backgrounds. Such vectors have been argued to be responsible for the observed matter/antimatter asymmetry in the Universe. From this perspective, having small remnants of such vector backgrounds today is a not so unrealistic possibility, given that the Universe may have undergone a phase transition at a certain temperature during an early era. Hence, it makes perfect sense to search for or bound such SME coefficients by precision atomic spectroscopy or other methods, such as EDMs, including comparison of the relevant properties of matter with antimatter, especially now that we have available man-made antimatter.

In the talk I also reviewed some of such tests, with direct interest to this conference, including a brief discussion on direct observations of T violation in entangled particle states, independently of CP properties. In addition I discussed a novel phenomenon that may characterise certain quantum gravity models, namely "intrinsic CPT violation" as a result of the fact that, due to the associated decoherence of matter propagating in a quantum space-time foam environment, the CPT operator is perturbatively ill-defined: although the anti particle exist, nevertheless the properties of the CPT operator when acting on entangled states of particles lead to modified EPR correlators. Such modifications imply a set of well-defined observables, which can be measured in current or upcoming facilities, such as Φ or B-factories.

The signatures of quantum-gravity induced decoherence in entangled states of mesons are rather unique, and in this sense they constitute "smoking-gun" evidence for this type of CPT Violation, if realised in Nature. The other important advantage of such searches is that they are virtually cost free, in the sense that the relevant tests can be performed in facilities that have already been or are to be built for other purposes at no extra cost, apart from minor modifications/adjustments in the relevant Monte-Carlo programmes to take proper account of these quantum-gravity effects. As a related topic to cases of CPTV decoherence, where the CPT operator is not well defined, I also discussed searches for violations of the Pauli equivalence principle, and more generally the spin statistics theorem.

Acknowledgements

I would like to thank the organisers of DISCRETE 2016 Symposium (November 28-December 3 2016, U. Warsaw, Poland), for their kind invitation to give a plenary talk at the conference. The work of N.E.M. is partially supported by STFC (UK) via the research grant ST/L000326/1.

References

[1] R. F. Streater and A. S. Wightman 2000, *PCT, spin and statistics, and all that* (Princeton, USA: Princeton Univ. Press) and references therein.
[2] A. D. Sakharov 1967, Pisma Zh. Eksp. Teor. Fiz. **5**, 32 [JETP Lett. **5**, 24 (1967)] [Sov. Phys. Usp. **34**, 392 (1991)] [Usp. Fiz. Nauk **161**, 61 (1991)].
[3] V. A. Kuzmin, V. A. Rubakov and M. E. Shaposhnikov 1985, Phys. Lett. B **155**, 36; M. B. Gavela, P. Hernandez, J. Orloff and O. Pene 1994, Mod. Phys. Lett. A **9**, 795 doi:10.1142/S0217732394000629 [hep-ph/9312215]; M. B. Gavela, P. Hernandez, J. Orloff, O. Pene and C. 1994 , Nucl. Phys. B **430**, 382 doi:10.1016/0550-3213(94)00410-2 [hep-ph/9406289].
[4] For a review see: W. Buchmuller, R. D. Peccei and T. Yanagida 2005, Ann. Rev. Nucl. Part. Sci. **55**, 311 [hep-ph/0502169], and references therein.
[5] D. N. Spergel *et al.* [WMAP Collaboration] 2003, Astrophys. J. Suppl. **148**, 175 [astro-ph/0302209].

P. A. R. Ade *et al.* [Planck Collaboration] 2014, Astron. Astrophys. **571**, A16 [arXiv:1303.5076 [astro-ph.CO]].

[6] R. N. Mohapatra and P. B. Pal 1998 [2004], World Sci. Lect. Notes Phys. **60**, 1 [World Sci. Lect. Notes Phys. **72**, 1 (2004)]. M. Trodden 1999, Rev. Mod. Phys. **71**, 1463 [hep-ph/9803479]; W. Buchmuller, arXiv:0710.5857 [hep-ph].

[7] M. C. Chen 2007, *TASI 2006 Lectures on Leptogenesis*, hep-ph/0703087 [HEP-PH]. A. Pilaftsis 2009, 2013, J. Phys. Conf. Ser. **171**, 012017 [arXiv:0904.1182 [hep-ph]]; J. Phys. Conf. Ser. **447**, 012007.

[8] M. Fukugita and T. Yanagida 1986, Phys. Lett. B **174**, 45.

[9] M. A. Luty 1992, Phys. Rev. D **45**, 455.

[10] P. Minkowski 1977, Phys. Lett. B **67**, 421; M. Gell-Mann, P. Ramond and R. Slansky 1979, in *Supergravity*, eds. D.Z. Freedman and P. van Nieuwenhuizen (North-Holland, Amsterdam); T. Yanagida 1979, in Proc. of the *Workshop on the Unified Theory and the Baryon Number in the Universe*, Tsukuba, Japan, eds. O. Sawada and A. Sugamoto; R. N. Mohapatra and G. Senjanovic 1980, Phys. Rev. Lett. **44**, 912; J. Schechter and J. W. F. Valle 1980, Phys. Rev. D **22**, 2227; G. Lazarides, Q. Shafi and C. Wetterich 1981, Nucl. Phys. B **181**, 287; doi:10.1016/0550-3213(81)90354-0

[11] M. C. Gonzalez-Garcia and M. Maltoni 2008, Phys. Rept. **460**, 1 (2008) [arXiv:0704.1800 [hep-ph]].

[12] T. Asaka, S. Blanchet and M. Shaposhnikov 2005, Phys. Lett. B **631**, 151; doi:10.1016/j.physletb.2005.09.070 [hep-ph/0503065]. For cosmological/astrophysical applications of this model see: A. Boyarsky, O. Ruchayskiy and M. Shaposhnikov 2009, Ann. Rev. Nucl. Part. Sci. **59**, 191 doi:10.1146/annurev.nucl.010909.083654 [arXiv:0901.0011 [hep-ph]]; M. Shaposhnikov, Prog. Theor. Phys. **122**, 185 (2009). doi:10.1143/PTP.122.185

[13] L. Canetti, M. Drewes, T. Frossard and M. Shaposhnikov, Phys. Rev. D **87**, 093006 (2013) doi:10.1103/PhysRevD.87.093006 [arXiv:1208.4607 [hep-ph]].

[14] C. R. Argüelles, N. E. Mavromatos, J. A. Rueda and R. Ruffini 2016 , JCAP **1604**, no. 04, 038 (2016) doi:10.1088/1475-7516/2016/04/038 [arXiv:1502.00136 [astro-ph.GA]]; N. E. Mavromatos, C. R. Argüelles, R. Ruffini and J. A. Rueda 2016, Int. J. Mod. Phys. D **26**, no. 03, 1730007; doi:10.1142/S0218271817300075, and references therein.

[15] O. W. Greenberg 2002, Phys. Rev. Lett. **89**, 231602 doi:10.1103/PhysRevLett.89.231602 [hep-ph/0201258].

[16] D. Colladay and V. A. Kostelecky 1998, Phys. Rev. D **58**, 116002 [hep-ph/9809521].

[17] M. Chaichian, K. Fujikawa and A. Tureanu 2012, 2013, Phys. Lett. B **712**, 115 (2012) doi:10.1016/j.physletb.2012.04.048 [arXiv:1203.0267 [hep-th]]; Eur. Phys. J. C **73**, no. 3, 2349 doi:10.1140/epjc/s10052-013-2349-2 [arXiv:1205.0152 [hep-th]].

[18] G. J. Milburn 2006, New J. Phys. **8**, 96 doi:10.1088/1367-2630/8/6/096 [gr-qc/0308021].

[19] R. M. Wald 1980, Phys. Rev. D **21**, 2742 doi:10.1103/PhysRevD.21.2742

[20] J. Bernabeu, N. E. Mavromatos and J. Papavassiliou 2004, Phys. Rev. Lett. **92**, 131601 doi:10.1103/PhysRevLett.92.131601 [hep-ph/0310180]; E. Alvarez, J. Bernabeu, N. E. Mavromatos, M. Nebot and J. Papavassiliou 2005, Phys. Lett. B **607**, 197 doi:10.1016/j.physletb.2004.12.032 [hep-ph/0410409]; J. Bernabeu, N. E. Mavromatos, J. Papavassiliou and A. Waldron-Lauda 2006, Nucl. Phys. B **744**, 180 doi:10.1016/j.nuclphysb.2006.03.028 [hep-ph/0506025].

[21] J. Ellis, N. E. Mavromatos and S. Sarkar 2013, Phys. Lett. B **725**, 407 doi:10.1016/j.physletb.2013.07.016 [arXiv:1304.5433 [gr-qc]].

[22] M. de Cesare, N. E. Mavromatos and S. Sarkar 2015, Eur. Phys. J. C **75**, no. 10, 514 doi:10.1140/epjc/s10052-015-3731-z [arXiv:1412.7077 [hep-ph]].

[23] O. Bertolami, D. Colladay, V. A. Kostelecky and R. Potting 1997, Phys. Lett. B **395**, 178 doi:10.1016/S0370-2693(97)00062-2 [hep-ph/9612437].

[24] H. Davoudiasl, R. Kitano, G. D. Kribs, H. Murayama and P. J. Steinhardt, Phys. Rev. Lett. **93**, 201301 (2004) doi:10.1103/PhysRevLett.93.201301 [hep-ph/0403019].

[25] B. Mukhopadhyay, Mod. Phys. Lett. A **20**, 2145 (2005) doi:10.1142/S0217732305017640 [astro-ph/0505460]. U. Debnath, B. Mukhopadhyay and N. Dadhich, Mod. Phys. Lett. A **21**, 399 (2006) doi:10.1142/S0217732306019542 [hep-ph/0510351]. G. Lambiase and S. Mohanty, Phys. Rev. D **84**, 023509 (2011) doi:10.1103/PhysRevD.84.023509 [arXiv:1107.1213 [hep-ph]]. G. Lambiase, Phys. Lett. B **642**, 9 (2006) doi:10.1016/j.physletb.2006.09.037 [hep-ph/0612212]. S. Capozziello and G. Lambiase, Frascati Phys. Ser. **58**, 17 (2014) [arXiv:1409.3370 [gr-qc]]. J. I. McDonald and G. M. Shore 2015, Phys. Lett. B **751**, 469 doi:10.1016/j.physletb.2015.10.075 [arXiv:1508.04119 [hep-ph]]. Phys. Lett. B **766**, 162 (2017) doi:10.1016/j.physletb.2016.11.057 [arXiv:1604.08213 [hep-ph]].

[26] E. V. Arbuzova and A. D. Dolgov 2017, arXiv:1702.07477 [gr-qc]; arXiv:1612.06206 [gr-qc].

[27] J. Bernabeu, N. E. Mavromatos and S. Sarkar 2006, Phys. Rev. D **74**, 045014 doi:10.1103/PhysRevD.74.045014 [hep-th/0606137]; N. E. Mavromatos and S. Sarkar 2013, Eur. Phys. J. C **73**, no. 3, 2359 doi:10.1140/epjc/s10052-013-2359-0 [arXiv:1211.0968 [hep-ph]].

[28] D. J. Gross and J. H. Sloan 1987, Nucl. Phys. B **291**, 41 doi:10.1016/0550-3213(87)90465-2 R. R. Metsaev and A. A. Tseytlin 1987, Nucl. Phys. B **293**, 385 doi:10.1016/0550-3213(87)90077-0.

[29] N. E. Mavromatos and A. Pilaftsis 2012, Phys. Rev. D **86**, 124038 doi:10.1103/PhysRevD.86.124038 [arXiv:1209.6387 [hep-ph]].

[30] E. W. Kolb and M. S. Turner 1990, *The Early Universe*, Front. Phys. **69**, 1-547.

[31] N. E. Mavromatos and S. Sarkar 2013, Eur. Phys. J. C **73**, no. 3, 2359 doi:10.1140/epjc/s10052-013-2359-0 [arXiv:1211.0968 [hep-ph]].

[32] J. R. Ellis, N. E. Mavromatos and M. Westmuckett 2004, 2005, Phys. Rev. D **70**, 044036 (2004) doi:10.1103/PhysRevD.70.044036 [gr-qc/0405066]; *ibid.* Phys. Rev. D **71**, 106006 doi:10.1103/PhysRevD.71.106006 [gr-qc/0501060]; J. R. Ellis, N. E. Mavromatos, D. V. Nanopoulos and M. Westmuckett 2006, Int. J. Mod. Phys. A **21**, 1379 (2006) doi:10.1142/S0217751X06028990 [gr-qc/0508105].

[33] J. Bernabeu, N. E. Mavromatos and S. Sarkar 2006, Phys. Rev. D **74**, 045014 doi:10.1103/PhysRevD.74.045014 [hep-th/0606137].

[34] N. E. Mavromatos, V. A. Mitsou, S. Sarkar and A. Vergou 2012, Eur. Phys. J. C **72**, 1956 doi:10.1140/epjc/s10052-012-1956-7 [arXiv:1012.4094 [hep-ph]].

[35] N. E. Mavromatos, M. Sakellariadou and M. F. Yusaf 2013, JCAP **1303**, 015 doi:10.1088/1475-7516/2013/03/015 [arXiv:1211.1726 [hep-th]]. T. Elghozi, N. E. Mavromatos, M. Sakellariadou and M. F. Yusaf 2016, JCAP **1602**, no. 02, 060 doi:10.1088/1475-7516/2016/02/060 [arXiv:1512.03331 [hep-th]].

[36] T. Li, N. E. Mavromatos, D. V. Nanopoulos and D. Xie 2009, Phys. Lett. B **679**, 407 doi:10.1016/j.physletb.2009.07.062 [arXiv:0903.1303 [hep-th]].

[37] G. G. Raffelt 2003, Proc. Int. Sch. Phys. Fermi **152**, 161 doi:10.3254/978-1-61499-008-6-161 [hep-ph/0208024].

[38] J. R. Ellis, N. E. Mavromatos, D. V. Nanopoulos and G. Volkov 2000, Gen. Rel. Grav. **32**, 1777 doi:10.1023/A:1001980530113 [gr-qc/9911055].

[39] I. I. Kogan, N. E. Mavromatos and J. F. Wheater, Phys. Lett. B **387**, 483 (1996) doi:10.1016/0370-2693(96)01067-2 [hep-th/9606102]; J. R. Ellis, N. E. Mavromatos and D. V. Nanopoulos, Int. J. Mod. Phys. A **13**, 1059 (1998) doi:10.1142/S0217751X98000470 [hep-th/9609238].

[40] J. Bernabeu, F. J. Botella and J. Roldan 1989, FTUV/89-35, IFIC/89-11, Proc. Blois Conference on CP Violations, pp. 389-400

[41] A. Di Domenico 2016, *Search for CPT and Lorentz-Symmetry Violation in Entangled Neutral Kaons*, arXiv:1608.00241 [hep-ph]; A. De Santis [KLOE-2 Collaboration and DAFNE Team] 2015, *DAΦNE and KLOE-2*, Phys. Scripta T **166**, 014015 doi:10.1088/0031-8949/2015/T166/014015 [arXiv:1503.06002 [physics.ins-det]] and references therein.

[42] J. P. Lees *et al.* [BaBar Collaboration] 2012, Phys. Rev. Lett. **109**, 211801 [arXiv:1207.5832 [hep-ex]]; G. Eigen [BaBar Collaboration] 2015, *Recent BABAR Results*, J. Phys. Conf. Ser. **631**, no. 1, 012033 doi:10.1088/1742-6596/631/1/012033 [arXiv:1503.02867 [hep-ex]] and references therein.

[43] V. A. Kostelecky and N. Russell 2011, *Data Tables for Lorentz and CPT Violation*, Rev. Mod. Phys. **83**, 11 doi:10.1103/RevModPhys.83.11 [arXiv:0801.0287 [hep-ph]].

[44] P. A. Bolokhov and M. Pospelov 2008, Phys. Rev. D **77**, 025022 [hep-ph/0703291 [HEP-PH]], and references therein.

[45] N. Russell 2002, hep-ph/0209251; M. Hori and J. Walz 2013, Prog. Part. Nucl. Phys. **72**, 206 (2013) doi:10.1016/j.ppnp.2013.02.004 [arXiv:1304.3721 [physics.atom-ph]], and references therein.

[46] B. R. Heckel, E. G. Adelberger, C. E. Cramer, T. S. Cook, S. Schlamminger and U. Schmidt 2008, Phys. Rev. D **78**, 092006 [arXiv:0808.2673 [hep-ex]].

[47] M. Hori *et al.* [ASACUSA Collaboration] 2011, Nature **475**, 484; doi:10.1038/nature10260; M. Hori *et al.* [ASACUSA Collaboration] 2016, Science **354**, no. 6312, 610 doi:10.1126/science.aaf6702

[48] P. A. Bolokhov, M. Pospelov and M. Romalis 2008, Phys. Rev. D **78**, 057702 [hep-ph/0609153] and references therein.

[49] O. G. Kharlanov and V. C. .Zhukovsky 2007, J. Math. Phys. **48**, 092302 [arXiv:0705.3306 [hep-th]].

[50] See, *e.g.*: A.D. Dolgov 2010, Phys. Atom. Nucl., **73**, 588, arXiv: 0903.4318, and references therein.

[51] F. Ambrosino *et al.* [KLOE Collaboration], Phys. Lett. B **642**, 315 (2006); A. Di Domenico 2016, arXiv:1608.00241 [hep-ph].

[52] M. C. Banuls and J. Bernabeu 2000, Nucl. Phys. B **590**, 19 [hep-ph/0005323]; J. Bernabeu, M. C. Banuls and F. Martinez-Vidal 2001, PoS HEP **2001**, 057 [hep-ph/0111073]; J. Bernabeu, F. Martinez-Vidal and P. Villanueva-Perez 2012, JHEP **1208**, 064 [arXiv:1203.0171 [hep-ph]]; See also: E. Alvarez and A. Szynkman 2008, Mod. Phys. Lett. A **23**, 2085 [hep-ph/0611370].

[53] J. Bernabu, F. J. Botella and M. Nebot 2016, JHEP **1606**, 100 doi:10.1007/JHEP06(2016)100 [arXiv:1605.03925 [hep-ph]].

[54] J. Bernabeu, A. Di Domenico and P. Villanueva-Perez 2013, Nucl. Phys. B **868**, 102 [arXiv:1208.0773 [hep-ph]].

[55] J. Bernabeu, F. J. Botella, N. E. Mavromatos and M. Nebot 2016, arXiv:1612.05652 [hep-ph].

[56] J. R. Ellis, J. S. Hagelin, D. V. Nanopoulos and M. Srednicki 1984, Nucl. Phys. B **241**, 381; J. R. Ellis, J. L. Lopez, N. E. Mavromatos and D. V. Nanopoulos 1996, Phys. Rev. D **53**, 3846. P. Huet and M. E. Peskin 1995, Nucl. Phys. B **434**, 3. For entangled states, the requirement of complete positivity implies a different parametrization for the foam effects (in some cases one may consider $\alpha = \gamma, \beta = 0$ in the parameterization of Ellis *et al.*): F. Benatti and R. Floreanini 1997, Nucl. Phys. B **488**, 335. The experiment can independently measure all three decoherence parameters α, β, γ of Ellis *et al.* and hence test the assumption of complete positivity, which notably may not be a property of quantum gravity.

[57] E. Alvarez, J. Bernabeu, N. E. Mavromatos, M. Nebot and J. Papavassiliou 2005, Phys. Lett. B **607**, 197 ; E. Alvarez, J. Bernabeu and M. Nebot 2006, JHEP **0611**, 087.

[58] J. Marton *et al.* 2017, arXiv:1703.01615 [quant-ph]; A. Pichler *et al.* 2016, J. Phys. Conf. Ser. **718**, no. 5, 052030 doi:10.1088/1742-6596/718/5/052030 [arXiv:1602.00898 [physics.ins-det]]; C. Curceanu *et al.* 2016, Found. Phys. **46**, 263 doi:10.1007/D10701-015-9923-4, 10.1007/s10701-015-9923-4 [arXiv:1601.06617 [quant-ph]]. H. Shi *et al.* [VIP Collaboration] 2016, J. Phys. Conf. Ser. **718**, no. 4, 042055 doi:10.1088/1742-6596/718/4/042055 [arXiv:1601.05828 [physics.ins-det]]. C. Curceanu *et al.* 2015, J. Phys. Conf. Ser. **631**, no. 1, 012068 doi:10.1088/1742-6596/631/1/012068

IOP Conf. Series: Journal of Physics: Conf. Series **873** (2017) 012007 doi:10.1088/1742-6596/873/1/012007

CERN-TH-2017-070

Spontaneous symmetry breaking in three-Higgs-doublet S_3-symmetric models

D Emmanuel-Costa[1,5] O M Ogreid[2], P Osland[3] and M N Rebelo[1,4]

[1] Centro de Física Teórica de Partículas – CFTP and Dept de Física Instituto Superior Técnico – IST, Universidade de Lisboa (UL), Av. Rovisco Pais, P-1049-001 Lisboa, Portugal
[2] Western Norway University of Applied Sciences, Postboks 7030, N-5020 Bergen, Norway
[3] Department of Physics and Technology, University of Bergen, Postboks 7803, N-5020 Bergen, Norway
[4] Theoretical Physics Department, CERN, CH-1211, Geneva 23, Switzerland

E-mail: `omo@hvl.no`, `Per.Osland@uib.no` and `rebelo@tecnico.ulisboa.pt`

Abstract. The talk summarises work done by the authors consisting of a detailed study of the possible vacua in models with three Higgs doublets with S_3 symmetry and without explicit CP violation. Different vacua require special regions of the parameter space which were analysed in our work. We establish the possibility of spontaneous CP violation in this framework and we also show which complex vacua conserve CP. In our work we discussed constraints from vacuum stability. The results presented here are relevant for model building.

1. Introduction

In the Standard Model of Particle Physics (SM) there is one Higgs doublet responsible for spontaneous electroweak symmetry breaking and for the mechanism that gives mass to fermions and to electroweak gauge bosons. The model predicts the existence of one Higgs boson. In 2012 a scalar boson was discovered at the LHC [1, 2] with properties consistent with those predicted by the SM. However there are good motivations to consider models with more than one Higgs doublet such as the possibility of having CP symmetry broken spontaneously [3] or new sources of CP violation. Supersymmetric models require two Higgs doublets. Furthermore, models with two Higgs doublets have a rich phenomenology with many interesting possible manifestations of physics beyond the SM [4, 5]. Extensions of the SM with more than one Higgs doublet are good candidates to explain some of the present flavour anomalies and to solve some of the puzzles left unanswered by the SM.

Models with more than one Higgs doublet can give substantial contributions to flavour changing neutral currents (FCNC). Current experimental bounds require these to be strongly suppressed. One possibility is to completely forbid Higgs mediated FCNC at tree level via a symmetry, as is the case in models with natural flavour conservation (NFC) [6, 7] where only one Higgs doublet is allowed to couple to each charge quark sector. In the case of two Higgs doublets this is achieved by means of a Z_2 symmetry and as a result neither spontaneous nor hard CP violation can occur in the Higgs sector. It is possible to have CP violation in the scalar sector,

[5] Affiliation valid till the end of 2016

with two Higgs doublets and NFC, with the Z_2 symmetry softly broken in the Higgs potential [8]. Three Higgs doublets and NFC with exact Z_2 symmetries allow for CP to be violated either explicitly [9] or spontaneously [10]. FCNC can also be eliminated at tree level assuming alignment of Yukawa couplings in flavour space [11], in this case complex Yukawa couplings give rise to new sources of CP violation. There have been several attempts at obtaining Yukawa alignment in several extensions of the SM [12, 13, 14, 15]. Another very interesting possibility to avoid the problem of having too large FCNC with two Higgs doublets, is to build models allowing for Higgs mediated FCNC which are under control by means of a symmetry that leads to couplings suppressed by small off-diagonal elements of the quark mixing matrix $V_{\rm CKM}$. The first models of this type, based on a symmetry, were proposed by Branco, Grimus and Lavoura (BGL) [16]. Implications of BGL-type models have been extensively analysed recently, in the light of the LHC results, for several different implementations [17, 18, 19, 20, 21, 22]. BGL-type models can have FCNC either in the up sector or in the down sector but not in both sectors at the same time. Recently, a generalisation of BGL-type models allowing for FCNC in both sectors was built [23]. BGL models models can also be extended to the case of three doublets [17].

Three-Higgs-doublet models may provide good dark matter candidates [24, 25, 26, 27, 28, 29]. One may also speculate that nature is such that three generations of fermions come with three Higgs doublets, which is, of course, an issue to be settled by experiment. As the number of doublets increases so does the complexity of the scalar potential and the number of free parameters in the theory [30]. Discrete symmetries play an important role in reducing this number and lead at the same time to testable predictions. Symmetries also play an important role in stabilising dark matter [31, 32, 33].

As mentioned above, CP can be spontaneously violated in models with three Higgs doublets with Z_2 symmetries [10]. Such is also the case for an S_3 symmetry [34, 35, 36].

This talk is based on the work done in Ref. [35] where the possible vacuum solutions were analysed and the possibility of having spontaneous CP violation in the context of three Higgs doublets, with an S_3 symmetry, was studied. In this work we concentrated our attention on the scalar potential. Several authors have considered implications of three-Higgs-doublet models with an S_3 symmetry for flavour physics (see, for example: [37, 38, 39, 40, 41, 42, 43, 44]).

2. Vacua of S_3-symmetric three-Higgs-doublet potential

The S_3 group is the permutation group involving three Higgs doublets ϕ_1, ϕ_2 and ϕ_3, identified as the defining representation and constituting a reducible triplet of Higgs doublets. The study of the S_3-symmetric three-Higgs-doublet potential can be done in two different frameworks, either in terms of the defining representation [45, 46], or in terms of the irreducible representations [47]. In what follows, we classify the vacua in terms of constraints on the potential. Vacua can also be classified in terms of their residual symmetries [48].

2.1. The scalar potential

The S_3 symmetric potential has a quadratic and a quartic part, which in terms of the defining representation can be written [45]:

$$V = V_2 + V_4 \tag{1}$$

$$V_2 = -\lambda \sum_i \phi_i^\dagger \phi_i + \frac{1}{2}\gamma \sum_{i<j} [\phi_i^\dagger \phi_j + \text{h.c.}], \tag{2a}$$

$$\begin{aligned} V_4 = {} & A \sum_i (\phi_i^\dagger \phi_i)^2 + \sum_{i<j} \{C(\phi_i^\dagger \phi_i)(\phi_j^\dagger \phi_j) + \overline{C}(\phi_i^\dagger \phi_j)(\phi_j^\dagger \phi_i) + \frac{1}{2}D[(\phi_i^\dagger \phi_j)^2 + \text{h.c.}]\} \\ & + \frac{1}{2}E_1 \sum_{i\neq j} [(\phi_i^\dagger \phi_i)(\phi_i^\dagger \phi_j) + \text{h.c.}] + \sum_{i\neq j\neq k\neq i, j<k} \{\frac{1}{2}E_2[(\phi_i^\dagger \phi_j)(\phi_k^\dagger \phi_i) + \text{h.c.}] \\ & + \frac{1}{2}E_3[(\phi_i^\dagger \phi_i)(\phi_k^\dagger \phi_j) + \text{h.c.}] + \frac{1}{2}E_4[(\phi_i^\dagger \phi_j)(\phi_i^\dagger \phi_k) + \text{h.c.}]\}. \end{aligned} \tag{2b}$$

Here all fields appear on an equal footing. This representation is not irreducible, it splits into two irreducible representations consisting of a singlet, h_S and a doublet of S_3 with components h_1 and h_2. The decomposition into these two irreducible representations is given by:

$$\begin{pmatrix} h_1 \\ h_2 \\ h_S \end{pmatrix} = \begin{pmatrix} \frac{1}{\sqrt{2}} & -\frac{1}{\sqrt{2}} & 0 \\ \frac{1}{\sqrt{6}} & \frac{1}{\sqrt{6}} & -\frac{2}{\sqrt{6}} \\ \frac{1}{\sqrt{3}} & \frac{1}{\sqrt{3}} & \frac{1}{\sqrt{3}} \end{pmatrix} \begin{pmatrix} \phi_1 \\ \phi_2 \\ \phi_3 \end{pmatrix} \tag{3}$$

This matrix has a striking similarity with the tribimaximal mixing matrix [49] which is very close to the observed leptonic mixing.

The scalar potential written in term of fields from irreducible representations has the form [47, 41, 50]

$$V_2 = \mu_0^2 h_S^\dagger h_S + \mu_1^2 (h_1^\dagger h_1 + h_2^\dagger h_2), \tag{4a}$$

$$\begin{aligned} V_4 = {} & \lambda_1 (h_1^\dagger h_1 + h_2^\dagger h_2)^2 + \lambda_2 (h_1^\dagger h_2 - h_2^\dagger h_1)^2 + \lambda_3 [(h_1^\dagger h_1 - h_2^\dagger h_2)^2 + (h_1^\dagger h_2 + h_2^\dagger h_1)^2] \\ & + \lambda_4 [(h_S^\dagger h_1)(h_1^\dagger h_2 + h_2^\dagger h_1) + (h_S^\dagger h_2)(h_1^\dagger h_1 - h_2^\dagger h_2) + \text{h.c.}] + \lambda_5 (h_S^\dagger h_S)(h_1^\dagger h_1 + h_2^\dagger h_2) \\ & + \lambda_6 [(h_S^\dagger h_1)(h_1^\dagger h_S) + (h_S^\dagger h_2)(h_2^\dagger h_S)] + \lambda_7 [(h_S^\dagger h_1)(h_S^\dagger h_1) + (h_S^\dagger h_2)(h_S^\dagger h_2) + \text{h.c.}] \\ & + \lambda_8 (h_S^\dagger h_S)^2. \end{aligned} \tag{4b}$$

In this form the potential has no symmetry for the interchange of h_1 and h_2 but there is a Z_2 symmetry of the form $h_1 \to -h_1$. There is an equivalent doublet representation which has also been used in the literature [40]:

$$\begin{pmatrix} \hat{\chi}_1 \\ \hat{\chi}_2 \end{pmatrix} = \frac{1}{\sqrt{2}} \begin{pmatrix} i & 1 \\ -i & 1 \end{pmatrix} \begin{pmatrix} h_1 \\ h_2 \end{pmatrix}, \tag{5}$$

where the symmetry appears as a symmetry for the interchange of the fields $\hat{\chi}_1$ and $\hat{\chi}_2$. Both expressions for the potential describe the same physics and the coefficients in the different frameworks are related through linear equations.

With the special choice of $\lambda_4 = 0$ the potential acquires a continuous $SO(2)$ symmetry defined by:

$$\begin{pmatrix} h_1' \\ h_2' \end{pmatrix} = \begin{pmatrix} \cos\theta & -\sin\theta \\ \sin\theta & \cos\theta \end{pmatrix} \begin{pmatrix} h_1 \\ h_2 \end{pmatrix}. \tag{6}$$

Spontaneous breaking of this symmetry leads to a massless scalar which of course must be avoided.

S_3 has three irreducible representations, a doublet, a singlet and a pseudosinglet, h_A. The latter has no direct translation into the initial fields ϕ_1, ϕ_2 and ϕ_3 and transforms under S_3 into

(-h_A). From a group theoretical point of view we can choose to write an S_3 symmetric potential in terms of the doublet and h_A. The new potential becomes:

$$V_2 = \mu_0^2 h_A^\dagger h_A + \mu_1^2(h_1^\dagger h_1 + h_2^\dagger h_2), \tag{7a}$$

$$\begin{aligned} V_4 = {} & \lambda_1(h_1^\dagger h_1 + h_2^\dagger h_2)^2 + \lambda_2(h_1^\dagger h_2 - h_2^\dagger h_1)^2 + \lambda_3[(h_1^\dagger h_1 - h_2^\dagger h_2)^2 + (h_1^\dagger h_2 + h_2^\dagger h_1)^2] \\ & + \lambda_4[(h_A^\dagger h_2)(h_1^\dagger h_2 + h_2^\dagger h_1) - (h_A^\dagger h_1)(h_1^\dagger h_1 - h_2^\dagger h_2) + \text{h.c.}] + \lambda_5(h_A^\dagger h_A)(h_1^\dagger h_1 + h_2^\dagger h_2) \\ & + \lambda_6[(h_A^\dagger h_1)(h_1^\dagger h_A) + (h_A^\dagger h_2)(h_2^\dagger h_A)] + \lambda_7[(h_A^\dagger h_1)(h_A^\dagger h_1) + (h_A^\dagger h_2)(h_A^\dagger h_2) + \text{h.c.}] \\ & + \lambda_8(h_A^\dagger h_A)^2, \end{aligned} \tag{7b}$$

which reduces to the same potential we had before with h_1 and h_2 interchanged. At this stage there is no new physics from this choice of representations. However this may change depending on how the couplings to the fermions are introduced.

In order to study the possibility of having spontaneous CP violation we start with a potential with real coefficients. This choice guarantees, without loss of generality, that the potential conserves CP. In this case we are left with ten independent parameters irrespective of the choice of representations. This potential does not fall into a CP conserving potential with irremovable complex parameters [51].

We use the following field notations for the decomposition of the $SU(2)$ Higgs doublets:

$$\phi_i = \begin{pmatrix} \varphi_i^+ \\ (\rho_i + \eta_i + i\chi_i)/\sqrt{2} \end{pmatrix}, \quad i = 1, 2, 3, \tag{8}$$

$$h_i = \begin{pmatrix} h_i^+ \\ (w_i + \tilde{\eta}_i + i\tilde{\chi}_i)/\sqrt{2} \end{pmatrix}, \quad i = 1, 2, \quad h_S = \begin{pmatrix} h_S^+ \\ (w_S + \tilde{\eta}_S + i\tilde{\chi}_S)/\sqrt{2} \end{pmatrix}. \tag{9}$$

2.2. Real vacuum solutions

Real vacuum solutions do not violate CP spontaneously. It is interesting to understand what are the possible real solutions for the vacuum. In this case one has to solve three minimisation conditions corresponding to the vanishing of the three relevant derivatives of the potential. In the irreducible framework these conditions can be solved in terms of μ_0^2 and μ_1^2 leading to [50]:

$$\mu_0^2 = \frac{1}{2w_S}\left[\lambda_4(w_2^2 - 3w_1^2)w_2 - (\lambda_5 + \lambda_6 + 2\lambda_7)(w_1^2 + w_2^2)w_S - 2\lambda_8 w_S^3\right], \tag{10a}$$

$$\mu_1^2 = -\frac{1}{2}\left[2(\lambda_1 + \lambda_3)(w_1^2 + w_2^2) + 6\lambda_4 w_2 w_S + (\lambda_5 + \lambda_6 + 2\lambda_7)w_S^2\right], \tag{10b}$$

$$\mu_1^2 = -\frac{1}{2}\left[2(\lambda_1 + \lambda_3)(w_1^2 + w_2^2) - 3\lambda_4(w_2^2 - w_1^2)\frac{w_S}{w_2} + (\lambda_5 + \lambda_6 + 2\lambda_7)w_S^2\right]. \tag{10c}$$

The first equation comes from the derivative of the potential with respect to w_S and the second and third from the derivatives with respect to w_1 and w_2. Eqs. (10b) and (10c) were obtained dividing by w_1 and w_2 respectively. Clearly, these two equations are not automatically consistent. There are several possible consistency cases:

- for $w_1 = 0$ the corresponding derivative is zero and there is no clash with the determination of μ_1^2 from Eq. (10c).
- otherwise, $\lambda_4(3w_2^2 - w_1^2)w_S = 0$ is required. This can be achieved in three different ways: $\lambda_4 = 0$ or $w_1 = \pm\sqrt{3}w_2$ or $w_S = 0$.
- for $w_S = 0$ a special condition arises from Eq. (10a): $\lambda_4 w_2(3w_1^2 - w_2^2) = 0$ so that in addition we must have $\lambda_4 = 0$ or $w_2 = \pm\sqrt{3}w_1$, or $w_2 = 0$.

Table 1. Possible real vacua (partly after Derman and Tsao [46]). This classification uses the notation R-X-y, where R refers to "real". The roman numeral X gives the number of constraints on the parameters of the potential that arise from solving the stationary-point equations. The letter y is used to distinguish different vev's that have the same X, and λ_a is defined in Eq. (11).

Vacuum	ρ_1, ρ_2, ρ_3	w_1, w_2, w_S	Comment
R-0	$0,0,0$	$0,0,0$	Not interesting
R-I-1	x,x,x	$0,0,w_S$	$\mu_0^2 = -\lambda_8 w_S^2$
R-I-2a	$x,-x,0$	$w,0,0$	$\mu_1^2 = -(\lambda_1+\lambda_3)\, w_1^2$
R-I-2b	$x,0,-x$	$w,\sqrt{3}w,0$	$\mu_1^2 = -\frac{4}{3}(\lambda_1+\lambda_3)\, w_2^2$
R-I-2c	$0,x,-x$	$w,-\sqrt{3}w,0$	$\mu_1^2 = -\frac{4}{3}(\lambda_1+\lambda_3)\, w_2^2$
R-II-1a	x,x,y	$0,w,w_S$	$\mu_0^2 = \frac{1}{2}\lambda_4\frac{w_2^3}{w_S} - \frac{1}{2}\lambda_a w_2^2 - \lambda_8 w_S^2$, $\mu_1^2 = -(\lambda_1+\lambda_3)\, w_2^2 + \frac{3}{2}\lambda_4 w_2 w_S - \frac{1}{2}\lambda_a w_S^2$
R-II-1b	x,y,x	$w,-w/\sqrt{3},w_S$	$\mu_0^2 = -4\lambda_4\frac{w_2^3}{w_S} - 2\lambda_a w_2^2 - \lambda_8 w_S^2$, $\mu_1^2 = -4(\lambda_1+\lambda_3)\, w_2^2 - 3\lambda_4 w_2 w_S - \frac{1}{2}\lambda_a w_S^2$
R-II-1c	y,x,x	$w,w/\sqrt{3},w_S$	$\mu_0^2 = -4\lambda_4\frac{w_2^3}{w_S} - 2\lambda_a w_2^2 - \lambda_8 w_S^2$, $\mu_1^2 = -4(\lambda_1+\lambda_3)\, w_2^2 - 3\lambda_4 w_2 w_S - \frac{1}{2}\lambda_a w_S^2$
R-II-2	$x,x,-2x$	$0,w,0$	$\mu_1^2 = -(\lambda_1+\lambda_3)\, w_2^2,\ \lambda_4 = 0$
R-II-3	$x,y,-x-y$	$w_1,w_2,0$	$\mu_1^2 = -(\lambda_1+\lambda_3)\,(w_1^2+w_2^2),\ \lambda_4 = 0$
R-III	ρ_1,ρ_2,ρ_3	w_1,w_2,w_S	$\mu_0^2 = -\frac{1}{2}\lambda_a(w_1^2+w_2^2) - \lambda_8 w_S^2$, $\mu_1^2 = -(\lambda_1+\lambda_3)\,(w_1^2+w_2^2) - \frac{1}{2}\lambda_a w_S^2$, $\lambda_4 = 0$

Derman and Tsao [46] analysed spontaneous symmetry breaking with real vacua taking also into account the residual symmetries. Their work was done in the reducible framework where the condition $\lambda_4 = 0$ corresponds to $4A - 2(C+\overline{C}+D) - E_1 + E_2 + E_3 + E_4 = 0$. This condition was obtained before by Derman [45] who considered it very unnatural, since in his context it was not clear that it was associated to an additional symmetry. With $\lambda_4 \neq 0$ there were only three possible real solutions [46]:

- (x,x,x) leaving S_3 unbroken and translating into the doublet-singlet notation as $(0,0,w_S)$; consistency condition: $w_1 = 0$ (also verifies $w_1 = \pm\sqrt{3}w_2$).
- (x,x,y) leaving a residual S_2 symmetry. In terms of the reducible representation any ordering of the vevs is equivalent, however, in the definition of the doublet of S_3 a special direction is chosen. As a result, different orderings correspond to different translations:
  (x,x,y) translates into $(0,w_2,w_S)$; consistency condition: $w_1 = 0$.
  (x,y,x) translates into $(w_1, -\frac{1}{\sqrt{3}}w_1, w_S)$; consistency condition: $w_1 = -\sqrt{3}w_2$.
  (y,x,x) translates into $(w_1, \frac{1}{\sqrt{3}}w_1, w_S)$; consistency condition: $w_1 = \sqrt{3}w_2$.
- $(x,y,z) = (x,-x,0)$ leaving a residual S_2 symmetry. This is the only possible real solution with all three vevs different from each other, unless one imposes $4A - 2(C+\overline{C}+D) - E_1 + E_2 + E_3 + E_4 = 0$ ($\lambda_4 = 0$). The translation into the irreducible representation is now:
  $(x,-x,0)$ translates into $(w_1 = \sqrt{2}x, 0, 0)$: consistency conditions: $w_S = 0$ together with $w_2 = 0$.
  $(x,0,-x)$ translates into $(w_1 = \frac{1}{\sqrt{2}}x, w_2 = \frac{\sqrt{3}}{\sqrt{2}}x, 0)$; consistency conditions: $w_S = 0$ together with $w_2 = \sqrt{3}w_1$.
  $(0,x,-x)$ translates into $(w_1 = -\frac{1}{\sqrt{2}}x, w_2 = \frac{\sqrt{3}}{\sqrt{2}}x, 0)$; consistency conditions: $w_S = 0$

IOP Conf. Series: Journal of Physics: Conf. Series **873** (2017) 012007 doi:10.1088/1742-6596/873/1/012007

together with $w_2 = -\sqrt{3}w_1$.

Table 1 summarises all the possible real solutions together with the constraints imposed on the parameters of the potential. The following abbreviation was introduced:

$$\lambda_a = \lambda_5 + \lambda_6 + 2\lambda_7. \tag{11}$$

Table 2. Complex vacua. Notation: $\epsilon = 1$ and -1 for C-III-d and C-III-e, respectively; $\xi = \sqrt{-3\sin 2\rho_1/\sin 2\rho_2}$, $\psi = \sqrt{[3+3\cos(\rho_2 - 2\rho_1)]/(2\cos\rho_2)}$. Due to the constraints the vacua labelled with an asterisk (*) are in fact real.

	IRF (Irreducible Rep.)	RRF (Reducible Rep.)
	w_1, w_2, w_S	ρ_1, ρ_2, ρ_3
C-I-a	$\hat{w}_1, \pm i\hat{w}_1, 0$	$x, xe^{\pm\frac{2\pi i}{3}}, xe^{\mp\frac{2\pi i}{3}}$
C-III-a	$0, \hat{w}_2 e^{i\sigma_2}, \hat{w}_S$	$y, y, xe^{i\tau}$
C-III-b	$\pm i\hat{w}_1, 0, \hat{w}_S$	$x+iy, x-iy, x$
C-III-c	$\hat{w}_1 e^{i\sigma_1}, \hat{w}_2 e^{i\sigma_2}, 0$	$xe^{i\rho} - \frac{y}{2}, -xe^{i\rho} - \frac{y}{2}, y$
C-III-d,e	$\pm i\hat{w}_1, \epsilon\hat{w}_2, \hat{w}_S$	$xe^{i\tau}, xe^{-i\tau}, y$
C-III-f	$\pm i\hat{w}_1, i\hat{w}_2, \hat{w}_S$	$re^{i\rho} \pm ix, re^{i\rho} \mp ix, \frac{3}{2}re^{-i\rho} - \frac{1}{2}re^{i\rho}$
C-III-g	$\pm i\hat{w}_1, -i\hat{w}_2, \hat{w}_S$	$re^{-i\rho} \pm ix, re^{-i\rho} \mp ix, \frac{3}{2}re^{i\rho} - \frac{1}{2}re^{-i\rho}$
C-III-h	$\sqrt{3}\hat{w}_2 e^{i\sigma_2}, \pm\hat{w}_2 e^{i\sigma_2}, \hat{w}_S$	$xe^{i\tau}, y, y$ $y, xe^{i\tau}, y$
C-III-i	$\sqrt{\frac{3(1+\tan^2\sigma_1)}{1+9\tan^2\sigma_1}}\hat{w}_2 e^{i\sigma_1}$, $\pm\hat{w}_2 e^{-i\arctan(3\tan\sigma_1)}, \hat{w}_S$	$x, ye^{i\tau}, ye^{-i\tau}$ $ye^{i\tau}, x, ye^{-i\tau}$
C-IV-a*	$\hat{w}_1 e^{i\sigma_1}, 0, \hat{w}_S$	$re^{i\rho} + x, -re^{i\rho} + x, x$
C-IV-b	$\hat{w}_1, \pm i\hat{w}_2, \hat{w}_S$	$re^{i\rho} + x, -re^{-i\rho} + x, -re^{i\rho} + re^{-i\rho} + x$
C-IV-c	$\sqrt{1+2\cos^2\sigma_2}\hat{w}_2$, $\hat{w}_2 e^{i\sigma_2}, \hat{w}_S$	$re^{i\rho} + r\sqrt{3(1+2\cos^2\rho)} + x$, $re^{i\rho} - r\sqrt{3(1+2\cos^2\rho)} + x, -2re^{i\rho} + x$
C-IV-d*	$\hat{w}_1 e^{i\sigma_1}, \pm\hat{w}_2 e^{i\sigma_1}, \hat{w}_S$	$r_1 e^{i\rho} + x, (r_2 - r_1)e^{i\rho} + x, -r_2 e^{i\rho} + x$
C-IV-e	$\sqrt{-\frac{\sin 2\sigma_2}{\sin 2\sigma_1}}\hat{w}_2 e^{i\sigma_1}$, $\hat{w}_2 e^{i\sigma_2}, \hat{w}_S$	$re^{i\rho_2} + re^{i\rho_1}\xi + x, re^{i\rho_2} - re^{i\rho_1}\xi + x$, $-2re^{i\rho_2} + x$
C-IV-f	$\sqrt{2 + \frac{\cos(\sigma_1 - 2\sigma_2)}{\cos\sigma_1}}\hat{w}_2 e^{i\sigma_1}$, $\hat{w}_2 e^{i\sigma_2}, \hat{w}_S$	$re^{i\rho_1} + re^{i\rho_2}\psi + x$, $re^{i\rho_1} - re^{i\rho_2}\psi + x, -2re^{i\rho_1} + x$
C-V*	$\hat{w}_1 e^{i\sigma_1}, \hat{w}_2 e^{i\sigma_2}, \hat{w}_S$	$xe^{i\tau_1}, ye^{i\tau_2}, z$

2.3. Complex vacuum solutions

In the discussion of possible complex vacua we now adopt a convention where w_S is real and non-negative and take

$$w_1 = \hat{w}_1 e^{i\sigma_1}, \quad w_2 = \hat{w}_2 e^{i\sigma_2}, \tag{12}$$

with the $\hat{w}_i$ also real and non-negative. With this convention w_S is also denoted by $\hat{w}_S$. A systematic analysis of possible solutions was performed in [35]. The results are summarised in Table 2. The list of the constraints on the potential that are consistent with each solution is not given here, it can be found in Ref. [35].

Several solutions require $\lambda_4 = 0$. This is not a new feature, it also happened in the context of real solutions. For $\lambda_4 = 0$ the potential acquires a continuous $SO(2)$ symmetry which can be broken spontaneously by the vacuum solutions, therefore, leading to a massless scalar. Massless scalars are ruled out by experiment. It is possible to avoid this problem by introducing soft breaking terms. The most general form for the V_2 part of the potential with soft breaking terms would be:

$$V_2 = \mu_0^2 h_S^\dagger h_S + \mu_1^2(h_1^\dagger h_1 + h_2^\dagger h_2) + \mu_2^2(h_1^\dagger h_1 - h_2^\dagger h_2) + \frac{1}{2}\nu^2(h_2^\dagger h_1 - h_1^\dagger h_2) \\ + \mu_3^2(h_S^\dagger h_1 - h_1^\dagger h_S) + \mu_4^2(h_S^\dagger h_2 - h_2^\dagger h_S). \tag{13}$$

However, soft breaking terms involving h_S and one h_i are not consistent with $\lambda_4 = 0$.

In Table 3 we collect all possible complex vacuum solutions indicating whether or not they require λ_4 equal to zero and whether or not they allow for spontaneous CP violation. One important conclusion from our analysis is that there are cases where CP can be violated spontaneously, however, no solution requiring $\lambda_4 = 0$ can lead to spontaneous CP violation. In order to confirm that CP could indeed be violated spontaneously we used a powerful tool based on CP-odd Higgs-basis-invariant conditions, verifying that there were indeed conditions that were violated. There are several such conditions which were especially built for the analysis of the Higgs potential [52, 53, 54, 55, 56]. In the next subsection we discuss spontaneous CP violation using a few illustrative examples.

Table 3. Spontaneous CP violation

Vacuum	λ_4	SCPV	Vacuum	λ_4	SCPV	Vacuum	λ_4	SCPV
C-I-a	X	no	C-III-f,g	0	no	C-IV-c	X	yes
C-III-a	X	yes	C-III-h	X	yes	C-IV-d	0	no
C-III-b	0	no	C-III-i	X	no	C-IV-e	0	no
C-III-c	0	no	C-IV-a	0	no	C-IV-f	X	yes
C-III-d,e	X	no	C-IV-b	0	no	C-V	0	no

2.4. Spontaneous CP violation

Spontaneous CP violation can only occur if the Lagrangian conserves CP but the vacuum does not. This can only happen when there is no transformation that can be identified with a CP transformation leaving both the Lagrangian and the vacuum invariant. Under a CP transformation a single Higgs doublet Φ transforms into its complex conjugate. In models with several Higgs doublets the most general CP transformation is given by:

$$\Phi_i \xrightarrow{\text{CP}} U_{ij}\Phi_j^*. \tag{14}$$

Here, U is a unitary matrix mixing different Higgs doublets and corresponds to a Higgs basis transformation[6]. Higgs basis transformations do not change the physics.

If all the coefficients of the potential are real the potential conserves CP explicitly and the above equation is verified for U the identity matrix. Checking for explicit CP invariance of a multi-Higgs potential may be a non-trivial task since Higgs basis transformations, in general, can transform couplings that are real in one Higgs basis into couplings that are complex in another basis. For this purpose CP-odd Higgs basis invariants are of great help [53, 54]. Once

[6] This transformation is often referred to as a "generalized" CP transformation, thus suggesting that there is also a "non-generalized" CP transformation. This is, of course, misleading.

it is known that a Lagrangian conserves CP it remains to check whether or not CP is violated spontaneously. It has been shown [57] that in order for the vacuum to conserve CP the following relation has to be obeyed:

$$U_{ij}\langle 0|\Phi_j|0\rangle^* = \langle 0|\Phi_i|0\rangle \tag{15}$$

with U now a unitary matrix corresponding to a symmetry of the Lagrangian. This relation is very powerful and allows to show that vacua that would at first sight violate CP are indeed CP conserving. This can be illustrated with a few examples taken from Table 3. For a full discussion see Ref. [35].

- Let us consider the vacuum identified as C-I-a, given by $\left(x, xe^{\pm\frac{2\pi i}{3}}, xe^{\mp\frac{2\pi i}{3}}\right)$ in the reducible representation. It is not possible to rephase the three Higgs doublets in such a way that the three vevs become real keeping at the same time the potential real. This is a vacuum solution with calculable non-trivial phases, fixed by the symmetry of the potential with no explicit dependence on the parameters of the potential. Such phases are called geometrical phases [57]. It was shown in Ref. [57] that this vacuum does not violate CP since Eq. (15) can be verified for U given by:

$$U = \begin{pmatrix} 1 & 0 & 0 \\ 0 & 0 & 1 \\ 0 & 1 & 0 \end{pmatrix}. \tag{16}$$

  This matrix makes use of the symmetry of the potential for the interchange of ϕ_2 and ϕ_3.
- Another interesting example is the C-III-c vacuum which is of the form $\left(\hat{w}_1 e^{i\sigma_1}, \hat{w}_2 e^{i\sigma_2}, 0\right)$ in the irreducible representation framework. It can also be written, without loss of generality, through an overall phase rotation, in the form $\left(\hat{w}_1 e^{i\sigma}, \hat{w}_2, 0\right)$. At first sight this vacuum looks like a CP violating vacuum, especially taking into consideration the fact that the moduli of w_1 and w_2 are different from each other. However, once again we can use Eq. (15) to show that this vacuum conserves CP. Notice that this solution requires $\lambda_4 = 0$ (see Table 3) and therefore there is an $SO(2)$ symmetry for the fields h_1 and h_2. With this knowledge one can build the necessary matrix U and Eq. (15) becomes:

$$e^{i(\delta_1+\delta_2)} \begin{pmatrix} \cos\theta & \sin\theta & 0 \\ -\sin\theta & \cos\theta & 0 \\ 0 & 0 & 1 \end{pmatrix} \begin{pmatrix} 0 & 1 & 0 \\ 1 & 0 & 0 \\ 0 & 0 & 1 \end{pmatrix} \begin{pmatrix} \cos\theta & -\sin\theta & 0 \\ \sin\theta & \cos\theta & 0 \\ 0 & 0 & 1 \end{pmatrix} \begin{pmatrix} \hat{w}_1 e^{i\sigma} \\ \hat{w}_2 \\ 0 \end{pmatrix}^* = \begin{pmatrix} \hat{w}_1 e^{i\sigma} \\ \hat{w}_2 \\ 0 \end{pmatrix}, \tag{17}$$

  or

$$e^{i(\delta_1+\delta_2)} \begin{pmatrix} \sin 2\theta & \cos 2\theta & 0 \\ \cos 2\theta & -\sin 2\theta & 0 \\ 0 & 0 & 1 \end{pmatrix} \begin{pmatrix} \hat{w}_1 e^{i\sigma} \\ \hat{w}_2 \\ 0 \end{pmatrix}^* = \begin{pmatrix} \hat{w}_1 e^{i\sigma} \\ \hat{w}_2 \\ 0 \end{pmatrix}. \tag{18}$$

  In this example the matrix U has several components:
  - an $SO(2)$ rotation of h_1 and h_2 by an angle θ, which should be chosen as:

$$\tan 2\theta = \frac{\hat{w}_1^2 - \hat{w}_2^2}{2\hat{w}_1\hat{w}_2\cos\sigma}. \tag{19}$$

  With this choice, the vevs of the new S_3 doublet fields acquire the same modulus and the new vacuum acquires the form $(ae^{i\delta_1}, ae^{i\delta_2}, 0)$,
  - an overall phase rotation of the three Higgs doublets by $\exp[-i(\delta_1+\delta_2)/2]$, so that now the first two vevs acquire symmetric phases: $(ae^{i\delta}, ae^{-i\delta}, 0)$,
  - finally we just need to use the symmetry for the interchange $h_1' \leftrightarrow h_2'$ in the S_3 doublet representation.

The last example illustrates how powerful the condition given by Eq. (15) can be, but at the same time it shows that, as complexity grows, it may be non-trivial, in cases where such a matrix exists, to build the necessary matrix U. In fact this may require special insight and there is the danger of missing it, in a CP conserving case. In Ref. [58] we propose an alternative simple method, which is very useful in such cases, and allows to detect or eliminate the possibility of having spontaneous CP violation in multi-Higgs models. The three tools, consisting of the use of CP-odd invariant conditions, the relation given by Eq. (15) and the simple method proposed in Ref. [58], combined together, provide a reliable procedure to determine whether or not a given Higgs potential violates CP spontaneously.

3. Conclusions

We have presented here a summary of the work done in Ref. [35]. We have focused on some important features of three-Higgs-doublet models with an S_3 symmetry with emphasis on the discussion of spontaneous CP violation. Some aspects which were dealt with in the paper were not included in this short presentation. We refer the reader to the original work for a more detailed discussion of these aspects and for other topics such as ideas about constraining the potential by the vevs, relations among complex and real vacua, and a discussion on positivity beyond the necessary conditions given by Das and Dey [50] following the approach of Refs. [59, 24] (see also [60]). Models with multi-Higgs doublets such as those discussed in our work are very interesting and can in principle provide answers for several open questions. In particular they can provide viable dark matter candidates. These and other questions such as ways of generating realistic fermion masses and mixing in this context or looking for viable models with spontaneous CP violation are still challenging despite the fact that a lot of work has been already done along these lines. These questions are very timely due to the potential for being tested at the LHC.

Acknowledgements

MNR thanks the local organising committee of the Symposium Discrete 2016 for the stimulating scientific atmosphere and the warm hospitality in Warsaw, and acknowledges financial support from the National Science Centre, Poland, the HARMONIA project under contract UMO-2015/18/M/ST2/00518 (2016-2019) to participate in the Symposium and in the 1st Harmonia Meeting that took place immediately after the Symposium. Special thanks go to Maria Krawczyk. The work of MNR is partially supported by Fundação para a Ciência e a Tecnologia (FCT, Portugal) through the projects CERN/FIS-NUC/0010/2015, CFTP-FCT Unit 777 (UID/FIS/00777/2013) which are partially funded through POCTI (FEDER), COMPETE, QREN and EU.

References

[1] Aad G *et al.* [ATLAS Collaboration], 2012 "Observation of a new particle in the search for the Standard Model Higgs boson with the ATLAS detector at the LHC," Phys. Lett. B **716** 1 [arXiv:1207.7214 [hep-ex]].
[2] Chatrchyan S *et al.* [CMS Collaboration], 2012 "Observation of a new boson at a mass of 125 GeV with the CMS experiment at the LHC," Phys. Lett. B **716** 30 [arXiv:1207.7235 [hep-ex]].
[3] Lee T D, 1973 "A Theory of Spontaneous T Violation," Phys. Rev. D **8** 1226.
[4] Gunion J F, Haber H E, Kane G L and Dawson S, 2000 "The Higgs Hunter's Guide," Front. Phys. **80** 1.
[5] Branco G C, Ferreira P M, Lavoura L, Rebelo M N, Sher M and Silva J P, 2012 "Theory and phenomenology of two-Higgs-doublet models," Phys. Rept. **516** 1 [arXiv:1106.0034 [hep-ph]].
[6] Glashow S L and Weinberg S, 1977 "Natural Conservation Laws for Neutral Currents," Phys. Rev. D **15** 1958.
[7] Paschos E A, 1977 "Diagonal Neutral Currents," Phys. Rev. D **15** 1966.
[8] Branco G C and Rebelo M N, 1985 "The Higgs Mass in a Model With Two Scalar Doublets and Spontaneous CP Violation," Phys. Lett. **160B** 117.
[9] Weinberg S, 1976 "Gauge Theory of CP Violation," Phys. Rev. Lett. **37** 657.

[10] Branco G C, 1980 "Spontaneous CP Nonconservation and Natural Flavor Conservation: A Minimal Model," Phys. Rev. D **22** 2901.
[11] Pich A and Tuzon P, 2009 "Yukawa Alignment in the Two-Higgs-Doublet Model," Phys. Rev. D **80** 091702 [arXiv:0908.1554 [hep-ph]].
[12] Serodio H, 2011 "Yukawa Alignment in a Multi Higgs Doublet Model: An effective approach," Phys. Lett. B **700** 133 [arXiv:1104.2545 [hep-ph]].
[13] De Medeiros Varzielas I, 2011 "Family symmetries and alignment in multi-Higgs doublet models," Phys. Lett. B **701** 597 [arXiv:1104.2601 [hep-ph]].
[14] Celis A, Fuentes-Martn J and Serôdio H, 2014 "Effective Aligned 2HDM with a DFSZ-like invisible axion," Phys. Lett. B **737** 185 [arXiv:1407.0971 [hep-ph]].
[15] Botella F J, Branco G C, Coutinho A M, Rebelo M N and Silva-Marcos J I, 2015 "Natural Quasi-Alignment with two Higgs Doublets and RGE Stability," Eur. Phys. J. C **75** 286 [arXiv:1501.07435 [hep-ph]].
[16] Branco G C, Grimus W and Lavoura L, 1996 "Relating the scalar flavor changing neutral couplings to the CKM matrix," Phys. Lett. B **380** 119 [hep-ph/9601383].
[17] Botella F J, Branco G C and Rebelo M N, 2010 "Minimal Flavour Violation and Multi-Higgs Models," Phys. Lett. B **687** 194 [arXiv:0911.1753 [hep-ph]].
[18] Botella F J, Branco G C, Nebot M and Rebelo M N, 2011 "Two-Higgs Leptonic Minimal Flavour Violation," JHEP **1110** 037 [arXiv:1102.0520 [hep-ph]].
[19] Bhattacharyya G, Das D, Pal P B and Rebelo M N, 2013 "Scalar sector properties of two-Higgs-doublet models with a global U(1) symmetry," JHEP **1310** 081 [arXiv:1308.4297 [hep-ph]].
[20] Botella F J, Branco G C, Carmona A, Nebot M, Pedro L and Rebelo M N, 2014 "Physical Constraints on a Class of Two-Higgs Doublet Models with FCNC at tree level," JHEP **1407** 078 [arXiv:1401.6147 [hep-ph]].
[21] Bhattacharyya G, Das D and Kundu A, 2014 "Feasibility of light scalars in a class of two-Higgs-doublet models and their decay signatures," Phys. Rev. D **89** 095029 [arXiv:1402.0364 [hep-ph]].
[22] Botella F J, Branco G C, Nebot M and Rebelo M N, 2016 "Flavour Changing Higgs Couplings in a Class of Two Higgs Doublet Models," Eur. Phys. J. C **76** no.3, 161 [arXiv:1508.05101 [hep-ph]].
[23] Alves J M, Botella F J, Branco G C, Cornet-Gomez F and Nebot M, 2017 "Controlled Flavour Changing Neutral Couplings in Two Higgs Doublet Models," arXiv:1703.03796 [hep-ph].
[24] Grzadkowski B, Ogreid O M and Osland P, 2009 "Natural Multi-Higgs Model with Dark Matter and CP Violation," Phys. Rev. D **80** 055013 [arXiv:0904.2173 [hep-ph]].
[25] Grzadkowski B, Ogreid O M, Osland P, Pukhov A and Purmohammadi M, 2011 "Exploring the CP-Violating Inert-Doublet Model," JHEP **1106** 003 [arXiv:1012.4680 [hep-ph]].
[26] Keus V, King S F, Moretti S and Sokolowska S, 2014 "Dark Matter with Two Inert Doublets plus One Higgs Doublet," JHEP **1411** 016 [arXiv:1407.7859 [hep-ph]].
[27] Machado A C B and Pleitez V, 2016 "A model with two inert scalar doublets," Annals Phys. **364** 53 [arXiv:1205.0995 [hep-ph]].
[28] Fortes E C F S, Machado A C B, Montaño J and Pleitez V, 2015 "Scalar dark matter candidates in a two inert Higgs doublet model," J. Phys. G **42** 105003 [arXiv:1407.4749 [hep-ph]].
[29] Fortes E C F S, Machado A C B, Montaño J and Pleitez V, 2015 "Prediction of $h \to \gamma Z$ from $h \to \gamma\gamma$ at LHC for the IMDS$_3$ Model," J. Phys. G **42** 115001 [arXiv:1408.0780 [hep-ph]].
[30] Olaussen K, Osland P and Solberg M A, 2011 "Symmetry and Mass Degeneration in Multi-Higgs-Doublet Models," JHEP **1107** 020 [arXiv:1007.1424 [hep-ph]].
[31] Ma E, 2006 "Verifiable radiative seesaw mechanism of neutrino mass and dark matter," Phys. Rev. D **73** 077301 [hep-ph/0601225].
[32] Barbieri R, Hall L J and Rychkov V S, 2006 "Improved naturalness with a heavy Higgs: An Alternative road to LHC physics," Phys. Rev. D **74** 015007 [hep-ph/0603188].
[33] Lopez Honorez L, Nezri E, Oliver J F and Tytgat M H G, 2007 "The Inert Doublet Model: An Archetype for Dark Matter," JCAP **0702** 028 [hep-ph/0612275].
[34] Barradas-Guevara E, Félix-Beltrán O and Rodríguez-Jáuregui E, 2015 "CP breaking in $S(3)$ flavoured Higgs model," arXiv:1507.05180 [hep-ph].
[35] Emmanuel-Costa D, Ogreid O M, Osland P and Rebelo M N, 2016 "Spontaneous symmetry breaking in the S_3-symmetric scalar sector," JHEP **1602** 154 Erratum: [JHEP **1608** 169] [arXiv:1601.04654 [hep-ph]].
[36] Barradas-Guevara E, Félix-Beltrán O and Rodríguez-Jáuregui E, 2016 "Higgs sector with spontaneous CP violation in $S(3)$ Standard Model," JNPMSRA, Vol. 4 [arXiv:1606.07773 [hep-ph]].
[37] Pakvasa S and Sugawara H, 1978 "Discrete Symmetry and Cabibbo Angle," Phys. Lett. **73B** 61.
[38] Ma E, 2000 "Permutation symmetry for neutrino and charged lepton mass matrices," Phys. Rev. D **61** 033012 [hep-ph/9909249].
[39] Araki T, Kubo K and Paschos E A, 2006 "S(3) flavor symmetry and leptogenesis," Eur. Phys. J. C **45** 465 [hep-ph/0502164].

[40] Bhattacharyya G, Leser P and Pas H, 2011 "Exotic Higgs boson decay modes as a harbinger of S_3 flavor symmetry," Phys. Rev. D **83** 011701 doi:10.1103/PhysRevD.83.011701 [arXiv:1006.5597 [hep-ph]].
[41] Teshima T, 2012 "Higgs potential in S_3 invariant model for quark/lepton mass and mixing," Phys. Rev. D **85** 105013 [arXiv:1202.4528 [hep-ph]].
[42] Canales F G, Mondragón A, Mondragón M, Saldaña Salazar U J and Velasco-Sevilla L, 2013 "Fermion mixing in an S_3 model with three Higgs doublets," J. Phys. Conf. Ser. **447** 012053.
[43] Das D, Dey U K and Pal P B, 2016 "S_3 symmetry and the quark mixing matrix," Phys. Lett. B **753** 315 [arXiv:1507.06509 [hep-ph]].
[44] Cruz A A and Mondragón M, "Neutrino masses, mixing, and leptogenesis in an S3 model," arXiv:1701.07929 [hep-ph].
[45] Derman E, 1979 "Flavor Unification, τ Decay and b Decay Within the Six Quark Six Lepton Weinberg-Salam Model," Phys. Rev. D **19** 317.
[46] Derman E and Tsao H S, 1979 "SU(2) X U(1) X S(n) Flavor Dynamics and a Bound on the Number of Flavors," Phys. Rev. D **20** 1207.
[47] Kubo J, Okada H and Sakamaki F, 2004 "Higgs potential in minimal S(3) invariant extension of the standard model," Phys. Rev. D **70** 036007 [hep-ph/0402089].
[48] Ivanov I P and Nishi C C, 2015 "Symmetry breaking patterns in 3HDM," JHEP **1501** 021 [arXiv:1410.6139 [hep-ph]].
[49] Harrison P F, Perkins D H and Scott W G, 2002 "Tri-bimaximal mixing and the neutrino oscillation data," Phys. Lett. B **530** 167 [hep-ph/0202074].
[50] Das D and Dey U K, 2014 "Analysis of an extended scalar sector with S_3 symmetry," Phys. Rev. D **89** 095025 Erratum: [Phys. Rev. D **91** (2015) no.3, 039905] [arXiv:1404.2491 [hep-ph]].
[51] Ivanov I P and Silva J P, 2016 "CP-conserving multi-Higgs model with irremovable complex coefficients," Phys. Rev. D **93** 095014 [arXiv:1512.09276 [hep-ph]].
[52] Lavoura L and Silva J P, 1994 "Fundamental CP violating quantities in a SU(2) x U(1) model with many Higgs doublets," Phys. Rev. D **50** 4619 [hep-ph/9404276].
[53] Branco G C, Rebelo M N and Silva-Marcos J I, 2005 "CP-odd invariants in models with several Higgs doublets," Phys. Lett. B **614** 187 [hep-ph/0502118].
[54] Gunion J F and Haber H E, 2005 "Conditions for CP-violation in the general two-Higgs-doublet model," Phys. Rev. D **72** 095002 [hep-ph/0506227].
[55] de Medeiros Varzielas I, King S F, Luhn C and Neder T, 2016 "CP-odd invariants for multi-Higgs models: applications with discrete symmetry," Phys. Rev. D **94** 056007 [arXiv:1603.06942 [hep-ph]].
[56] Branco G C, Lavoura L and Silva J P, 1999 *CP Violation,* Int. Ser. Monogr. Phys. **103** 1.
[57] Branco G C, Gerard J M and Grimus W, 1984 "Geometrical T Violation," Phys. Lett. B **136** 383.
[58] Ogreid O M, Osland P and Rebelo M N, 2017 "A Simple Method to detect spontaneous CP Violation in multi-Higgs models," arXiv:1701.04768 [hep-ph].
[59] El Kaffas A W, Khater W, Ogreid O M and Osland P, 2007 "Consistency of the two Higgs doublet model and CP violation in top production at the LHC," Nucl. Phys. B **775** 45 [hep-ph/0605142].
[60] Emmanuel-Costa D, Felix-Beltran O, Mondragon M and Rodriguez-Jauregui E, 2007 "Stability of the tree-level vacuum in a minimal S(3) extension of the standard model," AIP Conf. Proc. **917** 390.

IOP Conf. Series: Journal of Physics: Conf. Series **873** (2017) 012008 doi:10.1088/1742-6596/873/1/012008

Natural Alignment in the Two Higgs Doublet Model

P. S. Bhupal Dev[1], **Apostolos Pilaftsis**[2]

[1]Department of Physics and McDonnell Center for the Space Sciences, Washington University, St. Louis, MO 63130, USA

[2]Consortium for Fundamental Physics, School of Physics and Astronomy, University of Manchester, Manchester M13 9PL, United Kingdom.

E-mail: bdev@wustl.edu, apostolos.pilaftsis@manchester.ac.uk

Abstract. As the LHC Higgs data persistently suggest the couplings of the observed 125 GeV Higgs boson to be consistent with the Standard Model (SM) expectations, any extended Higgs sector must lead to the so-called SM *alignment limit*, where one of the Higgs bosons behaves exactly like that of the SM. In the context of the Two Higgs Doublet Model (2HDM), this alignment is often associated with either decoupling of the heavy Higgs sector or accidental cancellations in the 2HDM potential. We present a novel symmetry justification for 'natural' alignment without necessarily decoupling or fine-tuning. We show that there exist only *three* different symmetry realizations of the natural alignment scenario in 2HDM. We identify the 2HDM parameter space satisfying the natural alignment condition up to the Planck scale. We also analyze new collider signals for the heavy Higgs sector in the natural alignment limit, which dominantly lead to third-generation quarks in the final state and can serve as a useful observational tool during the Run-II phase of the LHC.

1. Introduction

The discovery of a Higgs boson in the Run-I phase of the LHC [1] provides the first experimental evidence for the Higgs mechanism [2] as the standard theory of electroweak symmetry breaking (EWSB). As more data are being collected at the LHC, the coupling measurements of the observed Higgs boson seem to be very close to the Standard Model (SM) predictions [3]. The constraints deduced from the Higgs signal strength data severely limit the form of a possible heavy scalar sector in the observable sub-TeV range, as predicted by various well-motivated new-physics scenarios, such as supersymmetry.

Here we consider the Two Higgs Doublet Model (2HDM) [4], where the SM Higgs doublet is supplemented by another isodoublet with hypercharge $Y = 1$. In the doublet field space $\Phi_{1,2}$, where $\Phi_i = (\phi_i^+, \phi_i^0)^{\mathsf{T}}$, the general 2HDM potential reads

$$\begin{aligned} V \;=\; & -\mu_1^2(\Phi_1^\dagger\Phi_1) - \mu_2^2(\Phi_2^\dagger\Phi_2) - \Big[m_{12}^2(\Phi_1^\dagger\Phi_2) + \text{H.c.}\Big] \\ & + \lambda_1(\Phi_1^\dagger\Phi_1)^2 + \lambda_2(\Phi_2^\dagger\Phi_2)^2 + \lambda_3(\Phi_1^\dagger\Phi_1)(\Phi_2^\dagger\Phi_2) + \lambda_4(\Phi_1^\dagger\Phi_2)(\Phi_2^\dagger\Phi_1) \\ & + \left[\frac{\lambda_5}{2}(\Phi_1^\dagger\Phi_2)^2 + \lambda_6(\Phi_1^\dagger\Phi_1)(\Phi_1^\dagger\Phi_2) + \lambda_7(\Phi_1^\dagger\Phi_2)(\Phi_2^\dagger\Phi_2) + \text{H.c.}\right], \end{aligned} \qquad (1)$$

which contains *four* real mass parameters $\mu_{1,2}^2$, $\text{Re}(m_{12}^2)$, $\text{Im}(m_{12}^2)$, and *ten* real quartic couplings $\lambda_{1,2,3,4}$, $\text{Re}(\lambda_{5,6,7})$, and $\text{Im}(\lambda_{5,6,7})$. Thus, the vacuum structure of the general 2HDM can be quite rich [5], as compared to the SM.

The quark-sector Yukawa Lagrangian in the general 2HDM is given by

$$-\mathcal{L}_Y^q \;=\; \bar{Q}_L(h_1^d\Phi_1 + h_2^d\Phi_2)d_R \;+\; \bar{Q}_L(h_1^u\widetilde{\Phi}_1 + h_2^u\widetilde{\Phi}_2)u_R \,, \tag{2}$$

where $\widetilde{\Phi}_i = i\sigma^2\Phi_i^*$ (σ^2 being the second Pauli matrix) are the isospin conjugates of Φ_i, $Q_L = (u_L, d_L)^\mathsf{T}$ are the $SU(2)_L$ quark doublets and u_R, d_R are right-handed quark singlets. To avoid potentially large flavor-changing neutral current processes at the tree level induced by the Yukawa interactions in (2), one imposes a discrete Z_2 symmetry [4] under which

$$\Phi_1 \to -\Phi_1, \quad \Phi_2 \to \Phi_2, \quad u_{Ra} \to u_{Ra}, \quad d_{Ra} \to d_{Ra} \;\text{ or }\; d_{Ra} \to -d_{Ra} \,, \tag{3}$$

($a = 1, 2, 3$ being the quark family index) so that only Φ_2 gives mass to up-type quarks, and only Φ_1 or only Φ_2 gives mass to down-type quarks. The Z_2 symmetry (3) is satisfied by four discrete choices of tree-level Yukawa couplings between the Higgs doublets and SM fermions, which are known as the Type I, II, X (lepton-specific) and Y (flipped) 2HDMs [4]. Global fits to the LHC Higgs data (see e.g., [6, 7]) suggest that all four types of 2HDMs with natural flavor conservation are constrained to lie close to the so-called SM *alignment limit* [8–12], where the mass eigenbasis of the CP-even scalar sector aligns with the SM gauge eigenbasis.

The SM alignment is often associated with the decoupling limit, in which all the non-standard Higgs bosons are assumed to be much heavier than the electroweak scale so that the lightest CP-even scalar behaves just like the SM Higgs boson. The alignment limit can also be achieved, without decoupling [9, 10, 13], but for small $\tan\beta$ values, this is usually attributed to accidental cancellations in the 2HDM potential. We present a novel symmetry argument to naturally justify the alignment limit [11], independently of the kinematic parameters of the theory, such as the heavy Higgs mass and $\tan\beta$. In particular, we show that there exist only *three* possible symmetry realizations of the scalar potential which predict natural alignment. A striking outcome of this analysis is that the heavy Higgs sector is predicted to be *quasi-degenerate*, apart from being *gaugephobic*, which is a generic feature in the alignment limit. Moreover, the current experimental constraints force the heavy Higgs sector to lie above the top-quark threshold. Thus, the dominant collider signal for this sector involves final states with third-generation quarks. We make a detailed study of some of these signals, which can be useful for the heavy Higgs searches in the ongoing Run-II phase of the LHC.

2. Natural Alignment Condition

For simplicity, we consider the CP-conserving 2HDM, but our results can be easily generalized to the CP-violating case. After EWSB by the vacuum expectation values (VEVs) $v_{1,2}$ of the two scalar fields $\Phi_{1,2}$, there are five physical scalar mass eigenstates: two CP-even (h, H), one CP-odd (a) and two charged $(h^\pm)$ scalars. The corresponding mass eigenvalues are given by [14]

$$M_{h^\pm}^2 \;=\; \frac{m_{12}^2}{s_\beta c_\beta} - \frac{v^2}{2}(\lambda_4+\lambda_5) + \frac{v^2}{2s_\beta c_\beta}\left(\lambda_6 c_\beta^2 + \lambda_7 s_\beta^2\right), \qquad M_a^2 \;=\; M_{h^\pm}^2 + \frac{v^2}{2}(\lambda_4-\lambda_5), \tag{4a}$$

$$M_H^2 \;=\; \frac{1}{2}\left[(A+B) - \sqrt{(A-B)^2+4C^2}\right], \qquad M_h^2 \;=\; \frac{1}{2}\left[(A+B) + \sqrt{(A-B)^2+4C^2}\right], \tag{4b}$$

where we have used the short-hand notations $c_\beta \equiv \cos\beta$ and $s_\beta \equiv \sin\beta$ with $\tan\beta = v_2/v_1$ and

$$A \;=\; M_a^2 s_\beta^2 + v^2\left(2\lambda_1 c_\beta^2 + \lambda_5 s_\beta^2 + 2\lambda_6 s_\beta c_\beta\right), \tag{5a}$$

$$B \;=\; M_a^2 c_\beta^2 + v^2\left(2\lambda_2 s_\beta^2 + \lambda_5 c_\beta^2 + 2\lambda_7 s_\beta c_\beta\right), \tag{5b}$$

$$C \;=\; -M_a^2 s_\beta c_\beta + v^2\left(\lambda_{34} s_\beta c_\beta + \lambda_6 c_\beta^2 + \lambda_7 s_\beta^2\right). \tag{5c}$$

with $\lambda_{34} = \lambda_3 + \lambda_4$. The mixing between the mass eigenstates in the CP-odd and charged sectors is governed by the angle β, whereas in the CP-even sector, it is governed by the angle $\alpha = (1/2)\tan^{-1}[2C/(A-B)]$. The SM Higgs field can be identified as the linear combination

$$H_{\rm SM} \;=\; H\cos(\beta-\alpha) + h\sin(\beta-\alpha)\;. \qquad (6)$$

Thus, the couplings of h and H to the SM gauge bosons $V = W^{\pm}, Z$ with respect to the SM Higgs couplings $g_{H_{\rm SM}VV}$ will be respectively $\sin(\beta-\alpha)$ and $\cos(\beta-\alpha)$. The *SM alignment limit* is defined as $\alpha \to \beta$ (or $\alpha \to \beta - \pi/2$) when H (h) couples to vector bosons exactly as in the SM, whereas h (H) becomes *gaugephobic*. For concreteness, we will take the alignment limit as $\alpha \to \beta$.

To derive the alignment condition, we rewrite the CP-even scalar mass matrix as

$$M_S^2 \;=\; \begin{pmatrix} A & C \\ C & B \end{pmatrix} \;=\; \begin{pmatrix} c_\beta & -s_\beta \\ s_\beta & c_\beta \end{pmatrix}\begin{pmatrix} \widehat{A} & \widehat{C} \\ \widehat{C} & \widehat{B} \end{pmatrix}\begin{pmatrix} c_\beta & s_\beta \\ -s_\beta & c_\beta \end{pmatrix}, \qquad (7)$$

$$\begin{aligned} \text{where}\quad \widehat{A} &= 2v^2\Big[c_\beta^4\lambda_1 + s_\beta^2 c_\beta^2\lambda_{345} + s_\beta^4\lambda_2 + 2s_\beta c_\beta\Big(c_\beta^2\lambda_6 + s_\beta^2\lambda_7\Big)\Big]\,, \\ \widehat{B} &= M_a^2 + \lambda_5 v^2 + 2v^2\Big[s_\beta^2 c_\beta^2\Big(\lambda_1+\lambda_2-\lambda_{345}\Big) - s_\beta c_\beta\Big(c_\beta^2 - s_\beta^2\Big)\Big(\lambda_6-\lambda_7\Big)\Big]\,, \qquad (8) \\ \widehat{C} &= v^2\Big[s_\beta^3 c_\beta\Big(2\lambda_2-\lambda_{345}\Big) - c_\beta^3 s_\beta\Big(2\lambda_1-\lambda_{345}\Big) + c_\beta^2\Big(1-4s_\beta^2\Big)\lambda_6 + s_\beta^2\Big(4c_\beta^2-1\Big)\lambda_7\Big], \end{aligned}$$

and we have used the short-hand notation $\lambda_{345} \equiv \lambda_3 + \lambda_4 + \lambda_5$. Evidently, the SM alignment limit $\alpha \to \beta$ is obtained when $\widehat{C} = 0$ in (7) [9]. This yields the quartic equation

$$\lambda_7\tan^4\beta - (2\lambda_2-\lambda_{345})\tan^3\beta + 3(\lambda_6-\lambda_7)\tan^2\beta + (2\lambda_1-\lambda_{345})\tan\beta - \lambda_6 \;=\; 0\,. \qquad (9)$$

For *natural* alignment, (9) should be satisfied for *any* value of $\tan\beta$, which requires the coefficients of the polynomial in $\tan\beta$ to vanish identically [11]. This implies

$$2\lambda_1 \;=\; 2\lambda_2 \;=\; \lambda_{345}\,, \qquad \lambda_6 \;=\; \lambda_7 \;=\; 0\,. \qquad (10)$$

In particular, for $\lambda_6 = \lambda_7 = 0$ as in the Z_2-symmetric 2HDMs, (9) has a solution

$$\tan^2\beta \;=\; \frac{2\lambda_1-\lambda_{345}}{2\lambda_2-\lambda_{345}} \;>\; 0\,, \qquad (11)$$

independent of M_a. After some algebra, the simple solution (11) to our *natural alignment condition* (9) can be shown to be equivalent to that derived in [10].

3. Symmetry Classifications of the 2HDM Potential

To identify all accidental symmetries of the 2HDM potential (1), it is convenient to work in the *bilinear scalar field formalism* [15] by introducing an 8-dimensional complex multiplet $\mathbf{\Phi} \equiv (\Phi_1, \Phi_2, \widetilde{\Phi}_1, \widetilde{\Phi}_2)^{\sf T}$ [5, 16, 17]. In terms of the $\mathbf{\Phi}$-multiplet, we define a *null* 6-dimensional Lorentz vector $R^A \equiv \mathbf{\Phi}^\dagger\Sigma^A\mathbf{\Phi}$, where $A = 0, 1, ..., 5$ and the six 8×8-dimensional matrices Σ^A may be expressed in terms of the three Pauli matrices $\sigma^{1,2,3}$ and the identity matrix $\mathbf{1}_{2\times 2} \equiv \sigma^0$, as follows:

$$\begin{aligned} \Sigma^{0,1,3} &= \frac{1}{2}\sigma^0\otimes\sigma^{0,1,3}\otimes\sigma^0, \qquad & \Sigma^2 &= \frac{1}{2}\sigma^3\otimes\sigma^2\otimes\sigma^0, \\ \Sigma^4 &= -\frac{1}{2}\sigma^2\otimes\sigma^2\otimes\sigma^0, \qquad & \Sigma^5 &= -\frac{1}{2}\sigma^1\otimes\sigma^2\otimes\sigma^0. \end{aligned} \qquad (12)$$

IOP Conf. Series: Journal of Physics: Conf. Series **873** (2017) 012008 doi:10.1088/1742-6596/873/1/012008

symmetry	μ_1^2	μ_2^2	m_{12}^2	λ_1	λ_2	λ_3	λ_4	$\mathrm{Re}(\lambda_5)$	$\lambda_6 = \lambda_7$
$Z_2 \times O(2)$	-	-	Real	-	-	-	-	-	Real
$(Z_2)^2 \times SO(2)$	-	-	0	-	-	-	-	-	0
$(Z_2)^3 \times O(2)$	-	μ_1^2	0	-	λ_1	-	-	-	0
$O(2) \times O(2)$	-	-	0	-	-	-	-	0	0
$Z_2 \times [O(2)]^2$	-	μ_1^2	0	-	λ_1	-	-	$2\lambda_1 - \lambda_{34}$	0
$O(3) \times O(2)$	-	μ_1^2	0	-	λ_1	-	$2\lambda_1 - \lambda_3$	0	0
$SO(3)$	-	-	Real	-	-	-	-	λ_4	Real
$Z_2 \times O(3)$	-	μ_1^2	Real	-	λ_1	-	-	λ_4	Real
$(Z_2)^2 \times SO(3)$	-	μ_1^2	0	-	λ_1	-	-	$\pm\lambda_4$	0
$O(2) \times O(3)$	-	μ_1^2	0	-	λ_1	$2\lambda_1$	-	0	0
$SO(4)$	-	-	0	-	-	-	0	0	0
$Z_2 \times O(4)$	-	μ_1^2	0	-	λ_1	-	0	0	0
$SO(5)$	-	μ_1^2	0	-	λ_1	$2\lambda_1$	0	0	0

Table 1: Relations between the parameters of the $U(1)_Y$-invariant 2HDM potential (1) for the 13 accidental symmetries [17] in a diagonally reduced basis, where $\mathrm{Im}(\lambda_5) = 0$ and $\lambda_6 = \lambda_7$. The symmetries in blue satisfy the natural alignment condition (10).

Note that the bilinear field space spanned by the 6-vector R^A realizes an *orthochronous* $SO(1,5)$ symmetry group [5,17].

In terms of the null-vector R^A, the 2HDM potential (1) takes on a simple quadratic form:

$$V = -\frac{1}{2} M_A R^A + \frac{1}{4} L_{AB} R^A R^B , \tag{13}$$

where M_A and L_{AB} are $SO(1,5)$ constant 'tensors' that depend on the mass parameters and quartic couplings given in (1) and their explicit forms may be found in [17,18]. Requiring that the $SU(2)_L$ gauge-kinetic term of the $\mathbf{\Phi}$-multiplet remains canonical restricts the allowed set of rotations from SO(1,5) to SO(5), where only the spatial components R^I (with $I = 1, ..., 5$) transform and the zeroth component R^0 remains invariant. Consequently, in the absence of the hypercharge gauge coupling and fermion Yukawa couplings, the maximal symmetry group of the 2HDM is $G^R_{\mathrm{2HDM}} = SO(5)$. Including all proper, improper and semi-simple subgroups of SO(5), the accidental symmetries for the 2HDM potential were completely classified in [5,17], as shown in Table 1. Here we have used a diagonally reduced basis [19], where $\mathrm{Im}(\lambda_5) = 0$ and $\lambda_6 = \lambda_7$, thus reducing the number of independent quartic couplings to seven. Each of the symmetries listed in Table 1 leads to certain constraints on the mass and/or coupling parameters. From Table 1, we find that there are *only* three symmetries, namely $Z_2 \times [O(2)]^2$, $O(3) \times O(2)$, and SO(5), which satisfy the natural alignment condition given by (10).[1] In the next two sections, we analyze each of these three realizations of the SM alignment. The simplest case based on the SO(5) group has already been discussed in details in Refs. [11,23]. Our preliminary results reported for the other two cases are taken from our upcoming publication [24].

[1] In Type-I 2HDM, there exists an additional possibility of realizing an exact Z_2 symmetry [20] which leads to an exact alignment, i.e. in the context of the so-called inert 2HDM [21]. In general, for nHDM with $m < n$ inert scalar doublets, there are still three continuous alignment symmetries in the field space of the non-inert sector [22].

4. Maximally Symmetric 2HDM

From Table 1, we see that the maximal symmetry group in the bilinear field space is SO(5), in which case the parameters of the 2HDM potential (1) satisfy the following relations:

$$\begin{aligned} \mu_1^2 &= \mu_2^2\,, \quad m_{12}^2 \;=\; 0\,, \\ \lambda_2 &= \lambda_1\,, \quad \lambda_3 \;=\; 2\lambda_1\,, \quad \lambda_4 \;=\; \mathrm{Re}(\lambda_5) \;=\; \lambda_6 \;=\; \lambda_7 \;=\; 0\,, \end{aligned} \tag{14}$$

Thus, the 2HDM potential (1) is parameterized by just a *single* mass parameter $\mu_1^2 = \mu_2^2 \equiv \mu^2$ and a *single* quartic coupling $\lambda_1 = \lambda_2 = \lambda_3/2 \equiv \lambda$, as in the SM:

$$V \;=\; -\,\mu^2\left(|\Phi_1|^2 + |\Phi_2|^2\right) \;+\; \lambda\left(|\Phi_1|^2 + |\Phi_2|^2\right)^2 \;=\; -\,\frac{\mu^2}{2}\,\mathbf{\Phi}^\dagger\mathbf{\Phi} \;+\; \frac{\lambda}{4}\left(\mathbf{\Phi}^\dagger\mathbf{\Phi}\right)^2. \tag{15}$$

Given the isomorphism of the Lie algebras SO(5) $\sim$ Sp(4), the maximal symmetry group of the 2HDM in the original $\mathbf{\Phi}$-field space is $\mathrm{G}^{\mathbf{\Phi}}_{\mathrm{2HDM}} = [\mathrm{Sp}(4)/\mathrm{Z}_2] \times \mathrm{SU}(2)_L$ [11, 17] in the custodial symmetry limit of vanishing g' and fermion Yukawa couplings.

4.1. Scalar Spectrum

Using the parameter relations given by (14), we find from (4a) and (4b) that in the MS-2HDM, the CP-even scalar H has mass $M_H^2 = 2\lambda_2 v^2$, whilst the remaining four scalar fields, denoted hereafter as h, a and $h^\pm$, are massless. This is a consequence of the Goldstone theorem, since after electroweak symmetry breaking, $\mathrm{SO(5)} \xrightarrow{\langle\Phi_{1,2}\rangle \neq 0} \mathrm{SO(4)}$. Thus, we identify H as the SM-like Higgs boson with the mixing angle $\alpha = \beta$ [cf. (6)], i.e. the SM alignment limit can be naturally attributed to the SO(5) symmetry of the theory.

In the exact SO(5)-symmetric limit, the scalar spectrum of the MS-2HDM is experimentally unacceptable. This is because the four massless pseudo-Goldstone particles, viz. h, a and $h^\pm$, have sizeable couplings to the SM Z and $W^\pm$ bosons, and could induce additional decay channels, such as $Z \to ha$ and $W^\pm \to h^\pm h$, which are experimentally excluded [30]. However, as we will see below, the SO(5) symmetry may be violated predominantly by RG effects due to g' and third-generation Yukawa couplings, as well as by soft SO(5)-breaking mass parameters, thereby lifting the masses of these pseudo-Goldstone particles to be consistent with the experimental constraints.

4.2. RG and Soft Breaking Effects

To calculate the RG and soft-breaking effects in a technically natural manner, we assume that the SO(5) symmetry is realized at some high scale μ_X much above the electroweak scale. The physical mass spectrum at the electroweak scale is then obtained by the RG evolution of the 2HDM parameters given by (1). Using the two-loop RG equations (RGEs) given in Ref. [11], we first examine the deviation of the Higgs spectrum from the SO(5)-symmetric limit due to g' and Yukawa coupling effects, in the absence of the soft-breaking term. This is illustrated in Figure 1 for a typical choice of parameters in the Type-II realization of the 2HDM. We find that the RG-induced g' effects only lift the charged Higgs mass $M_{h^\pm}$, while the corresponding Yukawa coupling effects also lift slightly the mass of the non-SM CP-even pseudo-Goldstone boson h. However, they still leave the CP-odd scalar a massless, which can be identified as a $\mathrm{U(1)_{PQ}}$ axion [31].

Therefore, g' and Yukawa coupling effects are *not* sufficient to yield a viable Higgs spectrum at the weak scale, starting from a SO(5)-invariant boundary condition (14) at some high scale μ_X. To minimally circumvent this problem, we include soft SO(5)-breaking effects, by assuming a non-zero soft-breaking term $\mathrm{Re}(m_{12}^2)$. In the SO(5)-symmetric limit for the scalar quartic

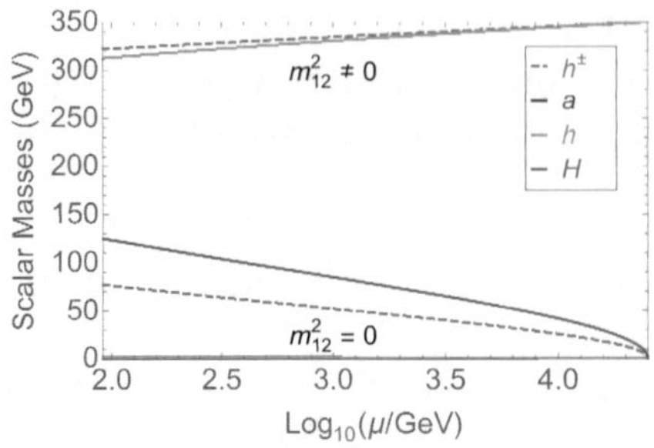

Figure 1: The scalar spectrum in the MS-2HDM without and with soft-breaking effects.

couplings, but with $\mathrm{Re}(m_{12}^2) \neq 0$, we obtain the following mass spectrum [cf. (4a) and (4b)]:

$$M_H^2 = 2\lambda_2 v^2 , \qquad M_h^2 = M_a^2 = M_{h^\pm}^2 = \frac{\mathrm{Re}(m_{12}^2)}{s_\beta c_\beta} , \tag{16}$$

as well as an equality between the CP-even and CP-odd mixing angles: $\alpha = \beta$, thus predicting an *exact* alignment for the SM-like Higgs boson H, simultaneously with an experimentally allowed heavy Higgs spectra (cf. Figure 1 for $m_{12}^2 \neq 0$ case). Note that in the alignment limit, the heavy Higgs sector is exactly degenerate [cf. (16)] at the SO(5) symmetry-breaking scale, and at the low-energy scale, this degeneracy is mildly broken by the RG effects. Thus, we obtain a *quasi-degenerate* heavy Higgs spectrum, which is a unique prediction of the MS-2HDM, valid even in the non-decoupling limit, and can be used to distinguish this model from other 2HDM scenarios.

4.3. Misalignment Predictions

As discussed in Section 4.2, there will be some deviation from the alignment limit in the low-energy Higgs spectrum of the MS-2HDM due to RG and soft-breaking effects. By requiring that the mass and couplings of the SM-like Higgs boson H are consistent with the LHC Higgs data [3], we derive predictions for the remaining scalar spectrum and compare them with the existing (in)direct limits on the heavy Higgs sector. We use the constraints in the $(\tan\beta,\ \beta-\alpha)$ plane derived from a recent global fit for the Type-II 2HDM [32], and require that for a given set of SO(5) boundary conditions $\{\mu_X, \tan\beta(\mu_X), \lambda(\mu_X)\}$, the RG-evolved 2HDM parameters at the weak scale must satisfy these alignment constraints on the lightest CP-even Higgs boson sector. This puts stringent constraints on the MS-2HDM parameter space, as shown in Figure 2 by the blue shaded region. In the red shaded region, there is no viable solution to the RGEs. We ensure that the remaining allowed (white) region satisfies the necessary theoretical constraints, i.e. positivity and vacuum stability of the Higgs potential, and perturbativity of the Higgs self-couplings [4]. From Figure 2, we find that there exists an *upper* limit of $\mu_X \lesssim 10^9$ GeV on the SO(5)-breaking scale of the 2HDM potential, beyond which an ultraviolet completion of the theory must be invoked. The situation can be alleviated with the other two natural alignment scenarios listed in Table 1 and this will be the subject of the next section.

For the allowed parameter space of our MS-2HDM as shown in Figure 2, we obtain concrete predictions for the remaining Higgs spectrum. In particular, the alignment condition imposes a *lower* bound on the soft breaking parameter $\mathrm{Re}(m_{12}^2)$, and hence, on the heavy Higgs spectrum. A comparison of the global fit limit on the charged Higgs-boson mass as a function of $\tan\beta$ [32] with our predicted limits from the alignment condition in the MS-2HDM for a typical value of the boundary scale $\mu_X = 3 \times 10^4$ GeV is shown in Figure 3 (left panel). It is clear that the alignment limits are stronger than the global fit limits, except in the very small and very large $\tan\beta$ regimes.

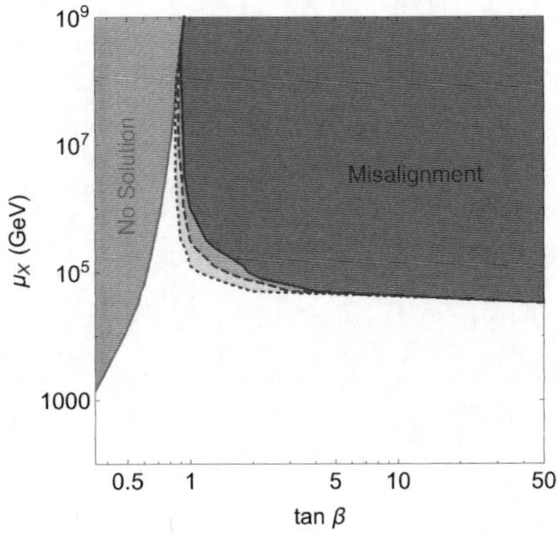

Figure 2: The 1σ (dotted), 2σ (dashed) and 3σ (solid) exclusion contours (blue shaded region) from the alignment constraints in MS-2HDM. The red shaded region is theoretically excluded.

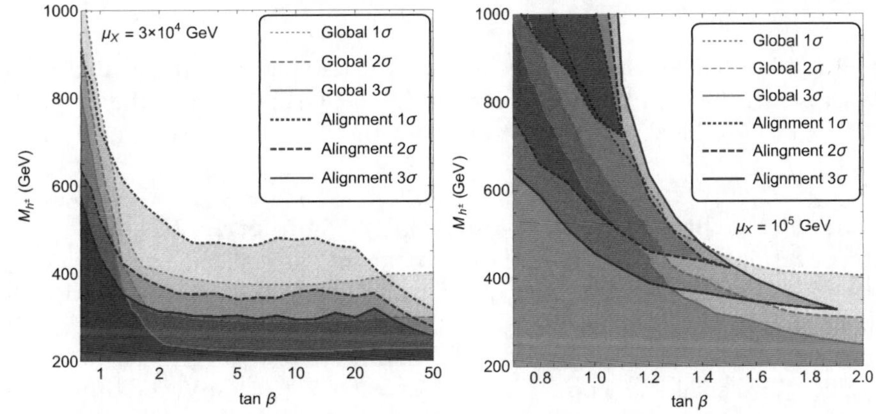

Figure 3: *Left:* The 1σ (dotted), 2σ (dashed) and 3σ (solid) *lower* limits on the charged Higgs mass obtained from the alignment condition (blue lines) in the MS-2HDM with $\mu_X = 3 \times 10^4$ GeV. *Right:* The 1σ (dotted), 2σ (dashed) and 3σ (solid) *allowed* regions from the alignment condition (blue lines) for $\mu_X = 10^5$ GeV. For comparison, the corresponding lower limits from a global fit are also shown (red lines).

From Figure 2, we note that for $\mu_X \gtrsim 10^5$ GeV, phenomenologically acceptable alignment is not possible in the MS-2HDM for large $\tan\beta$ *and* large m_{12}^2. Therefore, we also get an *upper* bound on the charged Higgs-boson mass $M_{h^\pm}$ from the misalignment condition, depending on $\tan\beta$. This is illustrated in Figure 3 (right panel) for $\mu_X = 10^5$ GeV.

Similar alignment constraints are obtained for the heavy neutral pseudo-Goldstone bosons h and a, which are predicted to be quasi-degenerate with the charged Higgs boson $h^\pm$ in the MS-2HDM [cf. (16)]. The current experimental limits on the heavy neutral Higgs sector [30] are weaker than the alignment constraints in this case. Thus, the MS-2HDM scenario provides a natural reason for the absence of a heavy Higgs signal below the top-quark threshold, and this has important consequences for the heavy Higgs searches in the Run-II phase of the LHC, as discussed in Sec. 6.

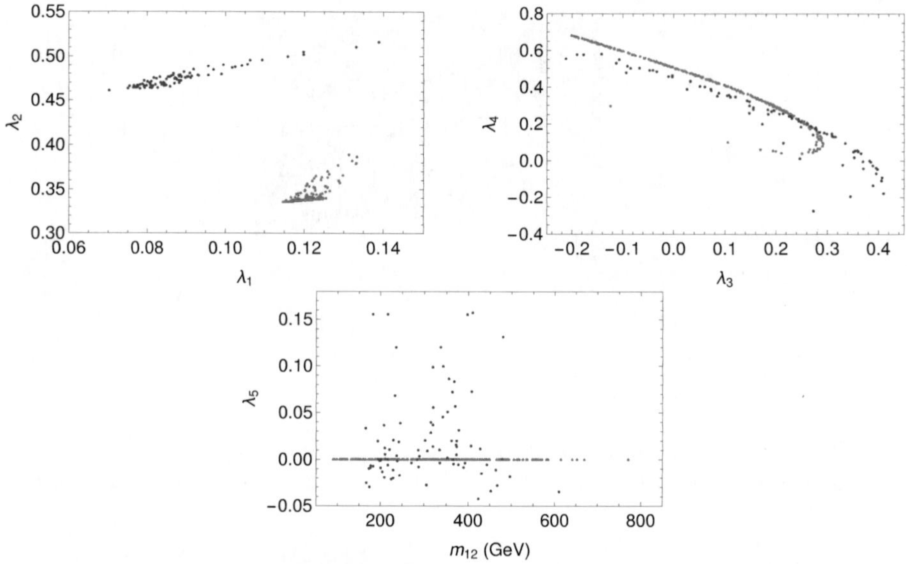

Figure 4: The 2HDM parameter space at the EW scale satisfying the natural alignment condition up to the Planck scale. The red points are for the $O(3) \times O(2)$ case and the blue points are for the $Z_2 \times [O(2)]^2$ case. We have shown the results for the $\tan\beta = 2$ case.

5. Planck-scale Perturbative Next-to-maximally Symmetric 2HDM

As discussed in Section , the MS-2HDM scenario cannot be realized beyond $\mu_X \gtrsim 10^9$ GeV. However, as we show in this section, the other two natural alignment scenarios discussed in Table 1, namely, those based on the $Z_2 \times [O(2)]^2$ and $O(3) \times O(2)$ symmetry of the 2HDM potential, can be realized all the way up to the Planck scale, assuming no other intermediate new physics scale. To see this, we choose $\mu_X = M_{\rm Pl} = 1.2 \times 10^{19}$ GeV and apply the boundary conditions given in Table 1 for these two cases. For the $Z_2 \times [O(2)]^2$ symmetry, this leaves three quartic couplings free, namely, $\lambda_{1,3,4}$, whereas for the $O(3) \times O(2)$ case, only two are free, namely, $\lambda_{1,3}$. We also allow for a non-zero m_{12}^2 to get a realistic scalar mass spectrum. Using the one-loop RGEs given in Ref. [11], we then study the top-down evolution of the 2HDM parameter space up to the weak scale and examine those points satisfying the vacuum stability, perturbativity and the LHC Higgs data, including the SM Higgs mass and the alignment constraints. Our numerical results are given in Figure 4 for the $\lambda_1 - \lambda_2$, $\lambda_3 - \lambda_4$, and $\lambda_5 - m_{12}$ parameter space at the weak scale, respectively. We have shown the results for the $\tan\beta = 2$ case, and the corresponding results for higher t_β values can be found in Ref. [24].

The corresponding scalar mass spectra are shown in Figure 5, where in the x-axis, we have plotted the heavy neutral Higgs mass $m_h \simeq m_a$ and on the y-axis, the charged Higgs mass $m_{h^\pm}$. It is clear that as in the MS-2HDM case, the heavy Higgs spectrum is quasi-degenerate. Thus, we conclude that the near-degeneracy of the heavy Higgs spectrum is a generic prediction of the natural alignment scenario, irrespective of the underlying symmetry of the 2HDM potential. This is a key result that will be important for the collider signal analysis in the next section.

6. Collider Signatures in the Alignment Limit

In the alignment limit, the couplings of the lightest CP-even Higgs boson are exactly similar to the SM Higgs couplings, while the heavy CP-even Higgs boson is gaugephobic. Therefore, two of the relevant Higgs production mechanisms at the LHC, namely, the vector boson fusion and Higgsstrahlung processes are suppressed for the heavy neutral Higgs sector. As a consequence,

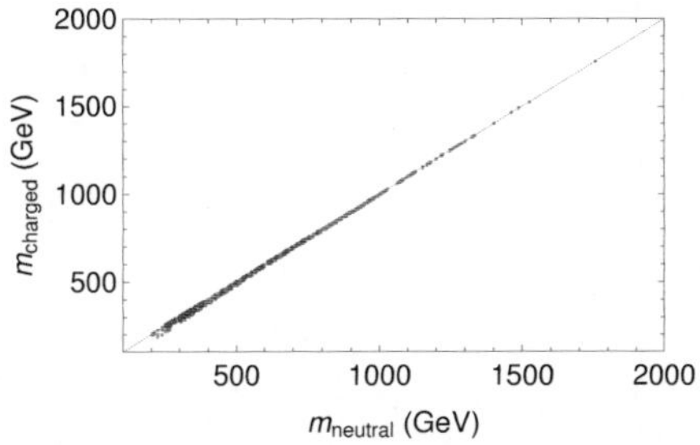

Figure 5: The scalar mass spectrum at the EW scale satisfying the natural alignment condition up to the Planck scale. The red points are for the $O(3) \times O(2)$ case and the blue points are for the $Z_2 \times [O(2)]^2$ case. The gray line shows the exact degeneracy. We have shown the results for the $\tan\beta = 2$ case.

the only relevant production channels to probe the neutral Higgs sector of the MS-2HDM are the gluon-gluon fusion and $t\bar{t}h$ ($b\bar{b}h$) associated production mechanisms at low (high) $\tan\beta$. For the charged Higgs sector of the MS-2HDM, the dominant production mode is the associated production process: $gg \to \bar{t}bh^+ + t\bar{b}h^-$, irrespective of $\tan\beta$.

Similarly, for the decay modes of the heavy neutral Higgs bosons in the MS-2HDM, the $t\bar{t}$ ($b\bar{b}$) channel is the dominant one for low (high) $\tan\beta$ values, whereas for the charged Higgs boson $h^{+(-)}$, the $t\bar{b}(\bar{t}b)$ mode is the dominant one for any $\tan\beta$. Thus, the heavy Higgs sector of the MS-2HDM can be effectively probed at the LHC through the final states involving third-generation quarks.

6.1. Charged Higgs Signal

The most promising channel at the LHC for the charged Higgs boson in the MS-2HDM is

$$gg \;\to\; \bar{t}bh^+ + t\bar{b}h^- \;\to\; t\bar{t}b\bar{b}\,. \tag{17}$$

Experimentally, this is a challenging mode due to large QCD backgrounds and the non-trivial event topology, involving at least four b-jets [25]. Nevertheless, a recent CMS study [33] has presented for the first time a realistic analysis of this process, in the leptonic decay mode of the W's coming from top decays:

$$gg \;\to\; h^{\pm}tb \;\to\; (\ell\nu_\ell bb)(\ell'\nu_{\ell'}b)b \tag{18}$$

(ℓ, ℓ' beings electrons or muons). Using the $\sqrt{s} = 8$ TeV LHC data, they have derived 95% CL upper limits on the production cross section $\sigma(gg \to h^{\pm}tb)$ times the branching ratio $\mathrm{BR}(h^{\pm} \to tb)$ as a function of the charged Higgs mass, as shown in Figure 6. In the same Figure, we show the corresponding predictions at $\sqrt{s} = 14$ TeV LHC in the Type-II MS-2HDM for some representative values of $\tan\beta$. The cross section predictions were obtained at leading order (LO) by implementing the 2HDM in `MadGraph5_aMC@NLO` [34] and using the `NNPDF2.3` PDF sets [35]. A comparison of these cross sections with the CMS limit suggests that the run-II phase of the LHC might be able to probe the low $\tan\beta$ region of the MS-2HDM parameter space using the process (17). Note that the production cross section $\sigma(gg \to \bar{t}bh^+)$ decreases rapidly with increasing $\tan\beta$ due to the Yukawa coupling suppression, even though $\mathrm{BR}(h^+ \to t\bar{b})$ remains close to 100%. Therefore, this channel is only effective for low $\tan\beta$ values.

In order to make a rough estimate of the $\sqrt{s} = 14$ TeV LHC sensitivity to the charged Higgs signal (17) in the MS-2HDM, we perform a parton level simulation of the signal and background events using `MadGraph5` [34]. For the event reconstruction, we use some basic selection cuts

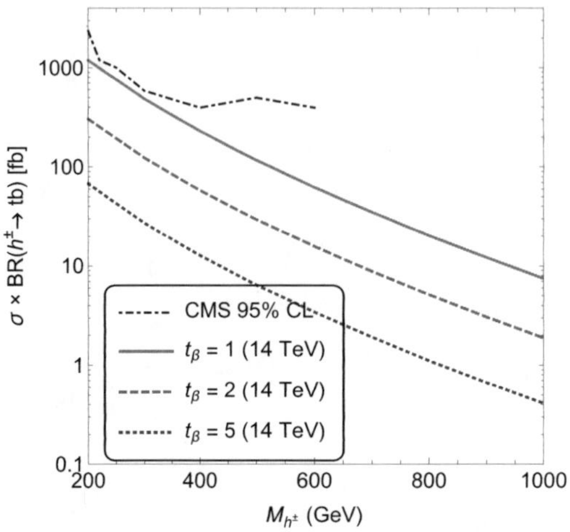

Figure 6: Predictions for the cross section of the process (17) in the Type-II MS-2HDM at $\sqrt{s} = 14$ TeV LHC for various values of $\tan\beta$. For comparison, we have also shown the current 95% CL CMS upper limit from the $\sqrt{s} = 8$ TeV data [33].

on the transverse momentum, pseudo-rapidity and dilepton invariant mass, following the CMS analysis [33]:

$$\begin{aligned} p_T^\ell \;&>\; 20\text{ GeV}, \quad |\eta^\ell| < 2.5, \quad p_T^j \;>\; 30\text{ GeV}, \quad |\eta^j| < 2.4, \quad \not{E}_T > 40\text{ GeV} \\ \Delta R^{\ell\ell} &> 0.4, \quad \Delta R^{\ell j} > 0.4, \quad M_{\ell\ell} > 12\text{ GeV}, \quad |M_{\ell\ell} - M_Z| > 10\text{ GeV}. \end{aligned} \tag{19}$$

Jets are reconstructed using the anti-k_T clustering algorithm [26] with a distance parameter of 0.5. Since four b-jets are expected in the final state, at least two b-tagged jets are required in the signal events, and we assume the b-tagging efficiency for each of them to be 70%.

The inclusive SM cross section for $pp \to t\bar{t}b\bar{b} + X$ is $\sim$ 18 pb at NLO, with roughly 30% uncertainty due to higher order QCD corrections [37]. Most of the QCD background for the $4b + 2\ell + \not{E}_T$ final state given by (18) can be reduced significantly by reconstructing at least one top-quark. The remaining irreducible background due to SM $t\bar{t}b\bar{b}$ production can be suppressed with respect to the signal by reconstructing the charged Higgs boson mass, once a valid signal region is defined, e.g. in terms of an observed excess of events at the LHC in future. For the semi-leptonic decay mode of top-quarks as in (18), one cannot directly use an invariant mass observable to infer $M_{h^\pm}$, as both the neutrinos in the final state give rise to missing momentum. A useful quantity in this case is the M_{T2} variable [38], defined as

$$M_{T2} \;=\; \min_{\left\{\not{\mathbf{p}}_{T_\mathrm{a}} + \not{\mathbf{p}}_{T_\mathrm{b}} = \not{\mathbf{p}}_T\right\}} \left[\max\left\{m_{T_\mathrm{a}}, m_{T_\mathrm{b}}\right\}\right], \tag{20}$$

where $\{\mathrm{a}\}, \{\mathrm{b}\}$ stand for the two sets of particles in the final state, each containing a neutrino with part of the missing transverse momentum ($\not{\mathbf{p}}_{T_{\mathrm{a,b}}}$). Minimization over all possible sums of these two momenta gives the observed missing transverse momentum $\not{\mathbf{p}}_T$, whose magnitude is the same as $\not{E}_T$ in our specific case. In (20), m_{T_i} (with i =a,b) is the usual transverse mass variable for the system $\{i\}$, defined as

$$m_{T_i}^2 \;=\; \left(\sum_{\text{visible}} E_{T_i} + \not{E}_{T_i}\right)^2 - \left(\sum_{\text{visible}} \mathbf{p}_{T_i} + \not{\mathbf{p}}_{T_i}\right)^2. \tag{21}$$

IOP Conf. Series: Journal of Physics: Conf. Series **873** (2017) 012008 doi:10.1088/1742-6596/873/1/012008

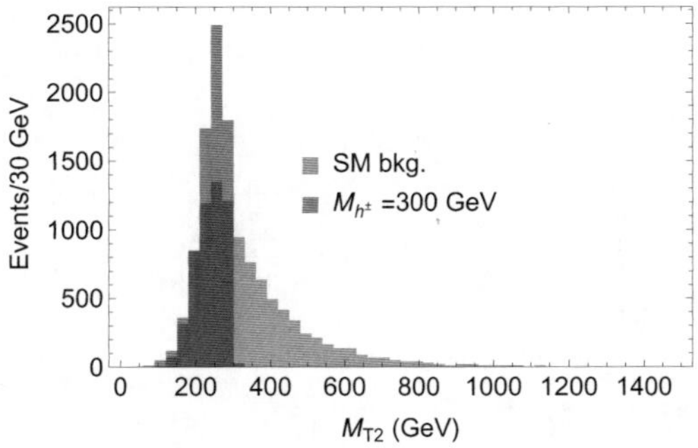

Figure 7: An illustration of the charged Higgs boson mass reconstruction using the M_{T2} variable. The irreducible SM background distribution is also shown for comparison.

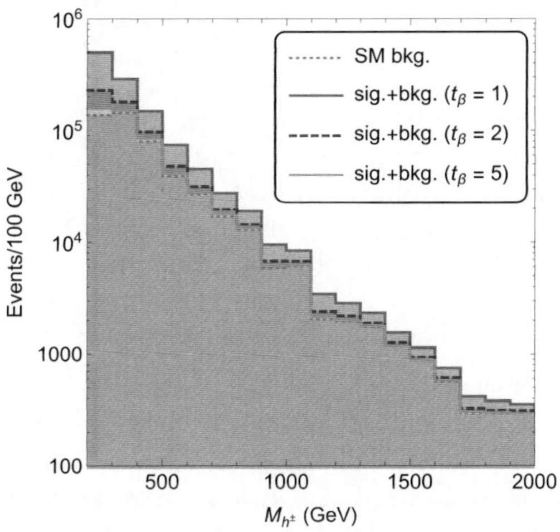

Figure 8: Predicted number of events for the $t\bar{t}b\bar{b}$ signal from the charged pseudo-Goldstone boson in the MS-2HDM at $\sqrt{s}$ = 14 TeV LHC with 300 fb^{-1} integrated luminosity. The irreducible SM background (red shaded) is controlled by assuming an efficient mass reconstruction technique [11].

For the correct combination of the final state particles in (18), i.e. for $\{\text{a}\} = (\ell\nu_\ell bb)$ and $\{\text{b}\} = (\ell'\nu_{\ell'} bb)$ in (20), the maximum value of M_{T2} represents the charged Higgs boson mass, with the M_{T2} distribution smoothly dropping to zero at this point. This is illustrated in Figure 7 for a typical choice of $M_{h^\pm} = 300$ GeV. For comparison, we also show the M_{T2} distribution for the SM background, which obviously does not have a sharp endpoint. Thus, for a given hypothesized signal region defined in terms of an excess due to $M_{h^\pm}$, we may impose an additional cut on $M_{T2} \leq M_{h^\pm}$ to enhance the signal (18) over the irreducible SM background.

Assuming that the charged Higgs boson mass can be reconstructed efficiently, we present an estimate of the signal to background ratio for the charged Higgs signal given by (17) at $\sqrt{s} = 14$ TeV LHC with 300 fb^{-1} for some typical values of $\tan\beta$ in Figure 8. Since the mass of the charged Higgs boson is a priori unknown, we vary the charged Higgs mass, and for each value of $M_{h^\pm}$, we assume that it can be reconstructed around its actual value within 30 GeV uncertainty.

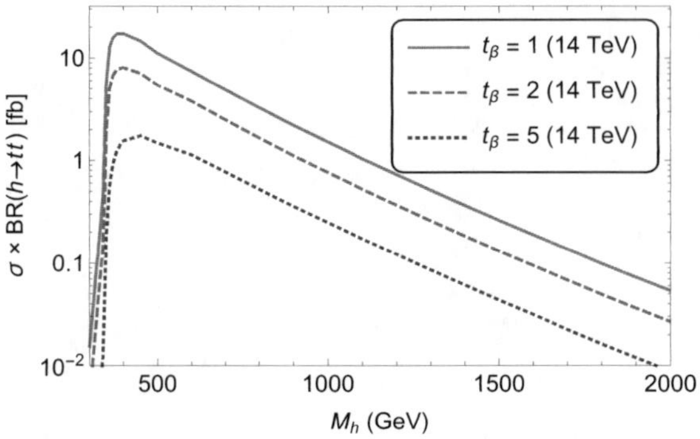

Figure 9: Predictions for the cross section of the process (22) in the Type-II MS-2HDM at $\sqrt{s} = 14$ TeV LHC for various values of $\tan\beta$.

6.2. Heavy Neutral Higgs Signal

So far there have been no direct searches for heavy neutral Higgs bosons involving $t\bar{t}$ and/or $b\bar{b}$ final states, mainly due to the challenges associated with uncertainties in the jet energy scales and the combinatorics arising from complicated multiparticle final states in a busy QCD environment. Nevertheless, these channels become pronounced in the MS-2HDM scenario, and hence, we have made a preliminary attempt to study them in [11] (see also [27]). In particular, we focus on the search channel

$$gg \;\to\; t\bar{t}h \;\to\; t\bar{t}t\bar{t}\,. \tag{22}$$

Such four top final states have been proposed before in the context of other exotic searches at the LHC (see e.g. [28]). However, their relevance for heavy Higgs searches have not been explored so far. We note here that the existing 95% CL experimental upper limit on the four top production cross section is 59 fb from ATLAS [29] and 32 fb from CMS [40], whereas the SM prediction for the inclusive cross section of the process $pp \to t\bar{t}t\bar{t} + X$ is about 10-15 fb [41].

To get a rough estimate of the signal to background ratio for our four-top signal (22), we perform a parton-level simulation of the signal and background events at LO in QCD using `MadGraph5_aMC@NLO` [34] with `NNPDF2.3` PDF sets [35]. For the inclusive SM cross section for the four-top final state at $\sqrt{s} = 14$ TeV LHC, we obtain 11.85 fb, whereas our proposed four-top signal cross sections are found to be comparable or smaller depending on M_h and $\tan\beta$, as shown in Figure 9. However, since we expect one of the $t\bar{t}$ pairs coming from an on-shell h decay to have an invariant mass around M_h, we can use this information to significantly boost the signal over the irreducible SM background. Note that all the predicted cross sections shown in Figure 9 are well below the current experimental upper bound [40].

Depending on the W decay mode from $t \to Wb$, there are 35 final states for four top decays. According to a recent ATLAS analysis [42], the experimentally favoured channel is the semi-leptonic/hadronic final state with two same-sign isolated leptons. Although the branching fraction for this topology (4.19%) is smaller than most of the other channels, the presence of two same-sign leptons in the final state allows us to reduce the large QCD background substantially, including that due to the SM production of $t\bar{t}b\bar{b}$+jets [42]. Therefore, we will only consider the following decay chain in our preliminary analysis:

$$gg \;\to\; t\bar{t}h \;\to\; (t\bar{t})(t\bar{t}) \;\to\; \Big((\ell^{\pm}\nu_\ell b)(jjb)\Big)\Big((\ell'^{\pm}\nu_{\ell'} b)(jjb)\Big)\,. \tag{23}$$

For event reconstruction, we will use the same selection cuts as in (19), and in addition, following [42], we require the scalar sum of the p_T of all leptons and jets (defined as H_T) to exceed 350 GeV.

IOP Conf. Series: Journal of Physics: Conf. Series **873** (2017) 012008 doi:10.1088/1742-6596/873/1/012008

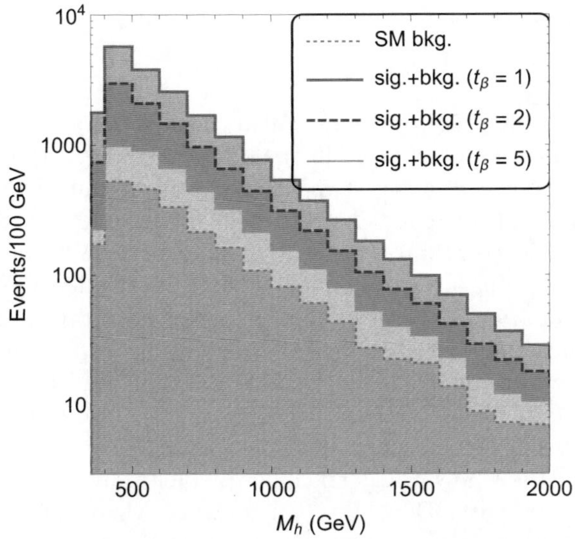

Figure 10: Predicted number of events for the $t\bar{t}t\bar{t}$ signal from the neutral pseudo-Goldstone boson in the MS-2HDM at $\sqrt{s}$ = 14 TeV LHC with 300 fb^{-1} integrated luminosity. The results are shown for three different values of $\tan\beta$ =1 (green solid), 2 (blue dashed) and 5 (orange solid). The SM background (red dotted) is controlled by assuming an efficient mass reconstruction technique, as outlined in the text.

As in the charged Higgs boson case (cf. Figure 7), the heavy Higgs mass can be reconstructed from the signal given by (23) using the M_{T2} endpoint technique, and therefore, an additional selection cut on $M_{T2} \leq M_h$ can be used to enhance the signal over the irreducible background. Our simulation results for the predicted number of signal and background events for the process (23) at $\sqrt{s} = 14$ TeV LHC with 300 fb^{-1} luminosity are shown in Figure 10. The signal events are shown for three representative values of $\tan\beta$. Here we vary the a priori unknown heavy Higgs mass, and for each value of M_h, we assume that it can be reconstructed around its actual value within 30 GeV uncertainty. From this preliminary analysis, we find that the $t\bar{t}t\bar{t}$ channel provides the most promising collider signal to probe the heavy Higgs sector in the MS-2HDM at low values of $\tan\beta \lesssim 5$.

The above analysis is also applicable for the CP-odd Higgs boson a, which has similar production cross sections and $t\bar{t}$ branching fractions as the CP-even Higgs h. However, the $t\bar{t}h(a)$ production cross section as well as the $h(a) \to t\bar{t}$ branching ratio decreases with increasing $\tan\beta$. This is due to the fact that the $ht\bar{t}$ coupling in the alignment limit is $\cos\alpha/\sin\beta \sim \cot\beta$, which is same as the $at\bar{t}$ coupling. Thus, the high $\tan\beta$ region of the MS-2HDM cannot be searched via the $t\bar{t}t\bar{t}$ channel proposed above, and one needs to consider the channels involving down-sector Yukawa couplings, e.g. $b\bar{b}b\bar{b}$ and $b\bar{b}\tau^+\tau^-$ [25]. It is also worth commenting here that the simpler process $pp \to h/a \to t\bar{t}$ $(b\bar{b})$ at low (high) $\tan\beta$ suffers from a huge SM $t\bar{t}$ $(b\bar{b})$ QCD background, even after imposing an $M_{t\bar{t}\ (b\bar{b})}$ cut. Some parton-level studies of this signal in the context of MSSM have been performed in [43].

We should clarify that the results obtained in this section are valid only at the parton level. In a realistic detector environment, the sharp features of the signal [see e.g., Figure 7] used to derive the sensitivity reach in Figures 8 and 10 may not survive, and therefore, the signal-to-background ratio might get somewhat reduced than that shown here. A detailed detector-level analysis of these signals, including realistic top reconstruction efficiencies and smearing effects, as well as possible interference effects between the charged and neutral Higgs signals and with

IOP Conf. Series: Journal of Physics: Conf. Series **873** (2017) 012008 doi:10.1088/1742-6596/873/1/012008

the SM background, is currently being pursued in a separate dedicated study [44].

7. Conclusions

We provide a symmetry justification of the so-called SM alignment limit, independently of the heavy Higgs spectrum and the value of $\tan\beta$ in the 2HDM. We show that in the Type-II 2HDM, there exist *only* three different symmetry realizations, which could lead to the SM alignment by satisfying the natural alignment condition (10) for *any* value of $\tan\beta$. In the context of the Maximally Symmetric 2HDM based on the SO(5) group, we demonstrate how small deviations from this alignment limit are naturally induced by RG effects due to the hypercharge gauge coupling g' and third generation Yukawa couplings, which also break the custodial symmetry of the theory. In addition, a non-zero soft SO(5)-breaking mass parameter is required to yield a viable Higgs spectrum consistent with the existing experimental constraints. Employing the current Higgs signal strength data from the LHC, which disfavour large deviations from the alignment limit, we derive important constraints on the 2HDM parameter space. In particular, we predict lower limits on the heavy Higgs spectrum, which prevail the present global fit limits in a wide range of parameter space. Depending on the scale where the maximal symmetry could be realized in nature, we also obtain an upper limit on the heavy Higgs masses in certain cases, which could be probed during the run-II phase of the LHC. Finally, we have studied the collider signatures of the heavy Higgs sector in the alignment limit beyond the top-quark threshold. We find that the final states involving third-generation quark final states can become a valuable observational tool to directly probe the heavy Higgs sector of the 2HDM in the alignment limit for low values of $\tan\beta$.

Acknowledgements

This work was supported by the Lancaster-Manchester-Sheffield Consortium for Fundamental Physics under STFC grant ST/L000520/1. P.S.B.D. would like to acknowledge the local hospitality provided by the CFTP and IST, Lisbon where part of these proceedings was written.

References

[1] G. Aad *et al.* [ATLAS Collaboration], Phys. Lett. B **716**, 1 (2012); S. Chatrchyan *et al.* [CMS Collaboration], Phys. Lett. B **716**, 30 (2012).
[2] P. W. Higgs, Phys. Rev. Lett. **13**, 508 (1964); F. Englert and R. Brout, Phys. Rev. Lett. **13**, 321 (1964); G. S. Guralnik, C. R. Hagen and T. W. B. Kibble, Phys. Rev. Lett. **13**, 585 (1964).
[3] G. Aad *et al.* [ATLAS and CMS Collaborations], JHEP **1608**, 045 (2016).
[4] G. C. Branco *et al.*, Phys. Rept. **516**, 1 (2012); I. P. Ivanov, arXiv:1702.03776 [hep-ph].
[5] R. A. Battye, G. D. Brawn and A. Pilaftsis, JHEP **1108**, 020 (2011).
[6] ATLAS collaboration, ATLAS-CONF-2014-010; V. Khachatryan *et al.* [CMS Collaboration], Phys. Rev. D **90**, 112013 (2014).
[7] A. Celis, V. Ilisie and A. Pich, JHEP **1307**, 053 (2013); C. W. Chiang and K. Yagyu, JHEP **1307**, 160 (2013); C.-Y. Chen, S. Dawson and M. Sher, Phys. Rev. D **88**, 015018 (2013); N. Craig, J. Galloway and S. Thomas, arXiv:1305.2424 [hep-ph]; K. Cheung, J. S. Lee and P.-Y. Tseng, JHEP **1401**, 085 (2014); L. Wang and X.-F. Han, JHEP **1411**, 085 (2014); B. Dumont, J. F. Gunion, Y. Jiang and S. Kraml, Phys. Rev. D **90**, 035021 (2014); S. Kanemura, K. Tsumura, K. Yagyu and H. Yokoya, Phys. Rev. D **90**, 075001 (2014); B. Grinstein and P. Uttayarat, JHEP **1306**, 094 (2013) [Erratum-ibid. **1309**, 110 (2013)]; A. Broggio, E. J. Chun, M. Passera, K. M. Patel and S. K. Vempati, JHEP **1411**, 058 (2014); N. Craig, F. D'Eramo, P. Draper, S. Thomas and H. Zhang, JHEP **1506**, 137 (2015); D. Chowdhury and O. Eberhardt, JHEP **1511**, 052 (2015); X. F. Han and L. Wang, arXiv:1701.02678 [hep-ph].
[8] H. Georgi and D. V. Nanopoulos, Phys. Lett. B **82**, 95 (1979).
[9] J. F. Gunion and H. E. Haber, Phys. Rev. D **67**, 075019 (2003).
[10] M. Carena, I. Low, N. R. Shah and C. E. M. Wagner, JHEP **1404**, 015 (2014); M. Carena, H. E. Haber, I. Low, N. R. Shah and C. E. M. Wagner, Phys. Rev. D **91**, 035003 (2015).
[11] P. S. B. Dev and A. Pilaftsis, JHEP **1412**, 024 (2014).

IOP Conf. Series: Journal of Physics: Conf. Series **873** (2017) 012008 doi:10.1088/1742-6596/873/1/012008

[12] D. Das and I. Saha, Phys. Rev. D **91**, 095024 (2015); J. Bernon, J. F. Gunion, H. E. Haber, Y. Jiang and S. Kraml, Phys. Rev. D **92**, 075004 (2015); H. E. Haber and O. Stål, Eur. Phys. J. C **75**, no. 10, 491 (2015).
[13] P. H. Chankowski, T. Farris, B. Grzadkowski, J. F. Gunion, J. Kalinowski and M. Krawczyk, Phys. Lett. B **496**, 195 (2000); I. F. Ginzburg and M. Krawczyk, Phys. Rev. D **72**, 115013 (2005); A. Delgado, G. Nardini and M. Quiros, JHEP **1307**, 054 (2013); G. Bhattacharyya and D. Das, Phys. Rev. D **91**, 015005 (2015).
[14] H. E. Haber and R. Hempfling, Phys. Rev. D **48**, 4280 (1993); A. Pilaftsis and C. E. M. Wagner, Nucl. Phys. B **553**, 3 (1999).
[15] M. Maniatis, A. von Manteuffel, O. Nachtmann and F. Nagel, Eur. Phys. J. C **48**, 805 (2006); C. C. Nishi, Phys. Rev. D **74**, 036003 (2006) [Erratum *ibid.* **76**, 119901 (2007)].
[16] C. C. Nishi, Phys. Rev. D **83**, 095005 (2011).
[17] A. Pilaftsis, Phys. Lett. B **706**, 465 (2012).
[18] M. Maniatis, A. von Manteuffel and O. Nachtmann, Eur. Phys. J. C **57**, 719 (2008). I. P. Ivanov, Phys. Rev. D **77**, 015017 (2008); C. C. Nishi, Phys. Rev. D **77**, 055009 (2008).
[19] J. F. Gunion and H. E. Haber, Phys. Rev. D **72**, 095002 (2005); M. Maniatis and O. Nachtmann, JHEP **1111**, 151 (2011).
[20] N. G. Deshpande and E. Ma, Phys. Rev. D **18**, 2574 (1978).
[21] R. Barbieri, L. J. Hall and V. S. Rychkov, Phys. Rev. D **74**, 015007 (2006).
[22] A. Pilaftsis, Phys. Rev. D **93**, no. 7, 075012 (2016).
[23] P. S. B. Dev and A. Pilaftsis, J. Phys. Conf. Ser. **631**, no. 1, 012030 (2015); PoS PLANCK **2015**, 105 (2015).
[24] N. Chakrabarty, P. S. B. Dev, B. Mukhopadhyaya and A. Pilaftsis, *in preparation.*
[25] D. de Florian *et al.* [LHC Higgs Cross Section Working Group], arXiv:1610.07922 [hep-ph].
[26] M. Cacciari, G. P. Salam and G. Soyez, JHEP **0804**, 063 (2008).
[27] S. Gori, I. W. Kim, N. R. Shah and K. M. Zurek, Phys. Rev. D **93**, no. 7, 075038 (2016).
[28] M. Spira and J. D. Wells, Nucl. Phys. B **523**, 3 (1998); A. Pomarol and J. Serra, Phys. Rev. D **78**, 074026 (2008); K. Kumar, T. M. P. Tait and R. Vega-Morales, JHEP **0905**, 022 (2009); S. Jung and J. D. Wells, JHEP **1011**, 001 (2010); G. L. Kane, E. Kuflik, R. Lu and L. T. Wang, Phys. Rev. D **84**, 095004 (2011); G. Cacciapaglia, R. Chierici, A. Deandrea, L. Panizzi, S. Perries and S. Tosi, JHEP **1110**, 042 (2011); E. Alvarez, D. A. Faroughy, J. F. Kamenik, R. Morales and A. Szynkman, Nucl. Phys. B **915**, 19 (2017).
[29] ATLAS Collaboration, ATLAS-CONF-2013-051.
[30] C. Patrignani *et al.* [Particle Data Group], Chin. Phys. C **40**, no. 10, 100001 (2016).
[31] R. D. Peccei and H. R. Quinn, Phys. Rev. Lett. **38**, 1440 (1977).
[32] O. Eberhardt, U. Nierste and M. Wiebusch, JHEP **1307**, 118 (2013); J. Baglio, O. Eberhardt, U. Nierste and M. Wiebusch, Phys. Rev. D **90**, 015008 (2014).
[33] CMS Collaboration, CMS-PAS-HIG-13-026.
[34] J. Alwall *et al.*, JHEP **1407**, 079 (2014).
[35] R. D. Ball *et al.*, Nucl. Phys. B **867**, 244 (2013).
[36] N. Craig, F. D'Eramo, P. Draper, S. Thomas and H. Zhang, JHEP **1506**, 137 (2015).
[37] G. Bevilacqua, M. Czakon, C. G. Papadopoulos, R. Pittau and M. Worek, JHEP **0909**, 109 (2009); M. Flechl, R. Klees, M. Kramer, M. Spira and M. Ubiali, Phys. Rev. D **91**, 075015 (2015).
[38] C. G. Lester and D. J. Summers, Phys. Lett. B **463**, 99 (1999).
[39] S. Kanemura, H. Yokoya and Y. J. Zheng, Nucl. Phys. B **898**, 286 (2015); Y. P. Kuang and L. H. Xia, Phys. Lett. B **747**, 193 (2015); N. Chen, J. Li and Y. Liu, arXiv:1509.03848 [hep-ph].
[40] V. Khachatryan *et al.* [CMS Collaboration], JHEP **1411**, 154 (2014); G. Aad *et al.* [ATLAS Collaboration], JHEP **1508**, 105 (2015).
[41] G. Bevilacqua and M. Worek, JHEP **1207**, 111 (2012).
[42] D. Paredes [ATLAS Collaboration], PhD Thesis, Université Blaise Pascal (2013), CERN-THESIS-2013-202; J. Keaveney [CMS Collaboration], arXiv:1412.4641 [hep-ex].
[43] S. Moretti and D. A. Ross, Phys. Lett. B **712**, 245 (2012); A. Djouadi, L. Maiani, A. Polosa, J. Quevillon and V. Riquer, JHEP **1506**, 168 (2015); B. Bhattacherjee, A. Chakraborty and A. Choudhury, Phys. Rev. D **92**, no. 9, 093007 (2015); M. Carena and Z. Liu, JHEP **1611**, 159 (2016).
[44] A. Choudhury, P. S. B. Dev, W. Klemm, E. Orgill, Y. Peters, A. Pilaftsis, *work in progress.*

IOP Conf. Series: Journal of Physics: Conf. Series **873** (2017) 012009 doi:10.1088/1742-6596/873/1/012009

Tests of Lorentz and CPT symmetry with hadrons and nuclei

J P Noordmans[1]

[1] CENTRA, Departamento de Física, Universidade do Algarve, 8005-139 Faro, Portugal

E-mail: jnoordmans@ualg.pt

Abstract. We apply chiral-perturbation-theory techniques to the QCD sector of the Lorentz and CPT violating standard-model extension. We derive the effective Lagrangian in terms of pions and nucleons for a selected set of dimension-five operators involving quarks and gluons. This derivation is based on chiral-symmetry properties of the operators, as well as on their behaviour under C,P, and T transformations. We consider the power counting rules and apply the heavy-baryon approach to account for the large nucleon mass. Having obtained the relevant Lorentz-violating contributions to the pion-nucleon Lagrangian, we proceed to derive the particle and anti-particle Hamiltonian, from which we obtain the Lorentz-violating contribution to comagnetometer experiments. This allows us to place stringent limits on some of the parameters. For some other parameters we find that the best bounds will come from nucleon-nucleon interactions, and we derive the relevant nucleon-nucleon potential. These considerations imply possible new opportunities for spin-precession experiments involving for example the deuteron.

1. Introduction

The unification of general relativity with quantum mechanics is one of the main outstanding problems in theoretical physics today. Many different solutions to this problem have been proposed. However, none of these have proven to be the definitive answer yet. One of the main difficulties is the energy scale at which the relevant theories become distinguishable from what should be their low-energy limits: general relativity and the standard model of particle physics. The generally accepted paradigm is that these theories should become unified around an energy scale of about the Planck mass: $m_P \simeq 10^{19}\,\mathrm{GeV}$. The relevant deviations from the conventional physics should thus become apparent at this energy scale, which is far out of reach of any present-day experiments.

One possible solution to this problem can be found in the study of violations of spacetime symmetries. Many theories that attempt to unify the standard model of particle physics with general relativity modify the notion of spacetime at Planck-scale energies. Consequently, the fate of spacetime symmetries such as Lorentz and CPT symmetry becomes an open issue. Tests of these symmetries might therefore provide valuable insights about a possible theory of quantum gravity. Moreover, it turns out that very sensitive experimental tests are possible for signals that are characteristic for the breaking of Lorentz and/or CPT symmetry. Therefore, tiny remnants of such high-energy symmetry breakings might be detectable at presently attainable energies. This high sensitivity, combined with the fact that Lorentz- and CPT-symmetry-violating signals (or their absence) might guide us in the quest for a theory of quantum gravity, has motivated a large amount of research in this area over the last two decades.

This research is facilitated by an effective-field-theory approach to Lorentz and CPT violation, called the standard-model extension (SME) [1], which we will discuss in more detail in the next section. One of the merits of the SME is its generality. It includes all Lorentz-violating effects that can be incorporated into a local quantum field theory. Additionally, it allows for a classification of the different Lorentz-violating operators and for quantitative bounds on the corresponding parameters for Lorentz violation. Therefore, one can compare different (potential) experiments and their sensitivity and identify possible sectors in the parameter space for Lorentz violation that might have been tested relatively poorly.

One of the sectors where limits are relatively weak, is the QCD sector of the SME [2]. The reason for this is obvious: the nonperturbative nature of QCD at energies below 1 GeV. Although very precise bounds on effective Lorentz-violating coefficients for protons, neutrons, and other hadrons exist, the corresponding limits on the underlying quark and gluon parameters are hard to obtain. To remedy this situation, we recently started to apply the machinery of chiral perturbation theory to the QCD sector of the SME [3]. Further work along the same lines can be found in Refs. [4, 5]. Using the effective operators in terms of the relevant degrees of freedom at low energies – the pion and nucleon fields – it becomes possible to put stringent constraints on quark and gluon parameters for Lorentz violation.

2. A Lorentz noninvariant effective field theory

It is not hard to argue that the best way to approach the problem of investigating Lorentz violation in theoretical particle physics today, is by using an effective-field-theory approach. A general construction of such an effective field theory was described in Ref. [1]. Similar, albeit more restricted approaches, were taken in Refs. [6, 7]. The resulting framework is called the standard-model extension (SME) and it has been succesfully used many times to describe and parametrize Lorentz- and CPT-violating effects, as well as to obtain quantitative bounds on many of its coefficients for Lorentz- and CPT violation. In Ref. [2] a complete and up-to-date overview of the experimental limits can be found.

Although it is possible to conceive of other methods to describe Lorentz- and CPT-symmetry breaking, the SME has several important advantages [8]. First of all, it is general, in the sense that all possible Lorentz-violating effects are included in the theory, as long as these effects can be described in the context of a local quantum field theory. Secondly, we can describe the SME as being realistic, since it is a perturbation of the conventional standard model of particle physics and thus includes all the known and well-established particle physics. Finally, the SME is coordinate invariant, as it should be, since we do not want the physics to depend on the coordinates we use.

The operators in the Lagrangian of the SME are build from the conventional standard-model quantum fields that represent all the known elementary particles and transform under the Lorentz group in the conventional way. The fields continue to be representations of the standard-model gauge group $SU(3)_c \times SU(2)_L \times U(1)_Y$ and the Lagrangian is invariant under the corresponding gauge transformations. The only restrictions on the form of the operators that are relaxed, is those of Lorentz and CPT symmetry. Operationally this means that Lorentz indices are no longer contracted to each other, but instead can be contracted to coefficients that parametrize the Lorentz and CPT violation. The value of these coefficients presumably originates from some unknown high-energy theory that incorporates (spontaneous) Lorentz-symmetry breaking.

A typical example of an SME operator is

$$\mathcal{L}^{\rm LV} \supset \langle T \rangle_{\mu\nu} \, \bar{\psi} \gamma^\mu \partial^\nu \psi \, , \tag{1}$$

where $\langle T\rangle_{\mu\nu}$ is the Lorentz-violating coefficient. The Lorentz-violating coefficients are generally taken to have fixed background values and do not transform under so-called 'particle Lorentz transformations'. These are transformations that effect only the quantities associated to the particles, like their momentum and spin, i.e. they are transformations of the particle fields only. Particle Lorentz transformations can be viewed as corresponding to physical rearrangements within the system one is trying to describe. They should be contrasted with 'observer Lorentz transformations', which affect both the particle fields and the Lorentz-violating coefficients. These transformations are just redefinitions of the coordinate system and as such should not influence the physics. Therefore, all SME operators are invariant under observer Lorentz transformations, as clearly exemplified by Eq. (1).

The Lagrangian of the SME in principle contains all local and gauge-invariant Lorentz-violating operators. This is an infinite set of operators, starting at a mass dimensionality three. On phenomenological grounds the coefficients of the operators are assumed to be suppressed by one or more powers of a large mass scale that corresponds to the scale of Lorentz-symmetry breaking (in many cases this scale is taken to be the Planck mass). This leads to a slight conundrum for the minimal standard-model extension (mSME), which contains only operators with a mass dimension of three or four. Naturalness arguments suggest that the coefficients of these operators are proportional the large mass scale or to unity, respectively. This obviously is in sharp contrast with existing experimental limits. One way out of this is to conjecture the existence of some additional symmetry that forbids the offensive operators. This symmetry might be broken at some intermediate energy scale M_{IR}, making low-energy values of the coefficients proportional to powers of M_{IR}/M_P. For example, it seems that in a minimal supersymmetric model the lowest dimension for Lorentz-violating operators is five [9], such that at low energy the symmetry-breaking operators are suppressed by at least one power of the high-energy scale.

Although the SME in principle contains all possible operators, the explicit identification and classification of all the operators has only been performed up to mass dimension five. The power-counting renormalizable operators of dimension three and four are described in Ref. [1], while the authors of Ref. [10] identify all mass-dimension five operators that are subject to certain 'UV-safety constraints', which prevent these operators from mixing with the lower-dimensional operators. For QED, all operators have been identified up to dimension six [11], while for kinetic operators the program of classifying all operators of arbitrary dimensions has been finished for photons, Dirac fermions, and neutrinos [12].

As mentioned in the introduction, in particular the QCD sector of the SME is poorly constrained, due to the nonperturbative nature of QCD. In Ref. [3] we started to remedy this situation, by applying chiral perturbation theory techniques to the Lorentz-violating operators (see Refs.[4] and [5] for more work in this direction). In Ref. [3] we focused on a set of dimension five quark-gluon operators, which are relevant at an energy scale of about 1 GeV. They were first written down in Ref. [10] and are given by

$$\mathcal{L}^{\rm LV} = \sum_{q=u,d} [C^q_{\mu\nu\rho}\bar{q}\gamma^\mu G^{\rho\nu}q + D^q_{\mu\nu\rho}\bar{q}\gamma^\mu\gamma^5 G^{\rho\nu}q] + H_{\mu\nu\rho}\,{\rm Tr}\left(G^{\mu\lambda}D^\nu\tilde{G}^\rho{}_\lambda\right)\,. \tag{2}$$

The sum in this equation runs over the lightest two quark flavors, the fields q are the quark fields and $G^{\mu\nu} = t^a G^{a,\mu\nu}$ ($t^a = \frac{1}{2}\lambda^a$, $a = 1,\ldots 8$, where λ^a are the Gell-Mann matrices, are the generators of the $SU(3)$ color group). All three operators violate Lorentz symmetry as well as CPT symmetry. This symmetry breaking is parametrized by the coefficients $C^q_{\mu\nu\rho}$, $D^q_{\mu\nu\rho}$, and $H_{\mu\nu\rho}$. In addition to being antisymmetric in their last two Lorentz indices, the $C^{\mu\nu\rho}$ and $D^{\mu\nu\rho}$ coefficients are symmetric in their first and last Lorentz indices, while $H^{\mu\nu\rho}$ is fully symmetric, i.e.

$$X^q_{\mu\nu\rho} = X^q_{\rho\nu\mu} = -X^q_{\mu\rho\nu} \,, \tag{3}$$

$$H_{\mu\nu\rho} = H_{\mu\rho\nu} = H_{\nu\mu\rho} \,, \tag{4}$$

with $X \in \{C, D\}$. These symmetry constraints follow from the UV-safety considerations, which are discussed in Ref. [10]. The properties of the coefficients under the action of C, P, and T operations are summarized in table 1.

There are many more operators that can be considered in the context of chiral perturbation theory, however these are left for future considerations (see for example Refs. [4] and [5]). Also, we consider only the lightest two quark flavors. In principle the strange quark can be included and our work can be extended to $SU(3)$ chiral perturbation theory (e.g. see Ref. [5]).

3. The Lorentz-violating chiral-perturbation-theory Lagrangian

At energies below 1 GeV, the operators in Eq. (2) induce interactions between the relevant degrees of freedom, i.e. the pions, nucleons, and photons. Chiral perturbation theory [13, 14] (see e.g. Refs. [15, 16] for reviews) can be used to derive the form of the operators that describe these interactions. In Ref. [3] we constructed the low-energy effective Lagrangian that corresponds to the Lorentz-violating operators in Eq. (2) in the $SO(4)$ approach to chiral perturbation theory. This approach is based on the observation that the massless (and Lorentz-symmetric) two-flavor QCD Lagrangian possesses an $SU(2)_L \times SU(2)_R \sim SO(4)$ symmetry under chiral transformations of the quark fields. Moreover, the structure of the baryon and light-meson spectrum implies that the axial part of this symmetry is spontaneously broken, down to its $SO(3)$ isospin subgroup. This happens at the chiral breaking scale: $\Lambda_\chi \simeq 1$ GeV. The pseudoscalar Goldstone bosons associated with this spontaneous symmetry breaking are identified with the pions. The small nonzero mass of the pions originates from the fact that chiral symmetry is not an exact symmetry of the QCD Lagrangian, due to the (small) quark masses.

To be able to construct a Lagrangian which has the correct symmetry properties, one constructs chiral- and gauge-covariant derivatives of the pion and nucleon fields. The covariant derivative for the pion field is given by

$$(D_\mu \boldsymbol{\pi})_a = D^{-1}(\partial_\mu \delta_{ab} + eA_\mu \epsilon_{3ab})\pi_b \,, \tag{5}$$

where $\boldsymbol{\pi}$ is the pion triplet, A_μ is the photon field, $e > 0$ is the proton charge, $D = 1 + \boldsymbol{\pi}^2/F_\pi^2$, $F_\pi \simeq \Lambda_\chi/(2\pi) \simeq 185$ GeV is the pion decay constant, and a, b are isospin indices. For the nucleon field the covariant derivative is

$$\mathcal{D}_\mu N = \left(\partial_\mu + \frac{i}{F_\pi^2}\boldsymbol{\tau}\cdot\boldsymbol{\pi}\times D_\mu\boldsymbol{\pi} + \frac{ie}{2}A_\mu(1+\tau_3)\right) N \,, \tag{6}$$

where $N = (p, n)^T$ is the nucleon doublet and $\boldsymbol{\tau}$ are the Pauli isospin matrices. In the chiral invariant Lagrangian that can now be constructed, pion fields can only appear with one or more derivatives acting on them. Consequently all such operators can be ordered in terms of powers of the small quantity p/Λ_χ, where $p \ll \Lambda_\chi$ is a momentum scale of the order of the pion mass. For the nucleon field on the other hand, the derivative does not give rise to a small quantity. In fact, a time derivative acting on the nucleon field is of order $m_N/\Lambda_\chi = \mathcal{O}(1)$. A way to deal with this is described in Section 4. For now we just consider only the operators with the smallest number of derivatives on either pion or nucleon fields and order all interactions by their chiral index, defined by

$$\Delta = d + f/2 - 2 \,, \tag{7}$$

where d is the number of (covariant) derivatives and f is the number of nucleon fields.

Table 1. The sign that follows from the action of C, P, or T on the Lorentz-violating operators in Eq. (2). As usual $(-1)^\mu = 1$ if $\mu = 0$ and $(-1)^\mu = -1$ otherwise.

	$C_{\mu\nu\rho}$	$D_{\mu\nu\rho}$	$H_{\mu\nu\rho}$
P	$(-1)^\mu(-1)^\nu(-1)^\rho$	$-(-1)^\mu(-1)^\nu(-1)^\rho$	$-(-1)^\mu(-1)^\nu(-1)^\rho$
T	$-(-1)^\mu(-1)^\nu(-1)^\rho$	$-(-1)^\mu(-1)^\nu(-1)^\rho$	$(-1)^\mu(-1)^\nu(-1)^\rho$
C	$+1$	-1	$+1$

Operators that break chiral symmetry can be incorporated in the formalism as well. The way the operators, relevant above 1 GeV, transform under chiral transformations, determines the form of the chiral-breaking operators in the effective Lagrangian below 1 GeV. These operators will generally contain pion fields without derivatives, but are proportional to some small symmetry-breaking parameter. The most important example of this is the breaking of chiral symmetry by the quark masses, which induces the pion mass at low energy.

In a similar fashion we constructed the chiral effective operators corresponding to the Lorentz-violating dimension five quark-gluon operators, given in Eq. (2). The first two operators in Eq. (2) can be split into a chiral invariant part, proportional to $X^+_{\mu\nu\rho} = X^u_{\mu\nu\rho} + X^d_{\mu\nu\rho}$, with $X \in \{C, D\}$, and a chiral noninvariant part, proportional to $X^-_{\mu\nu\rho} = X^u_{\mu\nu\rho} - X^d_{\mu\nu\rho}$. The final operator in Eq. (2) is trivially invariant under chiral transformations, since it does not contain any quark fields.

Using the chiral properties of the operators in Eq. (2), together with their C, P, and T characteristics, given in table 1, it is possible to construct the low-energy effective operators in terms of the relevant degrees of freedom, i.e. the pions and the nucleons (see Ref. [3] for details). The set of lowest-order relativistic operators that is non-redundant under the equations of motion, is given by

$$\begin{aligned}\mathcal{L}_\chi^{\mathrm{LV}} &= \frac{i}{m_N}\bar{N}\left(\tilde{C}^+_{\mu\nu\rho} + \tilde{C}^-_{\mu\nu\rho}\left[\tau_3 - \frac{2}{F_\pi^2 D}\left(\boldsymbol{\pi}^2\tau_3 - \pi_3\boldsymbol{\tau}\cdot\boldsymbol{\pi}\right)\right]\right)\sigma^{\nu\rho}\mathcal{D}^\mu N \\ &+\frac{1}{m_N^2}\tilde{H}_{\mu\nu\rho}\bar{N}\gamma^\mu\gamma^5\mathcal{D}^\nu\mathcal{D}^\rho N + \frac{i}{m_N F_\pi D}\tilde{D}^-_{\mu\nu\rho}\bar{N}(\boldsymbol{\tau}\times\boldsymbol{\pi})_3\sigma^{\nu\rho}\mathcal{D}^\mu N + \\ &+\frac{1}{m_N^2 F_\pi}\breve{D}^+_{\mu\nu\rho\alpha\beta}\bar{N}(\boldsymbol{\tau}\cdot D^\mu\boldsymbol{\pi})\sigma^{\nu\rho}\mathcal{D}^\alpha\mathcal{D}^\beta N + \mathrm{H.c.}\,,\end{aligned} \tag{8}$$

where we included a factor of $1/m_N$ for each covariant nucleon derivative, to keep the time derivatives from spuriously lowering the chiral index of the operators. The parameter $\breve{D}^+$ is given by

$$\breve{D}^+_{\mu\rho\nu\alpha\beta} = \tilde{D}^{+,1}_{\mu\rho\nu}\eta_{\alpha\beta} + \tilde{D}^{+,2}_{\alpha\nu\rho}\eta_{\mu\beta} + \tilde{D}^{+,3}_{\alpha[\beta\rho]}\eta_{\mu\nu}\,, \tag{9}$$

where square brackets mean antisymmetrization over the enclosed indices. In Eqs. (8) and (9), the Lorentz-violating coefficients at the hadronic level are denoted by a tilde. This represents the fact that they are proportional to the Lorentz-violating coefficients in Eq. (2), but additionally contain an unknown low-energy constant, i.e. $\tilde{C}^\pm_{\mu\nu\rho} = c^\pm C^\pm_{\mu\nu\rho}$, $\tilde{D}^-_{\mu\nu\rho} = d^- D^-_{\mu\nu\rho}$, $\tilde{D}^{+,i}_{\mu\nu\rho} = d_i^+ D^+_{\mu\nu\rho}$, and $\tilde{H}^{\mu\nu\rho} = hH^{\mu\nu\rho}$, with $c^\pm$, d^-, d_i^+, and h the low-energy constants. The value of these low-energy constants cannot be determined using symmetry arguments, although chiral symmetry does imply that $d^- = 2c^-$. In principle the values of the low-energy constants can be calculated using lattice QCD, however for the Lorentz-violating operators under consideration this has not been done. Therefore we use naive dimensional analysis (NDA) [17] to get an estimate of the value of the low-energy constants at the order-of-magnitude level. NDA gives that $c^{\pm}, d^{-} = \mathcal{O}(\Lambda_\chi F_\pi)$, $d_i^{+} = \mathcal{O}(F_\pi)$, and $h = \mathcal{O}(\Lambda_\chi^2)$. The operators in Eq. (8) also induce pure-pion operators, however these are all of higher order. Moreover, one can write down additional nucleon-pion operators, that have the correct symmetry properties. However, it can be shown that these are redundant [3].

As expected, we see that the chiral invariant operators (coupled to $C^{+}_{\mu\nu\rho}$ and $H_{\mu\nu\rho}$) only induce nucleon operators without pion fields at lowest order. The chiral noninvariant operators induce kinetic nucleon terms, as well as pion-nucleon interaction terms, which are related by chiral symmetry. Also, in contrast to the $C^{\pm}_{\mu\nu\rho}$ and $H_{\mu\nu\rho}$ parameters, the $D^{\pm}_{\mu\nu\rho}$ coefficient does not introduce any contribution to the nucleon two-point function. This is important, since it turns out that the strongest constraints come from properties of the free nucleon field. Consequently, the bounds on $D^{\pm}_{\mu\nu\rho}$ will be much weaker than those on $C^{\pm}_{\mu\nu\rho}$ and $H_{\mu\nu\rho}$.

4. Heavy baryon chiral perturbation theory

Since an effective field theory in principle contains an infinite number of terms, to be useful there has to be some way to order the effective operators, based on the expected size of their contribution to observables. As mentioned, this ordering is performed based on the chiral index of the operators, defined in Eq. (7). The estimated size of Feynman diagrams is then determined by their chiral order, which, for diagrams with only one baryon in the initial and final state, is given by

$$\nu = 2N_L + 1 + \sum_i \Delta_i \, , \tag{10}$$

where N_L is the number of independent loops and the sum runs over all the interactions that contribute to the relevant diagram. As mentioned, this is based on the assumption that all energy scales in the problem are small compared to the chiral breaking scale, i.e. that we can use the expansion parameter p/Λ_χ. However, when doing loop calculations involving nucleons, the nucleon mass constitutes an extra scale in the problem, which is not small, in fact $m_N/\Lambda_\chi = \mathcal{O}(1)$. Loop calculations performed with dimensional regularization receive contributions from loop momenta of order m_N. This upsets the assumed power counting. This holds for the Lorentz-violating effective operators in Eq. (8) as well. We therefore apply the framework of heavy-baryon chiral perturbation theory [18]. It consists of introducing heavy-nucleon fields with a fixed velocity v, given by

$$N_v = \frac{1+\not{v}}{2} e^{i m_N v^\mu x_\mu} N \, , \tag{11}$$

where $p^\mu = m_N v^\mu + k^\mu$, with k^μ a small residual momentum. Derivatives acting on the heavy-nucleon field will now be proportional to this small residual momentum. Moreover, the propagator of the heavy-nucleon field will no longer contain the nucleon mass, such that the results of loop integrals scale with powers of p/Λ_χ. Conventionally, in heavy-baryon chiral perturbation theory, the operators are written in terms of the nucleon velocity v^μ and the covariant spin vector S^μ, with $S^\mu = \frac{i}{2}\gamma^5\sigma^{\mu\nu}v_\nu$. In the heavy-baryon formalism the Lagrangian in Eq. (8) becomes

$$\begin{aligned}
\mathcal{L}_\chi^{\rm HB} &= 4\left(\epsilon^{\mu\nu\alpha\beta}\tilde{C}^{+}_{\rho\alpha\beta} - \tilde{H}^{\mu\nu}{}_{\rho}\right) v^\rho v_\nu \bar{N} S_\mu N \\
&\quad + 4\epsilon^{\mu\nu\alpha\beta}\tilde{C}^{-}_{\rho\alpha\beta} v_\nu v^\rho \bar{N}\left[\tau_3 - \frac{2}{F_\pi^2 D}\left(\boldsymbol{\pi}^2\tau_3 - \pi_3 \boldsymbol{\tau}\cdot\boldsymbol{\pi}\right)\right] S_\mu N \\
&\quad + \frac{4}{F_\pi D}\epsilon^{\mu\nu\alpha\beta}\tilde{D}^{-}_{\rho\alpha\beta} v^\rho v_\nu \bar{N}(\boldsymbol{\tau}\times\boldsymbol{\pi})_3 S_\mu N + 4\check{D}^{+}_{\mu\nu\rho\alpha\beta}\epsilon^{\nu\rho\lambda\kappa} v_\lambda v^\alpha v^\beta \bar{N}(\boldsymbol{\tau}\cdot D^\mu \boldsymbol{\pi}) S_\kappa N
\end{aligned} \tag{12}$$

IOP Conf. Series: Journal of Physics: Conf. Series **873** (2017) 012009 doi:10.1088/1742-6596/873/1/012009

Notice that $\tilde{C}^{+}_{\mu\nu\rho}$ and $\tilde{H}_{\mu\nu\rho}$ are coupled to an identical operator. However, in principle they can still be experimentally distinguished by using the symmetry properties of their Lorentz indices.

5. Particle and antiparticle Hamiltonian

As we mentioned before, the best constraints on the present set of Lorentz-violating coefficients come from kinetic nucleon terms. They are set by clock-comparison experiments. To analyse the effect of nucleon Lorentz violation on such experiments it is convenient to have a block-diagonalized form of the relevant Hamiltonian. Such a form can be obtained by performing a Foldy-Wouthuysen transformation on the nucleon fields. When the Hamiltonian is in the block-diagonalized form, the equations for the particle and the anti-particle become decoupled.

A comparable decoupling is achieved in the heavy-baryon approach, discussed in the previous section. In that approach a nonrelativistic expansion in p/m_N is employed. Using a Foldy-Wouthuysen transformation, in some cases the diagonalization can be achieved exactly, however in many cases the off-diagonal parts are of some higher order in a small quantity, which can also be p/m_N.

For the kinetic nucleon operators in Eq. (8) one can perform a Foldy-Wouthuysen transformation such that the off-diagonal parts of the Hamiltonian are of higher order in the Lorentz-violating coefficients and in the electromagnetic fields. These higher-order terms can safely be ignored and we obtain a relativistic expression for the block-diagonal Hamiltonian. One subtlety involves the higher-order time derivatives in the operators in Eq. (8). To deal with this complication, we use the approach of Ref. [12], i.e. we first diagonalize the Hamiltonian and then substitute $i\partial^0 \to \sqrt{\vec{p}^2 + m_N^2}$ for the fermion and $i\partial^0 \to -\sqrt{\vec{p}^2 + m_N^2}$ for the antifermion. Any terms that we miss in this way are of higher order in the Lorentz-violating coefficients.

For a particle of species w, with $w \in \{p, n\}$, the resulting Hamiltonian has a 2×2 upper left block $h_{w,+}$, which we associate with the particle, and a 2×2 lower right block $h_{w,-}$, pertaining to the antiparticle. These particle/antiparticle Hamiltonians can be written as

$$h_{w,\pm} = h_{w,0} \pm \delta h_w \, , \tag{13}$$

where $h_{w,0}$ is the conventional particle/antiparticle Hamiltonian, while the Lorentz-violating contribution is given by

$$\delta h_w = -2\gamma \left[\boldsymbol{\sigma} \cdot \bar{\boldsymbol{\xi}}_w - \gamma \boldsymbol{\sigma} \cdot \boldsymbol{\beta} \left(\bar{\xi}^0_w - \frac{\gamma}{\gamma+1} \boldsymbol{\beta} \cdot \bar{\boldsymbol{\xi}}_w \right) \right] \, , \tag{14}$$

where $\boldsymbol{\beta} = \boldsymbol{p}/E$ is the (anti)particle velocity, γ is the relativistic boost factor, and

$$\bar{\xi}^\mu_w = \xi^{\mu\nu\rho}_w \beta_\nu \beta_\rho = \left[\tilde{H}^{\mu\nu\rho} - \epsilon^{\mu\nu\alpha\beta} (\tilde{C}^w)^\rho{}_{\alpha\beta} \right] \beta_\nu \beta_\rho \, , \tag{15}$$

with $\beta = (1, \boldsymbol{\beta})$, $\tilde{C}^p_{\rho\alpha\beta} = \tilde{C}^+_{\rho\alpha\beta} + \tilde{C}^-_{\rho\alpha\beta}$, and $\tilde{C}^n_{\rho\alpha\beta} = \tilde{C}^+_{\rho\alpha\beta} - \tilde{C}^-_{\rho\alpha\beta}$. Comparing Eq. (15) to the expression in Eq. (12), we see that the same combination of $C^{\pm}_{\mu\nu\rho}$ and $H_{\mu\nu\rho}$ coefficients appears. This combination, which is symmetric in ν and ρ, is the one to which free nucleon experiments are sensitive. It has 32 independent components. A subset of these can be constrained using clock-comparison experiments, as we will illustrate in the following.

6. Constraints

In Ref. [3] we derived bounds on $C_{\mu\nu\rho}$ and $H_{\mu\nu\rho}$ from comagnetometer experiments. In finding the particular components to which the experiments are sensitive, we followed the approach of Ref. [19]. We will sketch this approach here briefly and refer to Ref. [3] or [19] for more details.

IOP Conf. Series: Journal of Physics: Conf. Series **873** (2017) 012009 doi:10.1088/1742-6596/873/1/012009

Essentially, one calculates the shift of the rotational energy levels of an atom, ion, or nucleus W, that follows from the Lorentz-violating Hamiltonian described in the previous section. The part of the Hamiltonian for W that is linear in the Lorentz-violating coefficients, is given by

$$h'_W = \sum_w \sum_{N=1}^{N_{w,W}} \delta h_{w,N} \ , \tag{16}$$

where $\delta h_{w,N}$ is the Hamiltonian for the N-th particle of species w, described in the previous section. The second sum runs over all particles of species w that are present in W.

This Hamiltonian induces a shift in the energy levels corresponding to

$$\delta E(F, M_F) = \langle F, M_F | h'_W | F, M_F \rangle \ , \tag{17}$$

where $|F, M_F\rangle$ is the state of W with spin angular momentum F and projection M_F. The Lorentz-violating contribution to a transition frequency $F, M_F \to F', M'_F$ can then be calculated by $\delta\omega = \delta E(F, M_F) - \delta E(F', M'_F)$

What the contribution of different (components of) Lorentz-violating coefficients to Eq. (17) looks like depends on the rotational properties of those coefficients. For the Lorentz-violating coefficients that are presently relevant the Lorentz-violating shift can be written as

$$\delta E(F, M_F) = \widetilde{M}^1_F E^W_1 + \widetilde{M}^2_F E^W_2 + \widetilde{M}^3_F E^W_3 \ , \tag{18}$$

where $\widetilde{M}^n_F$ $(n = 1, 2, 3)$ follow from ratios of Clebsch-Gordan coefficients, while E^W_n represent contributions from spherical tensors of rank n in the original Hamiltonian. Because of the triangle inequality for angular momenta, E^W_n requires W to have an angular momentum of at least $n/2$ to be nonvanishing. In Ref. [3] we calculated E^W_1, E^W_2, and E^W_3 for the Lorentz-violating coefficients under consideration, in the nonrelativistic limit. We found that the dominant contribution is part of E^W_1, which constitutes a dipole contribution to the energy. The resulting energy shift is given by

$$\delta E(F, M_F) = -\frac{2M_F}{F} \sum_w \sum_{N=1}^{N_w} \left[H^{300} + (\tilde{C}^w)^{012} - (\tilde{C}^w)^{021} \right] \langle [\sigma_3]_{w,N} \rangle \ , \tag{19}$$

Here, $\langle [\sigma_3]_{w,N} \rangle$ is an unknown matrix element of the third spin component, defined in the 'stretched' state $|F, F\rangle$ (the 3-axis is the quantization axis). One has to adopt some nuclear-structure model to calculate the values of such matrix elements. Besides Eq. (19), there are other Lorentz-violating contributions to the energy shift, which contain different linear combinations of the Lorentz-violating coefficients. However, these are suppressed by factors of p^2/m_N^2 with respect to the one in Eq. (19).

The best bounds on the Lorentz-violating dipole contribution in Eq. (19) come from a ^{3}He/^{129}Xe comagnetometer [20] for the X and Y direction and from a ^{199}Hg/^{133}Cs comagnetometer [21] for the Z direction. Here the X, Y, and Z direction are defined in the Sun-centered inertial reference frame [2]. These experiments allow us to put bounds on components of $\xi^{\mu\nu\rho}$ in the order of 10^{-33}–10^{-30} GeV^{-1}. Additional bounds can be derived by considering effects due to the nonzero velocity of the Earth. Such effects are ignored in most cases, since they are suppressed by powers of the Earth's velocity with respect to the Sun, i.e. powers of $\beta_\oplus \simeq 10^{-4}$. However, due to the high precision of the comagnetometer experiments, in this case relevant bounds can be derived from such effects. An analysis of such boost-dependent signals was performed in Ref. [22].

Using the experiments and analyses of Refs. [20, 21, 22], we are able to bound 7 linear combinations of components of the Lorentz-violating tensors [3]. These linear combinations

Table 2. Order-of-magnitude bounds on the LV tensor components defined in Eqs. (2).

Tensor component	Limit in GeV^{-1}
H_{TTX}, H_{TTY}, $C^q_{T[XZ]}$, $C^q_{T[YZ]}$	$< 10^{-33}$
H_{TTZ}	$< 10^{-30}$
$C^q_{T[XY]}$	$< 10^{-29}$
H_{TXX}, H_{TXY}, H_{TYY}	$< 10^{-28}$
H_{TTT}, H_{TXZ}, H_{TYZ}, $C^q_{T[TZ]}$, $C^q_{X[YZ]}$, $C^q_{Y[XZ]}$, $C^q_{Y[YZ]}$	$< 10^{-27}$
$C^q_{T[TX]}$, $C^q_{T[TY]}$, $C^q_{T[TZ]}$, $C^q_{X[XZ]}$, $C^q_{Z[XZ]}$, $C^q_{Z[YZ]}$	$< 10^{-26}$

involve the low-energy constants described below Eq. (8). Therefore, bounds on the actual coefficients are order-of-magnitude bounds until we can do better than NDA estimates for the low-energy constants. Moreover, we assume that there are no large (unexpected) cancellations between the different coefficients, such that the 7 experimental bounds can be interpreted as bounds on the separate components of the Lorentz-violating tensors. The resulting order-of-magnitude limits are collected in table 2.

7. Spin-precession equation

An alternative way of bounding Lorentz-violating coefficients such as the ones in Eq. (8) is by considering storage-ring experiments that measure spin-precession frequencies. To interpret such experiments, we derived relevant Lorentz-violating spin-precession equation that corresponds to Eq. (13). Using the Heisenberg equation of motion to derive the time evolution of the spin operator, we find that the Lorentz-violating contribution to the spin-precession frequency $\boldsymbol{\omega}_s$ of particle species w is given by

$$\mp\frac{\delta\boldsymbol{\omega}_{s,w}}{4} = -\gamma\bar{\boldsymbol{\xi}}_w + \gamma^2\boldsymbol{\beta}\left(\bar{\xi}^0_w - \frac{\gamma}{\gamma+1}\boldsymbol{\beta}\cdot\bar{\boldsymbol{\xi}}_w\right), \tag{20}$$

where γ and $\boldsymbol{\beta}$ are the Lorentz boost factor and particle velocity, respectively. The Lorentz-violating coefficient $\bar{\xi}^\mu$ is defined in Eq. (15) and the upper (lower) sign applies to particles (antiparticles).

In Ref. [3], we used Eq. (20) to derive a bound $|\xi_p^{ZTT}| < 2.7\times10^{-21}$ GeV by comparing proton and antiproton spin-precession frequencies. These were measured in two double Penning-trap experiments, performed in Mainz [23] and at CERN [24]. Although the sensitivity of these experiments is much lower than the comagnetometer experiments, described in the previous section, they have the added advantage that they do not need to rely on a nuclear model to determine the nuclear matrix elements.

Also, storage-ring experiments have the potential to bound additional (boost-dependent) components of the Lorentz-violating coefficients. In particular an experiment with deuterons is expected to be benificial, since it will have sensitivity to the $D^\pm_{\mu\nu\rho}$ coefficients that do not contribute to single-nucleon observables. We will discuss this in more detail in the next section.

8. The $D^\pm_{\mu\nu\rho}$ parameter

As can be seen from Eq. (8), the Lorentz-violating coefficient $D^\pm_{\mu\nu\rho}$ does not contribute to any nucleon two-point function, due to its symmetry properties. As such it does not contribute to free nucleon properties and thus cannot be bounded to the same precission as $C^\pm_{\mu\nu\rho}$ and $H_{\mu\nu\rho}$, using comagnetometer experiments.

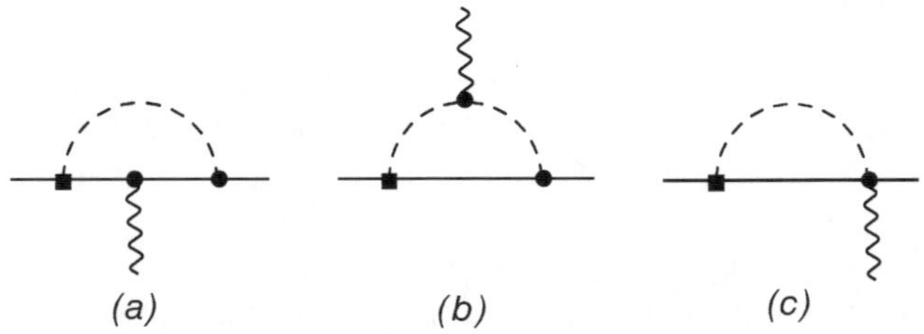

Figure 1. Contribution of $D^-_{\mu\nu\rho}$ to the nucleon electromagnetic form factor. The squares denote the pion-nucleon vertex from Eq. (12), while the circles denote leading-order vertices from the usual chiral perturbation Lagrangian.

In Ref. [3], we therefore calculated the pion-loop contribution of $D^\pm_{\mu\nu\rho}$ to the nucleon electromagnetic form factor (see Fig. 1). The result for the Lorentz-violating current that follows from the loop calculation is given by

$$I^\mu(q) = \left(F^+_{\nu\rho\sigma}(Q^2) + F^-_{\nu\rho\sigma}(Q^2)\tau_3\right) v^\nu v^\mu q^{[\sigma} S^{\rho]} + \cdots , \tag{21}$$

where the dots represent terms that are not relevant for on-shell photons. The function $F^+_{\nu\rho\sigma}(Q^2)$ ($F^-_{\nu\rho\sigma}(Q^2)$) represents a isoscalar (isovector) form factor, proportional to the Lorentz-violating parameters. For on-shell photons, the loop contribution to the isovector form-factor $F^-_{\nu\rho\sigma}$ turns out to vanish, while the loop calculation does give a nonzero isoscalar form factor. After renormalization, the loop contribution to the form factors can be written as

$$F^+_{\nu\rho\sigma}(Q^2=0) = \tilde{D}^-_{\nu\rho\sigma} \frac{8eg_A}{(2\pi F_\pi)^2} \ln\frac{m_N^2}{m_\pi^2} + \cdots , \tag{22}$$

where the dots represent contributions from other low-energy effective operators. We see that the chiral loop contribution is slightly enhanced by the factor $\ln\frac{m_N^2}{m_\pi^2}$. This result exemplifies how chiral perturbation theory allows for precision calculations for Lorentz-violating hadronic observables.

The problem with using the electromagnetic nucleon form factor to bound Lorentz-violating coefficients is the fact that the presence of physical electromagnetic fields suppresses the effects by many order of magnitudes. The suppression factor with respect to the dominant Lorentz-violating effects due to C and H will be in the order of $eF^{\mu\nu}/\Lambda_\chi^2$. For a magnetic field of 1 tesla this is about 10^{-16}. We thus estimate that the bounds on $D^-_{\mu\nu\rho}$ from comagnetometer experiments are at best in the order of $\mathcal{O}(10^{-17})$ GeV.

A more promising option is the contribution of $D^\pm_{\mu\nu\rho}$ to nucleon-nucleon interactions. Although it is not possible to write a two-point function for spin-$\frac{1}{2}$ particles involving $D^\pm_{\mu\nu\rho}$, this is not necessarily true for higher-spin particles, such as the deuteron. In Ref. [3] we thus calculated the Lorentz-violating nucleon-nucleon potential that follows from one-pion exchange between nucleons in the context of chiral EFT (see Refs. [25, 26, 27] for reviews). In the relevant pion-exchange diagram, one of the two interaction vertices is a Lorentz-violating vertex from Eq. (12). The resulting potential is given by

$$\begin{aligned} V_{\rm LV} &= -\left(\epsilon^{ijk}\tilde{D}^-_{0ij}\right)\frac{2ig_A}{F_\pi^2}(\boldsymbol{\tau}_1\times\boldsymbol{\tau}_2)_3\frac{(\boldsymbol{\sigma}_1\cdot\boldsymbol{k})\sigma_2^k + (\boldsymbol{\sigma}_2\cdot\boldsymbol{k})\sigma_1^k}{\boldsymbol{k}^2+m_\pi^2} \\ &\quad -\left(\epsilon^{jkl}\breve{D}^+_{ijk00}\right)\frac{4g_A}{F_\pi}\boldsymbol{\tau}_1\cdot\boldsymbol{\tau}_2\frac{(\sigma_1^l\sigma_2^m + \sigma_1^m\sigma_2^l)k^ik^m}{\boldsymbol{k}^2+m_\pi^2} , \end{aligned} \tag{23}$$

where $\boldsymbol{\sigma}_{1,2}$ ($\boldsymbol{\tau}_{1,2}$) are the spin (isospin) operators of nucleon 1 and 2 and the momentum transfer $\boldsymbol{k} = \boldsymbol{p} - \boldsymbol{p}'$ flows from nucleon 1 to nucleon 2; $\boldsymbol{p}$ and $\boldsymbol{p}'$ are the relative momenta of the incoming and outgoing nucleon pair in the center-of-mass frame. The Latin indices $i, j, k, \ldots$ denote spatial directions. There are other contributions from the D-parameter to the potential, which can be found in Ref. [3].

The detailed study of the effects of Eq. (23) are left for future work. However, it seems clear that it will have an effect on for example the spin-precession frequency of the deuteron as well as on other light-nucleon systems or clock-comparison experiments. For such systems a bound on $D^{\pm}_{\mu\nu\rho}$ that is a significant improvement over the current bounds seems feasible.

9. Conclusion and outlook

In Ref. [3], we started the program of deriving the hadronic low-energy effective Lagrangian for Lorentz-violation, based on chiral-symmetry considerations and induced by operators of the Lorentz-violating standard-model extension. Here, we limited the discussion to a selected set of dimension-five quark-gluon operators, which obey a set of UV-safety conditions [10]. This work has been extended to several different operators in other work [4, 5].

We applied a Foldy-Wouthuysen transformation to the low-energy effective Lagrangian to obtain a decoupled particle/antiparticle Hamiltonian and derived the resulting shift in the rotational energy levels of atoms. Using these results we derived strict bounds on components of two of the three Lorentz-violating coefficients, from comagnetometer experiments [20, 21]. The resulting limits are collected in table 2. Comparing these to the expected size of the dimension-five operators, i.e. $1/m_P \simeq \mathcal{O}(10^{-19})\,\mathrm{GeV}^{-1}$, we see that many of the components of $C_{\mu\nu\rho}$ and $H_{\mu\nu\rho}$ have bounds that are well beyond the Planck scale.

An important result is that the $D_{\mu\nu\rho}$ coefficients do not contribute to the nucleon two-point function and therefore cannot be limited by the same methods that provided the stringent bounds on $C_{\mu\nu\rho}$ and $H_{\mu\nu\rho}$. In Ref. [5] we found that this also holds for a particular pure-gluon parameter in the mSME and we expect there will be more such parameters in the SME Lagrangian.

We calculated pion-loop contributions to the nucleon electromagnetic form factor in terms of the $D_{\mu\nu\rho}$ coefficient. However, due to the supression that follows from the necessity of the presence of a physical electromagnetic field, the resulting bounds on components of $D_{\mu\nu\rho}$ are in the order of $\mathcal{O}(10^{-17})\,\mathrm{GeV}^{-1}$ at best. This is not yet at the expected level of $1/m_P$ and improvements of this bound are thus desirable.

Such improvement might come from considering effects of the one-pion exchange nucleon-nucleon potential in Eq. (23). The effects of the NN interaction in nuclei could provide much better bounds than the electromagnetic interactions. Especially the spin precession of the deuteron is promising in this respect and for example storage-ring experiments might be able to place stringent constraints [28].

Acknowledgments

The author acknowledges the financial support of the Portuguese Foundation for Science and Technology (FCT) under grant SFRH/BPD/101403/2014 and program POPH/FSE.

References

[1] Colladay D and Kostelecký V A 1997 *Phys. Rev.* D **55** 6760
Colladay D and Kostelecký V A 1998 *Phys. Rev.* **58** 116002
[2] Kostelecký V A and Russell N 2016 *Preprint 0801.0287v10 [hep-ph]*
[3] Noordmans J P, de Vries J and Timmermans R G E 2016 *Phys. Rev.* C **94** 025502
[4] Kamand R, Altschul B and Schindler M R 2017 *Phys. Rev.* D **95** 056005
[5] Noordmans J P 2017 *Preprint 1701.04334 [hep-ph]*
[6] Carroll S M, Field G B and Jackiw W 1990 *Phys. Rev.* D **41** 1231
[7] Coleman S R and Glashow S L 1999 *Phys. Rev.* D **59** 116008

IOP Conf. Series: Journal of Physics: Conf. Series **873** (2017) 012009 doi:10.1088/1742-6596/873/1/012009

[8] Kostelecký V A 2008 *Preprint 0802.0581 [gr-qc]*
[9] Groot Nibbelink S and Pospelov M 2005 *Phys. Rev. Lett.* **94** 081601
Bolokhov P A, Groot Nibbelink S and Pospelov M 2005 *Phys. Rev.* D **72** 015013
Mattingly D 2008 *Preprint 0802.1561 [gr-qc]*
[10] Bolokhov P A and Pospelov M 2008 *Phys. Rev.* D **77** 025022
[11] Ding Y and Kostelecký V A 2016 *Phys. Rev.* D **94** 056008
[12] Kostelecký V A and Mewes M 2009 *Phys. Rev.* D **80** 015020
Kostelecký V A and Mewes M 2012 *Phys. Rev.* D **85** 096005
Kostelecký V A and Mewes M 2013 *Phys. Rev.* D **88** 096006
[13] Weinberg S 1979 *Physica* A **96** 327
[14] Gasser J and Leutwyler H 1984 *Ann. Phys.* **158** 142
[15] Weinberg S 1996, *The Quantum Theory of Fields* vol 2 (Cambridge: Cambridge University Press)
[16] Bernard V and Meißner U -G 2007 *Ann. Rev. Nucl. Part. Sci.* **57** 33
Scherer S and Schindler M R 2012 *Lect. Notes Phys.* **830** 1
[17] Manohar A V and Georgi H 1984 *Nucl. Phys.* B **234** 189
Georgi H and Randall L 1986 *Nucl. Phys.* B **276** 241
[18] Jenkins E E and Manohar A V 1991 *Phys. Lett.* B **255** 558
Bernard V, Kaiser N, Kambor J and Meißner U -G 1992 *Nucl. Phys.* B **388** 315
[19] Kostelecký V A and Lane C D 1999 *Phys. Rev.* D **60** 116010
[20] Allmendinger F *et al.* 2014 *Phys. Rev. Lett.* **112** 110801
[21] Peck S K *et al.* 2012 *Phys. Rev.* A **86** 012109
[22] Canè F *et al.* 2004 *Phys. Rev. Lett.* **93** 230801
[23] Mooser A *et al.* 2014 *Nature* **509** 596
[24] DiSciacca J *et al.* 2013 *Phys. Rev. Lett.* **110** 130801
[25] Bedaque P F and van Kolck U 2002 *Ann. Rev. Nucl. Part. Sci.* **52** 339
[26] Epelbaum E, Hammer H W and Meißner U -G 2009 *Rev. Mod. Phys.* **81** 1773
[27] de Vries J and Meißner U -G 2016 *Int. J. Mod. Phys.* E **25** 1641008
[28] Eversmann D *et al.* (JEDI Collaboration) 2015 *Phys. Rev. Lett.* **115** 094801

The MoEDAL experiment at the LHC: status and results

Vasiliki A Mitsou
on behalf of the MoEDAL Collaboration
Instituto de Física Corpuscular (IFIC), CSIC – Universitat de València, C/ Catedrático José Beltrán 2, E-46980 Paterna (Valencia), Spain
E-mail: vasiliki.mitsou@ific.uv.es

Abstract. The MoEDAL experiment at the LHC is optimised to detect highly ionising particles such as magnetic monopoles, dyons and (multiply) electrically charged stable massive particles predicted in a number of theoretical scenarios. MoEDAL, deployed in the LHCb cavern, combines passive nuclear track detectors with magnetic monopole trapping volumes (MMTs), while spallation-product backgrounds are being monitored with an array of MediPix pixel detectors. An introduction to the detector concept and its physics reach, complementary to that of the large general purpose LHC experiments ATLAS and CMS, will be given. Emphasis is given to the recent MoEDAL results at 13 TeV, where the null results from a search for magnetic monopoles in MMTs exposed in 2015 LHC collisions set the world-best limits on particles with magnetic charges more than 1.5 Dirac charge. The potential to search for heavy, long-lived supersymmetric electrically-charged particles is also discussed.

1. Introduction

MoEDAL (Monopole and Exotics Detector at the LHC) [1, 2], the 7th experiment at the Large Hadron Collider (LHC) [3], was approved by the CERN Research Board in 2010. It is designed to search for manifestations of new physics through highly-ionising particles in a manner complementary to ATLAS and CMS [4]. The most important motivation for the MoEDAL experiment is to pursue the quest for magnetic monopoles and dyons at LHC energies. Nonetheless the experiment is also designed to search for any massive, stable or long-lived, slow-moving particles [5] with single or multiple electric charges arising in many scenarios of physics beyond the Standard Model (SM). A selection of the physics goals and their relevance to the MoEDAL experiment are described here and elsewhere [6]. For an extended and detailed account of the MoEDAL discovery potential, the reader is referred to the *MoEDAL Physics Review* [7]. Emphasis is given here on recent MoEDAL results, based on the exposure of magnetic monopole trapping volumes to 7-TeV and 8-TeV proton-proton collisions.

The structure of this paper is as follows. Section 2 provides a brief description of the MoEDAL detector. Magnetic monopoles and monopolia are briefly discussed in Section 3, whilst Section 4 presents the MoEDAL results on monopole searches. Section 5.1 is dedicated to supersymmetric models predicting massive (meta)stable states. Scenarios with doubly-charged Higgs bosons and their observability in MoEDAL are highlighted in Section 5.2. Highly-ionising exotic structures in models with extra spatial dimensions, namely microscopic black holes and D-matter, relevant to MoEDAL are briefly mentioned in Sections 5.3 and Sections 5.4, respectively. The paper concludes with a summary and an outlook in Section 6.

2. The MoEDAL detector

The MoEDAL detector [2] is deployed around the intersection region at Point 8 of the LHC in the LHCb experiment Vertex Locator (VELO) [8] cavern. A three-dimensional depiction of the MoEDAL experiment is presented in Fig. 1. It is a unique and largely passive LHC detector comprised of four sub-detector systems.

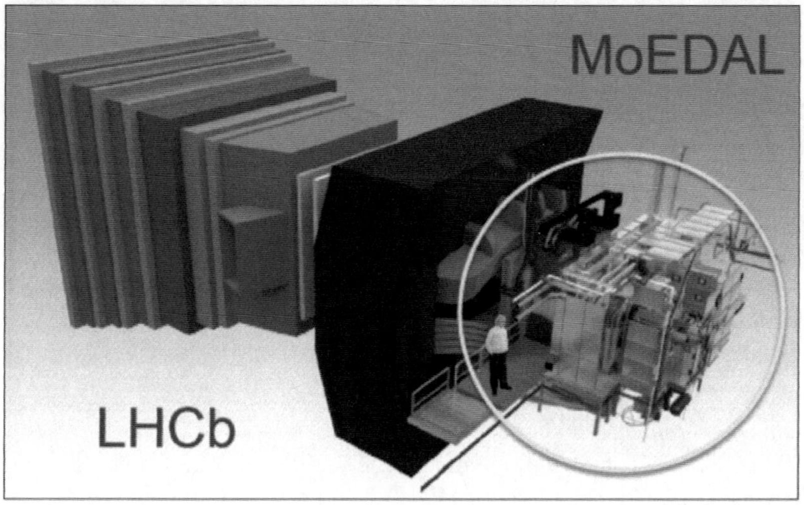

Figure 1. A three-dimensional schematic view of the MoEDAL detector (on the right) around the LHCb VELO region at Point 8 of the LHC.

2.1. Low-threshold nuclear track detectors

The main sub-detector system is made of a large array of CR39®, Makrofol® and Lexan® nuclear track detector (NTD) stacks surrounding the intersection area. The passage of a highly-ionising particle through the plastic detector is marked by an invisible damage zone along the trajectory. The damage zone is revealed as a cone-shaped etch-pit when the plastic detector is etched using a hot sodium hydroxide solution. Then the sheets of plastics are scanned looking for aligned etch pits in multiple sheets. The MoEDAL NTDs have a threshold of $Z/\beta \sim 5$, where Z is the charge and $\beta = v/c$ the velocity of the incident particle. In proton-proton collision running, the only source of known particles that are highly ionising enough to leave a track in MoEDAL NTDs are spallation products with range that is typically much less than the thickness of one sheet of the NTD stack. In that case the ionising signature will be that of a very low-energy electrically-charged *stopped* particle. This signature is distinct to that of a *penetrating* electrically or magnetically charged particle that will usually traverse every sheet in a MoEDAL NTD stack, accurately demarcating a track that points back to the collision point with a resolution of ~ 1 cm. The part of the Run-2 NTD deployment which rests on top of the LHCb VELO is visible in Fig. 2. This is the closest possible location to the interaction point and represents a novelty of this run with respect to earlier installations during Run-1.

Figure 2. Part of the Run 2 NTD deployment on top of the LHCb VELO.

Figure 3. The VHCC between RICH1 and TT installed for Run 2.

2.2. Very high-charge catcher

Another new feature of the Run-2 deployment is the installation of a high-threshold NTD array ($Z/\beta \sim 50$): the Very High Charge Catcher (VHCC). The VHCC sub-detector, consisting of two flexible low-mass stacks of Makrofol® in an aluminium foil envelope, is deployed in the forward acceptance of the LHCb experiment between the LHCb RICH1 detector and the Trigger Tracker (TT), as shown in Fig. 3. It is the only NTD (partly) covering the forward region, adding only $\sim 0.5\%$ to the LHCb material budget while enhancing considerably the overall geometrical coverage of MoEDAL NTDs.

2.3. Magnetic trappers

A unique feature of the MoEDAL detector is the use of paramagnetic magnetic monopole trappers (MMTs) to capture electrically- and magnetically-charged highly-ionising particles. Such volumes installed in IP8 for the 2015 proton-proton collisions is shown in Fig. 4. The aluminium absorbers of MMTs are subject to an analysis looking for magnetically-charged particles at a remote SQUID magnetometer facility [9, 10]. The search for the decays of long-lived electrically charged particles that are stopped in the trapping detectors will subsequently be carried out at a remote underground facility.

A trapping detector prototype was exposed to 8 TeV proton-proton collisions for an integrated luminosity of 0.75 fb^{-1} in 2012. It comprised an aluminium volume consisting of 11 boxes each containing 18 cylindrical rods of 60 cm length and 2.5 cm diameter. For the 2015 run at 13 TeV, the MMT was upgraded to an array consisting of 672 square aluminium rods with dimension $19 \times 2.5 \times 2.5$ cm^3 for a total mass of 222 kg in 14 stacked boxes that were placed 1.62 m from the IP8 LHC interaction point under the beam pipe on the side opposite to the LHCb detector. The results for both aforementioned configurations and energies, interpreted in terms of monopole mass and magnetic charge, are presented in Section 4.

Figure 4. Deployment of the MMT for the LHC Run 2.

Figure 5. Run 2 deployment of TimePix chips in MoEDAL.

2.4. TimePix radiation monitors

The only non-passive MoEDAL sub-detector system comprises an array of TimePix pixel device arrays (256 × 256 square pixels with a pitch of 55 μm) distributed throughout the MoEDAL cavern at IP8, forming a real-time radiation monitoring system of highly-ionising beam-related backgrounds. A photo of its readout setup for the 2015 installations is shown in Fig. 5. Each pixel of the innovative TimePix chip comprises a preamplifier, a discriminator with threshold adjustment, synchronisation logic and a 14-bit counter. The operation of TimePix in time-over-threshold mode allows a 3D mapping of the charge spreading effect in the whole volume of the silicon sensor, thus differentiating between different types of particles species from mixed radiation fields and measuring their energy deposition [11].

3. Magnetic monopoles

The MoEDAL detector is designed to fully exploit the energy-loss mechanisms of magnetically charged particles [12–15] in order to optimise its potential to discover these messengers of new physics. There are various theoretical scenarios in which magnetic charge would be produced at the LHC [7]: (light) 't Hooft-Polyakov monopoles [14, 16], electroweak monopoles [17–19], global monopoles [20–24] and monopolium [13, 25–27]. Magnetic monopoles that carry a non-zero magnetic charge and dyons possessing both magnetic and electric charge are among the most fascinating hypothetical particles. Even though there is no generally acknowledged empirical evidence for their existence, there are strong theoretical reasons to believe that they do exist, and they are predicted by many theories including grand unified theories and superstring theory [28, 29].

The theoretical motivation behind the introduction of magnetic monopoles is the symmetrisation of the Maxwell's equations and the explanation of the charge quantisation [12].

Dirac showed that the mere existence of a monopole in the universe could offer an explanation of the discrete nature of the electric charge, leading to the Dirac Quantisation Condition (DQC),

$$\alpha\, g = \frac{N}{2} e, \quad N = 1, 2, ..., \tag{1}$$

where e is the electron charge, $\alpha = \frac{e^2}{4\pi\hbar\, c\varepsilon_0} = \frac{1}{137}$ is the fine structure constant (at zero energy, as appropriate to the fact that the quantisation condition of Dirac pertains to long (infrared) distances from the centre of the monopole), ε_0 is the vacuum permittivity, and g is the monopole magnetic charge. In Dirac's formulation, magnetic monopoles are assumed to exist as point-like particles and quantum mechanical consistency conditions lead to Eq. (1), establishing the value of their magnetic charge. Although monopoles symmetrise Maxwell's equations in form, there is a numerical asymmetry arising from the DQC, namely that the basic magnetic charge is much larger than the smallest electric charge. A magnetic monopole with a single Dirac charge ($g_{\rm D}$) has an equivalent electric charge of $\beta(137e/2)$. Thus for a relativistic monopole the energy loss is around 4,700 times (68.5^2) that of a minimum-ionising electrically-charged particle. The monopole mass remains a free parameter of the theory.

A possible explanation for the lack of experimental confirmation of monopoles is Dirac's proposal [12, 13, 25] that monopoles are not seen freely because they form a bound state called *monopolium* [26, 27, 30] being confined by strong magnetic forces. Monopolium is a neutral state, hence it is difficult to detect directly at a collider detector, although its decay into two photons would give a rather clear signal for the ATLAS and CMS detectors [31, 32], which however would not be visible in the MoEDAL detector. Nevertheless according to a novel proposal [33], the LHC radiation detector systems can be used to turn the LHC itself into a new physics search machine by detecting final-state protons $pp \to pXp$ exiting the LHC beam vacuum chamber at locations determined by their fractional momentum losses. Such technique would be appealing for detecting monopolia. Furthermore the monopolium might break up in the medium of MoEDAL into highly-ionising dyons, which subsequently can be detected in MoEDAL [7]. Moreover its decay via photon emission would produce a peculiar trajectory in the medium, should the decaying states are also magnetic multipoles [7].

4. Searches for monopoles in MoEDAL

The high ionisation of slow-moving magnetic monopoles and dyons, implies quite characteristic trajectories when such particles interact with the MoEDAL NTDs, which can be revealed during the etching process [2,7]. In addition, the high magnetic charge of a monopole (which is expected to be at least one Dirac charge $g_D = 68.5e$ (*cf.* Eq. (1)) implies a strong magnetic dipole moment, which in turn may result in a strong binding of the monopole with the $^{27}_{13}$Al nuclei of the aluminium MoEDAL MMTs. In such a case, the presence of a monopole trapped in an aluminium bar of an MMT would de detected through the existence of a persistent current, defined as the difference between the currents in the SQUID of a magnetometer before and after the passage of the bar through the sensing coil.

In the context of MoEDAL searches, two configurations of MMTs have been used at two different LHC c.m. energies as described in Section 2.3. No magnetic charge exceeding $0.5g_{\rm D}$ was detected in any of the exposed samples when passed through the ETH Zurich SQUID facility, allowing limits to be placed on monopole production. Model-independent cross-section limits have been obtained in fiducial regions of monopole energy and direction for $1g_{\rm D} \leq |g| \leq 6g_{\rm D}$ with the 8-TeV analysis [34]. Model-dependent cross-section limits are obtained for Drell-Yan (DY) pair production of spin-1/2 and spin-0 monopoles for $1g_{\rm D} \leq |g| \leq 5g_{\rm D}$ at 13 TeV [35], as shown in Fig. 6. Caution, however, should be exerted here in the sense that the non-perturbative nature of the large magnetic Dirac charge of the monopole invalidate any perturbative treatment based on Drell-Yan calculations of the pertinent cross sections and hence any result based on the latter is only indicative, due to the lack of any other concrete theoretical treatment.

The weaker limits for $|g| = g_{\rm D}$ displayed in Fig. 6 when compared to higher charges are mostly due to loss of acceptance from monopoles punching through the trapping volume. For higher charges, monopoles ranging out before reaching the trapping volume decrease the acceptance for DY monopoles with increasing charge and reaches below 0.1% for a charge of $6g_{\rm D}$. The spin dependence is solely due to the different event kinematics: more central and more energetic monopoles for spin 0.

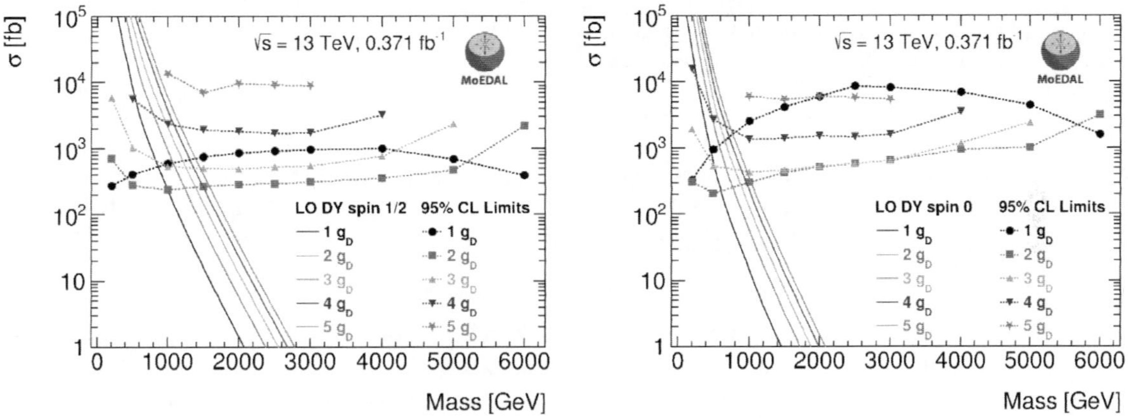

Figure 6. Cross-section upper limits at 95% confidence level for DY monopole production as a function of mass for spin-1/2 (left) and for spin-0 monopoles (right). The various line styles correspond to different monopole charges. The solid lines represent DY cross-section calculations at leading order. From Ref. [35].

Under the assumption of Drell-Yan cross sections, mass limits are derived for $g_{\rm D} \leq |g| \leq 4g_{\rm D}$ at the LHC, complementing previous results from ATLAS Collaboration [36, 37], which placed limits for monopoles with magnetic charge $|g| \leq 1.5g_{\rm D}$ (c.f. Fig. 7). The ATLAS bounds are better that the MoEDAL ones for $|g| = 1g_{\rm D}$ due to the higher luminosity delivered in ATLAS and the loss of acceptance in MoEDAL for small magnetic charges. On the other hand, higher charges are difficult to be probed in ATLAS due to the limitations of the electromagnetic-calorimeter-based level-1 trigger deployed for such searches. A comparison of the limits on monopole production cross sections set by other colliders with those set by MoEDAL is presented in Ref. [29], while general limits including searches in cosmic radiation are reviewed in Ref. [38].

5. Beyond magnetic monopoles

5.1. Electrically-charged long-lived particles in supersymmetry

Supersymmetry (SUSY) is an extension of the Standard Model which assigns to each SM field a superpartner field with a spin differing by a half unit. SUSY provides elegant solutions to several open issues in the SM, such as the hierarchy problem, the identity of dark matter, and the grand unification. SUSY scenarios propose a number of massive slowly moving electrically charged particles. If they are sufficiently long-lived to travel a distance of at least $\mathcal{O}(1{\rm m})$ before decaying and their $Z/\beta \gtrsim 0.5$, then they will be detected in the MoEDAL NTDs. No highly-charged particles are expected in such a theory, but there are several scenarios in which supersymmetry may yield massive, long-lived particles that could have electric charges ± 1, potentially detectable in MoEDAL if they are produced with low velocities ($\beta \lesssim 0.2$) .

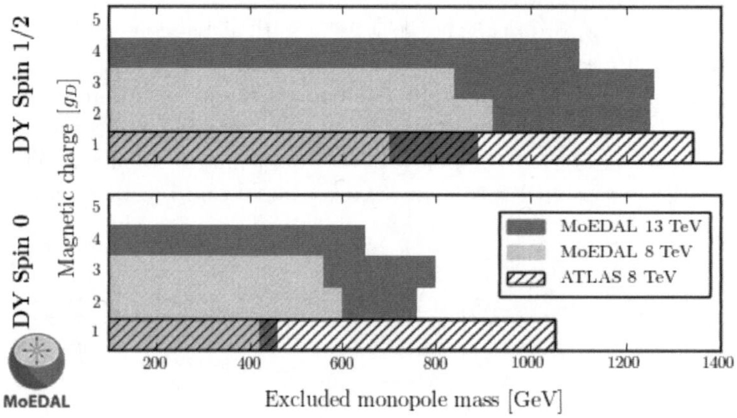

Figure 7. Excluded monopole masses for DY production for spin-1/2 (top) and spin-0 (bottom) monopoles. The MoEDAL results obtained at 8 TeV (yellow, light grey) [34] and 13 TeV (red, dark grey) [35] are superimposed on the ATLAS 8-TeV limits (hatched area) [37].

The lightest supersymmetric particle (LSP) is stable in models where R parity is conserved. The LSP should have no strong or electromagnetic interactions, for otherwise it would bind to conventional matter and be detectable in anomalous heavy nuclei [39]. Possible weakly-interacting neutral candidates in the Minimal Supersymmetric Standard Model (MSSM) include the sneutrino, which has been excluded by LEP and direct searches, the lightest neutralino $\tilde{\chi}_1^0$ (a mixture of spartners of the Z, H and γ) and the gravitino $\tilde{G}$.

5.1.1. Supersymmetric scenarios with R-parity violation Several scenarios featuring metastable charged sparticles might be detectable in MoEDAL. One such scenario is that R parity *may not be exact*, since there is no exact local symmetry associated with either L or B, and hence no fundamental reason why they should be conserved. One could consider various ways in which L and/or B could be violated in such a way that R is violated, as represented by the following superpotential terms:

$$W_{RV} = \lambda''_{ijk}\bar{U}_i\bar{D}_j\bar{D}_k + \lambda'_{ijk}L_iQ_j\bar{D}_k + \lambda_{ijk}L_iL_j\bar{E}_k + \mu_iL_iH, \tag{2}$$

where $Q_i, \bar{U}_i, \bar{D}_i, L_i$ and $\bar{E}_i$ denote chiral superfields corresponding to quark doublets, antiquarks, lepton doublets and antileptons, respectively, with i, j, k generation indices. The simultaneous presence of terms of the first and third type in Eq. (2), namely λ and λ'', is severely restricted by lower limits on the proton lifetime, but other combinations are less restricted. The trilinear couplings in Eq. (2) generate sparticle decays such as $\tilde{q} \to \bar{q}\bar{q}$ or $q\ell$, or $\tilde{\ell} \to \ell\ell$, whereas the bilinear couplings in Eq. (2) generate Higgs-slepton mixing and thereby also $\tilde{q} \to q\ell$ and $\tilde{\ell} \to \ell\ell$ decays [40]. For a nominal sparticle mass ~ 1 TeV, the lifetime for such decays would exceed a few nanoseconds for $\lambda, \lambda', \lambda'' < 10^{-8}$.

If R parity is broken, the LSP would be unstable, and might be charged and/or coloured. In the former case, it might be detectable directly at the LHC as a massive slowly-moving charged particle. In the latter case, the LSP would bind with light quarks and/or gluons to make colour-singlet states, the so-called *R-hadrons*, and any charged state could again be detectable as a massive slowly-moving charged particle. If $\lambda \neq 0$, the prospective experimental signature would be similar to a stau next-to-lightest sparticle (NLSP) case to be discussed later. On the other hand, if λ' or $\lambda'' \neq 0$, the prospective experimental signature would be similar to a stop NLSP case, yielding the possibility of charge-changing interactions while passing through matter. This could yield a metastable charged particle, created whilst passing through the material surrounding the intersection point, that would be detected by MoEDAL.

5.1.2. Metastable lepton NLSP in the CMSSM with a neutralino LSP However, even if R parity *is* exact, the NLSP may be long lived. This would occur, for example, if the LSP is the gravitino, or if the mass difference between the NLSP and the neutralino LSP is small, offering more scenarios for long-lived charged sparticles. In *neutralino dark matter* scenarios based on the constrained MSSM (CMSSM), for instance, the most natural candidate for the NLSP is the lighter stau slepton $\tilde{\tau}_1$ [41], which could be long lived if $m_{\tilde{\tau}_1} - m_{\tilde{\chi}_1^0}$ is small. There are several regions of the CMSSM parameter space that are compatible with the constraints imposed by unsuccessful searches for sparticles at the LHC, as well as the discovered Higgs boson mass. These include a strip in the focus-point region where the relic density of the LSP is brought down into the range allowed by cosmology because of its relatively large Higgsino component, a region where the relic density is controlled by rapid annihilation through direct-channel heavy Higgs resonances, and a strip where the relic LSP density is reduced by coannihilations with near-degenerate staus and other sleptons. It was found in a global analysis that the two latter possibilities are favoured [42].

In the coannihilation region of the CMSSM, the lighter $\tilde{\tau}_1$ is expected to be the lightest slepton [41], and the $\tilde{\tau}_1 - \tilde{\chi}_1^0$ mass difference may well be smaller than m_τ: indeed, this is required at large LSP masses. In this case, the dominant stau decays for $m_{\tilde{\tau}_1} - m_{\tilde{\chi}_1^0} > 160$ MeV are expected to be into three particles: $\tilde{\chi}_1^0 \nu \pi$ or $\tilde{\chi}_1^0 \nu \rho$. If $m_{\tilde{\tau}_1} - m_{\tilde{\chi}_1^0} < 1.2$ GeV, the $\tilde{\tau}_1$ lifetime is calculated to be so long, in excess of ~ 100 ns, that it is likely to escape the detector before decaying, and hence would be detectable as a massive, slowly-moving charged particle [43, 44]. The relevance of such scenarios while considering cosmological constraints is demonstrated in Fig. 8. Even is lepton-flavour violating couplings are allowed, the long lifetime of the staus remains [44].

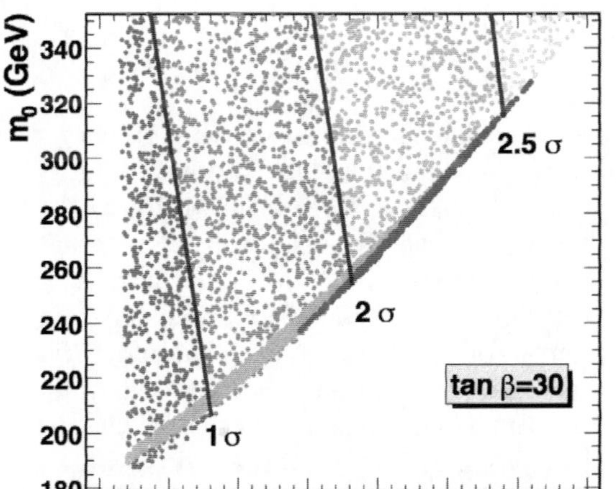

Figure 8. Allowed parameter regions in $M_{1/2}-m_0$ plane fixing $A_0 = 600$ GeV and $\tan\beta = 30$. The red (dark) narrow band is consistent region with dark matter abundance and $\delta m < m_\tau$ and the yellow (light) narrow band is that with $\delta m > m_\tau$. The green regions are inconsistent with the dark matter abundance, and in the white (excluded) area the LSP is the stau. The favoured regions of the muon anomalous magnetic moment at 1σ, 2σ and 2.5σ confidence level are indicated by solid lines. From Ref. [44].

5.1.3. Metastable sleptons in gravitino LSP scenarios On the other hand, in *gravitino dark matter* scenarios with more general options for the pattern of supersymmetry breaking, other options appear quite naturally, including the lighter selectron or smuon, or a sneutrino [45], or the lighter stop squark $\tilde{t}_1$ [46]. If the gravitino $\tilde{G}$ is the LSP, the decay rate of a slepton NLSP is given by

$$\Gamma(\tilde{\ell} \to \tilde{G}\ell) = \frac{1}{48\pi M_*^2} \frac{m_{\tilde{\ell}}^5}{M_{\tilde{G}}^2} \left[1 - \frac{M_{\tilde{G}}^2}{m_{\tilde{\ell}}^2}\right]^4, \tag{3}$$

where M_* is the Planck scale. Since M_* is much larger than the electroweak scale, the NLSP lifetime is naturally very long.

Gravitino (or axino) LSP with a long-lived charged stau may arise in gauge mediation and minimal supergravity models [47]. Large part of the parameter space potentially attractive for long-lived slepton searches with MoEDAL are compatible with cosmological constraints on the dark-matter abundance in superweakly interacting massive particle scenarios [48].

5.1.4. Long-lived gluinos in split supersymmetry The above discussion has been in the context of the CMSSM and similar scenarios where all the supersymmetric partners of Standard Model particles have masses in the TeV range. Another scenario is "split supersymmetry", in which the supersymmetric partners of quarks and leptons are very heavy, of a scale m_s, whilst the supersymmetric partners of SM bosons are relatively light [49]. In such a case, the gluino could have a mass in the TeV range and hence be accessible to the LHC, but would have a very long lifetime:

$$\tau \approx 8 \left(\frac{m_s}{10^9 \text{ GeV}}\right)^4 \left(\frac{1 \text{ TeV}}{m_{\tilde{g}}}\right)^5 \text{ s}. \tag{4}$$

Long-lived gluinos would form long-lived gluino R-hadrons including gluino-gluon *(gluinoball)* combinations, gluino-$q\bar{q}$ *(mesino)* combinations and gluino-qqq *(baryino)* combinations. The heavier gluino hadrons would be expected to decay into the lightest species, which would be metastable, with a lifetime given by Eq. (4), and it is possible that this metastable gluino hadron could be charged.

In the same way as stop hadrons, gluino hadrons may flip charge through conventional strong interactions as they pass through matter, and it is possible that one may pass through most of a conventional LHC tracking detector undetected in a neutral state before converting into a metastable charged state that could be detected by MoEDAL.

5.1.5. Experimental considerations There are several considerations supporting the complementary aspects of MoEDAL w.r.t. ATLAS and CMS when discussing the observability of (meta-)stable massive electrically-charged particles. Most of them stem from MoEDAL being "time-agnostic" due to the passive nature of its detectors. Therefore signal from very slowly moving particles will not be lost due to arriving in several consecutive bunch crossings. ATLAS and CMS, on the other hand, perform triggered-based analyses relying either on triggering on accompanying "objects", e.g. missing transverse energy, or by developing and deploying specialised triggers. In both cases the efficiency may lower and in the former the probed parameter space may be reduced. MoEDAL is mainly limited by the geometrical acceptance of the detectors, especially the MMTs, and by the requirement of passing the Z/β threshold of NTDs. In general ATLAS and CMS have demonstrated to cover high-velocities $\beta \gtrsim 0.2$, while MoEDAL is sensitive to lower ones $\beta \lesssim 0.2$.

When discussing the detection of particles stopped *(trapped)* in material that they may decay later, different possibilities are explored. CMS and ATLAS look in empty bunch crossings for decays of trapped particles into jets. MoEDAL MMTs may be monitored in a underground/basement laboratory for tracks arising from such decays. The background in the latter case, coming from cosmic rays, should be easier to control and assess. The probed lifetimes should be larger due to the unlimited monitoring time.

5.2. Doubly-charged Higgs bosons

Doubly-charged particles appear in many theoretical scenarios beyond the SM. For example, doubly-charged scalar states, usually termed doubly-charged Higgs fields, appear in left-right symmetric models [50–52] and in see-saw models for neutrino masses with Higgs triplets. A number of models encompasses additional symmetries and extend the SM Higgs sector by introducing doubly-charged Higgs bosons. A representative example of such a model is the L-R Symmetric Model (LRSM) [50–52], proposed to amend the fact that the weak-interaction couplings are strictly left handed by extending the gauge group of the SM so as to include a right-handed sector. The simplest realisation is an LRSM [50,51] that postulates a right-handed version of the weak interaction, whose gauge symmetry is spontaneously broken at high mass scale, leading to the parity-violating SM. This model naturally accommodates recent data on neutrino oscillations and the existence of small neutrino masses. The model generally requires Higgs triplets containing doubly-charged Higgs bosons ($H^{\pm\pm}$) Δ_R^{++} and Δ_L^{++}, which could be light in the minimal supersymmetric left-right model [53].

Single production of a doubly-charged Higgs boson at the LHC proceeds via vector boson fusion, or through the fusion of a singly-charged Higgs boson with either a $W^{\pm}$ or another singly-charged Higgs boson. The amplitudes of the $W_L W_L$ and $W_R W_R$ vector boson fusion processes are proportional to $v_{L,R}$, the vacuum expectation values of the neutral members of the scalar triplets of the LRSM. For the case of Δ_R^{++} production, the vector boson fusion process dominates. Pair production of doubly-charged Higgs bosons is also possible via a Drell-Yan process, with γ, Z or Z_R exchanged in the s-channel, but at a high kinematic price since substantial energy is required to produce two heavy particles. In the case of Δ_L^{++}, double production may nevertheless be the only possibility if v_L is very small or vanishing.

The decay of a doubly-charged Higgs boson can proceed via several channels. The dilepton signature leads to the (experimentally clean) final state $q\bar{q} \to \Delta_L^{++}\Delta_L^{--} \to 4\ell$. However as long as the triplet vacuum expectation value, v_Δ, is much larger than 10^{-4} GeV, the doubly-charged Higgs decay predominantly into a pair of same-sign W bosons. For very small Yukawa couplings $H_{\ell\ell} \lesssim 10^{-8}$, the doubly-charged Higgs boson can be quasi-stable [54]. In Fig. 9, the partial decay width of the doubly charged Higgs boson into a W boson pair is shown as a function of its mass. For $v_\Delta \gg 10^{-4}$ GeV, this partial width is roughly equal to the total width of the doubly charged Higgs boson. In the case of long lifetimes, slowly moving pseudo-stable Higgs bosons could be detected in the MoEDAL NTDs. For example with CR39, one could detect doubly-charged Higgs particles moving with speeds less than around $\beta \simeq 0.4$.

5.3. Black hole remnants in large extra dimensions

Over the last decades, models based on compactified extra spatial dimensions (ED) have been proposed in order to explain the large gap between the electroweak (EW) and the Planck scale of $M_{\rm EW}/M_{\rm PL} \approx 10^{-17}$. The four main scenarios relevant for searches at LHC the Arkani-Hamed-Dimopoulos-Dvali (ADD) model of large extra dimensions [55], the Randall-Sundrum (RS) model of warped extra dimensions [56], TeV^{-1}-sized extra dimensions [57], and the Universal Extra Dimensions (UED) model [58].

The existence of extra spatial dimensions [55, 56] and a sufficiently small fundamental scale of gravity open up the possibility that microscopic black holes be produced and detected [59–64] at the LHC. Once produced, the black holes will undergo an evaporation process categorised in three stages [59, 60]: the *balding phase*, the actual *evaporation phase*, and finally the Planck phase. It is generally assumed that the black hole will decay completely to some last few SM particles. However, another intriguing possibility is that the remaining energy is carried away by a stable remnant.

The prospect of microscopic black hole production at the LHC within the framework of models with large extra dimensions has been studied in Ref. [55]. Black holes produced at the

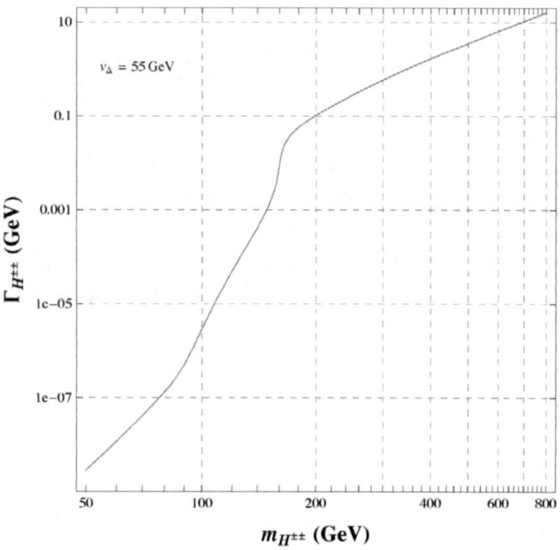

Figure 9. Partial decay width of $H^{\pm\pm} \to W^\pm W^\pm$ as a function of $m_{H^{\pm\pm}}$ for $v_\Delta = 55$ GeV. From Ref. [54].

LHC are expected to decay with an average multiplicity of $\sim 10-25$ into SM particles, most of which will be charged, though the details of the multiplicity distribution depend on the number of extra dimensions [65]. After the black holes have evaporated off enough energy to reach the remnant mass, some will have accumulated a net electric charge. According to purely statistical considerations, the probability for being left with highly-charged black hole remnants drops fast with the deviation from the average. The largest fraction of the black holes should have charges ± 1 or zero, although a smaller but non-negligible fraction would be multiply charged.

The fraction of charged black-hole remnants has been estimated [65,66] using the PYTHIA event generator [67] and the CHARYBDIS program [64]. It was assumed that the effective temperature of the black hole drops towards zero for a finite remnant mass, M_R. The value of M_R does not noticeably affect the investigated charge distribution, as it results from the very general statistical distribution of the charge of the emitted particles.

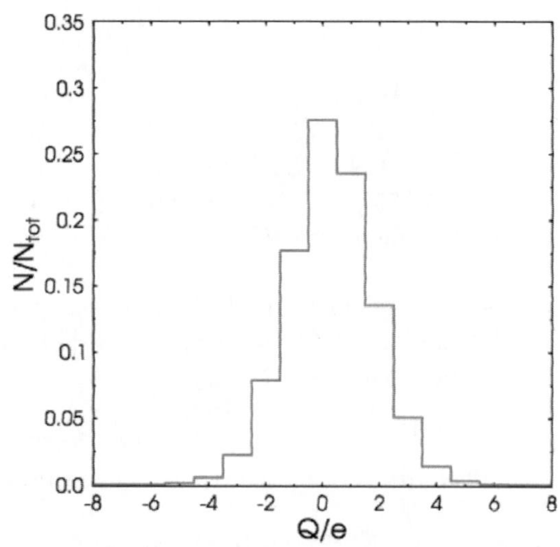

Figure 10. The distribution of black-hole remnant charges in proton-proton interactions at $\sqrt{s} = 14$ TeV calculated with the PYTHIA event generator [67] and the CHARYBDIS program [64]. From Ref. [66].

Thus, independent of the underlying quantum-gravitational assumption leading to the remnant formation, it was found that about 30% of the remnants are neutral, whereas $\sim 40\%$ would be singly-charged black holes, and the remaining $\sim$30% of remnants would be multiply-charged. The distribution of the remnant charges obtained is shown in Fig. 10. The black hole remnants considered here are heavy, with masses of a TeV or more. A significant fraction of the black-hole remnants produced would have a Z/β of greater than five, high enough to register in the CR39 NTDs forming the LT-NTD sub-detector of MoEDAL.

5.4. D-matter

Some versions of string theory include higher-dimensional "domain-wall"-like membrane *(brane)* structures in space-time, called *D-branes.* In some cases the bulk is "punctured" by lower-dimensional D-brane defects, which are either point-like or have their longitudinal dimensions compactified [68]. From a low-energy observer's perspective, such structures would effectively appear to be point-like *D-particles.* The latter have dynamical degrees of freedom, thus they can be treated as quantum excitations above the vacuum [68, 69] collectively referred to as *D-matter.* D-matter states are non-perturbative stringy objects with masses of order $m_D \sim M_s/g_s$, where g_s is the string coupling, typically of order one so that the observed gauge and gravitational couplings is reproduced. Hence, the D-matter states could be light enough to be phenomenologically relevant at the LHC.

Depending on their type, D-branes could carry integral or torsion (discrete) charges with the lightest D-particle (LDP) being stable. Therefore the LDPs are possible candidates for cold dark matter [69]. D-particles are solitonic non-perturbative objects in the string/brane theory. As discussed in the relevant literature [69], there are similarities and differences between D-particles and magnetic monopoles with non-trivial cosmological implications [68, 70–72]. An important difference is that they could have *perturbative* couplings, with no magnetic charge in general. Nonetheless, in the context of brane-inspired gauge theories, brane states with magnetic charges can be constructed, which would manifest themselves in MoEDAL in a manner similar to magnetic monopoles.

Non-magnetically-charged D-matter, on the other hand, could be produced at colliders and also produce interesting signals of direct relevance to the MoEDAL experiment. For instance, excited states of D-matter ($D^\star$) can be electrically-charged. For typical string couplings of phenomenological relevance, the first few massive levels may be accessible to the LHC. Depending on the details of the microscopic model considered, and the way the SM is embedded, such massive charged states can be relatively long-lived, and could likewise be detectable with MoEDAL. D-matter/antimatter pairs can be produced [70, 71] by the decay of intermediate off-shell Z-bosons, as shown in Fig. 11.

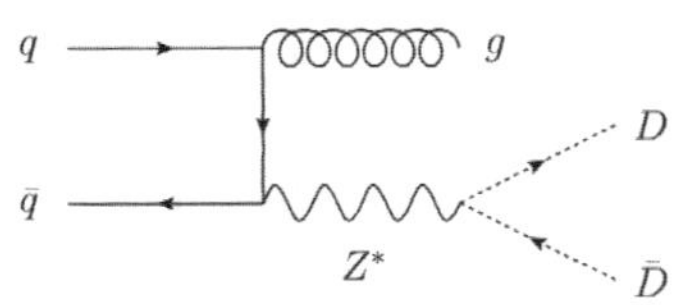

Figure 11. An example of parton-level diagrams for production of D-particles by $q\bar{q}$ collisions in a generic D-matter low-energy model [71].

6. Summary and outlook

MoEDAL is going to extend considerably the LHC reach in the search for (meta)stable highly ionising particles. The latter are predicted in a variety of theoretical models and include:

magnetic monopoles, SUSY stable (or, rather, long-lived) spartners, quirks, strangelets, Q-balls, fractionally-charged massive particles, etc [7]. Such particles can be light enough to be producible at the LHC energies (see e.g. Q-balls in the context of some SUSY or brane models [73]). In this paper we have described searches for only a subset of those particles, due to lack of space. Specifically, we discussed monopoles, partners in some SUSY models, doubly charged Higgs bosons, black hole remnants in models with large extra spatial dimensions, as well as some (more exotic) scenarios on D-matter which characterises some brane models.

The MoEDAL design is optimised to probe precisely all such long lived states, unlike the other LHC experiments [4]. Furthermore it combines different detector technologies: plastic nuclear track detectors (NTDs), trapping volumes and pixel sensors [2]. The first physics results, pertaining to magnetic monopole trapping detectors, obtained with LHC Run 1 data [34], and the corresponding analysis at 13 TeV has been published recently [35]. The MoEDAL Collaboration is preparing new analyses with more Run 2 data, with other detectors (NTDs) and with a large variety of interpretations involving not only magnetic but also electric charges.

Acknowledgments

The author is grateful to the DISCRETE2016 Symposium organisers for the kind invitation to present this talk. She acknowledges support by the Spanish Ministry of Economy and Competitiveness (MINECO) under the project FPA2015-65652-C4-1-R, by the Generalitat Valenciana through the project PROMETEO II/2013-017, by the Spanish National Research Council (CSIC) under the CT Incorporation Program 201650I002 and by the Severo Ochoa Excellence Centre Project SEV 2014-0398.

References

[1] For general information on the MoEDAL experiment, see: http://moedal.web.cern.ch/
[2] MoEDAL Collaboration, Technical Design Report of the MoEDAL Experiment *CERN Preprint* CERN-LHC-2009-006, MoEDAL-TDR-1.1 (2009), and references therein.
[3] L. Evans and P. Bryant, JINST **3**, S08001 (2008).
[4] A. De Roeck, A. Katre, P. Mermod, D. Milstead and T. Sloan, Eur. Phys. J. C **72**, 1985 (2012).
[5] M. Fairbairn, A. C. Kraan, D. A. Milstead, T. Sjostrand, P. Z. Skands and T. Sloan, Phys. Rept. **438**, 1 (2007); S. Burdin, M. Fairbairn, P. Mermod, D. Milstead, J. Pinfold, T. Sloan and W. Taylor, Phys. Rept. **582**, 1 (2015).
[6] V. A. Mitsou [MoEDAL Collaboration], EPJ Web Conf. **95**, 04042 (2015); PoS **EPS-HEP2015**, 109 (2015); N. E. Mavromatos and V. A. Mitsou [MoEDAL Collaboration], "Physics reach of MoEDAL at LHC: magnetic monopoles, supersymmetry and beyond," EPJ Web Conf. *to appear* [arXiv:1612.07012 [hep-ph]].
[7] B. Acharya *et al.* [MoEDAL Collaboration], Int. J. Mod. Phys. A **29**, 1430050 (2014).
[8] A. A. Alves, Jr. *et al.* [LHCb Collaboration], JINST **3**, S08005 (2008).
[9] M. D. Joergensen, A. De Roeck, H.-P. Hachler, A. Hirt, A. Katre, P. Mermod, D. Milstead and T. Sloan, "Searching for magnetic monopoles trapped in accelerator material at the Large Hadron Collider," arXiv:1206.6793 [physics.ins-det] (2012).
[10] A. De Roeck, H. P. Hächler, A. M. Hirt, M.-D. Joergensen, A. Katre, P. Mermod, D. Milstead and T. Sloan, Eur. Phys. J. C **72**, 2212 (2012).
[11] A. Sopczak *et al.*, IEEE Trans. Nucl. Sci. **62**, 3225 (2015).
[12] P. A. M. Dirac, Proc. Roy. Soc. Lond. A **133**, 60 (1931).
[13] P. A. M. Dirac, Phys. Rev. **74**, 817 (1948).
[14] G. 't Hooft, Nucl. Phys. B **79**, 276 (1974); A. M. Polyakov, JETP Lett. **20**, 194 (1974) [Pisma Zh. Eksp. Teor. Fiz. **20**, 430 (1974)].
[15] B. Julia and A. Zee, Phys. Rev. D **11**, 2227 (1975); Y. Nambu, Nucl. Phys. B **130**, 505 (1977); E. Witten, Phys. Lett. **86B**, 283 (1979); G. Lazarides, M. Magg and Q. Shafi, Phys. Lett. **97B**, 87 (1980); R. d. Sorkin, Phys. Rev. Lett. **51**, 87 (1983); D. J. Gross and M. J. Perry, Nucl. Phys. B **226**, 29 (1983); J. S. Schwinger, Science **165**, 757 (1969); J. Preskill, Ann. Rev. Nucl. Part. Sci. **34**, 461 (1984); A. Achucarro and T. Vachaspati, Phys. Rept. **327**, 347 (2000) [Phys. Rept. **327**, 427 (2000)]; T. W. Kephart, C. A. Lee and Q. Shafi, JHEP **0701**, 088 (2007); D. G. Pak, P. M. Zhang and L. P. Zou, Int. J. Mod. Phys. A **30**, no. 27, 1550164 (2015); A. Rajantie, JHEP **0601**, 088 (2006).
[16] V. Vento and V. S. Mantovani, "On the magnetic monopole mass," arXiv:1306.4213 [hep-ph] (2013).

[17] Y. M. Cho and D. Maison, Phys. Lett. B **391**, 360 (1997); W. S. Bae and Y. M. Cho, J. Korean Phys. Soc. **46**, 791 (2005).
[18] Y. M. Cho, K. Kimm and J. H. Yoon, Mod. Phys. Lett. A **31**, no. 09, 1650053 (2016); Phys. Lett. B **761**, 203 (2016).
[19] J. Ellis, N. E. Mavromatos and T. You, Phys. Lett. B **756**, 29 (2016); J. Ellis, V. Sanz and T. You, JHEP **1407**, 036 (2014).
[20] M. Barriola and A. Vilenkin, Phys. Rev. Lett. **63**, 341 (1989).
[21] A. S. Goldhaber, Phys. Rev. Lett. **63**, 2158 (1989); S. H. Rhie and D. P. Bennett, Phys. Rev. Lett. **67**, 1173 (1991); L. Perivolaropoulos, Nucl. Phys. B **375**, 665 (1992); G. W. Gibbons, M. E. Ortiz, F. Ruiz Ruiz and T. M. Samols, Nucl. Phys. B **385**, 127 (1992); M. Hindmarsh, Nucl. Phys. B **392**, 461 (1993); G. Arreaga, I. Cho and J. Guven, Phys. Rev. D **62**, 043520 (2000); A. Achucarro and J. Urrestilla, Phys. Rev. Lett. **85**, 3091 (2000); R. Gregory and C. Santos, Class. Quant. Grav. **20**, 21 (2003); S. B. Gudnason and J. Evslin, Phys. Rev. D **92**, no. 4, 045044 (2015); A. Achucarro, B. Hartmann and J. Urrestilla, JHEP **0507**, 006 (2005).
[22] A. K. Drukier and S. Nussinov, Phys. Rev. Lett. **49**, 102 (1982).
[23] P. O. Mazur and J. Papavassiliou, Phys. Rev. D **44**, 1317 (1991).
[24] N. E. Mavromatos and S. Sarkar, "Magnetic Monopoles from Global Monopoles in the presence of Kalb-Ramond Torsion," arXiv:1607.01315 [hep-th] (2016).
[25] Y. B. Zeldovich and M. Y. Khlopov, Phys. Lett. **79B**, 239 (1978).
[26] C. T. Hill, Nucl. Phys. B **224**, 469 (1983).
[27] V. K. Dubrovich, Grav. Cosmol. Suppl. **8N1**, 122 (2002).
[28] A. Rajantie, Contemp. Phys. **53**, 195 (2012).
[29] A. Rajantie, Phys. Today **69**, no. 10, 40 (2016).
[30] L. N. Epele, H. Fanchiotti, C. A. Garcia Canal and V. Vento, Eur. Phys. J. C **56**, 87 (2008); L. N. Epele, H. Fanchiotti, C. A. G. Canal and V. Vento, Eur. Phys. J. C **62**, 587 (2009).
[31] L. N. Epele, H. Fanchiotti, C. A. G. Canal, V. A. Mitsou and V. Vento, Eur. Phys. J. Plus **127**, 60 (2012).
[32] L. N. Epele, H. Fanchiotti, C. A. G. Canal, V. A. Mitsou and V. Vento, "Can the 750 GeV enhancement be a signal of light magnetic monopoles?," arXiv:1607.05592 [hep-ph] (2016).
[33] M. Kalliokoski, J. W. Lämsä, M. Mieskolainen and R. Orava, "Turning the LHC Ring into a New Physics Search Machine," arXiv:1604.05778 [hep-ex] (2016).
[34] B. Acharya *et al.* [MoEDAL Collaboration], JHEP **1608**, 067 (2016).
[35] B. Acharya *et al.* [MoEDAL Collaboration], Phys. Rev. Lett. **118**, 061801 (2017).
[36] G. Aad *et al.* [ATLAS Collaboration], Phys. Rev. Lett. **109**, 261803 (2012).
[37] G. Aad *et al.* [ATLAS Collaboration], Phys. Rev. D **93**, 052009 (2016).
[38] L. Patrizii and M. Spurio, Ann. Rev. Nucl. Part. Sci. **65**, 279 (2015).
[39] H. Goldberg, Phys. Rev. Lett. **50**, 1419 (1983) Erratum: [Phys. Rev. Lett. **103**, 099905 (2009)]; J. R. Ellis, J. S. Hagelin, D. V. Nanopoulos, K. A. Olive and M. Srednicki, Nucl. Phys. B **238**, 453 (1984).
[40] V. A. Mitsou, PoS **PLANCK 2015**, 085 (2015).
[41] J. R. Ellis, K. A. Olive, Y. Santoso and V. C. Spanos, Phys. Lett. B **565**, 176 (2003).
[42] O. Buchmueller *et al.*, Eur. Phys. J. C **72**, 2243 (2012); Eur. Phys. J. C **74**, 2922 (2014).
[43] T. Jittoh, J. Sato, T. Shimomura and M. Yamanaka, Phys. Rev. D **73**, 055009 (2006) Erratum: [Phys. Rev. D **87**, 019901 (2013)].
[44] S. Kaneko, J. Sato, T. Shimomura, O. Vives and M. Yamanaka, Phys. Rev. D **78**, 116013 (2008) Erratum: [Phys. Rev. D **87**, 039904 (2013)].
[45] J. R. Ellis, K. A. Olive and Y. Santoso, JHEP **0810**, 005 (2008).
[46] J. L. Diaz-Cruz, J. R. Ellis, K. A. Olive and Y. Santoso, JHEP **0705**, 003 (2007).
[47] K. Hamaguchi, M. M. Nojiri and A. de Roeck, JHEP **0703**, 046 (2007).
[48] J. L. Feng, S. Iwamoto, Y. Shadmi and S. Tarem, JHEP **1512**, 166 (2015).
[49] N. Arkani-Hamed and S. Dimopoulos, JHEP **0506**, 073 (2005); G. F. Giudice and A. Romanino, Nucl. Phys. B **699**, 65 (2004) Erratum: [Nucl. Phys. B **706**, 487 (2005)].
[50] J. C. Pati and A. Salam, Phys. Rev. D **10**, 275 (1974) Erratum: [Phys. Rev. D **11**, 703 (1975)].
[51] R. N. Mohapatra and J. C. Pati, Phys. Rev. D **11**, 566 (1975).
[52] G. Senjanovic and R. N. Mohapatra, Phys. Rev. D **12**, 1502 (1975); R. N. Mohapatra and G. Senjanovic, Phys. Rev. D **23**, 165 (1981).
[53] C. S. Aulakh, A. Melfo and G. Senjanovic, Phys. Rev. D **57**, 4174 (1998); Z. Chacko and R. N. Mohapatra, Phys. Rev. D **58**, 015003 (1998); C. S. Aulakh, K. Benakli and G. Senjanovic, Phys. Rev. Lett. **79**, 2188 (1997).
[54] C. W. Chiang, T. Nomura and K. Tsumura, Phys. Rev. D **85**, 095023 (2012).
[55] N. Arkani-Hamed, S. Dimopoulos and G. R. Dvali, Phys. Lett. B **429**, 263 (1998); Phys. Rev. D **59**, 086004

(1999); I. Antoniadis, N. Arkani-Hamed, S. Dimopoulos and G. R. Dvali, Phys. Lett. B **436**, 257 (1998).
[56] L. Randall and R. Sundrum, Phys. Rev. Lett. **83**, 4690 (1999).
[57] I. Antoniadis, Phys. Lett. B **246**, 377 (1990); I. Antoniadis and K. Benakli, Phys. Lett. B **326**, 69 (1994); I. Antoniadis, K. Benakli and M. Quiros, Phys. Lett. B **331**, 313 (1994).
[58] T. Appelquist, H. C. Cheng and B. A. Dobrescu, Phys. Rev. D **64**, 035002 (2001).
[59] S. B. Giddings and S. D. Thomas, Phys. Rev. D **65**, 056010 (2002).
[60] W. Fischler, "A Model for high-energy scattering in quantum gravity," arXiv:hep-th/9906038 (1999).
[61] P. C. Argyres, S. Dimopoulos and J. March-Russell, Phys. Lett. B **441**, 96 (1998); S. Dimopoulos and G. L. Landsberg, Phys. Rev. Lett. **87**, 161602 (2001); G. L. Alberghi, R. Casadio and A. Tronconi, J. Phys. G **34**, 767 (2007); M. Cavaglia, R. Godang, L. Cremaldi and D. Summers, Comput. Phys. Commun. **177**, 506 (2007); D. C. Dai, G. Starkman, D. Stojkovic, C. Issever, E. Rizvi and J. Tseng, Phys. Rev. D **77**, 076007 (2008).
[62] R. Casadio and B. Harms, Int. J. Mod. Phys. A **17**, 4635 (2002).
[63] M. Cavaglia, Int. J. Mod. Phys. A **18**, 1843 (2003); P. Kanti, Int. J. Mod. Phys. A **19**, 4899 (2004).
[64] C. M. Harris, P. Richardson and B. R. Webber, JHEP **0308**, 033 (2003).
[65] B. Koch, M. Bleicher and H. Stoecker, J. Phys. G **34**, S535 (2007).
[66] S. Hossenfelder, B. Koch and M. Bleicher, "Trapping black hole remnants," hep-ph/0507140 (2005).
[67] T. Sjostrand, S. Mrenna and P. Z. Skands, JHEP **0605**, 026 (2006).
[68] J. R. Ellis, N. E. Mavromatos and D. V. Nanopoulos, Gen. Rel. Grav. **32**, 943 (2000); J. R. Ellis, N. E. Mavromatos and D. V. Nanopoulos, Phys. Lett. B **665**, 412 (2008); J. R. Ellis, N. E. Mavromatos and M. Westmuckett, Phys. Rev. D **70**, 044036 (2004); *ibid.* **71**, 106006 (2005).
[69] G. Shiu and L. T. Wang, Phys. Rev. D **69**, 126007 (2004).
[70] N. E. Mavromatos, S. Sarkar and A. Vergou, Phys. Lett. B **696**, 300 (2011).
[71] N. E. Mavromatos, V. A. Mitsou, S. Sarkar and A. Vergou, Eur. Phys. J. C **72**, 1956 (2012).
[72] See, e.g., E. Witten, "Comments on string theory," arXiv:hep-th/0212247 (2002), and references therein.
[73] S. Kasuya and M. Kawasaki, Phys. Rev. Lett. **85**, 2677 (2000); S. Kasuya, M. Kawasaki and T. T. Yanagida, PTEP **2015**, no. 5, 053B02 (2015); S. Abel and A. Kehagias, JHEP **1511**, 096 (2015).

IOP Conf. Series: Journal of Physics: Conf. Series **873** (2017) 012011 doi:10.1088/1742-6596/873/1/012011

The discrete charm of flavour and CP violation

G C Branco and M N Rebelo

Centro de Física Teórica de Partículas – CFTP and Dept de Física Instituto Superior Técnico – IST, Universidade de Lisboa (UL), Av. Rovisco Pais, P-1049-001 Lisboa, Portugal

E-mail: gbranco@tecnico.ulisboa.pt and rebelo@tecnico.ulisboa.pt

Abstract. We point out that in the Standard Model (SM) there is no explanation why $|V_{23}|^2 + |V_{13}|^2$ is of order 10^{-3}. A framework is described for explaining this small mixing, involving the introduction of vector-like quarks. A symmetry is introduced so that at a first stage $V_{CKM} = \mathbb{1}$ and only the third generation of quarks acquires a mass. It is shown that when interactions of vector like quarks are taken into account a realistic quark mass spectrum is generated together with a correct V_{CKM} matrix.

1. Introduction

The closely related questions of Flavour and CP Violation are two of the major open problems in Particle Physics. Discrete symmetries have played a major role in the development of Particle Physics. The symmetries C, P, CP and T are all violated in Nature. Yet, they are very important, since they are conserved by electromagnetic and strong interactions. Only CPT is not violated, as far as we know today. Parity is explicitly violated in the Standard Model (SM), but in the SM CP can be violated explicitly in the Yukawa sector provided that there are three or more fermion generations. Spontaneous CP violation, can only occur if the scalar sector of the SM is extended. The first model of spontaneous CP violation was proposed by Lee [1] in the context of the SM extended with one additional Higgs doublet, in order to accommodate the experimental evidence for CP violation in the quark sector at a time when it was not yet known that there were more than two generations of quarks. Models with two Higgs doublets have a very rich phenomenology [2, 3] and they provide very interesting candidates for physics beyond the SM, which have the potential of being tested at the LHC. There is, by now, clear experimental evidence that the Cabibbo-Kobayashi-Maskawa (CKM) matrix is complex [4]. Therefore, in any model of spontaneous CP violation, with the breaking of CP arising from a CP violating vacuum, it must be possible to generate a complex CKM matrix from the vacuum phase, in order to conform to experiment. This makes it non-trivial to have viable models of spontaneous CP violation. The simplest realistic model of spontaneous CP violation, involves the addition to the SM of a vector-like isosinglet quark and a complex singlet scalar [5, 6, 7]. Models with several Higgs doublets have potentially large Higgs mediated flavour changing neutral currents. Experimental data show that the latter are extremely suppressed thus implying the need for a mechanism to avoid, in a natural way, sizeable effects. One option is to use a symmetry that eliminates these currents at tree level, this is the natural flavour conservation paradigm [8, 9]. In alternative it is possible to use a symmetry that leads to tree level effects suppressed by small entries of V_{CKM} [10]. These are the so-called BGL models which have been extended later-on

[11, 12] and where the suppression conforms with the experimental evidence.

At present, all experimental results on quark mixing and CP violation in the quark sector, are in essential agreement with experiment. However, there is plenty of room for New Physics. CP violation has profound implications for cosmology, since CP breaking is one of the required ingredients [13] in order to generate the observed Baryon Asymmetry of the Universe (BAU). By now, it is clear that in the SM it is not possible to generate sufficient BAU. One of the reasons for this is the smallness of CP violation in the SM. This is a clear indication that there should be additional sources of CP violation, beyond the one present in the SM. The great challenge is to find out where are the new sources of CP violation and how they could be detected. In the quark sector, an interesting possibility may arise if vector-like quarks are added to the spectrum of the SM [6, 14, 15, 16, 17, 18, 19, 20, 21, 22, 23, 24, 25, 26, 27, 28, 29, 30, 31]. These quarks can have bare masses in the Lagrangian, since their mass terms are invariant under the gauge symmetry. They may also receive their mass from the couplings to SU(2) singlet scalars, once these acquire a vacuum expectation value (vev). In any case, they mix with the standard quarks, and the 3×3 unitarity of the CKM matrix can be violated. These violations of 3×3 unitarity lead in general to Z-mediated Flavour-Changing-Neutral-Currents (FCNC) at tree level, which are naturally suppressed. In spite of this suppression, these FCNC, like $B_d - \overline{B}_d$ mixing, $B_s - \overline{B}_s$ mixing and $D^0 - \bar{D}^0$ mixing [32], can contribute to processes which in the SM only occur at loop level.

Another possible source of CP breaking is leptonic CP violation. It is well known that in the SM neutrinos are strictly massless, so the discovery of neutrino oscillations, implying that at least two of the neutrinos have mass, is a clear experimental evidence of Physics Beyond the SM. Let us assume that one adds three right-handed neutrinos to the SM. Then the most general renormalizable Lagrangian must include Majorana-type bare mass terms for the right-handed neutrinos, together with Dirac neutrino masses, leading to the seesaw mechanism [33, 34, 35, 36, 37] of type one, providing an elegant explanation for the smallness of the neutrino masses. In this scenario, there are three types of CP violation which may arise: CP violation at low energies of Dirac type, CP violation at low energies of Majorana type and CP violation at high energy which provides a very attractive scenario of Baryogenesis through Leptogenesis [38]. At low energies one may detect leptonic CP violation of Dirac type through neutrino oscillations, which is one of the major tasks for future experiments. Majorana CP phases are difficult to detect at low energies but they affect double-beta decay. An important question is whether there is a relationship between leptonic CP violation at low energy and CP violation entering in Leptogenesis. It has been shown that in general such a relationship does not exist [39, 40] unless one introduces leptonic flavour symmetries.

2. Comparing the Quark and Leptonic Sectors

In the SM, the flavour structure of Yukawa couplings is not constrained by the gauge symmetry which implies that the Yukawa matrices are 3×3 general complex matrices. This leads to the existence of 36 parameters in the quark flavour sector. It is clear that there is a large redundancy, since in the quark flavour sector we have only ten physical parameters, namely the six quark masses plus the four physical parameters of the CKM matrix. In the lepton sector, one encounters an entirely analogous situation in the case of massive Dirac neutrinos. In the case of Majorana neutrinos, one has two extra CP violating phases which reflect the Majorana character of neutrinos. This large flavour redundancy is due to the freedom that one has to make weak basis (WB) transformations which, for the quark sector, can be written:

$$d_L^0 \to U' d_L^0, \quad u_L^0 \to U' u_L^0, \quad d_R^0 \to V' d_R^0 \quad u_R^0 \to W' u_R^0 \tag{1}$$

As previously mentioned, the Yukawa matrices are arbitrary complex matrices. However, one can make a weak basis transformation so that both Y_u, Y_d become Hermitian matrices. In the context of specific anzätze with texture zeros it is common to consider Hermitian matrices (see, for example, [41]) yet, this need not be the case [42]. The large redundancy in Yukawa couplings renders it specially difficult to extract from the data some flavour symmetry, in a bottom up approach. Even if such symmetry exists, in what WB would the symmetry be apparent? There is a specially convenient WB where one of the quark mass matrices (M_d or M_u) is diagonal real and the other matrix is Hermitian. It can be readily seen that in this basis the number of free parameters in the quark mass matrices equals the number of physical measurable parameters, namely the six quark masses and the four physical parameters in V_{CKM}.

The discovery of neutrino oscillations providing experimental evidence for non-vanishing neutrino masses, brought further intriguing aspects to the Flavour Problem. Why are the pattern of fermion mixings so different in the quark and lepton sectors? Do we understand why lepton mixing is large? Do we understand why quark mixing is small? An important difference between the quark and lepton sectors has to do with experiment. In the quark sector the V_{CKM} matrix is overdetermined, in the sense that once we measure V_{us}, V_{ub} and V_{cb} through strange particle and B-meson decays, there is a large amount of physical quantities like $B_d - \overline{B}_d$, $B_s - \overline{B}_s$ mixings, the strength of CP violation in the kaon sector as well as CP violation in the B sector, with the measurement of the angles β and γ, which within the SM, have to be fitted through a single parameter, namely the CP violating phase entering in the parametrization of V_{CKM}. This is to be contrasted to the situation encountered in the lepton sector.

Let us assume that neutrinos are Majorana particles and consider the effective neutrino mass matrix at low energies. The leptonic mass terms are then of the form:

$$\mathcal{L}_{\text{mass}} = -\frac{1}{2}\nu_L^{0\,T} C^{-1} m_\nu \nu_L^0 - \overline{\ell_L^0} m_\ell \ell_R^0 + \text{h.c.}\ . \tag{2}$$

The neutrino mass matrix is a complex symmetric matrix, with six real parameters and three phases. If we work in the WB where the charged lepton mass matrix is diagonal and real, the number of independent parameters of the neutrino mass matrix is equal to the number of physical quantities derived from it, to wit, three neutrino masses, and six parameters - three mixing angles and three CP violating phases - appearing in the leptonic mixing matrix. It has been pointed out that it is not possible to completely determine the neutrino mass matrix from viable experimental measurements [43]. This motivated the authors of Ref. [43] to consider a restricted class of neutrino mass matrices with zero textures consistent with all experimental data and to explore their phenomenological implications. The imposition of these textures is based on the underlying idea that they are linked to a flavour symmetry. The effective neutrino mass matrix written in Eq. (2) must be generated through physics beyond the SM. The seesaw framework is a simple and elegant possibility. The authors of Ref. [44] studied seesaw realisations of the texture zeros proposed in [43]. Implications for CP violation in the leptonic sector at high energies, and in particular for leptogenesis, of these seesaw realizations were analysed in [45]. With two right-handed neutrinos only, zero textures allow to establish a connecting link between the sign of the baryon asymmetry of the universe and the sign of the CP violating parameter which will be measured in low energy neutrino oscillations [46]. With three right-handed neutrinos these signs cease to be correlated [47] but a connection still exists. In this respect the Casas and Ibarra parametrisation [48] is very useful to show that zeros in the Dirac type neutrino mass matrix, in the basis where the charged lepton mass matrix is diagonal, lead to orthogonality relations such that for some of the textures the variables relevant for leptogenesis can be fully determined in terms of low energy parameters and heavy neutrino masses [49, 50, 51]. An alternative way of reducing the number of free parameters in the Lagrangian is by means of discrete symmetries, such as for example an A_4 symmetry [52, 53] (for a review see [54]), leading, once again, to definite predictions. In general without a flavour model the three CP violating phases appearing at low energies and the three phases relevant for high energy leptonic CP violation, and in particular Leptogenesis, are not related [39, 40].

3. The Origin of CP Violation

In the SM with three or more generations, CP violation may originate at the Lagrangian level with complex Yukawa couplings, as it has been pointed out by Kobayashi and Maskawa [55]. The most elegant way to show this consists of separating the Lagrangian of the SM in two parts:

$$\mathcal{L} = \mathcal{L}_{CP} + \mathcal{L}_{\text{remaining}} \tag{3}$$

$\mathcal{L}_{CP}$ consists of the gauge part of the Lagrangian which necessarily conserves CP. One can then consider the most general CP transformation which leaves $\mathcal{L}_{CP}$ invariant. Applying this CP transformation to $\mathcal{L}_{\text{remaining}}$ one can easily derive that the necessary condition for CP invariance for any number of fermion generations is [56]:

$$\text{Tr}\,[H_u, H_d]^3 = 0 \tag{4}$$

The condition of Eq. (4) is automatically satisfied for one or two generations. For three generations Eq. (4) is a necessary and sufficient condition for CP invariance and it can be expressed in terms of quark masses and mixing:

$$\begin{aligned}\text{tr}\,[H_u, H_d]^3 = 6i \quad & (m_t^2 - m_c^2) \quad (m_t^2 - m_u^2) \quad (m_c^2 - m_u^2) \times \\ & \times (m_b^2 - m_s^2) \quad (m_b^2 - m_d^2) \quad (m_s^2 - m_d^2) \quad \text{Im}Q_{uscb}\end{aligned} \tag{5}$$

where Q stands for a rephasing invariant quartet of V_{CKM}, defined by $Q_{\alpha i \beta j} \equiv V_{\alpha i} V_{\beta j} V_{\alpha j}^* V_{\beta i}^*$ $(\alpha \neq \beta,\, i \neq j)$.

Another scenario for the generation of CP violation is the one put forward by Lee [1] about the same time that the KM mechanism [55] was proposed. The idea is that CP is a good symmetry of the Lagrangian but the vacuum is not CP invariant.

In general a realistic model of Spontaneous CP violation has to fulfil the following conditions.

i) The model should have a natural suppression mechanism for FCNC both in the scalar and vector sectors.

ii) The vacuum CP violating phase phase should be able to generate a complex CKM matrix. This requirement arises from the fact that experimentally there is clear evidence that the CKM matrix is non-trivially complex even if one allows for the presence of New Physics.

The minimal viable model of spontaneous CP violation involves the following extension of the SM:

i) Introduction of at least one vector-like quark with charge either -1/3 or +2/3.

ii) Introduction of one complex scalar field, singlet under the gauge group, which we denote S.

With the above mentioned addition to the SM, one achieves spontaneous CP violation through a non-trivial phase in the vev of the scalar isosinglet S. For definiteness, let us assume that we have introduced a down type vector-like quark. This leads to a 4×4 quark mass matrix in the down sector and the quark mixing matrix becomes a 3×4 matrix, leading to deviations of 3×3 unitarity in the CKM matrix connecting standard quarks and also to Z-mediated FCNC in the down sector. One of the nice features of this class of models is the fact that both deviations of 3×3 unitarity and tree level FCNC are naturally suppressed by the ratio of masses m^2/M^2 of standard quarks and of the heavy isosinglet quark. Recall that since the mass term of the vector-like quark is gauge invariant, its value can naturally be much larger than the electroweak scale.

There is mixing between standard quarks and the vector-like quark which in turn generates at low energies a complex 3×3 effective down quark mixing. Upon diagonalisation of the quark mass matrices, this leads to a complex 3×3 CKM matrix, in agreement with experiment. As mentioned above there are naturally suppressed FCNC at tree level which in spite of this suppression can contribute to $B_d - \overline{B}_d$ and $B_s - \overline{B}_s$ mixings.

4. Generating a Viable Model of Quark Mixing with Vector-Like Quarks

In the SM the Yukawa couplings Y_u, Y_d, responsible for generating the up and down quark mass matrices, are two independent complex matrices. As a result, we show next that there is no reason for having the CKM matrix close to the identity, even if one takes into account the strong hierarchy of quark masses. In order to prove this result we shall consider the extreme chiral limit, where only the top and the bottom quarks have a mass.

4.1. The Extreme Chiral Limit

In the EC limit the up and down quark mass matrices are two rank one matrices which can be written as :

$$M_d = U_L^{d\dagger}\ \mathrm{diag}(0,0,m_b)\ U_R^d, \qquad M_u = U_L^{u\dagger}\ \mathrm{diag}(0,0,m_t)\ U_R^u \tag{6}$$

Note that no generality is lost in writing in Eq. (6) the quark masses in a given ordering, taking into account that a permutation which would change these positions could always be incorporated in the matrices $U_{L,R}^{d,u}$.

Keeping in mind that at this stage the first two generations are massless, one can make an arbitrary redefinition of the massless quarks through a unitary transformation with the following form:

$$W_{u,d} = \begin{bmatrix} X_{u,d} & 0 \\ 0 & 1 \end{bmatrix} \tag{7}$$

where $X_{u,d}$ are 2×2 unitary matrices. One can then show that the quark mixing matrix V_{CKM} is an orthogonal matrix mixing only the third and second quark generation, parametrised by an angle α which is entirely arbitrary in the EC limit. It is clear that in the SM the smallness of $|V_{23}|^2 + |V_{13}|^2$ has no relation with the fact that the quark mass ratios are small. The experimentally observed smallness of $|V_{23}|^2 + |V_{13}|^2$ may be an indication of the need of a flavour symmetry.

5. Obtaining Small Mixing Through a Symmetry

Let us introduce the following Z_6 flavour symmetry in the SM:

$$\begin{array}{lllll} Q_{L1}^0 \to e^{i\tau}\ Q_{L1}^0 & Q_{L2}^0 \to e^{-2i\tau}\ Q_{L2}^0 & Q_{L3}^0 \to e^{-i\tau}\ Q_{L3}^0 & & \\ d_{R1}^0 \to e^{-i\tau} d_{R1}^0 & d_{R2}^0 \to e^{-i\tau} d_{R2}^0 & d_{R3}^0 \to e^{-2i\tau} d_{R3}^0 & & \\ u_{R1}^0 \to e^{i\tau} u_{R1}^0 & u_{R2}^0 \to e^{i\tau} u_{R3}^0 & u_{R3}^0 \to u_{R3}^0; & \Phi \to e^{i\tau}\Phi; & \tau = \dfrac{2\pi}{6} \end{array} \tag{8}$$

where the Q_{Lj}^0 stand for the left-handed quark doublets, d_{Rj}^0 and u_{Rj}^0 denote right-handed quark singlets and Φ is the scalar doublet. The Yukawa interactions can be written:

$$\mathcal{L}_Y = \left[-\overline{Q}_{Li}^0\ \Phi\, Y_d\ d_{Rj}^0 - \overline{Q}_{Li}^0\ \tilde{\Phi}\, Y_u\ u_{Rj}^0 \right] + \mathrm{h.c.}, \tag{9}$$

IOP Conf. Series: Journal of Physics: Conf. Series **873** (2017) 012011 doi:10.1088/1742-6596/873/1/012011

As a result of the Z_6 symmetry, the Yukawa couplings have the following flavour structure:

$$Y_d = \begin{bmatrix} 0 & 0 & 0 \\ 0 & 0 & 0 \\ 0 & 0 & \times \end{bmatrix}, \qquad Y_u = \begin{bmatrix} 0 & 0 & 0 \\ 0 & 0 & 0 \\ 0 & 0 & \times \end{bmatrix} \tag{10}$$

which leads to $V_{CKM} = \mathbb{1}$, at this stage.

5.1. Generation of Quark Masses for the First Two Families

We show next that one may generate masses for the first two generations, through the introduction of vector-like quarks. For definiteness, we introduce three down $(D^0_{Li}, D^0_{Ri})$and three up (U^0_{Li}, U^0_{Ri}) vector-like quarks which are singlets under the gauge group. The Yukawa interactions can now be written:

$$\mathcal{L}_{\rm Y} = \left[-\overline{Q}^0_{Li}\, \Phi\, (Y_d)_{i\alpha}\, d^0_{R\alpha} - \overline{Q}^0_{Li} \tilde{\Phi}\, (Y_u)_{i\beta}\, u^0_{R\beta} \right] + \text{h.c.}, \tag{11}$$

where the index i runs from 1 to 3 while the indices α and β run from 1 to 6. We introduce the following generic bare mass terms

$$\mathcal{L}_{b.m.} = [-\, \overline{D}^0_{Lj}(\eta_d)_{j\alpha}\, d^0_{R\alpha} - \overline{U}^0_{Lk}(\eta_u)_{k\beta}\, u^0_{R\beta}] + \text{h.c.} \tag{12}$$

The Z_6 symmetry is extended to the full Lagrangian, and we further introduce a complex scalar singlet denoted S. Soft-breaking terms are introduced in the scalar sector to avoid a massless Goldstone boson and also to solve the domain-wall problem. The couplings of the scalar singlet to the quark singlets can be written:

$$\mathcal{L}_{\rm g} = [-\, \overline{D}^0_{Lj}[(g_d)_{j\alpha} S + (g'_d)_{j\alpha} S^*]\, d^0_{R\alpha} - \overline{U}^0_{Lk}[(g_u)_{k\beta} S + (g'_u)_{k\beta} S^*]\, u^0_{R\beta}] + \text{h.c.} \tag{13}$$

After gauge symmetry breaking one obtains:

$$\mathcal{L}_{\rm M} = \left[-\frac{v}{\sqrt{2}}\, \overline{d}^0_{Li}(Y_d)_{i\alpha}\, d^0_{R\alpha} - \frac{v}{\sqrt{2}} \overline{u}^0_{Li}(Y_u)_{i\alpha}\, u^0_{R\alpha} - \overline{D}^0_{Li}(\mu_d)_{i\alpha}\, d^0_{R\alpha} - \overline{U}^0_{Li}(\mu_u)_{i\alpha}\, u^0_{R\alpha} \right] + \text{h.c.} \tag{14}$$

In a more compact form one can write:

$$\mathcal{L}_{\rm M} = -\left(\overline{d}^0_L\ \overline{D}^0_L \right) \mathcal{M}_d \begin{pmatrix} d^0_R \\ D^0_R \end{pmatrix} - \left(\overline{u}^0_L\ \overline{U}^0_L \right) \mathcal{M}_u \begin{pmatrix} u^0_R \\ U^0_R \end{pmatrix} \tag{15}$$

where $\mathcal{M}_d$, $\mathcal{M}_u$ are 6×6 matrices

$$\mathcal{M}_d = \begin{pmatrix} m_d & \omega_d \\ X_d & M_d \end{pmatrix} \quad \mathcal{M}_u = \begin{pmatrix} m_u & \omega_u \\ X_u & M_u \end{pmatrix} \tag{16}$$

These mass matrices can be diagonalised through unitary transformations:

$$\begin{aligned} &\begin{pmatrix} d^0_L \\ D^0_L \end{pmatrix} = \begin{pmatrix} A_{dL} \\ B_{dL} \end{pmatrix} (\, d_L\,) \equiv \mathcal{U}^d_L d_L \qquad \begin{pmatrix} u^0_L \\ U^0_L \end{pmatrix} = \begin{pmatrix} A_{uL} \\ B_{uL} \end{pmatrix} (\, u_L\,) \equiv \mathcal{U}^u_L u_L \\ &\begin{pmatrix} d^0_R \\ D^0_R \end{pmatrix} \equiv \mathcal{U}^d_R d_R \qquad \begin{pmatrix} u^0_R \\ U^0_R \end{pmatrix} \equiv \mathcal{U}^u_R u_R \end{aligned} \tag{17}$$

One can then write the charged currents as:

$$\mathcal{L}_{\rm W} = -\frac{g}{\sqrt{2}}\, (\overline{u}^0_L \gamma^\mu d^0_L)\, \mathbf{W}^+_\mu + \text{h.c.} = -\frac{g}{\sqrt{2}}\, (\overline{u}_L V \gamma^\mu d_L)\, \mathbf{W}^+_\mu + \text{h.c.} \tag{18}$$

with $V = A^\dagger_{uL} A_{dL}$. The neutral gauge couplings are:

$$\mathcal{L}_Z = \frac{g}{\cos\theta_W} Z_\mu \left[\frac{1}{2} \left(\overline{u}_L W^u \gamma^\mu u_L - \overline{d}_L W^d \gamma^\mu u_L \right) - \sin^2\theta_W \left(\frac{2}{3} \overline{u} \gamma^\mu u - \frac{1}{3} \overline{d} \gamma^\mu d \right) \right] \tag{19}$$

with $W^d = V^\dagger V$ and $W^u = V V^\dagger$. In the quark mass eigenstate basis the SM-like Higgs couplings, and the couplings of the would-be Goldstone bosons G^+, G^0 can be written:

$$\begin{aligned} \mathcal{L}_{\rm Mh} &= -\tfrac{\sqrt{2}G^+}{v} \left[\overline{u}_L V \mathcal{D}_d d_R - \overline{u}_R \mathcal{D}_u V d_L \right] \\ &-i \tfrac{G^0}{v} \left[\overline{d}_L W^d \mathcal{D}_d d_R - \overline{u}_L W^u \mathcal{D}_u u_R \right] - \tfrac{h}{v} \left[\overline{d}_L W^d \mathcal{D}_d d_R + \overline{u}_L W^u \mathcal{D}_u u_R \right] + \text{h.c.} \end{aligned} \tag{20}$$

The vector-like quarks transform in the following way under Z_6:

$$\begin{array}{llll} D^0_{L1} \to e^{-3i\tau}\, D^0_{L1} & D^0_{L2} \to e^{-2i\tau}\, D^0_{L2} & D^0_{L3} \to e^{-i\tau}\, D^0_{L3} & \\ D^0_{R1} \to e^{-2i\tau}\, D^0_{R1} & D^0_{R2} \to e^{-3i\tau}\, D^0_{R2} & D^0_{R3} \to D^0_{R3} & \\ U^0_{L1} \to e^{-i\tau}\, U^0_{L1} & U^0_{L2} \to U^0_{L2} & U^0_{L3} \to e^{i\tau}\, U^0_{L3} & \\ U^0_{R1} \to U^0_{R1} & U^0_{R2} \to e^{-i\tau}\, U^0_{R2} & U^0_{R3} \to e^{2i\tau}\, U^0_{R3}; & S \to e^{i\tau}\, S; \quad \tau = \frac{2\pi}{6} \end{array} \tag{21}$$

Diagonalisation of the quark mass matrices is performed through:

$$\mathcal{U}^{d\dagger}_L \mathcal{M}_d\, \mathcal{U}^d_R = \mathcal{D}_d \equiv \text{diag}(d_d, D_d) \tag{22}$$

with:

$$\mathcal{U}_L = \begin{pmatrix} K & R \\ S & T \end{pmatrix} \tag{23}$$

Unitarity of $\mathcal{U}_L$ implies $KK^\dagger = \mathbb{1} - RR^\dagger$ as well as $K^\dagger K = \mathbb{1} - S^\dagger S$. Deviations of 3×3 unitarity are small since R and S are suppressed by the ratio m/M. The matrices K_d, K_u can be computed from an effective Hermitian squared matrix given by:

$$\mathcal{H}_{eff} = (mm^\dagger + \omega\omega^\dagger) - (mX^\dagger + \omega M^\dagger)(XX^\dagger + MM^\dagger)^{-1}(Xm^\dagger + M\omega^\dagger) \tag{24}$$

To an excellent approximation K is the matrix that diagonalises $\mathcal{H}_{eff}$:

$$K^{-1} \mathcal{H}_{eff} K = d^2 \tag{25}$$

We assume that the 6×6 matrix $\mathcal{M}_d$ has a Froggatt-Nielsen structure [57]:

$$\mathcal{M}_d = \mu \cdot \begin{pmatrix} 0 & 0 & 0 & 0 & 0 & \lambda^2 z \\ 0 & 0 & 0 & 0 & \lambda y & 0 \\ 0 & 0 & 1 & \lambda x & 0 & 0 \\ 0 & 0 & \frac{C_1}{\lambda} & \frac{D_1}{\lambda^2} & 0 & 0 \\ \frac{A_2}{\lambda} & \frac{B_2}{\lambda} & \frac{C_2}{\lambda} & 0 & \frac{D_2}{\lambda^2} & 0 \\ \frac{A_3}{\lambda} & 0 & \frac{C_3}{\lambda} & 0 & 0 & \frac{D_3}{\lambda^2} \end{pmatrix}, \tag{26}$$

with $\mu \approx m_b$. This leads to the following structure for $\mathcal{H}_{eff}$:

$$\mathcal{H}_{eff} \sim \begin{pmatrix} \lambda^6 r z^2 & \lambda^5 u y z & -\lambda^3 c_3 z \\ \lambda^5 u^* y z & \lambda^4 r' y^2 & -\lambda^2 c_2 y \\ -\lambda^3 c_3^* z & -\lambda^2 c_2^* y & 1 - \lambda^2 \hat{r} \end{pmatrix} \tag{27}$$

with:

$$\begin{aligned} r &= |a_3|^2 + |c_3|^2 \\ r' &= |a_2|^2 + |b_2|^2 + |c_2|^2 \\ \hat{r} &= |c_1|^2 + |c_2|^2 + |c_3|^2 + (c_1 + c_1^*)x \\ u &= a_2^* a_3 + c_2^* c_3 \\ a_i = \frac{A_i}{D_i}, \quad b_i = \frac{B_i}{D_i}, \quad c_i &= \frac{C_i}{D_i} \end{aligned} \tag{28}$$

It can be shown that a realistic V_{CKM} is generated with:

$$|V_{cb}| \approx K_{23}\lambda^2 \qquad |V_{ub}| \approx K_{13}\lambda^3 \tag{29}$$

where the K_{ij} are of order one, namely $K_{23} = c_2 y$ and $K_{13} = c_3 z$.

The general FCNC structure was analysed in detail [58] for this model, including loop-induced FCNC. The decay channels of the vector-like quarks were also studied, together with the possibility of discovering them at the LHC.

6. Conclusions

We emphasise a flavour fine-tuning problem present in the SM, which results from the fact that there is no reason to have a V_{CKM} matrix close to the identity, even taking into account the strong hierarchy of quark masses. We describe a solution to this fine-tuning problem which involves the introduction of a flavour symmetry and the introduction of vector-like quarks of charges (-1/3) and (2/3), together with a complex singlet scalar. Prior to the introduction of the vector-like quarks V_{CKM} equals the identity, but in the presence of vector-like quarks, a correct quark mass spectrum and a realistic CKM mixing matrix is obtained. We also point out that in realistic models, some of the vector-like quarks are at the reach of second run of LHC.

Acknowledgements

The authors thank the local organising committee of the Symposium Discrete 2016 for the stimulating scientific atmosphere and the warm hospitality in Warsaw, and acknowledge financial support from the National Science Centre, Poland, the HARMONIA project under contract UMO-2015/18/M/ST2/00518 (2016-2019) to participate in the Symposium and in the 1st Harmonia Meeting that took place immediately after the Symposium. Special thanks go to Maria Krawczyk. Our work is partially supported by Fundação para a Ciência e a Tecnologia (FCT, Portugal) through the projects CERN/FIS-NUC/0010/2015, CFTP-FCT Unit 777 (UID/FIS/00777/2013) which are partially funded through POCTI (FEDER), COMPETE, QREN and EU.

References

[1] Lee T D, 1973 "A Theory of Spontaneous T Violation," Phys. Rev. D **8** 1226.
[2] Gunion J F, Haber H E, Kane G L and Dawson S, 2000 "The Higgs Hunter's Guide," Front. Phys. **80** 1.
[3] Branco G C, Ferreira P M, Lavoura L, Rebelo M N, Sher M and Silva J P, 2012 "Theory and phenomenology of two-Higgs-doublet models," Phys. Rept. **516** 1 [arXiv:1106.0034 [hep-ph]].

[4] Botella F J, Branco G C, Nebot M and Rebelo M N, 2005 "New physics and evidence for a complex CKM," Nucl. Phys. B **725** 155 [hep-ph/0502133].

[5] Bento L and Branco G C, 1990 "Generation of a K-M phase from spontaneous CP breaking at high-energy scale," Phys. Lett. B **245** 599.

[6] Bento L, Branco G C and Parada P A, 1991 "A Minimal model with natural suppression of strong CP violation," Phys. Lett. B **267** 95.

[7] Branco G C, Parada P A and Rebelo M N, 2003 "A Common origin for all CP violations," hep-ph/0307119.

[8] Glashow S L and Weinberg S, 1977 "Natural Conservation Laws for Neutral Currents," Phys. Rev. D **15** 1958.

[9] Paschos E A, 1977 "Diagonal Neutral Currents," Phys. Rev. D **15** 1966.

[10] Branco G C, Grimus W and Lavoura L, 1996 "Relating the scalar flavor changing neutral couplings to the CKM matrix," Phys. Lett. B **380** 119 [hep-ph/9601383].

[11] Botella F J, Branco G C and Rebelo M N, 2010 "Minimal Flavour Violation and Multi-Higgs Models," Phys. Lett. B **687** 194 [arXiv:0911.1753 [hep-ph]].

[12] Botella F J, Branco G C, Nebot M and Rebelo M N, 2011 "Two-Higgs Leptonic Minimal Flavour Violation," JHEP **1110** 037 [arXiv:1102.0520 [hep-ph]].

[13] Sakharov A D, 1967 "Violation of CP Invariance, c Asymmetry, and Baryon Asymmetry of the Universe," Pisma Zh. Eksp. Teor. Fiz. **5** 32 [JETP Lett. **5** (1967) 24] [Sov. Phys. Usp. **34** (1991) 392] [Usp. Fiz. Nauk **161** (1991) 61].

[14] del Aguila F, Chase M K and Cortes J, 1986 "Vector-like fermion contributions to ," Nucl. Phys. B **271** 61.

[15] Branco G C and Lavoura L, 1986 "On the Addition of Vector Like Quarks to the Standard Model," Nucl. Phys. B **278** 738.

[16] Lavoura L and Silva J P, 1993 "Bounds on the mixing of the down type quarks with vector - like singlet quarks," Phys. Rev. D **47** 1117.

[17] Branco G C, Morozumi T, Parada P A, and Rebelo M N, 1993 "CP asymmetries in B0 decays in the presence of flavor changing neutral currents," Phys. Rev. D **48** 1167.

[18] Barenboim G, Botella F J, Branco G C and Vives O, 1997 "How sensitive to FCNC can B0 CP asymmetries be?," Phys. Lett. B **422** 277 [hep-ph/9709369].

[19] Barenboim G, Botella F J and Vives O, 2001 "Tree level FCNC in the B system: From CP asymmetries to rare decays," Phys. Rev. D **64** 015007 [hep-ph/0012197].

[20] Barenboim G, Botella F J and Vives O, 2001 "Constraining models with vector - like fermions from FCNC in K and B physics," Nucl. Phys. B **613** 285 [hep-ph/0105306].

[21] Aguilar-Saavedra J A, 2004 "Effects of mixing with quark singlets," Phys. Rev. D **67** (2003) 035003 Erratum: [Phys. Rev. D **69** (2004) 099901] [hep-ph/0210112].

[22] Aguilar-Saavedra J A, Botella F J, Branco G C and Nebot M, 2005 "The Size of chi = arg (-V(ts) V*(tb) V*(cs) V(cb)) and physics beyond the standard model," Nucl. Phys. B **706** 204 [hep-ph/0406151].

[23] Botella F J, Branco G C and Nebot M, 2009 "Small violations of unitarity, the phase in $B_s^0 - \bar{B}_s^O$ and visible $t \to cZ$ decays at the LHC," Phys. Rev. D **79** 096009 [arXiv:0805.3995 [hep-ph]].

[24] Higuchi K and Yamamoto K, 2010 "Flavor-changing interactions with singlet quarks and their implications for the LHC," Phys. Rev. D **81** 015009 [arXiv:0911.1175 [hep-ph]].

[25] Botella F J, Branco G C and Nebot M, 2012 "The Hunt for New Physics in the Flavour Sector with up vector-like quarks," JHEP **1212** 040 [arXiv:1207.4440 [hep-ph]].

[26] Fajfer S, Greljo A, Kamenik J F and Mustac I, 2013 "Light Higgs and Vector-like Quarks without Prejudice," JHEP **1307** 155 [arXiv:1304.4219 [hep-ph]].

[27] Aguilar-Saavedra J A, Benbrik R, Heinemeyer S and Pérez-Victoria M, 2013 "Handbook of vectorlike quarks: Mixing and single production," Phys. Rev. D **88** 094010 [arXiv:1306.0572 [hep-ph]].

[28] Ellis S A R, Godbole R M, Gopalakrishna S and Wells J D, 2014 "Survey of vector-like fermion extensions of the Standard Model and their phenomenological implications," JHEP **1409** 130 [arXiv:1404.4398 [hep-ph]].

[29] Alok A K, Banerjee S, Kumar D, Sankar S U and London D, 2015 "New-physics signals of a model with a vector-singlet up-type quark," Phys. Rev. D **92** 013002 [arXiv:1504.00517 [hep-ph]].

[30] Ishiwata K, Ligeti Z and Wise M B, 2015 "New Vector-Like Fermions and Flavor Physics," JHEP **1510** 027 [arXiv:1506.03484 [hep-ph]].

[31] Bobeth C, Buras A J, Celis A and Jung M, 2016 "Patterns of Flavour Violation in Models with Vector-Like Quarks," arXiv:1609.04783 [hep-ph].

[32] Branco G C, P. A. Parada and Rebelo M N, 1995 "D0 - anti-D0 mixing in the presence of isosinglet quarks," Phys. Rev. D **52** 4217 [hep-ph/9501347].

[33] Minkowski P, 1077 "$\mu \to e\gamma$ at a Rate of One Out of 10^9 Muon Decays?," Phys. Lett. **67B** 421.

[34] Yanagida T, 1979 KEK. in Proc. of the Workshop on Unified Theory and Baryon Number in the Universe March.

[35] Glashow S L, 1980 "Quarks and Leptons - Plenum. in Cargèse", Lvy M et al. (eds.) New York, p. 707.
[36] Gell-Mann M, Ramond P and Slansky R, 1979 "Complex Spinors and Unified Theories," Conf. Proc. C **790927** 315 [arXiv:1306.4669 [hep-th]].
[37] Mohapatra R N and Senjanovic G, 1980 "Neutrino Mass and Spontaneous Parity Violation," Phys. Rev. Lett. **44** 912.
[38] Fukugita M and Yanagida T, 1986 "Baryogenesis Without Grand Unification," Phys. Lett. B **174** (1986) 45.
[39] Branco G C, T. Morozumi, B. M. Nobre and Rebelo M N, 2001 "A Bridge between CP violation at low-energies and leptogenesis," Nucl. Phys. B **617** 475 [hep-ph/0107164].
[40] Rebelo M N, 2002 "Leptogenesis without CP violation at low-energies," Phys. Rev. D **67** 013008 [hep-ph/0207236].
[41] Ramond P, Roberts R G and Ross G G, 1993 "Stitching the Yukawa quilt," Nucl. Phys. B **406** 19 [hep-ph/9303320].
[42] Branco G C and Silva-Marcos J I, 1994 "NonHermitian Yukawa couplings?," Phys. Lett. B **331** 390.
[43] Frampton P H, Glashow S L and Marfatia D, 2002 "Zeroes of the neutrino mass matrix," Phys. Lett. B **536** 79 [hep-ph/0201008].
[44] Kageyama A, Kaneko S, Shimoyama N and Tanimoto M, 2002 "Seesaw realization of the texture zeros in the neutrino mass matrix," Phys. Lett. B **538** 96 [hep-ph/0204291].
[45] Kaneko S and Tanimoto M, 2003 "Neutrino mass matrix with two zeros and leptogenesis," Phys. Lett. B **551** 127 [hep-ph/0210155].
[46] Frampton P H, Glashow S L and Yanagida T, 2002 "Cosmological sign of neutrino CP violation," Phys. Lett. B **548** 119 [hep-ph/0208157].
[47] Branco G C, Gonzalez Felipe R, Joaquim F R, Masina I, Rebelo M N and Savoy C A, 2003 "Minimal scenarios for leptogenesis and CP violation," Phys. Rev. D **67** 073025 [hep-ph/0211001].
[48] Casas J A and Ibarra A, 2001 "Oscillating neutrinos and muon —¿ e, gamma," Nucl. Phys. B **618** 171 [hep-ph/0103065].
[49] Branco G C, Rebelo M N and Silva-Marcos J I, 2006 "Leptogenesis, Yukawa textures and weak basis invariants," Phys. Lett. B **633** 345 [hep-ph/0510412].
[50] Branco G C, Emmanuel-Costa D, Rebelo M N and Roy P, 2008 "Four Zero Neutrino Yukawa Textures in the Minimal Seesaw Framework," Phys. Rev. D **77** 053011 [arXiv:0712.0774 [hep-ph]].
[51] Björkeroth F, de Anda F J, de Medeiros Varzielas I and King S F, 2015 "Leptogenesis in minimal predictive seesaw models," JHEP **1510** 104 [arXiv:1505.05504 [hep-ph]].
[52] Hagedorn C, Molinaro E and Petcov S T, 2009 "Majorana Phases and Leptogenesis in See-Saw Models with A(4) Symmetry," JHEP **0909** 115 [arXiv:0908.0240 [hep-ph]].
[53] Branco G C, Gonzalez Felipe R, Rebelo M N and Serodio H, 2009 "Resonant leptogenesis and tribimaximal leptonic mixing with A(4) symmetry," Phys. Rev. D **79** 093008 [arXiv:0904.3076 [hep-ph]].
[54] Altarelli G and Feruglio F, 2010 "Discrete Flavor Symmetries and Models of Neutrino Mixing," Rev. Mod. Phys. **82** 2701 [arXiv:1002.0211 [hep-ph]].
[55] Kobayashi M and Maskawa T, 1973 "CP Violation in the Renormalizable Theory of Weak Interaction," Prog. Theor. Phys. **49** 652.
[56] Bernabeu J, Branco G C and Gronau M, 1986 "Cp Restrictions On Quark Mass Matrices," Phys. Lett. **169B** 243.
[57] Froggatt C D and Nielsen H B, 1979 "Hierarchy of Quark Masses, Cabibbo Angles and CP Violation," Nucl. Phys. B **147** 277.
[58] Botella F J Branco, G C, Nebot M, Rebelo M N and Silva-Marcos J I, 2016 "Vector-like Quarks at the Origin of Light Quark Masses and Mixing," arXiv:1610.03018 [hep-ph].

IOP Conf. Series: Journal of Physics: Conf. Series **873** (2017) 012012 doi:10.1088/1742-6596/873/1/012012

Gravitational waves

Andrzej Trautman
Faculty of Physics, University of Warsaw
E-mail: Andrzej.Trautman@fuw.edu.pl

Abstract. Historical remarks on early theoretical work on the subject. Very early on, Einstein introduced the notion of gravitational waves, but later became convinced that they did not exist as a physical phenomenon. Exact solutions of Einstein's equations representing waves were found by a number of authors, contributing to their final acceptance as part of physics.

The recent observation of gravitational waves emitted by a system of two orbiting black holes [1] increased the interest in the phenomenon of gravitational radiation. This short note recalls some of the early theoretical work on this subject.

The first two papers on gravitational waves were by Albert Einstein [2] [3]. There was a small computational error in the first one.

In the linear approximation, the metric tensor is written as $g_{\mu\nu} = \eta_{\mu\nu} + h_{\mu\nu}$ and the following coordinate conditions, analogous to the Lorentz condition, have been used by Einstein:

$$(h^{\mu\nu} - \tfrac{1}{2}\eta^{\mu\nu}\eta_{\rho\sigma}h^{\rho\sigma})_{,\nu} = 0.$$

They are the linear form of the harmonic coordinates, later introduced by V.A. Fock,

$$(\sqrt{|g|}g^{\mu\nu})_{,\nu} = 0 \qquad \text{or} \qquad \Box x^{\mu} = 0.$$

The linearized Einstein field equations reduce to

$$\Box h^{\mu\nu} = \text{const.}\, T^{\mu\nu}.$$

By solving them, Einstein found plane gravitational waves and radiation from a bounded source – all of this was valid in the linear, weak field approximation to the field equations.

In the 1920s there appeared a paper [4] by the mathematician H. W. Brinkmann on exact plane gravitational waves (that name was not used by the author). At that time the paper was not noticed by physicists.

In the 1930s, probably under the influence of work on cosmology, Einstein recognizes the importance of using full – rather than linearized – equations. With Nathan Rosen he attempted to find exact plane waves. They obtained solutions of $R_{\mu\nu} = 0$ exhibiting singularities and, in 1936, they sent to the *Physical Review* a manuscript entitled *Do gravitational waves exist?*, concluding that they do not.

The referee – thanks to Daniel Kennefick [5] we now know that was H.P. Robertson – pointed out that the singularities may be due to the choice of coordinates and suggested revisions.

Einstein wrote to the editor of *Physical Review*: "We sent you the paper for publication and had not authorized you to show it to specialists before it is printed... I prefer to publish the paper elsewhere." Indeed, there is a paper [6] by Einstein and Rosen "On gravitational waves", published in *Journal of the Franklin Institute*, **223** (1937) 43–54, where they accepted Robertson suggestion and interpreted the solution as representing cylindrical waves.

Leopold Infeld, a Polish physicist, was with Einstein in Princeton in 1936-38. With Banesh Hofmann they then created a new approximate method (EIH) of finding the equations of motion of gravitating bodies. Infeld did not participate in the Einstein–Rosen collaboration, but under Einstein's influence, became convinced that gravitational waves did not exist. After the war, he returned to Poland and created the Institute of Theoretical Physics in Warsaw. Jerzy Plebański was his first collaborator here; in 1954 he gave a course of theoretical physics at Politechnika Warszawska and invited Andrzej Trautman (AT), one of the students, to join Infeld's group as a Ph. D. student. He suggested gravitational waves as the subject, warning that Infeld did not believe in their existence. In 1957, using Sommerfeld's outgoing radiation conditions, adapted to the case of gravitation, AT showed that an isolated system could radiate energy and that the curvature tensor was then of the form

$$N/r + O(r^{-2})$$

where N is a tensor of the same, simple form, as it appears for plane waves.

When AT has shown his results to Felix Pirani, who was then visiting Warsaw, Felix invited him to King's College, London, to give lectures (1958). Ivor Robinson (IR) and AT met there for the first time.

In 1958, there appeared a paper [7] by H. Bondi, F. Pirani and I. Robinson giving a new account of plane gravitational waves, free of singularities, making an explicit use of the 5-dimensional group of symmetries of those waves.

Robinson introduced the notion of congruences of (foliations by) null geodesics without shear; later they were characterized (by IR and AT) by a null vector field k in space-time M such that the flow generated by k preserves the conformal geometry of the screen bundle

$$\mathcal{K}^{\perp}/\mathcal{K} \to M, \qquad \mathcal{K}_p = \mathbb{R}k_p.$$

Many new gravitational fields were found using the requirement that they admit such a shear-free flow. The most important among them was the field of a general, rotating, black hole discovered by Roy Kerr (1963) [8].

IR and AT found [9] a new class of solutions of Einstein equations, containing gravitational waves with spherical wave fronts, carrying information in the sense of depending on arbitrary functions of $u = t - r$. They are given in coordinates (x, y, u, r) with the function $p(x, y, u)$ as

$$2\,\mathrm{d}r\,\mathrm{d}u + (K - 2p^{-1}\frac{\partial p}{\partial u}r - \frac{2m}{r})\,\mathrm{d}u^2 - r^2p^{-2}(\mathrm{d}x^2 + \mathrm{d}y^2),$$

where

$$K = \Delta \log p, \qquad \Delta = p^2(\frac{\partial^2}{\partial x^2} + \frac{\partial^2}{\partial y^2}),$$

is the Gaussian curvature of the surface $r = 1,\ u = \text{const.}$.

The field equations $R_{\mu\nu} = 0$ reduce to

$$12m\frac{\partial}{\partial u}\log p + \Delta\Delta \log p = 0.$$

If $m = 0$, then the dependence of p on u is unrestricted.

There is 'peeling off' of the curvature exhibiting its Petrov types,

$$\text{Riemann tensor} = \frac{N}{r} + \frac{III}{r^2} + \frac{D}{r^3}.$$

The Schwarzschild metric is among those solutions; in this case the Riemann tensor is D/r^3.

There is a clear electromagnetic analog of the N/r solution: plane e.m. reflected from a perfectly conducting paraboloid of revolution.

A discrete aside. Spinors that appear throughout physics have a 'discrete aspect'. The Pythagorean equation $x^2+y^2=z^2$ can be interpreted as a condition for the vector $(x,y,z)\in\mathbb{Z}^3$ to be *null* w.r.t. a scalar product of signature $(2,1)$. Euclid's solution

$$x=p^2-q^2,\quad y=2pq,\quad z=p^2+q^2,$$

written as

$$2\begin{pmatrix}p\\q\end{pmatrix}\begin{pmatrix}p & q\end{pmatrix}=\begin{pmatrix}x+z & y\\ y & z-x\end{pmatrix}$$

reads now as the equality

$$(\text{spinor})^2=\text{null vector}.$$

There is a homomorphism of groups
$\rho:\mathrm{SL}_2(\mathbb{Z})\to\mathrm{SO}_{2,1}$ given by

$$\rho(A)X=AXA^t,\quad X=\begin{pmatrix}x+z & y\\ y & z-x\end{pmatrix},$$

where now $(x,y,z)\in\mathbb{R}^3$ and A^t = transpose of the matrix $A\in\mathrm{SL}_2(\mathbb{Z})$.

One shows that the group $\mathrm{SL}_2(\mathbb{Z})$ acts transitively on the set of integer-valued spinors with relatively prime components. The group $\rho(\mathrm{SL}_2(\mathbb{Z}))$ contains matrices with half-integer elements, e.g.

$$\rho\begin{pmatrix}1 & 1\\ 0 & 1\end{pmatrix}=\begin{pmatrix}\frac{1}{2} & 1 & \frac{1}{2}\\ -1 & 1 & 1\\ -\frac{1}{2} & 1 & \frac{3}{2}\end{pmatrix}.$$

References

[1] B. P. Abbott *et al.* Phys. Rev. Lett. **116**, 061102 (2016), doi
[2] A. Einstein, Sitzungsber. Preuss. Akad. Wiss. Berlin (Math. Phys.) **1916**, 688 (1916),
[3] A. Einstein, Sitzungsber. Preuss. Akad. Wiss. Berlin (Math. Phys.) **1918**, 154 (1918),
[4] H. W. Brinkmann, Math. Ann. **94**, 119 (1925), doi:10.1007/BF01208647
[5] D. Kennefick, Phys. Today **58**, 43 (2005), doi: 10.1063/1.2117822
[6] A. Einstein and N. Rosen, J. Franklin Inst. **223**, 43 (1937), doi:10.1016/S0016-0032(37)90583-0,
[7] H. Bondi, F. A. E. Pirani and I. Robinson, Proc. Roy. Soc. Lond. A **251**, 519 (1959). doi:10.1098/rspa.1959.0124
[8] R. P. Kerr, Phys. Rev. Lett. **11**, 237 (1963). doi:10.1103/PhysRevLett.11.237
[9] I. Robinson and A. Trautman, Phys. Rev. Lett. **4**, 431 (1960). doi:10.1103/PhysRevLett.4.431,

Fifth Symposium on Prospects in the Physics of Discrete Symmetries IOP Publishing
IOP Conf. Series: Journal of Physics: Conf. Series **873** (2017) 012013 doi:10.1088/1742-6596/873/1/012013

First detections of gravitational waves from binary black holes

Michał Bejger

Nicolaus Copernicus Astronomical Center, Polish Academy of Sciences, ul. Bartycka 18, 00-716 Warsaw, Poland

E-mail: bejger@camk.edu.pl

Abstract. Recent direct detections of gravitational waves from coalescing binary black holes systems herald a new era in the observational astronomy, as well as in experimental verifications of the theories of gravity. I will present the principles of detection of gravitational waves, current state-of-art laser interferometric detectors (Advanced LIGO and Advanced Virgo), and the most promising astrophysical sources of gravitational waves.

1. Introduction

Before I begin, I would like to express my pleasure to be here and to present these exciting new results on the special session of the DISCRETE 2016 Symposium and the Leopold Infeld Colloquium. As a collaborator of Albert Einstein himself, prof. Infeld was, as you know, intimately involved in the early gravitational-wave research.

According to Einstein's general theory of relativity, published in 1915, the force of gravity is an illusion. Massive bodies seem to attract each other because they move in a curved spacetime on the "most straight" trajectories, geodesics. As phrased by John Archibald Wheeler, there is a reciprocal relation between massive objects and the geometry of spacetime: "mass tells spacetime how to curve, and spacetime tells mass how to move". This has to be contrasted with the usual Newtonian viewpoint on the mechanics of moving bodies - any observer can always tell the time and space coordinate of any given event, because objects move in absolute time and within absolute space. Because of the relation between the geometry and objects acting in it, phenomena thought to be stable (everlasting, not changing in time) are no longer such. For example, the classic example and success of the Newtonian theory - the binary system - is doomed to change its parameters and finally seize to exist. In the story of general relativity, in addition to two bodies comprising the binary there is a third actor, taking the energy from the system - the spacetime.

Very shortly after announcing the general theory of relativity, Albert Einstein realized that a linearised version of his equations resembles the wave equation [1]. The solution is interpreted as a short-wavelength, time-varying curvature deformation propagating with the speed of light on an otherwise slowly-varying, large-scale curvature background (a gravitational-wave "ripple" propagating through the four-dimensional spacetime); from the point of view of a metric tensor, it represents a small perturbation of a stationary background metric. Linear approximation corresponds to the waves propagating in the far-field limit. An intuitive picture was provided by Kip Thorne in [2]. Far-field limit for gravitational waves is there illustrated with an orange:

the background (slow-varying) curvature radius is the radius of the orange, and the small-scale dimples on the surface of the orange are the gravitational waves. By exploiting the gauge freedom of the theory one may show that the solution has features similar to electromagnetic waves: it is a transverse wave which may be polarized (has two independent polarizations).

Over the next 40 years, during which Einstein changed his mind to argue against their genuineness, a controversy persisted over the true nature of gravitational waves. One of the fundamental concepts of general relativity is that the results should be coordinate independent: one should be able to choose a system of coordinates and recover the same results. Einstein became convinced that gravitational waves are an artifact of a poor choice of coordinate system (Leopold Infeld was involved in these works as Einstein's assistant while working with him on the problem of relativistic binary system motion in the Institute for Advanced Study in Princeton).

Only in the late 50s and early 60s the works of Felix Pirani, [3], Herman Bondi [4], Ivor Robertson and Andrzej Trautman [5] convincingly showed that gravitational waves are indeed physical phenomena that carry and deposit energy. Anecdotally, a simple argument called "the sticky bead argument" was supplied anonymously by Richard Feynman under a pseudonym Mr. Smith at a conference at Chapel Hill, North Carolina in 1957. Feynman's gravitational-wave detector would consists of "two beads sliding freely (but with a small amount of friction) on a rigid rod. As the wave passes over the rod, atomic forces hold the length of the rod fixed, but the proper distance between the two beads oscillates. Thus, the beads rub against the rod, dissipating heat" (from the foreword of Kip Thorne and John Preskill in [6]). Leopold Infeld was also finally convinced of the reality of gravitational waves by one of his student collaborators, Róża Michalska-Trautman [7, 8].

The idea to use the light (beam detectors) for the gravitational-wave detection was first suggested in 1956 by Felix Pirani. Critical insight of the late 1950s convinced the experimentalists that it is worth investing some effort into building a device able to detect gravitational waves. In 1962 M. E. Gertsenshtein and V. I. Pustovit were the first to suggest an optical interferometer as a detecting device [9]. From the mid-1960s Joseph Weber pursued the idea of a *resonant bar* detector - a gravitational wave passing through an isolated mass of known characteristic frequency would induce tiny vibrations that may in principle be detected. The concept of laser interferometry was independently re-discovered by Rainer Weiss in the 1970s, which ultimately led to the LIGO project.

Before going to describe the current state of detectors and the first detections, I would also like to mention an early deep insight of Bohdan Paczyński, who already in 1967 understood the impact of gravitational wave emission on the fate of tight binary systems of stars [10]: the WZ Sge and AM CVn couldn't exist in their present state if not for the gravitational-wave emission. In the 1974 the first relativistic binary neutron-star system was discovered by Russell Hulse and Joseph Taylor [11]; it was shown without doubt that it is losing orbital energy in concordance with the emission of gravitational waves in the general relativity [12, 13]. These early observations provided much needed, *indirect* support for the existence of gravitational waves.

2. The era of detections

Recent first direct detections of gravitational waves with the two LIGO detectors [14, 15] create an unprecedented opportunity for studying the Universe through a novel, never before explored channel of spacetime fluctuations. Gravitational wave astronomy is often compared to 'listening to' rather than 'looking at' the skies. By design which is motivated by the choice of potential sources, the ground-based gravitational wave detectors of Advanced LIGO [16] and Advanced Virgo [17] are sensitive in the range of frequencies similar to the audible range of human ears - between 10 Hz and a few kHz. As in the case of an ear, a solitary laser interferometric detector is practically omnidirectional (has a poor angular resolution), and has no imaging capabilities.

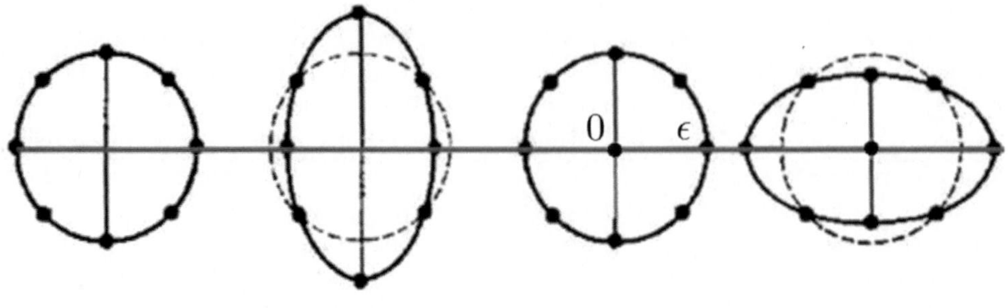

Figure 1. Proper distance change in the ring of test particles.

It registers a coherent signal emitted by a bulk movement of large, rapidly-moving masses. Once emitted, gravitational waves are weakly coupled to the surrounding matter and propagate freely without scattering. This has to be contrasted with the electromagnetic emission which originates at the microscopic level, is strongly coupled to the surroundings and often reprocessed; it carries a reliable information from the last scattering surface only. It seems therefore that gravitational wave detectors are the perfect counterpart to the electromagnetic observatories as they may provide us with information impossible to obtain by other means.

A realistic wave phenomenon (and not, say, a coordinate artifact) must be capable of transmitting energy from the source to infinity. If the amplitude of an exemplary isotropic field at a radial distance r from the source is $h(r)$, then the flux of energy over a spherical surface at r is $\mathcal{F}(r) \propto h^2(r)$, and the total emitted power (the luminosity) is $\mathcal{L}(r) \propto 4\pi r^2 h^2(r)$. Since the energy has to be conserved, the amplitude $h(r)$ falls like $1/r$, irrespectively of the multipole character of the source (the lowest radiating multipole in the gravity theory is the quadrupole distribution, because for an isolated system a time-changing monopole would correspond to the violation of the mass-energy conservation, and a time-changing dipole would violate the momentum conservation law).

Gravitational waves are related to the changes in the spacetime distance (the proper time interval), therefore they cannot be detected by a local measurement - one has to compare the spacetime positions of remote events [3]. The difficulty of the experiment is then described by a question: how to measure space-time distances, if the ruler also changes length? Fortunately, there is one aspect of the world that is constant at all times: the speed of light.

Consider the proper distance between free-falling particles initially forming a ring (see Fig. 1). Assuming a metric tensor composed of two parts, $g_{\mu\nu} = \eta_{\mu\nu} + h_{\mu\nu}$, where the $h_{\mu\nu}$ is denoting the gravitational-wave distortion, and taking a coordinate system in which for two test particles, both initially at rest, one at $x = 0$ and the other at $x = \epsilon$ the proper distance between them is

$$\Delta s = \int |g_{\mu\nu} dx^\mu dx^\nu|^{1/2} \to \int_0^\epsilon |g_{xx}|^{1/2} \approx \epsilon \sqrt{g_{xx}\,(x=0)}. \tag{1}$$

By taking $g_{xx}\,(x=0) = \eta_{xx} + h_{xx}\,(x=0)$ and keeping only the linear terms one gets

$$\Delta s \approx \epsilon \left(1 + \frac{1}{2} h_{xx}(x=0)\right), \tag{2}$$

which is, in general, time-varying. This simple argument can be made robust by solving the geodesic deviation equation.

This detection principle in the case of the laser interferometric detector is to measure the difference of the relative change in its perpendicular arms' lengths L_x and L_y, $\delta L_x - \delta L_y = \Delta L/L$, by measuring the interference pattern in the output port located at the apex of the device (or, equivalently, to measure the difference of round-trip times of a laser light circulating in the arms of the detector). Due to the quadrupolar nature of a gravitational wave, shortening of one arm corresponds to elongation of the other. This change of lengths is reflected in varying paths that the laser light has to cross before the interference. The dimensionless gravitational-wave amplitude $h = \Delta L/L$ (the "strain") is proportional to the amount of outgoing laser light. The fact that the directly-measurable quantity is the amplitude $h \propto 1/r$, not the energy of the wave as in the electromagnetic antennæ, has a direct consequence for the reach of the observing device: one-order-of-magnitude sensitivity improvement corresponds to one-order-of-magnitude growth of distance reach r, as opposed to the factor of $\sqrt{10}$ in the electromagnetic observations (consequently, the volume of space grows like r^3 in case of gravitational-wave observations, encompassing hundreds of thousands of galaxies for a distance reach of the order of hundreds of Mpc, see [18]).

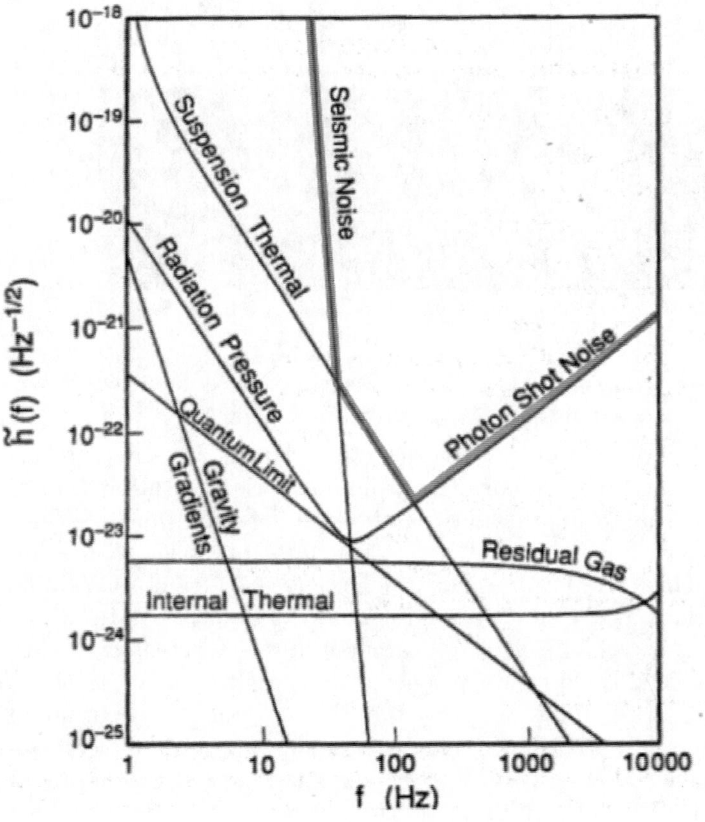

Figure 2. The sensitivity curve of the Initial LIGO detector (from the 1989 LIGO proposal to NSF[2]). The main sources of noise is the ground movement (seismic noise) for low frequencies, thermal movement of mirrors and their suspensions for frequencies about 100 Hz, and photon-shot noise related to the quantum nature of the laser light for high frequencies. For comparison see inset (b) in Fig. 5 for the LIGO O1 observational campaign sensitivity curve.

Among promising sources of gravitational radiation are all asymmetric collapses and explosions (supernovæ), rotating deformed stars (gravitational-wave 'pulsars' of continuous and transient nature), and tight binary systems of e.g., neutron stars and black holes. In the following we will focus on the binary systems, since their properties make them the analogues to standard candles of traditional astronomy. The fittingly descriptive term "standard siren" was first used in [19] in the context of gravitational waves from super-massive binary black holes as a target for the planned spaceborne LISA detector [20]. The idea of using well-understood signals to infer the

[2] `https://dcc.ligo.org/LIGO-M890001/public`

IOP Conf. Series: Journal of Physics: Conf. Series **873** (2017) 012013 doi:10.1088/1742-6596/873/1/012013

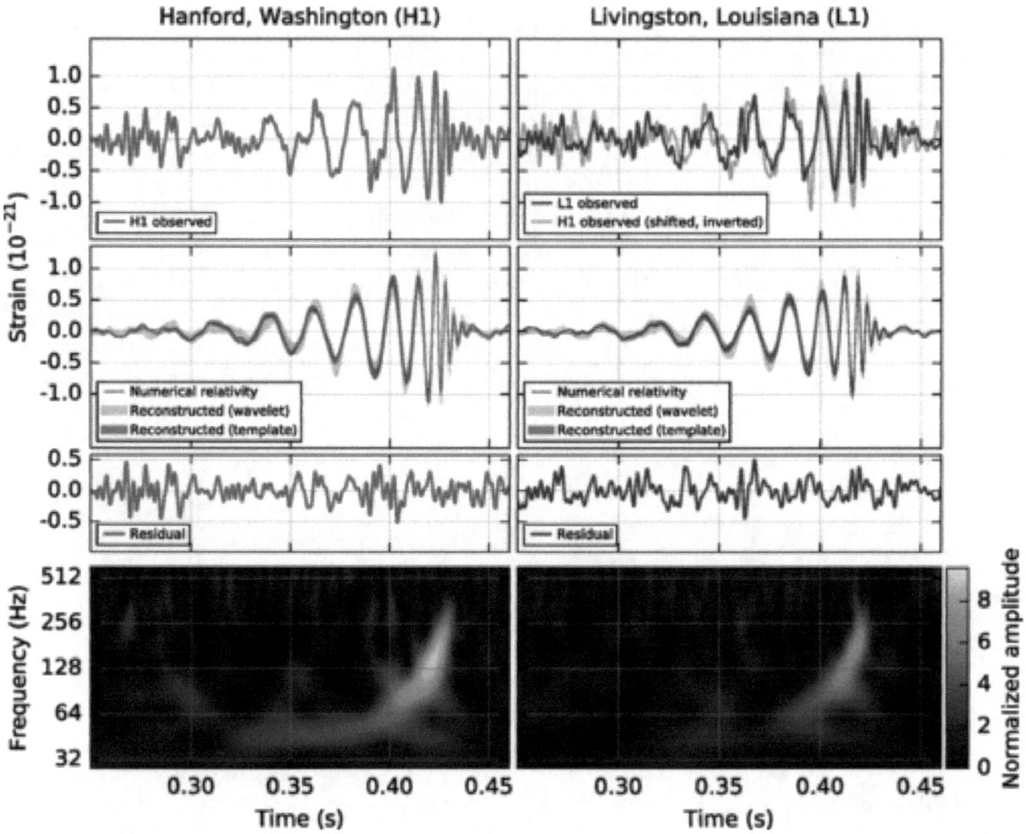

Figure 3. The gravitational-wave event GW150914 observed by the LIGO Hanford (H1, left column panels) and Livingston (L1, right column panels) detectors. Times are shown relative to September 14, 2015 at 09:50:45 UTC. For visualization, all time series are filtered with a 35-350 Hz bandpass filter to suppress large fluctuations outside the detectors' most sensitive frequency band, and band-reject filters to remove the strong instrumental spectral lines seen in the Fig. 3 spectra. *Top row, left:* H1 strain. *Top row, right:* L1 strain. GW150914 arrived first at L1 and $6.9^{+0.5}_{-0.4}$ ms later at H1; for a visual comparison, the H1 data are also shown, shifted in time by this amount and inverted (to account for the detectors' relative orientations). *Second row:* Gravitational-wave strain projected onto each detector in the 35-350 Hz band. Solid lines show a numerical relativity waveform for a system with parameters consistent with those recovered from GW150914. Shaded areas show 90% credible regions for two independent waveform reconstructions. *Third row:* Residuals after subtracting the filtered numerical relativity waveform from the filtered detector time series. Bottom row:A time-frequency representation of the strain data, showing the signal frequency increasing over time (the characteristic chirp). Reprinted from [14] under the terms of the Creative Commons Attribution 3.0 License.

distance and constrain the cosmological parameters was however proposed much earlier [21, 22]. Here we will focus on the inspiraling and merging binary systems.

Magnitudes of the gravitational-wave strain h and the luminosity $\mathcal{L}$ may be estimated using dimensional analysis and Newtonian physics. As the waves are generated by the accelerated movement of masses and the mass distribution should be quadrupolar, one may assume that h

is proportional to a second time derivative of the quadrupole moment $I_{ij} = \int \rho(\mathbf{x}) x_i x_j d^3x$ for some matter distribution $\rho(\mathbf{x})$. For a binary composed of masses m_1 and m_2, orbiting the center of mass at a separation a with the orbital angular velocity ω, h is proportional to the system's moment of inertia μa^2 and to ω^2, as well as inversely proportional to the distance, $h \propto \mu a^2 \omega^2 / r$, where $\mu = m_1 m_2 / M$ is the reduced mass, and $M = m_1 + m_2$ the total mass. In order to recover the dimensionless h, the characteristic constants of the problem, G and c, are used to obtain

$$h \simeq \frac{G}{c^4}\frac{1}{r}\mu a^2 \omega^2 = \frac{G^{5/3}}{c^4}\frac{1}{r}\mu M^{2/3}\omega^{2/3} \qquad \left(h_{ij} = \frac{2G}{c^4 r}\ddot{I}_{ij} \right), \tag{3}$$

with the use of Kepler's third law ($GM = a^3\omega^2$) in the second equation. The expression in brackets represents the strain tensor h_{ij} in the non-relativistic *quadrupole approximation* [23]. Similarly, the luminosity $\mathcal{L}$ (the rate of energy loss in gravitational waves, integrated over a sphere at a distance r) should be proportional to $h^2 r^2$ and some power of ω. From dimensional analysis one has

$$\mathcal{L} = \frac{dE_{GW}}{dt} \propto \frac{G}{c^5} h^2 \omega^2 \propto \frac{G}{c^5}\mu^2 a^4 \omega^6 \qquad \left(\mathcal{L} = \frac{c^3}{16G\pi} \iint \langle \dot{h}_{ij}\dot{h}^{ij} \rangle \, dS = \frac{G}{5c^5}\langle \dddot{I}_{ij} \dddot{I}^{ij} \rangle \right), \tag{4}$$

with the proportionality factor of 32/5. Again, the expression in brackets refers to the quadrupole approximation; the angle brackets denote averaging over the orbital period. Waves leave the system at the expense of its orbital energy $E_{orb} = -Gm_1m_2/(2a)$. Using the time derivative of the third Kepler's third law, $\dot{a} = -2a\dot{\omega}/(3\omega)$, one gets the evolution of the orbital frequency driven by the gravitational-wave emission:

$$\frac{dE_{orb}}{dt} \equiv \frac{Gm_1m_2}{2a^2}\dot{a} = -\frac{dE_{GW}}{dt} \quad \Longrightarrow \quad \dot{\omega} = \frac{96}{5}\frac{\omega^{11/3}}{c^5} G^{5/3}\mathcal{M}^{5/3}. \tag{5}$$

The system changes by increasing its orbital frequency; at the same time the strain amplitude h of emitted waves also grows. Orbital frequency is related in a straightforward manner to the gravitational-wave frequency f_{GW}: from the geometry of the problem it is evident that the frequency of radiation is predominantly at twice the orbital frequency, $f_{GW} = \omega/\pi$. This characteristic frequency-amplitude evolution is called the *chirp*, by the similarity to birds' sounds, and the characteristic function of component masses,

$$\mathcal{M} = \left(\mu^3 M^2\right)^{1/5} = \frac{c^3}{G}\left(\frac{5}{96}\pi^{-8/3} f_{GW}^{-11/3} \dot{f}_{GW}\right)^{3/5} \tag{6}$$

is called the *chirp mass*. It is a quantity measured *directly* from the recorded values of f_{GW} and $\dot{f}_{GW}$. By combining the equations for $\dot{f}_{GW}$ and h, one recovers the distance to the source r. The distance, as a function of the frequency and amplitude parameters is, similarly to $\mathcal{M}$, *directly* measured by the detector:

$$r = \frac{5}{96\pi^2}\frac{c}{h}\frac{\dot{f}_{GW}}{f_{GW}^3} = 512 \frac{1}{h_{21}} \left(\frac{0.01\ \mathrm{s}}{\tau}\right) \left(\frac{100\ \mathrm{Hz}}{f_{GW}}\right)^2 \mathrm{Mpc}, \tag{7}$$

where h_{21} denotes the strain in the units of 10^{-21} and $\tau = f_{GW}/\dot{f}_{GW}$ denotes the rate of change of the gravitational-wave chirp frequency. Note that within the simple Newtonian approximation presented here (at the leading order of the post-Newtonian expansion) the product $h\tau$ is independent of the components' masses [24]. The simplified analysis presented above does not take into account the full post-Newtonian waveform, polarization information, network of detectors analysis etc., but is intended to demonstrate that binary systems are indeed truly extraordinary "standard sirens". Their observations provide absolute, physical distances directly, without the need for a calibration or a 'distance ladder'. Among them, binary black hole systems occupy a special position. In the framework of general relativity, binary black holes waveforms are independent from astrophysical assumptions about the systems' intrinsic parameters and their environment. At cosmological scales the distance r has a true meaning of the *luminosity distance*. However, since the vacuum (black-hole) solutions in general relativity are scale-free, the measurements of their waveforms alone cannot determine the source's redshift. The parameters measured by the detector are related to the rest-frame parameters by the redshift z: $f_{GW} = f^{rf}_{GW}/(1+z)$, $\tau = \tau^{rf}(1+z)$, $\mathcal{M} = \mathcal{M}^{rf}(1+z)$. Independent measurements of the redshift which would facilitate the cosmographic studies of the large-scale Universe requires the collaboration with the electromagnetic observers i.e., the *multi-messenger astronomy*. This may be obtained by assessing the redshift of the galaxy hosting the binary by detecting the electromagnetic counterpart of the event [25, 26] or by performing statistical study for galaxies' redshifts correlated with the position of the signal's host galaxy [27]. The omnidirectional nature of a solitary detector is mitigated by simultaneous data analysis from at least three ground-based detectors in order to perform the triangulation of the source position, hence the crucial need for the LIGO-Virgo collaboration, which will be in the future enhanced by the LIGO-India detector, and the KAGRA detector in Japan [28].

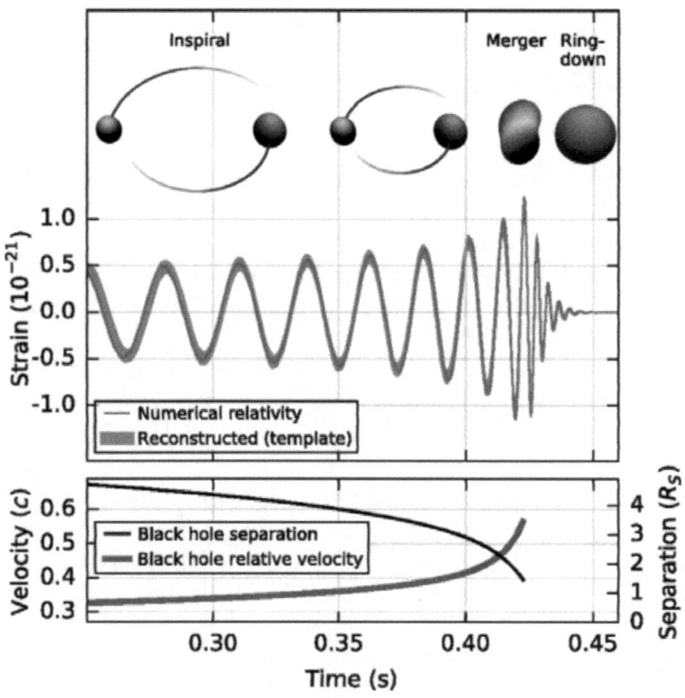

Figure 4. *Top:* Estimated gravitational-wave strain amplitude from GW150914 projected onto the H1 detector. This shows the full bandwidth of the waveforms, without the filtering used for Fig. 3. The inset images show numerical relativity models of the black hole horizons as the black holes coalesce. *Bottom:* The Keplerian effective black hole separation in units of Schwarzschild radii $R_s = 2GM/c^2$, and the effective relative velocity given by the post-Newtonian parameter $v/c = (GM\pi f_{GW})^{1/3}/c$, where f_{GW} is the gravitational-wave frequency obtained from numerical relativity calculations and M is the total mass of the system. Reprinted from [14] under the terms of the Creative Commons Attribution 3.0 License.

3. GW150914

On the 14th of September 2015, both LIGO detectors (first Livingston, Hanford 7 ms later) registered a signal featured in Fig. 3. The false alarm probability (probability of this signal to be spurious and occurring in pure noise data) was estimated to be less than 1 in 5 million, which

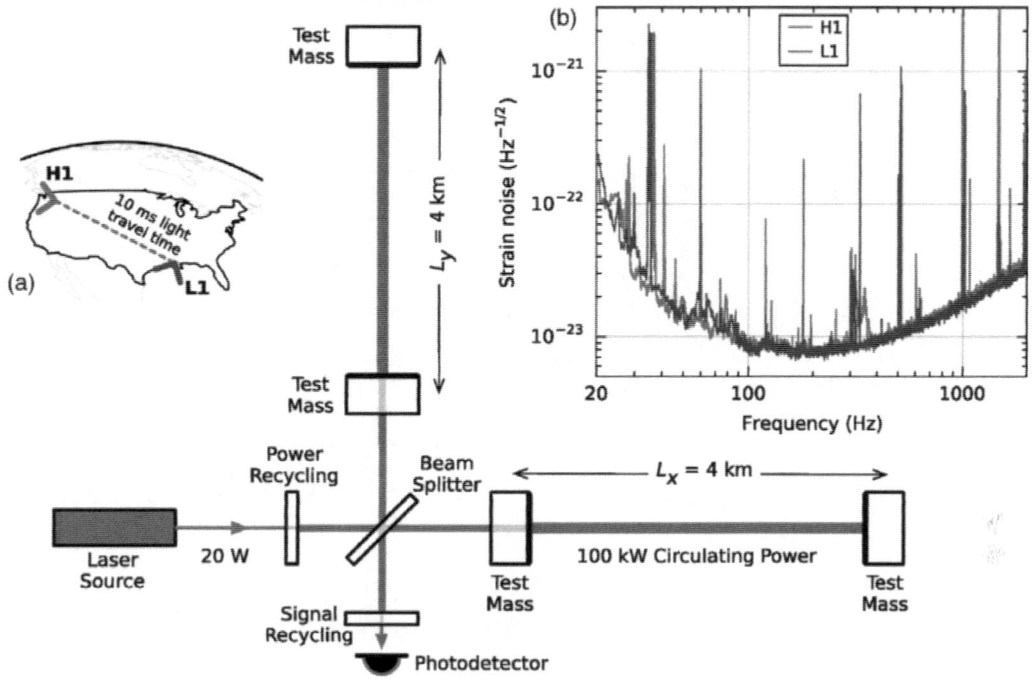

Figure 5. Simplified diagram of an Advanced LIGO detector (not to scale). A gravitational wave propagating orthogonally to the detector plane and linearly polarized parallel to the 4-km optical cavities will have the effect of lengthening one 4-km arm and shortening the other during one half-cycle of the wave; these length changes are reversed during the other half-cycle. The output photo-detector records these differential cavity length variations. While a detector's directional response is maximal for this case, it is still significant for most other angles of incidence or polarizations (gravitational waves propagate freely through the Earth). Inset (a): Location and orientation of the LIGO detectors at Hanford, WA (H1) and Livingston, LA (L1). Inset (b): The instrument noise for each detector near the time of the signal detection; this is an amplitude spectral density, expressed in terms of equivalent gravitational-wave strain amplitude. The sensitivity is limited by photon shot noise at frequencies above 150 Hz, and by a superposition of other noise sources at lower frequencies. Narrow-band features include calibration lines (33-38, 330, and 1080 Hz), vibrational modes of suspension fibers (500 Hz and harmonics), and 60 Hz electric power grid harmonics. Reprinted from [14] under the terms of the Creative Commons Attribution 3.0 License.

corresponds to a false alarm rate of less than 1 in 200000 years. The source frame component black-hole masses were estimated, by selecting a best-fit model in an offline comparison with detailed numerical relativity simulations, to $m_1 = 36^{+5}_{-4}\, M_\odot$ and $m_2 = 29^{+4}_{-4}\, M_\odot$. The final black hole mass is $M_f = 62^{+4}_{-4}\, M_\odot$, and its spin was estimated to be $a = 0.67^{+0.05}_{-0.07}$. The event happened at a distance of 410^{+160}_{-180} Mpc (the central value corresponds to 1 billion 300 million light years), and a redshift of $z = 0.09^{+0.03}_{-0.04}$, assuming standard cosmology. The uncertainties define the 90% credible intervals. This first ever direction of gravitational waves shows its potential for making truly long-range cosmological observations.

The whole event took less than 0.2 s in the sensitivity band of the detector, that is for frequencies higher than about 30 Hz. Final orbital velocity of the system components was larger than 0.5 c (see Fig. 4). Total energy emitted in gravitational waves during the lifetime

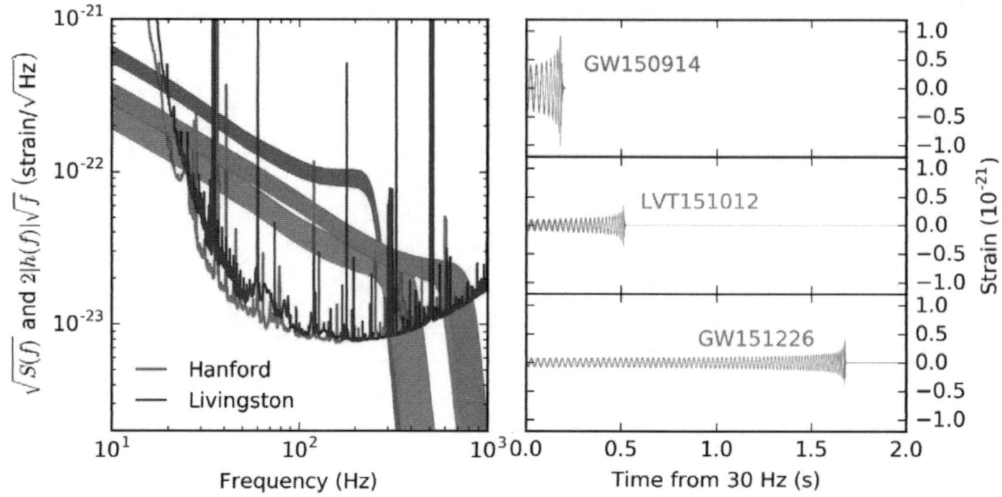

Figure 6. *Left panel:* Amplitude spectral density of the total strain noise of the H1 and L1 detectors, $\sqrt{S(f)}$, in units of strain per $\sqrt{\text{Hz}}$, and the recovered signals of GW150914, GW151226, and LVT151012 plotted so that the relative amplitudes can be related to the signal-to-noise of the signal (see Eq. 9). *Right panel:* Time evolution of the recovered signals from when they enter the detectors' sensitive band at 30 Hz. Both figures show the 90% credible regions of the LIGO Hanford signal reconstructions from a coherent Bayesian analysis using a non-precessing spin waveform model. Reprinted from [30] under the terms of the Creative Commons Attribution 3.0 License.

of the system was estimated to be equal to $3^{+0.5}_{-0.5}\ M_{\odot}c^2$. Peak "brightness" of the event, which happened while the two black holes were already merging was estimated to be $3.6^{+0.5}_{-0.4} \times 10^{49}$ Joule/s. This amazing efficiency of transmitting the energy into spacetime means that this event was the most energetic phenomenon observed by humanity to date; the brightness exceeded the electromagnetic brightness of all the 10^{22} stars in the observed Universe. From the point of view of energy production, it equals $200^{+30}_{-30}\ \mathrm{M}_{\odot}\mathrm{c}^2/\mathrm{s}$.

Why do we think it was a binary black hole system? From the gravitational-wave amplitude dependence for a binary system,

$$h \propto \mathcal{M}^{5/3} \times f_{GW}^{2/3} \times r^{-1}, \tag{8}$$

the function of masses, the chirp mass $\mathcal{M}$, is known function of the observed frequency f_{GW} (see Eq. 6). From higher order post-Newtonian corrections one gets the mass ratio m_2/m_1 and the components of the black-hole spins. Measured $\mathcal{M}$ is about 30 $M_{\odot}$, which means that the sum of masses is about $M = 70\ M_{\odot}$ (one can obtain this quick estimate assuming $m_1 = m_2$, then $M = 2^{6/5}\mathcal{M}$). In the sensitivity band of the detector 8 last orbits of the system were recorded until a limiting frequency of about 150 Hz (corresponding to the orbital frequency of 75 Hz).

In order to explain these observations one could argue that the observations were linked to a binary neutron star system. Double neutron star system is compact enough to exist at the observed orbital frequencies, but it's too light to be reconciled with the measured $\mathcal{M}$ and the total mass. Another possibility would be a mixed black hole-neutron star binary - here one could obtain the correct total mass, but the final orbital frequency at which the neutron star plunges into the black hole is much lower than the observed one. The basic physics of the binary black hole merger GW150914 is described in [29].

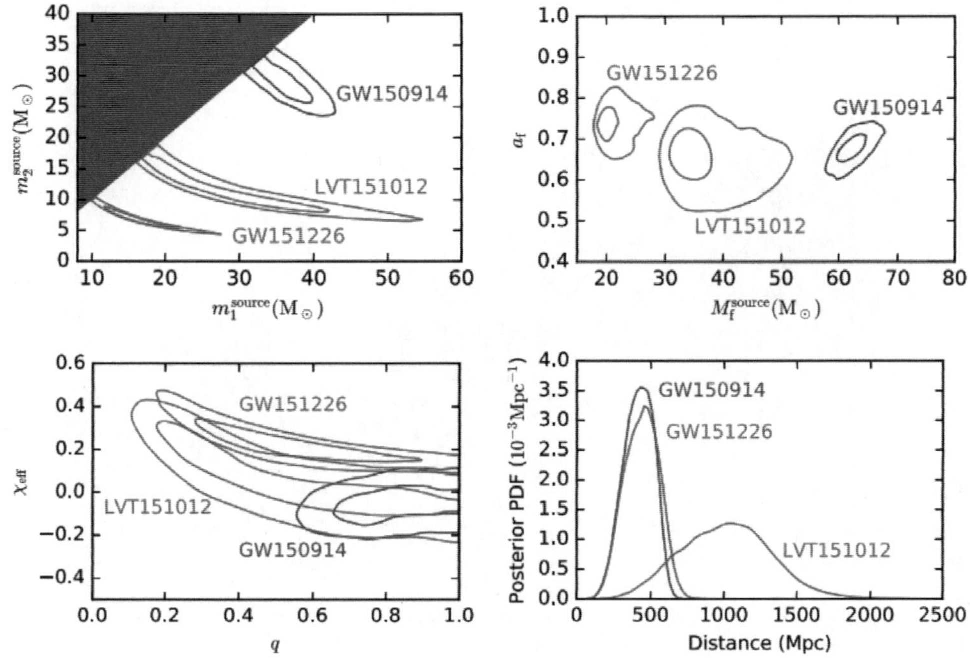

Figure 7. Posterior probability densities of the masses, spins, and distance to the three events GW150914, LVT151012, and GW151226. For the two-dimensional distributions, the contours show 50% and 90% credible regions. *Top left panel:* Component masses m_1 and m_2 in the source frame. *Top right panel:* The mass and dimensionless spin magnitude of the final black holes. *Bottom left panel:* The effective spin and mass ratios of the binary components. *Bottom right panel:* The luminosity distance to the three events. Reprinted from [30] under the terms of the Creative Commons Attribution 3.0 License.

4. O1 detections

During the Advanced LIGO O1 observational campaign two other binary black hole mergers were recorded. GW151226, with a significance of greater than 5σ over the observing period. A third possible signal, LVT151012, was also identified, with substantially lower significance and with an 87% probability of being of astrophysical origin [30], hence the number of detections is sometimes described as "two and a half". Figure 6 shows the detected signal wave trains in comparison with the sensitivity curve of detectors. The signal to noise ratio ρ is defined as

$$\rho^2 = \int_0^\infty \left(\frac{2|\tilde{h}(f)|\sqrt{f}}{\sqrt{S(f)}} \right)^2 d\ln(f), \tag{9}$$

where the $\tilde{h}(f)$ is the best match template signal waveform and $S(f)$ is the amplitude spectral density of the data stream. From this form it is evident that the signal-to-noise ratio is the area between the template waveform $\tilde{h}(f)$ and the sensitivity curve; for GW150914 $\rho \simeq 24$, whereas for GW151226 $\rho \simeq 13$, and in case of LVT151012 $\rho \simeq 10$.

The estimated parameters of the two ("and a half") O1 detections are presented in Fig. 7. The spins a_i, the component of spin angular momentum S_i aligned with the direction of the orbital angular momentum $\hat{L}$, and the effective spin parameters $\chi_{\rm eff}$ are defined as follows:

IOP Conf. Series: Journal of Physics: Conf. Series **873** (2017) 012013 doi:10.1088/1742-6596/873/1/012013

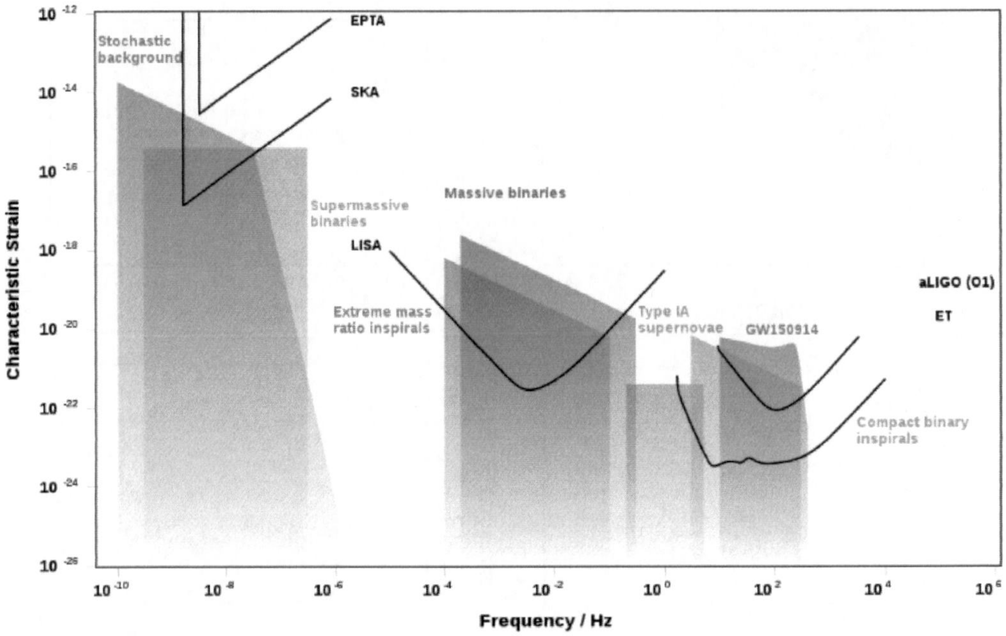

Figure 8. A comparison of Earth-based detectors' frequency sensitivity range with the proposed space-borne detector LISA [31] and pulsar timing array projects (here EPTA, `http://www.epta.eu.org`). ET denotes the planned third generation underground 10 km arm interferometer (`http://www.et-gw.eu`), and SKA is the Square Kilometer Array (`skatelescope.org`).

$$a_i = \frac{c}{Gm_i^2} S_i, \quad \chi_i = \frac{c}{Gm_i^2} S_i \cdot \hat{L}, \quad \chi_{\text{eff}} = \frac{\chi_1 m_1 + \chi_2 m_2}{m_1 + m_2}. \tag{10}$$

Measured masses draw the mass function of stellar-mass black holes much different than the one inferred from the observations within our Galaxy. Especially, the high masses of GW150914 suggest low-metallicity environment in which high mass progenitors of supernovæ lose less mass during their evolution and are able to produce high black-hole remnants.

5. Summary

We have recently witnessed first direct detections of gravitational waves, first observations of near-horizon dynamics, new method for measuring BH masses and spins, first observations of a BH binary system, "brightest" phenomenons ever recorded in history and new ways to test gravity theories in the strong-field regime. Up to date the three O1 detections are in concordance with general relativity. The inferred merger rate for stellar mass binary black hole systems was estimated to be in the range of $9 - 240$ Gpc^{-3} yr^{-1}.

Astrophysics of gravitational waves is qualitatively different from anything we were studying so far. Observations of objects dark in electromagnetic waves, and joint observations of gravitational and electromagnetic waves (the multi-messenger astronomy) will allow to probe the details of previously unaccessible phenomena. For example, stellar-mass black hole binary inspirals similar to GW150914 and GW151226 would be visible at lower frequencies [32] by a proposed space-borne detector LISA [31] (see also Fig. 8). The onset of the gravitational-wave astronomy heralds a completely new way to study the Universe.

6. Acknowledgements

This work was partially supported by the Polish National Science Centre grant no. UMO-2014/14/M/ST9/00707.

References

[1] Einstein A 1916 *Sitzungsberichte der Königlich Preußischen Akademie der Wissenschaften (Berlin), Seite 688-696.*

[2] Thorne K S 1987 *Gravitational radiation.* pp 330–458

[3] Pirani F A E 1956 *Acta Physica Polonica* **15** 389–405

[4] Bondi H 1957 Nature **179** 1072–3

[5] Robinson I and Trautman A 1960 *Physical Review Letters* **4** 431–2

[6] Feynman R P, Morinigo F B and Wagner W G 1995 *Feynman lectures on gravitation* (Reading, Addison-Wesley)

[7] Infeld L and Michalska-Trautman R 1969 *Annals of Physics* **55** 561–75

[8] Infeld L and Michalska-Trautman R 1969 *Annals of Physics* **55** 576–86

[9] Gertsenshtein M E and Pustovoit V I 1962 *Soviet Journal of Experimental and Theoretical Physics* **43** 605–7

[10] Paczyński B 1967 Acta Astron. **17** 287

[11] Hulse R A and Taylor J H 1975 ApJ **195** L51–3

[12] Weisberg J M, Taylor J H and Fowler L A 1981 *Scientific American* **245** 74–82

[13] Taylor J H and Weisberg J M 1982 ApJ **253** 908–20

[14] Abbott B P, Abbott R, Abbott T D, Abernathy M R *et al.* 2016 Phys. Rev. Lett. **116** 061102 (*Preprint* 1602.03837)

[15] Abbott B P, Abbott R, Abbott T D, Abernathy M R *et al.* 2016 Phys. Rev. Lett. **116** 241103 (*Preprint* 1606.04855)

[16] Aasi J, Abbott B P, Abbott R, Abbott T *et al.* 2015 Class. Quant. Grav. **32** 074001 (*Preprint* 1411.4547)

[17] Acernese F, Agathos M, Agatsuma K, Aisa D *et al.* 2015 Class. Quant. Grav. **32** 024001 (*Preprint* 1408.3978)

[18] Abbott B P, Abbott R, Abbott T D, Abernathy M R *et al.* 2016 *Living Reviews in Relativity* **19** (*Preprint* 1304.0670)

[19] Holz D E and Hughes S A 2005 ApJ **629** 15–22 (*Preprint* astro-ph/0504616)

[20] Danzmann K *et al.* (LISA Study Team) 1996 *Class. Quantum Grav.* **13** A247–50

[21] Schutz B F 1986 Nature **323** 310

[22] Schutz B F 2002 Lighthouses of Gravitational Wave Astronomy *Lighthouses of the Universe: The Most Luminous Celestial Objects and Their Use for Cosmology* ed Gilfanov M, Sunyeav R and Churazov E p 207 (*Preprint* gr-qc/0111095)

[23] Einstein A 1918 *Sitzungsberichte der Königlich Preußischen Akademie der Wissenschaften (Berlin), Seite 154-167.*

[24] Królak A and Schutz B F 1987 *Gen. Relativ. Gravit.* **19** 1163–71

[25] Bloom J S, Holz D E, Hughes S A, Menou K *et al.* 2009 (*Preprint* 0902.1527)

[26] Singer L P, Chen H Y, Holz D E, Farr W M *et al.* 2016 ApJ **829** L15 (*Preprint* 1603.07333)

[27] MacLeod C L and Hogan C J 2008 Phys. Rev. D **77** 043512 (*Preprint* 0712.0618)

[28] Aso Y, Michimura Y, Somiya K, Ando M *et al.* 2013 Phys. Rev. D **88** 043007 (*Preprint* 1306.6747)

[29] Abbott B P, Abbott R, Abbott T D Abernathy M R *et al.* 2016 (*Preprint* 1608.01940)

IOP Conf. Series: Journal of Physics: Conf. Series **873** (2017) 012013 doi:10.1088/1742-6596/873/1/012013

[30] Abbott B P, Abbott R, Abbott T D, Abernathy M R *et al.* 2016 *Physical Review X* **6** 041015 (*Preprint* 1606.04856)

[31] Amaro-Seoane P, Audley H, Babak S, Baker J *et al.* 2017 (*Preprint* 1702.00786)

[32] Sesana A 2016 *Physical Review Letters* **116** 231102 (*Preprint* 1602.06951)

IOP Conf. Series: Journal of Physics: Conf. Series **873** (2017) 012014 doi:10.1088/1742-6596/873/1/012014

Supergravity at 40: Reflections and Perspectives

Sergio Ferrara

Theoretical Physics Department, CERN, CH - 1211 Geneva 23, SWITZERLAND
INFN - Laboratori Nazionali di Frascati, Via Enrich Fermi 40, I-00044 Frascati, ITALY
Department of Physics and Astronomy and Mani L. Bhaumik Institute for Theoretical Physics, U.C.L.A., Los Angeles CA 90095-1547, USA

E-mail: sergio.ferrara@cern.ch

Augusto Sagnotti

Scuola Normale Superiore and INFN, Piazza dei Cavalieri 7 I-56126 Pisa, ITALY

E-mail: sagnotti@sns.it

Abstract. The fortieth anniversary of the original construction of Supergravity provides an opportunity to combine some reminiscences of its early days with an assessment of its impact on the quest for a quantum theory of gravity.

1. Introduction

The year 2016 marked the fortieth anniversary of the discovery of Supergravity (SGR) [1], an extension of Einstein's General Relativity [2] (GR) where Supersymmetry, promoted to a gauge symmetry, accompanies general coordinate transformations. Supersymmetry, whose first realization in four–dimensional Quantum Field Theory was introduced by Julius Wess and Bruno Zumino in [3, 4], extends the very notion of spacetime, adjoining to the Poincaré group of translations and Lorentz rotations new symmetries that change the Statistics (Bose-Einstein vs Fermi-Dirac) of particles and fields. A peculiar mathematical structure, called "super-algebra", achieves this goal while circumventing classic no-go theorems that constrain attempts to unify space-time symmetries (connected to mass and spin) with internal ones (connected to charges of various types) [5] [1]. Supergravity implies the existence of a new type of elementary quantum of gravitational origin, a spin 3/2 particle called gravitino.

An exact Supersymmetry would require the existence of super-partners in the Standard Model of Electroweak and Strong interactions and for the gravitational field, but it would also imply mass degeneracies between the known particles and their super-partners. This option has been ruled out, over the years, by several experiments, and therefore Supersymmetry can be at best broken, with super-partner masses that seem to lie beyond the TeV energy region currently explored at the CERN LHC. In Supergravity one would expect that the breaking be

[1] Supersymmetry was inspired by "dual resonance models", an early version of String Theory pioneered by Gabriele Veneziano [6] and extended by André Neveu, Pierre Ramond and John Schwarz [7]. Earlier work done in France by Jean-Loup Gervais and Benji Sakita [8], and in the Soviet Union by Yuri Golfand and Evgeny Likhtman [9] and by Dmitry Volkov and Vladimir Akulov [10], had anticipated some salient features.

spontaneous, as in the Brout-Englert-Higgs (BEH) mechanism of the standard Model [11], which was remarkably confirmed by the 2012 discovery of a Higgs particle [12].

Supersymmetry would have dramatic consequences. It would affect the subatomic world, via supersymmetric extensions of the Standard Model (MSSM [13] and alike), but also large-scale phenomena and the cosmological evolution of our Universe [14]. Supergravity has the potential to provide important clues for dark matter, dark energy and inflation [15], and for the links between the corresponding breaking scale and the one that ought to have superseded it and presumably still characterizes the present epoch. The recent discovery of gravitational waves from black-hole (BH) mergers is a stunning confirmation of GR [16], the gauge theory of the gravitational field, and one can dream of future revelations of its spin-two quantum, the graviton. The gravitino ought to acquire mass via a supersymmetric version of the BEH mechanism, whose details would also control other mass splittings of crucial importance for super-particle searches (squarks, gluinos, sleptons).

Let us now describe some key steps in the development of Supergravity, with an eye to achievements and difficulties of this endeavor and to its impact on different fields.

2. The Early Times

The first instance of Supergravity was built in the spring of 1976 by Daniel Freedman, Peter van Nieuwenhuizen and one of us (S.F.) [17], in a collaboration that had started in the fall of 1975 in Paris, at École Normale Supérieure. The construction relied on the vierbein formulation of General Relativity and on the Noether method, an iterative procedure that would result in the non-linear Yang-Mills or Einstein-Hilbert action principles if applied to gauge theories or gravity. Inconsistencies, if present, would have led to obstructions that no further modifications could have overcome. Shortly thereafter, Stanley Deser and Bruno Zumino recovered the result in a simpler and elegant way [18], extending the first-order (Palatini) formalism of General Relativity [2].

These original developments are well captured by eq. (1) below, where we display the Lagrangian of $N = 1$ Supergravity in four dimensions in the "1.5 order" formalism [1], the torsion equation for the gravitino field ψ_μ and the supersymmetry transformations for the vierbein field e_a^μ and ψ_μ (in "mostly plus" signature, as in other examples below) [3]:

$$\begin{aligned} \mathcal{S} &= \frac{1}{2k^2}\int d^4x\, e\left[\, e_a^\mu\, e_b^\nu\, R_{\mu\nu}{}^{ab}(\omega) \;-\; \overline{\psi}_\mu\, \gamma^{\mu\nu\rho}\, D_\nu(\omega)\,\psi_\rho\right] , \\ \frac{\delta\,\mathcal{S}}{\delta\,\omega} &= 0 \longrightarrow D_\mu\, e_\nu^a \;-\; D_\nu\, e_\mu^a \;=\; \frac{1}{2}\,\overline{\psi}_\mu\,\gamma^a\,\psi_\nu\,, \\ \delta\, e_\mu^a &= \frac{1}{2}\,\overline{\epsilon}\,\gamma^a\,\psi_\mu\,, \qquad \delta\,\psi_\mu \;=\; D_\mu\,\epsilon\,. \end{aligned} \tag{1}$$

Further simplifications of the procedure emerged once its full significance was better appreciated. A mixed formalism was eventually adopted, where it became far simpler to track unwanted terms, and during a Summer Institute held at École Normale in August 1976 the Noether procedure led to the first matter couplings [22], which opened the way to a host of more complicated examples. Moreover, the "spinning string" [7], or String Theory as it is now called [23], was connected to space-time Supersymmetry via a Gliozzi-Scherk-Olive (or GSO) projection [24]. A first extended version of four–dimensional Supergravity, involving two gravitinos, came to light shortly thereafter [25].

[2] Alternative approaches were soon developed, including Supergravity as the gauge theory of the anti-deSitter group [19], Supergravity on a group manifold [20] and Supergravity from broken superconformal symmetry [21].

[3] For convenience, we use in all equations the conventions set out by Daniel Freedman and Antoine Van Proeyen in [1]. Notice that their definition of $\overline{\psi}_\mu$, say, differs from the usual one, since it includes an imaginary factor i.

Table 1. The helicity content of supergravity multiplets in four dimensions. The factors indicate the multiplicities, and the overall multiplets for $N = 7$ and $N = 8$ coincide.

N	Helicity Content
1	$\left[(2), \left(\frac{3}{2}\right)\right]$
2	$\left[(2), 2\left(\frac{3}{2}\right), (1)\right]$
3	$\left[(2), 3\left(\frac{3}{2}\right), 3(1), \left(\frac{1}{2}\right)\right]$
4	$\left[(2), 4\left(\frac{3}{2}\right), 6(1), 4\left(\frac{1}{2}\right), 2(0)\right]$
5	$\left[(2), 5\left(\frac{3}{2}\right), 10(1), 11\left(\frac{1}{2}\right), 10(0)\right]$
6	$\left[(2), 6\left(\frac{3}{2}\right), 16(1), 26\left(\frac{1}{2}\right), 30(0)\right]$
8	$\left[(2), 8\left(\frac{3}{2}\right), 28(1), 56\left(\frac{1}{2}\right), 70(0)\right]$

The low-energy spectra that emerged from the GSO projection pointed to yet unknown ten-dimensional versions of Supergravity, including the counterparts of several gravitinos [26], and to a four-dimensional supersymmetric Yang-Mills theory (SYM) invariant under four distinct supersymmetries [24, 27].

When S. F. visited Caltech in the Fall of 1976, he became aware that Murray Gell-Mann had worked out many consequences of Supersymmetry, including upper bounds on the number of gravitinos and on the gauge symmetries allowed, in principle, in all instances of "pure" Supergravity, where all particles would be connected to the graviton [28] (see Table 2). Gell-Mann had realized, in particular, that the largest theory would include eight gravitinos, and would allow for a maximal gauge group, $SO(8)$, which would not suffice to accommodate the $SU(3) \times SU(2) \times U(1)$ gauge symmetry of the Standard Model.

3. The Golden Age

The following years, 1977 and 1978, were most performing, and a widespread enthusiasm drew into the field many new adepts. Important developments followed readily, including the discovery of "minimal" formulations where $N = 1$ Supersymmetry is manifest [29, 30]. These involve extra (non-propagating) auxiliary degrees of freedom, which result into equal numbers of Bose and Fermi fields and particles. Finding these "off-shell" formulations has proved very hard beyond $N = 1$, but in this case they have allowed systematic investigations of the spontaneous breaking of local Supersymmetry and precise characterizations of scalar geometries. This is fortunate, since only $N = 1$ Supersymmetry is directly compatible with the chiral (parity-violating) interactions of the Standard Model. Auxiliary field formulations proved important also to understand higher–derivative extensions of Supergravity [31], including the supersymmetric extension [32] of Starobinsky's model of inflation [33]. More recently, they were instrumental in supersymmetric localization techniques, where curved backgrounds providing infrared regulators are captured by diverse auxiliary–field configurations [34].

Complete matter couplings for $N = 1, 2$ in four dimensions were thus constructed by the early 1980s [35, 36], and some key features of the general $N = 1$ case can be neatly summarized as in eq. (2) below:

$$\mathcal{S} \;=\; \int d^4x\, e \left[\frac{1}{2\,k^2}\, e_a^\mu\, e_b^\nu\, R_{\mu\nu}{}^{ab}(\omega) \;-\; \partial_i\, \partial_{\bar{j}}\, \mathcal{G}\, D_\mu z^i\, D_\nu \bar{z}^{\bar{j}}\, g^{\mu\nu} \;-\; V(z,\bar{z}) \;+\; \ldots \right],$$
$$V \;=\; e^{\mathcal{G}} \Big[\mathcal{G}_i\, \mathcal{G}_{\bar{j}}\, \left(\mathcal{G}^{-1}\right)^{i\bar{j}} \;-\; 3 \Big], \qquad \mathcal{G} \;=\; K \;+\; \log|W|^2 \,. \tag{2}$$

The main ingredients of the construction are the Kahler potential $K(z^i, \overline{z}^{\bar{i}})$ and the superpotential $W(z^i)$, which depend on the scalar fields and enter the theory via the invariant combination $\mathcal{G}$.

A key step in the development of the theory had to do with the maximal model promptly identified by Gell-Mann. The maximal $N = 8$ Supergravity was derived in 1978 by Eugene Cremmer and Bernard Julia [37] from their previous, remarkable construction with Joel Scherk, of the unique Supergravity in eleven dimensions [38]. Its key features are summarized in

$$\mathcal{S} \;=\; \frac{1}{2\,k^2} \int d^{11}x\, e \left[e_a^\mu\, e_b^\nu\, R_{\mu\nu}{}^{ab}(\omega) \;-\; \overline{\psi}_\mu\, \gamma^{\mu\nu\rho}\, D_\nu \left(\frac{\omega + \hat{\omega}}{2}\right) \psi_\rho \right.$$
$$-\; \frac{1}{24}\, F_{\mu\nu\rho\sigma}\, F^{\mu\nu\rho\sigma} \;-\; \frac{\sqrt{2}}{192} \left(\overline{\psi}_\mu\, \gamma^{\mu\nu\alpha\beta\gamma\delta}\, \psi_\nu \;+\; 12\, \overline{\psi}^\alpha\, \gamma^{\gamma\delta}\, \psi^\beta\right) \left(F_{\alpha\beta\gamma\delta} \;+\; \widehat{F}_{\alpha\beta\gamma\delta}\right)$$
$$\left. -\; \frac{2\sqrt{2}}{(144)^2}\, \epsilon^{\alpha_1 \ldots \alpha_4 \beta_1 \ldots \beta_4 \mu\nu\rho}\, F_{\alpha_1 \ldots \alpha_4}\, F_{\beta_1 \ldots \beta_4}\, C_{\mu\nu\rho} \right], \tag{3}$$
$$\widehat{\omega}_{\mu\, ab} \;=\; \omega_{\mu\, ab} \;+\; \frac{1}{8}\, \overline{\psi}_\alpha\, \gamma_{\mu ab}{}^{\alpha\beta}\, \psi_\beta\,, \qquad \widehat{F}_{\mu\nu\rho\sigma} \;=\; F_{\mu\nu\rho\sigma} \;+\; \frac{3\sqrt{2}}{2}\, \overline{\psi}_{[\mu}\, \gamma_{\nu\rho}\, \psi_{\sigma]}\,,$$

where ω solves its field equation, F is the curl of three-form C and "hats" denote supercovariant quantities.

Curiously, at most seven additional spatial dimensions are indeed allowed in Supergravity [39], in contrast with General Relativity. In general, extra dimensions beyond our space time could exist, and yet be inaccessible to our senses, if they were curled into tiny internal spaces. This is the spirit of the Kaluza-Klein (KK) scenario that first linked, in the 1920s, higher-dimensional GR and Electromagnetism [40].

The first key ingredient of the $N = 8$ construction was the geometrical nature of the scalar interactions, which result from an analogue of the pion model involving the 70 different fields present in the maximal theory, associated to the $E_{7(7)}/SU(8)$ coset. The second was a set of generalized electric-magnetic dualities, which extend the manifest symmetry of the vacuum Maxwell equations under the interchange of electric and magnetic fields and had already surfaced in simpler models [41]. For instance, in the $N = 8$ model, the $E_{7(7)}$ group acts on the 56 "electric" and "magnetic" field strengths as a generalized electric–magnetic duality [37]. Hidden (infinite–dimensional) symmetries extending it have been widely explored in recent years, following [42].

4. Supergravity and Particle Physics

The MSSM [13] and other supersymmetric extensions of the Standard Model were introduced and widely investigated by Pierre Fayet, Savas Dimopoulos, Howard Georgi and others [13], relying heavily on the soft-breaking terms proposed by Luciano Girardello and Marc Grisaru in [43]. These low–dimensional couplings are not supersymmetric, and were introduced to overcome restrictions accompanying the spontaneous breaking of rigid Supersymmetry. A glimpse of the modifications induced by Supergravity is captured by the super–trace formula, here restricted for simplicity to chiral multiplets [44, 35],

$$\frac{1}{2}\, \mathrm{Str}\, M^2 \Big|_{V_{z_i}=0, V=0} \;\equiv\; \frac{1}{2} \sum_J (-1)^{2\,J}\, (2\,J \;+\; 1)\, m_J^2 \;=\; e^{\mathcal{G}} \Big[N \;-\; 1 \;+\; \mathcal{G}^i\, \mathcal{G}^{\bar{j}}\, \mathcal{R}_{i\bar{j}} \Big], \tag{4}$$

IOP Conf. Series: Journal of Physics: Conf. Series **873** (2017) 012014 doi:10.1088/1742-6596/873/1/012014

Table 2. Standard field multiplets in $D = 10, 11$. The IIA model is directly related via KK to eleven dimensions. The IIB model contains doublets of gravitini, spinors and two-forms, and chiral four-form, D^+, with self–dual field strength, and two scalars parametrizing the $SL(2,R)/U(1)$ coset. The field content of the $(1,0)$ SGR model is a truncation of the others. The L, R suffixes indicate the chirality of Fermi fields.

Model	Field Content
D=11 SGR	$\left(e_\mu^a, \psi_\mu, C_{\mu\nu\rho}\right)$
D=10 IIA, or (1,1) SGR	$\left(e_\mu^a, \psi_{\mu\,L}, \psi_{\mu\,R}, \lambda_L, \lambda_R, C_{\mu\nu\rho}, B_{\mu\nu}, A_\mu, \phi\right)$
D=10 IIB, or (2,0) SGR	$\left(e_\mu^a, \psi_{\mu\,L}^i, \lambda_R^i, D^+{}_{\mu\nu\rho\sigma}, B_{\mu\nu}^i, \phi, \phi'\right) \quad (i = 1, 2)$
D=10, (1,0) SGR	$\left(e_\mu^a, \psi_{\mu\,L}, \lambda_R, B_{\mu\nu}, \phi\right)$
D=10, (1,0) SYM	(A_μ, λ_L)

where N is the number of scalar multiplets, which emerges in the $m_{3/2}/m_{Pl} \to 0$ limit. For the sake of comparison, in renormalizable models of rigid Supersymmetry the $r.h.s.$ would vanish, reflecting patterns where scalars pair, in mass, around fermions.

Supergravity provided a rationale for the emergence of soft–breaking terms as low–energy relics of the super–Higgs mechanism. This was shown by Riccardo Barbieri, Carlos Savoy and one of us (S.F.), and independently by Lawrence Hall, Joseph Lykken and Steven Weinberg, and by Richard Arnowitt, Ali Chamseddine and Pran Nath [45].

No-scale Supergravity, a theory with a naturally vanishing cosmological constant, was built in [46], and was readily applied to Physics beyond the Standard Model.

All these developments made it possible to derive parameter spaces of masses and couplings for supersymmetric extensions of the Standard Model, which can be explored in experimental searches for new physics, as was done at LEP, at Tevatron and elsewhere in the past and is currently done extensively at the LHC collider [4].

"Split Supersymmetry" [47] was advocated in recent years to obtain large mass separations among MSSM super partners while maintaining a number of attractive features of Supersymmetry, at the price of a fine–tuned cosmological constant [48], and was further developed by the same authors and many others.

5. Supergravity and String Theory

The ultraviolet behavior of Supergravity theories was vigorously investigated soon after the original discovery. As in gravity [49], no divergences were found in the one-loop S-matrix of "pure" models involving solely the gravitational multiplet [50]. Symmetry arguments soon pushed them at least to the three-loop order [51, 52], and thus beyond GR, where divergences begin to show up at two loops [53]. In subsequent years refined symmetry arguments and explicit computations relying on novel methods spurred by these studies have proceeded hand-in-hand, under the drive of Zvi Bern, Lance Dixon, David Kosower and others, revealing further, unexpected cancelations of divergences [54]. The case of $N = 8$ Supergravity remains unsettled, and some authors still envisage the possibility that this maximal theory be finite to all orders. In particular, the double–copy structure relating $N = 4$ Yang–Mills to $N = 8$ Supergravity might provide a clue to its actual divergence structure [55], and recently played a role in the identification of an unexpected symmetry [56], the dual superconformal symmetry, whose lessons could reverberate on Supergravity. Moreover, some subtleties that were well appreciated only recently might lead to deeper insights into the whole scenario [57].

Supergravity allows in general continuous deformations that combine gauged internal symmetries and the emergence of scalar potentials. Explicit constructions soon clarified [58, 59], however, that these bring along negative vacuum energies, and thus maximally (super)symmetric anti-de Sitter, or AdS, vacua. This is the case, in particular, for the $N = 8$ model of Bernard de Wit and Hermann Nicolai [60], which realized the full $SO(8)$ gauge symmetry that we have already mentioned but does not admit a Minkowski vacuum. In a KK compactification from eleven dimensions, a gauged $SO(8)$ symmetry would be inherited from a seven-dimensional internal sphere [61]. A number of enticing variants were readily explored [62].

The GSO projection opened the way to connecting String Theory to gravity along lines that were foreseen, in the mid 1970s, by Scherk and Schwarz [63], and independently by Tamiaki Yoneya [64]. All ten-dimensional versions of Supergravity (and corresponding strings) were constructed by the 1980s [65, 66], but a widespread activity in this direction only started in 1984, when Michael Green and Schwarz discovered that gauge and gravitational [67] anomalies cancel, unexpectedly, in all versions of ten-dimensional supersymmetric String Theory (or Superstring Theory, for brief) [68]. Anomalies are quantum violations of classical symmetries that are very troublesome when they affect gauge interactions. Their cancellation is a fundamental consistency condition, which is automatically granted in the Standard Model by its known particle content [69].

The allowed gauge groups in ten dimensions are $SO(32)$ and $E_8 \times E_8$ in closed Heterotic superstrings [70], whose discovery followed closely the anomaly cancellation, and $SO(32)$ in the Type-I theory involving both open and closed superstrings [71], akin respectively to segments and circles. Here $E_8 \times E_8$ denotes two copies of the largest exceptional Lie algebra in the Cartan classification, and both $SO(32)$ and $E_8 \times E_8$ contain the $SU(3) \times SU(2) \times U(1)$ gauge symmetry of the Standard Model as a small subgroup. At low energies, all these theories reduce to minimal Supergravity coupled to a Yang-Mills multiplet in ten dimensions.

Unprecedented avenues thus opened up for linking ten-dimensional strings to the chiral interactions of Particle Physics, in enticing scenarios that are free, by construction, of the ultraviolet problems of gravity. One might well say that Supergravity started officially, with this 1984 "first superstring revolution", a second life as a low-energy manifestation of String Theory. It actually led to reconsider the very notion of anomaly, since in the Green-Schwarz mechanism some of its gauge potentials remove part of it. The original construction involved a single antisymmetric tensor potential $B_{\mu\nu}$ (to be contrasted with the electro-magnetic vector potential A_μ), but more complicated cases that have emerged later rest on several fields of this and similar types.

The anomaly cancellation mechanism quickly resulted in definite KK scenarios granting String Theory a four-dimensional interpretation [72]. It pointed to a specific class of six-dimensional internal manifolds leading to chiral spectra with $N = 1$ Supersymmetry in four dimensions that were widely studied in Mathematics, Calabi-Yau spaces. These lead naturally, in this context, to a grand-unification gauge group E_6, which was known to connect to the Standard Model with right-handed neutrinos. All thus converged into an intriguing dictionary between the resulting low-energy dynamics and the topology of these complicated manifolds, or if you will between four-dimensional Supergravity and Algebraic Geometry. A remarkable property of these spaces, with deep physical implications, is the so-called "mirror symmetry" [73], which exchanges manifolds while swapping their Hodge numbers. This symmetry allows one to compute non–perturbative couplings in terms of classical ones, and finds a rationale in the C-map of the four–dimensional low–energy Supergravity [74], where $N = 2$ vector multiplets and hypermultiplets are interchanged. The structure of these multiplets rests on "special geometry", whose mathematical significance was first stressed in [75]. No-scale Supergravity [46] was linked to Calabi-Yau compactifications in [76, 77]. Actually, strings behave properly also on singular (orbifold) limits that compare to smooth Calabi-Yau spaces like tetrahedra to spheres [78]. These limits afford more complete descriptions, and have led to deep insights into the structure of String Theory, opening the way to early direct constructions of four–dimensional heterotic string spectra [79]

Other (type II) theories with maximal Supersymmetry exist in ten dimensions, are also free of anomalies and allow similar compactifications, but for a while they seemed totally unrelated to Particle Physics.

6. Branes and M–Theory

The early 1990s were marked by many detailed studies of classical solutions of Supergravity [80]. These generalize BH's, and form fields are their key new ingredient. In four dimensions, electric charges source a vector potential A_μ, whose Lorentz index reflects somehow their "world lines". In a similar fashion, strings source antisymmetric potentials $B_{\mu\nu}$, while other potentials present in ten-dimensional Supergravity are related to extended objects of higher dimensionality, generically dubbed "p-branes". Here p is the number of spatial dimensions of these extended objects ($p = 0$ for particles, $p = 1$ for strings, $p = 2$ for membranes, and so on). Supergravity cannot forego the presence of a wide class of extended objects, with which String Theory would appear at odds, insofar as it focuses solely on strings. This was strongly advocated, over the years, by a number of scientists, and most notably by Michael Duff [81] and Paul Townsend [82].

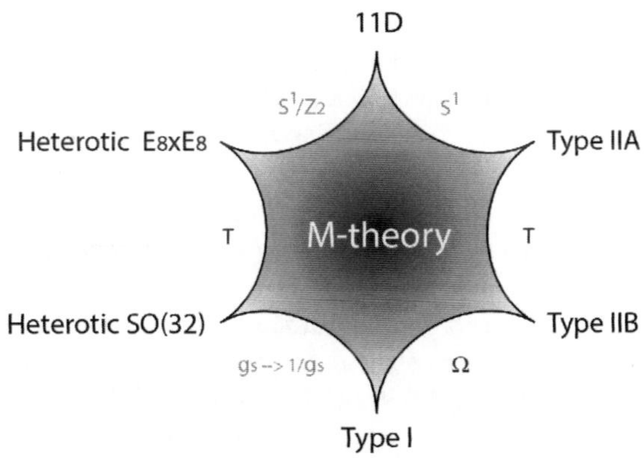

Figure 1. Duality links among the five ten–dimensional supersymmetric versions of String Theory and eleven–dimensional Supergravity.

The "second superstring revolution" is associated to the mid-1990s, and to the resolution of this apparent dichotomy when it became widely recognized that String Theory, after all, cannot be merely a theory of strings. Behind it, as we now concede, lies indeed a far more complicated soup of strings and more general p-branes. The novel ingredient of this wider picture was a class of p-branes that show up in string perturbation theory in the presence of boundaries, the D-branes whose role was clarified by Joseph Polchinski [83], but the (electric-magnetic) dualities of the low-energy Supergravity were again a main tool. Strong–weak coupling dualities in String Theory were actually first advocated by Anna-Maria Font, Dieter Lust, Luis Ibanez and Fernando Quevedo in [84], and were inspired by continuous symmetries of the low–energy

Supergravity. The end result is an awesome unified picture, which was largely due to Edward Witten [85] and is usually referred to as M-theory.

When strings combine with other extended objects, even the very notion of space time is blurred, to the extent that ten and eleven dimensions emerge from singular corners of this correspondence. One of these is the eleven-dimensional Supergravity. These results are usually summarized via the hexagon–like diagram of Fig. 1, whose sides reflect different duality links, three of which were deeply inspired by Supergravity [86, 85, 87]. The others had already surfaced in the late Eighties: they are beyond its reach but find their rationale in T-duality, a peculiar string correspondence between large and small KK radii [88], and in the Ω orientifold link between type-IIB and type-I strings proposed by one of us (A. S.) [89]. Supergravity thus started officially a third, parallel life, as a probe into the elusive inner workings of String Theory. [4]

7. Supergravity and the AdS/CFT Correspondence

The late 1990s witnessed the emergence of a peculiar duality that has had a huge impact on the literature. This is the AdS/CFT correspondence [91], which links "holographically" gravity and gauge theories [5]. It was originally conjectured by Juan Maldacena, and its most explored form concerns String Theory compactified on a special KK manifold, $AdS_5 \times S_5$, the direct product of a five-dimensional anti-de Sitter space and a five-dimensional internal sphere. AdS_5 has a boundary at spatial infinity, which is identified with a four-dimensional Minkowski space, and the correspondence posits the equivalence between a weakly coupled String Theory in the bulk and a strongly coupled field theory, the $N = 4$ supersymmetric Yang-Mills theory that we have already met in a ten–dimensional guise, on this boundary. This surprising correspondence, which brought AdS Supergravity to the forefront, also vindicates some intriguing ideas that had been around since Jacob Bekenstein connected the black-hole entropy to its area [93]. Supergravity thus started officially, with this "third superstring revolution", one more parallel life as an unprecedented tool for exploring non-perturbative features of gauge theories.

In $N = 2$ Supergravity supersymmetric black holes give rise to universal (duality–invariant) area formulas for their Bekenstein-Hawking Entropy, which generalize the Reissner–Nordstrom entropy formula for charged dyons and rest on the (square modulus) of the central charge computed at its extremum in moduli space [94]. The name "attractors" was coined since, independently of the initial conditions at large distances, the scalar trajectories in moduli space approach the same extremal points, which coincide with the values at the BH Horizon. This explains, in particular, why the ADM mass is a continuous moduli dependent function while the entropy is quantized in terms of BH charges. In the large–charge limit, Andrew Strominger and Cumrun Vafa associated to these types of results for extremal BH's a microscopic String Theory counting [95]. The actual nature of the microstates, however, is still a debated issue [96].

8. Conclusions and Perspectives

The last two decades have witnessed a multitude of applications of AdS/CFT outside its original realm. These have touched upon fluid dynamics, the quark-gluon plasma, and more recently Condensed Matter Physics, providing a number of useful insights on strongly coupled matter systems. Perhaps more unexpectedly, AdS/CFT duality has stimulated work related to scattering amplitudes, which may also shed light on the old issue of the ultraviolet behavior of supergravity, but the reverse program of gaining information about gravity from gauge dynamics has proved harder. Above all, however, there is a pressing need to shed light on the geometrical

[4] Recently, double field theory [90] has emerged as an enticing intermediate framework to accommodate T-duality within Field Theory.

[5] There were a number of notable anticipations of aspects of the correspondence, including some contributions in [66, 92].

principles and the deep symmetries underlying String Theory, which have proved elusive over the years.

The interplay between Particle Physics and Cosmology is a natural arena to explore consequences of Supergravity. Recent experiments probing the Cosmic Microwave Background, and in particular the results of the Planck mission [97], have lent some definite support to inflationary models of the Early Universe. An elusive particle, the inflaton, should have driven this primordial acceleration, and while our current grasp of String Theory does not allow a detailed analysis of the problem, Supergravity can provide fundamental clues on this and the subsequent particle physics epochs.

Applications of Supergravity to Cosmology have attracted an increasing interest in recent years [98, 32]. Supersymmetry is broken in a de Sitter like inflationary phase, where one typically encounters more fields than would be needed for the early Universe, and moreover some familiar scenarios tend to be plagued by instabilities. The novel ingredient that appears to get around these problems is non-linear supersymmetry [99], whose foundations lie in the prescient 1973 work of Volkov and Akulov [10] and on the technique of constrained superfields [100]. Non-linear supersymmetry arises when some super-partners are exceedingly massive, seems to play an intriguing role in String Theory [101] and connects naturally to the KKLT scenario [102]. The current lack of signals for supersymmetry at the LHC makes one wonder whether it might also hold a prominent place in an eventual picture of Particle Physics. This resonates with the idea of "split supersymmetry" [47], which allows for large mass separations among superpartners and can be accommodated in Supergravity at the price of reconsidering hierarchy issues.

Crossing the current frontiers appears to require a deeper understanding of broken Supersymmetry in Supergravity and in String Theory. Broken Supersymmetry made an early entry in String Theory via orbifold realizations of the Scherk-Schwarz mechanism [103] in models of oriented closed strings [104]. The resulting spectra are closely connected to special versions of gauged Supergravity, whose AdS versions play also a central role in the AdS/CFT correspondence, and afford interesting generalizations in String Theory in the presence of branes [105], even in the presence of internal magnetic fields [108]. These constructions involve the Born–Infeld action [106], which was originally recovered from open strings in [107]. Branes were actually instrumental in Supersymmetry in another context, since they provided early clues on mechanisms for its partial breaking [109], and this phenomenon is connected again, in the non–linear limit, to extensions of the Born–Infeld theory [110]. Finally, in recent years, gauged Supergravity found a geometrical framework within generalized geometry [111], a setting that is related to the flux compactifications of [77], while the embedding tensor of [112, 59] captures their algebraic foundations.

In breaking Supersymmetry, one is confronted with important conceptual challenges: the resulting vacua are deeply affected, in general, by quantum fluctuations, and this reverberates on old conundrums related to dark energy and the cosmological constant. There are even signs that this type of investigations could shed light on the backbone of String Theory, and we are confident that Supergravity will lead us farther once more. Finally, Supergravity may have something to say about the dark matter in the Universe, since gravitini or other light superpartners dubbed neutralinos might perhaps account for it.

Acknowledgments

This overview was based in part on the talk delivered by S. F. at the "Infeld Colloquium and Discrete", in Warsaw, on December 1 2016, and on a joint CERN Courier article [113]. It is focussed on topics that are closer to the research interests of the authors, who apologize in advance if they have missed some important references. S. F. was supported in part by the CERN TH-Department and by INFN (IS CSN4-GSS-PI). AS was supported in part by Scuola Normale, by INFN (IS CSN4-GSS-PI) and by APC-U. Paris VII, and would like to thank APC-U. Paris VII for the kind hospitality. We are grateful to Carlo Angelantonj, Bianca Letizia Cerchiai, Emilian Dudas and Marine Samsonyan, who read the manuscript and offered constructive suggestions.

References

[1] For a recent review see:
Freedman D Z and Van Proeyen A 2012 *Supergravity* (Cambridge Univ. Press).
For an early review see:
Van Nieuwenhuizen P 1981 Phys. Rept. **68** 189
[2] For reviews see:
Misner C W, Thorne K S and Wheeler J A 1970 *Gravitation* (Freeman);
Weinberg S 1972 *Gravitation and Cosmology : Principles and Applications of the General Theory of Relativity* (Wiley)
Hawking S W and Ellis G F R 1973 *The Large Scale Structure of Space-Time* (Cambridge Univ. Press)
Wald R M 1984 *General Relativity* (Chicago Univ. Press)
[3] Wess J and Zumino B 1974 Phys. Lett. **49B** 52, 1974 Nucl. Phys. B **70** 39
Salam A and Strathdee J A 1974 Nucl. Phys. B **76** 477
Ferrara S and Zumino B 1974 Nucl. Phys. B **79** 413
Salam A and Strathdee J A 1974 Phys. Lett. **51B** 353
Fayet P and Iliopoulos J 1974 Phys. Lett. B **51** 461
[4] For a review of early results see:
Fayet P and Ferrara S 1977 Phys. Rept. **32** 249
For a collection of early papers see: Ferrara S 1987 *Supersymmetry* (North-Holland, World Scientific)
More recent reviews are:
West P C 1990 *Introduction to supersymmetry and supergravity* (World Scientific)
Wess J and Bagger J 1992 *Supersymmetry and supergravity* (Princeton Univ. Press)
Weinberg S 1995 *The quantum theory of fields. Vol. 3: Supersymmetry* (Cambridge Univ. Press)
[5] Coleman S R and Mandula J 1967 Phys. Rev. **159** 1251
Haag R, Lopuszanski J T and Sohnius M 1975 Nucl. Phys. B **88** 257
[6] Veneziano G 1968 Nuovo Cim. A **57** 190
[7] Neveu A and Schwarz J H 1971 Nucl. Phys. B **31** 86
Ramond P 1971 Phys. Rev. D **3** 2415
[8] Gervais J L and Sakita B 1971 Nucl. Phys. B **34** 632
[9] Golfand Y A and Likhtman E P 1971 JETP Lett. **13** 323 [Pisma Zh. Eksp. Teor. Fiz. **13** 452]
[10] Volkov D V and Akulov V P 1973 Phys. Lett. **46B** 109
[11] Englert F and Brout R 1964 Phys. Rev. Lett. **13** 321
Higgs P W 1964 Phys. Lett. **12** 132, Phys. Rev. Lett. **13** 508, 1986 Phys. Rev. **145** 1156
[12] Aad G *et al.* [ATLAS Collaboration] 2012 Phys. Lett. B **716** 1 [arXiv:1207.7214 [hep-ex]]
Chatrchyan S *et al.* [CMS Collaboration] 2012 Phys. Lett. B **716** 30 [arXiv:1207.7235 [hep-ex]]
[13] Fayet P 1977 Phys. Lett. **69B** 489
Dimopoulos S and Georgi H 1981 Nucl. Phys. B **193** 150
and references therein
For a review see:
Nilles H P 1984 Phys. Rept. **110** 1
[14] For reviews see
Mukhanov V 2005 *Physical foundations of cosmology* (Cambridge Univ. Press)
Weinberg S *Cosmology* 2008 (Oxford Univ. Press)
Lyth D H and Liddle A R 2009 *The primordial density perturbation: Cosmology, inflation and the origin of structure* (Cambridge Univ. Press)
Gorbunov D S and Rubakov V A 2011 *Introduction to the theory of the early Universe: Cosmological perturbations and inflationary theory* (World Scientific)
[15] Starobinsky A A 1980 Phys. Lett. **B 91** 99
Kazanas D 1980 Astrophys. J. **241** L59
Sato K 1981 Phys. Lett. B **99** 66
Guth A H 1981 Phys. Rev. **D 23** 347
Mukhanov V F and Chibisov G V 1981 JETP Lett. **33** 532 [Pisma Zh. Eksp. Teor. Fiz. **33** 549]
Linde A D 1982 Phys. Lett. **B 108** 389
Albrecht A and Steinhardt P J 1982 Phys. Rev. Lett. **48** 1220

Linde A D 1983 Phys. Lett. **B129** 177
For reviews see:
Bartolo N, Komatsu E, Matarrese S and Riotto A 2004 Phys. Rept. **402** 103 [astro-ph/0406398]
Martin J, Ringeval C and Vennin V 2014 Phys. Dark Univ. **5-6** 75 [arXiv:1303.3787 [astro-ph.CO]]
[16] Abbott B P *et al.* [LIGO Scientific and Virgo Collaborations] 2016 Phys. Rev. Lett. **116** no.6 061102 [arXiv:1602.03837 [gr-qc]]
[17] Freedman D Z, van Nieuwenhuizen P and Ferrara S 1976 Phys. Rev. D **13** 3214
[18] Deser S and Zumino B 1976 Phys. Lett. **62B** 335
[19] MacDowell S W and Mansouri F 1977 Phys. Rev. Lett. **38** 739 Erratum: [Phys. Rev. Lett. **38** 1376]
[20] Ne'eman Y and Regge T 1978 Phys. Lett. **74B** 54
For a review see:
Castellani L, D'Auria R and Fré P 1981 *Supergravity and superstrings*, 3 vols (World Scientific)
[21] Kaku M, Townsend P K and van Nieuwenhuizen P 1977 Phys. Lett. **69B** 304
[22] Ferrara S, Freedman D Z, van Nieuwenhuizen P, Breitenlohner P, Gliozzi F and Scherk J 1977 Phys. Rev. D **15** 1013
Ferrara S, Gliozzi F, Scherk J and van Nieuwenhuizen P 1976 Nucl. Phys. B **117** 333
[23] For reviews see:
Green M M, Schwarz J H and Witten E 1987 *Superstring Theory*, 2 vols., (Cambridge Univ. Press)
Polchinski J 1998 *String theory*, 2 vols. (Cambridge Univ. Press)
Johnson C V 2003 *D-branes* (Cambridge Univ. Press)
Zwiebach B 2004 *A first course in string theory* (Cambridge Univ. Press)
Becker K, Becker M and Schwarz J H 2007 *String theory and M-theory: A modern introduction* (Cambridge Univ. Press)
Kiritsis E 2007 *String theory in a nutshell* (Princeton Univ. Press)
West P 2012 *Introduction to strings and branes* (Cambridge Univ. Press)
[24] Gliozzi F, Scherk J and Olive D I 1977 Nucl. Phys. B **122** 253
[25] Ferrara S and van Nieuwenhuizen P 1976 Phys. Rev. Lett. **37** 1669
[26] Cremmer E, Scherk J and Ferrara S 1978 Phys. Lett. **74B** 61
[27] Brink L, Schwarz J H and Scherk J 1977 Nucl. Phys. B **121** 77
[28] Gell-Mann M, Ramond P and Slansky R 1978 Rev. Mod. Phys. **50** 721
[29] Ferrara S and van Nieuwenhuizen P 1978 Phys. Lett. **74B** 333
Stelle K S and West P C 1978 Phys. Lett. **74B** 330
[30] Sohnius M F and West P C 1981 Phys. Lett. **105B** 353, 1982 Nucl. Phys. B **198** 493
Ferrara S and Sabharwal S 1989 Annals Phys. **189** 318
[31] Ferrara S, Grisaru M T and van Nieuwenhuizen P 1978 Nucl. Phys. B **138** 430
Cecotti S 1987 Phys. Lett. B **190** 86
[32] Farakos F, Kehagias A and Riotto A 2013 Nucl. Phys. B **876** 187 [arXiv:1307.1137 [hep-th]]
Ferrara S, Kallosh R, Linde A and Porrati M 2013 Phys. Rev. D **88** no.8, 085038 [arXiv:1307.7696 [hep-th]]
[33] Starobinsky A A 1980 Phys. Lett. **91B** 99
[34] Pestun V 2012 Commun. Math. Phys. **313** 71 [arXiv:0712.2824 [hep-th]]
Festuccia G and Seiberg N 2011 JHEP **1106** 114 [arXiv:1105.0689 [hep-th]]
Benini F and Zaffaroni A arXiv:1605.06120 [hep-th]
[35] Cremmer E, Julia B, Scherk J, Ferrara S, Girardello L and van Nieuwenhuizen P 1979 Nucl. Phys. B **147** 105
Bagger J and Witten E 1982 Phys. Lett. **118B** 103
Cremmer E, Ferrara S, Girardello L and Van Proeyen A 1983 Nucl. Phys. B **212** 413
[36] Gunaydin M, Sierra G and Townsend P K 1984 Nucl. Phys. B **242** 244
Bagger J and Witten E 1983 Nucl. Phys. B **222** 1
de Wit B, Lauwers P G and Van Proeyen A 1985 Nucl. Phys. B **255** 569
Cremmer E, Kounnas C, Van Proeyen A, Derendinger J P, Ferrara S, de Wit B and Girardello L 1985 Nucl. Phys. B **250** 385
Andrianopoli L, Bertolini M, Ceresole A, D'Auria R, Ferrara S, Fré P and Magri T 1997 J. Geom. Phys. **23** 111 [hep-th/9605032]
For a review see:
Samtleben H 2008 Class. Quant. Grav. **25** 214002 [arXiv:0808.4076 [hep-th]]
For a review of harmonic Supersace see:
Galperin A S, Ivanov E A, Ogievetsky V I and Sokatchev E S 2001 *Harmonic Superspace* (Cambridge Univ. Press)

[37] Cremmer E and Julia B 1979 Nucl. Phys. B **159** 141
[38] Cremmer E, Julia B and Scherk J 1978 Phys. Lett. **76B** 409
[39] Nahm W 1978 Nucl. Phys. B **135** 149
[40] Kaluza T 1921 Sitzungsber. Preuss. Akad. Wiss. Berlin (Math. Phys.) **1921** 966
Klein O 1926 Z. Phys. **37** 895 [Surveys High Energ. Phys. **5** 241]
[41] Ferrara S, Scherk J and Zumino B 1977 Nucl. Phys. B **121** 393
Gaillard M K and Zumino B 1981 Nucl. Phys. B **193** 221
For a review see:
Aschieri P, Ferrara S and Zumino B 2008 Riv. Nuovo Cim. **31** 625 [arXiv:0807.4039 [hep-th]]
[42] West P C 2001 Class. Quant. Grav. **18** 4443 [hep-th/0104081]
Damour T, Henneaux M and Nicolai H 2002 Phys. Rev. Lett. **89** 221601 [hep-th/0207267]
[43] Girardello L and Grisaru M T 1982 Nucl. Phys. B **194** 65
[44] Ferrara S, Girardello L and Palumbo F 1979 Phys. Rev. D **20** 403
Grisaru M T, Rocek M and Karlhede A 1983 Phys. Lett. **120B** 110
[45] Barbieri R, Ferrara S and Savoy C A 1982 Phys. Lett. **119B** 343
Chamseddine A H, Arnowitt R L and Nath P 1982 Phys. Rev. Lett. **49** 970
Hall L J, Lykken J D and Weinberg S 1983 Phys. Rev. D **27** 2359
Giudice G F and Masiero A 1988 Phys. Lett. B **206** 480
[46] Cremmer E, Ferrara S, Kounnas C and Nanopoulos D V 1983 Phys. Lett. **133B** 61
Ellis J R, Lahanas A B, Nanopoulos D V and Tamvakis K 1984 Phys. Lett. **134B** 429
For a review see:
Lahanas A B andNanopoulos D V 1987 Phys. Rept. **145** 1
[47] Wells J D hep-ph/0306127
Arkani-Hamed N and Dimopoulos S 2005 JHEP **0506** 073 [hep-th/0405159]
Giudice G F and Romanino A 2004 Nucl. Phys. B **699** 65 Erratum: [2005 Nucl. Phys. B **706** 487] [hep-ph/0406088]
Arkani-Hamed N, Dimopoulos S, Giudice G F and Romanino A 2005 Nucl. Phys. B **709** 3 [hep-ph/0409232]
[48] For a classic review, see:
Weinberg S 1989 Rev. Mod. Phys. **61** 1
[49] 't Hooft G and Veltman M J G 1974 Ann. Inst. H. Poincare Phys. Theor. A **20** 69
[50] Grisaru M T, van Nieuwenhuizen P andVermaseren J A M 1976 Phys. Rev. Lett. **37** 1662
[51] Grisaru M T 1977 Phys. Lett. **66B** 75
E. Tomboulis 1977 Phys. Lett. **67B** 417
[52] Deser S, Kay J H and Stelle K S 1977 Phys. Rev. Lett. **38** 527 [arXiv:1506.03757 [hep-th]]
Kallosh R E 1981 Phys. Lett. **99B** 122
Grisaru M T and Siegel W 1981 Nucl. Phys. B **187** 149, 1982 Nucl. Phys. B **201** 292 Erratum: [Nucl. Phys. B **206** 496]
[53] Goroff M H and Sagnotti A 1985 Phys. Lett. **160B** 81, 1986 Nucl. Phys. B **266** 709
van de Ven A E M 1992 Nucl. Phys. B **378** 309
[54] For recent reviews see:
Stelle K S 2012 Int. J. Geom. Meth. Mod. Phys. **09** 1261013
Bern Z 2014 Mod. Phys. Lett. A **29** no.32, 1430036
For a comprehensive discussion of superfield methods see:
Gates S J, Grisaru M T, Rocek M and Siegel W 1983 *Superspace Or One Thousand and One Lessons in Supersymmetry* Front. Phys. **58** 1 [hep-th/0108200]
[55] Bern Z, Carrasco J J M and Johansson H 2010 Phys. Rev. Lett. **105** 061602 [arXiv:1004.0476 [hep-th]]
[56] Drummond J M, Henn J, Korchemsky G P and Sokatchev E 2010 Nucl. Phys. B **828** 317 [arXiv:0807.1095 [hep-th]]
[57] Bern Z, Cheung C, Chi H H, Davies S, Dixon L and Nohle J 2015 Phys. Rev. Lett. **115** no.21, 211301 [arXiv:1507.06118 [hep-th]]
Bern Z, Chi H H, Dixon L and Edison A arXiv:1701.02422 [hep-th]
[58] Freedman D Zand Das A K 1977 Nucl. Phys. B **120** 221
[59] For a recent review, see:
Trigiante M arXiv:1609.09745 [hep-th]
[60] de Wit B and Nicolai H 1982 Nucl. Phys. B **208** 323
[61] de Wit B and Nicolai H 1987 Nucl. Phys. B **281** 211
[62] Hull C M 1984 Phys. Lett. **142B** 39
Gunaydin M, Romans L J and Warner N P 1985 Phys. Lett. **154B** 268, 1986 Nucl. Phys. B **272** 598

[63] Scherk J and Schwarz J H 1974 Nucl. Phys. B **81** 118
[64] Yoneya T 1974 Prog. Theor. Phys. **51** 1907
[65] Green M B and Schwarz J H 1982 Phys. Lett. **109B** 444
Green M B, Schwarz J H and Brink L 1982 Nucl. Phys. B **198** 474
Schwarz J H and West P C 1983 Phys. Lett. **126B** 301
Howe P S and West P C 1984 Nucl. Phys. B **238** 181
Bergshoeff E, de Roo M, de Wit B and van Nieuwenhuizen P 1982 Nucl. Phys. B **195** 97
Chapline G F and Manton N S 1983 Phys. Lett. **120B** 105
Romans L J 1986 Phys. Lett. **169B** 374
[66] Schwarz J H 1983 Nucl. Phys. B **226** 269
[67] Alvarez-Gaume L and Witten E 1984 Nucl. Phys. B **234** 269
[68] Green M B and Schwarz J H 1984 Phys. Lett. **149B** 117
[69] Bouchiat C, Iliopoulos J and Meyer P 1972 Phys. Lett. **38B** 519
[70] Gross D J, Harvey J A, Martinec E J and Rohm R 1985 Phys. Rev. Lett. **54** 502, Nucl. Phys. B **256** 253, 1986 Nucl. Phys. B **267** 75
[71] Paton J E and Chan H M 1969 Nucl. Phys. B **10** 516
Schwarz J H 1982 Phys. Rept. **89** 223
Marcus N and Sagnotti A 1982 Phys. Lett. **119B** 97
[72] Candelas P, Horowitz G T, Strominger A and Witten E 1985 Nucl. Phys. B **258** 46
[73] For reviews see:
Yau S T *Mirror symmetry I* 1997 (AMS/IP studies in advanced mathematics. 9)
Greene B and Yau S T 1997 *Mirror symmetry II* (AMS/IP studies in advanced mathematics. 1)
[74] Cecotti S, Ferrara S and Girardello L 1989 Int. J. Mod. Phys. A **4** 2475
[75] Strominger A 1990 Commun. Math. Phys. **133** 163
For a recent review see:
Cecotti S 2015 *Supersymmetric Field Theories : Geometric Structures and Dualities* (Cambridge Univ. Press)
[76] Witten E 1985 Phys. Lett. **155B** 151
[77] Giddings S B, Kachru S and Polchinski J 2002 Phys. Rev. D **66** 106006 [hep-th/0105097]
[78] Dixon L J, Harvey J A, Vafa C and Witten E 1985 Nucl. Phys. B **261** 678, 1986 Nucl. Phys. B **274** 285
[79] Lerche W, Lust D and Schellekens A N 1987 Nucl. Phys. B **287** 477
Narain K S, Sarmadi M H and Vafa C 1987 Nucl. Phys. B **288** 551
Antoniadis I, Bachas C P and Kounnas C 1987 Nucl. Phys. B **289** 87
[80] For a review see:
Duff M J, Khuri R R and Lu J X 1995 Phys. Rept. **259** 213 [hep-th/9412184]
[81] Duff M J 1988 Class. Quant. Grav. **5** 189
[82] Bergshoeff E, Sezgin E and Townsend P K 1987 Phys. Lett. B **189** 75, 1988 Annals Phys. **185** 330
[83] Polchinski J 1995 Phys. Rev. Lett. **75** 4724 [hep-th/9510017]
For reviews see:
Polchinski J, Chaudhuri S and Johnson C V hep-th/9602052
Polchinski J hep-th/9611050
[84] Font A, Ibanez L E, Lust D and Quevedo F 1990 Phys. Lett. B **249** 35
[85] Witten E 1995 Nucl. Phys. B **443** 85 [hep-th/9503124]
[86] Hull C M and Townsend P K 1995 Nucl. Phys. B **438** 109 [hep-th/9410167]
Townsend P K 1995 Phys. Lett. B **350** 184 [hep-th/9501068]
[87] Horava P and Witten E 1996 Nucl. Phys. B **460** 506 [hep-th/9510209], 1996 Nucl. Phys. B **475** 94 [hep-th/9603142]
[88] Narain K S 1986 Phys. Lett. **169B** 41
Narain K S, Sarmadi M H and Witten E 1987 Nucl. Phys. B **279** 369
For a review see:
Giveon A, Porrati M and Rabinovici E 1994 Phys. Rept. **244** 77 [hep-th/9401139]
[89] Sagnotti A 1988 in Cargese '87, *Non-Perturbative Quantum Field Theory*, eds. G. Mack et al (Pergamon Press) arXiv:hep-th/0208020
Pradisi G and Sagnotti A 1989 Phys. Lett. **B 216** 59
Horava P 1989 Nucl. Phys. **B 327** 461, 1989 Phys. Lett. **B 231** 251
Bianchi M and Sagnotti A 1990 Phys. Lett. **B 247** 517, 1991 Nucl. Phys. **B 361** 519
Bianchi M, Pradisi G and Sagnotti A 1992 Nucl. Phys. **B 376** 365
Sagnotti A 1992 Phys. Lett. **B 294** 196 [arXiv:hep-th/9210127]
For reviews see:

Dudas E 2000 Class. Quant. Grav. **17** R41 [arXiv:hep-ph/0006190]
Angelantonj C and Sagnotti A 2002 Phys. Rept. **371** 1 [Erratum-ibid. 2003 **376** 339] [arXiv:hep-th/0204089]
[90] Siegel W 1993 Phys. Rev. D **47** 5453 [hep-th/9302036]
Hull C M and Zwiebach B 2009 JHEP **0909** 099 [arXiv:0904.4664 [hep-th]]
[91] Maldacena J M 1999 Int. J. Theor. Phys. **38** 1113 [1998 Adv. Theor. Math. Phys. **2** 231] [hep-th/9711200]
Witten E 1998 Adv. Theor. Math. Phys. **2** 253 [hep-th/9802150]
Gubser S S, Klebanov I R and Polyakov A M 1998 Phys. Lett. B **428** 105 [hep-th/9802109]
For a review see:
Aharony O, Gubser S S, Maldacena J M, Ooguri H and Oz Y 2000 Phys. Rept. **323** 183 [hep-th/9905111]
[92] Flato M and Fronsdal C 1978 Lett. Math. Phys. **2** 421
Angelopoulos E, Flato M, Fronsdal C and Sternheimer D 1981 Phys. Rev. D **23** 1278
Breitenlohner P and Freedman D Z 1982 Annals Phys. **144** 249
Gunaydin M and Marcus N 1985 Class. Quant. Grav. **2** L11
Kim H J, Romans L J and van Nieuwenhuizen P 1985 Phys. Rev. D **32** 389
[93] Bekenstein J D 1973 Phys. Rev. D **7** 2333, 1994 Phys. Rev. D **49** 1912 [gr-qc/9307035]
't Hooft G 1993 in *Salamfest* [gr-qc/9310026]
Susskind L 1995 J. Math. Phys. **36** 6377 [hep-th/9409089]
[94] Ferrara S, Kallosh R and Strominger A 1995 Phys. Rev. D **52** R5412 [hep-th/9508072]
Ferrara S and Kallosh R 1996 Phys. Rev. D **54** 1514 [hep-th/9602136]
Ferrara S, Gibbons G W and Kallosh R 1997 Nucl. Phys. B **500** 75 [hep-th/9702103]
Ooguri H, Strominger A and Vafa C 2004 Phys. Rev. D **70** 106007 [hep-th/0405146]
[95] Strominger A and Vafa C 1996 Phys. Lett. B **379** 99 [hep-th/9601029]
[96] For reviews see:
Mathur S D 2005 Fortsch. Phys. **53** 793 [hep-th/0502050]
Bena I and Warner N P 2008 Lect. Notes Phys. **755** 1 [hep-th/0701216]
[97] Ade P A R *et al.* [Planck Collaboration] 2016 Astron. Astrophys. **594** A13 [arXiv:1502.01589 [astro-ph.CO]], 2016 Astron. Astrophys. **594** A20 [arXiv:1502.02114 [astro-ph.CO]]
[98] Kallosh R, Linde A and Rube T 2011 Phys. Rev. D **83** 043507 [arXiv:1011.5945 [hep-th]]
Ellis J, Nanopoulos D V and Olive K A 2013 Phys. Rev. Lett. **111** 111301 Erratum: [2013 Phys. Rev. Lett. **111** no.12, 129902] [arXiv:1305.1247 [hep-th]]
[99] Antoniadis I, Dudas E, Ferrara S and Sagnotti A 2014 Phys. Lett. B **733** 32 [arXiv:1403.3269 [hep-th]]
Ferrara S, Kallosh R and Linde A 2014 JHEP **1410** 143 [arXiv:1408.4096 [hep-th]]
Dall'Agata G and Zwirner F 2014 JHEP **1412** 172 [arXiv:1411.2605 [hep-th]]
Galante M, Kallosh R, Linde A and Roest D 2015 Phys. Rev. Lett. **114** no.14, 141302 [arXiv:1412.3797 [hep-th]]
Ferrara S, Porrati M and Sagnotti A 2015 Phys. Lett. B **749** 589 [arXiv:1508.02939 [hep-th]]
Ferrara S, Kallosh R and Thaler J 2016 Phys. Rev. D **93** no.4, 043516 [arXiv:1512.00545 [hep-th]]
Carrasco J J M, Kallosh R and Linde A 2016 Phys. Rev. D **93** no.6, 061301 [arXiv:1512.00546 [hep-th]]
For a review see:
Ferrara S, Kehagias A and Sagnotti A 2016 Int. J. Mod. Phys. A **31** no.25, 1630044 [arXiv:1605.04791 [hep-th]]
[100] Rocek M 1978 Phys. Rev. Lett. **41** 451
Ivanov E A and Kapustnikov A A 1978 J. Phys. A **11** 2375
Lindstrom U and Rocek M 1979 Phys. Rev. D **19** 2300
Casalbuoni R, De Curtis S, Dominici D, Feruglio F and Gatto R 1989 Phys. Lett. B **220** 569
Komargodski Z and Seiberg N 2009 JHEP **0909** 066 [arXiv:0907.2441 [hep-th]], 2010 JHEP **1007** 017 [arXiv:1002.2228 [hep-th]]
Kuzenko S M and Tyler S J 2011 Phys. Lett. B **698** 319 [arXiv:1009.3298 [hep-th]]
[101] Sugimoto S 1999 Prog. Theor. Phys. **102** 685 [arXiv:hep-th/9905159]
Antoniadis I, Dudas E and Sagnotti A 1999 Phys. Lett. **B 464** 38 [arXiv:hep-th/9908023]
Angelantonj C 2000 Nucl. Phys. **B 566** 126 [arXiv:hep-th/9908064]
Aldazabal A and Uranga A M 1999 JHEP **9910** 024 [arXiv:hep-th/9908072]
Angelantonj C, Antoniadis I, D'Appollonio G, Dudas E and Sagnotti A 2000 Nucl. Phys. **B 572** 36 [arXiv:hep-th/9911081]
Dudas E and Mourad J 2001 Phys. Lett. B **514** 173 [hep-th/0012071]
Pradisi G and Riccioni F 2001 Nucl. Phys. B **615** 33 [hep-th/0107090]
[102] Kachru S, Kallosh R, Linde A D and Trivedi S P 2003 Phys. Rev. D **68** 046005 [hep-th/0301240]
Kachru S, Kallosh R, Linde A D, Maldacena J M, McAllister L P and Trivedi S P 2003 JCAP **0310** 013

[hep-th/0308055]
Kallosh R, Vercnocke B and Wrase T 2016 JHEP **1609** 063 [arXiv:1606.09245 [hep-th]]
[103] Scherk J and Schwarz J H 1979 Nucl. Phys. B **153** 61
[104] Rohm R 1984 Nucl. Phys. B **237** 553
Kounnas C and Porrati M 1988 Nucl. Phys. B **310** 355
Ferrara S, Kounnas C, Porrati M and Zwirner F 1989 Nucl. Phys. B **318** 75
Kounnas C and Rostand B 1990 Nucl. Phys. B **341** 641
Antoniadis I and Kounnas C 1991 Phys. Lett. B **261** 369
Kiritsis E and Kounnas C 1997 Nucl. Phys. B **503** 117 [hep-th/9703059]
[105] Antoniadis I, Dudas E and Sagnotti A 1999 Nucl. Phys. B **544** 469 [hep-th/9807011]
Antoniadis I, D'Appollonio G, Dudas E and Sagnotti A 1999 Nucl. Phys. B **553** 133 [hep-th/9812118]
[106] Born M and Infeld L 1934 Proc. Roy. Soc. Lond. A **144** 425
[107] Fradkin E S and Tseytlin A A 1985 Phys. Lett. **163B** 123
Abouelsaood A, Callan C G, Nappi C R and Yost S A 1987 Nucl. Phys. B **280** 599
[108] Bachas C hep-th/9503030
Bianchi M and Stanev Y S 1998 Nucl. Phys. B **523** 193 [hep-th/9711069]
Blumenhagen R, Goerlich L, Kors B and Lust D 2000 JHEP **0010** 006 [hep-th/0007024]
Angelantonj C, Antoniadis I, Dudas E and Sagnotti A 2000 Phys. Lett. B **489** 223 [hep-th/0007090]
Bianchi M and Trevigne E 2005 JHEP **0508** 034 [hep-th/0502147]
For a review see:
Blumenhagen R, Kors B, Lust D and Stieberger S 2007 Phys. Rept. **445** 1 [hep-th/0610327]
[109] Hughes J and Polchinski J 1986 Nucl. Phys. B **278** 147
Hughes J, Liu J and Polchinski J 1986 Phys. Lett. B **180** 370
Antoniadis I, Partouche H and Taylor T R 1996 Phys. Lett. B **372** 83 [hep-th/9512006]
[110] Deser S and Puzalowski R 1980 J. Phys. A **13** 2501
Cecotti S and Ferrara S 1987 Phys. Lett. B **187** 335
Bagger J and Galperin A 1997 Phys. Rev. D **55** 1091 [hep-th/9608177]
Rocek M and Tseytlin A A 1999 Phys. Rev. D **59** 106001 [hep-th/9811232]
Kuzenko S M and Theisen S 2001 Fortsch. Phys. **49** 273 [hep-th/0007231]
Ferrara S, Porrati M and Sagnotti A 2014 JHEP **1412** 065 [arXiv:1411.4954 [hep-th]]
Ferrara S, Porrati M, Sagnotti A, Stora R and Yeranyan A 2015 Fortsch. Phys. **63** 189 [arXiv:1412.3337 [hep-th]]
Cerchiai B L and Trigiante M 2016 JHEP **1610** 160 [arXiv:1609.07399 [hep-th]]
[111] For a review see:
Grana M 2006 Phys. Rept. **423** 91 [hep-th/0509003]
[112] de Wit B, Samtleben H and Trigiante M 2003 Nucl. Phys. B **655** 93 [hep-th/0212239]
See also:
Dall'Agata G and Inverso G 2012 Nucl. Phys. B **859** 70 [arXiv:1112.3345 [hep-th]], and references therein
[113] Ferrara S and Sagnotti A 2017 CERN Courier, vol. 57, n. 1, January/February

IOP Conf. Series: Journal of Physics: Conf. Series **873** (2017) 012015 doi:10.1088/1742-6596/873/1/012015

The NA62 experiment at CERN

Francesco Gonnella
On behalf of the NA62 collaboration[1]
University of Birmingham, Edgbaston, Birmingham B15 2TT, UK
E-mail: francesco.gonnella@cern.ch

Abstract. $K^+ \to \pi^+ \nu\bar{\nu}$ is one of the theoretically cleanest meson decay where to look for indirect effects of new physics complementary to LHC searches. The NA62 experiment at CERN SPS is designed to measure the branching ratio of this decay with 10% precision. NA62 took data in 2014 and 2015 and will accumulate kaon decays till the end of 2018. The quality of data acquired in view of the final measurement will be presented. The experiment is also collecting data to search for exotic processes, like Lepton Flavour and Number violation and Axions.

[1] The NA62 Collaboration: G. Aglieri Rinella, R. Aliberti, F. Ambrosino, R. Ammendola, B. Angelucci, A. Antonelli, G. Anzivino, R. Arcidiacono, I. Azhinenko, S. Balev, M. Barbanera, J. Bendotti, A. Biagioni, L. Bician, C. Biino, A. Bizzeti, T. Blazek, A. Blik, B. Bloch-Devaux, V. Bolotov, V. Bonaiuto, M. Boretto, M. Bragadireanu, D. Britton, G. Britvich, M.B. Brunetti, D. Bryman, F. Bucci, F. Butin, J. Calvo, E. Capitolo, C. Capoccia, T. Capussela, A. Cassese, A. Catinaccio, A. Cecchetti, A. Ceccucci, P. Cenci, V. Cerny, C. Cerri, B. Checcucci, O. Chikilev, S. Chiozzi, R. Ciaranfi, G. Collazuol, A. Conovaloff, P. Cooke, P. Cooper, G. Corradi, E. Cortina Gil, F. Costantini, F. Cotorobai, A. Cotta Ramusino, D. Coward, G. D'Agostini, J. Dainton, P. Dalpiaz, H. Danielsson, J. Degrange, N. De Simone, D. Di Filippo, L. Di Lella, S. Di Lorenzo, N. Dixon, N. Doble, B. Dobrich, V. Duk, V. Elsha, J. Engelfried, T. Enik, N. Estrada-Tristan, V. Falaleev, R. Fantechi, V. Fascianelli, L. Federici, S. Fedotov, A. Filippi, M. Fiorini, J. Fry, J. Fu, A. Fucci, L. Fulton, S. Gallorini, S. Galeotti, E. Gamberini, L. Gatignon, G. Georgiev, S. Ghinescu, A. Gianoli, M. Giorgi, S. Giudici, L. Glonti, A. Goncalves Martins, F. Gonnella, E. Goudzovski, R. Guida, E. Gushchin, F. Hahn, B. Hallgren, H. Heath, F. Herman, T. Husek, O. Hutanu, D. Hutchcroft, L. Iacobuzio, E. Iacopini, E. Imbergamo, O. Jamet, P. Jarron, E. Jones, T. Jones K. Kampf, J. Kaplon, V. Kekelidze, S. Kholodenko, G. Khoriauli, A. Khotyantsev, A. Khudyakov, Yu. Kiryushin, A. Kleimenova, K. Kleinknecht, A. Kluge, M. Koval, V. Kozhuharov, M. Krivda, Z. Kucerova, Yu. Kudenko, J. Kunze, G. Lamanna, G. Latino, C. Lazzeroni, G. Lehmann-Miotto, R. Lenci, M. Lenti, E. Leonardi, P. Lichard, R. Lietava, V. Likhacheva, L. Litov, R. Lollini, D. Lomidze, A. Lonardo, M. Lupi, N. Lurkin, K. McCormick, D. Madigozhin, G. Maire, C. Mandeiro, I. Mannelli, G. Mannocchi, A. Mapelli, F. Marchetto, R. Marchevski, S. Martellotti, P. Massarotti, K. Massri, P. Matak, E. Maurice, M. Medvedeva, A. Mefodev, E. Menichetti, E. Migliore, E. Minucci, M. Mirra, M. Misheva, N. Molokanova, J. Morant, M. Morel, M. Moulson, S. Movchan, D. Munday, M. Napolitano, I. Neri, F. Newson, J. Noël, A. Norton, M. Noy, G. Nuessle, T. Numao, V. Obraztsov, A. Ostankov, S. Padolski, R. Page, V. Palladino, G. Paoluzzi, C. Parkinson, E. Pedreschi, M. Pepe, F. Perez Gomez, M. Perrin-Terrin, L. Peruzzo, P. Petrov, F. Petrucci, R. Piandani, M. Piccini, D. Pietreanu, J. Pinzino, I. Polenkevich, L. Pontisso, Yu. Potrebenikov, D. Protopopescu, F. Raffaelli, M. Raggi, P. Riedler, A. Romano, P. Rubin, G. Ruggiero, V. Russo, V. Ryjov, A. Salamon, G. Salina, V. Samsonov, C. Santoni, G. Saracino, F. Sargeni, V. Semenov, A. Sergi, M. Serra, A. Shaikhiev, S. Shkarovskiy, I. Skillicorn, D. Soldi, A. Sotnikov, V. Sugonyaev, M. Sozzi, T. Spadaro, F. Spinella, R. Staley, A. Sturgess, P. Sutcliffe, N. Szilasi, D. Tagnani, S. Trilov, M. Valdata-Nappi, P. Valente, M. Vasile, T. Vassilieva, B. Velghe, M. Veltri, S. Venditti, P. Vicini, R. Volpe, M. Vormstein, H. Wahl, R. Wanke, P. Wertelaers, A. Winhart, R. Winston, B. Wrona, O. Yushchenko, M. Zamkovsky, A. Zinchenko.

IOP Conf. Series: Journal of Physics: Conf. Series **873** (2017) 012015 doi:10.1088/1742-6596/873/1/012015

1. Introduction

NA62 is a fixed-target experiment studying the decay products of a high-intensity kaon beam. The NA62 apparatus is located in the CERN North Area at the SPS extraction site. Its main goal is measuring the Branching Ratio of the ultra-rare Flavor Changing Neutral Current (FCNC) kaon decay

$$K^+ \to \pi^+ \nu\bar{\nu}$$

with a 10% accuracy [1].

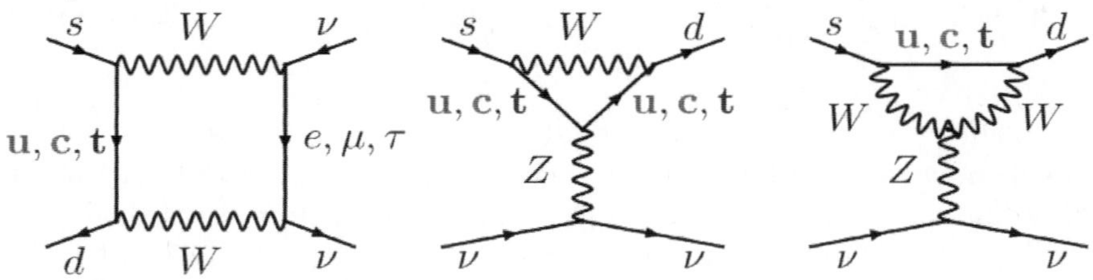

Figure 1. Flavor Changing Neutral Current (FCNC) loop processes, highly suppressed by GIM mechanism, but theoretically very clean [2].

This decay, together with its neutral partner $K_L \to \pi^0\nu\bar{\nu}$, is driven by FCNC (see figure 1), and its branching ratio (BR) is highly suppressed by the GIM mechanism. The BR is predicted in Standard Model with high precision, making these processes extremely sensitive to new physics.

The actual Standard Model prediction [2] for the branching ratio is

$$BR_{SM}(K^+ \to \pi^+\nu\bar{\nu}) = (8.4 \pm 1.0) \times 10^{-11}.$$

Should the experimental value be different from the prediction, new physics contributions have to be present.

On the other hand, this process is almost unexplored experimentally: the only existing measurement has been obtained combining datasets collected by Brookhaven National Laboratory (BNL) E787 and E949 experiments [3, 4]. The measured branching ratio, based on 7 candidates, is:

$$BR_{exp}(K^+ \to \pi^+\nu\bar{\nu}) = (17.3^{+11.5}_{-10.5} \times 10^{-11}).$$

2. The NA62 experimental apparatus

A schematic view of the NA62 apparatus [5] is shown in figure 2. NA62 exploits the 400 GeV/c proton beam delivered by the CERN SPS to create a secondary beam. The protons impinging on a beryllium target produce a hadron beam with a momentum 75 GeV/c and a rate of 750 MHz. The beam is composed of 71% pions, 23% protons, and 6% kaons.

2.1. Beam-particle identification and momentum measurement

Kaons are identified by a differential Cherenkov counter detector (CEDAR) equipped with photon detector, called KTAG. The KTAG detector combines a time resolution of ∼100ps and a kaon identification efficiency higher than 95%.

The kaon direction, momentum and time are measured by the GigaTracKer (GTK) detector. It is composed of three stations interspersed by two pairs of dipole magnets. Each station is made of 200 μm thick silicon pixel matrix ($300{\times}300 \mu m^2$) coupled with 100 μm thick readout chip for a total of 0.5 X_0 per station with time resolution better than 200 ps and momentum resolution 0.2%. The GTK is operated in a high particle rate environment of 750 MHz.

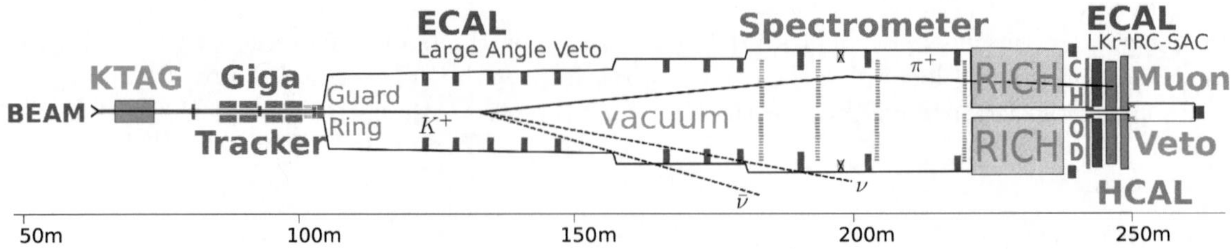

Figure 2. Schematic section of the NA62 experimental setup (indicative scale).

2.2. Final-state particle momentum measurement

The decay products momentum is measured by the Straw spectrometer (STRAW). This detector is composed of 4 stations made of 1792 straw tubes positioned in 4 views (U, V, X, Y). They are operated in vacuum and located on each side of a dipole magnet which produces a vertical field of 0.36 T. The dipole magnet, reused from the earlier NA48 experiment, is placed after the second chamber and provides a 270 MeV/c kick in the horizontal plane to charged-particle trajectory, allowing momentum measurement. The STRAW have momentum resolution better than 1%.

2.3. Final-state particle identification

A first pion/muon separation is obtained with the RICH detector, wich is a Cherenkov counter.

This detector is 17 m long, filled with Neon at atmospheric pressure and equipped with 2000 photomultipliers. An evacuated beam pipe run trough the detector to avoid beam interactions with the gas. The RICH is designed to operate in the momentum range of $15 \div 35$ GeV/c and it leads to an additional muon suppression factor of 10^2. The NA62 RICH is able to identify muons with a 99% efficiency. A time resolution of less than 100 ps ensures the time matching between primary and secondary tracks. Thanks to its high time resolution, the RICH is the reference time for the level-0 trigger system.

The Charged particle HODoscope (CHOD) consisting of 128 scintillator slabs arranged in 2 planes (horizontal and vertical) measures the track crossing time with $\sim$200 ps resolution. It also participates to the level-0 trigger as an an independent system used for trigger efficiency purposes.

2.4. Photon-veto system

The photon-veto system allows to reject decays with π^0 in the final state with an inefficiency better than 10^{-8}. The system covers the solid angle up to 50 mrad and it includes 4 different detectors (see figure 3):

- Large Angle Veto (LAV) covering the angle region $8.5 \div 50$ mrad;
- Liquid Kripton electromagnetic calorimeter (LKr), vetoing the region $1 \div 8.5$ mrad;
- Small Angle Calorimeter (SAC) and Inner Radius Calorimeter (IRC) covering the angle region < 1 mrad.

The Large Angle Veto is composed of 12 stations placed along the NA62 length. Each station is in turn composed of 4 or 5 layers of lead-glass blocks coupled to photomultipliers. The LKr, located just downstream of the RICH detector, is reused from the NA48 experiment. The IRC and SAC are lead-plastic shintillator calorimeters. The IRC is used to cover the small angle around the LKr central hole, while the SAC is located at the very end of the experiment, after a magnet bending away all the charged particle remaining in the beam.

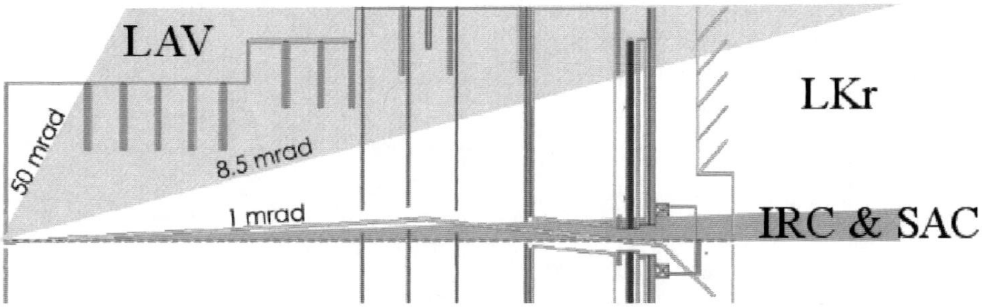

Figure 3. Angular coverage of the NA62 veto subsystems.

Upstream, inelastic interactions and muon halo are vetoed by the CHANTI detector placed after the third station of GTK (referred as Guard Ring in figure 2).

2.5. Hadronic calorimeters and muon veto
The iron/scintillator hadronic calorimeters provide further pion/muon separation. Together with RICH, they guarantee a separation inefficiency of 10^{-7}. Finally the fast muon veto detector (MUV) composed of an iron wall followed by 5 cm thick scintillators, coupled to PMTs, identifies muons. Muon Veto provides also signals for the trigger system.

3. NA62 trigger and data acquisition
In order to reduce the raw data rate, NA62 makes use of a 3-level trigger system consisting of:

- L0 hardware: from 10 MHz to 1 MHz
- L1 software: from 10 MHz to 100 kHz
- L2 software: from 100 kHz to 10 kHz

The hardware L0 trigger processor receives reduced information (trigger primitives) form a subset of detectors to issue the L0 trigger. The L0 logic is implemented on an FPGA, handling a maximum of ~10MHz of trigger primitives from CHOD, RICH, LAV12, Calorimeters, and MUV.

When the detector acquisition systems receive the L0 trigger, they transfer data from the readout electronics to a dedicated farm of computers. On the farm a 2-stage software based high-level Trigger (L1 and L2) runs with the purpose of reducing data rate down to a manageable level. This is done by rejecting as much as possible all the main decay modes of the K^+, being $K^+ \to \mu^+\nu$ (~60%) and $K^+ \to \pi^+\pi^0$ (~20%), while keeping a large efficiency for the $K^+ \to \pi^+\nu\bar{\nu}$ signal.

Assuming 10% signal acceptance, at least the observation of 10^{13} K^+ decays is required to reach the designed precision level. In this framework the fully functional experimental apparatus is expected to detect ~ 45 events per year.

4. Data-quality studies
The $K^+ \to \pi^+\nu\bar{\nu}$ decay is characterised by one track in the GTK making a vertex with one charged track identified as π^+ in the detector downstream. A key variable used to distinguish the signal from the background is the squared missing mass defined as:

$$m^2_{miss} \stackrel{def}{=} (p_{K^+} - p_{\pi^+})^2$$

where p_{K^+} and p_{π^+} are the four momenta of incoming kaon and outgoing pion respectively (the pion mass is assumed for the outgoing particle). It is possible to identify two regions in the m^2_{miss} distribution where the signal to background ratio is most favourable (see figure 4).

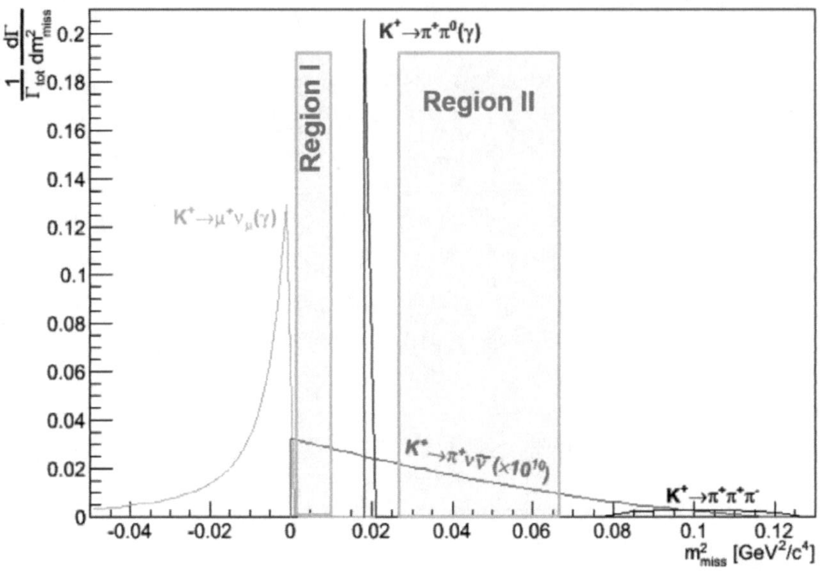

Figure 4. Theoretical m^2_{miss} distribution for signal and backgrounds from the main K^+ decay modes. In the figure the two regions where the signal/ratio is larger are highlighted.

In this section, some results coming from a study on data quality based on data collected in 2015 and relevant for the measurement of the $K^+ \to \pi^+ \nu \bar{\nu}$ decay are presented.

4.1. Kinematic results

A single outgoing track must form a vertex with the incoming kaon, required to be in the 65-metre long region between the last GTK station and the first STRAW chamber. A quality check on the vertex is based on the distance of closest approach (CDA) between the GTK hit location and the tracks reconstructed in the STRAW spectrometer. CDA must be less than 1.5 cm.

Further selection requirements for the signal are:

- The track projection needs to spatially intersect an energy deposits in the calorimeters and the corresponding hit in the CHOD.
- The associated CHOD hit must be in time with the KTAG candidate.

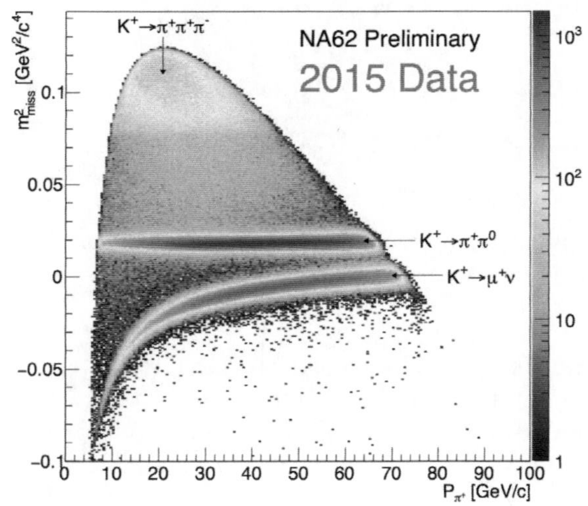

Figure 5. m^2_{miss} versus the momentum of the downstream track in single-track events confirmed by GTK and KTAG signals.

The squared missing mass of tracks that satisfy all those constraints are reported in figure 5 as a function of the track momentum measured with the STRAW spectrometer. In this plot, one can clearly distinguish the three main kaon decay channels: $K \to \pi^+\pi^+\pi^-$ on the top, $K \to \pi^+\pi^0$ as an horizontal band centred around the value of neutral pion mass, and $K \to \mu^+\nu$ at the bottom when the pion mass hypothesis results in a negative missing mass values. The plots shown in this paper, are based on datasets recorded in 2015 at around 3% of nominal intensity.

The KTAG can also be used in anti-coincidence with a Gigatracker track in order to select single track events not related to kaons. This technique allows to study the background generated by the non-kaon beam. Figure 6 shows the main sources of non-kaon tracks:

- decay from beam π^+,
- elastic scattering of beam particles in the material along the beam line (KTAG and Gigatracker stations),
- inelastic scatterings in the last GTK station.

The resolution of the $K^+ \to \pi^+\pi^0$ squared missing mass peak is essential to reduce the tails that can span across the two signal regions. In figure 7 the squared missing mass resolution is

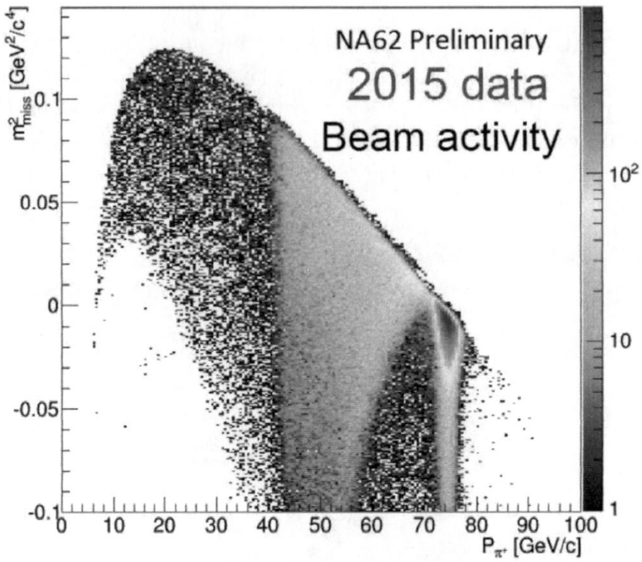

Figure 6. m^2_{miss} vs P_{π^+} momentum, under π^+ mass hypothesis. The downstream track was required to be in anti-coincidence with the kaon track in KTAG.

plotted as a function of momentum. The resolution measured is 1.2 $\times 10^{-3}$ GeV$^2/c^4$, close to the design value represented by the solid black line.

Figure 7 shows that the resolution increases by a factor of three if a coincidence with the GTK detector is not required.

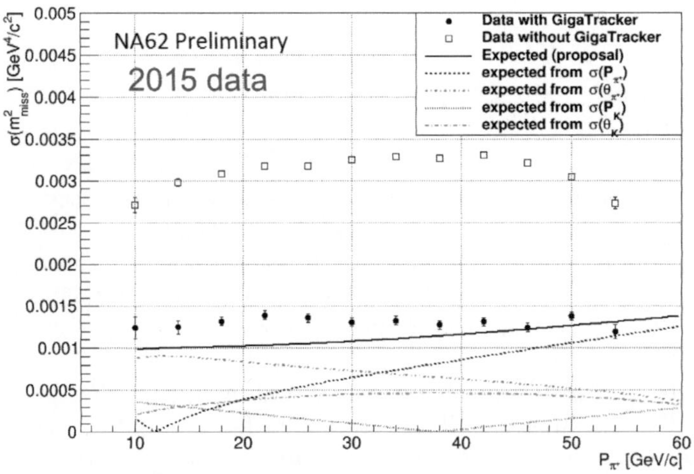

Figure 7. $\sigma(m^2_{miss})$ distribution obtained with individual kaon track measured with GTK (black points) and assuming kaon track aligned on the central beam axis (empty squares). The solid line shows the expected distribution evaluated as the sum of the various contributions coming from P_π, θ_π, P_K, θ_K shown by the different dashed lines.

4.2. Particle identification

The particle identification systems of NA62 are designed to separate π^+ from μ^+ and e^+ in order to guarantee at least 7 orders of magnitude suppression of $K^+ \to \mu^+\nu_\mu$ in addition to the kinematic rejection. The RICH is designed to contribute with a factor 100 to this rejection.

To study the performance, $K^+ \to \mu^+\nu_\mu$ and $K^+ \to \pi^+\pi^0$ decays were selected applying the kinematic constraints, Muon Veto, and calorimetric information. The positive tracks were selected to have a momentum between 15 and 35 Gev/c. Figure 8 shows that remaining muon contamination of 0.01 corresponds to a pion efficiency of 80%, close to the design target. This figures were improved in the 2016 run.

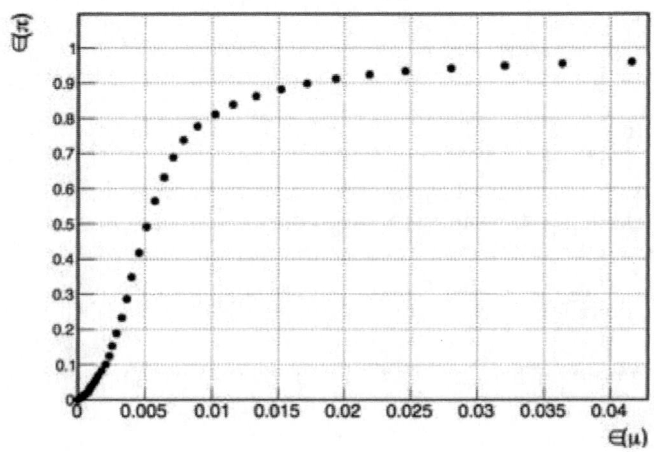

Figure 8. π^+ efficiency as a function of μ^+ remaining contamination measured by the RICH.

4.3. Photon veto

The photon veto system has to provide a suppression factor of 10^8 to reject π^0 from the decay: $K^+ \to \pi^+\pi^0$ ($\pi^0 \to \gamma\gamma$).

The rejection of such events is implemented by vetoing the events containing at least one photon in one of the electromagnetic calorimeters, LAV, LKr, IRC or SAC. The target suppression factor, is reached thanks to a photon detection inefficiency lower than 10^{-5} for energy greater than 10GeV. This is true for pions that have at least 35 GeV/c momentum. Figure 9 shows the inefficiency for the detection of π^0: when only the LKr calorimeter is used, when LKr and LAV are used, and when the full photon veto system is used. The measurement of the efficiency with 2015 data results is less than 10^{-6} at 90% C.L. but it is statistically limited. New measurements are ongoing on 2016 datasets.

5. NA62 physics besides $K^+ \to \pi^+\nu\overline{\nu}$

Thanks to the performance of the NA62 apparatus, physics beyond the $K^+ \to \pi^+\nu\overline{\nu}$ ultra-rare decay can be addressed. NA62 can significantly improve the existing limits on lepton flavour and number violating decays like $K^+ \to \pi^+\mu^\pm e^\mp$ or $K^+ \to \pi^+l^+l^-$. Experimentally π^0 physics can take advantage of the performances of the electromagnetic calorimeters and processes like $\pi^0 \to$ *invisible* or dark photon production can be investigated. Thanks to the quality of the kinematic reconstruction, searches for heavy neutral lepton (N) produced in $K^+ \to l^+N$ decays can be performed at NA62 and improve the present sensitivity.

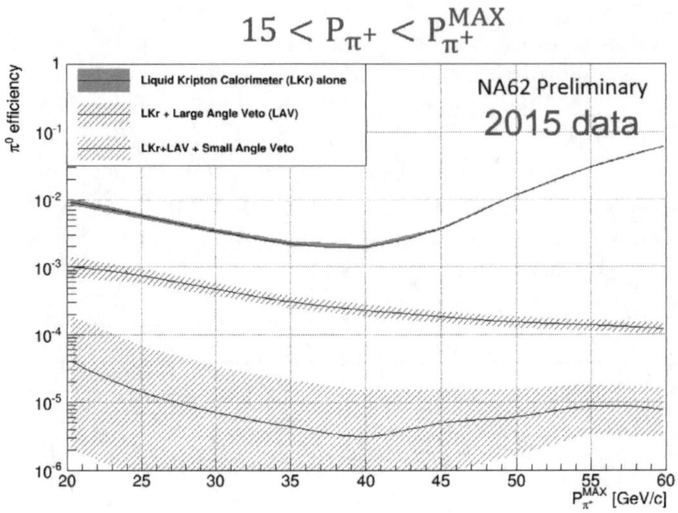

Figure 9. π^0 veto efficiency for different combinations of the calorimeters information.

The longitudinal scale of the apparatus opens the possibility to search for long-lived particles through their decays. Dark photon, heavy neutral leptons or axion-like particles produced at the target or in beam dump configurations. NA62 is already addressing part of the above physics programme simultaneously with the $K^+ \to \pi^+\nu\overline{\nu}$ measurement.

6. Conclusions and Prospects

$K^+ \to \pi^+\nu\overline{\nu}$ is one of the theoretically cleanest meson decay to look for indirect effects of new physics complementary to LHC searches. The NA62 experiment will measure the branching ratio of this decay with 10% precision. NA62 took data in 2014 and 2015. In 2016 the first NA62 physics run took place, collecting about 10^{12} kaon decays, data are currently being analysed. The data-quality studies show that the physics sensitivity for the $K^+ \to \pi^+\nu\overline{\nu}$ decay is in line with the design, allowing NA62 to mark CERNs return to the exploration of the Standard Model using high-intensity kaon beams.

References

[1] Proposal to Measure the Rare Decay $K^+ \to \pi^+\nu\overline{\nu}$ at the CERN SPS, CERN-SPSC-2005-013 (2005)
[2] Andrzej J. Buras, Journal of High Energy Physics **1511** 033 (2015)
[3] S. Adler et al. (E949 and E787 Collaborations), Phys. Rev. D **77**, 052003 (2008).
[4] S. Adler et al. (E949 and E787 Collaborations), Phys. Rev. D **79**, 092004 (2009).
[5] NA62-Collaboration, Technical Design Document , 092004 (2010).

IOP Conf. Series: Journal of Physics: Conf. Series **873** (2017) 012016 doi:10.1088/1742-6596/873/1/012016

Neutral pion form factor measurement by the NA62 experiment

Michal Zamkovsky

Speaker, Institute of Particle and Nuclear Physics, Charles University

E-mail: michal.zamkovsky@cern.ch

F Ambrosino, A Antonelli, G Anzivino, R Arcidiacono, W Baldini, S Balev, J R Batley, M Behler, S Bifani, C Biino, A Bizzeti, T Blazek, B Bloch-Devaux, G Bocquet, V Bolotov, F Bucci, N Cabibbo, M Calvetti, N Cartiglia, A Ceccucci, P Cenci, C Cerri, C Cheshkov, J B Chze, M Clemencic, G Collazuol, F Costantini, A Cotta Ramusino, D Coward, D Cundy, A Dabrowski, G DAgostini, P Dalpiaz, C Damiani, H Danielsson, M De Beer, G Dellacasa, J Derr, H Dibon, D Di Filippo, L DiLella, N Doble, V Duk, J Engelfried, K Eppard, V Falaleev, R Fantechi, M Fidecaro, L Fiorini, M Fiorini, T Fonseca Martin, P L Frabetti, A Fucci, S Gallorini, L Gatignon, E Gersabeck, A Gianoli, S Giudici, A Gonidec, E Goudzovski, S Goy Lopez, E Gushchin, B Hallgren, M Hita-Hochgesand, M Holder, P Hristov, E Iacopini, E Imbergamo, M Jeitler, G Kalmus, V Kekelidze, K Kleinknecht, M Koval, V Kozhuharov, W Kubischta, V Kurshetsov, G Lamanna, C Lazzeroni, M Lenti, E Leonardi, L Litov, N Lurkin, D Madigozhin, A Maier, I Mannelli, F Marchetto, G Marel, M Markytan, P Marouelli, M Martini, L Masetti, P Massarotti, E Mazzucato, A Michetti, I Mikulec, M Misheva, N Molokanova, E Monnier, U Moosbrugger, C Morales Morales, M Moulson, S Movchan, D J Munday, M Napolitano, A Nappi, G Neuhofer, A Norton, T Numao, V Obraztsov, V Palladino, M Patel, M Pepe, A Peters, F Petrucci, M C Petrucci, B Peyaud, R Piandani, M Piccini, G Pierazzini, I Polenkevich, I Popov, Y Potrebenikov, M Raggi, B Renk, F Retire, P Riedler, A Romano, P Rubin, G Ruggiero, A Salamon, G Saracino, M Savri, M Scarpa, V Semenov, A Sergi, M Serra, M Shieh, S Shkarovskiy, M W Slater, M Sozzi, T Spadaro, S Stoynev, E Swallow, M Szleper, M Valdata-Nappi, P Valente, B Vallage, M Velasco, M Veltri, S Venditti, M Wache, H Wahl, A Walker, R Wanke, L Widhalm, A Winhart, R Winston, M D Wood, S A Wotton, O Yushchenko, A Zinchenko and M Ziolkowski

Abstract. The NA62 experiment at CERN collected a large sample of charged kaon decays with a highly efficient trigger for decays into electrons in 2007. The kaon beam represents a source of tagged neutral pion decays in vacuum. A measurement of the electromagnetic transition form factor slope of the neutral pion in the time-like region from ~1 million fully reconstructed π^0 Dalitz decay is presented. The limits on dark photon production in π^0 decays from the earlier kaon experiment at CERN, NA48/2, are also reported.

Introduction

The NA48/2 experiment at the CERN SPS recorded a sample of 2×10^{11} charged kaon decays in 2003-04. The following NA62 experiment collected a ten times smaller sample of $K^\pm$ in 2007 and was using the NA48 detector [1] and a modified beam line [2]. This paper reports on the measurements of neutral pion physics on NA62 and NA48/2, namely the measurement of the slope of the π^0 electromagnetic transition form factor from a sample of $\pi^0 \to e^+e^-\gamma$ decays and the search for the dark photon in the $\pi^0 \to \gamma A'$ decay.

1. Experimental setup

Narrow momentum band $K^\pm$ beams with central momentum 60 (74) GeV/c, with $\delta P_K/P_K \sim 1\%$ (rms), were extracted from 400 GeV/c proton beam for NA48/2 (NA62) configuration. The kaon beams were focused at the first drift chamber with $\sim$ 10 mm transverse size. The beam flux with an intensity of $2.5 \cdot 10^7$ protons per pulse entered a decay region, where $\sim$ 18% of kaons decayed. The decay region was 113 m long, with a diameter from 1.92 m in the first 66 m increasing to 2.4 m in the last 47 m. It was contained in an evacuated cylindrical steel tank with a volume $\sim$ 120 m^3. The vacuum at a level $\leq 10^{-4}$ mbar avoided interactions of kaon decay products before detection. The downstream part of the vacuum tank was closed by convex hemispherical (0.8 mm thick, that corresponds to $3 \cdot 10^{-3} X_0$) Kevlar window with a radius of 1.3 m. The difference between the NA48/2 and NA62 detector setups was the presence of the KABES detector, which has been removed after NA48/2. In NA48/2 two simultaneous beams of K^+ and K^- were used, while in NA62 one beam was alternatively stopped. The fiducial region was followed by a magnetic spectrometer, consisting of four drift chambers (DCH), housed in a vessel filled with helium gas at nearly atmospheric pressure. Through the middle of DCH and following detector a beam pipe allowed the undecayed particles to continue their path in vacuum. Each DCH consisted of 8 octagonal planes with 256 grounded wires and was operated with a gas mixture of 50% Argon and 50% Ethan. The maximum drift time in a plane was 100 ns. The resolution on the spatial position of the reconstructed hit was 90 μm and the time resolution was about 700 ps. A spectrometer magnet was located between the chambers DCH2 and DCH3. It provided a momentum kick of 120 (257) MeV/c. The resolution of momentum measurement was: $\sigma_p/p = (1.02 \oplus 0.044 \cdot p)\%$ and $\sigma_p/p = (0.48 \oplus 0.009 \cdot p)$ in 2007, p in GeV/c. The spectrometer was followed by a scintillator hodoscope consisting of two planes segmented into horizontal and vertical strips achieving a very good $\sim$ 150 ps time resolution. Further downstream was a liquid krypton calorimeter (LKr) that measured the energy of electrons and photons. The transverse segmentation into 13248 2 cm $\times$ 2 cm projective cells and the 27 radiation length thickness resulted in an energy resolution $\sigma(E)/E = (3.2/\sqrt{E} \oplus 9.0/E \oplus 0.42)\%$ (E in GeV/c) and a transverse position resolution of 1.5 mm for 10 GeV showers. The calorimeter was followed by a muon system consisting of three scintillator planes, with an iron wall installed in front of each plane. A detailed description of the beam line, detectors and trigger is available in [1].

2. Theory

The Dalitz decay $\pi^0_D \to e^+e^-\gamma$ proceeds through the $\pi^0\gamma\gamma$ vertex with an off-shell photon. The commonly used independent kinematic variables x and y are defined in terms of particle four-momenta:

$$x = \left(\frac{M_{ee}}{m_{\pi^0}}\right)^2 = \frac{(p_{e^+} + p_{e^-})^2}{m^2_{\pi^0}}, \qquad y = \frac{2p_{\pi^0} \cdot (p_{e^+} - p_{e^-})}{m^2_{\pi^0}(1-x)}. \tag{1}$$

The x variable is the normalized invariant mass squared of the electron-positron pair, while y is related to the angles between the final state particle momenta. The kinematic limits on the variables are given by

$$r^2 \leq x \leq 1, \quad -\beta \leq y \leq \beta, \quad \text{where} \quad r = \frac{2m_e}{m_{\pi_0}} \quad \text{and} \quad \beta = \sqrt{1 - \frac{r^2}{x}}. \tag{2}$$

The normalized π^0_D leading order (LO) differential decay width reads:

$$\frac{1}{\Gamma(\pi^0_{2\gamma})}\frac{d^2\Gamma(\pi^0_D)}{dxdy} = \frac{\alpha}{4\pi}\frac{(1-x)^3}{x}(1 + y^2 + \frac{r^2}{x})|F(x)|^2. \tag{3}$$

In the allowed kinematic region for the π^0_D decay, $F(x)$ is expected to vary slowly and it is sufficient to use a linear approximation $F(x) = 1 + ax$, where a is the so-called π^0 *transition form factor* (TFF) slope parameter. The expected π^0 TFF slope value from Vector Meson Dominance model, dominated by the ρ and ω mesons, is $a \approx 0.03$. The theoretical models of π^0 TFF enter hadronic light-by-light scattering contribution to $(g-2)_\mu$. For a recent overview and references see [3]. The comparison of TFF slope predictions with a model independent measurement is an important test of the theory models.

2.1. Radiative Corrections

The radiation corrections are a crucial aspect of measuring the π^0 TFF, as their effect on the differential decay rate exceed that of the TFF. The first study of radiative corrections was done in [4] and extended in [5], where the following diagrams were considered:

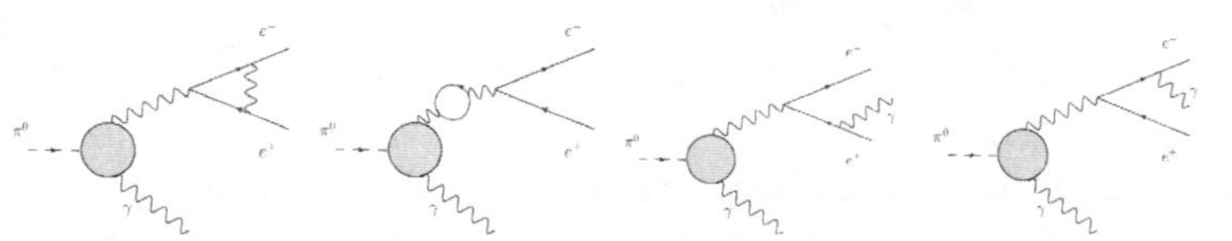

$\delta_{virt} + \delta_{BS}$ corrections.

The revisited and improved radiative corrections in [6] were largely triggered by and used for this analysis. They include additional second-order contributions to the internal bremsstrahlung:

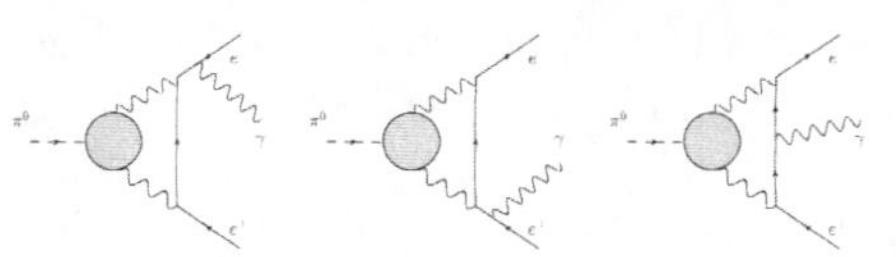

$\delta_{1\gamma\text{IR}}$ correction.

The contribution of the radiative corrections in the $(x,\ y)$ plane together with the separate contributions from different origins are shown in Figure 1. They enter the differential decay width through a function $\delta(x,y) = \delta_{virt} + \delta_{BS} + \delta_{1\gamma\text{IR}}$ in the next-to-leading-order (NLO):

$$\frac{d^2\Gamma}{dxdy} = \frac{d^2\Gamma^{LO}}{dxdy}(1 + \delta(x,y)), \qquad \delta(x,y) = \frac{d^2\Gamma^{NLO}}{dxdy}\Big/\frac{d^2\Gamma^{LO}}{dxdy}. \tag{4}$$

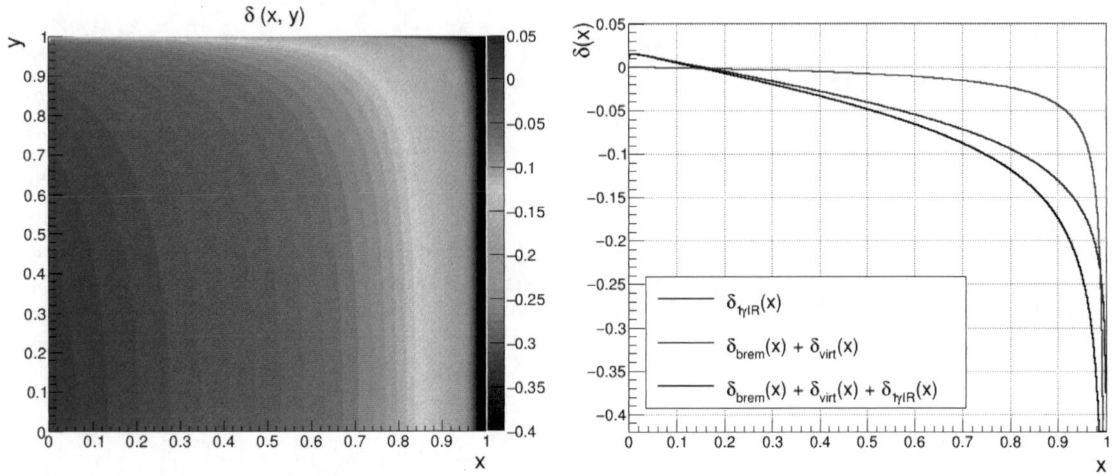

Figure 1. Left: Size of the radiative corrections to the π^0_D decay in the x, y plane. Right: Size of the radiative corrections versus x. Black line represents the total radiative correction size; blue corresponds to its component first computed in [5]; red shows the new contribution, introduced in [6].

As the π^0 TFF does not depend on the y Dalitz variable, the slope parameter is extracted from the reconstructed distribution of the x Dalitz variable. One can therefore integrate the π^0_D differential decay width in eq. 3 over y and obtain:

$$\frac{1}{\Gamma(\pi^0_{2\gamma})}\frac{d\Gamma(\pi^0_D)}{dx} = \frac{2\alpha}{3\pi}\frac{(1-x)^3}{x}(1+\frac{r^2}{2x})\sqrt{1-\frac{r^2}{x}}(1+\delta(x))(1+ax)^2, \tag{5}$$

where the TFF is described by the linear parametrization and the radiative corrections to the x distribution are included.

3. Event selection

The tagged π^0s are originating from $K^\pm \to \pi^\pm\pi^0$ decays, which branching fraction is $BR(K_{2\pi}) \approx 21\%$. In 2007, the data taking conditions were optimized for $e^\pm$ from $K^\pm \to e^\pm\nu_e$, which is favorable also for the $K^\pm \to \pi^\pm\pi^0$ with subsequent decay $\pi^0 \to \gamma e^+e^-$. During 2007 a large and clean sample of π^0_D decays was recorded, counting 1.05×10^6 events. The selection of signal events was based on the three-track topology, a photon cluster in LKr and full kinematic closure. The selection was restricted to area with Dalitz variable $x > 0.01$. Figure 2 shows the reconstructed and simulated invariant mass of π^0_D, and the distribution of the Dalitz x variable.

4. TFF measurement

The TFF was obtained by adjusting the simulation to the data x spectrum. A χ^2 fit to the reconstructed x distribution of the π^0_D event candidates distributed over equipopulated bins has been performed to extract the TFF slope value. The NA62 preliminary result on the π^0 TFF slope parameter is obtained as:

$$a = (3.70 \pm 0.53_{stat} \pm 0.36_{syst}) \times 10^{-2}.$$

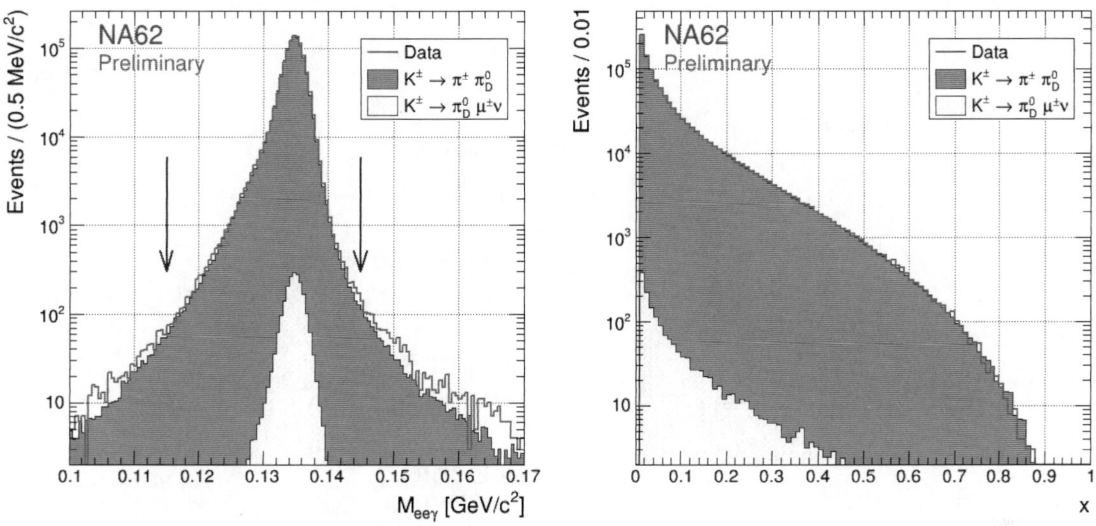

Figure 2. Left: Reconstructed and simulated invariant mass of π^0_D.
Right: Reconstructed x Dalitz variable for data and MC components. The TFF slope parameter value used to generate MC events was $a_{sim} = 0.032$.

The fit result is illustrated in Figure 3 where the effect of a positive TFF slope is clearly seen from the ratio of the data and MC distribution reweighted to a value $a = 0$. The horizontal positions of black markers correspond to the barycenters of the data distributed over 20 equipopulated bins. The red solid line represents the TFF function with the slope value equal to the fit central value. Red dashed lines correspond to the 1 σ band. The breakdown of the uncertainties is presented in Table 1. The value of the slope parameter can be compared to the various TFF theory expectations:

- $a = (2.90 \pm 0.50) \times 10^{-2}$, χPT [7];
- $a = (3.07 \pm 0.06) \times 10^{-2}$, dispersion theory [8];
- $a = (2.92 \pm 0.04) \times 10^{-2}$, two-hadron saturation [9].

The results from other experiments are shown in Figure 3.

5. Search for the dark photon

The idea of a new U(1) gauge symmetry with a very light gauge boson, called a Dark Photon (DP, usually denoted as A' boson) has been proposed in [10], [11]. The DP could mediate the annihilation of Dark Matter particles and explain the excess of positrons in cosmic rays [12] and also explain the muon gyromagnetic ratio $(g-2)$ measurement [13]. Dark photon presence could manifest itself in the NA48/2 data set by the process $\pi^0 \to \gamma A'$ decay followed by the prompt $A' \to e^+e^-$ decay. The signal decay chain has the same particles in the final state as the π^0 Dalitz decay. The DP is characterized by two a priori unknown parameters, the mass $m_{A'}$ and the mixing parameter ε^2. The expected branching fraction of π^0 into the DP is [14]

$$\mathcal{B}(\pi^0 \to \gamma A') = 2\varepsilon^2 \left(1 - \frac{m^2_{A'}}{m^2_{\pi^0}}\right)^3 \mathcal{B}(\pi^0 \to \gamma\gamma), \tag{6}$$

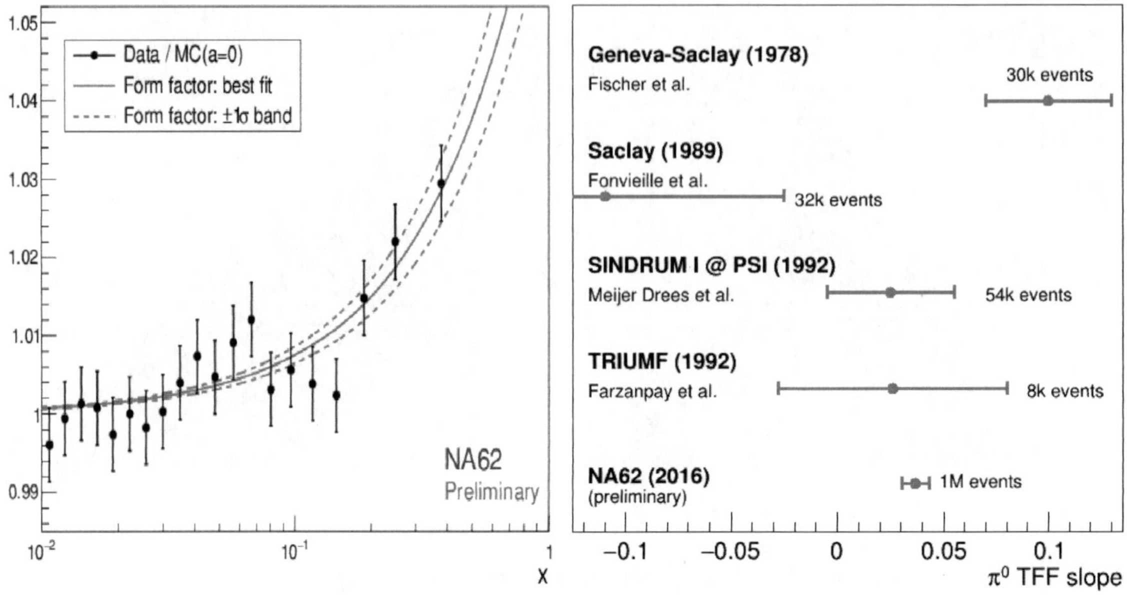

Figure 3. Left: Fit result illustration. Right: Results on the TFF slope from π^0_D measurements.

Table 1. Summary of uncertainty sources.

Source	$\delta a(\times 10^2)$
Statistical-Data	0.49
Statistical-MC	0.20
Beam momentum spectrum simulation	0.30
Spectrometer momentum scale	0.15
Spectrometer resolution	0.05
LKr non-linearity and energy scale	0.04
Particle mis-ID	0.08
Accidental background	0.08
Neglected π^0_D sources in MC	0.01

which is kinematically suppressed as $m_{A'}$ approaches m_{π^0}. In the DP mass range $2m_e < m_{A'} < m_{\pi^0}$ accessible in pion decays, the only allowed tree-level decay into SM fermions is $A' \to e^+e^-$, while the loop-induced SM decays ($A' \to 3\gamma$, $A' \to \nu\bar{\nu}$) are highly suppressed. Therefore, for a DP decaying only into SM particles, $\mathcal{B}(A' \to e^+e^-) \approx 1$, the expected total decay width is [14]

$$\Gamma_{A'} \approx \Gamma(A' \to e^+e^-) = \frac{1}{3}\alpha\varepsilon^2 m_{A'}\sqrt{1 - \frac{4m_e^2}{m_{A'}^2}}\left(1 + \frac{2m_e^2}{m_{A'}^2}\right). \tag{7}$$

It follows that, for $2m_e \ll m_{A'} < m_{\pi^0}$, the DP mean proper lifetime $\tau_{A'}$ satisfies the relation

$$c\tau_{A'} = hc/\Gamma_{A'} \approx 0.8\mu m \times \left(\frac{10^{-6}}{\varepsilon^2}\right) \times \left(\frac{100\text{MeV}/c^2}{m_{A'}}\right). \tag{8}$$

This analysis is performed assuming that the DP decays at the production point (prompt decay), which is valid for sufficiently large values of $m_{A'}$ and ε^2. Then the DP has the same production and signature as the Dalitz decay $\pi^0_D \to e^+e^-\gamma$, that represents an irreducible but well controlled background.

The result of the analysis of the full data set of NA48/2 including a total of 1.69×10^7 π^0_D reconstructed events is reported. The π^0_D events are coming from kaon decays $K^\pm \to \pi^\pm\pi^0$ ($K_{2\pi}$) and $K^\pm \to \pi^0\mu^\pm\nu$ ($K_{\mu 3}$), followed by the prompt $\pi^0 \to \gamma A'$, $A \to e^+e^-$ decay chain. These two exclusive selections apply the following criteria:
$K_{2\pi}$ selection:

- $|m_{\pi\gamma ee} - m_K| < 20$ MeV/c^2;
- $|m_{\gamma ee} - m_{\pi^0}| < 8$ MeV/c^2;
- no missing momentum.

$K_{\mu 3}$ selection:

- $m^2_{miss} = (P_K - P_\mu - P_{\pi^0})^2$ compatible with 0;
- $|m_{\gamma ee} - m_{\pi^0}| < 8$ MeV/c^2;
- missing total and transverse momentum.

The numbers of reconstructed events from $K_{2\pi}$ ($K_{\mu 3}$) are 1.38×10^7 (0.31×10^7), respectively. A search for resonances in the di-electron invariant mass spectrum has been performed in the mass range 9 MeV/$c^2 \le m_{A'} <$ 120 MeV/c^2. The lower boundary of the mass range is determined by the limited accuracy of the π^0_D background simulation at low e^+e^- mass. The obtained upper limits on the numbers of DP candidates and $\mathcal{B}(\pi^0 \to \gamma A')$ in each mass hypothesis considered are presented in Figure 4.

No signal was observed and upper limits on the DP mixing parameter ε^2 were improved in the DP mass range 9 MeV/$c^2 < m_{A'} <$ 70 MeV/c^2, see Figure 5. If dark photon couples to quarks and decays mainly to SM fermions, it is ruled out as explanation for the anomalous $(g-2)_\mu$ measurement.

Conclusion

A total of 1.05×10^6 π^0 Dalitz decays, originating from $K^\pm \to \pi^\pm\pi^0$ decays, have been fully reconstructed. A preliminary measurement of the π^0 transition form factor slope based on NA62 2007 data statistics was performed:

$$a = (3.70 \pm 0.53_{stat} \pm 0.36_{syst}) \times 10^{-2}.$$

The precision of the TFF measurement has been improved in the time-like momentum region. Final result is in preparation.

Improved limits (down to $\epsilon^2 = 2 \times 10^{-7}$) in the 9-70 MeV/$c^2$ mass range of the dark photon search in π^0 decays has been achieved by exploring NA48/2 data. The whole region favored by $(g-2)_\mu$ is excluded, assuming that DP decays into SM fermions only.

Acknowledgments

The speaker received support from project GA UK no. 404716.

IOP Conf. Series: Journal of Physics: Conf. Series **873** (2017) 012016 doi:10.1088/1742-6596/873/1/012016

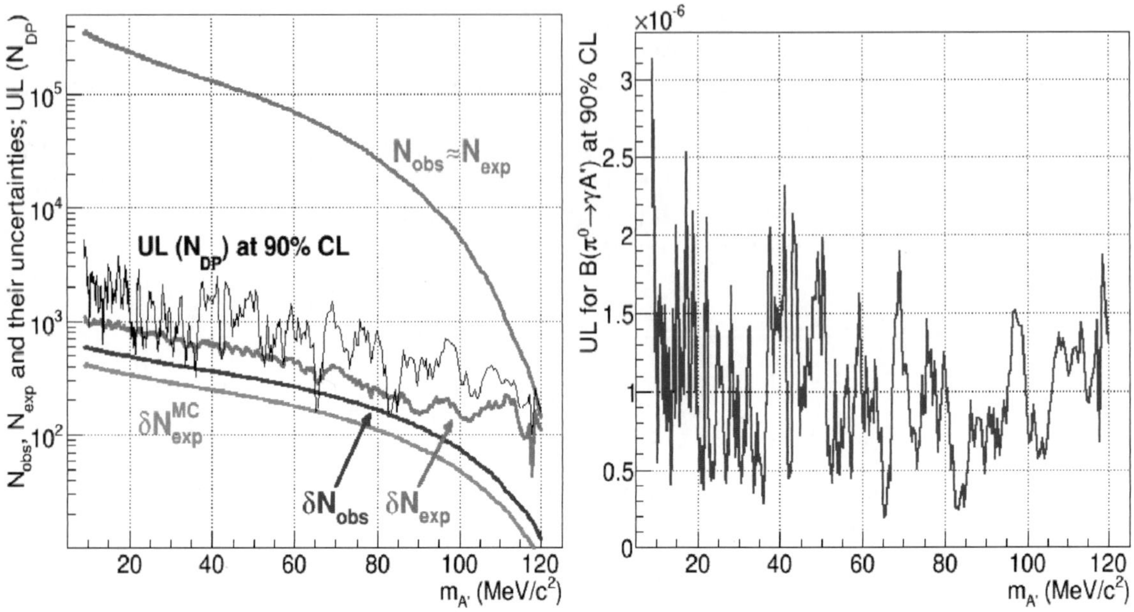

Figure 4. Left: Numbers of observed data events (N_{obs}) and expected π^0_D background events (N_{exp}) passing the joint DP selection (indistinguishable in a logarithmic scale), estimated uncertainties $\delta N_{obs} = \delta N_{exp}$ and δN_{exp}, and obtained upper limits at 90% CL on the numbers of DP candidates (N_{DP}) for each DP mass vale $m_{A'}$. The contribution to δN_{exp} from the MC statistical uncertainty is shown separately (δN^{MC}_{exp}). The remaining and dominant component is due to the statistical errors on the trigger efficiencies measured in the DP signal region. Right: Obtained upper limits on $\mathcal{B}(\pi^0 \to \gamma A')$ at 90% CL for each DP mass value $m_{A'}$.

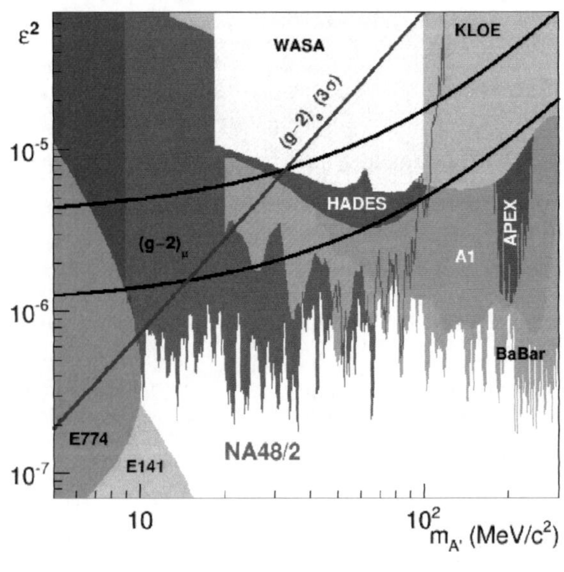

Figure 5. Upper limits at 90% CL on the mixing parameter ε^2 versus the DP mass $m_{A'}$ obtained from $\pi^0 \to \gamma A'$ search by NA48/2, compared to other published exclusion limits from meson decay, beam dump and e^+e^- collider experiments.Further details can be found in [15].

References

[1] V Fanti *et al*, *Nucl. Instrum. Meth.* A **574**, 433 (2007).
[2] J R Batley *et al*, *Eur. Phys. J.* C **52**, 875 (2007).
[3] A Nyffeler, arXiv:1602.03398 [hep-ph].

[4] B E Lautrup and J Smith, *Phys. Rev.* D **3**, 1122 (1971).
[5] Mikaelian and Smith, *Phys. Rev.* D **5**, (1972) 1763.
[6] T Husek *et al*, *Phys. Rev.* D **92**, 054027 (2015).
[7] K Kampf *et al*, *EPJ* C **46** (2006), 191.
[8] M Hoferichter *et al*,*EPJ* C **74**, (2014), 3180.
[9] T Husek *et al*, *EPJ* C **75**, (2015) 12, 586.
[10] P Fayet, *Phys.Lett.* B **95**, 285 (1980).
[11] C Boehm, P Fayet, and J Silk, *Phys.Rev.* D **69**, 101302 (2004).
[12] M Aguilar *et al*, *Phys.Rev.Lett.* **110**, 141102 (2013).
[13] M Pospelov, *Phys. Rev.* D **80**, (2009) 095002.
[14] B Batell, M Pospelov and A Ritz, *Phys. Rev.* D **80**, (2009) 095024.
[15] J R Batley *et al*, *Phys. Lett.* B **746**, (2015) 178.

IOP Conf. Series: Journal of Physics: Conf. Series **873** (2017) 012017 doi:10.1088/1742-6596/873/1/012017

Cherenkov-like emission of Z bosons

D Colladay[1], **J P Noordmans**[2] **and R Potting**[2]

[1] New College of Florida, 5800 Bay Shore Road, Sarasota, FL 34243, USA
[2] CENTRA and Department of Physics, FCT, University of the Algarve, Campus de Gambelas, 8005-139 Faro, Portugal

E-mail: rpotting@ualg.pt

Abstract. We study CPT and Lorentz violation in the electroweak gauge sector of the Standard Model in the context of the Standard-Model Extension (SME). In particular, we show that any non-zero value of a certain relevant Lorentz violation parameter that is thus far unbounded by experiment would imply that for sufficiently large energies one of the helicity modes of the Z boson should propagate with spacelike four-momentum and become stable against decay in vacuum. In this scenario, Cherenkov-like radiation of Z bosons by ultra-high-energy cosmic-ray protons becomes possible. We deduce a bound on the Lorentz violation parameter from the observational data on ultra-high energy cosmic rays.

1. Introduction

Lorentz invariance is a fundamental ingredient of both quantum field theory and General Relativity. Nevertheless, in the past two decades there has been a growing interest in the possibility that Lorentz symmetry may not be exact in Nature. On the theoretical side, a number of candidate theories for quantum gravity have been shown to involve Lorentz invariance violation as a possible effect. We may mention in particular string field theory [1] and loop quantum gravity [2]. More simple toy models involving spontaneous breaking of Lorentz symmetry that have been studied are bumblebee models [3] and lattice models where fermion bilinears carrying Lorentz indices acquire nonzero vacuum expectation values [4]. Another scenario involves cosmologically varying scalars [5] in which a scalar acquires a vacuum expectation value that has a (slow) variation as a function of spacetime. In other words, a preferred direction is selected through a non-zero value for the gradient of a scalar. A class of Lorentz-violating models that have received a lot of attention involve noncommutative geometry. Here the coordinates of spacetime are taken to be noncommuting quantities [6], $[x_\alpha, x_\beta] = i\frac{1}{\Lambda_{NC}^2}\theta_{\alpha\beta}$, where $\theta_{\alpha\beta}$ is a tensor-valued set of coefficients of $\mathcal{O}(1)$, while Λ_{NC} denotes the noncommutative energy scale. Noncommutative quantum field theories can be constructed by taking an ordinary quantum field theory and replacing the ordinary multiplication of fields with Moyal products. It is possible to re-express resulting noncommutative field theory in terms of a conventional one, by use of the so-called Seiberg-Witten map [7]. This yields a field theory with Lorentz-violating terms that are at least of mass dimension six. Finally, it is worthwhile to mention Hořava-Lifshitz gravity [8, 9] as a fundamentally Lorentz-noninvariant model, with Lorentz invariance arising at low energy as an emergent symmetry.

However, the most important reason for the recent interest in experimental testing of Lorentz and CPT symmetry has been the development of low-energy effective field theories with Lorentz-invariance violation, in particular de Standard-Model Extension (SME) [10]. The Lagrangian of the matter sector of this framework contains all Lorentz-violating gauge-invariant effective operators that can be build from the conventional Standard-Model fields, coupled to vector and tensor coefficients that parametrize the Lorentz violation. In fact, the SME also contains all CPT-violating operators, since in any local interacting quantum field theory CPT violation implies Lorentz violation [11]. The SME can thus be used to provide a general quantification of the exactness of Lorentz and CPT symmetry in the form of observational contraints on the Lorentz-violation coefficients [12].

A possible observational consequence of Lorentz violation, that can be addressed using astrophysical data, is vacuum Cherenkov radiation [13]. The Lorentz-violation coefficients can in some cases be interpreted as inducing a refractive index for the vacuum. The velocity of charged particles above some energy threshold might then exceed the phase velocity of light. This causes these particles to rapidly lose energy through photon emission. The mere observation of high-energy cosmic particles can then be used to constrain the Lorentz violation coefficients.

In this talk, we consider a process analogous to vacuum Cherenkov radiation, but with the emitted photon replaced by a Z boson. We assume the latter to obey a Lorentz- and CPT violating dispersion relation, originating from the superficially renormalizable part of the SME, called the minimal SME (mSME). In this case, the Lorentz violation originates from a Chern-Simons-like addition to the Standard-Model Lagrangian [14] and is captured by one four-vector: k^{μ}_{ZZ}. Such a theory has been shown to be consistently and covariantly quantizable, despite the presence of spacelike momenta, which are necessary for vacuum Cherenkov radiation to occur [15]. If k^{μ}_{ZZ} is timelike, the nonzero gauge-boson mass is an important ingredient that prevents the theory from containing imaginary energies. However, at the relevant energies the Z-boson mass can be considered small, allowing us to obtain stringent limits on the previously unconstrained Lorentz-violation coefficients k^{μ}_{ZZ}.

2. Vacuum Cherenkov radiation

One of the first SME terms ever to be considered (even before the formulation of the full SME [14]) is the Lorentz- and CPT-violating photon term

$$\mathcal{L}_{AF} = \frac{1}{2}(k_{AF})_{\kappa}\epsilon^{\kappa\lambda\mu\nu}A_{\lambda}F_{\mu\nu}\,. \tag{1}$$

Inclusion of this term leads to birefringence. Therefore, observations of cosmological sources with known polarizations permit searching for energy-dependent polarization changes, either from distant sources or from the cosmological microwave background radiation. As a result, the k_{AF} coefficients have been bound with great precision [14]:

$$|(k_{AF})^{\kappa}| \leq 10^{-43}\,\mathrm{GeV}\,. \tag{2}$$

Another interesting property of the term (1) is that it can provoke vacuum Cherenkov radiation [13]. This can happen because one of the photon helicities acquires a spacelike momentum, allowing for the possibility for a charged particle to move faster than the phase velocity of the photon.

Let us now consider, instead of the photon, a term analogous to (1) for the Z boson:

$$\mathcal{L}_{AF,Z} = \frac{1}{2}(k_{ZZ})_{\kappa}\epsilon^{\kappa\lambda\mu\nu}Z_{\lambda}Z_{\mu\nu}\;, \tag{3}$$

where $Z_{\mu\nu} = \partial_{\mu}Z_{\nu} - \partial_{\nu}Z_{\mu}$ and Z_{μ} represents the Z boson. The Lorentz violation four-vector coefficient k^{μ}_{ZZ} is real and can either be timelike, lightlike, or spacelike.

IOP Conf. Series: Journal of Physics: Conf. Series **873** (2017) 012017 doi:10.1088/1742-6596/873/1/012017

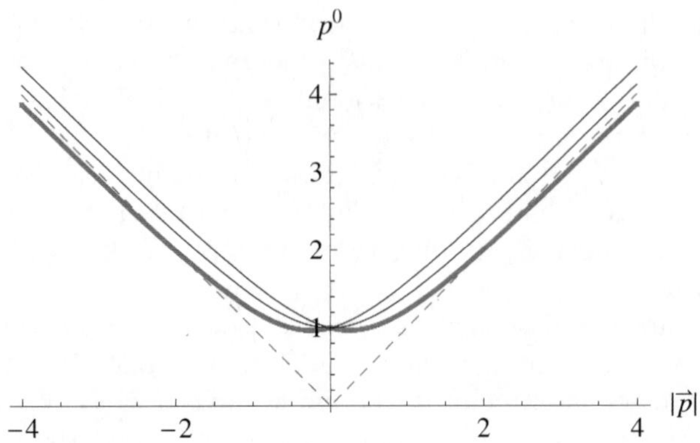

Figure 1. The dispersion relations of the Z boson subject to the inclusion the Lorentz-violating term (3) for purely timelike k^{μ}_{ZZ}, where we took $k^0_{ZZ} = \frac{M}{4}$, in arbitrary units. Indicated in red is the +-mode, which has spacelike four-momenta above an energy threshold.

These Z bosons obey the dispersion relations

$$\Lambda_0(p) = p^2 + M^2 = 0, \qquad \Lambda_\pm(p) \equiv p^2 - M^2 \pm 2\sqrt{(p \cdot k)^2 - p^2 k^2} = 0 \ , \tag{4}$$

where p^μ and M represent the momentum and the mass of the Z boson, while we dropped the subscript on k^{μ}_{ZZ} for simplicity. These dispersion relations are represented in Fig. 1 for the case of purely timelike k^{μ}_{ZZ}. As it turns out, the 0 and − gauge-boson polarization modes are timelike for any momentum. On the other hand, it follows from (4) that the gauge-boson momentum is spacelike for the + mode, if and only if

$$(p \cdot k)^2 > \frac{1}{4} M^4 \ . \tag{5}$$

This relation determines the energy threshold above which the Cherenkov-like process can take place. Since spacelike momenta are a necessary condition for the desired Cherenkov-like process, we will only consider Z bosons in the + polarization mode.

3. Cherenkov emission of Z bosons

We will first calculate the rate of Cherenkov emission of Lorentz-violating Z bosons with a spacelike momentum by a elementary Dirac fermion with mass m. Subsequently, we will consider the case of a proton.

First of all, we note that covariant quantization of free Z bosons in the presence of the term (3) can be performed in a fully consistent way, as described in Ref. [15]. Let us label

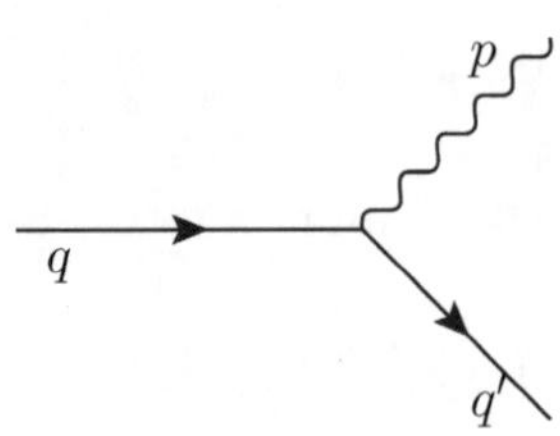

Figure 2. Relevant Feynman diagram for Z boson emission by incoming fermion.

the momentum of the incoming fermion as q, the emitted gauge boson as p, and the outgoing fermion as $q' = q - p$ (see Fig. 2). We will assume that the fermion obeys a conventional Lorentz-symmetric dispersion relation.

The rate at which the fermion loses energy by the Cherenkov-like emission of spacelike gauge bosons is given by the zeroth component of

$$\frac{dP^\mu}{dt} = \int p^\mu d\Gamma \,, \tag{6}$$

where $d\Gamma$ is the differential decay rate, which reads

$$d\Gamma = \frac{1}{2q^0}\frac{d^3p}{(2\pi)^3}\frac{1}{\Lambda'_+(p)}\frac{d^3q'}{(2\pi)^3}\frac{1}{2q'^0}\left(\frac{1}{2}\sum_{\text{spins}}|\mathcal{M}|^2\right)(2\pi)^4\delta^4(q-p-q') \,. \tag{7}$$

Here, the squared matrix element $|\mathcal{M}|^2$ is summed (averaged) over the final (initial) fermion spin. The unconventional factor $\Lambda'_+(p) = \partial\Lambda_+(p)/\partial p^0$ in the denominator defines a normalization in which the phase space and the matrix element are separately observer Lorentz invarian [15], i.e., invariant under simultaneous Lorentz transformations of the momenta and the Lorentz violation four-vector coefficient k^μ_{ZZ}. This is important, as traditional calculations in the literature all have divergent factors in the outgoing boson phase space factors when an observer frame is chosen in which the energy of the radiated boson goes to zero.

The matrix element that follows from the appropriate tree-level Feynman diagram is given by

$$i\mathcal{M} = i\bar{u}(q')\gamma^\mu(g_V + g_A\gamma^5)u(q)e^{(+)*}_\mu(p) \,, \tag{8}$$

where $u(q)$ and $u(q')$ are conventional Dirac spinors corresponding to the ingoing and outgoing particles (the case of antiparticles can be treated in an analogous way). The four-vector $e^{(+)}_\mu(p)$ is the gauge-boson polarization vector that corresponds to the dispersion relation in Eq. (4). (see [15] for explicit expressions). The coupling constants g_V and g_A, which multiply the vector and axial-vector current, respectively, depend on the process under consideration. For example, a charged lepton emitting a Z boson has tree-level values of $g_V = -g(1 - 4\sin^2\theta_w)/(4\cos\theta_w)$ and $g_A = -g/(4\cos\theta_w)$, with $g \simeq 0.65$ the $SU(2)$ coupling constant.

Using the fact that [15]

$$e^{(+)}_\mu(p)e^{(+)*}_\nu(p) = -\frac{1}{2}\eta_{\mu\nu} - \frac{p_\mu p_\nu k^2 + k_\mu k_\nu p^2 - (p_\mu k_\nu + p_\nu k_\mu)(p\cdot k)}{2((p\cdot k)^2 - p^2k^2)} + \frac{i\epsilon_{\mu\nu\alpha\beta}k^\alpha p^\beta}{2\sqrt{(p\cdot k)^2 - p^2k^2}} \,, \tag{9}$$

we find that the spin-averaged and spin-summed squared matrix element is given by

$$\frac{1}{2}\sum_{\text{spins}}|\mathcal{M}|^2 = \frac{-\xi}{2}\left[p^2\left(1 + 4X^2 - 8yX\right) - 4(1-2x)m^2\right] \,, \tag{10}$$

where $\xi = g_V^2 + g_A^2$, $x = g_V^2/\xi$, $y = g_A g_V/\xi$, and

$$X = \frac{k\cdot(q - \frac{1}{2}p)}{\sqrt{(p\cdot k)^2 - p^2k^2}} \,. \tag{11}$$

The rate of energy-momentum loss becomes

$$\frac{dP^\mu}{dt} = \frac{1}{8\pi^2 q^0}\int\frac{d^3p}{\Lambda'_+(p)}\theta(q^0 - p^0)\delta((q-p)^2 - m^2)p^\mu\left(\frac{1}{2}\sum_{\text{spins}}|\mathcal{M}|^2\right) \,. \tag{12}$$

To proceed, we note that dP^μ/dt transforms as a four-vector divided by q^0. Therefore, we make an ansatz in terms of all available four-vectors:

$$q^0 \frac{dP^\mu}{dt} = A(q,k)\, q^\mu + B(q,k)\, k^\mu \,, \tag{13}$$

where $A(q,k)$ and $B(q,k)$ are observer-Lorentz-invariant quantities that can depend on q^μ and k^μ. Contracting this expression with q^μ and k^μ gives two observer-Lorentz-invariant equations, which we can solve for $A(q,k)$ and $B(q,k)$ after having performed the integrations over $\vec{p}$, given in Eq. (12).

To achieve this, we make use of the observer Lorentz invariance of $A(q,k)$ and $B(q,k)$ by specializing to an observer frame that simplifies the calculation. For the cases that k^μ is timelike or spacelike, we go the frame where k^μ is purely timelike, i.e., $k = (k^0, \vec{0})$, or purely spacelike, i.e. $k = (0, \vec{k})$, respectively. If k^μ is lightlike, we do not need to specialize to a particular frame.

The calculation of the rate can now be done in any one of these special frames by explicitly performing the integral over $\vec{p}$ in (12). It turns out that, for small Lorentz-violation coefficient, the Z boson is emitted in a narrow forward cone:

$$\cos\theta_{pq} = 1 + \mathcal{O}(\kappa^2/M^2) \tag{14}$$

where θ_{pq} is the angle between the Z boson and the incoming fermion, while

$$\kappa = k^0 - \cos\theta_{pk}|\vec{k}| \tag{15}$$

is a value of the order of the Lorentz-violation coefficient. The θ function in (12) restricts the emitted Z boson momenta to be in a finite range of spacelike momenta. This in turn forces the incoming fermion to have a momentum above a threshold value

$$|\vec{q}|_{\text{th}} = \frac{M(M+2m)}{2|\kappa|} \tag{16}$$

in order that there is a nonzero value of the emission rate.

As it turns out, both $A(q,k)$ and $B(q,k)$ in Eq. (13) are of order M^2. Combined with Eq. (16) we thus find that the second term in Eq. (13) is of order κ^2/M^2 compared to the first term. We therefore neglect this second term and find that the rate of energy-momentum loss is given by

$$\frac{dP^\mu}{dt} = \frac{\xi M^2}{96\pi q^0}\theta(a-1) q^\mu F(a, \text{sgn}(\kappa)) \,. \tag{17}$$

where

$$F(a, \frac{m}{M}, \tilde{y}, \kappa) = (a-1)\left[8(1-\tilde{y}) + \left(\frac{5}{a} - \frac{1}{a^2}\right)(1+2\tilde{y})\right] - 12\ln a + \mathcal{O}\left(\frac{m}{M}, \frac{\kappa^2}{M^2}\right) \,. \tag{18}$$

Here $\tilde{y} = \text{sgn}(\kappa)y$, while the variable a is defined as the ratio of $|\vec{q}|$ to its threshold value, i.e., $a = |\vec{q}|/|\vec{q}|_{\text{th}}$. In a general observer frame and upto terms of order κ^2/M^2 and m^2/M^2, we can write this as

$$a = \frac{2|q\cdot k|}{M(M+2m)} \,. \tag{19}$$

The stepfunction demands that $a > 1$ for Eq. (17) to be nonvanishing. This is just the threshold condition for the initial fermion.

The zeroth component of the decay rate (17) can be viewed as a differential equation $dE(t)/dt = -W(E(t))$ in terms of the energy of the fermion. It can be solved explicitly in

IOP Conf. Series: Journal of Physics: Conf. Series **873** (2017) 012017 doi:10.1088/1742-6596/873/1/012017

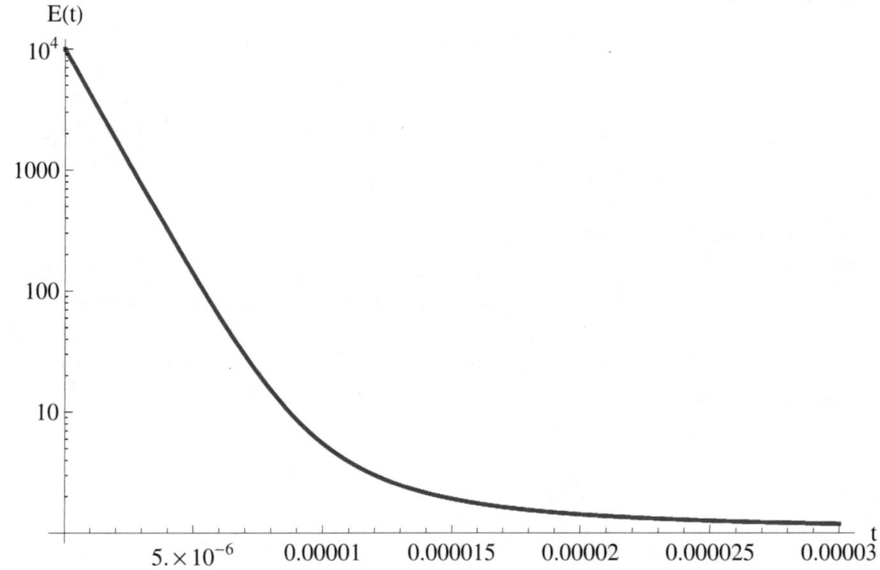

Figure 3. The ratio of the fermion energy to the threshold value as a function of time, for the parameter values: $M = 91\,\text{GeV}$, $m/M = 1/91$, $g_V = -0.0026$, $g_A = 0.0445$, $\kappa = 5\times10^{-15}\,\text{GeV}$.

the regimes $a \gg 1$ and close to threshold, $a - 1 \ll 1$. In the former case the decay is exponential, $E(t) = E_{\text{in}}e^{-t/t_\infty}$, with characteristic time $t_\infty = \frac{6\pi\hbar}{\xi|\kappa|(1-\tilde{y})}$. For $a-1 \ll 1$, $E(t)$ decays to the threshold energy E_{th} in finite time, with characteristic time constant $t_{\text{fin}} = \frac{8\pi\hbar}{\xi|\kappa|(1-2\tilde{y})}\left(\frac{M}{m}\right)^{3/2}$. Fig. 3 displays a typical decay curve for the ratio of the fermion energy to the threshold value as a function of time.

If the radiating particle is a composite fermion, the latter will disintegrate upon emission of a Z boson. In this case, it makes more sense to consider the decay rate $\Gamma = \int d\Gamma$, rather than the radiation rate calculated above. For a proton the decay rate can be written

$$\Gamma = \frac{1}{2q^0}\int \frac{d^3\vec{p}}{(2\pi)^3}\frac{4\pi}{\Lambda'_+(p)} e_\mu^{(+)}(p)e_\nu^{(+)*}(p)W^{\mu\nu}\,, \tag{20}$$

where $W^{\mu\nu}$ is the hadronic part, which can be calculated in the parton model. The calculation essentially amounts to calculating the decay rate of an elementary quark that caries a fraction x of the longitudonal proton momentum. We can thus use many of the results obtained for the elementary fermion rate. The final result for the decay rate is

$$\Gamma = \frac{|\kappa|}{16\pi}\sum_q \xi_q \int_0^1 (f_q(x) + \bar{f}_q(x))G_q(a,x)\theta(ax-1)\,dx\,. \tag{21}$$

Here $\xi_q = \frac{g^2}{2c_w^2}\left((T_q^3)^2 + 2Q_q s_w^2(Q_q s_w^2 - T_q^3)\right)$, with Q_q the quark charge and T_q^3 the third weak-isospin component. The functions $f_q(x)$ and $\bar{f}_q(x)$ are the parton distributions functions for the quarks and antiquarks of flavor q, respectively. They represent the chance of finding a quark with momentum xq inside the proton. The function $G_q(a,x)$ in (21) is given by

$$G_q(a,\frac{m}{M},y_q,\kappa,x) = (ax-1)\left(\frac{-7+6\tilde{y}_q(x)}{ax} - \frac{1-2\tilde{y}_q(x)}{(ax)^2}\right) - 4\left(1+\frac{1-2\tilde{y}_q(x)}{ax}\right)\ln a + \mathcal{O}\left(\frac{m}{M},\frac{\kappa^2}{M^2}\right) \tag{22}$$

with

$$\tilde{y}_q(x) = \frac{g^2 T_q^3 (2Q_q s_w^2 - T_q^3)}{4\xi_q c_w^2} \cdot \frac{f_q(x) - \bar{f}_q(x)}{f_q(x) + \bar{f}_q(x)} \,.$$

We see that (21) is essentially the sum over elementary-quark decay rates, weighted by the corresponding parton distribution functions.

4. Limits from ultra-high-energy cosmic rays

We can use the fact that a proton with an energy above threshold will disintegrate to use astrophysical data to limit k^μ. More precisely, such a proton cannot reach Earth if its mean free path L is much smaller than the distance from its source to Earth. Since many ultra-high-energy cosmic ray particles (UHECR) with energies above $57\,\mathrm{EeV} \equiv |\vec{q}|_{\mathrm{obs}}$ have been observed, more or less from all directions [16], it follows that

$$|\kappa| < \frac{M(M+2m)}{|\vec{q}|_{obs}} \approx 1.5 \times 10^{-7}\,\mathrm{GeV} \equiv |\kappa|_0 \tag{23}$$

We see from (21) that the mean lifetime of protons (in the Earth's frame) t_p is proportional to $|\kappa|^{-1}$, but it is enhanced by the minute values of the parton distribution functions for values of x close to one. A conservative (large) estimate gives a mean free path of

$$L \simeq ct_p \sim (\hbar c/|\kappa|_0) \times 10^{15} \sim 10^3\,\mathrm{km}\,. \tag{24}$$

It is clear that protons with an energy above this threshold will not be able to reach Earth from any viable UHECR source. This allows us to conclude that

$$|k^\mu_{ZZ}| < 1.5 \times 10^{-7}\,\mathrm{GeV} \tag{25}$$

as a bound on the components of k^μ_{ZZ}.

It is tempting to try and relate the coefficient k_{ZZ} to the $SU(2)$ and $U(1)$ gauge boson parameters k_1^μ and k_2^μ parametrizing all possible CPT violation in the $SU(2)\times U(1)$ gauge sector of the mSME [10]. However, it turns out that the latter generate, apart from the terms (3) and (1), a quadratic CPT-violating mixing term between the photon and the Z boson. The terms (3) and (1) by themselves are not invariant under general $SU(2) \times U(1)$ gauge transformations. However, a similar analysis as the one presented can be done in the case of a Lorentz- and CPT violation term for the W bosons, in which case no problematic mixing occurs [17].

Acknowledgments

This work is supported in part by the Portuguese Fundação para a Ciência e a Tecnologia (FCT) through projects UID/FIS/00099/2013 and SFRH/BPD/101403/2014 and program POPH/FSE.

References

[1] Kostelecky V A and Samuel C 1989 *Phys. Rev.* D **39** 683
Kostelecky V A and Potting R 1991 *Nucl. Phys.* B **359** 545
Kostelecky V A and Potting R 1996 *Phys. Lett.* B **381** 89
Hashimoto K and Murata M 2013 *PTEP* **2013** 043B01
[2] Ashtekar A 1986 *Phys. Rev. Lett.* **57**, 2244
Gambini R and Pullin J 1999 *Phys. Rev.* D **59** 124021
[3] Bluhm R and Kostelecky V A 2005 *Phys. Rev.* D **71** 065008;
[4] Bjorken J 2001 (*Preprint*: hep-th/0111196)
Kraus P and Tomboulis E T 2002 *Phys. Rev.* D **66** 045015
Chkareuli J L, Froggatt C D, Jejelava J G and Nielsen H B 2008 *Nucl. Phys.* B **796** 211

Tomboulis E T 2012 *Phys. Rev.* D **85** 024042
[5] Kostelecky V A, Lehnert R and Perry M J 2003 *Phys. Rev.* D **68** 123511
Arkani-Hamed N, Cheng H-C, Luty M A and Mukohyama S 2004 *JHEP* **0405** 074
[6] Connes A and Kreimer D 1998 *Commun. Math. Phys.* **199** 203
[7] Seiberg N and Witten E 1999 *JHEP* **9909** 032
Carroll S N, Harvey J A, Kostelecky V A, Lane C D and Okamoto T 2001 *Phys. Rev. Lett.* **87** 141601
[8] Horava P 2009 *Phys. Rev.* D **79** 084008
[9] Blas D, Pujolas O and Sibiryakov S 2010 *Phys. Rev. Lett.* **104** 181302
[10] Colladay D and Kostelecky V A 1997 *Phys. Rev.* D **55** 6760
Colladay D and Kostelecky V A 1998 *Phys. Rev.* D **58** 116002
[11] Greenberg O W 2002 *Phys. Rev. Lett.* **89** 231602
[12] Kostelecky V A and Russell N 2011 *Rev. Mod. Phys.* **83** 11 (*Preprint* arXiv:0801.0287v10 [hep-ph])
[13] Lehnert R and Potting R 2004 *Phys. Rev. Lett.* **93** 110402
Lehnert R and Potting R 2004 *Phys. Rev.* D **70** 125010 [*Erratum-ibid.* D **70** 129906]
Kaufhold K and Klinkhamer F R 2006 *Nucl. Phys.* B **734** 1
Altschul B 2010 *Phys. Rev.* D **82** 016002
[14] Carroll S M, Field G B and Jackiw R 1990 *Phys. Rev.* D **41** 1231
[15] Colladay D, McDonald P, Noordmans J P and Potting R 2017 *Phys. Rev. D* **95** 025025
[16] Aab A et al. [Pierre Auger Collaboration] 2015 *Astrophys. J.* **804** 15
Abu-Zayyad T et al. [Telescope Array Collaboration] 2013 *Astrophys. J.* **777** 88
[17] Colladay D, Noordmans J P and Potting R *work in progress.*

VIP-2 at LNGS: An experiment on the validity of the Pauli Exclusion Principle for electrons

J. Marton[1], S. Bartalucci[2], A. Bassi[3], M. Bazzi[2], S. Bertolucci[4], C. Berucci[1,2], M. Bragadireanu[2,3], M. Cargnelli[1], A. Clozza[2], C. Curceanu[2,3,9], L. De Paolis[2], S. Di Matteo[6], S. Donadi[4], J.-P. Egger[7], C. Guaraldo[2], M. Iliescu[2], M. Laubenstein[8], E. Milotti[4], A. Pichler[1], D. Pietreanu[2,3], K. Piscicchia[2,9], A. Scordo[2], H. Shi[2], D. Sirghi[2,3], F. Sirghi[2,3], L. Sperandio[2], O. Vazquez-Doce[10], E. Widmann[1] and J. Zmeskal[1]

[1] Stefan Meyer Institute for subatomic physics, Boltzmanngasse 3, 1090 Vienna, Austria
[2] INFN, Laboratori Nazionali di Frascati, CP 13, Via E. Fermi 40, I-00044, Frascati (Roma), Italy
[3] Horia HulubeiNational Institute of Physics and Nuclear Engineering, Str. Atomistilor no. 407, P.O. Box MG-6, Bucharest - Magurele, Romania
[4] Dipartimento di Fisica, Universit'a di Trieste and INFN Sezione di Trieste, Via Valerio, 2, I-34127 Trieste, Italy
[5] University and INFN Bologna, Via Irnerio 46, I-40126, Bologna, Italy
[6] Institut de Physique UMR CNRS-UR1 6251, Universite de Rennes1, F-35042 Rennes, France
[7] Institut de Physique, Universite de Neuchatel, 1 rue A.-L. Breguet, CH-2000 Neuchatel, Switzerland
[8] INFN, Laboratori Nazionali del Gran Sasso, S.S. 17/bis, I-67010 Assergi (AQ), Italy
[9] Museo Storico della Fisica e Centro Studi e Ricerche Enrico Fermi, Roma, Italy
[10] Excellence Cluster Universe, Technische Universitat Munchen, Garching, Germany

E-mail: `johann.marton@oeaw.ac.at`

Abstract. We are experimentally investigating possible violations of standard quantum mechanics predictions in the Gran Sasso underground laboratory in Italy. We test with high precision the Pauli Exclusion Principle and the collapse of the wave function (collapse models). We present our method of searching for possible small violations of the Pauli Exclusion Principle (PEP) for electrons, through the search for "anomalous" X-ray transitions in copper atoms. These transitions are produced by "new" electrons (brought inside the copper bar by circulating current) which can have the possibility to undergo Pauli-forbidden transition to the 1s level already occupied by two electrons. We describe the VIP2 (VIolation of the Pauli Exclusion Principle) experimental data taking at the Gran Sasso underground laboratories. The goal of VIP2 is to test the PEP for electrons in agreement with the Messiah-Greenberg superselection rule with unprecedented accuracy, down to a limit in the probability that PEP is violated at the level of 10^{-31}. We show preliminary experimental results and discuss implications of a possible violation.

1. Introduction

W. Pauli discovered the famous Exclusion Principle named after him which explained the periodic table of the elements [1, 2] (Nobel prize in 1945). The Pauli Exclusion Principle is one of the most important rules of nature and it has many consequences not only related to the periodic system of the elements but also to the stability of matter, the existence/stability of neutron stars and many other phenomena. We know that the Pauli Exclusion Principle (PEP) is itself a consequence of the spin-statistics theorem which divides nature in fermionic and bosonic systems. In spite of all efforts, no simple intuitive explanation for the PEP could be given - but several proofs of the Spin-statistics relation (Pauli exclusion principle) based on complicated arguments can be found in the literature [3, 4]. The proof by Lüders and Zumino [4] is based on a clear set of assumptions:

- Invariance with respect to the proper inhomogeneous Lorentz group
- Two operators of the same field at points separated by a spacelike interval either commute or anticommute (locality)
- The vacuum is the state of lowest energy
- The metric of the Hilbert space is positive definite
- The vacuum is not identically annihilated by a field

If at least one of these assumptions is invalid then a violation of the Pauli Principle would be possible. There are also theoretical attempts to accomplish PEP violations. Some recent theoretical studies can be found in refs. [5, 6].

2. Experimental tests of the Exclusion Principle

We know the PEP seems to be fulfilled to a high degree since no violations are found up-to-now. However, due to the outstanding importance of PEP in physics, experimental investigations were performed on many different systems: atomic transitions, nuclear transitions, nuclear reactions, anomalous atomic structure, anomalous nuclear structure, statistics of neutrinos, astrophysics and cosmology.

The different experimental approaches of PEP tests are based on various assumptions. According to S. Elliott [7] these experiments need to be distinguished in relation to the Messiah-Greenberg super-selection rule [8]. This rule states that the exchange symmetry of a steady state is constant in time. As a consequence, the symmetry of a quantum state can only change if a particle which is new to the system, interacts with the state. Some experiments investigating Pauli violation in stable states resulted in remarkable upper bounds for violation [9, 10, 11, 12]. However, there is the caveat that in these experimental cases the Messiah-Greenberg super-selection rule is not obeyed, meaning that one is testing an other fundamental rule, i.e. the stability of particles (e.g. electron decay [13]).

A pioneering experiment was performed by Ramberg and Snow [14] which searched for Pauli forbidden X-ray transitions in copper after introducing "new" electrons to the system. The concept is based on the assumption that an electric current running through a copper conductor represents a source of electrons which are "new" to the systems of copper atoms of the copper conductor. Thus one can search for Pauli-forbidden transitions in the copper atoms (see fig. 1). The transition energy of the PEP violating transition is shifted in energy due to the shielding by the "extra" electron in the 1s state. These shifted transition energies can be calculated using a multiconfiguration Dirac-Fock approach taking the relevant corrections (e.g. relativistic corrections) into account [15, 16]. Ramberg and Snow conducted the experiment in the basement of Fermilab and obtained the result

$$\beta^2/2 \leq 1.7 \times 10^{-26} \quad (1)$$

The quantity $\beta^2/2$ stands for the probability of a Pauli violating atomic transition and is de-facto standard in the literature.

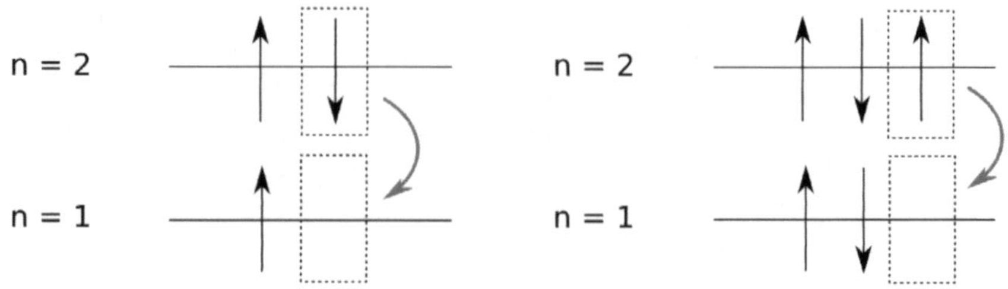

Figure 1. Transitions to the 1s ground state: Allowed transition 2p-1s (left) and Pauli forbidden transition to the fully occupied 1s state (right).

A critical point is the question how one can identify electrons in the conductor as "new" electrons to the atoms of the conductor. Indeed the de-coherence time (at room temperature) is very short - supporting the picture of "new" electrons (Edoardo Milotti, private communication, 2016).

2.1. VIP at LNGS

A much improved experiment VIP [17, 18] following the concept of Ramberg and Snow was set up in the underground laboratory LNGS in Gran Sasso/Italy (LNGS). VIP used charge coupled devices (CCDs) [19] as X-ray detectors with very good energy resolution, large area and high intrinsic efficiency. The CCDs were previously successfully employed in an experiment on kaonic atoms at LNF Frascati [20, 21]. The CCDs were positioned around a pure copper cylinder operated without and with up to 40 A current. The cosmic background in the LNGS laboratory is strongly suppressed ($\sim 10^{-6}$) due to the rock coverage. Additionally the setup was covered by massive lead shielding (see fig. 2).

Figure 2. The VIP experiment with the lead shielding mounted in operation at LNGS.

The VIP experiment took data alternating runs with and without current. It obtained for $\beta^2/2$ [22, 23]

Figure 3. This photo shows the interior of the VIP2 experiment. In the box the copper target, the SDDs and the plastic scintillators are mounted. An insulation vacuum ($\sim 10^{-5}$ mbar) inside the box is necessary to operate the SDDs at 100 K.

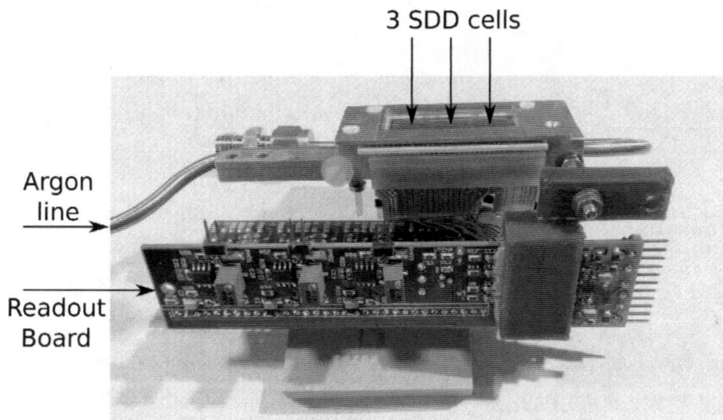

Figure 4. X-ray detector system of the VIP2 experiment. The 3 SDD cell detector is cooled by liquid argon to about 100 K and read out via the readout board.

$$\beta^2/2 \leq 4.7 \times 10^{-29} \tag{2}$$

Compared to the result of Ramberg-Snow it is an improvement by nearly 3 orders of magnitude.

2.2. VIP2 at LNGS

As a next step the experiment VIP2 with SDDs (Silicon Drift Detectors) as X-ray detectors was built and installed in LNGS. The experiment is designed for higher sensitivity by providing a larger X-ray detector solid angle, higher current and employing active shielding by plastic scintillators readout by silicon photomultipliers as background sensitive detectors. Due to the timing capability of SDDs the timing information of the SDD detectors and plastic scintillator signals can be used to additionally suppress background events.

3. Recent Results

The progress of the VIP2 experiment has been reported in [24, 25, 26, 27]. In 2016 we collected data in a time period of $\sim$70 days without current and $\sim$40 days with 100 A current. In fig. 5

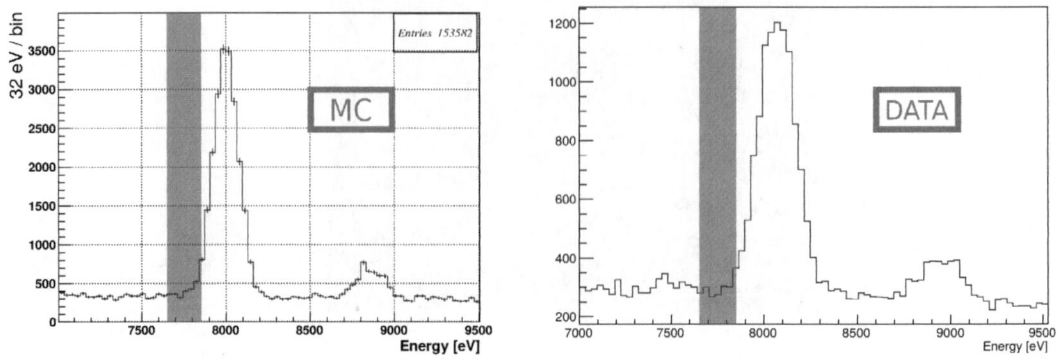

Figure 5. Comparison between 30 days of Monte Carlo simulation data (left) and 30 days of data measured at LNGS (right). The background in the region of interest (marked in red) differs only by around 30 %.

the Monte Carlo generated X-ray energy spectrum in the range 7-9.5 keV around the region of interest (marked in red) and the corresponding measured energy spectrum are displayed.

In order to compare our preliminary result we used the analysis technique of Ramberg and Snow [14]. The analysis of this data set leads to a preliminary upper limit for the probability that the PEP is violated for electrons in copper

$$\beta^2/2 \leq 1.4 \times 10^{-29} \tag{3}$$

It has to be emphasized that this preliminary result represents already the most stringent test of the PEP with no violation of the Messiah-Greenberg superselection rule.

4. Summary and Outlook

The experimental program for testing a possible PEP violation for electrons made great progress in 2016. The use of a new type of SDDs as X-ray detectors can further enhance the sensitivity by providing larger sensitive area. Furthermore, the cooling can be made more simple changing from liquid argon to Peltier cooling. Concerning the reduction of the X-ray background we will install a passive shielding with Teflon, lead and copper. Given a running time of 3 years and alternating measurement with and without current we expect either to lower the upper limit of PEP violation by about two orders of magnitude compared to the former VIP experiment or to discover the violation.

Acknowledgments

We thank H. Schneider, L. Stohwasser, and D. Stuckler from Stefan-Meyer-Institut for their fundamental contribution in designing and building the VIP2 setup. We acknowledge the very important assistance of the INFN-LNGS laboratory staff during all phases of preparation, installation and data taking. The support from the EU COST Action CA 15220 is gratefully acknowledged. We thank the Austrian Science Foundation (FWF) which supports the VIP2 project with the grants P25529-N20 and W1252-N27 (doctoral college particles and interactions) and Centro Fermi for the grant Problemi aperti nella meccania quantistica. Furthermore, this paper was made possible through the support of a grant from the Foundational Questions Institute, FOXi (Events as we see them: experimental test of the collapse models as a solution of the measurement problem) and a grant from the John Templeton Foundation (ID 581589).

The opinions expressed in this publication are those of the authors and do not necessarily reflect the views of the John Templeton Foundation.

References

[1] Pauli W 1925 *Zeitschrift für Physik* **31** 765–783 ISSN 0044-3328 URL `http://link.springer.com/10.1007/BF02980631`
[2] Pauli W 1946 Nobel prize lecture, www.nobelprize.org
[3] Pauli W 1940 *Phys. Rev.* **58** 716–722 URL `http://link.aps.org/doi/10.1103/PhysRev.58.716`
[4] Lüders G and Zumino B 1958 *Phys. Rev.* **110** 1450–1453 URL `http://link.aps.org/doi/10.1103/PhysRev.110.1450`
[5] Jackson M G 2008 *Physical Review D* **78** 126009 (*Preprint* 0809.2733) URL `http://dx.doi.org/10.1103/PhysRevD.78.126009`
[6] Balachandran A P, Joseph A and Padmanabhan P 2010 *Physical Review Letters* **105** 051601 ISSN 0031-9007 URL `http://link.aps.org/doi/10.1103/PhysRevLett.105.051601`
[7] Elliott S R, LaRoque B H, Gehman V M, Kidd M F and Chen M 2012 *Foundations of Physics* **42** 1015–1030 ISSN 1572-9516 URL `http://dx.doi.org/10.1007/s10701-012-9643-y`
[8] Messiah A M L and Greenberg O W 1964 *Physical Review* **136** B248—-B267 URL `http://link.aps.org/doi/10.1103/PhysRev.136.B248`
[9] Bernabei R, Belli P, Montecchia F, De Sanctis M, Di Nicolantonio W, Incicchitti A, Prosperi D, Bacci C, Dai C, Ding L, Kuang H and Ma J 1997 *Physics Letters B* **408** 439–444 ISSN 03702693 URL `http://linkinghub.elsevier.com/retrieve/pii/S0370269397008423`
[10] Bernabei R, Belli P, Cappella F, Cerulli R, Dai C J, D'Angelo A, He H L, Incicchitti A, Kuang H H, Ma X H, Montecchia F, Nozzoli F, Prosperi D, Sheng X D and Ye Z P 2010 *Journal of Physics: Conference Series* **202** 012039 URL `http://stacks.iop.org/1742-6596/202/i=1/a=012039`
[11] Borexino collaboration B and Back H O 2004 (*Preprint* 0406252) URL `http://arxiv.org/abs/hep-ph/0406252 http://dx.doi.org/10.1140/epjc/s2004-01991-1`
[12] Bellini G, Bonetti S, Buizza Avanzini M, Caccianiga B, D'Angelo D, Franco D, Giammarchi M, Lombardi P, Ludhova L, Meroni E, Miramonti L, Perasso L, Ranucci G, Re A, Suvorov Y, Benziger J, Cadonati L, Pocar A, Calaprice F, Chavarria A, Dalnoki-Veress F, Galbiati C, Goretti A, Ianni A, Leung M, Loer B, Saldanha R, Xu J, Carraro C, Davini S, Guardincerri E, Manuzio G, Pallavicini M, Perasso S, Risso P, Salvo C, Testera G, Zavatarelli S, De Kerret H, Kryn D, Obolensky M, Vignaud D, Derbin A, Muratova V, Etenko A, Litvinovich E, MacHulin I, Sabelnikov A, Skorokhvatov M, Sukhotin S, Fomenko K, Smirnov O, Sotnikov A, Zaimidoroga O, Gazzana S, Ghiano C, Ianni A, Korga G, Montanari D, Razeto A, Tartaglia R, Goeger-Neff M, Lewke T, Meindl Q, Oberauer L, Von Feilitzsch F, Winter Y, Wurm M, Grieb C, Hardy S, Joyce M, Manecki S, Papp L, Raghavan R S, Rountree D, Vogelaar R B, Maneschg W, Schönert S, Simgen H, Zuzel G, Misiaszek M, Wojcik M, Ortica F and Romani A 2010 *Physical Review C - Nuclear Physics* **81** 034317 ISSN 05562813 (*Preprint* 0911.0548) URL `http://link.aps.org/doi/10.1103/PhysRevC.81.034317`
[13] Elliott S R, Abgrall N, Arnquist I J, Avignone F T, Barabash A S, Bertrand F E, Bradley A W, Brudanin V, Busch M, Buuck M, Caldwell A S, Chan Y D, Christofferson C D, Chu P H, Cuesta C, Detwiler J A, Dunagan C, Efremenko Y, Ejiri H, Finnerty P S, Galindo-Uribarri A, Gilliss T, Giovanetti G K, Goett J, Green M P, Gruszko J, Guinn I S, Guiseppe V E, Henning R, Hoppe E W, Howard S, Howe M A, Jasinski B R, Keeter K J, Kidd M F, Konovalov S I, Kouzes R T, LaFerriere B D, Leon J, MacMullin J, Martin R D, Massarczyk R, Meijer S J, Mertens S, Orrell J L, O'Shaughnessy C, Poon A W P, Radford D C, Rager J, Rielage K, Robertson R G H, Romero-Romero E, Shanks B, Shirchenko M, Suriano A M, Tedeschi D, Trimble J E, Varner R L, Vasilyev S, Vetter K, Vorren K, White B R, Wilkerson J F, Wiseman C, Xu W, Yakushev E, Yu C H, Yumatov V and Zhitnikov I 2016 *The European Physical Journal C* **76** 619 ISSN 1434-6044 URL `http://link.springer.com/10.1140/epjc/s10052-016-4467-0`
[14] Ramberg E and Snow G A 1990 *Physics Letters B* **238** 438–441 ISSN 0370-2693 URL `http://www.sciencedirect.com/science/article/pii/037026939091762Z`
[15] Sperandio L 2008 *New experimental limit on the Pauli Exclusion Principle violation by electrons from the VIP experiment* Ph.D. thesis
[16] Matteo S D and Sperandio L 2005 VIP Technical Note IR - 4 Tech. rep.
[17] Collaboration V 2004 http://www.lnf.infn.it/esperimenti/vip URL `http://www.lnf.infn.it/esperimenti/vip`
[18] Bartalucci S, Bertolucci S, Bragadireanu M, Bucci C, Cargnelli M, Catitti M, Curceanu C, Di Matteo S,

Egger J H, Ferrari N, Fuhrmann H, Guaraldo C, Iliescu M, Ishiwatari T, Laubenstein M, Marton J, Milotti E, Pietreanu D, Ponta T, Sirghi D, Sirghi F, Sperandio L, Widmann E and Zmeskal J 2006 *Physics Letters B* **641** 18–22 ISSN 03702693 URL `http://linkinghub.elsevier.com/retrieve/pii/S0370269306009385`

[19] Egger J P, Chatellard D and Jeannet E 1993 *Muonic Atoms and Molecules* (Basel: Birkhäuser Basel) pp 331–344

[20] Beer G, Bragadireanu A M, Cargnelli M, Curceanu-Petrascu C, Egger J P, Fuhrmann H, Guaraldo C, Iliescu M, Ishiwatari T, Itahashi K, Iwasaki M, Kienle P, Koike T, Lauss B, Lucherini V, Ludhova L, Marton J, Mulhauser F, Ponta T, Schaller L A, Seki R, Sirghi D L, Sirghi F, Zmeskal J and Zmeskal J 2005 *Physical Review Letters* **94** 212302 ISSN 0031-9007 URL `http://link.aps.org/doi/10.1103/PhysRevLett.94.212302`

[21] Ishiwatari T, Beer G, Bragadireanu A, Cargnelli M, Curceanu (Petrascu) C, Egger J P, Fuhrmann H, Guaraldo C, Iliescu M, Itahashi K, Iwasaki M, Kienle P, Lauss B, Lucherini V, Ludhova L, Marton J, Mulhauser F, Ponta T, Schaller L, Sirghi D, Sirghi F, Strasser P and Zmeskal J 2006 *Nuclear Instruments and Methods in Physics Research Section A: Accelerators, Spectrometers, Detectors and Associated Equipment* **556** 509–515 ISSN 01689002

[22] Curceanu C, Bartalucci S, Bassi A, Bertolucci S, Berucci C, Bragadireanu A, Cargnelli M, Clozza A, De Paolis L, Matteo S, Donadi S, D'Uffizi A, Egger J P, Guaraldo C, Iliescu M, Laubenstein M, Marton J, Milotti E, Pichler A, Pietreanu D, Piscicchia K, Ponta T, Sbardella E, Scordo A, Shi H, Sirghi D, Sirghi F, Sperandio L, Doce O and Zmeskal J 2015 *Journal of Physics: Conference Series* **631** 012068 ISSN 17426596 17426588 URL `http://iopscience.iop.org/article/10.1088/1742-6596/631/1/012068/pdf`

[23] Piscicchia K, Curceanu C, Bartalucci S, Bassi A, Bertolucci S, Berucci C, Bragadireanu A, Cargnelli M, Clozza A, De Paolis L, Di Matteo S, Donadi S, D'Uffizi A, Egger J P, Guaraldo C, Iliescu M, Ishiwatari T, Laubenstein M, Marton J, Milotti E, Pietreanu D, Ponta T, Sbardella E, Scordo A, Shi H, Sirghi D, Sirghi F, Sperandio L, Vazquez Doce O and Zmeskal J 2015 *Acta Physica Polonica B* **46** 147 ISSN 0587-4254 (*Preprint* `1501.04462`) URL `http://arxiv.org/abs/1501.04462` `http://dx.doi.org/10.5506/APhysPolB.46.147` `http://www.actaphys.uj.edu.pl/vol46/abs/v46p0147`

[24] Pichler A, Bartalucci S, Bazzi M, Bertolucci S, Berucci C, Bragadireanu M, Cargnelli M, Clozza A, Curceanu C, De Paolis L, Di Matteo S, D'Uffizi A, Egger J P, Guaraldo C, Iliescu M, Ishiwatari T, Laubenstein M, Marton J, Milotti E, Pietreanu D, Piscicchia K, Ponta T, Sbardella E, Scordo A, Shi H, Sirghi D, Sirghi F, Sperandio L, Vazquez-Doce O, Widmann E and Zmeskal J 2016 *Journal of Physics: Conference Series* **718** 052030 (*Preprint* `1602.00898`) URL `http://iopscience.iop.org/article/10.1088/1742-6596/718/5/052030/meta`

[25] Shi H, Bazzi M, Beer G, Bellotti G, Berucci C, Bragadireanu A, Bosnar D, Cargnelli M, Curceanu C, Butt A, D'Uffizi A, Fiorini C, Ghio F, Guaraldo C, Hayano R, Iliescu M, Ishiwatari T, Iwasaki M, Levi Sandri P, Marton J, Okada S, Pietreanu D, Piscicchia K, Romero Vidal A, Sbardella E, Scordo A, Sirghi D, Sirghi F, Tatsuno H, Vazquez Doce O, Widmann E and Zmeskal J 2016 *EPJ Web of Conferences* **126** 04045 ISSN 2100-014X (*Preprint* `1601.02236`) URL `http://arxiv.org/abs/1601.02236` `http://dx.doi.org/10.1051/epjconf/201612604045` `http://www.epj-conferences.org/10.1051/epjconf/201612604045`

[26] Marton J, Bartalucci S, Bertolucci S, Berucci C, Bragadireanu M, Cargnelli M, Curceanu C, Clozza A, Matteo S, Egger J P, Guaraldo C, Iliescu M, Ishiwatari T, Laubenstein M, Milotti E, Pichler A, Pietreanu D, Piscicchia K, Ponta T, Scordo A, Shi H, Sirghi D, Sirghi F, Sperandio L, Doce O, Widmann E and Zmeskal J 2015 *Journal of Physics: Conference Series* **631** ISSN 17426596 17426588 URL `http://iopscience.iop.org/article/10.1088/1742-6596/631/1/012070/pdf`

[27] Curceanu C, Bertolucci S, Bassi A, Bazzi M, Bertolucci S, Berucci C, Bragadireanu A, Cargnelli M, Clozza A, De Paolis L, Di Matteo S, Donadi S, D'uffizi A, Egger J P, Guaraldo C, Iliescu M, Laubenstein M, Marton J, Milotti E, Pichler A, Pietreanu D, Piscicchia K, Ponta T, Scordo A, Shi H, Sirghi D, Sirghi F, Sperandio L, Doce O and Zmeskal J 2016 *International Journal of Quantum Information* **14** 1–10 ISSN 02197499 URL `http://www.worldscientific.com/doi/abs/10.1142/S0219749916400177`

Searches for lepton number violation and resonances in $K^{\pm} \to \pi\mu\mu$ decays at the NA48/2 experiment

Letizia Peruzzo [1]
Johannes Gutenberg University, Mainz, Germany
E-mail: lperuzzo@uni-mainz.de

Abstract. The NA48/2 experiment at CERN collected in 2003-2004 a large sample of charged kaon decays with multiple charged particles in the final state. A new upper limit on the rate of the lepton number violating decay $K^{\pm} \to \pi^{\mp}\mu^{\pm}\mu^{\pm}$ obtained from this sample is reported. Searches for two-body resonances in the $K^{\pm} \to \pi\mu\mu$ decays (including heavy neutral leptons and inflatons) in the accessible range of masses and lifetimes are presented.

1. Introduction

The massive nature of neutrinos has unambiguously been demonstrated by the observation of neutrino oscillations, although, within the Standard Model (SM), neutrinos are massless and right-handed neutrino states are not included. The possibility to include right-handed neutrinos is provided, for example, by the Neutrino Minimal Standard Model (νMSM) [1, 2], a SM extension in which sterile neutrinos mix with ordinary neutrinos. In the νMSM three massive right-handed neutrinos are introduced to explain neutrino oscillations, dark matter and the baryon asymmetry of the Universe. The lightest among the three massive neutrinos, with mass of $\mathcal{O}(1\ \text{keV}/c^2)$, is a dark matter candidate; the other two, with masses in the range of 100 MeV/c^2 to a few GeV/c^2, give masses to the SM neutrinos via the see-saw mechanism and introduce extra CP violating phases to explain the baryon asymmetry.
In a further extension of the νMSM a real scalar field is added to the theory in order to incorporate inflation and provide a common source for electroweak symmetry breaking and for right-handed neutrino masses [3].
The νMSM and its extension predict the existence of new particles which could be produced in kaon decays. In particular, a Majorana neutrino N_4 is produced on shell in $K^{\pm} \to N_4\mu^{\pm}$ and then decays to $N_4 \to \pi\mu$ in both the Lepton Number Violating decay (LNV) $K^{\pm} \to \pi^{\mp}\mu^{\pm}\mu^{\pm}$, which is forbidden in the SM, and in the Lepton Number Conserving decay (LNC) $K^{\pm} \to \pi^{\pm}\mu^{\pm}\mu^{\mp}$ [4]. An inflaton χ could be produced in the Lepton Number Conserving (LNC) $K^{\pm} \to \pi^{\pm}\chi$ decay with the subsequent decay $\chi \to \mu^{+}\mu^{-}$ [5, 6].
The large statistics of kaon decays with multiple charged particles in the final state collected in 2003–2004 by the NA48/2 experiment allows to search for the forbidden LNV $K^{\pm} \to \pi^{\mp}\mu^{\pm}\mu^{\pm}$ decay, as well as for two-body resonances in $K^{\pm} \to \pi\mu\mu$ decays. The search for the

[1] On behalf of the NA48/2 Collaboration:
Cambridge, CERN, Dubna, Chicago, Edinburgh, Ferrara, Firenze, Mainz, Northwestern, Perugia, Pisa, Saclay, Siegen, Torino, Wien

$K^{\pm} \to \pi^{\mp}\mu^{\pm}\mu^{\pm}$ together with the limits on the products of branching fractions $\mathcal{BR}(K^{\pm} \to \mu^{\pm}N_4)\mathcal{BR}(N_4 \to \pi\mu)$ and $\mathcal{BR}(K^{\pm} \to \pi^{\pm}\chi)\mathcal{BR}(\chi \to \mu^{+}\mu^{-})$ [7] are presented here.

2. The NA48/2 apparatus and data taking conditions

The NA48/2 experiment was located in the north area (NA) of the CERN SPS accelerator facility and was a successor of the NA48 experiment with almost the same detector set up. The primary goal of NA48/2 was the search of direct CP violation in $K^{\pm} \to \pi^{\pm}\pi^{+}\pi^{-}$ and $K^{\pm} \to \pi^{\pm}\pi^{0}\pi^{0}$ decays [4] with about 100 days of effective data taking in 2003–2004. The experiment used 400 GeV/c protons provided by the SPS which, hitting on a Beryllium target, produced a secondary hadronic beam. Thanks to a system of magnets and collimators only charged particles with momenta of (60 ± 3) GeV/c were selected and aligned with the longitudinal axis of the detector within 1 mm and with a transverse size of about 1 cm. The experiment was recording the decays of K^{+} and K^{-} mesons inside the fiducial decay region located in a 114 m long cylindrical vacuum tank. After the vacuum tank a thin Kevlar window separated the vacuum from a helium vessel at atmospheric pressure in which a magnetic spectrometer was located.
The spectrometer consisted of four drift chambers (DCH) and a dipole magnet, placed between the second and the third chamber, which provided a horizontal momentum kick of p_t = 120 MeV/c for charged particles. Each chamber had four different views and a spatial resolution of $\sigma_{x,y}$ = 90 μm. The nominal momentum resolution of the spectrometer was $\sigma_p/p = (1.02 \oplus 0.044\ p)\%$ where p is given in GeV/c. A scintillating hodoscope (HOD) followed the spectrometer. It was divided in four quadrants by a horizontal and a vertical plane of strip-shaped counters. The HOD was used in the trigger chain to provide a fast time measurement for charged particles with a 150 ps time resolution. A quasi homogeneous electromagnetic calorimeter filled with liquid krypton (LKr) and a depth of $27X_0$ was used both for photon detection and particle identification. It had an energy resolution of $\sigma_E/E = 0.032/\sqrt{E} \oplus 0.09/E \oplus 0.0042$ (E in GeV) corresponding to $\sigma_E/E = 0.94\%$ at 20 GeV. The particle identification was completed by a hadronic calorimeter and a muon veto system (MUV). The hadronic calorimeter consisted of alternated iron and scintillator planes, while the MUV of three plastic scintillator strips planes each of them preceded by a 80 cm iron wall. A more detailed description of the experiment can be found in Ref. [5].

3. Events selection

Two different samples are selected on the data: candidates for the Lepton Number Violating decay $K^{\pm} \to \pi^{\mp}\mu^{\pm}\mu^{\pm}$ ($K^{LNV}_{\pi\mu\mu}$) and Lepton Number Conserving $K^{\pm} \to \pi^{\pm}\mu^{+}\mu^{-}$ ($K^{LNC}_{\pi\mu\mu}$) events for the search of resonances. Due to the decay vertex resolution, the signature given by a particle X produced in $K \to \mu X$ (πX) and promptly decaying to $X \to \pi\mu$ ($X \to \mu^{+}\mu^{-}$) is indistinguishable from a real three-track decay. Therefore the trigger logic of three-track events (the main trigger of the experiment) can be used for the searches presented here.
The $K^{\pm} \to \pi^{\pm}\pi^{+}\pi^{-}$ ($K_{3\pi}$) decay is used as a normalization channel: the same three-track topology and the small difference in mass between π and μ allow a first order cancellation of systematic effects due to possible imperfect kaon beam description and detector and trigger inefficiencies.
The three-track event selection requires a total charge of the three tracks equal to one and a vertex inside the 98 m long fiducial decay region. For each track a momentum between 5 and 55 GeV/c is required and a total momentum for the three tracks inside the range 55-65 GeV/c, compatible with the beam momentum. A cut is applied to the total transverse momentum of the three tracks with respect to the beam direction, that is measured using the $K_{3\pi}$ sample ($p_t < 10$ MeV/c). The three tracks forming the vertex must be consistent in time with the trigger and within 10 ns from the average time of the three tracks. Each track is required to be inside the geometric acceptance of the DCH, HOD, LKr and MUV detectors. If more than one

vertex satisfies the requirements above only the one with the lowest fit χ^2 is taken into account. The $K_{\pi\mu\mu}$ signal candidates are selected applying particle identification and kinematic constraints. The three tracks have to consist of a $\pi^{\pm}$ candidate and two muons with the same/opposite sign for $K_{\pi\mu\mu}^{LNV}$ and $K_{\pi\mu\mu}^{LNC}$, respectively. The pion candidate is selected by requiring the ratio between the energy deposited in the LKr and the momentum measured in the spectrometer to be $E/p < 0.95$ and that no in-time associated hits are in the MUV. For muon identification $E/p < 0.2$ and associated hits in the first two planes of the MUV are required.

The signal region satisfies $|M_{\pi\mu\mu} - M_K| < 5(8)$ MeV/c^2 where $M_{\pi\mu\mu}$ is the invariant mass of the three tracks in the $\pi^{\mp}\mu^{\pm}\mu^{\pm}$ ($\pi^{\pm}\mu^{+}\mu^{-}$) hypothesis and M_K is the nominal $K^{\pm}$ mass [10]. The selected region corresponds to ± 2 and ± 3.2 times the resolution of $M_{\pi\mu\mu}$ respectively for $K_{\pi\mu\mu}^{LNV}$ and $K_{\pi\mu\mu}^{LNC}$, where $\sigma_{M(\pi\mu\mu)} = 2.5$ MeV/c^2. The different cuts on the mass resolution is due to the different background composition in the two samples. The data/Monte Carlo agreement is studied in the control region 456 MeV/c^2 $< M_{\pi\mu\mu} <$ 480 MeV/c^2.

For the $K_{3\pi}$ normalization channel the pion identification criteria used for the signal is applied only for the track with opposite electric charge with respect to the kaon and the invariant mass of the three pions must satisfy $|M_{3\pi} - M_K| < 5$ MeV/c^2 that corresponds to $\pm 3\sigma_{3\pi}$ ($\pm$ 5.1 MeV/c^2) where the three-pion mass resolution is 1.7 MeV/c^2. The number of $K^{\pm}$ decays in the fiducial region is measured to $N_K = 1.637 \times 10^{11}$ using the normalization channel $K_{3\pi}$ with $N_{3\pi} = 1.367 \times 10^7$ and the acceptance $A_{3\pi} = 14.96\%$.

The events passing the signal selection are shown in Figure 1 for both data and Monte Carlo simulation. Only one event is observed in the signal region for the $K_{\pi\mu\mu}^{LNV}$ sample while 3489 candidates are selected in the $K_{\pi\mu\mu}^{LNC}$ sample.

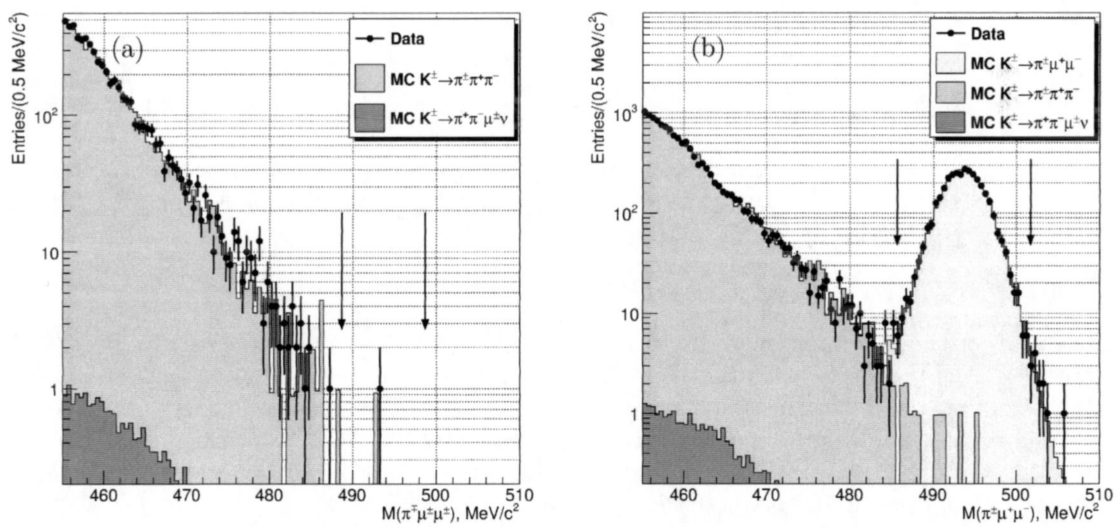

Figure 1. $M_{\pi\mu\mu}$ mass distribution for data and Monte Carlo simulation for events that pass the $K_{\pi\mu\mu}^{LNV}$ (a) and the $K_{\pi\mu\mu}^{LNC}$ (b) selection.

4. Upper limit on $\mathcal{BR}(K^{\pm} \to \pi^{\mp}\mu^{\pm}\mu^{\pm})$

Only one event remains after applying the signal selection. The background expectation is $N_{bkg} = 1.16 \pm 0.87_{stat.} \pm 0.12_{syst.}$. The background is evaluated with Monte Carlo simulation and is mostly composed of $K_{3\pi}$ events in which two pions decay into muons. Since no signal is observed an upper limit is set on $\mathcal{BR}(K^\pm \to \pi^\mp\mu^\pm\mu^\pm)$ applying the Rolke-Lopez method [11] to find the 90% confidence intervals for the case of a Poisson process in presence of multiple Poisson backgrounds with unknown mean:

$$\mathcal{BR}(K^\pm \to \pi^\mp\mu^\pm\mu^\pm) \;=\; \frac{N^{LNV}_{\pi\mu\mu}\cdot A_{3\pi}}{N_{3\pi}\cdot D\cdot A(K^{LNV}_{\pi\mu\mu})}\cdot \mathcal{BR}(K_{3\pi}) \;<\; 8.6\times 10^{-11},$$

where $D = 100$ is the trigger downscaling, $N^{LNV}_{\pi\mu\mu} < 2.92$ the upper limit on the signal events at 90% CL and $A(K^{LNV}_{\pi\mu\mu}) = (20.62 \pm 0.01)\%$ the signal acceptance evaluated with Monte Carlo simulation.

5. Search for two-body resonances

The search for the two-body resonances is performed on all events that survived $K^\pm \to \mu^\pm X(X \to \pi\mu)$ and $K^\pm \to \pi^\pm X(X \to \mu^+\mu^-)$ in the $K^{LNV}_{\pi\mu\mu}$ and $K^{LNC}_{\pi\mu\mu}$ selections respectively. Figure 2 shows the obtained signal acceptances as a functions of the resonance mass and lifetime.

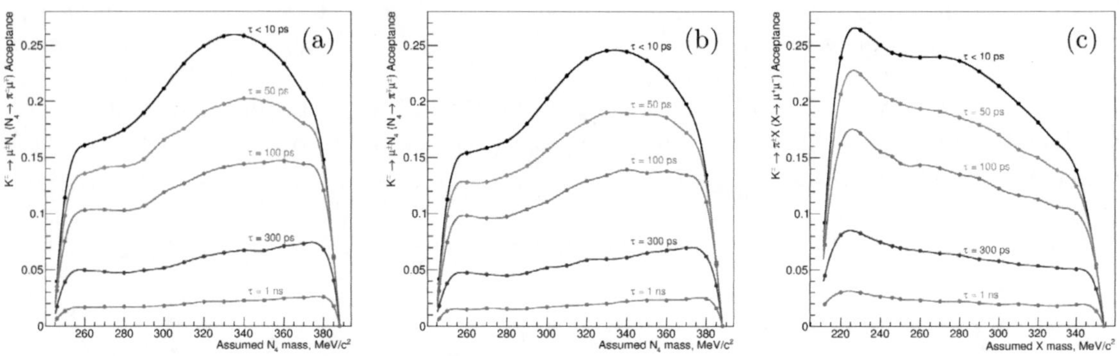

Figure 2. Signal acceptance as a function of the resonances mass and lifetime hypothesis. (a) $K^{LNV}_{\pi\mu\mu}$ selection for $K^\pm \to \mu^\pm N_4(N_4 \to \pi^\mp\mu^\pm)$; (b) $K^{LNC}_{\pi\mu\mu}$ selection for $K^\pm \to \mu^\pm N_4(N_4 \to \pi^\pm\mu^\mp)$; (c) $K^{LNC}_{\pi\mu\mu}$ selection for $K^\pm \to \pi^\pm\chi(\chi \to \mu^+\mu^-)$.

In the scan of the invariant mass $M_{ij}(ij = \pi\mu, \mu^+\mu^-)$ the mass step are given by $\sigma(M_{ij})/2$, where $\sigma(M_{ij})$ is the mass resolution, and the signal window by $\pm 2\sigma(M_{ij})$ at each mass M_{ij}. This means that the results obtained in neighbouring mass hypotheses are highly correlated, as the signal mass window is 8 times larger than the mass step.
In total, 284 and 267(280) mass hypotheses have been tested, covering the full kinematic range of the $M_{\pi\mu}(M_{\mu\mu})$ distributions for the $K^{LNV}_{\pi\mu\mu}$ and $K^{LNC}_{\pi\mu\mu}$ candidates.
In the $K^{LNC}_{\pi\mu\mu}$ selection a total of 3489 candidates is observed where the background contamination from $K_{3\pi}$ is estimated to be $(0.36 \pm 0.10)\%$ using Monte Carlo simulation. This level of purity allows to consider $K^{LNC}_{\pi\mu\mu}$ as the only background for resonance searches in that sample. For the search of two-body resonances also the Rolke-Lopez statistical method is used. The number of considered background events for the $K^{LNV}_{\pi\mu\mu}(K^{LNC}_{\pi\mu\mu})$ candidates is N=4 (N=1). The number of observed events and the obtained upper limits at 90% confidence level are shown in Figure 3.

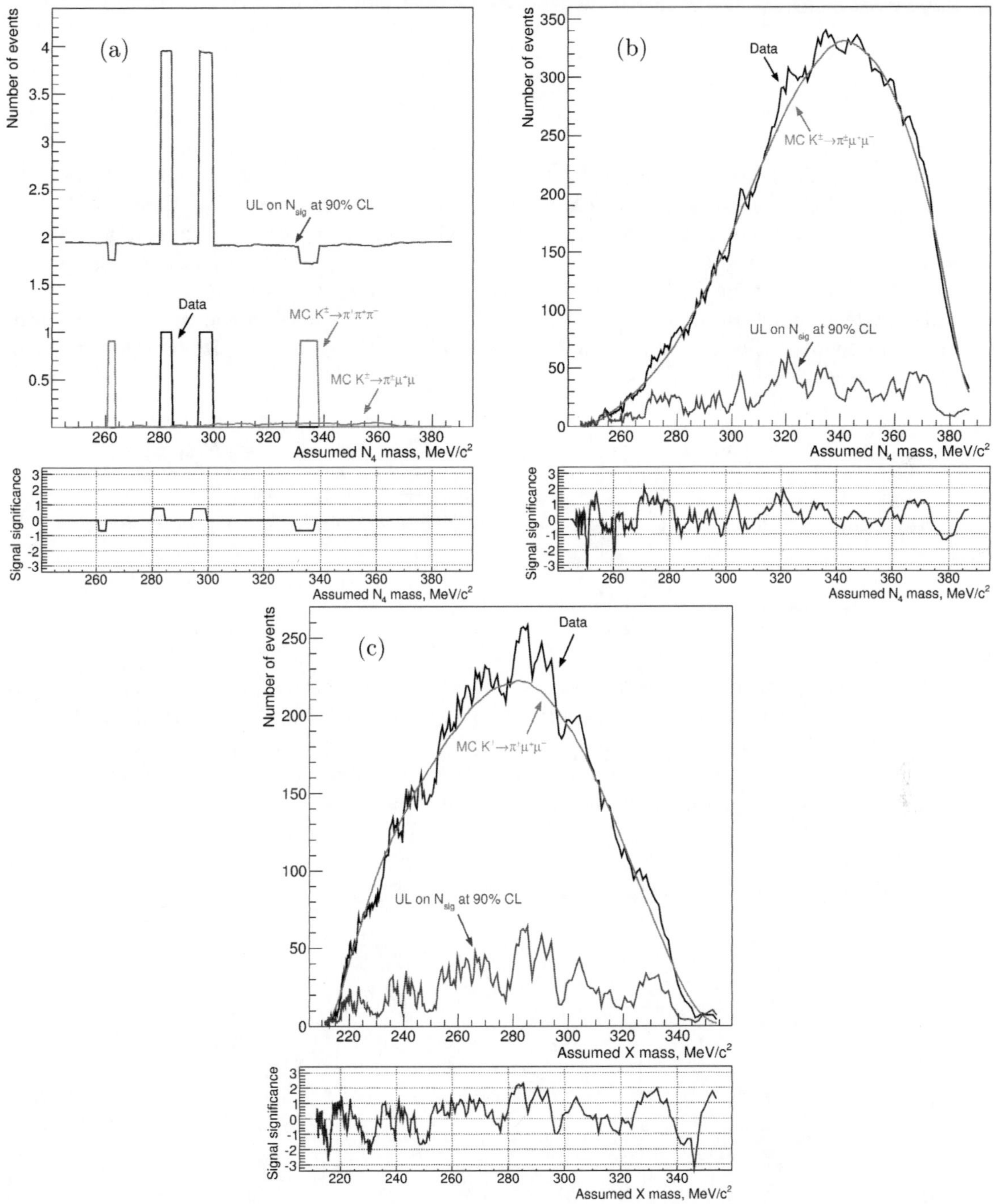

Figure 3. The numbers of observed data (black) and expected background events ($K^{\pm} \to \pi^{\pm}\pi^{+}\pi^{-}$ in green, $K^{\pm} \to \pi^{\pm}\mu^{+}\mu^{-}$ in red) are shown for (a) $M_{\pi\mu}$ with the $K^{LNV}_{\pi\mu\mu}$ selection; (b) $M_{\pi\mu}$ with the $K^{LNC}_{\pi\mu\mu}$ selection; (c) $M_{\mu\mu}$ with the $K^{LNC}_{\pi\mu\mu}$ selection (big plots). The light blue line represents the obtained upper limits at 90% CL on the numbers of signal candidates. In the separated figure the local significances (blue line) are shown for each resonance mass value.

For each mass hypothesis of the three resonance searches the local significance z is evaluated as

$$z = \frac{N_{obs} - N_{exp}}{\sqrt{\delta N_{obs}^2 + \delta N_{exp}^2}},$$

where N_{obs} and N_{exp} are the number of observed and expected background events with their uncertainties δN_{obs} and δN_{exp}. The results for z are also shown in Figure 3. Since the local significances never exceed 3 standard deviations no signal is observed, and upper limits are set on the product $\mathcal{BR}(K^\pm \to p_1 X)\mathcal{BR}(X \to p_2 p_3)$, with the three possibilities $p_1 p_2 p_3 = \mu^\pm\pi^\mp\mu^\pm, \mu^\pm\pi^\pm\mu^\mp, \pi^\pm\mu^+\mu^-$. The upper limits are calculated as a function of the resonance lifetime τ for each mass hypothesis m_i by using the values of the signal acceptances (these also depend on the resonances mass and lifetime) and the upper limits on the number of signal events in each mass hypothesis.

$$\mathcal{BR}(K^\pm \to p_1 X)\mathcal{BR}(X \to p_2 p_3)|_{m_i,\tau} = \frac{N_{sig}^i}{N_{3\pi} \cdot D} \cdot \frac{A_{3\pi}}{A_{\pi\mu\mu}(m_i, \tau)} \cdot \mathcal{BR}(K_{3\pi})$$

The obtained upper limits on the products $\mathcal{BR}(K^\pm \to p_1 X)\mathcal{BR}(X \to p_2 p_3)$ are shown in Figure 4.

6. Conclusions

The NA48/2 collaboration has performed searches for the lepton number violating $K^\pm \to \pi^\mp\mu^\pm\mu^\pm$ decay and resonances in $K^\pm \to \pi\mu\mu$ decays, using the data collected in 2003–2004. In those searches (LNV decay and resonances) no signals are observed.
For the LNV $K^\pm \to \pi^\mp\mu^\pm\mu^\pm$ decay an upper limit is set of $\mathcal{BR} < 8.6 \times 10^{-11}$ [7], which improves the previous limit [12] by more than one order of magnitude.
Upper limits are also set on the products $\mathcal{BR}(K^\pm \to \mu^\pm N_4)\mathcal{BR}(N_4 \to \pi^\mp\mu^\pm)$ and $\mathcal{BR}(K^\pm \to \pi^\pm\chi)\mathcal{BR}(\chi \to \mu^+\mu^-)$ as function of the resonance lifetimes and masses. These upper limits are in the range $10^{-10} - 10^{-9}$ for resonance lifetimes $\tau < 100$ ps.

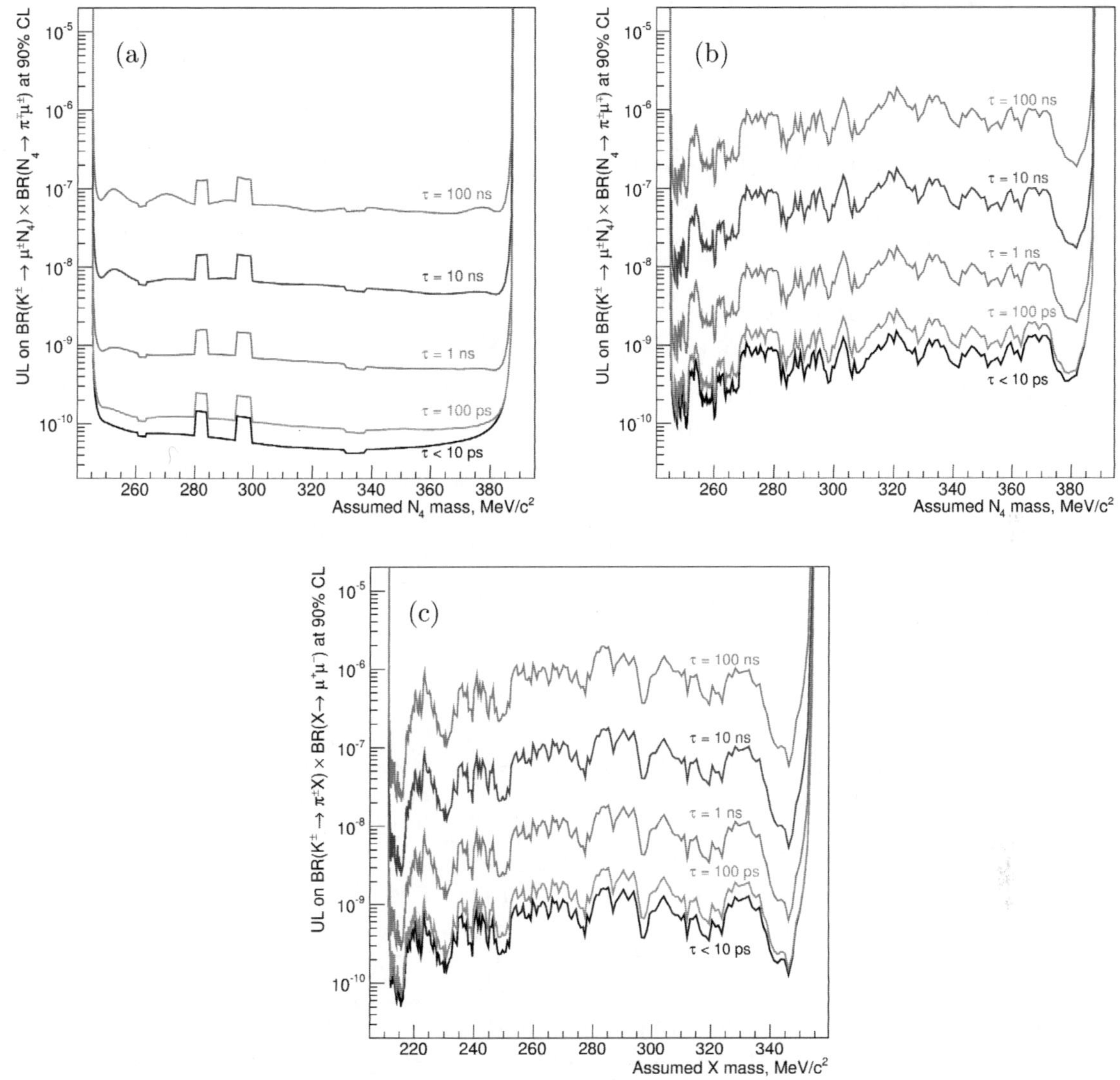

Figure 4. Obtained upper limits at 90% CL on the products of branching ratios as functions of the resonance mass and lifetime: (a) $\mathcal{BR}(K^\pm \to \mu^\pm N_4)\mathcal{BR}(N_4 \to \pi^\mp\mu^\pm)$; (b) $\mathcal{BR}(K^\pm \to \mu^\pm N_4)\mathcal{BR}(N_4 \to \pi^\pm\mu^\mp)$; (c) $\mathcal{BR}(K^\pm \to \pi^\pm\chi)\mathcal{BR}(\chi \to \mu^+\mu^-)$.

References

[1] Asaka T and Shaposhnikov M 2005 *Phys. Lett.* B**620** 17.
[2] Asaka T, Blanchet S and Shaposhnikov M 2005 *Phys. Lett.* B**631** 151.
[3] Shaposhnikov M and Tkachev I 2006 *Phys. Lett.* B**639** 414.
[4] Atre A, Han T, Pascoli S and Zhang B 2009 *JHEP* **0905** 030.
[5] Bezrukov F and Gorbunov D 2010 *JHEP* **1005** 010.
[6] Bezrukov F and Gorbunov D 2014 *Phys. Lett.* B**736** 494.
[7] Batley J R *et al* (NA48/2 collaboration), arXiv:1612.04723.
[8] Batley J R *et al* (NA48/2 collaboration) 2007 Search for direct CP violating charge asymmetries in $K^\pm \to \pi^\pm\pi^+\pi^-$ and $K^\pm \to \pi^\pm\pi^0\pi^0$ decays *Eur. Phys. J.* C**52** 875.
[9] Fanti V *et al* (NA48 collaboration) 2007 *Nucl. Instrum. Methods* A**574** 433.
[10] Patrignani C *et al* (Particle Data Group) 2016 *Chin. Phys.* C**40** 100001.
[11] Rolke W A and Lopez A M 2001 *Nucl. Instrum. Meth.* A**458** 745.
[12] Batley J R *et al.* (NA48/2 collaboration) 2011 *Phys. Lett.* B**697** 107.

IOP Conf. Series: Journal of Physics: Conf. Series **873** (2017) 012020 doi:10.1088/1742-6596/873/1/012020

Model independent measurement of the leptonic kaon decay $K^{\pm} \to \mu^{\pm}\nu_{\mu}e^{+}e^{-}$

Letizia Peruzzo [1]
Johannes Gutenberg University, Mainz, Germany
E-mail: lperuzzo@uni-mainz.de

Abstract. The NA48/2 experiment at CERN collected a very large sample of charged kaon decays into multiple final states. From this data sample we have reconstructed about 1500 events of the very rare decay $K^{\pm} \to \mu^{\pm}\nu_{\mu}e^{+}e^{-}$ over almost negligible background in the region with M($e^{+}e^{-}$) above 140 MeV, which is of great interest in Chiral Perturbation Theory. We present the M_{ee} spectrum and a model-independent measurement of the decay rate for this region.

1. Introduction

Quantum Chromodynamics (QCD), at low energies, is described by theories that can be tested using radiative decays of K mesons. Chiral Perturbation Theory (ChPT) is one of the frameworks available to describe QCD at low energy. In this theory QCD is an effective field theory with a cutoff scale Λ of $\mathcal{O}(1\text{ GeV})$, where for energies below the cutoff an expansion can be performed in p/Λ (chiral expansion). The study of radiative kaon decays expressed in chiral expansion at next-to-leading order can so be used to test ChPT.
The present work is focused on the study of the leptonic decay $K^{\pm} \to \mu^{\pm}\nu_{\mu}e^{+}e^{-}$ with data collected in 2003–2004 at the NA48/2 experiment at CERN.
Of particular interest is the kinematic region in which the invariant mass of the electron-positron pair is above the π^0 mass where the ChPT contribution increases the tree-level branching fraction by 70%. A precise description of the theoretical calculation of the branching fraction for the full phase space and for $M_{ee} \geq 140\text{ MeV}/c^2$ can be found in Ref. [1].
The amplitude is dominated by the Inner Bremsstrahlung of the muon (IB) in the final state which is a purely Quantum Electrodynamical (QED) process and can be exactly calculated within the Standard Model (SM). Going to next-to-leading order, the impact of ChPT on the branching fraction becomes more and more important especially for large M_{ee}, the so-called Structure Dependent (SD) amplitude. In Figure 1 the Feynman diagrams of both the IB and SD contributions are shown. The Monte Carlo simulation used for the analysis is based on Ref. [1], which includes the form factors computed up to $\mathcal{O}(p^4)$ in ChPT and no radiative corrections. The phase space is expressed in terms of the dimensionless variable $z = (M_{ee}/M_K)^2$. For $z \geq 0.1$ a sizeable part of the amplitude originates from the SD contribution (Figure 2).

[1] On behalf of the NA48/2 Collaboration:
Cambridge, CERN, Dubna, Chicago, Edinburgh, Ferrara, Firenze, Mainz, Northwestern, Perugia, Pisa, Saclay, Siegen, Torino, Wien

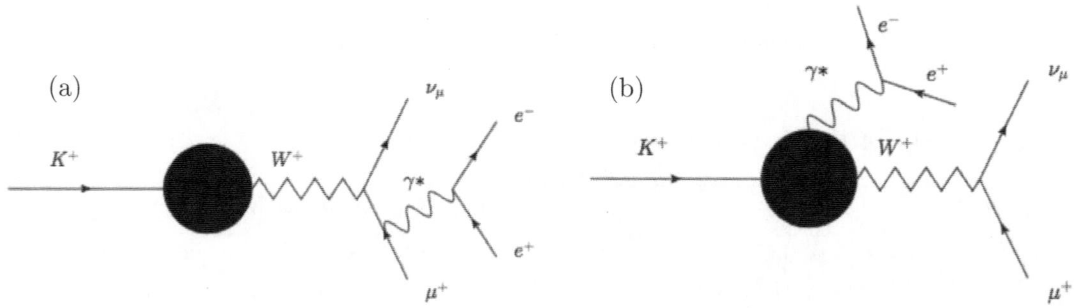

Figure 1. The Feynman diagrams of $K^\pm \to \mu^\pm \nu_\mu e^+ e^-$ proceed via a W boson and a virtual photon. The Inner Bremsstrahlung (IB) contribution from the muon in the final state (a) is a pure QED process, while the Structure Dependent (SD) contribution (b) is described in terms of ChPT.

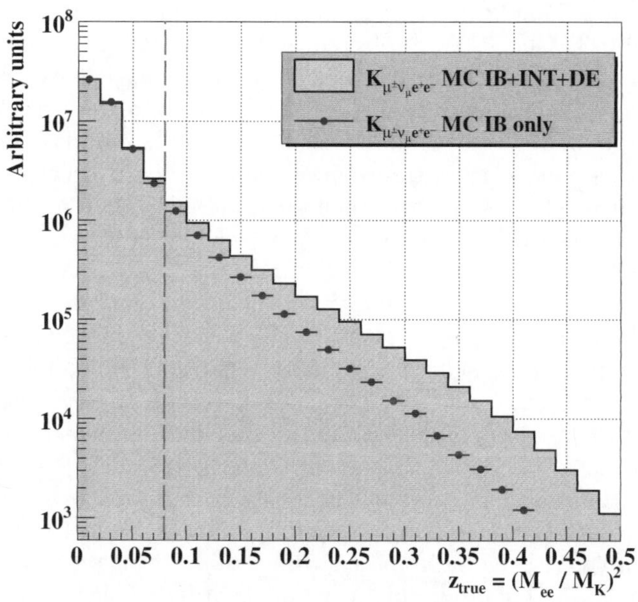

Figure 2. Simulated $z = (M_{ee}/M_K)^2$ distribution before applying any selection criteria. The Monte Carlo simulation is produced according to [1]. Two types of simulation are show here: the red dots correspond to a simulation with only the IB contribution while the light-blue solid plot includes all the ChPT form factors contributions, providing an enhancement at large z. The analysis is performed in the phase space $z > 0.08$ which means $M_{ee} \geq 140$ MeV/c^2 (right region with respect to red dashed vertical line).

The first measurement of $\mathcal{BR}(K^\pm \to \mu^\pm \nu_\mu e^+ e \,|M_{e^+e} \geq 140 \text{ MeV}/c^2) = (12.3 \pm 3.2)^{-8}$ was performed at CERN in 1976 [2] with 14 observed events. With a more stringent cut on the phase space the E865 experiment at BNL in 2002 measured $\mathcal{BR}(K^\pm \to \mu^\pm \nu_\mu e^+ e \,|M_{e^+e} \geq$

IOP Conf. Series: Journal of Physics: Conf. Series **873** (2017) 012020 doi:10.1088/1742-6596/873/1/012020

145 MeV/c^2) = $(7.06 \pm 0.31) \times 10^{-8}$ from 2679 observed events [3].
Table 1 summarizes the theoretical results obtained in Ref. [1].

Table 1. Theoretical prediction of $\mathcal{BR}(K^{\pm} \to \mu^{\pm}\nu_{\mu}e^{+}e^{-})$ [1].

	IB contribution	IB+SD contribution
Full phase space	2.49×10^{-5}	22.49×10^{-5}
$M_{e^+e^-} \geq 140\ MeV/c^2$	4.98×10^{-8}	8.51×10^{-8}

2. The NA48/2 apparatus and data taking conditions

The NA48/2 experiment was located in the north area (NA) of the CERN SPS accelerator facility and was a successor of the NA48 experiment with almost the same detector set up. The primary goal of NA48/2 was the search of direct CP violation in $K^{\pm} \to \pi^{\pm}\pi^{+}\pi^{-}$ and $K^{\pm} \to \pi^{\pm}\pi^{0}\pi^{0}$ decays [4] with about 100 days of effective data taking in 2003–2004. The experiment used 400 GeV/c protons provided by the SPS which, hitting on a Beryllium target, produced a secondary hadronic beam. Thanks to a system of magnets and collimators only charged particles with momenta of (60 ± 3) GeV/c were selected and aligned with the longitudinal axis of the detector within 1 mm and with a transverse size of about 1 cm. The experiment was recording the decays of K^{+} and K^{-} mesons inside the fiducial decay region located in a 114 m long cylindrical vacuum tank. After the vacuum tank a thin Kevlar window separated the vacuum from a helium vessel at atmospheric pressure in which a magnetic spectrometer was located.
The spectrometer consisted of four drift chambers (DCH) and a dipole magnet, placed between the second and the third chamber, which provided a horizontal momentum kick of p_t = 120 MeV/c for charged particles. Each chamber had four different views and a spatial resolution of $\sigma_{x,y}$ = 90 μm. The nominal momentum resolution of the spectrometer was $\sigma_p/p = (1.02 \oplus 0.044\ p)\%$ where p is given in GeV/c. A scintillating hodoscope (HOD) followed the spectrometer. It was divided in four quadrants by a horizontal and a vertical plane of strip-shaped counters. The HOD was used in the trigger chain to provide a fast time measurement for charged particles with a 150 ps time resolution. A quasi homogeneous electromagnetic calorimeter filled with liquid krypton (LKr) and a depth of $27X_0$ was used both for photon detection and particle identification. It had an energy resolution of $\sigma_E/E = 0.032/\sqrt{E} \oplus 0.09/E \oplus 0.0042$ (E in GeV) corresponding to $\sigma_E/E = 0.94\%$ at 20 GeV. The particle identification was completed by a hadronic calorimeter and a muon veto system (MUV). The hadronic calorimeter consisted of alternated iron and scintillator planes, while the MUV of three plastic scintillator strips planes each of them preceded by a 80 cm iron wall. A more detailed description of the experiment can be found in Ref. [5].

3. Events selection

The $K^{\pm} \to \mu^{\pm}\nu_{\mu}e^{+}e^{-}$ ($K_{\mu\nu ee}$) selection is based on the reconstruction of a three-track vertex. Two of the three tracks have to be identified as electron and positron while the other has to be compatible with a muon. The decay mode $K^{\pm} \to \pi^{\pm}\pi^{+}\pi^{-}$ ($K_{3\pi}$) has the same vertex topology and is used as a normalization channel for the $K_{\mu\nu ee}$ branching ratio measurement, leading to a first order cancellation of systematic effects. Using the $K_{3\pi}$ normalization channel, the total number of kaon decays in the fiducial volume is measured to be $N_K = (1.56 \pm 0.01) \times 10^{11}$.
In the $K_{\mu\nu ee}$ selection the three tracks have to form a vertex of good quality inside the fiducial volume with a total charge of $Q = \pm 1$ corresponding to the kaon charge. Each track has to be

inside the geometrical acceptance of the detectors and has to have a momentum between 3 and 50 GeV/c. The total momentum of the three tracks is requested to be less than 66 GeV/c in order to be consistent with the momentum of the beam kaons.
The particle identification of the electron-positron pair is based on the ratio between the total energy measured in the calorimeter over the momentum measured by the spectrometer (E/p) which has to be within the range 0.95-1.05 and on a linear discriminant method based on the shower shape. The good e/π separation of the discriminant provides an almost complete suppression of pions coming from $K_{3\pi}$ and $K^{\pm} \rightarrow \pi^{+}\pi^{-}e^{\pm}\nu_{e}$ (K_{e4}) background decays, that can mimic the signal if they are not really identified as pions. The muon is required to behave like a minimum ionising particle in the LKr ($E/p < 0.2$) and to have an in-time coincidence in the MUV detector.
Two further cuts are applied to the $K_{\mu\nu ee}$ signal:

(1) the invariant mass of the electron-positron is requested to be $M_{ee} \geq 140$ MeV/c^2 in order to reject decays with an $e^{+}e^{-}$ pair coming from a π^0 ($m_{\pi^0} = 135$ MeV/c^2) like $K^{\pm} \rightarrow \pi^{\pm}\pi^{0}$ or $K^{\pm} \rightarrow \pi^{0}\mu^{\pm}\nu_{\mu}$ decays followed by $\pi^{0} \rightarrow e^{+}e^{-}\gamma$.

(2) the invariant mass of the muon and neutrino is required to be above 170 MeV/c^2 in order to reject background events coming from $K^{\pm} \rightarrow \pi^{\pm}e^{+}e^{-}$ followed by $\pi^{\pm} \rightarrow \mu^{\pm}\nu_{\mu}$ decay. The cut decreases the signal acceptance by 11% but completely suppresses $K^{\pm} \rightarrow \pi^{\pm}e^{+}e^{-}$ events.

4. Background estimation

After the signal selection few irreducible background events remain in the signal region. The events contributing to this background arise from multiple pions in the final state, with one pion decaying into $\mu\nu$. Three different channels are contributing:

(1) $K^{\pm} \rightarrow \pi^{\pm}\pi^{0}\pi^{0}$ ($K_{3\pi^0}$) decays with $\pi^{\pm} \rightarrow \mu^{\pm}\nu_{\mu}$ and both π^0 undergo Dalitz decays $\pi^{0} \rightarrow e^{+}e^{-}\gamma$, with only two of the four electrons from different π^0's being detected.

(2) $K^{\pm} \rightarrow \pi^{\pm}\pi^{+}\pi^{-}$ ($K_{3\pi}$) decays where one pion decays into $\mu\nu$ and the other two are misidentified as electrons.

(3) $K^{\pm} \rightarrow \pi^{+}\pi^{-}e^{\pm}\nu_{e}$ (K_{e4}) decays in which one of the pions is misidentified as electron and the other either misidentified as muon or decaying into $\mu\nu$.

The background estimation is performed using data by selecting events with same sign electrons, while all other requirements remain unchanged.
Since the detector acceptances, trigger inefficiencies and misidentification probabilities do not depend on the electric charge the expected number of background events in the signal sample is equal to the observed number of events in the same sign sample multiplied by a corresponding combinatorial factor. For both $K_{3\pi^0}$ and $K_{3\pi}$ the combinatorial factor is 2, because there are two possibilities for having a opposite sign pair ($e^{+}e^{-}$), while for K_{e4} the corresponding factor is 1. Using this method the total number of observed background events is $N_{bgk} = 54 \pm 10(stat.) \pm 5(syst.)$.

5. Results

After the selection 1663 signal candidates are observed with a background contamination of 3% and a signal acceptance varying between 12 and 15% depending on M_{ee}.
Figure 3 and 4 show the surviving events for both data and Monte Carlo simulation after applying the selection criteria.

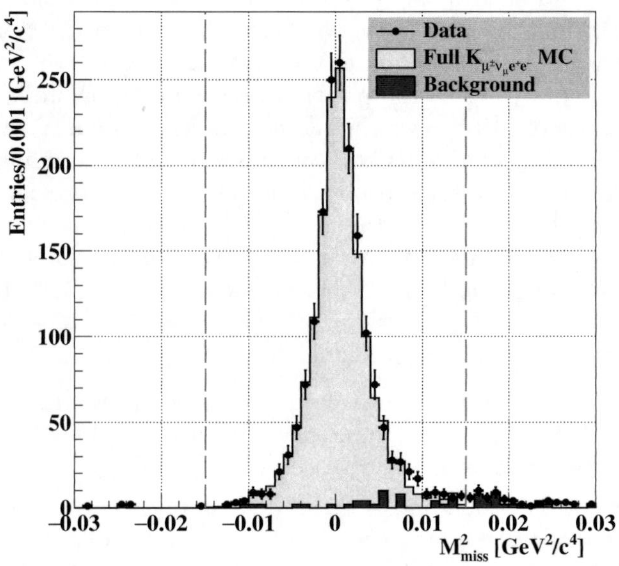

Figure 3. Missing mass $M^2_{miss} = (P_K - P_\mu - P_{e^+} - P_{e^-})^2$ after the full signal selection. The expected background contribution is represented by the dark-blue solid plot which dominates the right tail of the squared missing mass. The two dashed red vertical lines define the signal region.

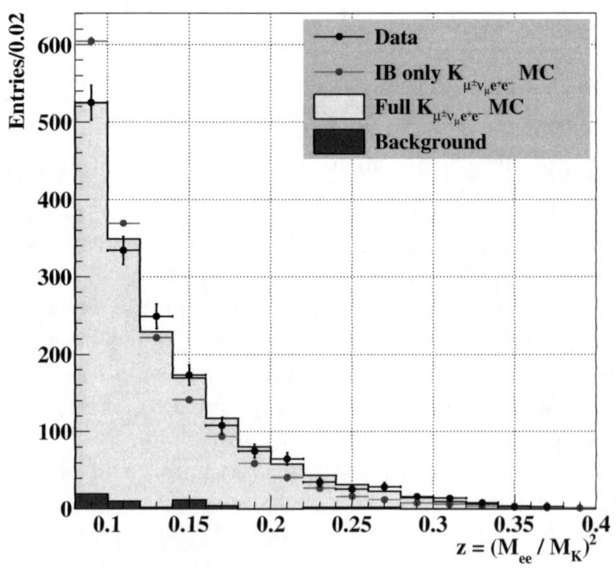

Figure 4. Distribution of $z = (M_{ee}/M_K)^2$ for both data (black points) and Monte Carlo simulation (IB+SD solid, only IB red points). Data are in good agreement with the shape predicted by ChPT.

In order to properly account for the variation of the acceptance the branching ratio $\mathcal{BR}_{(K_{\mu\nu ee})}$ is measured in 15 different bins of z as shown in Figure 5.

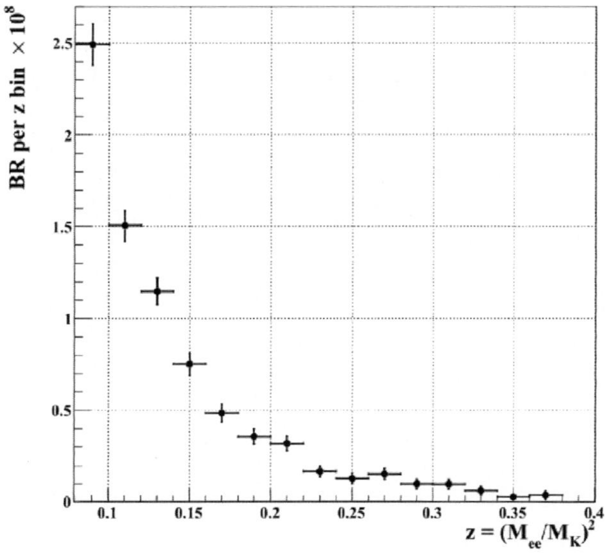

Figure 5. Branching ratio of the decay $K^{\pm} \to \mu^{\pm}\nu_{\mu}e^{+}e^{-}$ computed in 15 different bins of $z = (M_{ee}/M_K)^2$ in the interesting region $M_{ee} \geq 140$ MeV/c^2. The final result is the sum of each individual bin.

The total branching ratio is then the sum over all bins of z and is obtained to be

$$\mathcal{BR}(K^{\pm} \to \mu^{\pm}\nu_{\mu}e^{+}e\,|M_{e^{+}e^{-}} \geq 140\ MeV/c^2) = (7.81{\pm}0.21(stat.){\pm}0.08(syst.){\pm}0.06(ext.)){\times}10^{-8}$$

The relative systematic uncertainty is 1.2% and is dominated by the effect of the radiative corrections on the signal acceptance and the statistical uncertainty of the background estimation. The external error is due to the uncertainty on the $K_{3\pi}$ branching fraction. Table 2 summarizes all contributions to the total uncertainty.

6. Conclusion

A model independent measurement of the leptonic kaon decay $K^{\pm} \to \mu^{\pm}\nu e^{+}e^{-}$ has been presented, using data collected by the NA48/2 experiment in 2003–2004. A total number of 1663 events are observed in the signal region with an estimated background of 54 events. The branching ratio for the phase space of $M_{ee} \geq 140$ MeV/c^2 is measured to be

$$\mathcal{BR}(K^{\pm} \to \mu^{\pm}\nu_{\mu}e^{+}e^{-}) = (7.81 \pm 0.21(stat.) \pm 0.08(syst.) \pm 0.06(ext.)) \times 10^{-8}.$$

The measurement is statistically dominated and the total uncertainty is improved by a factor 1.5 with respect to the previous result [3]. For the first time also radiative corrections are included. The observed z spectrum is in agrement with the shape predicted by ChPT [1].

IOP Conf. Series: Journal of Physics: Conf. Series **873** (2017) 012020 doi:10.1088/1742-6596/873/1/012020

Table 2. Contributions to the uncertainty on $\mathcal{BR}(K^{\pm} \to \mu^{\pm}\nu_{\mu}e^{+}e^{-})$. The list includes statistical, systematic and external sources.

Source	$\Delta\mathcal{BR}/\mathcal{BR}[\times 10^2]$
Data statistics	2.54
Normalization channel statistics	0.02
Total statistical	2.54
Radiative corrections	0.70
Background statistics	0.62
Trigger efficiency	0.54
Background systematic	0.30
Muon ID efficiency	0.13
Signal channel acceptance	0.12
Electron ID uncertainty	0.04
Normalization channel acceptance	0.03
Total systematic	1.15
External uncertainty $\mathcal{BR}(K_{3\pi})$	0.72
Total uncertainty	2.88

References

[1] Bijnens J, Ecker G and Gasser J 1993 Radiative semileptonic Kaon decays *Nucl. Phys.* B**396** 81-118.
[2] Diamant-Berger A M *et al* 1976 Study of some rare decays of K^+ meson *Phys. Lett.* B**62** Issue **4** 485-490.
[3] Poblaguev A A *et al* 2002 Experimental study of the radiative decays $K^+ \to \mu^+\nu e^+ e$ and $K^+ \to e^+\nu e^+ e$ *Phys. Rev. Lett.* **89** 061803.
[4] Batley J R *et al* (NA48/2 collaboration) 2007 Search for direct CP violating charge asymmetries in $K^{\pm} \to \pi^{\pm}\pi^{+}\pi^{-}$ and $K^{\pm} \to \pi^{\pm}\pi^{0}\pi^{0}$ decays *Eur. Phys. J.* C**52** 875.
[5] Fanti V *et al* (NA48 collaboration) 2007 *Nucl. Instrum. Methods* A**574** 433.

IOP Conf. Series: Journal of Physics: Conf. Series **873** (2017) 012021 doi:10.1088/1742-6596/873/1/012021

The KLOE-2 experiment at DAΦNE

E. De Lucia on behalf of the KLOE-2 Collaboration
Laboratori Nazionali di Frascati dell'INFN, via E. Fermi 40, 00044 Frascati (RM), Italy
E-mail: erika.delucia@lnf.infn.it

Abstract. The KLOE-2 experiment at the INFN Laboratori Nazionali di Frascati (LNF) is currently taking data at the upgraded e^+e^- DAΦNE collider. Data taking campaign started in November 2014, reaching 2.4 fb^{-1} integrated luminosity by July 2016 and restarted in September 2016 with the aim of collecting more than 5 fb^{-1} in the next 1-1.5 years. An overview of the KLOE-2 experiment will be given including performance of the upgraded KLOE-2 detector, data taking conditions and data quality.

1. Introduction

KLOE-2 represents the continuation of KLOE [1] with a new physics program mainly focused on the study of K short, η and η' decays as well as on kaon interferometry, test of discrete symmetries, and search for physics beyond the Standard Model [2]. The new data taking campaign aiming to collect more than 5 fb^{-1} integrated luminosity, will allow in particular to perform CPT symmetry and quantum coherence tests using entangled neutral kaons with an unprecedented precision. The first data taking campaign started in November 2014, reaching 2.4 fb^{-1} integrated luminosity by July 2016 and restarted data taking in September 2016 with the aim of collecting more than 5 fb^{-1} in the next 1-1.5 years. Record performance in terms of $2x10^{32}$ $cm^{-2}s^{-1}$ peak luminosity and 13 pb^{-1} maximum daily integrated luminosity were achieved with the innovative crab-waist beam collision scheme, developed in Frascati, that will be employed in the upgrade of the B-factory currently under construction at the KEK Laboratory in Japan and is considered a valid option in several future projects.

2. KLOE-2 detector

The general purpose KLOE detector, composed by one of the biggest drift chamber [3] ever built surrounded by a lead-scintillating fiber Electromagnetic Calorimeter [4] among the best ones for energy and timing performance at low energies, undergone several upgrades. These include State-of-The-art cylindrical GEM detector, the Inner Tracker, LET and HET taggers for $\gamma-\gamma$ physics, and the CCALT and QCALT detectors, crystal and tile calorimeters positioned near the interaction point and along the beam-pipe respectively.

2.1. Inner Tracker

To improve its vertex reconstruction capabilities near the interaction region, KLOE-2 is the first high-energy experiment using the GEM technology with a cylindrical geometry, a novel idea that was developed at LNF exploiting the kapton properties to build a transparent and compact tracking system. This innovative tracking detector, the Inner Tracker (IT) has been inserted

IOP Conf. Series: Journal of Physics: Conf. Series **873** (2017) 012021 doi:10.1088/1742-6596/873/1/012021

in the free space between the beam pipe and the DC inner wall, at 25 cm from IP [5]. The IT is the first ever built and operated cylindrical GEM detector, with a total material budget below 2% of radiation length, allowing to minimize the multiple scattering of low- momentum tracks and the probability of photon conversions. The resolution on vertices close to the IP is expected to improve of about a factor 2. The IT is composed by four concentric Cylindrical GEM (CGEM) detectors, with radii from 13 cm, to preserve the K_S-K_L quantum interference, up to 20.5 cm, due to the constraint from DC inner wall. Each layer, with a total active length of 70 cm, is a triple-GEM detector with 5 concentric cylindrical electrodes: a cathode, to set the drift field, 3 GEM foils acting as multiplication stages, and an anode/readout plane. The anode plane is a multi-layer circuit which implements X-V stereo readout with a 650 μm pitch for a total amount of 30000 front-end channels. The IT is operated with an Ar/iC_4H_{10} 90/10 gas mixture at a nominal gain of 12000. The measured efficiency [6] with cosmic-ray muon tracks reconstructed by the DC is 98% for the single-view and 95% for the two-views. The IT operational parameters with colliding beams have been optimized as a function of the beam currents and background conditions, which are carefully monitored through the current measured on the IT layer. Alignment and calibration of this detector was never done before and represents one of the challenging activities of the experiment. DC excellent track reconstruction has been exploited for IT alignment and calibration. Several effects were studied and accounted for with cosmic-ray muons acquired with and without magnetic field and then Bhabha scattering events are used to check the evaluated corrections. After applying a first set of alignment and calibration parameters as obtained with cosmic-ray muons acquired with and without magnetic field, the preliminary IT-DC Bhabha tracking residual distributions show a resolution of ~350 μm. Fig.1 shows the residual distribution obtained for IT layer 2. Further alignment and calibration refinements are ongoing and will definitely improve these first results.

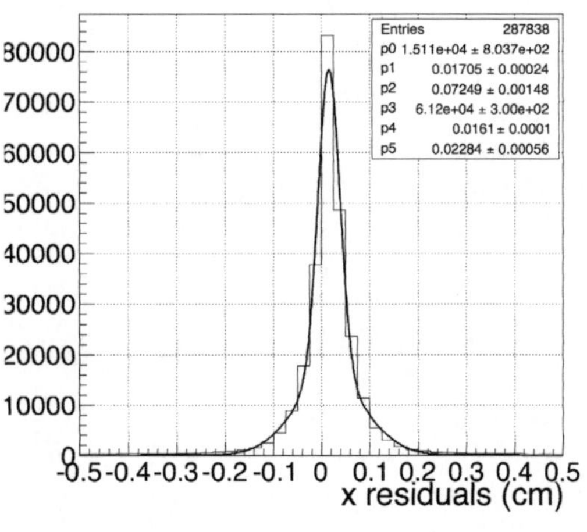

Figure 1. IT-DC residuals as obtained with Bhabha scattering events with Layer #2 and using first alignment and calibration parameters.

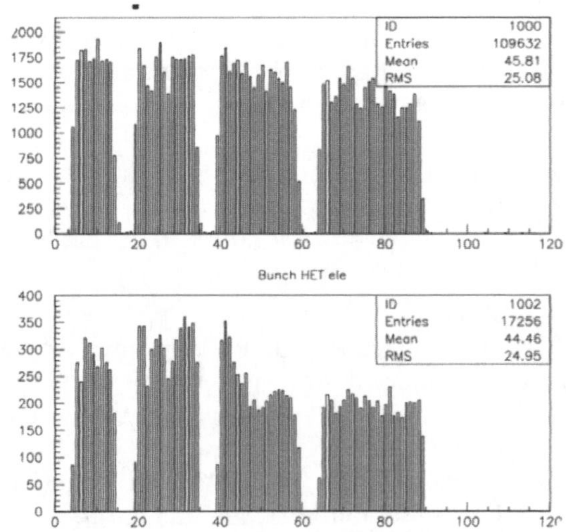

Figure 2. DAΦNE bunch structure as reproduced by HET taggers (top) and KLOE-2 detector (bottom)

2.2. $\gamma\gamma$-taggers

To study $\gamma - \gamma$ physics stations of electron-positron taggers have been installed: i) inside the KLOE-2 detector at 1 m from IP, the Low Energy Taggers (LETs)[7] and ii) 11 m from IP after the first bending dipole, the High Energy Taggers (HETs) [8]. The aim of these taggers is the detection of electrons and positrons scattered in $e^+e^- \to e^+e^-\gamma^*\gamma^* \to e^+e^-X$ reactions, which deviate from the equilibrium orbit during the propagation along the accelerator lattice with energy below 510 MeV.

The two identical LET stations tag electrons and positrons within the 160 < E < 400 MeV energy range. Each LET station consists in a 20 LYSO crystal matrix, each crystal read out by a Silicon photomultiplier (SiPM) with 1.5 x 1.5 cm^2 section and 12 cm length, pointing to the average direction of the arriving particles (11° with respect to the beam line). The two stations are rotated by an angle of 17° with respect to the horizontal plane in order to maximize the number of collected positrons and electrons. The energy resolution of the LET calorimeters is better than 10% for energies larger than 150 MeV, as measured at the Frascati Beam Test Facility (BTF) with electrons of energies between 50 and 450 MeV, and well matches the requirements. Equalization of the LET crystals response and time calibration have been performed selecting minimum ionizing particles (MIPs) as high-momentum tracks from cosmic-ray muons. The absolute energy scale calibration is performed with radiative Bhabha scattering events, with one photon and one lepton reconstructed in the KLOE-2 main detector and the other lepton detected in the LET.

HET stations are position detectors for measuring the deviation of electrons and positrons from the main beam orbit. Together with time information, the position measurement allows scattered electron and positron with energy grater than 420 MeV in $\gamma-\gamma$ processes to be tagged. The sensitive area of each HET station is a set of twenty-eight 3x5x6 mm^3 plastic scintillators which are placed at different distances from the beam-line and with an additional larger one used for coincidence purposes. Therefore the distance between the impinging particle and the beam can be measured knowing which scintillator has been fired. Energy and time resolutions are 2.5 MeV and 200 ps respectively. The HET operability at DAΦNE has been established with dedicated runs to characterize its stations in terms of time resolution and efficiency, and setting up first level calibration. After calibrations, the DAΦNE bunch structure has been successfully measured and nicely reproduced (Fig.2).

2.3. Low-angle calorimeters

Two new Quadrupole CALorimeters with Tiles, QCALTs [9], have been installed around the DAΦNE low-β quadrupoles, at both sides of the IP, with the goal of improving the acceptance for rejecting $K_L \to 3\pi^0$ background events in CP-violating $K_L \to 2\pi^0$ decays. Each calorimeter, 1 m long, consists in a dodecahedral structure, arranged as a sampling of 5 layers of 5 mm thick scintillator plates, alternated with 3.5 mm thick tungsten plates, for a total of $\sim$5X$_0$ thickness. The active part of each plane is divided into 20 tiles of 5x5 cm^2 area with 1 mm diameter wavelength shifter fibers embedded in circular grooves. Each fiber is then optically connected to a SiPM, for a total of 2400 channels. QCALT calorimeters have a $\sim$2 mm resolution along the beam axis and a time resoultion of $\sim$1 ns. Time calibration and channel-by-channel equalization is performed using cosmic-ray muon runs acquired with and without magnetic field. The KLOE calorimeter is used as timing reference. In order to enlarge the angular acceptance for particles coming from the IP from 20° to 10°, with the aim of improving multi-photon detection in rare decays such as $K_S \to \gamma\gamma$, $\eta \to \pi^0\gamma\gamma$ and $K_S \to 3\pi^0$, two identical Crystal CALorimeters with Timing (CCALTs) [9] have been installed very close to the IP, near the first focusing quadrupoles of DAΦNE. Each CCALT module is made of 4 aluminum shells, with projective geometry, containing 4 LYSO crystals readout by SiPM. Tests performed at the Frascati Beam Test Facility (BTF) allowed to measure an energy resolution better than 5% and a time resolution of about 49 ns (120 ps) at 100 MeV (500 MeV) energy. Similar performance are measured with cosmic-ray muons, used also for calibration and equalization of CCALT channels. Such detector provides also fast luminosity monitor to DAΦNE operations.

3. Data taking and Data quality

KLOE-2 data taking started in November 2014 and is progressing steadily. DAΦNE performance in terms of delivered luminosity and machine background conditions have constantly improved during this period, allowing the KLOE-2 detector to operate with increasingly higher beam currents and peak luminosity. By July 2016 DAΦNE delivered an integrated luminosity of 3.0 fb^{-1}, out of which 2.4 fb^{-1} were acquired by KLOE-2. Record performance in terms of 2x10^{32} $\text{cm}^{-2}\text{s}^{-1}$ peak luminosity and 13 pb^{-1} maximum daily integrated (11 pb^{-1} acquired) luminosity were achieved. DAΦNE uses a topping-up injection scheme, which allows to keep the average beam currents between 1.2÷1.6 A (electrons) and 0.7÷1.0 A (positrons). KLOE-2 detector provides instantaneous luminosity measurement, by counting large angle Bhabha scattering events. To safely operate all sub-detectors and for e^+e^- injection optimization, the information from three indicators of the machine background level is sent on-line to DAΦNE operators: the current measured on the first IT layer, the total current measured in the DC and the background level measured in the EMC end-caps. This allowed machine background conditions to be kept at an acceptable level and good data quality having at the same time good collider uptime and luminosity performance.

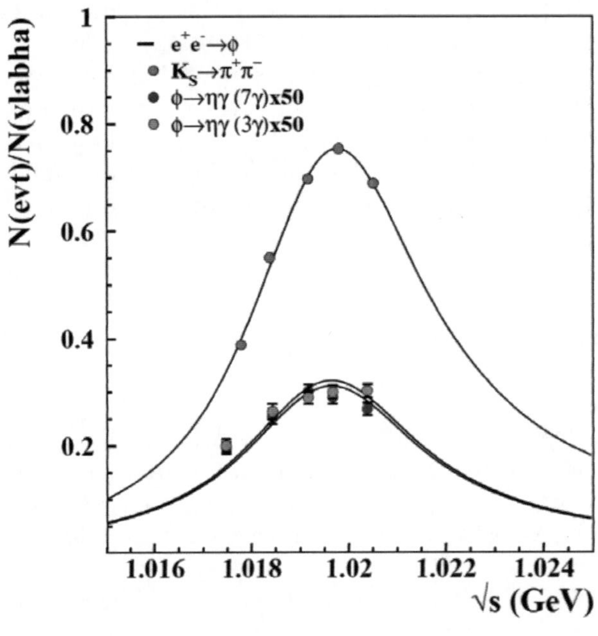

Figure 3. DAΦNE Energy scan around the *phi*-meson peak.

Data quality is monitored both on-line and off-line with beam parameters, center of mass energy, Interaction Point position and ϕ-meson momentum, and checking the stability of Event classification counters of several decays of interest. Benchmark physics channels have been selected and among these are $K_{S,L} \to \pi^+\pi^-$ and $\phi \to \eta\gamma$ ($\eta \to \gamma\gamma$ and $\eta \to 3\pi^0$). Looking at K_S lifetime and K_L, π^0 and η invariant mass distributions, an overall good agreement with previous KLOE data distributions is observed despite the increased machine background levels. An energy scan around the ϕ-meson peak has been performed to measure present beam energy scale, by shifting the central value of DAΦNE radiofrequency. Fig.3 shows the event rates in several ϕ-meson decay normalized to the large angle Bhabha scattering events used to measure luminosity. Distributions are nicely reproduced by ϕ-meson lineshape, including radiative corrections and beam energy spread effect. To run exactly at the ϕ-meson peak, DAΦNE center of mass energy has been shifted by +550 keV.

4. Conclusions

KLOE-2 data taking campaign is steadily progressing together with first physics analysis following the guidelines described in [2]. KLOE-2 experiment is fully operational and its new detectors are being successfully commissioned. KLOE-2 first data are being analyzed with benchmark physics channels.Good agreement with previous performance has been found. An integrated luminosity of more than 5 fb^{-1} is expected to be reached within one year. Meanwhile the KLOE-2 Collaboration is continuing the analysis of the KLOE dataset and producing interesting results both in Kaon and Hadron Physics, some of them presented at this same conference.

5. Acknowledgements

We warmly thank our former KLOE colleagues for the access to the data collected during the KLOE data taking campaign. We thank the DAΦNE team for their efforts in maintaining low background running conditions and their collaboration during all data taking. We want to thank our technical staff: G.F. Fortugno and F. Sborzacchi for their dedication in ensuring efficient operation of the KLOE computing facilities; M. Anelli for his continuous attention to the gas system and detector safety; A. Balla, M. Gatta, G. Corradi and G. Papalino for electronics maintenance; C. Piscitelli for his help during major maintenance periods. This work was supported in part by the Polish National Science Centre through the Grants No. 2013/08/M/ST2/00323, 2013/11/B/ST2/04245, 2014/14/E/ST2/00262, 2014/12/S/ST2/00459.

References

[1] Bossi F., De Lucia E., Franzini J. L., Miscetti S., Palutan M. and KLOE Collaboration, Nuovo Cimento, 30 (2008) 10.
[2] G. Amelino Camelia *et al.*, Physics with the KLOE-2 experiment at the upgraded DANE, *Eur. Phys. J.* C 68 619 (2010).
[3] M. Adinolfi *et al.*, Nucl. Inst. & Meth. **A 488** (2002) 364.
[4] M. Adinolfi *et al.*, Nucl. Inst. & Meth. **A 482** (2002) 51.
[5] A. Balla *et al.*, *Status of the cylindrical-GEM project for the KLOE-2 inner tracker*, Nucl. Inst. & Meth. **A 628** (2011) 18 and A. Balla *et al.*,*The cylindrical GEM detector for the KLOE-2 Inner Tracker*, JINST **9** C01014 (2014).
[6] Di Cicco A. and Morello G., Acta Phys. Pol. B, 46 (2015) 73.
[7] Babusci D. et al., Acta Phys. Pol. B, 46 (2015) 87.
[8] Babusci D. et al., Acta Phys. Pol. B, 46 (2015) 81.
[9] Happacher F. and Martini M., Acta Phys. Pol. B, 46 (2015) 87.

IOP Conf. Series: Journal of Physics: Conf. Series **873** (2017) 012022 doi:10.1088/1742-6596/873/1/012022

Belle achievements and Belle II prospects for CP violation

L Li Gioi, on behalf of the Belle and Belle II collaborations
Max Planck Institut für Physik, Föhringer Ring 6, 80805 Munich, Germany
E-mail: ligioi@mpp.mpg.de

Abstract. The Belle experiment has accomplished various CP violation measurements, in order to constrain the CKM Unitarity Triangle and to search for New Physics signatures, using a data sample of 772 million B-meson pairs. The Belle II experiment will extend research capabilities in this field, thanks to the 50 times larger integrated luminosity expected to be delivered by the SuperKEKB e^+e^- collider. The results from Belle and future prospects at Belle II are presented.

1. Introduction
Fifty years after its discovery in K^0 system [1], the violation of the CP symmetry is one of the open problems in particle physics. CP violation in the Standard Model arises from the presence of an irreducible phase in the Cabibbo-Kobayashi-Maskawa (CKM) [2, 3] quark mixing matrix. The unitarity condition of the CKM matrix can be written as a sum of three complex numbers equal to zero, which can be presented in a form of a triangle, called the "Unitarity Triangle", in the complex plane. The CP violation in the B meson system has been clearly established [4, 5] and, although there is good agreement with the expectations of the Standard Model, further measurements of CP violation in B decays are needed to over-constrain the Unitarity Triangle and search for New Physics effects.

2. SuperKEKB
The target instantaneous luminosity of SuperKEKB is $8 \cdot 10^{35}$ cm^{-2} s^{-1}, which is 40 times higher than the world record set by KEKB. The SuperKEKB collider is under commissioning. Most of the components of KEKB has been reused, however there are also many components that need to be modified or newly developed. The most important of them which have an impact on the Belle II physics program are:

Nano Beam. SuperKEKB is now based on the Nano Beam scheme, which was first proposed [6] for SuperB in Italy. The Nano Beam scheme follows the ideas of the final focus system envisaged for a future linear collider, transferred to a circular machine, where vertical transverse beam sizes of a few tens of nanometers are considered (with a few micrometers horizontally).

Reduced boost. In order to reach the target luminosity, the energy difference between the two beams has been reduced from 8 (e^-) and 3.5 (e^+) GeV of KEKB to 7 (e^-) and 4 (e^+) GeV of SuperKEKB. This makes more difficult to resolve the vertices of the two produced B mesons.

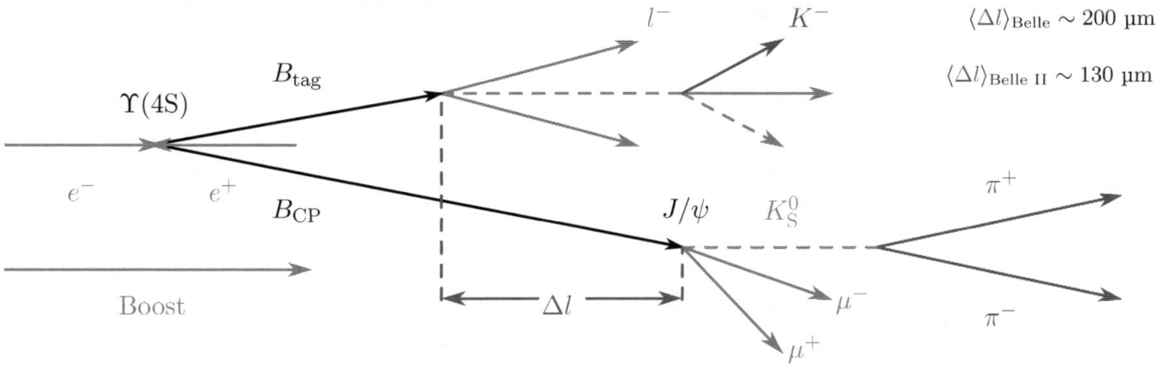

Figure 1. Schema of the time dependent analysis.

3. Belle II

The main design goals of Belle II demand to deal with much higher physics rates and significantly larger expected backgrounds relative to Belle, which implies a necessity of improvement in the overall physics performance. As the range of center of mass energies of SuperKEKB is basically the same of KEKB, the physics requirements of Belle II remain basically the same as those of Belle. This allows to reuse part of the Belle detector as the electromagnetic calorimeter, the solenoid, and the instrumented iron return yoke. However, the much more demanding physics rate and higher background together with the reduction the the energy asymmetry require an entirely new tracking, consisting of a Pixel Vertex Detector (PXD), a Silicon strip Vertex Detector (SVD) and a Central Drift Chamber (CDC); a new particle identification systems as well as a faster signal processing in the outer detectors.

3.1. The Pixel Vertex Detector (PXD)

The new two layers silicon pixel vertex detector [7], based on DEPFET [8] technology, has been designed to cope with the expected increase of background and to provide the same sensitivity of Belle in time dependent analysis in reduced boost environment.

The pixel vertex detector will be mounted directly on the beam pipe at a distance of 14 mm from the interaction point. With an expected single point resolution of 7.7 μm, it will provide an accurate measurement of the tracks position in order to precisely reconstruct the decay vertex of the short living particles.

4. Time dependent measurements

A clean laboratory to study the two-state system of B^0 and $\bar{B}^0$ mesons is provided by the $\Upsilon(4S)$ resonance which can be produced in an e^+e^- annihilation at the center of mass energy of 10.58 GeV. In SuperKEKB the beam energies to form the $\Upsilon(4S)$ are chosen to be asymmetric: 7 (e^-) and 4 (e^+) GeV, in order to translate the decay time into a decay length (Figure 1):

$$\Delta t = \Delta l/\beta\gamma c, \tag{1}$$

where l is the boost direction.

The $\Upsilon(4S)$ decays in an exclusively produced B meson pair: B^+B^- or $B^0\bar{B}^0$ with almost equal probability. In addition to the $\Upsilon(4S)$ there are contributions from non resonant background, consisting of lighter quarks (u, d, s, and c) and anti-quarks, and of lepton (e, μ, τ) pairs production. The relative rates of B mesons at the peak of the $\Upsilon(4S)$ to the lighter quarks component and to τ lepton are roughly 1:3:1.

IOP Conf. Series: Journal of Physics: Conf. Series **873** (2017) 012022 doi:10.1088/1742-6596/873/1/012022

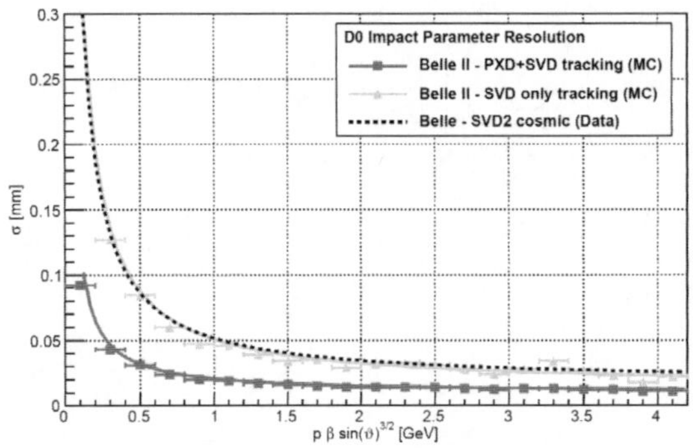

Figure 2. D0 resolution: Belle II PXD+SVD, Belle II SVD only and Belle SVD2 cosmic events for typical Belle/Belle II particle energies.

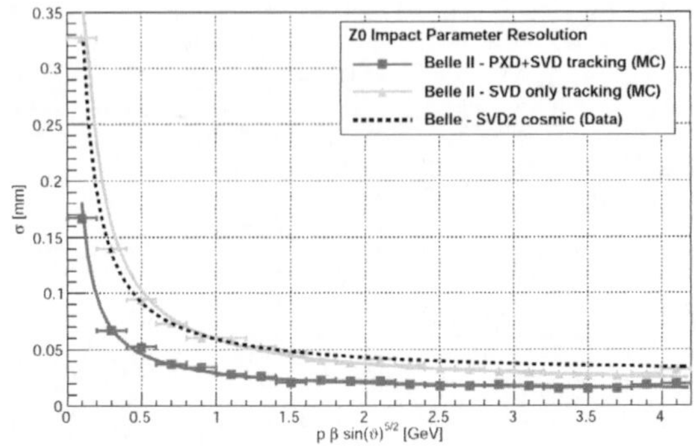

Figure 3. Z0 resolution: Belle II PXD+SVD, Belle II SVD only and Belle SVD2 cosmic events for typical Belle/Belle II particle energies.

The B^0 and $\bar{B^0}$ mesons produced by the decay of the $\Upsilon(4S)$ cannot oscillate independently before they decay, but are locked in a quantum-entangled state. Thus the identification, trough its decay, at any given time t, of one of the B mesons as a B^0 ($\bar{B^0}$) gives the information of the flavor of the other B meson: $\bar{B^0}$ (B^0) at the same time. The difference between matter and antimatter can then be studied using the shape of the distribution of the difference of the decay times of the two B mesons:

$$P(\Delta t, q) = \frac{e^{-|\Delta t|/\tau_B}}{4\tau_B}[1 + q(A_{CP}cos(\Delta m_d \Delta t) + S_{CP}sin(\Delta m_d \Delta t))]. \quad (2)$$

Here q represents the flavor of the B meson, τ_B the lifetime of the B^0, Δm_d the mass difference of the neutral B mass eigenstates and A_{CP} and S_{CP} the parameters related respectively to the direct and mixing induced CP violation.

5. Measurement of the Impact Parameter resolution

The impact parameters: D0 and Z0, defined as the projections of the distance from the point of closest approach to the origin, are a good measure of the overall performance of the tracking system and as such are used to find the optimal tracker configuration: they test the interplay between the geometry setup of individual sensitive detectors and physics effects. Figure 2 and 3 show the almost factor two expected improvement, compared with Belle, of the impact parameter resolution due to the PXD.

6. Vertex finding

The Belle II experiment has deployed two implementations of vertex fit: KFit [9], developed for the Belle experiment, and RAVE [10], the porting to a standalone package of the CMS vertexing libraries. In Belle II both KFit and RAVE are used for kinematic fits and RAVE is used for geometric fits.

6.1. Fully reconstructed particles

Particles decay vertex is usually identified performing kinematic vertex fits. In order to test the Belle II vertexing performance, the decay vertex of $J/\psi \to \mu^+\mu^-$ coming from the $B^0 \to J/\psi K_S$ decay mode is used. The vertex resolution in the direction of the beams is found to be 26 μm.

IOP Conf. Series: Journal of Physics: Conf. Series **873** (2017) 012022 doi:10.1088/1742-6596/873/1/012022

6.2. Tag side vertex fit

To be sensitive to time-dependent CP violating effects, the vertex resolution should be sufficiently good to resolve the oscillations of the neutral B mesons in the decay time distribution. The most important contribution to the Δt resolution comes from the vertex fit of the tag side. For this fit a selection of the tracks originating from the B vertex is required. In the decay tree of a B meson the tracks can be divided in three groups: tracks originating from the B decay, including the ones coming from decay vertices indistinguishable from the B (e.g. μ^+ and μ^- in $B^0 \to [J/\psi \to \mu^+\mu^-]K_S$); tracks originating from D mesons and tracks originating from K_S decays. The tag side vertex fit is performed using the RAVE Adaptive Vertex Fit [10] algorithm, using as input all tracks with at least one hit in the PXD, which are not used for the fully reconstructed B. Tracks originating from K_S decays are not considered. If the fit does not converge, also tracks with no PXD hits are used.

In order to reduce as much as possible the weight of the tracks originating from D mesons, the fit is constrained to a region defined by an ellipsoid around the boost direction, where the B has an higher probability to decay than a possible D meson. This constrain acts as a weight in the final χ^2. In the rare cases of non converging fit, the constrain is applied enlarging its size in the boost direction, becoming virtually equivalent to a cylinder, and the fit is performed again. The tag side vertex fit of correctly reconstructed B mesons returns a bias of $6\,\mu$m and a resolution of $53\,\mu$m, independently of the signal B decay mode. A total efficiency of 96% is obtained. This translates to a Δt resolution of 0.77 ps and a bias of -0.03 ps, providing a superior separation capability compared to Belle [11]. This exceeds the design parameters that request to maintain the Belle vertex separation capability in time dependent analysis in spite of the more difficult experimental environment: reduced boost and higher machine background.

7. Flavor Tagger

The goal of B-flavor tagging is to determine the flavor of a B meson (i.e. whether it contains a b or a $\bar{b}$ quark) at the time of its decay. At the B factories, flavor tagging is needed for most measurements of time-dependent CP asymmetries and B meson mixing. Flavor tagging relies on the fact that a large fraction of B mesons decay to a final state that is flavor-specific and, to good approximation, can only be formed either through the decay of a b quark, or through the decay of a $\bar{b}$ quark. Because of the large number of decay channels, full reconstruction of a sufficiently large number of flavor-specific B_{tag} decays is not feasible. Instead inclusive techniques are employed that make use of different flavor-specific signatures of B decays. For example, in semileptonic decays $B^0 \to D^* l^+ \nu_l$ the charge of the lepton unambiguously identifies the flavor of the decaying B meson. The Belle II flavor tagger proceeds in 3 steps or levels: TrackLevel, EventLevel and CombinerLevel. Each step relies on trained multivariate methods. At the end the dilution factor $D = (1-2w)$, with w the mistag probability, is extracted. The figure of merit for the performance of a tagging algorithm is the effective tagging efficiency $Q = \epsilon_{tag}(1-2w)^2$ where ϵ_{tag} denotes the fraction of events to which a flavor tag can be assigned. The definition of Q is motivated by the fact that CP and mixing asymmetries are diluted by wrong assignments of flavor tags. The statistical uncertainties on such asymmetry measurements generally scale approximately as $1/\sqrt{Q}$. Q can be maximized by grouping events into tagging categories according to their mistag probabilities w. The effective tagging efficiency is given by $Q = \sum_c \epsilon_c (1-2w_c)^2$. Using the same groups used by the Belle experiment, Belle II obtained $Q = 35\%$, sensibly better than the one (29%) obtained by Belle.

8. Measurement of sin(2β) with $b \to c\bar{c}s$ decay modes

These decays to CP eigenstates include B^0 decays to charmonium and a K_S or K_L. Belle has published a combined analysis of decay modes with CP -1 eigenvalue: $B^0 \to J/\psi K_S$, $B^0 \to \psi(2S)K_S$, $B^0 \to \chi_{c1}K_S$, and CP $+1$ eigenvalue: $B^0 \to J/\psi K_L$ [15]. The CP violation

Table 1. Belle II expected sensitivity on the CP parameters of $b \to c\bar{c}s$. Expected statistical, reducible systematic and non reducible systematic uncertainties are shown. An integrated luminosity of 50 ab^{-1} is assumed. Three cases are considered: 'No improvement', where Belle irreducible systematic uncertainties are assumed to not improve in Belle II; 'Vertex improvement', where an improvement of 50% is assumed for the systematic due to the vertex positions; 'Leptonic categories', where the analysis is performed using only the leptonic categories for flavor tagging [12].

	No improvement	Vertex improvement	Leptonic categories
S (50 ab^{-1})			
stat.	0.0027	0.0027	0.0048
syst. reducible	0.0026	0.0026	0.0026
syst. irreducible	0.0070	0.0036	0.0035
A (50 ab^{-1})			
stat.	0.0019	0.0019	0.0033
syst. reducible	0.0014	0.0014	0.0014
syst. irreducible	0.0106	0.0087	0.0035

parameters of the combined analysis:

$$\begin{aligned} S &= 0.667 \pm 0.023(\text{stat}) \pm 0.012(\text{syst}), \\ A &= 0.006 \pm 0.016(\text{stat}) \pm 0.012(\text{syst}), \end{aligned}$$

are still dominated by the statistical errors.

Systematic errors include uncertainties in the wrong-tag fractions, a possible fit bias, uncertainties in the signal fractions, the background Δt distribution, τ_{B^0} and Δm_d. All these depend on control samples and are expected to scale with the increased integrated luminosity in the same way as the statistical errors. The two remaining systematic errors, tag side interference and uncertainty due to the vertex reconstruction algorithm, do not scale with the integrated luminosity. A dedicated study is thus needed.

The simultaneous analysis of CP-odd and CP-even final states gives a combined tag side interference systematic smaller than the systematic in each individual mode. This term has an opposite sign for different CP eigenstates and produces, then, a partial cancellation in the combined result. In the Belle measurement [15] this term was evaluated by comparing the fit results with and without the tag-side interference term. This term is, however, well defined in the case of the $B^0 \to D^{*-}l^+\nu$ decay mode. The systematic error from tag side interference can then be reduced, in the measurement of the CP parameters in $b \to c\bar{c}s$, taking into account the tag-side interference term in the default fitter and assigning as error the tag-side interference term uncertainty. Another way of avoiding the effect of the tag-side interference is to use only leptonic categories for the flavor tagging. However, this would reduce the tagging efficiency by about a factor of three increasing the statistical error.

The new Belle II Pixel Vertex Detector improves the impact parameter determination compared to Belle. Together with new tracking, alignment and vertex algorithms, this translates to an improvement of almost a factor of two in the vertex resolution. A factor of two can then be reasonably assumed for the reduction of the systematic uncertainty due to the vertex reconstruction.

In Table 1 the Belle II expected sensitivity on the CP parameters of the combined $b \to c\bar{c}s$ analysis is presented. An experimental precision better than 1% is expected on S at Belle II.

Table 2. $B^0 \to \phi K_S$: sensitivity estimates for $S_{\phi K^0}$ and $A_{\phi K^0}$ parameters for 1 ab^{-1} and 5 ab^{-1} integrated luminosity [12].

Channel	ε_{reco}	yield	$\sigma(S_{\phi K^0})$	$\sigma(A_{\phi K^0})$
1 ab^{-1} lumi.:				
$\phi(K^+K^-)K_S(\pi^+\pi^-)$	35%	456	0.174	0.123
$\phi(K^+K^-)K_S(\pi^0\pi^0)$	25%	153	0.295	0.215
$\phi(\pi^+\pi^-\pi^0)K_S(\pi^+\pi^-)$	28%	109	0.338	0.252
K_S modes combination			0.135	0.098
$K_S + K_L$ modes combination			0.108	0.079
5 ab^{-1} lumi.:				
$\phi(K^+K^-)K_S(\pi^+\pi^-)$	35%	2280	0.078	0.055
$\phi(K^+K^-)K_S(\pi^0\pi^0)$	25%	765	0.132	0.096
$\phi(\pi^+\pi^-\pi^0)K_S(\pi^+\pi^-)$	28%	545	0.151	0.113
K_S modes combination			0.060	0.044
$K_S + K_L$ modes combination			0.048	0.035

9. Measurement of $\sin(2\beta)$ with $b \to q\bar{q}s$ ($q = u, d, s$) decay modes

The $b \to q\bar{q}s$ ($q = u, d, s$) decay modes are loop-dominated in the Standard Model, and hence may be more sensitive to new-physics effecs than the tree-dominated modes.

Belle II performed a full sensitivity study for the time dependent CP violation parameters in the penguin dominated modes $B^0 \to \phi K^0$ and $\eta' K^0$ [12]. Target integrated luminosity for these studies is taken to be 5 ab^{-1}. With this dataset, the measurements will still be dominated by statistical uncertainties. The statistical sensitivity of the analysis scales well with $1/\sqrt{\mathcal{L}}$ and no significant additional systematic uncertainties compared to those of $b \to c\bar{c}s$ is anticipated, so the method of Table 1 can be also applied to these cases.

9.1. $B^0 \to \phi K_S$

BaBar and Belle extracted the $B_d \to \phi K^0$ CP asymmetry parameters from time-dependent analysis of the $K^+K^-K^0$ final state [13, 14]. Belle II performed a sensitivity study using a quasi-two body approach, taking into account the decay modes: $\phi(K^+K^-)K_S(\pi^+\pi^-)$, $\phi(K^+K^-)K_S(\pi^0\pi^0)$, $\phi(\pi^+\pi^-\pi^0)K_S(\pi^+\pi^-)$, $\phi(K^+K^-)K_L$. The channel $\phi(\pi^+\pi^-\pi^0)K_S(\pi^+\pi^-)$ was never used before due to its very low significance with small integrated luminosity. For the decay modes with a K_L, sensitivities were extrapolated from previous BaBar and Belle results and no simulation was performed.

The sensitivity of the analysis using datasets of 1 and 5 ab^{-1} was estimated with toy Monte Carlo experiments. Table 2 summarizes the sensitivity estimates for the two integrated luminosity scenarios considered. The expected yield of ϕK_L^0 is based on previous BaBar and Belle analyses using the same Δt as resolution estimated in $\phi \to K^+K^-$ for Belle II.

9.2. $B^0 \to \eta' K_S$

The statistical uncertainties expected from this analysis was estimated using a set of toy Monte Carlo experiments. $S_{\eta' K_S} = 0.7$ and $A_{\eta' K_S} = 0$ were used as input parameters. The expected uncertainties for extracted $S_{\eta' K_S}$ and $A_{\eta' K_S}$ are summarized in Table 3. These results are comparable, for similar integrated luminosities, with BaBar and Belle published results.

For the decay mode $\eta'(\eta(2\gamma)\pi^{\pm})K_S^0(2\pi^0)$, the efficiency taken from similar study performed in Belle was used. The efficiency found in this sensitivity study is a factor of two lower, suffering from the poor π^0 reconstruction currently available. Improvements are expected before the

Table 3. $B^0 \to \eta' K_S$: estimated resolutions from toy MC studies for CP-violating S_f and A_f parameters for an integrated luminosity of 1 and 5 ab^{-1} for different channels [12].

Channel	yield	$\sigma(S_f)$	$\sigma(A_f)$	yield	$\sigma(S_f)$	$\sigma(A_f)$
		1 ab^{-1}			5 ab^{-1}	
$\eta(2\gamma)K^0_S(\pi^\pm)$	969	0.13	0.08	4840	0.06	0.04
$\eta(2\gamma)K^0_S(2\pi^0)$	215	0.27	0.17	1070	0.12	0.09
$\eta(3\pi)K^0_S(\pi^\pm)$	283	0.25	0.16	1415	0.11	0.08
$\rho(\pi^\pm)K^0_S(\pi^\pm)$	2100	0.06	0.07	10500	0.04	0.03
$\rho(\pi^\pm)K^0_S(2\pi^0)$	320	0.10	0.17	1600	0.10	0.07
K_S modes	3891	0.065	0.040	19500	0.028	0.021
K_L modes	1546	0.17	0.11	7730	0.08	0.05
$K_S + K_L$ modes	5437	0.060	0.038	27200	0.027	0.020

Belle II data taking. The sensitivity for the decay channel with $\eta'(\rho^0(\pi^+\pi^-)\gamma)$ has been estimated using the expected yield, based on Belle efficiency, and with the resolution found in the analysis for the $\eta'(\eta(2\gamma)\pi^\pm)K^0_S(2\pi^\pm)$ channel. The K_L modes were not analyzed yet, so Table 3 contains the extrapolation from the similar measurements done by Belle to the expected Belle II statistics.

10. BaBar + Belle $B^0 \to D_{CP}h^0$

A measurement of the time-dependent CP asymmetry of $B^0 \to D_{CP}h^0$ decays, has been performed by combining the final data samples collected at the $\Upsilon(4S)$ resonance by the BaBar and Belle experiments. Here h^0 is a π^0, η, or ω meson, and the neutral D meson is reconstructed in the CP eigenstates K^+K^-, $K_S\pi^0$, or $K_S\omega$. Since they contain only tree-level amplitudes, these decays can be used for the measurement of the angle β of the Unitarity Triangle and are not sensitive to most models of physics beyond the Standard Model. However, the measurement of the time-dependent CP violation enables testing of the measurements of $b \to c\bar{c}s$ transitions and provides a Standard Model reference for the beyond the Standard Model searches in the mixing-induced CP violation of $b \to s$ penguin-mediated B meson decays.

The CP asymmetry parameters have been measured excluding the hypothesis of no mixing-induced CP violation at the level of 5.4 standard deviations [16]:

$$\begin{aligned} -\eta_f S &= 0.66 \pm 0.10(\text{stat}) \pm 0.06(\text{syst}), \\ C &= -0.02 \pm 0.07(\text{stat}) \pm 0.03(\text{syst}). \end{aligned}$$

An experimental difficulty in the use of $B^0 \to D_{CP}h^0$ decays arises from low B and D meson branching fractions and low reconstruction efficiencies. However, with the high luminosity data sample expected to be collected by the Belle II experiment, an error on β of about 0.015 is expected.

11. Measurement of cos(2β) with $B^0 \to D^{(*)0}h^0$, $D^0 \to K_S\pi^+\pi^-$

Belle performed a measurement of the CP-violation parameter β obtained in a time-dependent analysis of $B^0 \to D^{(*)0}h^0$ decays followed by $D^0 \to K_S\pi^+\pi^-$ decay. A model-independent measurement is performed using the binned Dalitz plot technique. The three-body state $D^0 \to K_S\pi^+\pi^-$ is not a CP eigenstate. As a consequence, this state provides sensitivity to both $sin2\beta$ and $cos2\beta$. The following results are obtained [17]:

$$\begin{aligned} sin2\beta &= 0.43 \pm 0.27(\text{stat}) \pm 0.08(\text{syst}), \\ cos2\beta &= 1.06 \pm 0.33(\text{stat})^{+0.21}_{-0.15}(\text{syst}), \\ \beta &= 11.7^\circ \pm 7.8^\circ(\text{stat}) \pm 2.1^\circ(\text{syst}). \end{aligned}$$

The value $sin2\beta$ measured in $b \to c\bar{c}s$ transitions determines the value of $cos2\beta$, leading to two possible solutions in the $0^\circ \leq \beta < 180^\circ$ range. This measurement is inconsistent with the negative solution, corresponding to the value of 68.1° at the level of 5.1 standard deviations. It is in agreement with the positive solution, corresponding to the value of 21.9° within 1.3 standard deviations. Thus, this measurement clearly resolves the ambiguity in β inherent in the measurement of $sin2\beta$ using the $b \to c\bar{c}s$ transition.

12. Determination of α

The most precise way of determining the phase α is based on applying isospin symmetry to $B \to \pi\pi$, $\rho\rho$ decays [18].

The decays $B^+ \to \pi^+\pi^0$, $B^0 \to \pi^+\pi^-$, $B^0 \to \pi^0\pi^0$ including their charge-conjugates, and corresponding B decays involving longitudinally polarized ρ mesons, ρ_L , provide sufficient information to determine α. A complete study of these processes resolves discrete ambiguities in α.

Future measurements of $S(\pi^0\pi^0)$ at Belle II using photon conversion in the detector and the π^0 3-body decay [19] will distinguish between multiple solutions.

12.1. $B \to \rho\rho$

The Belle experiment has recently updated the measurement of the branching fraction and the longitudinal polarization fraction of $B^0 \to \rho^+\rho^-$ decays, together with the time-dependent CP violating parameters in decays into longitudinally polarized $\rho^+\rho^-$ pairs, obtaining [20]:

$$\begin{aligned} \mathcal{B}(B^0 \to \rho^+\rho^-) &= (28.3 \pm 1.5(\text{stat}) \pm 1.5(\text{syst})) \times 10^{-6}, \\ f_l &= 0.988 \pm 0.012(\text{stat}) \pm 0.023(\text{syst}), \\ S &= -0.13 \pm 0.15(\text{stat}) \pm 0.05(\text{syst}), \\ A &= 0.00 \pm 0.10(\text{stat}) \pm 0.06(\text{syst}). \end{aligned}$$

Performing an isospin analysis to constrain the Cabibbo-Kobayashi-Maskawa angle α, two solutions are obtained [20] with $\alpha = (93.4 \pm 10.6)^\circ$, being most compatible with other Standard-Model based fits to the data. Extrapolating the present Belle results to the integrated luminosity that will be collected by Belle II (50 ab^{-1}) and assuming no measurement of $S(\rho^0\rho^0)$, an error of about 3 degrees is expected on α [21].

12.2. $B \to \pi\pi$

Belle II is performing a sensitivity study of the time dependent analysis of $B^0 \to \pi^0\pi^0$. If the feasibility will be proved, the isospin analysis will distinguish between multiple solutions and an error of about 3 degrees is expected on α considering and integrated luminosity of 50 ab^{-1} [21].

As the measurement of the B^0 vertex is needed for the time dependent analysis, only events where one of the two π^0s decays to three bodies ($\pi^0 \to e^+e^-\gamma$) or when one of the four photons from the two neutral pions converts in the beam pipe or in the Pixel Vertex Detector can be used.

Figure 4 shows the Δt residuals obtained selecting both decay modes: the resolution is 1.18 ps and the bias is $-0.02\,\mu$m. This resolution is suitable for performing the time dependent analysis of $B^0 \to \pi^0\pi^0$; studies are ongoing for determining if the number of reconstructed events will be sufficient for performing the analysis.

IOP Conf. Series: Journal of Physics: Conf. Series **873** (2017) 012022 doi:10.1088/1742-6596/873/1/012022

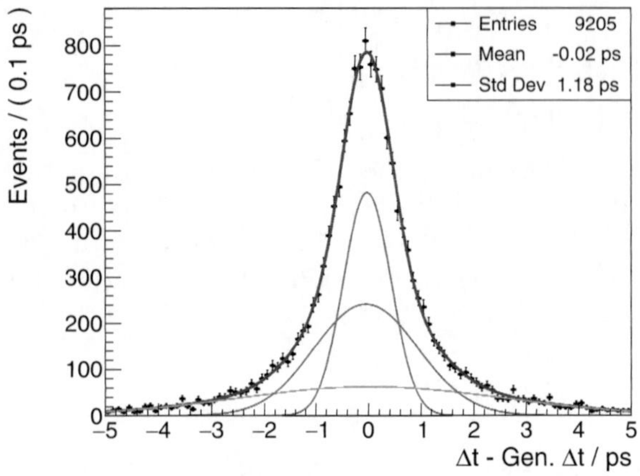

Figure 4. Δt residuals obtained selecting $B^0 \to \pi^0\pi^0$ events when one of the two π^0s decays to three bodies ($\pi^0 \to e^+e^-\gamma$) or when one of the four photons, from the two neutral pions, converts in the beam pipe or in the Pixel Vertex Detector.

13. $B^0 \to K_S\pi^0\gamma$

The radiative loop $b \to s\gamma$ processes have been extensively studied as a probe of New Physics beyond the Standard Model. In the Standard Model $b_L \to s_R\gamma_R$ is m_s/m_b suppressed compared to $b_R \to s_L\gamma_L$, if QCD interactions are switched off. In order to use this as a probe of New Physics, it is important to estimate reliably the QCD corrections to this expectations. At the same time, it is important to perform a precise time dependent analysis, as it gives an indirect measurement of the photon polarization. The $B^0 \to K^0_S\pi^0\gamma$ decay mode is promising because of its large branching fraction, $\mathcal{B}(B^0 \to K^0_S\pi^0\gamma) \sim 3.8 \cdot 10^{-5}$.

The experimental results of Belle [22]:

$$\begin{aligned} S_{K^0_S\pi^0\gamma} &= -0.10 \pm 0.31 \pm 0.07, \\ A_{K^0_S\pi^0\gamma} &= -0.20 \pm 0.20 \pm 0.06 \end{aligned}$$

are still dominated by statistical errors.

Belle II performed a complete sensitivity study of time dependent analysis of $B^0 \to K^0_S\pi^0\gamma$ decay mode. Expected errors on the CP parameters are shown, assuming different integrated luminosities, in Table 4. This mode will be used at Belle II as an important probe of New Physics.

Table 4. $B^0 \to K^0_S\pi^0\gamma$ error estimation on the CP parameters at different integrated luminosities.

Int. Lum. ab^{-1}	$\sigma(S)$	$\sigma(A)$
2	0.150	0.180
10	0.068	0.082
50	0.031	0.037

14. Summary

The experiments conducted with the BaBar and Belle detectors at the B-Factories PEP-II and KEKB have established the three-generation quark-mixing scheme within the Standard Model of particle physics. All present measurements from the B-Factories are in agreement with the

IOP Conf. Series: Journal of Physics: Conf. Series **873** (2017) 012022 doi:10.1088/1742-6596/873/1/012022

Standard Model, though some "tensions" exist. Still, there are important reasons to believe that New Physics beyond the Standard Model must exist. In this paper, the last Belle achievements and Belle II physics prospects for CP violation has been discussed. Belle II at the next generation flavor factory SuperKEKB will be used to search for New Physics in the flavor sector at the precision frontier.

References

[1] J. H. Christenson, J. W. Cronin, V. L. Fitch and R. Turlay *Phys. Rev. Lett.* **13**, (1964) 138.
[2] N. Cabibbo, *Phys. Rev. Lett.* **10** (1963) 531.
[3] M. Kobayashi and T. Maskawa, *Prog. Th. Phys.* **49** (1973) 652.
[4] B. Aubert *et al* (BaBar Collaboration) *Phys. Rev. Lett.*, **89** (2002) 201802.
[5] K. Abe *et al* (Belle Collaboration) *Phys. Rev.* D, **66** (2002) 071102.
[6] M. Bona *et al*, (SuperB) , *INFN/AE-07/2*, (2007).
[7] R.H. Richter *et al*, *Nuclear Instruments and Methods in Physics Research A* **511** (2003) 250.
[8] Z. Dolezal and C. Kiesling and C. Lacasta and H.-G. Moser, *The PXD Whitebook*, (2012).
[9] J. Tanaka, *Belle Note* 194.
[10] W. Waltenberger, W. Mitaroff, F. Moser, B. Pflugfelder and H. V. Riedel, *Journal of Physics: Conference Series* **119** (2008) 032037.
[11] T. Kawasaki *Simulation Study for Measurement of CP violation in B $\to$ charmonium + K_S at KEK B-factory*, Ph.D. Thesis.
[12] The Belle II collaboration and B2TiP theory community, *The Physics Prospects for Belle II*, to be published.
[13] J. P. Lees *et al* (BABAR Collaboration) *Phys. Rev.* D **85** (2012) 112010.
[14] Y. Nakahama *et al* (The Belle Collaboration) *Phys. Rev.* D **82** (2010) 073011.
[15] I. Adachi *et al* (The Belle Collaboration) *Phys. Rev. Lett.* **108** (2012), 171802.
[16] A. Abdesselam *et al* (BaBar Collaboration, Belle Collaboration) *Phys. Rev. Lett.* **115**(2015), 121604.
[17] V. Vorobyev *et al* (Belle Collaboration) *Phys. Rev.* D **94** (2016), 052004.
[18] M. Gronau and D. London, *Phys. Rev. Lett.* **65** (1990) 33813384.
[19] H. Ishino, M. Hazumi, M. Nakao, and T. Yoshikawa, *Preprint hep-ex/0703039 [HEP-EX]*.
[20] P. Vanhoefer *et al* (Belle Collaboration) *Phys. Rev.* D **93**, (2016) 032010
[21] P. Vanhoefer, presentation at the first B2TiP workshop, `https://kds.kek.jp/indico/event/15873/session/14/contribution/32/material/slides/0.pdf`.
[22] Y. Ushiroda *et al* (Belle Collaboration) *Phys. Rev.* **D** 74, (2006) 111104(R).

IOP Conf. Series: Journal of Physics: Conf. Series **873** (2017) 012023 doi:10.1088/1742-6596/873/1/012023

BABAR results on T and CPT symmetry in B^0-$\overline{B}^0$ mixing and in $B^0 \to c\overline{c}K^0$ decays

Klaus R. Schubert

Institut für Kernphysik, Johannes Gutenberg-Universität, Mainz, Germany

BABAR Collaboration

E-mail: `klaus.schubert@tu-dresden.de`

Abstract. The presentation reviews the results which have been obtained with the *BABAR* detector at the e^+e^- storage ring PEP-II at SLAC between 2004 and 2016 on T and CPT symmetry in B^0-$\overline{B}^0$ mixing and in the decay amplitudes of $B^0 \to c\overline{c}K^0$ decays. All results are based on the measured time dependence of two-decay events from entangled B^0-$\overline{B}^0$ pairs produced in decays of $\Upsilon(4S)$ mesons. Neither T nor CPT violation have been observed in B^0-$\overline{B}^0$ mixing, also no CPT violation in the $B^0 \to c\overline{c}K^0$ decay amplitudes.

1. Introduction

Solving the puzzle of two different K^+ mesons decaying into final states with opposite parity [1], Lee and Yang [2] proposed in 1956 that P symmetry is broken in weak interactions since there were no convincing tests of P conservation in weak processes. Soon later, two experimental groups concurrently proved that P is not only violated in K^+ decays, but also in the decay chain $\pi^+ \to \mu^+ \to e^+$ [3] and in β decays of ^{60}Co [4]. CP violation was discovered in 1964 in the decays $K^0 \to \pi^+\pi^-$ at late decay times [5]. In the following years, many authors asked if CPT could also be violated in Nature despite its validity in Lorentz-invariant QFT. CP violation implies that T or CPT or both are also violated. Bell and Steinberger [6] proposed in 1965 separate tests of T and CPT using a unitarity relation with the sum of CP violations in all K^0 decay modes. This relation could not yield numerical results before all essential inputs had been measured. The last essential input, the phase ϕ_{00} between the amplitudes for the decays of K^0_S and K^0_L into $\pi^0\pi^0$, was first determined in 1970 [7]. With this entry, the Bell-Steinberger unitarity analysis resulted in $\mathrm{Re}(\epsilon) = (1.7 \pm 0.3) \times 10^{-3}$ and $\mathrm{Im}(\delta) = (-0.3 \pm 0.4) \times 10^{-3}$ [8] immediately after the ϕ_{00} measurement. $\mathrm{Re}(\epsilon)$ describes T violation in K^0-$\overline{K}^0$ mixing, here established with $\sim 5\sigma$, and δ describes CPT violation therein, here compatible with zero. The most recent update of the analysis [9, 10] gives $\mathrm{Re}(\epsilon) = (161.1 \pm 0.5) \times 10^{-5}$ and $\mathrm{Im}(\delta) = (-0.7 \pm 1.4) \times 10^{-5}$.

In the 10 times heavier B^0-$\overline{B}^0$ system, large CP and T violation was observed in 2001 in $B^0 \to c\overline{c}K^0$ decays [11, 12, 13], up to very recently without a determination of CPT-violating parameters. No CP violation has been observed in B^0-$\overline{B}^0$ mixing so far. The present presentation discusses the contributions of the *BABAR* experiment to the determination of T- and CPT-violating parameters in mixing and decays of the $B^0(b\overline{d})$-$\overline{B}^0(\overline{b}d)$ system.

IOP Conf. Series: Journal of Physics: Conf. Series **873** (2017) 012023 doi:10.1088/1742-6596/873/1/012023

2. CPT and T symmetries in B^0-$\overline{B}^0$ mixing

Since mixing in the two-state system $\Psi = \psi_1 B^0 + \psi_2 \overline{B}^0$ proceeds on a similar time scale as weak decays, it is well described by the linear evolution equation

$$\mathrm{i}\,\frac{\partial}{\partial t}\begin{pmatrix}\psi_1\\ \psi_2\end{pmatrix} = \left[\begin{pmatrix} m_{11} & m_{12}\\ m_{12}^* & m_{22}\end{pmatrix} - \frac{\mathrm{i}}{2}\begin{pmatrix}\Gamma_{11} & \Gamma_{12}\\ \Gamma_{12}^* & \Gamma_{22}\end{pmatrix}\right]\begin{pmatrix}\psi_1\\ \psi_2\end{pmatrix}, \tag{1}$$

with 7 real parameters m_{11}, m_{22}, Γ_{11}, Γ_{22}, $|m_{12}|$, $|\Gamma_{12}|$, and $\phi(\Gamma_{12}/m_{12})$. As long as mixing is "slow", this description is independent of the dynamics which may include New Physics in addition to the Standard weak interaction; it is just the two-dimensional generalization of

$$\mathrm{i}\,\frac{\mathrm{d}\psi}{\mathrm{d}t} = (m - \mathrm{i}\Gamma/2)\psi \tag{2}$$

for a slowly decaying particle with wave-function $\psi(t)$. Two solutions of Eq. 1, B_H^0(heavy) and B_L^0(light), have an exponential decay law. In lowest order of z and $1-|q/p|$, as used throughout this presentation, they are given by

$$\begin{aligned} B_H^0(t) &= \mathrm{e}^{-\Gamma_H t/2 - \mathrm{i} m_H t}\left[p(1+\mathsf{z}/2)\,B^0 - q(1-\mathsf{z}/2)\,\overline{B}^0\right]/\sqrt{2},\\ B_L^0(t) &= \mathrm{e}^{-\Gamma_L t/2 - \mathrm{i} m_L t}\left[p(1-\mathsf{z}/2)\,B^0 + q(1+\mathsf{z}/2)\,\overline{B}^0\right]/\sqrt{2}, \end{aligned} \tag{3}$$

with 7 real observables m_H, $\Delta m = m_H - m_L$, Γ_H, $\Delta\Gamma = \Gamma_H - \Gamma_L$, $|q/p|$, Re(z), and Im(z). The 7 observables follow from the 7 parameters, e.g.,

$$\left|\frac{q}{p}\right| = 1 - \frac{2\,\mathrm{Im}(\Gamma_{12}/m_{12})}{4 + |\Gamma_{12}/m_{12}|^2}\,, \quad \mathsf{z} = \frac{(m_{11}-m_{22}) - \mathrm{i}(\Gamma_{11}-\Gamma_{22})/2}{\Delta m - \mathrm{i}\Delta\Gamma/2}\,. \tag{4}$$

In the $K^0 - \overline{K}^0$ system with the sign conventions $\Delta m = m(K_L) - m(K_S)$ and $\Delta\Gamma = \Gamma(K_L) - \Gamma(K_S) < 0$, the traditionally used observables are $\mathrm{Re}(\epsilon) = (1-|q/p|)/2$ and $\delta = -\mathsf{z}/2$.

From Eqs. 3, we obtain the transition rates as function of the evolution time t. With $\Gamma = (\Gamma_H + \Gamma_L)/2$ and $|\Delta\Gamma| \ll \Gamma$, they are

$$\begin{aligned} R(B^0 \to B^0) &= \mathrm{e}^{-\Gamma t}[1 + \cos(\Delta m t) - \mathrm{Re}(\mathsf{z})\Delta\Gamma t + 2\,\mathrm{Im}(\mathsf{z})\sin(\Delta m t)]/2\,,\\ R(\overline{B}^0 \to \overline{B}^0) &= \mathrm{e}^{-\Gamma t}[1 + \cos(\Delta m t) + \mathrm{Re}(\mathsf{z})\Delta\Gamma t - 2\,\mathrm{Im}(\mathsf{z})\sin(\Delta m t)]/2\,, \end{aligned} \tag{5}$$

$$\begin{aligned} R(B^0 \to \overline{B}^0) &= \mathrm{e}^{-\Gamma t}[1 - \cos(\Delta m t)]\cdot|q/p|^2/2\,,\\ R(\overline{B}^0 \to B^0) &= \mathrm{e}^{-\Gamma t}[1 - \cos(\Delta m t)]\cdot|p/q|^2/2\,. \end{aligned} \tag{6}$$

The rates in Eq. 5 depend only on z, they are not sensitive to T symmetry, and CPT symmetry requires equality of the two rates, i.e. $\mathsf{z} = 0$. The rates in Eq. 6 depend only on $|q/p|$, they are not sensitive to CPT symmetry, and T symmetry requires equal rates, i.e. $|q/p| = 1$. Since $\Delta\Gamma$ is unknown, the first rates determine up to now only Im(z), not Re(z). Transitions into states decaying into CP eigenstates like $c\bar{c}K_S$, $c\bar{c}K_L$ are also sensitive to Re(z) as shown later.

3. CPT and T symmetries in B^0 decays

If the decay amplitudes

$$A = \langle f|\mathcal{H}|B^0\rangle \text{ and } \overline{A} = \langle \overline{f}|\mathcal{H}|\overline{B}^0\rangle \tag{7}$$

IOP Conf. Series: Journal of Physics: Conf. Series **873** (2017) 012023 doi:10.1088/1742-6596/873/1/012023

have a singe weak phase, CPT symmetry requires [14]

$$|\overline{A}/A| = 1\,. \tag{8}$$

If the amplitudes are sums of amplitudes with different weak phases, we can have $|\overline{A}/A| \neq 1$ inspite of CPT symmetry ("direct CP violation" and "direct T violation") or $|\overline{A}/A| = 1$ inspite of CPT violation.

Tests of T symmetry require the interference of at least two contributing amplitudes to the same final state. Two early test examples are radiative decays of excited atoms and neutron decays. "Forbidden" atomic transitions proceed in E2 and M1 modes when selection rules exclude E1. In 1951, Lloyd [15] has shown that T symmetry requires $\mathrm{Im}(A_{M1}/A_{E2}) = 0$ for the observable ratio of the two transition amplitudes. In 1957, Jackson *et al* [16] have shown that the beta decay of polarized neutrons allows to measure the phase between the the axial-vector and vector coupling constants G_A and G_V. T symmetry requires this phase to be 0 or 180°, i. e. $\mathrm{Im}(G_A/G_V) = 0$.

The same symmetry principle applies to B^0 decays into CP eigenstates like $c\bar{c}K^0_S$ with eigenvalues $CP = -1$ and $c\bar{c}K^0_L$ with $CP = +1$, where $c\bar{c} = J/\psi,\ \psi'$ and other charmonium states. With

$$A = \langle c\bar{c}K^0|\mathcal{H}|B^0\rangle \text{ and } \overline{A} = \langle c\bar{c}\overline{K}^0|\mathcal{H}|\overline{B}^0\rangle\,, \tag{9}$$

the well-obeyed $\Delta S = \Delta b$ rule

$$\langle c\bar{c}\overline{K}^0|\mathcal{H}|B^0\rangle = \langle c\bar{c}K^0|\mathcal{H}|\overline{B}^0\rangle = 0\,, \tag{10}$$

and negligible CP violation in K^0-$\overline{K}^0$ mixing, we have

$$\begin{aligned} A_S = \langle c\bar{c}K^0_S|\mathcal{H}|B^0\rangle &= A/\sqrt{2}\,,\\ A_L = \langle c\bar{c}K^0_L|\mathcal{H}|B^0\rangle &= A/\sqrt{2}\,,\\ \overline{A}_S = \langle c\bar{c}K^0_S|\mathcal{H}|\overline{B}^0\rangle &= \overline{A}/\sqrt{2}\,,\\ \overline{A}_L = \langle c\bar{c}K^0_L|\mathcal{H}|\overline{B}^0\rangle &= -\overline{A}/\sqrt{2}\,. \end{aligned} \tag{11}$$

Because of mixing, an initial B^0 or $\overline{B}^0$ state decays into final states $c\bar{c}K^0_S$ or $c\bar{c}K^0_L$ with interfering amplitudes from both B^0 and $\overline{B}^0$ at the time of decay. T symmetry, therefore, requires a phase relation between A and $\overline{A}$. Since both states B^0 and $\overline{B}^0$ can be arbitrarily rephased, the phase of $\overline{A}/A$ is unobservable. However, the phase of the parameter

$$\lambda = \frac{q\overline{A}}{pA} \tag{12}$$

with q and p from Eqs. 3 is an observable, and T symmetry requires [17, 13, 18]

$$\mathrm{Im}(\lambda) = 0\,. \tag{13}$$

The observed CP and T violation in time-dependent B^0 decays with $\mathrm{Im}(\lambda) \neq 0$ is often called CP violation in the interplay of decay and mixing since it is not a property of the decay dynamics alone. It requires the observable phase of $(q/p) \times \overline{A}/A$ to be different from 0 and 180°.

4. Earlier *BABAR* results on T and CPT symmetry tests in B^0-$\overline{B}^0$ mixing

Neutral B mesons in the *BABAR* experiment are produced in the entangled two-particle state $(B^0\overline{B}^0 - \overline{B}^0 B^0)/\sqrt{2}$ from $\Upsilon(4S)$ decays. With a flavor-specific first decay into $\ell^- X$ ($\ell^+ X$) at time t_1, the remaining single-particle state is a B^0 ($\overline{B}^0$) at this time. The rates for flavor-specific decays into $\ell^- X$ ($\ell^+ X$) at time $t_2 = t_1 + t$ are given by Eqs. 6. The t-dependent difference of $\ell^-\ell^-$ and $\ell^+\ell^+$ pair rates determines, therefore, the T-symmetry parameter $|q/p|$. In 2006, *BABAR* analysed like-sign dilepton pairs from about 230×10^6 $\Upsilon(4S)$ decays [19], corresponding to about half of its finally collected data, resulting in

$$|q/p|\,(BABAR,\ \text{dileptons}) = 1 - (0.8 \pm 1.7 \pm 1.9) \times 10^{-3}\,. \tag{14}$$

In 2013, with about 425×10^6 $\Upsilon(4S)$ decays [20], *BABAR* found

$$|q/p|\,(BABAR,\ D^{*\pm}X^0\ell^{\pm}\ \text{and}\ K^{\pm}X) = 1 - (0.3 \pm 0.8 \pm 1.8) \times 10^{-3}\,, \tag{15}$$

where the flavor of one B-meson is identified by a charged lepton and the slow charged pion from a $D^{*\pm}$ decay, and the flavor of the other B by using the angular distribution of a charged kaon from sequential $B \to DX \to K^{\pm}X'$ decays.

These two results, together with results from Belle, CLEO, LHCb and D0, lead to the present world average [21] of

$$|q/p|\,(\text{HFAG 2016}) = 1 + (1,0 \pm 0.8) \times 10^{-3}\,, \tag{16}$$

with no observed T violation in B^0-$\overline{B}^0$ mixing.

In the same 2006 analysis determining $|q/p|$, *BABAR* [19] uses opposite-sign dilepton events for determining Im(z). These events have the rate $R_{-+} = R(B^0 \to B^0)$ in Eq. 5 when the ℓ^- decay occurs first, and $R_{+-} = R(\overline{B}^0 \to \overline{B}^0)$ when the ℓ^+ is first. The two rates $R_{-+}(t)$ with $t = t_+ - t_-$ and $R_{+-}(t)$ with $t = t_- - t_+$ have the property $R_{+-}(t) = R_{-+}(-t)$, required by the two-decay-time formula [22] as a consequence of entanglement independent of CPT symmetry. *BABAR*'s 2006 result [19] from opposite-sign dilepton events is

$$\text{Im(z)} = (-14 \pm 7 \pm 3) \times 10^{-3}\,. \tag{17}$$

As long as $\Delta\Gamma$ in B^0-$\overline{B}^0$ mixing remains unknown, opposite-sign dilepton events do not determine Re(z), see Eqs. 5. For completeness, I include in this Section a 2004 *BABAR* measurement of Re(z) [23] using decay pairs $\ell^{\pm}X$ from one neutral B and $c\bar{c}K$ from the other B with a data set of about about 90×10^6 $\Upsilon(4S)$ decays. The time-dependent rates for these decay pairs are derived in Section 6 where the same analysis is discussed with the final *BABAR* data set of about 470×10^6 $\Upsilon(4S)$ decays [24]. The 2004 result was

$$\text{Re(z)} = (-19 \pm 48 \pm 47) \times 10^{-3}\,. \tag{18}$$

5. The *BABAR* 2012 result on T violation in time-dependent $B^0 \to c\bar{c}K^0$ decays

The final result on CP violation in $B^0 \to c\bar{c}K^0$ decays has been published in 2009 [25] using about 470×10^6 $\Upsilon(4S)$ decays. Assuming CPT symmetry, the CP- and T-violating parameter Im(λ) in the definition of Eqs. 9 and 12 has been determined from a fit to the four time-dependent rates of the decay pairs ℓ^+X or ℓ^-X, $c\bar{c}K^0_S$ or $c\bar{c}K^0_L$, with the result

$$\text{Im}(\lambda) = 0.687 \pm 0.028 \pm 0.012\,. \tag{19}$$

The flavor-specific decays ℓ^+X and ℓ^-X include decays with high-momentum charged particles, charged kaons and charged pions from D^* decays in addition to high-momentum electrons and muons. The charmonium states $c\bar{c}$ include J/ψ, $\psi(2S)$, η_c and χ_{c1} in combination with K^0_S and only J/ψ in combination with K^0_L and $K^*(892)^0$. The K^* events with decays into $K^0_S\pi^0$ contribute to both the K^0_S and K^0_L samples with the eigenvalues $CP = -1$ and $CP = +1$ respectively; decay angular distributions are used for the separation of the two contributions.

Using essentially the same event selection, omitting $J/\psi\, K^*$ and $\eta_c\, K^0_S$ decays, and following a suggestion of J. Bernabéu [26, 27], *BABAR* [28] used the four time-dependent rates of the decay pairs ($\ell^- X$, $c\bar{c}K^0_L$), ($\ell^+ X$, $c\bar{c}K^0_L$), ($\ell^- X$, $c\bar{c}K^0_S$) and ($\ell^+ X$, $c\bar{c}K^0_S$) for a demonstration of "genuine" T violation in 2012. In contrast to testing a consequence of T-symmetry in the dynamics of a process, a genuine T-symmetry test [27] measures the detailed balance of two processes that are related by the exchange of initial and final state, i. e. the equality of the transition rates for $i \to f$ and $f \to i$. An example for a genuine T test in the strong interaction is the comparison of the cross sections for the nuclear reactions $^{24}Mg + \alpha \to {}^{27}Al + p$ and $^{27}Al + p \to {}^{24}Mg + \alpha$ by von Witsch *et al* in 1968 [29].

For the four decay pairs $\ell^\pm X$, $c\bar{c}K^0_{S.L}$, there are four independent rates, but for the clarity of demonstration, *BABAR* determined eight rates separating into $t(c\bar{c}K) - t(\ell X) > 0$ $t(c\bar{c}K) - t(\ell X) < 0$. The separated rates are related by the two-decay-time formula [22] following from entanglement of the B^0-$\overline{B}{}^0$ pair independent of any discrete symmetry, as already discussed for dilepton events in the text above Eq. 17. Table 1 shows the numbering of the eight chosen rates for the following discussion.

Table 1. Decay pairs in Ref. [28] for the measurement of the decay-time dependences with the coefficients $C_1 \cdots C_8$ and $S_1 \cdots S_8$. Rows 3 gives the initial state of the second B as prepared by the first decay, row 4 the final state as defined by the second decay of the decay pair.

i	1	2	3	4	5	6	7	8
1st decay	$\ell^- X$	$\ell^+ X$	$\ell^- X$	$\ell^+ X$	$c\bar{c}K_L$	$c\bar{c}K_L$	$c\bar{c}K_S$	$c\bar{c}K_S$
2nd decay	$c\bar{c}K_L$	$c\bar{c}K_L$	$c\bar{c}K_S$	$c\bar{c}K_S$	$\ell^- X$	$\ell^+ X$	$\ell^- X$	$\ell^+ X$
initial state	B^0	$\overline{B}{}^0$	B^0	$\overline{B}{}^0$	B_-	B_-	B_+	B_+
final state	B_+	B_+	B_-	B_-	$\overline{B}{}^0$	B^0	$\overline{B}{}^0$	B^0

Without assuming CPT symmetry in mixing, but assuming CPT symmetry in the amplitudes A for the decay $B^0 \to c\bar{c}\, K^0$ and $\overline{A}$ for $\overline{B}{}^0 \to \overline{K}{}^0$, i. e. $|\overline{A}/A| = 1$, the first decay of a B^0-$\overline{B}{}^0$ pair into $c\bar{c}K^0_S$ prepares a remaining single-B-meson state with $c\bar{c}K$ decays into only $c\bar{c}K^0_L$ at $t = 0$, called B_+. If the first decay is $c\bar{c}K^0_L$, the prepared state B_- decays into $c\bar{c}K$ states with only $c\bar{c}K^0_S$ at $t = 0$. In the same way, a first decay into $\ell^+ X$ ($\ell^- X$) prepares a remaining B in the state $\overline{B}{}^0$ (B^0). With $t = t(c\bar{c}K) - t(\ell X)$ for the first four and $t = t(\ell X) - t(c\bar{c}K)$ for the second four rates in Table 1, all eight measured rates are parametrized by

$$R_i(t) = N_i\, \mathrm{e}^{-\Gamma t}\, (1 + C_i\, \cos \Delta m\, t + S_i\, \sin \Delta m\, t)\,, \tag{20}$$

and the 16 coefficients $C_1 \cdots S_8$ are determined by eight fits including the resolution between the measured times $\tilde{t}$ and the true times t. The 3rd and 4th row in Table 1 give the initial state of the second B as prepared by the first decay and the final state as defined by the second decay of the decay pair. Detailed balance between the time-dependent transitions $B^0 \to B_+$ and $B_+ \to B^0$ requires $R_8(t) = R_1(t)$, i. e. $C_8 = C_1$ and $S_8 = S_1$, and the same for the other three exchanges of initial and final state. *BABAR* [28] finds that all C_i are compatible with zero, but the differences $S_8 - S_1$, $S_7 - S_2$, $S_6 - S_3$ and $S_5 - S_4$ are all different from zero with a combined significance of 14 standard deviations, strongly demonstrating T-symmetry violation in B^0-meson decays.

In the same publication [28], *BABAR* also demonstrates CPT symmetry by measuring the difference of the time-dependent transition rates for $B^0 \to B_+$ and $B_+ \to \overline{B}{}^0$ and the three other differences between CPT-related rates. All four differences are compatible with zero.

Since CPT asymmetry in the decay amplitudes has been assumed, this is a CPT test in B^0-$\overline{B}^0$ mixing. It is a qualitative test only; no values for Re(z) and Im(z) have been determined. This determination was performed by *BABAR* in 2016 [24] and is discussed in the next Section.

6. The *BABAR* 2016 results on CPT in B^0-$\overline{B}^0$ mixing and $B^0 \to c\bar{c}K^0$ decays

The eight time-dependent rates for the decays of entangled B^0-$\overline{B}^0$ pairs in Table 1 are sensitive to $|\lambda|$, Im(λ), Re(z) and Re(z), where z is defined in Eqs. 4 and λ in Eq. 12 with q/p from Eqs. 4 and A and $\overline{A}$ from Eqs. 9. Assuming

(i) $\Delta\Gamma = \Gamma_H - \Gamma_L = 0$,
(ii) absence of decays $B^0 \to c\bar{c}\overline{K}^0$ and $\overline{B}^0 \to c\bar{c}K^0$, and
(iii) negligible CP violation in K^0-$\overline{K}^0$ mixing,

the eight rates are of the form in Eq. 20 with $t = t(c\bar{c}K) - t(\ell X)$ for the first four and $t = t(\ell X) - t(c\bar{c}K)$ for the second four. In lowest order of the small quantities z, $|q/p| - 1$, and $|\lambda| - 1$, the coefficients $C_1 \cdots C_4$ and $S_1 \cdots S_4$ are given by

$$\begin{aligned}
C_1(B^0 \to c\bar{c}K_L) &= +(1-|\lambda|) - \mathrm{Re}(\lambda)\,\mathrm{Re}(\mathsf{z}) - \mathrm{Im}(\lambda)\,\mathrm{Im}(\mathsf{z})\,,\\
C_2(\overline{B}^0 \to c\bar{c}K_L) &= -(1-|\lambda|) + \mathrm{Re}(\lambda)\,\mathrm{Re}(\mathsf{z}) - \mathrm{Im}(\lambda)\,\mathrm{Im}(\mathsf{z})\,,\\
C_3(B^0 \to c\bar{c}K_S) &= +(1-|\lambda|) + \mathrm{Re}(\lambda)\,\mathrm{Re}(\mathsf{z}) + \mathrm{Im}(\lambda)\,\mathrm{Im}(\mathsf{z})\,,\\
C_4(\overline{B}^0 \to c\bar{c}K_S) &= -(1-|\lambda|) - \mathrm{Re}(\lambda)\,\mathrm{Re}(\mathsf{z}) + \mathrm{Im}(\lambda)\,\mathrm{Im}(\mathsf{z})\,,\\
S_1 &= +\mathrm{Im}(\lambda)/|\lambda| - \mathrm{Re}(\mathsf{z})\mathrm{Re}(\lambda)\mathrm{Im}(\lambda) + \mathrm{Im}(\mathsf{z})[\mathrm{Re}(\lambda)]^2\,,\\
S_2 &= -\mathrm{Im}(\lambda)/|\lambda| - \mathrm{Re}(\mathsf{z})\mathrm{Re}(\lambda)\mathrm{Im}(\lambda) - \mathrm{Im}(\mathsf{z})[\mathrm{Re}(\lambda)]^2\,,\\
S_3 &= -\mathrm{Im}(\lambda)/|\lambda| - \mathrm{Re}(\mathsf{z})\mathrm{Re}(\lambda)\mathrm{Im}(\lambda) + \mathrm{Im}(\mathsf{z})[\mathrm{Re}(\lambda)]^2\,,\\
S_4 &= +\mathrm{Im}(\lambda)/|\lambda| - \mathrm{Re}(\mathsf{z})\mathrm{Re}(\lambda)\mathrm{Im}(\lambda) - \mathrm{Im}(\mathsf{z})[\mathrm{Re}(\lambda)]^2\,.
\end{aligned} \tag{21}$$

Because of entanglement of the B^0-$\overline{B}^0$ pair and the two-decay-time formula, the coefficients for the decay pairs with $i = 5 \cdots 8$ follow Eq. 21 with $C_i = C_{i-4}$ and $S_i = -S_{i-4}$.

The 16 coefficients $C_1 \cdots S_8$ have been determined in 2012 in the T-violation paper of *BABAR* [28] together with their statistical and systematic covariance matrices. In the 2016 paper [24], these 16 results based on *BABAR*'s final data set with about 470×10^6 $\Upsilon(4S)$ decays are used to determine the four parameters $|\lambda|$, Im(λ), Re(z) and Re(z) without repeating reconstruction and selection of the $\ell^\pm X$, $c\bar{c}K_{S.L}$ decay pairs. The four parameters are obtained in an iterative linearized χ^2 fit of the parameters to the expressions in Eqs. 21. Since the expressions contain Re(z) only in products with Re(λ) and the fit parameters Im(λ) and $|\lambda|$ do not fix the sign of Re(λ), additional information is used [30] for taking Re(λ) negative. The χ^2 fit converges already in the second step and gives the results

$$\begin{aligned}
\mathrm{Im}(\lambda) = 0.689 \pm 0.034 \pm 0.019\ , &\quad |\lambda| = 0.999 \pm 0.023 \pm 0.017\\
\mathrm{Im}(\mathsf{z}) = 0.010 \pm 0.030 \pm 0.013\ , &\quad \mathrm{Re}(\mathsf{z}) = -0.065 \pm 0.028 \pm 0.014\,,
\end{aligned} \tag{22}$$

where the first uncertainties are statistical and the second uncertainties are systematic, as in all results throughout this report. The final results are independent of the assumption $\Delta\Gamma = 0$ as controlled by a "toy Monte Carlo" simulation varying $\Delta\Gamma$ within one standard deviation of the present world average [21]. Inserting the world average for $|q/p|$ in Eq. 16 into the definition of λ, we obtain

$$|\overline{A}/A| = 0.999 \pm 0.023 \pm 0.017\,. \tag{23}$$

7. Final remarks

The uncertainty on the T-violation result Im(λ) in Eqs. 22 is slightly larger than for *BABAR*'s final CP-violation result in Eq. 19, since that one is obtained from a slightly wider event selection and from assuming CPT symmetry. The three CPT-sensitive results Im(z), Re(z) and $|\overline{A}/A|$ are all in agreement with no CPT violation, assuming a single weak phase of the amplitude A. The Im(z) result has a larger uncertainty than *BABAR*'s dilepton result in Eq. 17 since that one is based on a larger number of decay pairs. But with the small and unknown value for $\Delta\Gamma$, dileptons are not sensitive to Re(z). The parameter Re(z) has been determined by *BABAR* and by Belle with comparable uncertainties; the 2012 result of Belle with 535×10^6 $\Upsilon(4S)$ events [31] is Re(z) $= 0.019 \pm 0.037 \pm 0.033$. To my knowledge, there are no additional measurements of Re(z) in the B^0-$\overline{B}^0$ system. LHCb gives only a result for the B_s-$\overline{B}_s$ system [32], Re(z, B_s) $= -0.022 \pm 0.033 \pm 0.005$, together with Im(z, B_s) $= 0.004 \pm 0.011 \pm 0.002$.

Acknowledgments

I would like to thank the organizers, especially J Bernabéu and A. Di Domenico, for inviting me to present this review at DISCRETE2016. I also thank all members of the *BABAR* Collaboration for numerous discussions on tests of discrete symmetries during the last 20 years. In addition, I especially thank T. Ruf and H.-J. Gerber (CPLEAR Collaboration) for extended discussions on T-symmetry violation and its parametrizations.

References

[1] Dalitz R H 1953 *Phil. Mag.* **44**, 1068
[2] Lee T D and Yang C N 1956 *Phys. Rev.* **104**, 254
[3] Garwin R L, Lederman L M and Weinrich M 1957 *Phys. Rev.* **105**, 1415
[4] Wu C S *et al* 1957 *Phys. Rev.* **105**, 1413
[5] Christenson J H, Cronin J W, Fitch V L and Turlay R 1964 *Phys. Rev. Lett.* **13**, 138
[6] Bell J S and Steinberger J 1966 in Walsh T R *et al* eds., *Oxford International Conference on Elementary Particles* Didcot Chilton, p. 195
[7] Chollet J C *et al* 1970 *Phys. Lett.* B **31**, 658
[8] Schubert K R *et al* 1970 *Phys. Lett.* B **31**, 662
[9] Antonelli M and D'Ambrosio G 2013 *CPT* invariance tests in neutral Kaon decay, review in ref. [10]
[10] Patrignani C *et al* (Particle Data Group) 2016 *Chin. Phys.* C, **40**, 100001
[11] Aubert B *et al* (*BABAR* Collaboration) 2001 *Phys. Rev. Lett.* **87**, 091801
[12] Abe K *et al* (Belle Collaboration) 2001 *Phys. Rev. Lett.* **87**, 091802
[13] Fidecaro M, Gerber H-J and Ruf T 2015 *Eur. Phys. J.* C **75**:73
[14] Lee T D, Oehme R and Yang C N 1957 *Phys. Rev.* **106**, 340
[15] Lloyd S P 1951 *Phys. Rev.* **81**, 161
[16] Jackson J D, Treiman S B and Wyld H W Jr. 1957 *Phys. Rev.* **106**, 517
[17] Enz C P and Lewis R R 1965 *Helv. Phys. Acta* **38**, 860
[18] Applebaum E *et al* 2014 *Phys. Rev.* D **89**,076011
[19] Aubert B *et al* (*BABAR* Collaboration) 2006 *Phys. Rev. Lett.* **96**, 251802
[20] Lees J P *et al* (*BABAR* Collaboration) 2013 *Phys. Rev. Lett.* **111**, 101802
[21] Amhis Y *et al* (Heavy Flavor Averaging Group) 2016 arXiv:1612.07233v1 [hep-ex]
[22] Lipkin H J 1968 *Phys. Rev.* **176**, 1715
[23] Aubert B *et al* (*BABAR* Collaboration) 2004 *Phys. Rev.* D **70**, 012007, inserting Re(λ) $= -0.73$
[24] Lees J P *et al* (*BABAR* Collaboration) 2016 *Phys. Rev.* D **94**, 011101
[25] Aubert B *et al* (*BABAR* Collaboration) 2009 *Phys. Rev.* D **79**, 072009
[26] Bañuls M C and Bernabéu J 1999 *Phys. Lett.* B **464**, 117
[27] Bernabéu J 2016 these Proceedings
[28] Lees J P *et al* (*BABAR* Collaboration) 2012 *Phys. Rev. Lett.* **109**, 211801
[29] von Witsch W, Richter A and von Brentano P 1968 *Phys. Rev.* **169**, 923
[30] See Ref. [24] and citations 19-22 therein.
[31] Higuchi T *et al* (Belle Collaboration) 2012 *Phys. Rev.* D **85**, 071105
[32] Aaij R *et al* (LHCb Collaboration) 2016 *Phys. Rev. Lett.* **116**, 241601

IOP Conf. Series: Journal of Physics: Conf. Series **873** (2017) 012024 doi:10.1088/1742-6596/873/1/012024

Entangled B_d mesons and genuine T, CP and CPT asymmetry parameters

Miguel Nebot[1]

[1] Centro de Física Teórica de Partículas (CFTP), Instituto Superior Técnico (IST), U. de Lisboa (UL), Av. Rovisco Pais, P-1049-001 Lisboa, Portugal.

E-mail: miguel.r.nebot.gomez@tecnico.ulisboa.pt

Abstract. We establish the precise connection between the theoretical T, CP, CPT asymmetries, in terms of transition probabilities between the neutral meson B_d states, and the experimental asymmetries, in terms of the double decay rate intensities for Flavour-CP eigenstate decay products in a B-Factory of entangled states. Genuine asymmetry parameters in the time distribution of the asymmetries are identified and their measurability analysed, disentangling genuine and possible fake terms. The nine asymmetry parameters – three different observables for each one of the three symmetries – are expressed in terms of the ingredients of the Weisskopf-Wigner dynamical description of the entangled B_d-meson states and we obtain a global fit to their values from the BaBar Collaboration experimental results. The possible fake terms are all compatible with zero. The information content of the nine asymmetry parameters is indeed different. The non-vanishing $\Delta\mathscr{S}_c^{\mathrm{T}} = -0.687 \pm 0.020$ and $\Delta\mathscr{S}_c^{\mathrm{CP}} = -0.680 \pm 0.021$ are impressive separate direct evidence of Time-Reversal-Violation and CP-Violation in these transitions (and compatible with Standard Model expectations). A 2σ effect for the $\mathrm{Re}(\theta)$ parameter responsible of CPT-Violation appears; interpreted as an upper limit, it leads to $|M_{\bar{B}^0\bar{B}^0} - M_{B^0B^0}| < 4.0 \times 10^{-5}$ eV at 95% C.L. for the diagonal flavour terms of the mass matrix. Finally, we consider scenarios where, in the presence of quantum gravity fluctuations (space-time foam), the CPT operator may be ill-defined. Its perturbative treatment leads to a modification of the Einstein-Podolsky-Rosen correlation of the neutral meson system by adding an Entanglement-weakening term of the wrong exchange symmetry, the ω-effect. The analysis is extended to identify how to probe the complex ω when the connection between the Intensities for the two time-ordered decays (f, g) and (g, f) is lost (f flavour and g CP eigenstate decay channels), and how the ω-effect is disentangled from CPT violation in the evolution Hamiltonian.

1. Introduction

The BaBar Collaboration demonstrated direct evidence of Time Reversal Violation in the time evolution of the B_d^0–$\bar{B}_d^0$ meson system [1], independent of CP Violation or CPT Invariance. The result does not dependent on a particular dynamical description of the B_d^0–$\bar{B}_d^0$ system: it is established in terms of asymmetries of observable transition rates. The essential quantum mechanical ingredients involved are (i) the entanglement of the B_d^0–$\bar{B}_d^0$ pair in a B-Factory before the first decay, (ii) the use of decays as filtering measurements to (a) prepare the initial, and (b) detection the final, B meson states participating in the transition, (iii) the time dependence of the double decay rates. The concept behind this approach originates in [2, 3] while the application to an actual experimental analysis was addressed in [4], avoiding, in particular, the need of the T-reversal of the decay (since, as emphasized by L.Wolfenstein [5], the T-reverse of a decaying state is not a physical state). The considered transitions are between Flavour and CP eigenstate decay products. Orthogonality between the meson states filtered by both types of decay products (the Flavour and CP "Tags" [6]), is well defined under certain conditions [4, 7, 8]; this opens the possibility of constructing Genuine Asymmetries without contaminating fake terms. The eight different Flavour $\leftrightarrows$ CP transitions provide separate and independent asymmetries for the T, CP and CPT transformations. Our main objectives are [9]: (1) provide the precise connection between theoretical asymmetries (in terms of transition probabilities for the meson states), and experimental asymmetries (in terms of double decay intensities), giving *genuine asymmetry parameters* for model-independent T, CP, CPT time dependent asymmetries; (2) analysing these asymmetry parameters within the Weisskopf-Wigner Approach (WWA) [10, 11] for the description of the dynamics/time evolution of the B_d^0–$\bar{B}_d^0$ system. In the following, we recall some generalities of the B_d^0–$\bar{B}_d^0$ effective hamiltonian, the evolution of the initial entangled state and the double decay rate intensities. In section 3, the conditions for the channels used in the Babar analysis of [1] to be truly appropriate for time reversal genuine asymmetries are addressed. Section 4 analyses the experimental asymmetries of [1] in detail, focusing on the connection with section 2. The reconstruction of complete genuine asymmetries beyond the BaBar ratios is shown in section 5. The contamination of genuine T and CPT asymmetries from deviations in the conditions discussed in section 3 is discussed and quantified in section 6. A summary of the most relevant results from a global fit to the Babar data is included in 7. Section 8 is devoted to a condensed discussion of the sensitivity to the ω-effect and to present selected results derived when the previous analysis is extended through the inclusion of ω.

2. Entanglement, Time evolution and double decay rates

The effective Hamiltonian for the two meson system B_d^0–$\bar{B}_d^0$ is $\mathbf{H} = \mathbf{M} - i\mathbf{\Gamma}/2$ with $\mathbf{M}$ and $\mathbf{\Gamma}$ 2×2 hermitian matrices, respectively the hermitian and the antihermitian parts of $\mathbf{H}$. In the notation of [12] the eigenvalues and eigenvectors are

$$\mu_{H,L} = M_{H,L} - \frac{i}{2}\Gamma_{H,L}; \qquad \mathbf{H}|B_H\rangle = \mu_H|B_H\rangle; \qquad |B_H\rangle = p_H|B_d^0\rangle + q_H|\bar{B}_d^0\rangle; \tag{1}$$

$$\mathbf{H}|B_L\rangle = \mu_L|B_L\rangle; \qquad |B_L\rangle = p_L|B_d^0\rangle - q_L|\bar{B}_d^0\rangle. \tag{2}$$

In general, with $\mathbf{H}$ not a normal operator, $[\mathbf{M}, \mathbf{\Gamma}] \neq 0$: then, the states (1) and (2) are not orthogonal. The averages and differences of masses and widths[1] are more convenient to use:

$$\mu = \frac{M_H + M_L}{2} - \frac{i}{2}\frac{\Gamma_H + \Gamma_L}{2} \equiv M - \frac{i}{2}\Gamma, \tag{3}$$

$$\Delta\mu = M_H - M_L - \frac{i}{2}(\Gamma_H - \Gamma_L) \equiv \Delta M - \frac{i}{2}\Delta\Gamma, \tag{4}$$

together with the complex parameters θ and q/p:

$$\frac{q_H}{p_H} = \frac{q}{p}\sqrt{\frac{1+\theta}{1-\theta}}, \qquad \frac{q_L}{p_L} = \frac{q}{p}\sqrt{\frac{1-\theta}{1+\theta}}, \tag{5}$$

and with

$$\delta = \frac{1 - |q/p|^2}{1 + |q/p|^2}. \tag{6}$$

[1] Subindices "H" and "L" label the "heavy" and "light" states, then $\Delta M > 0$ and the sign of $\Delta\Gamma$ is not a matter of convention, it is not fixed.

From $\mathbf{H}$, $\theta = (\mathbf{H}_{22} - \mathbf{H}_{11})/\Delta\mu$, $(q/p)^2 = \mathbf{H}_{21}/\mathbf{H}_{12}$; θ is a CP and CPT violating complex parameter; δ violates CP and T. Interestingly, $\mathbf{H}$ can be written in terms of physical parameters [13] (except for the phase of q/p, which is convention dependent),

$$\mathbf{H} = \begin{pmatrix} \mu - \frac{\Delta\mu}{2}\theta & \frac{p}{q}\frac{\Delta\mu}{2}\sqrt{1-\theta^2} \\ \frac{q}{p}\frac{\Delta\mu}{2}\sqrt{1-\theta^2} & \mu + \frac{\Delta\mu}{2}\theta \end{pmatrix}. \tag{7}$$

In a B factory, operating at the $\Upsilon(4S)$ peak, the initial two-meson state is EPR [14] entangled,

$$|\Psi_0\rangle = \frac{1}{\sqrt{2}}\Big(|B_d^0\rangle|\bar{B}_d^0\rangle - |\bar{B}_d^0\rangle|B_d^0\rangle\Big) = \frac{1}{\sqrt{2}(p_L q_H + p_H q_L)}\Big(|B_L\rangle|B_H\rangle - |B_H\rangle|B_L\rangle\Big), \tag{8}$$

which retains its entangled antisymmetric character in terms of $\mathbf{H}$ eigenstates. For the decay of the first state into $|f\rangle$ at time t_0, and then the second state into $|g\rangle$ at time $t+t_0$, the transition amplitude is

$$\langle f, t_0; g, t+t_0|T|\Psi_0\rangle = \frac{e^{-i(\mu_H+\mu_L)t_0}}{\sqrt{2}(p_L q_H + p_H q_L)}\Big(e^{-i\mu_H t}\mathcal{A}_f^L\mathcal{A}_g^H - e^{-i\mu_L t}\mathcal{A}_f^H\mathcal{A}_g^L\Big), \tag{9}$$

with $\mathcal{A}_f^{H,L} \equiv \langle f|T|B_{H,L}\rangle$ the decay amplitudes of the eigenstates into the final state f. The double decay rate $I(f,g;t)$ is obtained squaring, and integrating over t_0:

$$I(f,g;t) = \frac{e^{-\Gamma t}}{4\Gamma|p_L q_H + p_H q_L|^2}\left|e^{i\Delta M t/2}e^{\Delta\Gamma t/4}\mathcal{A}_f^H\mathcal{A}_g^L - e^{-i\Delta M t/2}e^{-\Delta\Gamma t/4}\mathcal{A}_f^L\mathcal{A}_g^H\right|^2. \tag{10}$$

Equation (10) makes transparent the following expected symmetry property: up to the global exponential decay factor $e^{-\Gamma t}$, the combined transformations $t \to -t$ and $f \rightleftarrows g$ give the identity. Expansion of the t dependence with the approximation $\Delta\Gamma = 0$, (valid for the B_d^0 states) gives

$$I(f,g;t) = e^{-\Gamma t}\frac{\langle\Gamma_f\rangle\langle\Gamma_g\rangle}{\Gamma}\Big\{\mathscr{C}_h[f,g] + \mathscr{C}_c[f,g]\cos(\Delta M\,t) + \mathscr{S}_c[f,g]\sin(\Delta M\,t)\Big\}, \tag{11}$$

with $\langle\Gamma_f\rangle$ defined after eq. (13). Therefore, with the initial entangled state in eq. (8) and the evolution in eq. (9), the following symmetry properties follow:

$$\mathscr{C}_h[f,g] = \mathscr{C}_h[g,f], \quad \mathscr{C}_c[f,g] = \mathscr{C}_c[g,f], \quad \mathscr{S}_c[f,g] = -\mathscr{S}_c[g,f]. \tag{12}$$

They are central in assessing the independent observables present in the Babar measurements. For the expressions to follow in the next sections, we define as usual the parameters of mixing times decay amplitudes

$$\lambda_f \equiv \frac{q}{p}\frac{\bar{A}_f}{A_f}, \quad C_f \equiv \frac{1-|\lambda_f|^2}{1+|\lambda_f|^2}, \quad S_f \equiv \frac{2\mathrm{Im}\,(\lambda_f)}{1+|\lambda_f|^2}, \quad R_f \equiv \frac{2\mathrm{Re}\,(\lambda_f)}{1+|\lambda_f|^2}, \tag{13}$$

where $\langle f|T|B_d^0\rangle \equiv A_f$, $\langle f|T|\bar{B}_d^0\rangle \equiv \bar{A}_f$ and $\langle\Gamma_f\rangle = \frac{1}{2}(|A_f|^2 + |\bar{A}_f|^2)$. Furthermore, for flavour specific channels $f = \ell^\pm + X$ ($f = \ell^\pm$ for short in the following), and assuming no wrong lepton charge sign decays, $C_{\ell^\pm} = \pm 1$, $R_{\ell^\pm} = S_{\ell^\pm} = 0$. It is also convenient to introduce the reduced intensity $\hat{\mathrm{I}}(f,g;t)$,

$$\hat{\mathrm{I}}(f,g;t) \equiv \frac{\Gamma}{\langle\Gamma_f\rangle\langle\Gamma_g\rangle}I(f,g;t) = e^{-\Gamma t}\Big\{\mathscr{C}_h[f,g] + \mathscr{C}_c[f,g]\cos(\Delta M\,t) + \mathscr{S}_c[f,g]\sin(\Delta M\,t)\Big\}. \tag{14}$$

3. Conditions for motion reversal asymmetries

The original proposal of [2,3] to observe T violation independently of CP violation, implemented by BaBar [1] after [4], contained three ingredients:
I - Time reversal in the B^0_d–$\bar{B}^0_d$ Hilbert space: define a reference transition $P_1 \to P_2(t)$ among meson states, and compare with the reversed transition $P_2 \to P_1(t)$. The probability that an initially prepared state P_1, evolved to $P_1(t)$, behaves like a P_2, is:

$$P_{12}(t) = |\langle P_2|U(t,0)|P_1\rangle|^2 \,. \tag{15}$$

The proposed T violating asymmetry is

$$P_{12}(t) - P_{21}(t)\,. \tag{16}$$

II - Beyond the use of $P_1, P_2 = B^0_d, \bar{B}^0_d$ states. If the transitions $B^0_d \rightleftarrows \bar{B}^0_d$ are used, the asymmetry is not independent of CP: by construction, it is both CP and T violating (and very small since it comes from the parameter δ). A new reference transition $B^0_d \to B_+$ was introduced, to be compared with $B_+ \to B^0_d$. In a decay channel with well-defined $CP = +$ and negligible CP violation, the reference transition can be accessed by looking for decay events f_1 where a B meson decays to a self-tagging channel of $\bar{B}^0_d$ and the other B meson decays later to a CP eigenstate $f_{CP=+}$ decay where one can neglect CP violation. Then, the main problem is how to measure the reverse transition.
III - The entangled character of the initial state is crucial to (i) connect double decay rates with meson transition rates and (ii) identify the reverse transition. If we assume that observing a $f_{CP=-}$ one filters in that side a B_-, then, due to the entanglement, we tag the orthogonal state to B_- in the opposite side. This state should be a B_+. In general, from eq. (8) we can say that, if at time t_1 we observe in one side the decay product f, the (still living) meson at time t_1 is tagged as *the state that does not decay into* f, $|B_{\not\to f}\rangle$,

$$|B_{\not\to f}\rangle = \frac{1}{\sqrt{|A_f|^2 + |\bar{A}_f|^2}} \left(\bar{A}_f|B^0_d\rangle - A_f|\bar{B}^0_d\rangle\right). \tag{17}$$

The orthogonal state $\langle B^\perp_{\not\to f}|B_{\not\to f}\rangle = 0$, the one filtered by a decay f, is

$$|B^\perp_{\not\to f}\rangle = \frac{1}{\sqrt{|A_f|^2 + |\bar{A}_f|^2}} \left(A^*_f|B^0_d\rangle + \bar{A}^*_f|\bar{B}^0_d\rangle\right). \tag{18}$$

The *filtering identity* [8,15] defines the precise meaning of the last statement:

$$\left|\langle B^\perp_{\not\to f}|B_1\rangle\right|^2 = \frac{|\langle f|T|B_1\rangle|^2}{|A_f|^2 + |\bar{A}_f|^2}. \tag{19}$$

Notice that, if $B_1 = B_{\not\to g}(t)$, this quantity is the reduced intensity $\hat{\mathrm{I}}(g, f; t)$ in eq. (14):

$$\hat{\mathrm{I}}(g, f; t) = \frac{|\langle f|T|B_{\not\to g}(t)\rangle|^2}{|A_f|^2 + |\bar{A}_f|^2} = \left|\langle B^\perp_{\not\to f}|B_{\not\to g}(t)\rangle\right|^2 . \tag{20}$$

Eq. (20) is, therefore, the precise connection between meson transition probabilities and double decay rates. By measuring $\hat{\mathrm{I}}(f_1, f_2; t)$, – we use the shorthand notation (f_1, f_2) in the following to refer to the considered first and second decays – we study probabilities $P_{12}(t)$ for transitions between meson states (B_1, B_2) given by

$$|B_1\rangle = |B_{\not\to f_1}\rangle\,, \quad |B_2\rangle = |B^\perp_{\not\to f_2}\rangle\,, \tag{21}$$

IOP Conf. Series: Journal of Physics: Conf. Series **873** (2017) 012024 doi:10.1088/1742-6596/873/1/012024

i.e. transition probabilities for $(B_1, B_2) = (B_{\nrightarrow f_1}, B^{\perp}_{\nrightarrow f_2})$. To compare with $P_{21}(t)$ one needs the reverse transition $(B^{\perp}_{\nrightarrow f_2}, B_{\nrightarrow f_1})$, but the applied filtering and tagging methods do not give this transition. Two new decay channels f_1' and f_2' in (f_2', f_1') give the transition $(B_{\nrightarrow f_2'}, B^{\perp}_{\nrightarrow f_1'})$. Then, provided they fulfill

$$|B_{\nrightarrow f_i'}\rangle = |B^{\perp}_{\nrightarrow f_i}\rangle, \tag{22}$$

this new transition (f_2', f_1') gives the reversed meson transition. For flavour specific decay channels, with no wrong lepton charge sign decays, $|B_d^0\rangle = |B_{\nrightarrow \ell^-}\rangle$ and $|\bar{B}_d^0\rangle = |B_{\nrightarrow \ell^+}\rangle$, the identity is obviously $|\bar{B}_d^0\rangle = |(B_d^0)^{\perp}\rangle$: if $f_1 = X\ell^+\nu_\ell$, then $f_1' = X'\ell^-\bar{\nu}_\ell$ ($f_1 = \ell^+$ and $f_1' = \ell^-$ for short). For the CP channel, equations (17), (18) and (22), give the condition these channels should satisfy:

$$\lambda_{f_2}\lambda^*_{f_2'} = -\left|\frac{q}{p}\right|^2 . \tag{23}$$

Following eq. (23), $f_2 = J/\psi K_+$ and $f_2' = J/\psi K_-$ were proposed in [2–4] and used by BaBar in [1], where $K_\pm$ are the neutral kaon states filtered by the CP eigenstate decay channels. Consequently, the states $B_\mp$ are well defined and given by equation (18) for each of the two decay channels. From now on, we use K_S for K_+ and K_L for K_- as it is an accurate approximation up to CP violation in the kaon system. Considering that $\lambda_{J/\psi K_S} \equiv \lambda_{K_S} \sim \left|\frac{q}{p}\right| e^{-i2\beta}$ and $\lambda_{J/\psi K_L} \equiv \lambda_{K_L} \sim -\left|\frac{q}{p}\right| e^{-i2\beta}$, we use the general parameterisation

$$\lambda_{K_S} = \left|\frac{q}{p}\right| \rho\,(1+\epsilon_\rho)\,e^{-i(2\beta+\epsilon_\beta)}\,, \quad \lambda_{K_L} = -\left|\frac{q}{p}\right| \frac{1}{\rho}\,(1+\epsilon_\rho)\,e^{-i(2\beta-\epsilon_\beta)}\,, \tag{24}$$

in terms of the real parameters $\{\rho, \beta, \epsilon_\rho, \epsilon_\beta\}$, which allows us to control deviations from condition (23). By properly comparing double decay rates corresponding to two channels, one from $\{\ell^+, \ell^-\}$ and the other from $\{J/\psi K_S, J/\psi K_L\}$ (K_S and K_L for short in the following), we will therefore measure genuine time-reverse processes provided

$$\epsilon_\rho = 0\,, \quad \epsilon_\beta = 0\,. \tag{25}$$

Deviations from (25) contaminate time reversal asymmetries and should be conveniently subtracted out. Notice that eq. (23) is fulfilled even if $\rho \neq 1$. Equation (23) guarantees that the considered channels allow to truly compare the transition $P_2 \to P_1(t)$ with the reversed transition $P_1 \to P_2(t)$. Nevertheless, in order to ensure that this motion reversal asymmetry is truly a time reversal asymmetry, one needs decay channels f such that in the limit of T invariance, $S_f = 0$ [8, 15]. For CP eigenstates, T invariance implies $S_f = 0$ provided there is no CPT violation in the corresponding decay amplitude, in accordance with [7]. This is equivalent to no CP violation in the decay, in the T invariant limit, giving, in addition to eq. (25), the condition $\rho = 1$. We then conclude that we should perform the data analysis with arbitrary ρ, ϵ_ρ and ϵ_β and that any deviation from

$$\rho = 1\,,\ \epsilon_\rho = 0\,,\ \epsilon_\beta = 0\,, \tag{26}$$

is a source of fake T violation. Notice that in the absence of CP violation in the decays filtering $B_\pm$, these states would be orthogonal, implying eq. (26), the orthogonality condition (23) would be automatically satisfied. Finally, it is convenient to clarify that without wrong flavour decays in $B_d^0 \to J/\psi K^0$ and $\bar{B}_d^0 \to J/\psi \bar{K}^0$, $\lambda_{K_S} + \lambda_{K_L} = 0$ [16], implying

$$\rho = 1\,,\ \epsilon_\beta = 0\,, \tag{27}$$

clearly showing full compatibility among eq. (23) and the absence of wrong flavour decays. In terms of C_{K_S}, C_{K_L}, S_{K_S}, S_{K_L}, R_{K_S} and R_{K_L} (eq. (13)), no wrong flavour decays imply $C_{K_S} - C_{K_L} = 0$, $S_{K_S} + S_{K_L} = 0$ and $R_{K_S} + R_{K_L} = 0$; if, in addition, we impose eq. (23), $C_{K_S} = C_{K_L} = \delta$.

4. Independent asymmetries at BaBar

To reduce dependences on detection efficiencies in the different channels, in [1], instead of measuring $\mathscr{C}_h[f,g]$, $\mathscr{C}_c[f,g]$ and $\mathscr{S}_c[f,g]$ in eq. (11) or eq. (14), the BaBar collaboration fixed the normalization of the constant term and used the decay intensity

$$\mathbf{g}_{f,g}(t) \propto e^{-\Gamma t}\{1 + \mathrm{C}[f,g]\cos(\Delta M\, t) + \mathrm{S}[f,g]\sin(\Delta M\, t)\}\,, \tag{28}$$

and thus two quantities,

$$\mathrm{C}[f,g] = \frac{\mathscr{C}_c[f,g]}{\mathscr{C}_h[f,g]}, \quad \mathrm{S}[f,g] = \frac{\mathscr{S}_c[f,g]}{\mathscr{C}_h[f,g]}, \tag{29}$$

are measured for each pair (f,g). According to eq. (12), they verify

$$\mathrm{C}[f,g] = \mathrm{C}[g,f], \quad \mathrm{S}[f,g] = -\mathrm{S}[g,f]\,. \tag{30}$$

Genuine discrete asymmetries can be constructed combining one flavour specific channel and one CP channel: starting from one reference transition, one can generate another three by means of T, CP and CPT transformations. Because of eq. (30), the four transitions $\bar{B}^0_d \to B_-$, $B_- \to \bar{B}^0_d$, $B^0_d \to B_-$ and $B_- \to B^0_d$ saturate all the independent parameters measurable with one flavour specific and one CP decays. Table 1 shows the different meson state transitions, the corresponding decay channels, and the effect of the discrete symmetry transformations. Only eight parameters are independent: they are the $\mathrm{C}[f,g]$ and $\mathrm{S}[f,g]$ corresponding to the decays (ℓ^+, K_S), (K_L, ℓ^-), (ℓ^-, K_S) and (K_L, ℓ^+). BaBar has, of course, at least two independent ways of measuring the same parameter by means of the time-ordering of the decays. This operation is not a symmetry transformation from the left to the right-hand side of Table 1. Only *six* independent asymmetries can be constructed out of the eight independent parameters,

Table 1. Double decay channels, the associated filtered meson states and their transformed transitions under the three discrete symmetries.

	Transition	$\mathbf{g}_{f,g}(t)$	$\mathbf{g}_{g,f}(t)$	Transition	
Reference	$\bar{B}^0_d \to B_-$	(ℓ^+, K_S)	(K_S, ℓ^+)	$B_+ \to B^0_d$	Reference
T-transformed	$B_- \to \bar{B}^0_d$	(K_L, ℓ^-)	(ℓ^-, K_L)	$B^0_d \to B_-$	T-transformed
CP-transformed	$B^0_d \to B_-$	(ℓ^-, K_S)	(K_S, ℓ^-)	$B_+ \to \bar{B}^0_d$	CP-transformed
CPT-transformed	$B_- \to B^0_d$	(K_L, ℓ^+)	(ℓ^+, K_L)	$\bar{B}^0_d \to B_+$	CPT-transformed

corresponding to three time dependent asymmetries

$$A_{\mathrm{T}}(t) = \mathbf{g}_{K_L,\ell^-}(t) - \mathbf{g}_{\ell^+,K_S}(t)\,, \tag{31}$$

$$A_{\mathrm{CP}}(t) = \mathbf{g}_{\ell^-,K_S}(t) - \mathbf{g}_{\ell^+,K_S}(t)\,, \tag{32}$$

$$A_{\mathrm{CPT}}(t) = \mathbf{g}_{K_L,\ell^+}(t) - \mathbf{g}_{\ell^+,K_S}(t)\,, \tag{33}$$

IOP Conf. Series: Journal of Physics: Conf. Series **873** (2017) 012024 doi:10.1088/1742-6596/873/1/012024

which can be explicitely expanded as

$$A_{\rm S}(t) = e^{-\Gamma t}\left\{\Delta C_{\rm S}[\ell^+, K_S]\cos(\Delta M\, t) + \Delta S_{\rm S}[\ell^+, K_S]\sin(\Delta M\, t)\right\}, \ {\rm S} = {\rm T}, {\rm CP}, {\rm CPT}, \quad (34)$$

where

$$\Delta S_{\rm T}^+ \equiv \Delta S_{\rm T}[\ell^+, K_S] = {\rm S}[K_L, \ell^-] - {\rm S}[\ell^+, K_S], \quad (35)$$

$$\Delta S_{\rm CP}^+ \equiv \Delta S_{\rm CP}[\ell^+, K_S] = {\rm S}[\ell^-, K_S] - {\rm S}[\ell^+, K_S], \quad (36)$$

$$\Delta S_{\rm CPT}^+ \equiv \Delta S_{\rm CPT}[\ell^+, K_S] = {\rm S}[K_L, \ell^+] - {\rm S}[\ell^+, K_S], \quad (37)$$

and the corresponding ones for the $\cos(\Delta M\, t)$ terms, are the six independent asymmetries that can be constructed (with the same notation of reference [1] for easy comparison). To illustrate the difference among asymmetries which would be equivalent in a CPT invariant world, one can write (expanding to linear order in ${\rm Re}\,(\theta)$, ${\rm Im}\,(\theta)$):

$$\Delta S_{\rm T}^+ \simeq S_{K_S} - S_{K_L} - {\rm Re}\,(\theta)\,(S_{K_S}R_{K_S} + S_{K_L}R_{K_L}) + {\rm Im}\,(\theta)\,(S_{K_S}^2 - S_{K_L}^2 + C_{K_S} + C_{K_L}), \quad (38)$$

$$\Delta S_{\rm CP}^+ \simeq 2S_{K_S} + 2{\rm Im}\,(\theta)\,(S_{K_S}^2 - 1), \quad (39)$$

No matter whether CPT Violation is expected to be small, conceptually it is very important to emphasize that $\Delta S_{\rm T}^+ \neq \Delta S_{\rm CP}^+$ for several reasons. For $\Delta S_{\rm T}^+$ to be truly T violating, eq. (26) should be fulfilled. Then, the dominant term in eqs. (38)–(39) should be equal: $S_{K_S} - S_{K_L} = 2S_{K_S}$. But, in general, $\Delta S_{\rm T}^+$ and $\Delta S_{\rm CP}^+$ differ by terms that are CPT violating and CP invariant in $\Delta S_{\rm T}^+$, and by terms that are CPT violating and T invariant in $\Delta S_{\rm CP}^+$. Only the pieces that do not depend on θ are identical. See [9] for an extended discussion including $\Delta C_{\rm T}^+ \neq \Delta C_{\rm CP}^+$ and the CPT asymmetries.

5. Genuine asymmetry parameters

Although the reduced intensity $\hat{\rm I}(f, g; t)$ involves three coefficients $\mathscr{C}_h$, $\mathscr{C}_c$ and $\mathscr{S}_c$, the BaBar analysis [1] focused on the ratios ${\rm C} = \mathscr{C}_c/\mathscr{C}_h$ and ${\rm S} = \mathscr{S}_c/\mathscr{C}_h$ in eq. (29). From the experimental point of view those ratios might be more appropriate, but the theoretical side, access to the three independent coefficients would be more desirable: for instance, while an asymmetry in the ratios does imply a symmetry violation, no asymmetry in the ratios may nevertheless come from asymmetries in both the numerator and the denominator. Obtaining the three independent coefficients $\mathscr{C}_h$, $\mathscr{C}_c$ and $\mathscr{S}_c$ for each pair of decay channels might be particularly interesting for asymmetries in the ratios with values that are, within uncertainties, compatible with zero, like e.g. CPT asymmetries. Fortunately, using input information for $|q/p|$ or, equivalently δ, it can be achieved. We have

$$\mathscr{C}_h[\ell^\pm, K_{S,L}] + \mathscr{C}_c[\ell^\pm, K_{S,L}] = \frac{(1 \pm \delta)(1 \mp C_{K_{S,L}})}{2(1 - \delta C_{K_{S,L}})} = \mathscr{C}_h[\ell^\pm, K_{S,L}]\left(1 + {\rm C}[\ell^\pm, K_{S,L}]\right). \quad (40)$$

Equation (40) is interpreted in the following way: while ${\rm C}[\ell^\pm, K_{S,L}]$ and $C_{K_{S,L}}$ will be constrained or extracted from data, including information on δ, we can compute $\mathscr{C}_h[\ell^\pm, K_{S,L}]$, and thus $\mathscr{C}_c[\ell^\pm, K_{S,L}]$ and $\mathscr{S}_c[\ell^\pm, K_{S,L}]$ separately. One can then build T, CP and CPT complete time-dependent asymmetries analog to eqs. (31), (32) and (33),

$$\mathscr{A}_{\rm S}(t) = e^{-\Gamma t}\left\{\Delta\mathscr{C}_h^{\rm S} + \Delta\mathscr{C}_c^{\rm S}\cos(\Delta M\, t) + \Delta\mathscr{S}_c^{\rm S}\sin(\Delta M\, t)\right\}, \qquad {\rm S} = {\rm T}, {\rm CP}, {\rm CPT}. \quad (41)$$

We refer to $\Delta\mathscr{C}_h^{\rm S}$, $\Delta\mathscr{C}_c^{\rm S}$ and $\Delta\mathscr{S}_c^{\rm S}$ in these asymmetries as "genuine asymmetry parameters" since they collect the full time-dependent difference of probabilities in transitions among meson states given in eq. (16). For detailed expressions expanded up to linear order in θ and δ and a detailed discussion of the physical interpretation of (40) we refer to [9].

IOP Conf. Series: Journal of Physics: Conf. Series **873** (2017) 012024 doi:10.1088/1742-6596/873/1/012024

6. Genuine T-reverse and fake asymmetries

In section 3 we have discussed how asymmetries like eqs. (31) and (33) are "contaminated" through contributions which are not truly T-violating (this also applies to the genuine asymmetry parameters from section 5). It occurs when the conditions in eq. (26) are not fulfilled. The challenge is to disentangle *fake* effects in T and CPT asymmetries due to deviations from the requirements of eq. (26). The reasoning is illustrated using, for example, the asymmetry $\Delta\mathscr{S}_c^{\mathrm{T}}$. First, we remind that in terms of all parameters – δ, ρ, β, ϵ_ρ and ϵ_β in eq. (24), plus the complex θ parameter –, $\Delta\mathscr{S}_c^{\mathrm{T}}$ is simply a function $\Delta\mathscr{S}_c^{\mathrm{T}}(\rho,\beta,\epsilon_\rho,\epsilon_\beta,\delta,\theta)$. $\Delta\mathscr{S}_c^{\mathrm{T}}$ would be a *true* T-violation asymmetry if $\epsilon_\rho = \epsilon_\beta = 0$ and $\rho = 1$ (eq. (26)). It is then possible to do the following separation, at each point in parameter space, when performing a fit to the BaBar observables:

$$\Delta\mathscr{S}_c^{\mathrm{T}}(\rho,\beta,\epsilon_\rho,\epsilon_\beta,\delta,\theta) = \left[\Delta\mathscr{S}_c^{\mathrm{T}}(\rho,\beta,\epsilon_\rho,\epsilon_\beta,\delta,\theta) - \Delta\mathscr{S}_c^{\mathrm{T}}(1,\beta,0,0,\delta,\theta)\right] + \Delta\mathscr{S}_c^{\mathrm{T}}(1,\beta,0,0,\delta,\theta)\,. \tag{42}$$

The term within square brackets, $\Delta\mathscr{S}_c^{\mathrm{T}}(\rho,\beta,\epsilon_\rho,\epsilon_\beta,\delta,\theta) - \Delta\mathscr{S}_c^{\mathrm{T}}(1,\beta,0,0,\delta,\theta)$ has exactly the desired properties for the *fake* contribution: independently of β, δ and θ, it vanishes when the conditions eqs. (23) and (26) are fulfilled. Then, the last term, $\Delta\mathscr{S}_c^{\mathrm{T}}(1,\beta,0,0,\delta,\theta)$ is the truly T-violating contribution, the *genuine T-reverse* one. It is then possible to quantify the amounts of fake and genuine T-reverse contributions to T and CPT asymmetries like the BaBar ones ΔS_{T}^+, ΔC_{T}^+, $\Delta S_{\mathrm{CPT}}^+$, $\Delta C_{\mathrm{CPT}}^+$, and also, of course, to the T and CPT genuine asymmetry parameters involving the individual $\mathscr{C}_h$, $\mathscr{C}_c$ and $\mathscr{S}_c$ coefficients. In terms of δ, ρ, β, ϵ_ρ and ϵ_β, the genuine T-reverse asymmetries are obtained for

$$\begin{Bmatrix} C_{K_S} \\ C_{K_L} \end{Bmatrix} \to \delta, \quad \begin{Bmatrix} S_{K_S} \\ -S_{K_L} \end{Bmatrix} \to -\sqrt{1-\delta^2}\sin 2\beta, \quad \begin{Bmatrix} R_{K_S} \\ -R_{K_L} \end{Bmatrix} \to \sqrt{1-\delta^2}\cos 2\beta\,. \tag{43}$$

7. Results

With the information on the $\mathrm{C}[\ell^\pm, K_{S,L}]$ and $\mathrm{S}[\ell^\pm, K_{S,L}]$ coefficients provided in [1], including full covariance information and separate statistical and systematic uncertainties, supplemented with information on $|q/p|$ from [17], we perform a fit in terms of $\{\mathrm{Re}\,(\theta)\,, \mathrm{Im}\,(\theta)\,, \delta, \rho, \beta, \epsilon_\rho, \epsilon_\beta\}$. We can also address a more restricted situation where no wrong flavour decays (i.e. $\Delta F = \Delta Q$) are allowed in $B_d^0, \bar{B}_d^0 \to J/\Psi K_{S,L}$, that is imposing $\lambda_{K_S} + \lambda_{K_L} = 0$: in terms of the previous set of parameters, that means setting $\rho = 1$ and $\epsilon_\beta = 0$. For a details of the fit procedure, more detailed results, tables and plots, we refer to [9], and focus here on a few significant aspects. Starting with the CPT violating θ parameter, from the fits,

$$\begin{Bmatrix} \mathrm{Re}\,(\theta) = \pm(5.92 \pm 3.03) \times 10^{-2} \\ \mathrm{Im}\,(\theta) = (0.22 \pm 1.90) \times 10^{-2} \end{Bmatrix}$$

$$\text{and} \begin{Bmatrix} \mathrm{Re}\,(\theta) = \pm(3.92 \pm 1.43) \times 10^{-2} \\ \mathrm{Im}\,(\theta) = (-0.22 \pm 1.64) \times 10^{-2} \end{Bmatrix} \text{with } \lambda_{K_S} + \lambda_{K_L} = 0, \tag{44}$$

which improve significantly on the uncertainty of the real part quoted by the Particle Data Group (PDG) in [17], based on BaBar [18, 19] and Belle [20] results. For similar results based on the samed Babar data, see [21, 22]. Figure 1 shows the result of the fit for the imaginary vs. real part of θ. To illustrate the difference between the genuine T-reverse and CP asymmetry parameters figures 2(a), 2(b), show true T-reverse asymmetries versus CP asymmetries for ΔS and for the genuine asymmetry coefficients $\Delta\mathscr{S}_c$ and $\Delta\mathscr{C}_c$. The dashed diagonal line would correspond to strict equality among both observables. In figures 2(c), 2(d), we show genuine T-reverse vs. fake contributions for ΔS_{T}^+ and for the genuine asymmetry parameter $\Delta\mathscr{S}_c^{\mathrm{T}}$. This is particularly relevant for the ΔS_{T}^+ BaBar asymmetry since sizable fake contributions could have weakened the evidence for the time reversal violation observation independent of CP. It is clear that the T-fake contributions to ΔS_{T}^+ and $\Delta\mathscr{S}_c^{\mathrm{T}}$ are below the percent level.

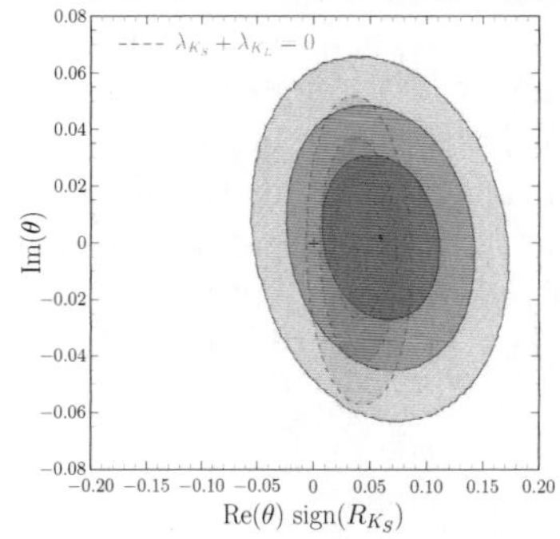

Figure 1. $\mathrm{Im}\,(\theta)$ vs. $\mathrm{Re}\,(\theta)\mathrm{sign}(R_{K_S})$ in the full fit (blue regions, solid contours), and in the fit with $\lambda_{K_S} + \lambda_{K_L} = 0$ (red regions, dashed contours); dark to light regions correspond, respectively, to two-dimensional 68%, 95% and 99% C.L. here and in Fig. 2.

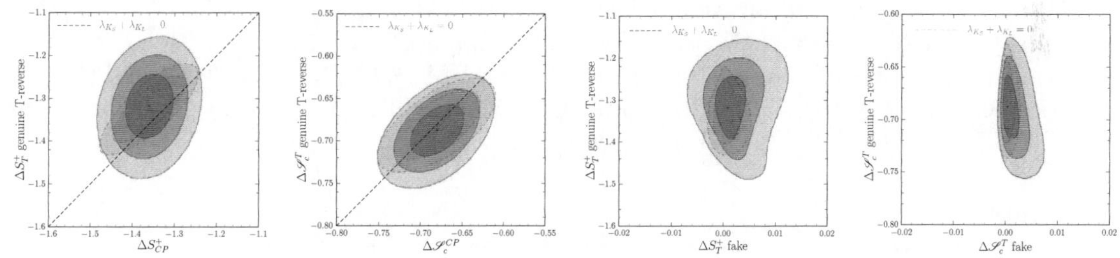

(a) ΔS. (b) $\Delta\mathscr{S}_c$. (c) $\Delta S_{\mathrm{T}}^{+}$. (d) $\Delta\mathscr{S}_c^{\mathrm{T}}$.

Figure 2. Genuine T-reverse vs. CP asymmetries and genuine T-reverse vs. fake contributions.

8. ω effect

Quantum gravity or in general deviations from any of the three assumptions – Lorentz invariance, locality of interactions and unitary – of the CPT theorem, may lead to (independent) violations of CPT and, accordingly, to an ill-defined CPT quantum mechanical operator [23–25]. The CPT breaking associated to ill-defined particle-antiparticle states modifies the EPR correlation and produces the ω-effect [26–28]. Treating it perturbatively, i.e. still talking the language of B_d^0, $\bar{B}_d^0$, the perturbed two-particle state will contain a component of the "wrong" symmetry at the instant of their production by the decay of $\Upsilon(4S)$: instead of eq. (8) it would read

$$|\Psi_0\rangle \propto |B_d^0\rangle|\bar{B}_d^0\rangle - |\bar{B}_d^0\rangle|B_d^0\rangle + \omega\big[|B_d^0\rangle|\bar{B}_d^0\rangle + |\bar{B}_d^0\rangle|B_d^0\rangle\big], \qquad (45)$$

where $\omega = |\omega|e^{i\Omega}$ is a complex CPT-breaking parameter [26,27], associated with the non-identical particle nature of the neutral meson and antimeson states. The presence of an ω-effect weakens the entanglement of the initial state (45), as follows from the fact that when $\omega = \pm 1$ the state simply reduces to a product state, whilst the state is fully entangled when $\omega = 0$. Furthermore, the coefficients in the general time-dependent intensity (11) are modified,

$$\mathscr{C}_h[f,g] \to \mathscr{C}_h[f,g], \qquad \mathscr{C}_c[f,g] \to \mathscr{C}_c[f,g], \qquad \mathscr{S}_c[f,g] \to \mathscr{S}_c[f,g], \qquad (46)$$

where now, contrary to eq. (12),

$$\mathscr{C}_h[f,g] \neq \mathscr{C}_h[g,f], \qquad \mathscr{C}_c[f,g] \neq \mathscr{C}_c[g,f], \qquad \mathscr{S}_c[f,g] \neq -\mathscr{S}_c[g,f]. \qquad (47)$$

The redundancy due to the different time orderings in the Babar data for the determination of $\mathscr{C}_h[f,g]$, $\mathscr{C}_c[f,g]$ and $\mathscr{S}_c[f,g]$ does now provide sensitivity to the presence of ω: $\mathrm{C}[\ell^\pm,g]-\mathrm{C}[g,\ell^\pm]$ and $\mathrm{S}[\ell^\pm,g]+\mathrm{S}[g,\ell^\pm]$ are linear in both the real and imaginary parts of ω. With these basic ingredients, one can develop a full analysis along the lines of section 7, but including ω. For detailed discussions and results, we refer the interested reader to reference [29], from which we only mention here that one can indeed extract or bound both the real and imaginary parts of ω: for the first time for $\mathrm{Im}(\omega)$, with less precision than analyses focusing only on flavour specific channels in the case of $\mathrm{Re}(\omega)$ [28].

Conclusions

We have discussed genuine T, CP and CPT asymmetry parameters at B-factories, the separation of fake contributions to T and CPT violating observables and obtained from a fit to existing data, improved bounds on the real part of the CPT violating B_d^0–$\bar{B}_d^0$ mixing parameter θ.

Acknowledgments

The author thanks the organization of the Discrete 2016 Symposium for the excellent development of the conference. MN also acknowledges support from *Fundação para a Ciência e a Tecnologia* (FCT, Portugal) through postdoctoral grant SFRH/BPD/112999/2015 and through the project CFTP-FCT Unit 777 (UID/FIS/00777/2013) which are partially funded through POCTI (FEDER), COMPETE, QREN and EU.

Bibliography

[1] Lees J P *et al.* (BaBar) 2012 *Phys. Rev. Lett.* **109** 211801 (*Preprint* 1207.5832)
[2] Bañuls M and Bernabéu J 1999 *Phys.Lett.* **B464** 117–122 (*Preprint* hep-ph/9908353)
[3] Bañuls M and Bernabéu J 2000 *Nucl.Phys.* **B590** 19–36 (*Preprint* hep-ph/0005323)
[4] Bernabéu J, Martínez-Vidal F and Villanueva-Pérez P 2012 *JHEP* **1208** 064 (*Preprint* 1203.0171)
[5] Wolfenstein L 1999 *Phys.Rev.Lett.* **83** 911–912, *Int.J.Mod.Phys.* **E8** 501–511
[6] Bañuls M C and Bernabéu J 1999 *JHEP* **06** 032 (*Preprint* hep-ph/9807430)
[7] Applebaum E, Efrati A, Grossman Y, Nir Y and Soreq Y 2014 *Phys. Rev.* **D89** 076011 (*Preprint* 1312.4164)
[8] Bernabéu J, Botella F J and Nebot M 2014 *Phys. Lett.* **B728** 95–98 (*Preprint* 1309.0439)
[9] Bernabéu J, Botella F J and Nebot M 2016 *JHEP* **06** 100 (*Preprint* 1605.03925)
[10] Weisskopf V and Wigner E 1930 *Z.Phys.* **65** 18–29
[11] Lee T, Oehme R and Yang C N 1957 *Phys.Rev.* **106** 340–345
[12] Branco G C, Lavoura L and Silva J P *CP Violation, Oxford University Press*
[13] Silva J P 2000 *Phys. Rev.* **D62** 116008 (*Preprint* hep-ph/0007075)
[14] Einstein A, Podolsky B and Rosen N 1935 *Phys. Rev.* **47** 777–780
[15] Bernabéu J, Botella F J and Nebot M 2015 *J. Phys. Conf. Ser.* **631** 012017
[16] Grossman Y, Kagan A L and Ligeti Z 2002 *Phys. Lett.* **B538** 327–334 (*Preprint* hep-ph/0204212)
[17] Olive K A *et al.* (Particle Data Group) 2014 *Chin. Phys.* **C38** 090001
[18] Aubert B *et al.* (BaBar) 2004 *Phys. Rev.* **D70** 012007 (*Preprint* hep-ex/0403002)
[19] Aubert B *et al.* (BaBar Collaboration) 2006 *Phys.Rev.Lett.* **96** 251802 (*Preprint* hep-ex/0603053)
[20] Higuchi T *et al.* 2012 *Phys. Rev.* **D85** 071105 (*Preprint* 1203.0930)
[21] Lees J P *et al.* (BaBar) 2016 *Phys. Rev.* **D94** 011101 (*Preprint* 1605.04545)
[22] Schubert K *these proceedings*
[23] Wald R M 1980 *Phys. Rev.* **D21** 2742–2755
[24] Ellis J R, Mavromatos N E and Nanopoulos D V 1996 *Int. J. Mod. Phys.* **A11** 1489–1508 (*Preprint* hep-th/9212057)
[25] Ellis J R, Lopez J L, Mavromatos N E and Nanopoulos D V 1996 *Phys. Rev.* **D53** 3846–3870 (*Preprint* hep-ph/9505340)
[26] Bernabéu J, Mavromatos N E and Papavassiliou J 2004 *Phys. Rev. Lett.* **92** 131601 (*Preprint* hep-ph/0310180)
[27] Álvarez E, Bernabéu J, Mavromatos N, Nebot M and Papavassiliou J 2005 *Phys.Lett.* **B607** 197–203 (*Preprint* hep-ph/0410409)
[28] Álvarez E, Bernabéu J and Nebot M 2006 *JHEP* **0611** 087 (*Preprint* hep-ph/0605211)
[29] Bernabeu J, Botella F J, Mavromatos N E and Nebot M 2016 (*Preprint* 1612.05652)

IOP Conf. Series: Journal of Physics: Conf. Series **873** (2017) 012025 doi:10.1088/1742-6596/873/1/012025

Discrete symmetries and QM studies with entangled neutral kaons at KLOE-2

Aleksander Gajos on behalf of the KLOE-2 Collaboration

Marian Smoluchowski Institute of Physics, Jagiellonian University
Łojasiewicza 11, 30-348 Cracow, Poland

E-mail: aleksander.gajos@uj.edu.pl

Abstract. The long history of kaon physics results produced by KLOE is being continued at the upgraded KLOE-2 detector. Entangled neutral kaon pairs produced at DAΦNE are a unique tool to probe discrete symmetries and QM basic principles at the utmost precision. The status of the newest ongoing analyses using the most refined analysis tools will be presented and discussed:(i) search for decoherence and CPT violation effects in the $\phi \to \mathrm{K_S K_L} \to \pi^+\pi^-\pi^+\pi^-$ decay,(ii) test of CP and CPT symmetries in $\mathrm{K_S}$ semileptonic decays,(iii) test of time reversal and CPT in transitions in $\phi \to \mathrm{K_S K_L} \to \pi e \nu, 3\pi^0, 2\pi$ decays,(iv) study of the $\mathrm{K_S} \to \pi^+\pi^-\pi^0$ decay.

1. Introduction

The test of fundamental discrete symmetries with the neutral K meson system has a long and rich history. The properties of this system are directly related to the CP, T and CPT symmetries and provide the potential of designing new tests and to search for violation effects. Such potential can be further enhanced by using the quantum entanglement of neutral kaons produced in certain decays, thus opening a whole new field of quantum interferometry studies. The KLOE experiment, being the only facility which can study entangled neutral kaon pairs, is one of the most important contributors to the state of knowledge of kaon physics and related discrete symmetries violation. Event though a lot of focus in the field of discrete symmetry studies in the recent years has been shifted towards heavier neutral meson systems such as B and D, the KLOE experiment strives to prove that the potential of neutral kaon physics is far from being fully exploited.

The KLOE (K LOng Experiment) detector located at Laboratori Nazionali di Frascati, Italy operates at the DAΦNE electron-positron collider, which provides e^+e^- collisions with center-of-mas energy set at the ϕ resonance mass peak. The decays of ϕ into charged and neutral kaon pairs recorded by KLOE provide an opportunity to extensively study kaon physics, inevitably related to fundamental discrete symmetries. A unique property of KLOE and DAΦNE is the production of neutral kaon pairs in a quantum-entangled state, which opens a way to probe discrete symmetries by means of quantum interferometry.

Operation of the KLOE detector yielded a dataset with an integrated luminosity of about 2.5 fb^{-1}, corresponding to 10^{10} of produced ϕ mesons. This data have been used to test discrete symmetries in the neutral kaon systems using diverse processes and techniques bringing multiple results which are reported on in this work. Moreover, improvement of some of the results

IOP Conf. Series: Journal of Physics: Conf. Series **873** (2017) 012025 doi:10.1088/1742-6596/873/1/012025

is possible by ongoing refined data analyses whose progress is presented. Finally, the recent upgrade of the detector and the new measurements presently performed by the KLOE-2 setup show perspectives to improve almost all of the KLOE results related to discrete symmetries. The expected improvements to be brought by KLOE-2 are also discussed in this work.

2. Search for CPT violation effects with the $\phi \to \mathrm{K_S K_L} \to \pi^+\pi^-\pi^+\pi^-$ process

One of the possible indirect probes of the CPT symmetry violation is a search for Lorentz noninvariance. The Standard Model Extension (SME) and the Greenberg Anti-CPT theorem [1, 2] state that CPT violation should appear together with Lorentz invariance breaking which may be manifested as a direction-dependent modulation of the usual δ_K parameter in neutral K meson system. This modulation can be parametrized as:

$$\delta_K \simeq i\, sin\phi_{SW} e^{i\phi_{SW}} \gamma_K (\Delta a_0 - \vec{\beta_K} \Delta \vec{a})/\Delta m, \tag{1}$$

where ϕ_{SW} is the superweak phase, γ_K and $\vec{\beta_K}$ are the Lorentz factor and velocity of a kaon in the laboratory frame and a_μ denote coefficients of the SME lagrangian part. It is thus possible to extract a_μ from a measurement of the δ_K dependence on the kaon momentum direction.

At KLOE, the pairs of neutral K mesons are produced in an entangled state for which the double decay rate for the kaons decaying into final states f_1 and f_2 in their proper times differing by $\Delta\tau$ reads:

$$I_{f_1 f_2}(\Delta\tau) = C_{12} e^{-\Gamma|\Delta\tau|} \left[|\eta_1|^2 e^{\frac{\Delta\Gamma}{2}\Delta\tau} + |\eta_2|^2 e^{-\frac{\Delta\Gamma}{2}\Delta\tau} - 2\Re e\left(\eta_1 \eta_2^* e^{-i\Delta m \Delta\tau}\right)\right], \tag{2}$$

where $\Delta\Gamma = \Gamma_S - \Gamma_L$, $\Gamma = (\Gamma_S + \Gamma_L)/2$ and

$$\eta_i = \frac{\langle f_i|T|\mathrm{K_L}\rangle}{\langle f_i|T|\mathrm{K_S}\rangle}, \quad i = 1, 2. \tag{3}$$

In fact, if two identical final states are chosen e.g. as $\pi^+\pi^-$, the aforementioned CPT violation effects can lead to a discrepancy between the η_1 and η_2 parameters due to different momenta of the two kaons. Their momentum dependence would appear in the following form:

$$\eta_1 \simeq \epsilon_K - \delta_K(\vec{p_1}), \tag{4}$$
$$\eta_2 \simeq \epsilon_K - \delta_K(\vec{p_2}), \tag{5}$$

where ϵ_K is the usual parameter of the neutral kaon system related to T and CP violation and δ_K depends on the kaon momentum. Therefore, a measurement of the double decay rate for the $\phi \to \mathrm{K_S K_L} \to \pi^+\pi^-\pi^+\pi^-$ as a function of kaons momenta configuration in a given reference frame allows for extraction of the direction-dependent modulation of δ_K and, as seen from equation 1, for a measurement of the Δa_μ parameters of the quark sector of the Standard Model Extension.

As the frame of fixed stars (sidereal frame) is natural choice of a reference frame to observe the modulation, the set of $\phi \to \mathrm{K_S K_L} \to \pi^+\pi^-\pi^+\pi^-$ events recorded by KLOE was divided into two subsamples of kaon emission angle in the laboratory ($\vec{p_1} \cdot \vec{p_\phi} \lessgtr 0$ where p_1 is the momentum of the kaon with a larger z momentum component of the kaons pair) and four bins of sidereal time corresponding to varying location of the laboratory w.r.t. the sidereal frame due to Earth rotation. The double decay rates $I_{\pi^+\pi^-,\pi^+\pi^-}(\Delta\tau)$ for thus obtained 8 subsamples were simultaneously fit in order to extract the Δa_μ parameters. The $I_{\pi^+\pi^-,\pi^+\pi^-}(\Delta\tau)$ distributions and the fit results are shown in figure 1.

Table 1 presents values Δa_μ SME parameters measured by KLOE. These results, reaching the sensitivity at the level of 10^{-18} GeV expected for the kaon sector, presently are the most precise

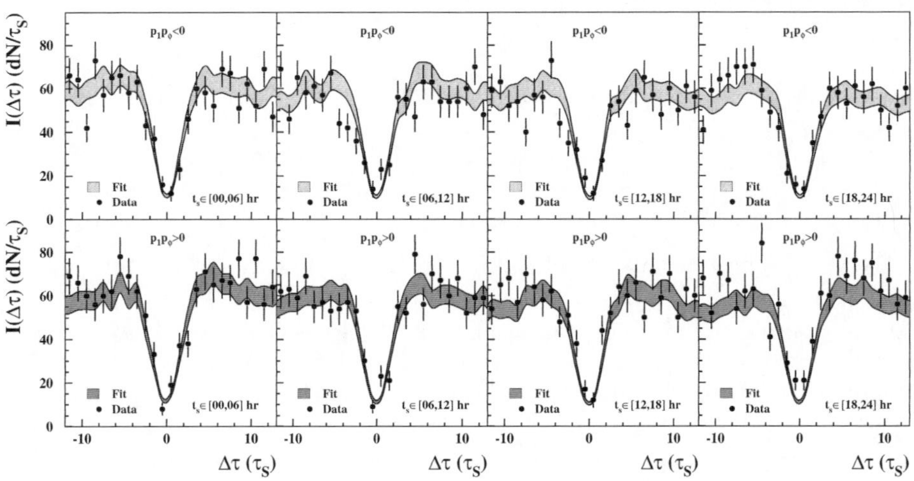

Figure 1. Double decay rates for the $\phi \rightarrow K_S K_L \rightarrow \pi^+\pi^-\pi^+\pi^-$ process (points) obtained for two subsamples of kaon emission angle (top and bottom row) and 4 bins of sidereal time (columns). Bands denote results of the fit. Figure adapted from [3].

measurement for the quark sector of the SME. It is worth noting that these results are dominated by statistical uncertainty, leaving room for improvement by the KLOE-2 detector which aims to collect a data sample of at least 5 fb^{-1}. On top of the larger integrated luminosity, KLOE-2 is also expected ro reduce the systematic error thanks to the improvement of the tracking and vertexing resolutions provided by the new inner tracking detector [4].

Table 1. Values of the SME lagrangian parameters measured by the KLOE experiment [3].

Parameter	KLOE measurement
Δa_0	$(-6.0 \pm 7.7_{stat} \pm 3.1_{syst}) \times 10^{-18}$ GeV
Δa_X	$(0.9 \pm 1.5_{stat} \pm 0.6_{syst}) \times 10^{-18}$ GeV
Δa_Y	$(-2.0 \pm 1.5_{stat} \pm 0.5_{syst}) \times 10^{-18}$ GeV
Δa_Z	$(3.1 \pm 1.7_{stat} \pm 0.6_{syst}) \times 10^{-18}$ GeV

3. Search for quantum decoherence with $\phi \rightarrow K_S K_L \rightarrow \pi^+\pi^-\pi^+\pi^-$

The process involving decays of both entangled kaons into a final state with two charged pions can also be studied to test the basic principles of quantum mechanics by searching for possible decoherence of the kaons state. According to quantum mechanics, the entangled state of the two kaons produced in a ϕ meson decay prohibits their simultaneous decay into an identical final state. This effect is caused by the completely destructive interference between states with identical kaon decay times, which results in a dip in the double decay intensity $I_{\pi^+\pi^-,\pi^+\pi^-}(\Delta\tau)$ for $\Delta\tau \approx 0$ (visible e.g. in the distributions shown in figure 1). However, according to the Furry hypothesis of "spontaneous factorization" [5], a decoherence effect might take place immediately after the $\phi \rightarrow K^0\bar{K}^0$ decay, leading to factorization of the kaons into a non-entangled state. This hypothetical effect violating the quantum mechanics principles can be parametrized by an ζ

IOP Conf. Series: Journal of Physics: Conf. Series **873** (2017) 012025 doi:10.1088/1742-6596/873/1/012025

parameter accounting for weakening of the interference as follows:

$$I(\pi^+\pi^-,\pi^+\pi^-,\Delta t) = \frac{N}{2}\Big[|\langle\pi^+\pi^-,\pi^+\pi^-|\mathrm{K^0\bar{K}^0}(\Delta t)\rangle|^2 + |\langle\pi^+\pi^-,\pi^+\pi^-|\mathrm{\bar{K}^0K^0}(\Delta t)\rangle|^2 - (1-\zeta_{0\bar{0}})\cdot 2\Re\left(\langle\pi^+\pi^-,\pi^+\pi^-|\mathrm{K^0\bar{K}^0}(\Delta t)\rangle\langle\pi^+\pi^-,\pi^+\pi^-|\mathrm{\bar{K}^0K^0}(\Delta t)\rangle^*\right)\Big]. \quad (6)$$

The $\zeta_{0\bar{0}}$ is the decoherence parameter defined in the $\{\mathrm{K^0},\mathrm{\bar{K}^0}\}$ basis and a similar ζ_{SL} parameter may be defined in the $\{\mathrm{K_S},\mathrm{K_L}\}$ basis.

At KLOE, a sample of 1.7 fb^{-1} was used to select $e^+e^- \to \phi \to \mathrm{K_SK_L} \to \pi^+\pi^-\pi^+\pi^-$ events and to determine the $I_{\pi^+\pi^-,\pi^+\pi^-}(|\Delta\tau|)$ double decay rate as a function of the time difference between kaon decays. The residual background originated mostly from non-resonant $e^+e^- \to \pi^+\pi^-\pi^+\pi^-$. The resulting $I_{\pi^+\pi^-,\pi^+\pi^-}(|\Delta\tau|)$ spectrum shown in figure 2 was fitted with the intensity function given by equation 6, including the effects of resolution and of background contamination, which account for the reduction of visible destructive interference pattern in the region around $|\Delta\tau| \approx 0$. With the fit (also shown in figure 2) the following results on the decoherence parameters were obtained [6]:

$$\zeta_{0\bar{0}} = (1.4 \pm 9.5_{stat} \pm 3.8_{syst}) \times 10^{-7}, \quad (7)$$
$$\zeta_{SL} = (0.3 \pm 1.8_{stat} \pm 0.6_{syst}) \times 10^{-2}. \quad (8)$$

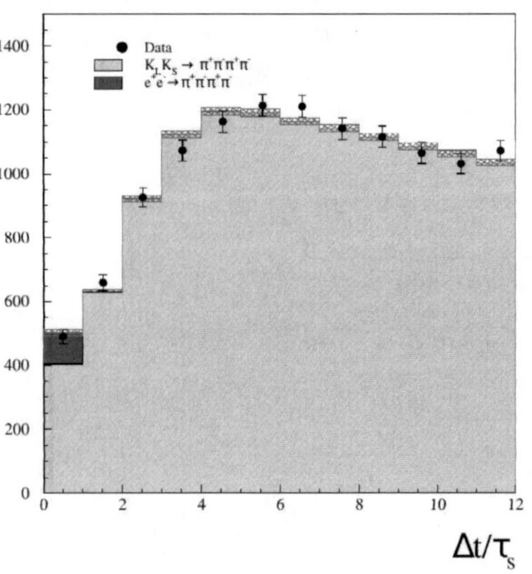

Figure 2. The double decay rate $I_{\pi^+\pi^-,\pi^+\pi^-}(|\Delta\tau|)$ for $e^+e^- \to \phi \to \mathrm{K_SK_L} \to \pi^+\pi^-\pi^+\pi^-$ (black points) and the result of a fit including effects of resolution and background contamination (solid histogram). Red hatched area denotes uncertainty related to the efficiency correction. Figure adapted from [6].

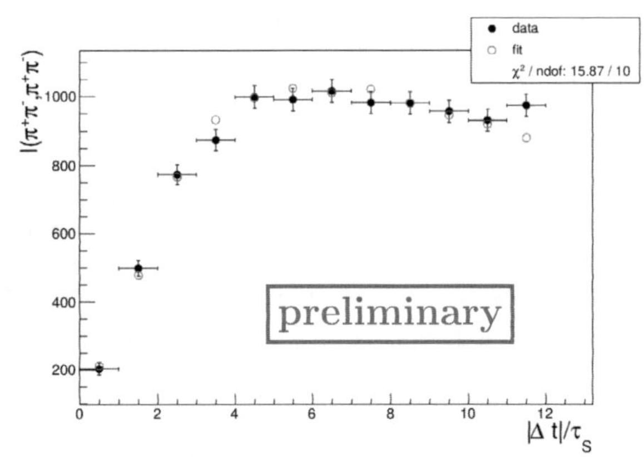

Figure 3. Preliminary double decay rate for $\phi \to \mathrm{K_SK_L} \to \pi^+\pi^-\pi^+\pi^-$ obtained from KLOE data with refined event selection (black circles). Hollow red circles denote the fit result.

The above results are compatible with the predictions of quantum mechanics, showing no indications of decoherence of the kaons' states. The achieved sensitivity on the $\zeta_{0\bar{0}}$ parameter at the level of 10^{-6} is especially good due to the CP suppression in the used decay channel and improves by several orders of magnitude the previous measurement with neutral kaons [7] and a result obtained with B mesons [8]. However, the sensitivity reached by KLOE in this measurement is still limited by the presence of residual background in the interference region (see figure 2). Recently, improved techniques of selection of the $\phi \to K_S K_L \to \pi^+\pi^-\pi^+\pi^-$ events have been devised and applied to KLOE data, yielding a cleaner sample with the same dataset of 1.7 fb^{-1} used in the previous study. The new event selection criteria are similar to those used in the search for CPT and Lorentz invariance violation [3] described in section 2. At present, a re-analysis of KLOE data is in progress and a preliminary $I_{\pi^+\pi^-,\pi^+\pi^-}(\Delta\tau)$ distribution is shown in figure 3 where a more pronounced region of destructive interference ($\Delta\tau \approx 0$) is clearly visible (compare to figure 2). This improved analysis is expected to bring a substantial reduction of the uncertainties on $\zeta_{0\bar{0}}$ and ζ_{SL}.

Moreover, a further improvement of sensitivity on the decoherence parameters is expected from the KLOE-2 experiment, both due to the larger statistical sample and to the use of the inner tracker detector [4], which will improve the tracking performances for $\pi^+\pi^-$ events thus allowing to reduce by a factor of 3 the resolution spread in the $\Delta\tau \approx 0$ region of the double kaon decay rate $I_{\pi^+\pi^-,\pi^+\pi^-}$.

4. Tests of CP and CPT in K_S semileptonic decays

The possible charge asymmetry in semileptonic decays of neutral K mesons, i.e. asymmetry between the $K_{S,L} \to \pi^+e^-\bar{\nu}_e$ and $K_{S,L} \to \pi^-e^+\nu_e$ decays, is closely related to fundamental discrete symmetries as its value may be expressed in terms of the following CP and CPT violating parameters:

$$A_{S,L} = \frac{\Gamma(K_{S,L} \to \pi^-e^+\nu_e) - \Gamma(K_{S,L} \to \pi^+e^-\bar{\nu}_e)}{\Gamma(K_{S,L} \to \pi^-e^+\nu_e) + \Gamma(K_{S,L} \to \pi^+e^-\bar{\nu}_e)} = 2\Big[Re(\epsilon_K) \pm Re(\delta_K) - Re(y) \pm Re(x_-)\Big]. \quad (9)$$

In the above expression, ϵ_K and δ_K are the usual parameters for the kaon system corresponding respectively to CP and CPT violation whereas y describes CPT violation in $K^0(\bar{K^0}) \to \pi e\nu$ decays and x_- is a function of semileptonic kaon decay amplitudes which vanishes if the $\Delta S = \Delta Q$ rule holds. Therefore, if CPT invariance is assumed, the semileptonic asymmetries for K_L and K_S should both reduce to the same value, related to CP violation and expected to be at the level of 10^{-3}:

$$A_S = A_L = 2Re(\epsilon_K) \approx 3 \cdot 10^{-3}. \quad (10)$$

Consequently, measurements of the semileptonic asymmetries in the decays of both short and long-lived neutral kaons not only provide a probe of CP violation but also allow for a test of the CPT symmetry through a comparison between A_S and A_L.

The most precise results available to date comprise a measurement of $A_L = (3.332 \pm 0.058_{stat} \pm 0.047_{syst}) \times 10^{-3}$ by the KTeV experiment [9] and a determination of A_S performed by KLOE using a data sample of 0.41 fb^{-1} with the following result [10]:

$$A_S = (1.5 \pm 9.6_{stat} \pm 2.9_{syst}) \times 10^{-3}. \quad (11)$$

Due to a significant difference of semileptonic decay rates for K_S and K_L, the uncertainty on the A_S asymmetry is the limiting factor of a possible CPT test and its reduction is crucial. The uncertainty on KLOE result is dominated by the statistical error which can be improved by using a larger data sample available from KLOE. Therefore presently the whole KLOE dataset of 1.7 fb^{-1} is subject to an improved analysis aiming at a new determination of the semileptonic asymmetry of the K_S.

In this analysis, K_S decays at KLOE are tagged by an interaction of K_L in the electromagnetic calorimeter and, subsequently, decays involving charged pion and electron/positron tracks are

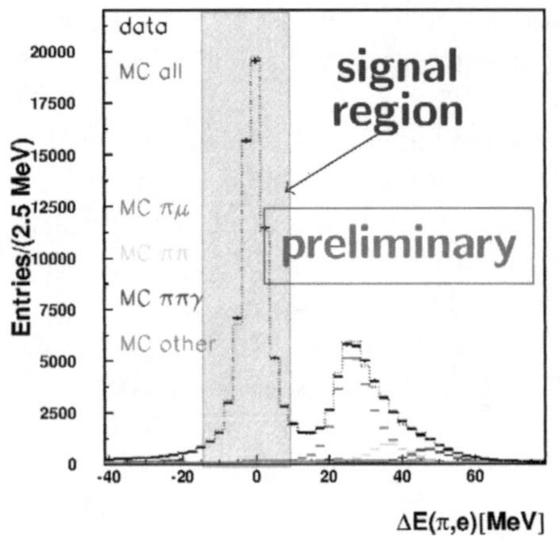

Figure 4. Distribution of the difference between missing energy and missing momentum for selected events. Shaded area denotes the signal region where semileptonic K_S decays with electron and positron are counted. Data is presented in black histogram, total MC is the dotted line and colored histograms show MC-based shapes for major background components used for background normalization and subtraction.

identified based on the particles' time of flight from the decay vertex to the calorimeter. In order to count the events with $K_S \to \pi^+ e^- \bar{\nu}_e$ and $K_S \to \pi^- e^+ \nu_e$, the spectrum of the difference between the missing energy and the missing momentum of the K_S decay $\Delta E(\pi, e)$ is built up, assuming that the two reconstructed tracks are an electron and a pion: as shown in figure 4 the semileptonic decays with an undetected neutrino result in a pronounced peak around zero. A certain amount of background mostly from $K_S \to \pi^+\pi^-(\gamma)(\to \pi\mu\nu)$ is allowed in the final $\Delta E(\pi, e)$ distribution as MC-simulated shapes for the major background components are normalized to the data and subtracted from the events counted in the signal region (see figure 4). More details of this ongoing analysis can be found in reference [11]. As the analyzed data sample is 4 times larger, an twofold reduction of the statistical error is expected. Moreover, the KLOE-2 experiment arises prospects of measuring A_S with the uncertainty at the level of 3×10^{-3}.

5. Test of time-reversal and CPT symmetries in neutral kaon transitions with $\phi \to K_S K_L \to \pi e \nu, 3\pi^0, 2\pi$ decays

Neutral meson pairs in a quantum entangled state provide a unique opportunity to test fundamental symmetries by means of a comparison between a certain transition of the neutral meson state and a process obtained through conjugation of the transition e.g. with the T or CPT operator. The symmetry under reversal in time (T) is a notable case where the processes compared in a direct test are obtained simply by an exchange of initial and final states. Such a test is feasible with the system of neutral K mesons by observing transitions between kaon states in the strangeness basis $\{K^0, \bar{K^0}\}$ and eigenstates of the CP operator corresponding to eigenvalues of +1 and −1, i.e. $\{K_+ \approx K_S, K_- \approx K_L\}$[1] [12].

In this test, the final state of a certain kaon transition is identified in the $\{K^0, \bar{K^0}\}$ basis by observation of its decay through a semileptonic channel with an electron or positron, where the charge of the lepton identifies the strangeness of the decaying kaon according to the $\Delta S = \Delta Q$ rule. Similarly, decays into a hadronic final state with $\pi^+\pi^-$ (CP=+1) or $3\pi^0$ (CP=-1) define the kaon state in the CP basis $\{K_+, K_-\}$. The identification of the initial state of a kaon transition must be performed before the decay occurs, and it is only possible by exploiting the quantum entanglement of neutral kaon pairs produced in a ϕ decay. An observation of the earlier between the two kaon decays provides information about the state of its partner particle being in an orthogonal state at the same moment of time.

Out of four possible transitions which may be compared with their T-conjugates, two can be

[1] While K_L and K_S are not exact CP eigenstates, it was shown that kaons mixing due to CP violation has a negligible effect on the T symmetry test [12]

used at KLOE-2 to determine with high statistics the asymmetry ratios dependent on the time difference Δt between the two kaon decays (refer to [12] for a detailed study). The probabilities entering these ratios are estimated by double decay rates $I(f_1, f_2; \Delta t)$ corresponding to processes characterized by a pair of kaon decays into specific final states f_1 and f_2 at times separated by Δt:

$$R_2(\Delta t) \quad = \frac{P[\mathrm{K}^0(0) \to \mathrm{K}_-(\Delta t)]}{P[\mathrm{K}_-(0) \to \mathrm{K}^0(\Delta t)]} \sim \frac{\mathrm{I}(\pi^+ e^- \bar{\nu}, 3\pi^0; \Delta t)}{\mathrm{I}(\pi\pi, \pi^- e^+ \nu; \Delta t)} \tag{12}$$

$$R_4(\Delta t) \quad = \frac{P[\bar{\mathrm{K}}^0(0) \to \mathrm{K}_-(\Delta t)]}{P[\mathrm{K}_-(0) \to \mathrm{K}^0(\Delta t)]} \sim \frac{\mathrm{I}(\pi^- e^+ \nu, 3\pi^0; \Delta t)}{\mathrm{I}(\pi\pi, \pi^+ e^- \bar{\nu}; \Delta t)} \tag{13}$$

It was shown that in the asymptotic region of large time differences ($\Delta t \gg \tau_S$ where τ_S is the lifetime of $\mathrm{K_S}$) the deviation of $R_2(\Delta t)$ and $R_4(\Delta t)$ from unity should be proportional to the real part of ϵ_K which is a T violating parameter of the neutral kaon system [12].

It is worth noting that although several experiments exist with access to entangled B mesons and such time-reversal symmetry test has been implemented at the BaBar experiment [13], KLOE-2 is the only existing experimental setup with entangled kaons, and represents the unique possibility to perform T symmetry tests in the neutral kaon system. With its goal integrated luminosity, a statistically significant test of time reversal symmetry is expected with a sensitivity at the level of 10^{-3}.

Moreover, the aforementioned concept of direct discrete symmetry test in transitions can be extended to the CPT symmetry by an appropriate choice of the compared processes. Thus, a similar set of time-dependent ratios sensitive to CPT violation is measurable at KLOE-2 [14]:

$$R_{2,CPT}(\Delta t) \quad = \frac{\mathrm{P}[\mathrm{K}^0(0) \to \mathrm{K}_-(\Delta t)]}{\mathrm{P}[\mathrm{K}_-(0) \to \overline{\mathrm{K}}^0(\Delta t)]}, \tag{14}$$

$$R_{4,CPT}(\Delta t) \quad = \frac{\mathrm{P}[\overline{\mathrm{K}}^0(0) \to \mathrm{K}_-(\Delta t)]}{\mathrm{P}[\mathrm{K}_-(0) \to \mathrm{K}^0(\Delta t)]}. \tag{15}$$

Additionally, in case of the CPT test, the above ratios may be combined into a double ratio which is a very robust CPT-violation observable in the $\Delta t \gg \tau_S$ limit:

$$\frac{R^{exp}_{2,CPT}(\Delta t \gg \tau_S)}{R^{exp}_{4,CPT}(\Delta t \gg \tau_S)} = 1 - 8\Re\delta_K - 8\Re x_-. \tag{16}$$

Any deviation from unity in this observable is produced by a genuine CPT noninvariance effect without the need for $\Delta S = \Delta Q$ rule assumption and with a negligible impact of CP violation [14]. The value of this CPT-violation observable has never been measured before. Therefore, determination of $R_{2,CPT}(\Delta t)$ and $R_{4,CPT}(\Delta t)$ establishes a test of CPT symmetry complementary to the one described in section 4 and is one of the goals of the KLOE-2 experiment along with the time-reversal symmetry test.

The experimental realization of both aforementioned tests requires selection and reconstruction of two classes of events: $\mathrm{K_S K_L} \to \pi^+\pi^-\pi e\nu$ and $\mathrm{K_S K_L} \to \pi e\nu\, 3\pi^0$. Methods to select the signal and to reconstruct the kaon decay times with sufficient resolution are already being devised using both KLOE dataset and KLOE-2 data collected so far. Among the kaon decays of interest, $\mathrm{K_L} \to 3\pi^0 \to 6\gamma$ involving no charged particles is of special importance as its reconstruction must be based solely on information from the electromagnetic calorimeter. A dedicated reconstruction algorithm has been prepared in order to allow for analytical calculation of the $\mathrm{K_L}$ decay location and time. Is is based on the trilateration technique also used in GPS positioning and was tested to provide a resolution of $\mathrm{K_L}$ decay time of about 2 τ_S [15]. A preliminary spectrum of the double decay rate $\mathrm{I}(\pi e\nu, 3\pi^0; \Delta t)$ at its present stage of event selection and obtained with the dedicated $\mathrm{K_L} \to 3\pi^0 \to 6\gamma$ reconstruction is shown in figure 5. The major residual background component is presently due to $\mathrm{K_S} \to \pi^+\pi^-$ and methods to purify the sample are being elaborated.

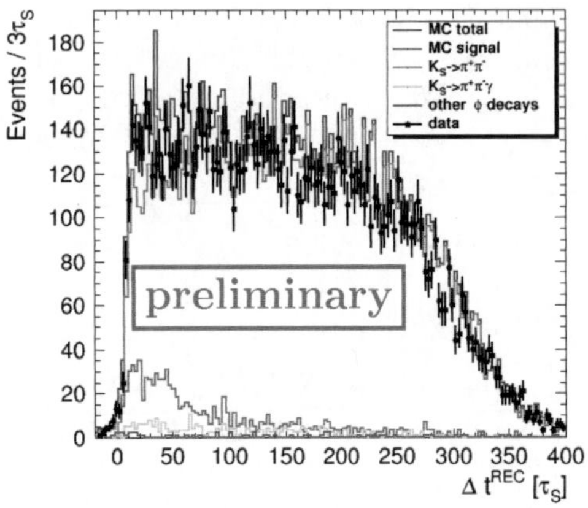

Figure 5. Raw distribution of double decay rate $I(\pi e\nu, 3\pi^0; \Delta t)$ for $K_S K_L \to \pi e \nu\, 3\pi^0$ events as a function of time difference between kaon decays. The spectrum was obtained with KLOE data (black points) and corresponding MC simulations (histograms). The major residual background component is presently due to $K_S \to \pi^+\pi^-$.

6. Search for CPT violation with rare K_S decays

Another possible approach to study the mechanism of CP violation is to search for CP-forbidden rare decays of neutral K mesons, e.g. decay of K_S into three pions. The $K_S \to 3\pi^0$ decay is a pure CP-violating process whereas $K_S \to \pi^+\pi^-\pi^0$ violates CP for I=1 or 3 in the final state.

CP violation in the system of neutral kaons is usually expressed in terms of the following amplitude ratios (for K_S decays):

$$\eta_{+-0} = \frac{\langle \pi^+\pi^-\pi^0 | H | K_S \rangle}{\langle \pi^+\pi^-\pi^0 | H | K_L \rangle} = \epsilon_K + \epsilon'_{+-0}, \quad (17)$$

$$\eta_{000} = \frac{\langle \pi^0\pi^0\pi^0 | H | K_S \rangle}{\langle \pi^0\pi^0\pi^0 | H | K_L \rangle} = \epsilon_K + \epsilon'_{000}, \quad (18)$$

where ϵ_K is the usual parameter of CP violation in mixing and ϵ'_{+-0} and ϵ'_{000} describe CP violation in decay.

Although the $K_S \to 3\pi^0$ decay has never been observed to date, the KLOE experiment has set the most precise upper limit on its branching ratio using a dataset of 1.7 fb^{-1}. The presence of short lived neutral kaons was tagged by detecting the associated K_L interaction in the calorimeter, similarly to the case of the charge asymmetry measurement described in section 4. The 6 photon candidates from $K_S \to 3\pi^0$ are then reconstructed based on calorimeter information and a kinematic fit is applied. Finallly, a discriminant analysis based on the kinematical difference between the signal and the $K_S \to 2\pi^0$ overwhelming background is performed. For further details of the analysis the reader is referred to [16].

The upper limit obtained by KLOE [16] amounts to:

$$\mathrm{BR}(K_S \to 3\pi^0) < 2.6 \times 10^{-8}, \quad (19)$$

which yields the limit $|\eta_{000}| < 0.0088$ at 90% confidence level. Further improvement is expected from the KLOE-2 experiment which is going to collect at least 5 fb^{-1} of data allowing to push its sensitivity to $\mathrm{BR}(K_S \to \pi^0\pi^0\pi^0)$ below 10^{-8}.

The same KLOE dataset used to study $K_S \to 3\pi^0$ is presently being explored in a direct search for K_S decays into the $\pi^+\pi^-\pi^0$ final state. Current knowledge of the branching ratio for this process is an average of three indirect measurements and amounts to $\mathrm{BR}(K_S \to \pi^+\pi^-\pi^0) = 3.5^{+1.1}_{-0.9} \times 10^{-7}$ [17] with uncertainty at the level of 30%. The ongoing analysis of KLOE data shows prospects to improve this result by a direct measurement with relative uncertainty below 20%.

7. Summary and perspectives

The KLOE experiment in the past years already produced important contributions to a variety of discrete symmetry studies, such as the search for Lorentz invariance and, consequently,

CPT symmetry tests, CP and CPT testing with charge asymmetry in semileptonic kaon decays, searches for deviation from quantum mechanics principles manifested by decoherence of entangled kaons' state and searches for rare kaon decays. Despite that, the data collected by KLOE still did not reveal all its secrets and currently two enhanced re-analyses as well as a new search for rare $K_S \to \pi^+\pi^-\pi^0$ are in progress to further improve the sensitivity to discrete symmetries' violation.

At the same time, the KLOE-2 detector is half way through its new datataking campaign with novel subdetectors and larger integrated luminosity. These two factors are expected to allow for an improvement of all of the KLOE measurements reported in this work.

Acknowledgments

We warmly thank our former KLOE colleagues for the access to the data collected during the KLOE data taking campaign. We thank the DAΦNE team for their efforts in maintaining low background running conditions and their collaboration during all data taking. We want to thank our technical staff: G.F. Fortugno and F. Sborzacchi for their dedication in ensuring efficient operation of the KLOE computing facilities; M. Anelli for his continuous attention to the gas system and detector safety; A. Balla, M. Gatta, G. Corradi and G. Papalino for electronics maintenance; M. Santoni, G. Paoluzzi and R. Rosellini for general detector support; C. Piscitelli for his help during major maintenance periods. This work was supported in part by the EU Integrated Infrastructure Initiative Hadron Physics Project under contract number RII3-CT- 2004-506078; by the European Commission under the 7th Framework Programme through the 'Research Infrastructures' action of the 'Capacities' Programme, Call: FP7-INFRASTRUCTURES-2008-1, Grant Agreement No. 227431; by the Polish National Science Centre through the Grants No. 2011/03/N/ST2/02652, 2013/08/M/ST2/00323, 2013/11/B/ST2/04245, 2014/14/E/ST2/00262, 2014/12/S/ST2/00459.

References

[1] Kostelecky V A 2001 *Phys. Rev.* **D64** 076001 (*Preprint* hep-ph/0104120)
[2] Greenberg O W 2002 *Phys. Rev. Lett.* **89** 231602 (*Preprint* hep-ph/0201258)
[3] Babusci D *et al.* (KLOE-2) 2014 *Phys. Lett.* **B730** 89–94 (*Preprint* hep-ex/1312.6818)
[4] Balla A *et al.* 2011 *Nucl. Instrum. Meth.* **A628** 194–198 (*Preprint* physics.ins-det/1003.3770)
[5] Furry W H 1936 *Phys. Rev.* **49**(5) 393–399
[6] Di Domenico A (KLOE-2) 2009 *Journal of Physics: Conference Series* **171** 012008
[7] Bertlmann R A, Grimus W and Hiesmayr B C 1999 *Phys. Rev. D* **60**(11) 114032
[8] Go A *et al.* (Belle Collaboration) 2007 *Phys. Rev. Lett.* **99**(13) 131802
[9] Alavi-Harati A *et al.* (KTeV) 2002 *Phys. Rev. Lett.* **88** 181601 (*Preprint* hep-ex/0202016)
[10] Ambrosino F *et al.* (KLOE) 2006 *Phys. Lett.* **B636** 173–182 (*Preprint* hep-ex/0601026)
[11] Kamińska D (KLOE-2) 2015 *J. Phys. Conf. Ser.* **631** 012042 (*Preprint* hep-ex/1509.01125)
[12] Bernabeu J, Di Domenico A and Villanueva-Perez P 2013 *Nucl. Phys.* **B868** 102–119 (*Preprint* hep-ph/1208.0773)
[13] Lees J P *et al.* (BaBar) 2012 *Phys. Rev. Lett.* **109** 211801 (*Preprint* hep-ex/1207.5832)
[14] Bernabeu J, Di Domenico A and Villanueva-Perez P 2015 *JHEP* **10** 139 (*Preprint* hep-ph/1509.02000)
[15] Gajos A (KLOE-2) 2015 *Acta Phys. Polon.* **B46** 13–18 (*Preprint* hep-ex/1501.04801)
[16] Babusci D *et al.* (KLOE) 2013 *Phys. Lett.* **B723** 54–60 (*Preprint* hep-ex/1301.7623)
[17] Patrignani C *et al.* (Particle Data Group) 2016 *Chin. Phys.* **C40** 100001

IOP Conf. Series: Journal of Physics: Conf. Series **873** (2017) 012026 doi:10.1088/1742-6596/873/1/012026

An Overview of BESIII recent results

Yadi Wang

Helmholtz-Institut Mainz, Staudingerweg 18, Mainz, DE

E-mail: yadiwang@uni-mainz.de

Abstract. Working in the center-of-mass energy region from 2 to 5 GeV, BESIII has published many interesting and important results in a broad range of physics topics, including charmonium and charm physics, hadron studies, determination of the tau mass, R measurements, and investigations of the still-mysterious XYZ particles. This paper reviews a selection of results from BESIII.

1. Introduction

The third generation of Beijing Spectrometer experiments (BESIII) operating at the second generation of Beijing Electron Positron Collider (BEPCII), has been run for eight years. The BESIII detector has consisted a single small-celled helium-based MDC, a plastic scintillator TOF system, a CsI(Tl) electromagnetic calorimeter, a 1.0 T superconducting magnet, and a muon counter with 9 resistive plate chamber (RPC) laryers in the barrel part and 8 in the end-cap portions interleaved in the steel of the flux return yoke. More details of the BESIII detector can be found in Ref. [1].

From the first event in 2008, BESIII has data sets at many Center-of-mass (CM) energies between 2 to 5 GeV, and very large data sets at the J/ψ (1.3 billion events), $\psi(2S)(2S)$ (0.45 billion), and $\psi(3770)$ (2.9 fb^{-1}). These are the world's largest exclusive charmonium data sets and allow for many precision measurements. In the following, we reported the results from: $D_{(s)}$, Λ_c, XYZ particles, light hadron progress, proton form factor and τ mass measurement, new physics.

2. Charmed meson Physics

At the $\psi(3770)$ resonance or 4.009 GeV, $D\bar{D}$ or D_sD_s pairs are produced with no other accompanying particles. Thereby the data sets at $\psi(3770)$ and 4.009 GeV can provide extremely clean and pure charmed-meson signals. The quantum coherence of the two produced D mesons provides opportunities to measure $D^0 - \bar{D}^0$ mixing parameters, determine strong phases, and search for CP violation. In addition, two elements of Cabibbo-Kobayashi-Maskawa (CKM) matrix which parameterizes the mixing between the quark flavors in weak interaction can be precisely measured in $D_{(s)}$ decays. Precise measurements of these elements are very important in testing the Standard Model (SM) and searching for New Physics (NP) beyond the SM. Any improved measurement of these elements would be the important input for precision test of the SM.

The following results in this section are based on the tagged technique and 2.93 fb^{-1} data at $\sqrt{s} = 3.773$ and 482 pb^{-1} data at $\sqrt{s} = 4.009$ GeV. (Throughout this paper, the inclusion of charge conjugate channels is implied.) Since the $D_{(s)}\bar{D}_{(s)}$ are pair produced just above threshold, the identification of one $D_{(s)}$ meson from a subset of tracks in an event guarantees that the remaining tracks have originated from the decay of the recoiling $\bar{D}_{(s)}$. Once a tag is found, the recoil tracks can be examined for the decay mode of interest. Frequently two variables, beam-constrained mass M_{BC} and energy difference ΔE, are used to identify the signals, defined as: $M_{BC} = \sqrt{E^2_{beam} - |\vec{p}_{D_{(s)}}|^2}$ and $\Delta E = E_{D_{(s)}} - E_{beam}$, where $E_{D_{(s)}}$ and $\vec{p}_{D_{(s)}}$ are the energies and momenta of the $D_{(s)}$ decay products in the center-of-mass system of the $\psi(3770)$ or 4.009 GeV. For true $D_{(s)}$ candidates, ΔE will be consistent with zero, and M_{BC} will be consistent with the $D_{(s)}$ mass.

2.1. Leptonic decays of $D^+_{(s)}$

The $D^+_{(s)}$ meson can decay into $l^+\nu_l$ (where $l = e$, μ or τ) via annihilation mediated by a virtual W^+ boson. In the SM, the decay width is given by

$$\Gamma(D^+_{(s)} \to l^+\nu_l) = \frac{G_F^2}{8\pi} f^2_{D^+_{(s)}} m_l^2 m_{D^+_{(s)}} \left(1 - \frac{m_l^2}{m^2_{D^+_{(s)}}}\right)^2 |V_{cd(s)}|^2, \tag{1}$$

where G_F is the Fermi coupling constant, $V_{cd(s)}$ is the $c \to d(s)$ Cabibbo-Kobayashi-Maskawa (CKM) matrix element, m_l is the lepton mass, and $m_{D^+_{(s)}}$ is the mass of $D^+_{(s)}$. All of the strong interaction effects between the two initial-state quarks are absorbed into the decay constant $f_{D^+_{(s)}}$. The decay constant is a very fundamental parameter that can be used to test our knowledge of hadronic dynamics. Moreover, the precise measurement of f_{D^+} can validate theoretical calculations of f_B, which is a parameter of critical importance for B physics.

For the analysis of $D^+ \to \mu^+\nu_\mu$, nine hadronic modes $K^+\pi^-\pi^-$, $K_s\pi^-$, K_sK^-, $K^+K^-\pi^-$, $K^+\pi^-\pi^-\pi^0$, $\pi^+\pi^-\pi^-$, $K_s\pi^-\pi^0$, $K^+\pi^-\pi^-\pi^-\pi^+$ and $K_s\pi^-\pi^-\pi^+$ are used for tagging D^-. The signal of D^- is clearly observed in M_{BC} spectra. A maximum likelihood fit to the M_{BC} spectra gives the number of the observed D^- events. The $D^+ \to \mu^+\nu_\mu$ decays are selected on the recoil side by requiring no extra tracks except a muon. The information of the neutrino is inferred from the variable of a missing mass. After subtracting the backgrounds and correcting for the reconstruction efficiency, the branching fraction is measured to be $\mathcal{B}(D^+ \to \mu^+\nu_\mu) = (3.71 \pm 0.19 \pm 0.06) \times 10^{-4}$. With the PDG value of D^+ lifetime, m_D, m_μ and $|V_{cd}|$, f_D is determined to be $(203.2 \pm 5.3 \pm 1.8)$ MeV. Alternatively, with the value of f_{D^+} from PDG, the $|V_{cd}|$ is calculated to be $(0.2210 \pm 0.0058 \pm 0.0047)$ [2].

For the D_s leptonic decays, nine hadronic modes of D_s^- to K_sK^-, $K^+K^-\pi^-$, $K^+K^-\pi^-\pi^0$, $K_sK^+\pi^-\pi^-$, $\pi^+\pi^-\pi^-$, $\pi^-\eta$, $\pi^-\pi^0\eta$, $\pi^-\eta'(\eta' \to \pi^+\pi^-\eta)$ and $\pi^-\eta'(\eta' \to \rho^0\gamma)$ are used. On the recoil side, $D^- \to \mu^+\nu_{mu}$ and $\tau^+\nu_\tau$ are selected by requiring there is only one charged track to be identified as μ or τ. Fitting on the missing mass distribution, the number of events are extracted and the corresponding branching fractions are calculated as $\mathcal{B}(D_s^+ \to \mu^+\nu_{tau}) = (0.495 \pm 0.067 \pm 0.026)\%$, and $\mathcal{B}(D_s^+ \to \tau^+\nu_\tau) = (4.83 \pm 0.65 \pm 0.26)\%$. The decay constant is then extracted as $241.0 \pm 16.3 \pm 6.5$ MeV [3].

2.2. Semileptonic decays of D

The semileptonic charm meson decay process is a good laboratory for both studying the quark-mixing mechanism and testing theoretical techniques developed for calculating the hadronic matrix element. In the SM, neglecting the lepton mass, the differential decay rate for $D^+ \to Pe^+\nu_e$ ($P = K^-, \pi^-, \bar{K}^0$ or π^0) is given by

$$\frac{d\Gamma}{dq^2} = X\frac{G_F^2}{24\pi^3}|V_{cd(s)}|^2 p^3 |f_+(q^2)|^2, \tag{2}$$

where p is the momentum of the pseudoscalar meson P in the rest frame of the D meson, and $f_+(q^2)$ is the form factor representing the hadronic form factors of the hadronic week current that depends on the square of the four-momentum transfer $q = p_{D_{(s)}} - p_P$. In Eq. 2, X is a factor due to isospin, which equals to 1 for $D^0 \to K^- e^+ \mu_e$, $D^0 \to \pi^- e^+ \mu_e$, $D^+ \to \bar{K}^0 e^+ \mu_e$ and 1/2 for $D^+ \to \pi^0 e^+ \mu_e$.

By tagging $\bar{D}^0$ decays to $K^+\pi^-$, $K^+\pi^-\pi^0$, $K^+\pi^+\pi^-\pi^-$, $K^+\pi^+\pi^+\pi^-\pi^-\pi^0$ and $K^+\pi^-\pi^0\pi^0$, the events of D^0 decays $K^- e^+ \mu_e$ and $\pi^- e^+ \mu_e$ are selected on the recoil side. By analyzing these events, the branching fraction are measured to be $\mathcal{B}(D^0 \to K^- e^+ \mu_e) = (3.505 \pm 0.14 \pm 0.033)\%$ and $\mathcal{B}(D^0 \to \pi^- e^+ \mu_e) = (0.295 \pm 0.004 \pm 0.003)\%$. The differential decay rates are measured in q^2 bins precisely. The product $f_+(0)|V_{cs(d)}|$ are extracted from the fit to the differential decay rates with different models. In conjunction with the values of $|V_{cs(d)}|$ determined from a global SM fit, the form factor is determined as $f_+^{D\to K}(0) = 0.7368 \pm 0.0026 \pm 0.0036$, $f_+^{D\to\pi}(0) = 0.6372 \pm 0.0080 \pm 0.0044$. Alternatively, the $|V_{cs(d)}|$ is extracted using the recent lattice QCD results of form factors: $V_{cs} = 0.9601 \pm 0.0033 \pm 0.0047 \pm 0.0239_{\rm LQCD}$ and $|V_{cd}| = 0.2155 \pm 0.0027 \pm 0.0014 \pm 0.0094_{\rm LQCD}$ [4]. With these precise results, the PDG group issued a new world weighted average value for $|V_{cd}|$ which is $\sim 48\%$ improved in precision. The PDG checked the unitarity of the CKM matrix, and indeed found about 2σ tension with three-generation unitarity [5, 6]. Similarly, $D^+ \to K^0 e^+ \nu_e$ and $D^+ \to \pi^0 e^+ \nu_e$ are analyzed. The numerical results are consistent with the charged decay modes.

3. Charmed baryon physics

Unlike the charmed meson decays, charmed baryon decays suffer from great uncertainties experimentally. Thus, the improved measurement of the charmed baryon Λ_c decays is of crucial importance for that they could both provide plentiful information for the dynamics in the charm sector and improved input for beauty physics. Utilizing the 567 pb^{-1} data at $\sqrt{s} = 4.599$ GeV which is just above the $\Lambda_c^+\bar{\Lambda}_c^-$ threshold, BESIII performed a measurement of the absolute branching fraction for 12 Cabibbo-favored Λ_c^+ hadronic decay modes [7]. With the single tag and double tag technique, the absolute branching fractions of Λ_c have been measured with 3-6 times better precision than the world average values. For the golden mode $\mathcal{B}(\Lambda_c^+ \to pK^-\pi^+)$, our result is about 2σ lower than Belle result.

The measurement of Λ_c decaying to neutron involved final states is very rare. It has been argured that the isospin symmetry works well in the charged baryon section. The relative branching fraction of the neutron-involving decay to the proton-involving decay provides an important observable in testing isospin symmetry in Λ_c decays. It will also help to explore the final state interactions. A measurement of $\Lambda_c \to n\bar{K}_0\pi^+$ is finished at BESIII. With 11 single tag modes, a clean Λ_c^+ sample is selected. The signal of $\Lambda_c^+ \to nK_s\pi^+$ is get by a simultaneous fit on the mass spectrum of M^2_{miss} (the mass of neutron) and $M_{\pi^+\pi^-}$ (the mass of K_s) and the mass spectrums from events of Λ_c^- sideband, shown in Fig. 1 (the left four plots). The branching fraction is measured as $\mathcal{B}(\Lambda_c^+ \to nK_s\pi^+) = (1.82 \pm 0.23 \pm 0.11)\%$.

The branching fraction for $\Lambda_c^+ \to \Lambda e^+ \nu_e$ is the benchmark for those of other Λ_c semileptonic channels. The absolute branching fraction of it is measured in 11 tag modes. From the U_{miss} distribution of neutrino which is shown in Fig. 1 (b), the number of events are extracted and the branching fraction is then determined as $(3.63 \pm 0.38 \pm 0.20)\%$ which improves the precision of the world average value by more than twofold [8].

4. *XYZ* particles

From 2013, many fruitful results from BESIII about the XYZ have extracted the most interest in particle physics. Up to now, four charged Z_c states with probable the corresponding neutral states have been found. Their masses, widths, and original decay channels are listed in Table 1.

IOP Conf. Series: Journal of Physics: Conf. Series **873** (2017) 012026 doi:10.1088/1742-6596/873/1/012026

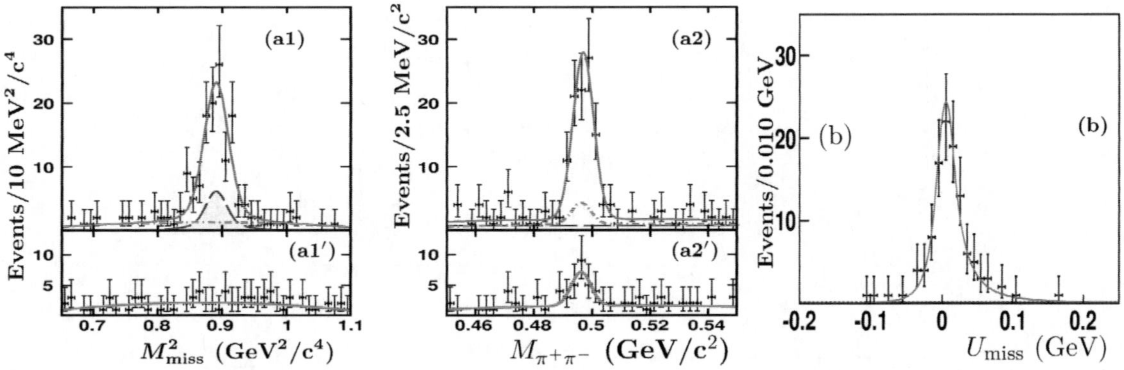

Figure 1. (a) For the $\Lambda_c \to nK_s\pi$ analysis: the simultaneous fit to M^2_{miss} and $M_{\pi^+\pi^-}$ of events in (a1, a2) the $\bar{\Lambda}_c^-$ signal region and (a1′, a2′) sideband regions. (b)For the $\Lambda_c \to \Lambda e\nu_e$ analysis: fit to the U_{miss} distribution within the Λ signal region.

Table 1. The summary of the found Z_c states.

Particles	channel	Mass (MeV/c^2)	Width (MeV)
$Z_c^{\pm}(3900)$	$e^+e^- \to \pi^+\pi^- J/\psi$	$3899.0 \pm 3.6 \pm 4.9$	$46 \pm 10 \pm 20$
$Z_c^0(3900)$	$e^+e^- \to \pi^0\pi^0 J/\psi$	3894.8 ± 2.3	29.6 ± 8.2
$Z_c^{\pm}(3885)$	$e^+e^- \to \pi(D^\star D)^{\pm}$	$3882.2 \pm 1.1 \pm 1.5$	$26.5 \pm 1.7 \pm 2.1$
$Z_c^0(3885)$	$e^+e^- \to \pi(D^\star D)^0$	$3885.7 \pm 5.7 \pm 8.4$	$35 \pm 12 \pm 15$
$Z_c^{\pm}(4020)$	$e^+e^- \to \pi^+\pi^- h_c$	$4022.9 \pm 0.8 \pm 2.7$	$7.9 \pm 2.7 \pm 2.6$
$Z_c^0(4020)$	$e^+e^- \to \pi^0\pi^0 h_c$	$4023.9 \pm 2.2 \pm 3.8$	Fixed at $Z_c^{\pm}(4020)$
$Z_c^{\pm}(4025)$	$e^+e^- \to \pi(D^\star D^\star)^{\pm}$	$4026.3 \pm 2.6 \pm 3.7$	$24.8 \pm 5.6 \pm 7.7$
$Z_c^0(4025)$	$e^+e^- \to \pi(D^\star D^\star)^0$	$4025.5 \pm 4.7 \pm 3.1$	$23.0 \pm 6.0 \pm 1.0$

For the Y state, the first to be discovered is $Y(4260)$, which is originally seen as a peak in the $e^+e^- \to \pi^+\pi^- J/\psi$ cross section. New results from BESIII, however, show that the $Y(4260)$ is not a simple peak. Using both a small number of high-statistics data points and a large number of low-statistics data points, the cross section of $e^+e^- \to \pi^+\pi^- J/\psi$ is renewed, shown in Fig. 2 (a,b). The peak that was formerly known as the $Y(4260)$ can be better described as a combination of two peaks, one with a mass of $4220.0 \pm 3.1 \pm 1.4$ MeV/c^2 and width of $44.1 \pm 4.1 \pm 2.0$ MeV and one with a mass of $4320.0 \pm 10.4 \pm 7.0$ MeV/c^2 and width of $101.4^{+25.3}_{-19.7} \pm 10.2$ MeV [9].

Similary, the $e^+e^- \to \pi^+\pi^- h_c$ cross section is much more complex than previously suspected [10]. The cross section can also be described as two peaks, shown in Fig. 2 (c), one with a mass of $4218.4 \pm 4.0 \pm 0.9$ MeV/c^2 and width of $66.0 \pm 9.0 \pm 0.4$ MeV and one with a mass of $4391.6 \pm 6.3 \pm 1.0$ MeV/c^2 and width of $139.5 \pm 16.1 \pm 0.6$ MeV. The parameters of the lighter peak agree with the parameters of the lighter peak in $\pi^+\pi^- J/\psi$.

An analysis with preliminary result on the cross section of $e^+e^- \to \pi^+\pi^-\psi(2S)$ is also made. The new measurements are in agreement with those from Belle and BaBar, which were used to determine the paramters of the $Y(4360)$.

Another well-known mystery in the charmonium spectrum is the nature of the $X(3872)$. Using high-statistics data above 4.0 GeV, BESIII observed the process $e^+e^- \to \gamma X(3872), X(3872) \to$

IOP Conf. Series: Journal of Physics: Conf. Series **873** (2017) 012026 doi:10.1088/1742-6596/873/1/012026

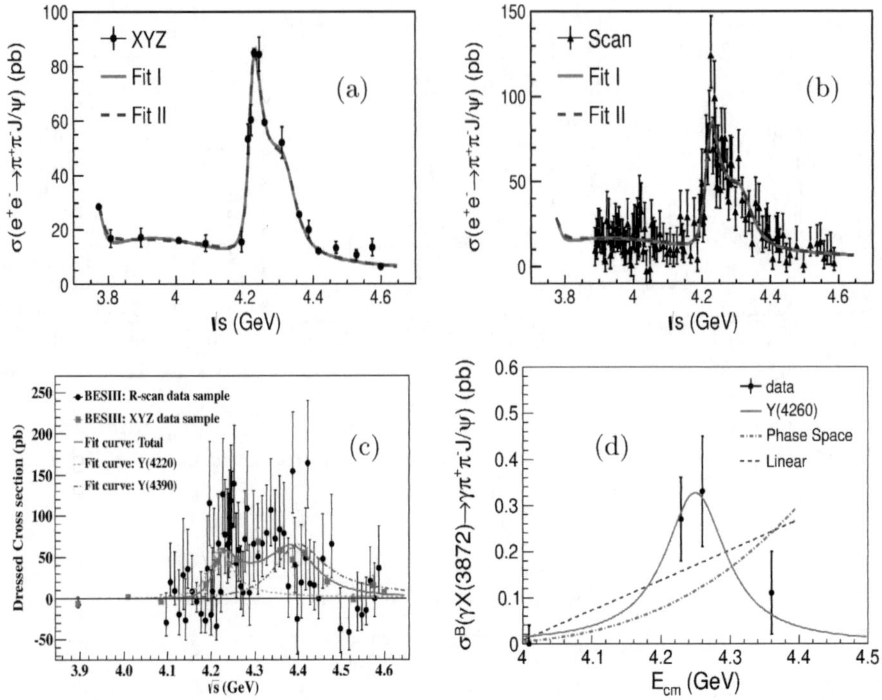

Figure 2. The (a) and (b) plots are for the measured cross section of $e^+e^- \to \pi^+\pi^- J/\psi$ using high-statistics data point (left) and a larger number of low-statistics data points. The plot of (c) is the measured cross section of $e^+e^- \to \pi^+\pi^- h_c$. The plot of (d) is the measured cross section of $e^+e^-\gamma X(3872) \to \gamma\pi^+\pi^- J/\psi$.

$\pi^+\pi^- J/\psi$ [11]. The cross section shows a shape that is consistent with a peak between 4.2 and 4.3 GeV, shown in Fig. 2 (d), which may indicate a connection between the $X(3872)$ and the Y states. More data is needed to resolve this issue, and find more connections among X, Y and Z states.

5. The $X(1835)$

Since its discovery, the $X(1835)$ was subject of different speculations about its nature. The possible interpretations include $p\bar{p}$ bound state, glueball, radia excitation of η' meason, and so on. The lowest lying pseudoscalar glueball meson predicted by the Lattice QCD is expected to have a mass around 2.3 GeV/c^2. Its decay dynamics suggests that it may have properties in common with the η_c, of which one of the strongest decay channel is $\pi^+\pi^-\eta'$. The $X(1835)$ was found in the $\pi^+\pi^-\eta'$ mass distribution at BESII, and confirmed at BESIII. The $X(1835)$ is also observed in the $\eta K_s K_s$ channel in J/ψ radiative decays, where its spin parity was determined as $J^P = 0^-$ by a partial wave analysis [12]. At the same time, an anomalously strong enhancement at the $p\bar{p}$ mass threshold, dubbed as $X(p\bar{p})$, was first observed by BESII in $J/\psi \to \gamma p\bar{p}$, and then confirmed at BESIII and CLEO. The quantum numbers were subsequently determined as $J^P = 0^-$ by BESIII.

With the 1.3 billion data of J/ψ, a significant abrupt change in slope of the $X(1835) \to \pi^+\pi^-\eta'$ line shape is observed. This indicates a connection between $X(1835)$ and $X(p\bar{p})$. The invariant mass spectrum lineshape around 1.85 GeV/c^2 might be characterized by two models.

The first one takes the advantage of the Flatté formula to describe the opening of an additional decay mode, and a resonance at $X(1920)$ with a significance of 5.7σ has be included, shown in Fig. 3 (a). The second model is based on the coherent sum of two Breit-Wigner amplitudes, the $X(1835)$ and a narrow $X(1870)$, and the fit plot is shown in Fig. 3 (b). Both models support the existence of a $p\bar{p}$ moleculelike or bound state. More data are needed to further study the structure around 1.85 GeV$/c^2$, and the line shapes from other decay modes are also needed.

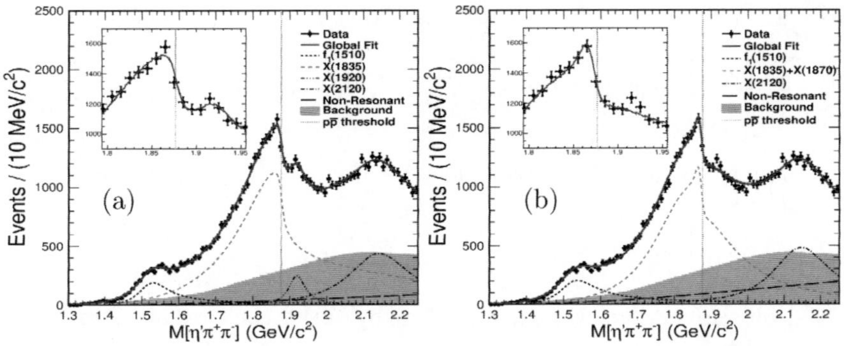

Figure 3. Fit results of the mass spectrum of $\pi^+\pi^-\eta'$ from J/ψ radiative decay with (a) the Flatté formula and the $X(1920)$ resonance and (b) a coherent sum of two Breit-Wigner amplitudes.

6. Baryon form factors

The baryon form factors (FFs) provide fundamental information on the baryon structure, giving crucial tests also to models of hadron internal structure in general. It describes the modifications of the point-like photon-hadron vertex due to the internal structure and dynamic of hadrons. Taking into account the symmetry properties of the EM interaction under Charge (C), Parity (P) and Time (T) transformations, the structure of any non-point-like particle of spin S is parameterized in terms of $(2S+1)$ FFs. For baryons with $\frac{1}{2}$ spin are described by electro FF (G_E) and magnetic FF (G_M). In the time-like region, like $e^+e^- \to p\bar{p}$ process, the G_E and G_M, or $R = \frac{G_E}{G_M}$ and G_M can be measured from the differential born cross section:

$$\frac{d\sigma_{Born}}{d\Omega} = \frac{\alpha^2\beta\mathcal{C}}{4q^2}\left[(1+\cos^2\theta)|G_M|^2 + \frac{1}{\tau}\sin^2\theta|G_E|^2\right] \quad (3)$$

The Born cross section can be written as:

$$\sigma_{Born} = \frac{4\pi\alpha^2\beta\mathcal{C}}{3q^2}\left[|G_M|^2 + \frac{1}{2\tau}|G_E|^2\right] \quad (4)$$

The factor $\mathcal{C}$ is a correction factor to take into account the coulomb interaction between the outgoing charged baryons.

For the proton FFs, the direct production of $p\bar{p}$ from e^+e^- annihilation or the production from ISR events from e^+e^- can both serve for the measurement. With the data collected at 2012, the proton FFs have been measured, and the cross section of $e^+e^- \to p\bar{p}$ and R were both consistent with Babar result. Recently, by means of ISR technique, the cross section and FFs are measured in a continuous q^2 region from the $p\bar{p}$ threshold with the large XYZ data sets. The

IOP Conf. Series: Journal of Physics: Conf. Series **873** (2017) 012026 doi:10.1088/1742-6596/873/1/012026

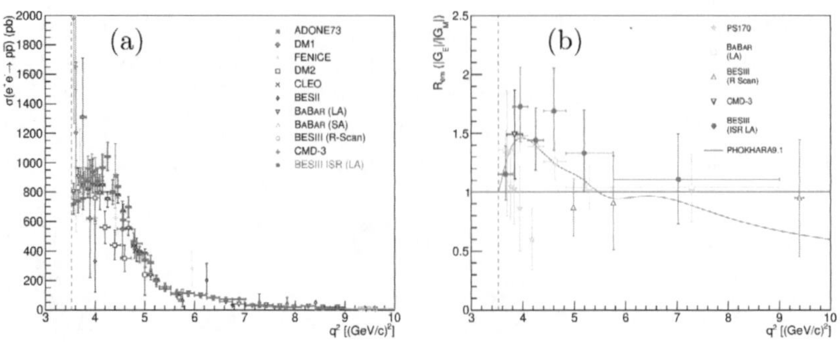

Figure 4. Preliminary results of the cross section of $e^+e^- \to p\bar{p}$ (a) and the R of proton (b) from ISR analysis.

dressed cross section and R are shown in Fig. 4. Obviously, the precisions in cross section and R are comparable with BaBar results.

In 2014 and 2015, an unprecedented amount of off-resonance data have been taken within the CMS energy range 2-3 GeV and above the threshold of the $\Lambda_c\overline{\Lambda}_c$. These data enable a great leap forward regarding baryon structure. The precision of proton form factors will reach 10% level, and the precision of R and G_M will be similar as that in the space-like region. For the other baryons (n, Λ, Σ, Λ_c), their form factors will be measured for the first time and the result is promising.

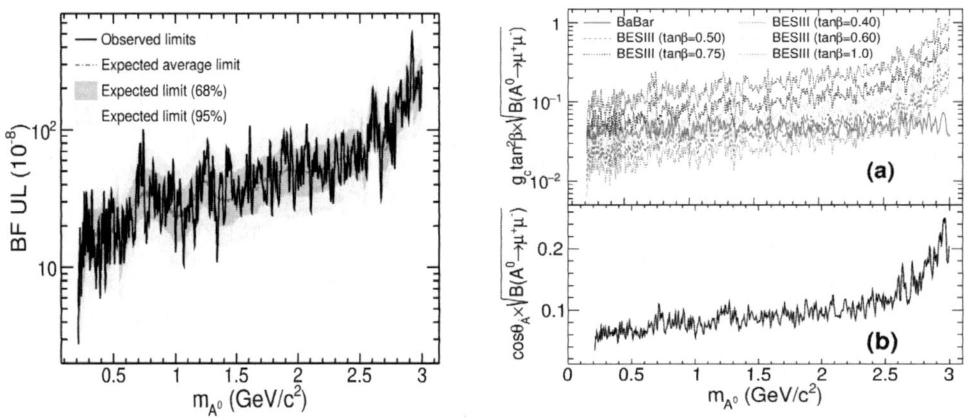

Figure 5. (Left) Upper limit of the branching fraction $\mathcal{B}((J/\psi \to \gamma A^0)) \times \mathcal{B}(A^0 \to \mu^+\mu^-)$. (Right) [top] The 90% C.L. upper limit of $g_b(= g_c \tan^2\beta) \times \sqrt{\mathcal{B}(A^0 \to \mu^+\mu^-)}$ and [bottom] $\cos\theta_A(= |sqrtg_b g_c| \times \sqrt{\mathcal{B}(A^0 \to \mu^+\mu^-)})$.

7. New physics

It is a very important part of BESIII project to search for new physics which can explain many unsolved mysteries or puzzles. New physics ideas or theories, even some of which are agains with traditional SM, are put forward as the extensions of SM. The next-to-minimal supersymmetric Standard Model (NMSSM) has a rich Higgs sector containing three CP-even, two CP-odd, and two charge Higgs bosons. The mass of the lightest CP-odd Higgs boson, A^0, maybe less than twice the mass of the charmed quark, and can decay to a muon pair from charmonium radiative decay.

BESIII has searched the light Higgs boson firstly through 106 million $\psi(2S)$ event. Non-resonant background can be reduced very effectively due to the intermediate production of J/ψ. Upper limit of number of A^0 signal is set at the 90% confidence level and corresponded upper limit of the branching fraction of $\mathcal{B}(\psi(2S) \to \gamma A^0) \times \mathcal{B}(A^0 \to \mu^+\mu^-)$ is from 4×10^{-7} to 2.1×10^{-5} [13]. With 225 million J/ψ data, A^0 is searched for again directly from J/ψ radiative decay [14]. This analysis gives a factor of five times improvement result for the upper limit of $\mathcal{B}(J/\psi \to \gamma A^0) \times \mathcal{B}(A^0 \to \mu^+\mu^-)$ ranging between 2.8×10^{-8} and 4.953×10^{-6}, shown in Fig. 5 (left). In this work, the coupling paramter g_b is also computed, shown in Fig. 5 (right). The new result is complementary and better than BaBar in the low mass region for $\tan\beta \le 0.6$. Combining BaBar results, the measured $\cos\theta_A \times \mathcal{B}(A^0 \to \mu^+\mu^-)$ result constrains the A^0 is mostly a singlet.

References

[1] M. Ablikim *et al.* [BESIII Collaboration] 2010 *Nucl. Instrum. Meth.* A **614** 345
[2] M. Ablikim *et al.* [BESIII Collaboration] 2014 *Phys. Rev.* D **89** 051104
[3] M. Ablikim *et al.* [BESIII Collaboration] 2016 *Phys. Rev.* D **94** 072004
[4] M. Ablikim *et al.* [BESIII Collaboration] 2015 *Phys. Rev.* D **92** 072012
[5] C. Patrignani *et al.* [Particle Data Group] 2016 *Chin. Phy.* C **40** 100001
[6] G. Rong, Y. Fang, H.L. Ma and J.Y. Zhao 2015 *Phys. Lett.* B **743** 315
[7] M. Ablikim *et al.* [BESIII Collaboration] 2016 *Phys. Rev. Lett.* **116** 052001
[8] M. Ablikim *et al.* [BESIII Collaboration] 2015 *Phys. Rev. Lett.* **115** 221805
[9] M. Ablikim *et al.* [BESIII Collaboration] *arXiv:* 1611.01317
[10] M. Ablikim *et al.* [BESIII Collaboration] *arXiv:* 1610.07044
[11] M. Ablikim *et al.* [BESIII Collaboration] 2014 *Phys. Rev. Lett.* **112** 092001
[12] M. Ablikim *et al.* [BESIII Collaboration] 2015 *Phys. Rev. Lett.* **115** 091803
[13] M. Ablikim *et al.* [BESIII Collaboration] 2012 *Phys. Rev.* D **85** 092012
[14] M. Ablikim *et al.* [BESIII Collaboration] 2016 *Phys. Rev.* D **93** 052005

IOP Conf. Series: Journal of Physics: Conf. Series **873** (2017) 012027 doi:10.1088/1742-6596/873/1/012027

Leptogenesis with GeV-Scale Right-Handed Neutrinos

Marco Drewes, Björn Garbrecht, Juraj Klarić and Dario Gueter
Physik-Department T70, Technische Universität München, James-Franck-Straße, 85748 Garching, Germany
E-mail: marco.drewes@tum.de, garbrecht@tum.de, dario.gueter@tum.de, juraj.klaric@tum.de

Abstract. We review the relation between leptogenesis and the discrete symmetry of charge-parity conjugation. The requirement of respecting the theorem of combined charge-parity-time reversal invariance at the level of the kinetic equations describing leptogenesis in the early Universe poses an interesting challenge that may most efficiently be addressed by the use of closed-time-path techniques. A byproduct of these methods is an accurate and unified description of leptogenesis from oscillations of right-handed neutrinos that applies to the regime of ultraheavy as well as GeV-scale RHNs. For the latter scenario, we discuss the mechanism on the example of the osicllatory and overdamped parametric regimes and very briefly comment on the prospect of experimental tests.

1. Introduction

The 2016 edition of the DISCRETE symposium in Warsaw takes place fifty years after Andrei D. Sakharov famously noted that the cosmic matter-antimatter asymmetry can be explained through the violation of discrete symmetries in conjunction with non-equilibrium thermodynamics [1]. The since sought-after, but yet unknown mechanism of creating the asymmetry has been called *baryogenesis*, and necessary ingredients for this process are often paraphrased as the celebrated *Sakharov conditions*:

- baryon number B violation,
- charge C and charge-parity CP violation,
- interactions out of thermal equilibrium.

Baryon-number violation is necessary to create $B \neq 0$ starting from initial conditions with $B \equiv 0$, C violation for the Universe to treat particles different from antiparticles and CP violation to avoid unwanted cancellations between fermionic degrees of freedom of left and right chirality.

The out-of-equilibrium condition is perhaps the most defining for the field of baryogenesis, and it is useful to look at it from different angles. To name two of these, one may note that an equilibrium state with an equal number of particles and antiparticles maximizes the entropy, while a state with more particles than antiparticles has a smaller entropy. The reduction of entropy due to a baryon asymmetry must therefore be compensated or overcompensated by an increase in entropy carried by other degrees of freedom, which can be achieved by a thermodynamically irreversible change of the state variables, *e.g.* through the expansion and

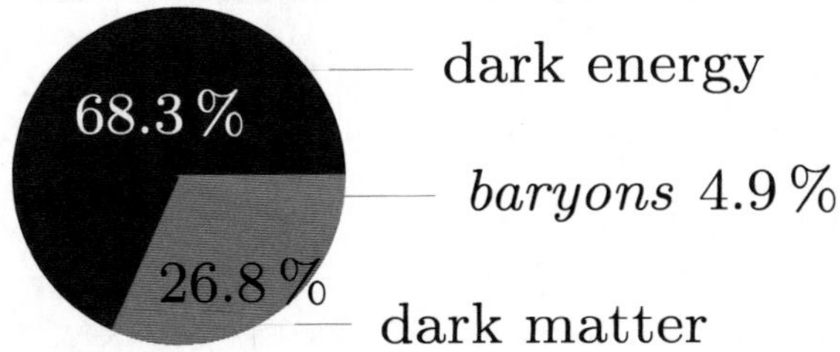

Figure 1. Contributions to the energy density of the Universe.

cooling of the Universe. Alternatively, one may argue that due to the combined invariance of quantum field theory under charge, parity and time reversal (CPT), an equilibrium state (which is time-reversal invariant) cannot support a non-vanishing (*i.e.* C and CP violating) baryon number.

By the year 1966, a number of theoretical and experimental insights have had come together setting the initial point for the problem of baryogenesis: the formulation of the Dirac equation [2] together with the discovery of the positron [3], the CPT-invarince theorem [4, 5], the observation of parity [6] and CP violation [7] in weak interactions and finally the detection of the cosmic microwave background (CMB) as the smoking gun of the hot big bang [8]. Modern surveys of the CMB also give the best measurement of the asymmetry [9] (*cf.* Figure 1), which is in excellent agreement wit big bang nucleosynthesis (BBN) [10]. The accurate understanding that has been achieved of the CMB and of BBN is a stunning success of applying particle physics well beyond laboratory experiments to the early Universe, and it motivates to tackle unresolved problems such as dark matter and baryogenesis.

2. Non-Equilibrium and Neutrinos

The non-equilibrium conditions may be realized in a large number of ways. To avoid reliance on peculiar initial conditions, there is a certain appeal to scenarios where the Universe is initially very close to equilibrium and a more pronounced deviation occurs subsequently during the expansion process. According to these rules of the game, baryogenesis from first-order phase transitions, such as electroweak baryogenesis, and baryogenesis from out-of-equilibrium decays of non-relativistic particles, such as ultraheavy sterile neutrinos, are of principal interest. We also find it interesting to relax above rule a little and to allow for a deviation of equilibrium due to very weakly coupled degrees of freedom that are not occupied initially and only populated at a later time due to subsequent equilibration. An example for this latter category is leptogenesis from GeV-scale sterile neutrinos. In this section, we discuss the parametric hints that point to an involvement of sterile neutrinos in baryogenesis.

Key observables that have recently been undergoing rapid experimental improvement are the neutrino mass and mixing parameters that can be inferred from a combination of the solar neutrino and atmospheric neutrino fluxes as well as from oscillations of reactor neutrinos and in muon neutrino beams. We denote by m_i the mass of a light neutrino, where $i = 1, 2, 3$, $\Delta m^2_{ij} = (m_i - m_j)^2$ and $m_1 < m_2 < m_3$ for the so-called normal hierarchy (NH), whereas $m_3 < m_1 < m_2$ for inverted hierarchy (IH). For the present discussion, we quote the mass differences $\Delta m^2_{21} = 7.50 \times 10^{-5}\mathrm{eV}^2$, $\Delta m^2_{31} = 2.457 \times 10^{-3}\mathrm{eV}^2$ for NH, (the best fit values for IH only differ by very little) summarized in Ref. [12] as well as the upper cosmological bound on the sum of the masses [9] $\sum_i m_i < 0.23\mathrm{eV}$.

These mass scales are way below those within the Standard Model (SM) and call for an explanation, where perhaps the most popular is the type I seesaw mechanism. It introduces n_N

hypothetical right-handed neutrinos (RHNs) N with a Majorana mass matrix M (which we can take to be diagonal through field redefinitions). The RHNs couple to SM leptons $\ell = (\nu, e_{\rm L})$ and Higgs doublets ϕ, where $\langle|\phi|\rangle = v = 174\,\text{GeV}$:

$$\mathcal{L}_{\rm SM} \to \mathcal{L}_{\rm SM} + \frac{1}{2}\bar{N}_i^c(\mathrm{i}\partial\!\!\!/ - M_{ij})N_j - Y_{ia}^*\bar{\ell}_a\phi^\dagger N_i - Y_{ia}\bar{N}_i\phi\ell_a\,, \tag{1}$$

with $a = e, \mu, \tau$. The RHNs and the SM leptons mix through the Yukawa couplings Y, and their masses can be arranged in terms of a $(3 + n_N) \times (3 + n_N)$ matrix:

$$\frac{1}{2}(\bar{\nu}\,\bar{N}^c)\underbrace{\begin{pmatrix} 0 & m_D \\ m_D & M \end{pmatrix}}_{=:\mathcal{M}}\begin{pmatrix} \nu^c \\ N \end{pmatrix}, \tag{2}$$

where $m_D = Y^\dagger v$. Assuming, $||M|| \gg ||m||$, it is easy to block diagonalize

$$\begin{pmatrix} m & 0 \\ 0 & M \end{pmatrix} = U\mathcal{M}U^\dagger\,, \quad U = \begin{pmatrix} 1 - \theta\theta^\dagger & \theta \\ -\theta^\dagger & 1 - \theta^\dagger\theta \end{pmatrix}, \quad \theta = Y^\dagger v M^{-1}\,. \tag{3}$$

This leads to a Majorana mass matrix for *light* neutrinos $m = v^2 Y^\dagger M^{-1} Y^*$ (with the eigenvalues m_i), diagonalized by U_ν, as well as one for *heavy* neutrinos $M_N = M + \frac{1}{2}\left(\theta^\dagger\theta M + M\theta^T\theta^*\right) + \mathcal{O}(\theta^3)$, diagonalized by U_N, where

$$\nu^{\rm light} = U_\nu^\dagger\left((\mathbb{1} - \frac{1}{2}\theta\theta^\dagger)\nu - \theta N^c\right), \tag{4a}$$

$$\nu^{\rm heavy} = U_N^\dagger\left((\mathbb{1} - \frac{1}{2}\theta^T\theta^*)N - \theta^T\nu^c\right). \tag{4b}$$

The RHNs are their own antiparticles and can decay either $N_i \to \ell\phi$ or $N_i \to \bar{\ell}\phi^\dagger$ and therefore violate lepton number L.

3. Out-of Equilibrium Decays, Asymmetries and Real Intermediate States

Now due to the approximate conformal invariance, a gas of thermalized and relativistic (*i.e.* of mass much below the temperature T) particles remains in equilibrium as the Universe expands. Another way of stating this is that the distribution function of a non-interacting massless particle in an expanding box remains of the Bose-Einstein or Fermi-Dirac form as the volume is changed, while this is not true for a massive species. Therefore, a deviation from thermal equilibrium will occur when the mass of a particular species becomes appreciable compared to the temperature. As we discuss below, ideally, the temperature when equilibration occurs coincides with the temperature where the particle becomes non-relativistic.

While in Sakharov's seminal work, the heavy particle responsible for the asymmetry bears the name "Maximon", in the arena of leptogenesis, the RHN is the gladiator [11]. The simplest meaningful network of kinetic equations describing leptogenesis may be stated as

$$\frac{dn_{Ni}}{dt} + 3Hn_{Ni} = \Gamma_i(n_{Ni} - n_{Ni}^{\rm eq})\,, \tag{5a}$$

$$\frac{dn_L}{dt} + 3Hn_L = \varepsilon\Gamma_i(n_{Ni} - n_{Ni}^{\rm eq}) - Wn_L\,, \tag{5b}$$

where

$$\Gamma_i = \frac{|Y_i|^2}{8\pi}M_{ii}\,, \quad W = \frac{\Gamma_i}{4}\left(\frac{M_{ii}}{T}\right)^{\frac{3}{2}} \mathrm{e}^{-\frac{M_{ii}}{T}}$$

IOP Conf. Series: Journal of Physics: Conf. Series **873** (2017) 012027 doi:10.1088/1742-6596/873/1/012027

are the decay rate of RHN N_i and the washout rate, respectively. Further, we have introduced here the Hubble expansion rate H, the number density n_X of a species X (L denoting leptons) and the decay asymmetry

$$\varepsilon = (\Gamma_{Ni\to\ell H} - \Gamma_{Ni\to\bar{\ell}H^*})/(\Gamma_{Ni\to\ell H} + \Gamma_{Ni\to\bar{\ell}H^*})\,. \tag{6}$$

(Γ_i is here taken to be the tree-level, CP-symmetric decay raye, while $\Gamma_{Ni\to\ell H}$ and $\Gamma_{Ni\to\bar{\ell}H^*}$ include CP violating contributions that typically arise at next-to-leading order.)

The coincidence between the equilibration temperature and the transition to the non-relativisic regime occurs when $\Gamma_i \sim H$ for $T \sim M_{ii}$, *i.e.* if $[YY^\dagger]_{ii}/(8\pi) \sim T^2/m_{\rm Pl} \sim M_{ii}^2/m_{\rm Pl}$. For smaller interaction strengths Γ_i, the processes that produce the lepton asymmetry n_L in Eq. (5b) will be suppressed, whereas for larger interactions, the system will be maintained close to equilibrium due to the right-hand side of Eq. (5a), what suppresses the outcome, such that above coincidence corresponds to ideal circumstances for creating a large lepton asymmetry. For simplicity, we now drop the flavour indices, such that the Majorana mass for light neutrinos reads $m \sim Y^2 v^2/M$. Using this to substitute the heavy neutrino mass M in the previous relations, we obtain that

$$m \sim 8\pi v^2/m_{\rm Pl} \sim 0.1\,{\rm meV}\,. \tag{7}$$

This is somewhat below the observed mass scale that we have quoted in Section 2 but yet remarkably close given the fact that the masses of the known fermions span over many orders of magnitude. Moreover, ignoring the flavour structure, this hints to Γ being moderately larger than H for $M \sim T$, implying that the RHNs are moderately close to equilibrium when they become non-relativistic. This situation is referred to as *strong washout*. To summarize this point, intriguingly, the neutrino mass scale points to the possibility that the RHNs of the type-I seesaw mechanism may play a key role in baryogenesis. Strong washout has the extra appeal that the outcome of leptogenesis is largely independent of the initial conditions, and there are fairly accurate analytic methods of solving Eqs.(10) in that regime [13, 14].

The decay asymmetry ε can be calculated by standard means from interfering tree-level with loop amplitudes for the process $N_i \to \ell H$ as well as for its CP conjugate. These interferences can be represented by the blue cuts in Figure 2. As stated above, we denote the CP symmetric tree-level result for the decay rate by Γ_i. Then,

$$\Gamma_{Ni\to\ell\phi} = \Gamma_i(1+\varepsilon) = \Gamma_{\bar{\ell}\phi^*\to Ni}\,, \tag{8}$$

where the latter equality is a direct consequence of the CPT theorem. Likewise, the charge-parity conjugate processes are related through conjugation of the couplings,

$$\Gamma_{Ni\to\bar{\ell}\phi^*} = \Gamma_i(1-\varepsilon) = \Gamma_{\ell\phi\to Ni}\,. \tag{9}$$

These relations suggest that if, for definiteness, $\varepsilon > 0$, then processes generating a positive asymmetry, *i.e.* those that create leptons or destroy antileptons, are faster than those that generate a negative asymmetry. Substituting this without further ado as rates into Boltzmann equations, this would neither be consistent with Eq. (5b) (where a process creating leptons of a definite charge and its inverse have the same rate) nor with the CPT theorem, that bars the creation of an asymmetry in equilibrium.

This difficulty is often addressed by including lepton number violating $2 \leftrightarrow 2$ processes, *i.e.* $\ell H \leftrightarrow \bar{\ell}H^*$ and by subtracting from these *partly* the contributions from the on-shell exchange of RHNs, namely precisely in a manner that complies with the CPT theorem, arguing that these contributions are already included within the $1 \leftrightarrow 2$ processes [15]. This procedure is known as real intermediate state (RIS) subtraction.

IOP Conf. Series: Journal of Physics: Conf. Series **873** (2017) 012027 doi:10.1088/1742-6596/873/1/012027

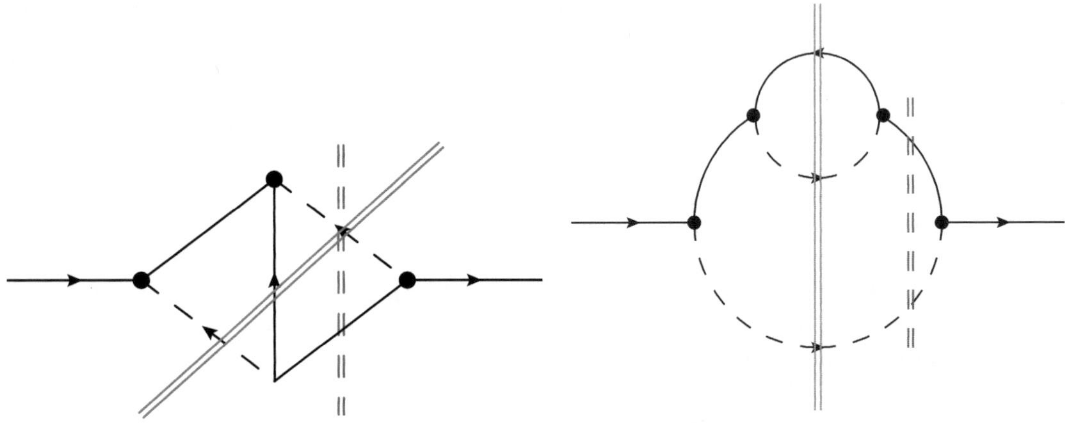

Figure 2. One loop contributions to the asymmetry in decays and inverse decays. RHNs are represented by black solid lines without arrows, leptons by solid lines with arrows and Higgs bosons by dashed lines with arrows. The double lines represent cuts, where the dashed one leads to interferences between tree and loop $1 \leftrightarrow 2$ processes, while the solid ones to interferences between scattering processes. In those contributions relevant for generating a CP asymmetry, scattering proceeds via an on-shell RHN.

Instead of substituting S-matrix elements into Boltzmann equations, a complementary and somewhat more systematic and direct way of addressing this problem is to directly solve for the time evolution of the operators of interest, *e.g.* the lepton charge density. The unitary time evolution follows from the Hamiltonian but a perturbative expansion becomes rather involved quickly. The task is greatly simplified in a functional approach, where the perturbation expansion can be organized within simple Feynman diagrammatic rules. This calculus is known as the Schwinger-Keldysh or Closed-Time-Path (CTP) formalism [16, 17]. The key difference to scattering problems is that for S-matrix elements, one does calculations in terms of time-ordered Green functions and in particular propagators, while in the CTP framework, diagrams are constructed from a more general basis of propagators (for example time-ordered Feynman propagators and Wightman functions without time ordering) that contain both spectral and statistical information.

An introduction to the derivation of kinetic equations from the CTP formalism in relativistic field theory is given in Ref. [18]. The kinetic equations descend from Schwinger-Dyson equations, and the various scattering processes appear in the self-energy terms. By construction, these processes are counted *inclusively*, such that the CPT theorem is respected. The propagators pertaining to different boundary conditions that appear in the CTP approach can be associated with certain cuts, that in turn lead to the CP-violating contribution. This is illustrated in Figure 2: Besides the interferences of tree and loop amplitudes, we encounter also interferences between $2 \leftrightarrow 2$ scattering amplitudes. The CP-violating contributions from the latter interferences come precisely from the region of momentum integration where one of the RHNs propagates on shell. This is shown for leptogenesis in Refs. [19, 20], and for a model with scalar particles in Refs. [21, 22]. The CP cuts that appear in fully inclusive calculations on soft leptogenesis, where substantial cancellations occur, are discussed in Ref. [23], where also numerous calculational details on handling CP asymmetries in the CTP approach are presented.

IOP Conf. Series: Journal of Physics: Conf. Series **873** (2017) 012027 doi:10.1088/1742-6596/873/1/012027

4. Resonant Leptogenesis and from Oscillations of GeV-Scale Sterile Neutrinos

Now suppose that the width $k^0\Gamma_i$ is larger or of the same order than the squared mass splitting $(M_{ii} - M_{jj})^2$, where k^0 is the typical energy of the RHN, *i.e.* $k^0 \approx M_{ii}$ in the non-relativistic case whereas $k^0 \sim T$ in the relativistic case. For the mass splitting to be comparable to the width, a mass degeneracy is therefore required in the non-relativistic regime, what leads to the typical scenarios of resonant leptogenesis [24, 25, 26, 27]. (Note here that even when $(M_{ii} - M_{jj})^2 \ll M_{ii}^2$ it is still possible that $(M_{ii} - M_{jj})^2 \gg k^0\Gamma_i$ such that no resummation of the RHN propagators is necessary). In the relativistic regime, a mass degeneracy is not necessary. This can typically lead to successful leptogenesis with RHNs around the GeV scale [28, 29]. (For specific configurations of the flavoured couplings in the seesaw mechanism, also scenarios with RHNs above the electroweak scale are viable [30].)

Clearly, when the squared mass splitting is smaller than the width, the expansion of the RHN propagator through the insertion of a wave function correction, as indicated in Figure 2, cannot be performed. Instead, the wave function corrections need to be resummed, what is automatically achieved in the CTP formulation of the Schwinger-Dyson equations. To illustrate this point explicitly, we note the kinetic equation for the deviation $\delta f_{Nh}(\mathbf{k})$ of the RHN distribution from equilibrium [32, 33],

$$\delta f'_{Nh} + \frac{a^2(\eta)}{2k^0}\mathrm{i}[M^2, \delta f_{Nh}] + f_N^{\mathrm{eq}\prime} = -2\left\{\mathrm{Re}[Y^*Y^t]\frac{k\cdot\hat{\Sigma}_N^{\mathcal{A}}}{k^0} - \mathrm{i}h\mathrm{Im}[Y^*Y^t]\frac{\tilde{k}\cdot\hat{\Sigma}_N^{\mathcal{A}}}{k^0}, \delta f_{Nh}\right\}. \tag{10}$$

This follows from the Schwinger-Dyson equations (or, more specifically to the CTP framework, the Kadanoff-Baym equations) from integrating out k^0 (*i.e.* taking the zeroth moment) and taking an appropriate trace over the spinor structure. Furthermore, in above equation, $\hat{\Sigma}_N^{\mathcal{A}}$ is the reduced (*i.e.* stripped of Yukawa couplings that appear explicitly in the equation) spectral (cut part) of the self energy, $a(\eta), \eta$ are the scale factor of the expanding Universe and the conformal time ($\prime \equiv d/d\eta$), h denotes helicity, and $\tilde{k} = (|\mathbf{k}|, |k^0|\mathbf{k}/|\mathbf{k}|)$. We have suppressed the flavour indices of $\delta f_{Nij\,h}$ and of the heavy neutrino mass matrix M_{ij} as well as on the Yukawa couplings Y_{ia}. This equation implicitly resums all wave-function insertions of $\hat{\Sigma}_N^{\mathcal{A}}$. The off-diagonal components of $\delta f_{Nij\,h}$ are crucial because these also appear in the kinetic equation for the left-handed SM leptons, where they lead to the creation of the CP asymmetry.

For non-relativistic neutrinos, one often encounters the situation that one can neglect the term $\delta f'_{Nh}$ because all eigenvalues of this ordinary and linear differential equation that follow from the commutator and anticommutator terms are larger than the rate of change $f_N^{\mathrm{eq}\prime}/f_N^{\mathrm{eq}}$ due to the Hubble expansion. (f_N^{eq} denotes the equilibrium Fermi-Dirac distribution of the RHNs.) In this case (which can include situations where $(M_{ii} - M_{jj})^2 \ll k^0\Gamma_i$), one can derive a quasi-static solution for $\delta f'_{Nh}$ that is more general than the well-known results but that agrees with these when $(M_{ii} - M_{jj})^2 \gg k^0\Gamma_i$ [33, 34]. In particular, these solutions apply throughout the strong washout regime. In case there are eigenvalues that are smaller than $f_N^{\mathrm{eq}\prime}/f_N^{\mathrm{eq}}$, which may happen outside the strong washout regime, it is necessary to track the full time-dependence of δf_{Nh}. In summary, making use of Eq. (10), it is possible to make accurate predictions for resonant leptogenesis for all ratios between $(M_{ii} - M_{jj})^2$ and $k^0\Gamma_i$. (We disagree with Ref. [40] that argues that there are additional contributions from the "mixing" of RHNs, that would result in a factor two enhancement of the asymmetry for typical parameter points within the strong washout regime.)

Now for GeV-scale RHNs, that are relativistic at the relevant temperatures above the electroweak phase transition (where baryon-minus-lepton number violating sphaleron transitions freeze out) the production rate of the individual RHNs may or may not be below the Hubble rate. While there exist different suitable analytic approximations for various regions in parameter space, the situation is more involved compared to the strong washout regime in standard resonant

IOP Conf. Series: Journal of Physics: Conf. Series **873** (2017) 012027 doi:10.1088/1742-6596/873/1/012027

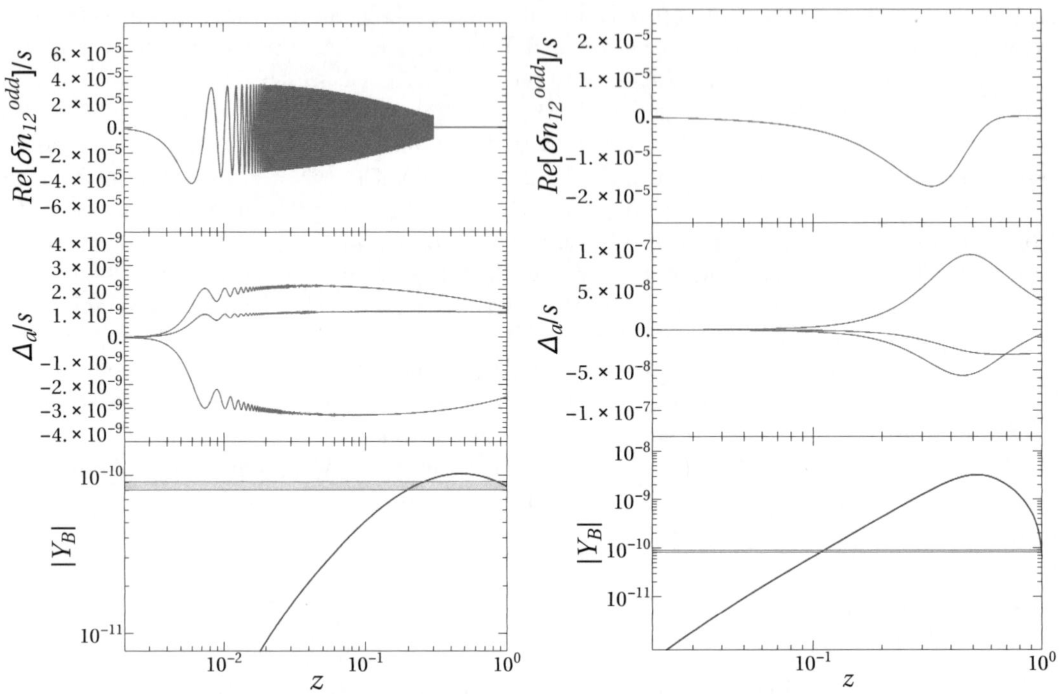

Figure 3. Examples for the oscillatory (left) and the overdamped regime of leptogenesis from GeV-scale RHNs. The helicity-odd number density of RHNs (*i.e.* the phase-space integral of $\delta f_{N+\frac{1}{2}} - \delta f_{N-\frac{1}{2}}$) is denoted by δn, $\Delta_a = B/3 - L_a$ (B being the baryon number density and L_a the number density of SM leptons of flavour $a = e, \mu, \tau$), $Y_B = B/s$, s the entropy density and $z = T_{\rm EW}/T$ with $T_{\rm EW}$ the temperature of the electroweak phase transition. The green band indicates the range for the baryon asymmetry that is in agreement with current observations.

leptogenesis. To illustrate the different parametric regions, we present in Figure 3 the *oscillatory* and the *overdamped* regimes [35].

In the oscillatory regime, $|M_{ii} - M_{jj}| \gg T_{\rm EW}|$ and $(M_{ii} - M_{jj})^2 \gg T\Gamma_i$, such that many oscillations can take place before the electroweak phase transition. The Yukawa couplings are small compared to the overdamped regime, such that the baryon asymmetry only suffers a small amount of washout prior to the electroweak phase transition. In the overdamped regime $(M_{ii} - M_{jj})^2 \gg T\Gamma_i$ does not hold (for at least one of the N_i), *i.e.* there is a smaller mass splitting and larger Yukawa couplings. Interestingly, this does not necessarily have to correspond to fine tuning but can also be motivated by an approximate conservation of lepton number [36, 37, 38, 39]. We can observe that larger Yukawa couplings also lead to a stronger washout of the asymmetry toward the electroweak phase transition.

What can be seen from both panels of Figure 3 is that at early times, the baryon asymmetry Y_B is very small. This is because the asymmetry generated initially is purely flavoured, such that summing the asymmetries over the active lepton flavours yields a vanishing result. Only at a later time toward the electroweak phase transition, an appreciable amount of asymmetry is transferred to the RHNs (where it takes the form of a helicity asymmetry), such that a non-vanishing net lepton number is present in the SM leptons. Sphaleron processes then lead to a non-vanishing baryon asymmetry Y_B.

As direct and indirect signals of RHNs are proportional to these quantities, the active-sterile mixing angles θ, Eq. (3), given present and aniticpated constraints from the PMNS matrix and

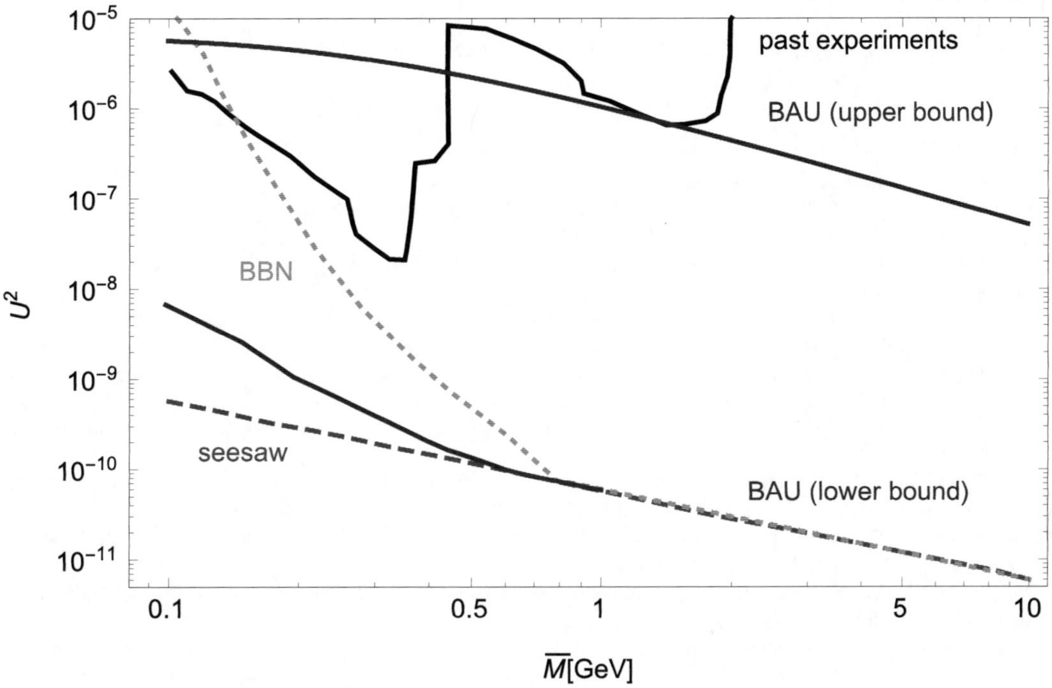

Figure 4. Between the two solid blue lines, there is the allowed range of the total active sterile mixing U^2 over the average mass $(M_1 + M_2)/2$ in a scenario with $n_N = 2$ RHNs, assuming normal hierarchy. There is a lower bound on the mixing from the requirement of explaining the light neutrino masses through the seesaw mechanism, and an additional one from big bang nucleosynthesis (BBN).

the light neutrino masses, are of phenomenological relevance. Recent work has discussed this topic, also concerning the different flavour effects in some detail [35, 41, 42, 43, 44]. For brevity, we here just present in Figure 4 the flavour-summed quantity $U^2 = \mathrm{tr}\theta^\dagger\theta$ that characterizes the total active-sterile mixing. Along the upper limit, leptogenesis is viable in the overdamped regime that allows for large mixing angles whereas the lower bound occurs at the transition between both regimes, where the baryon asymmetry turns out maximal for a given mixing angle (through tuning the mass degeneracy).

It can be anticipated that the experimental bounds will improve considerably in the future. Relevant in the lower mass range are NA62 and the proposed SHiP facility, whereas heavier RHNs may be particularly well accessible at future lepton colliders (see *e.g.* [45, 46, 47, 48]).

5. Conclusions

Closed-time-path techniques are the most powerful tool to meet the requirements of the CPT theorem when formulating kinetic equations for the dynamical emergence of the matter-antimatter asymmetry in the early Universe. In particular, they lead to a unified formulation of CP violation from mixing of RHNs throughout the entire parameter space of the type-I seesaw mechanism. This includes resonant leptogenesis from ultraheavy RHNs as well as leptogenesis from GeV-scale RHNs. In the latter case, it is possible that RHNs responsible for leptogenesis are accessible by present and proposed experiments. As this requires large active-sterile mixing, leptogenesis would then likely proceed in a regime of overdamped oscillations.

Acknowledgements
This research was supported by the DFG cluster of excellence "Origin and Structure of the Universe".

References
[1] A. D. Sakharov, "Violation of CP Invariance, C Asymmetry, and Baryon Asymmetry of the Universe," Pisma Zh. Eksp. Teor. Fiz. **5** (1967) 32 [JETP Lett. **5** (1967) 24] [Sov. Phys. Usp. **34** (1991) 392] [Usp. Fiz. Nauk **161** (1991) 61]. doi:10.1070/PU1991v034n05ABEH002497
[2] P. A. M. Dirac, "The quantum theory of the electron," Proc. Roy. Soc. Lond. A **117** (1928) 610. doi:10.1098/rspa.1928.0023
[3] C. D. Anderson, "The Positive Electron," Phys. Rev. **43** (1932) 491. doi:10.1103/PhysRev.43.491
[4] G. Luders, "On the Equivalence of Invariance under Time Reversal and under Particle-Antiparticle Conjugation for Relativistic Field Theories," Kong. Dan. Vid. Sel. Mat. Fys. Med. **28N5** (1954) 1.
[5] W. Pauli, "Exclusion principle, Lorentz group and reflexion of space-time and charge", chapter in "Niels Bohr and the Development of Physics", (eds. Pauli, W. and Rosenfeld, L. and Weisskopf, V.), 30, Pergamon Press (1955).
[6] C. S. Wu, E. Ambler, R. W. Hayward, D. D. Hoppes and R. P. Hudson, "Experimental Test of Parity Conservation in Beta Decay," Phys. Rev. **105** (1957) 1413. doi:10.1103/PhysRev.105.1413
[7] J. H. Christenson, J. W. Cronin, V. L. Fitch and R. Turlay, "Evidence for the 2π Decay of the K_2^0 Meson," Phys. Rev. Lett. **13** (1964) 138. doi:10.1103/PhysRevLett.13.138
[8] A. A. Penzias and R. W. Wilson, "A Measurement of excess antenna temperature at 4080-Mc/s," Astrophys. J. **142** (1965) 419. doi:10.1086/148307
[9] P. A. R. Ade *et al.* [Planck Collaboration], "Planck 2015 results. XIII. Cosmological parameters," Astron. Astrophys. **594** (2016) A13 doi:10.1051/0004-6361/201525830 [arXiv:1502.01589 [astro-ph.CO]].
[10] see, *e.g.*, G. Steigman, "Primordial Nucleosynthesis in the Precision Cosmology Era," Ann. Rev. Nucl. Part. Sci. **57** (2007) 463 doi:10.1146/annurev.nucl.56.080805.140437 [arXiv:0712.1100 [astro-ph]].
[11] M. Fukugita and T. Yanagida, "Baryogenesis Without Grand Unification," Phys. Lett. B **174** (1986) 45. doi:10.1016/0370-2693(86)91126-3
[12] M. C. Gonzalez-Garcia, M. Maltoni and T. Schwetz, "Updated fit to three neutrino mixing: status of leptonic CP violation," JHEP **1411** (2014) 052 doi:10.1007/JHEP11(2014)052 [arXiv:1409.5439 [hep-ph]].
[13] E. W. Kolb and M. S. Turner, "Grand Unified Theories and the Origin of the Baryon Asymmetry," Ann. Rev. Nucl. Part. Sci. **33** (1983) 645. doi:10.1146/annurev.ns.33.120183.003241
[14] W. Buchmuller, P. Di Bari and M. Plumacher, "Leptogenesis for pedestrians," Annals Phys. **315** (2005) 305 doi:10.1016/j.aop.2004.02.003 [hep-ph/0401240].
[15] E. W. Kolb and S. Wolfram, "Baryon Number Generation in the Early Universe," Nucl. Phys. B **172** (1980) 224 Erratum: [Nucl. Phys. B **195** (1982) 542]. doi:10.1016/0550-3213(80)90167-4, 10.1016/0550-3213(82)90012-8
[16] J. S. Schwinger, "Brownian motion of a quantum oscillator," J. Math. Phys. **2** (1961) 407. doi:10.1063/1.1703727
[17] L. V. Keldysh, "Diagram technique for nonequilibrium processes," Zh. Eksp. Teor. Fiz. **47** (1964) 1515 [Sov. Phys. JETP **20** (1965) 1018].
[18] E. Calzetta and B. L. Hu, "Nonequilibrium Quantum Fields: Closed Time Path Effective Action, Wigner Function and Boltzmann Equation," Phys. Rev. D **37** (1988) 2878. doi:10.1103/PhysRevD.37.2878
[19] M. Beneke, B. Garbrecht, M. Herranen and P. Schwaller, "Finite Number Density Corrections to Leptogenesis," Nucl. Phys. B **838** (2010) 1 doi:10.1016/j.nuclphysb.2010.05.003 [arXiv:1002.1326 [hep-ph]].
[20] A. Anisimov, W. Buchmller, M. Drewes and S. Mendizabal, "Quantum Leptogenesis I," Annals Phys. **326** (2011) 1998 Erratum: [Annals Phys. **338** (2011) 376] doi:10.1016/j.aop.2011.02.002, 10.1016/j.aop.2013.05.00 [arXiv:1012.5821 [hep-ph]].
[21] M. Garny, A. Hohenegger, A. Kartavtsev and M. Lindner, "Systematic approach to leptogenesis in nonequilibrium QFT: Vertex contribution to the CP-violating parameter," Phys. Rev. D **80** (2009) 125027 doi:10.1103/PhysRevD.80.125027 [arXiv:0909.1559 [hep-ph]].
[22] M. Garny, A. Hohenegger, A. Kartavtsev and M. Lindner, "Systematic approach to leptogenesis in nonequilibrium QFT: Self-energy contribution to the CP-violating parameter," Phys. Rev. D **81** (2010) 085027 doi:10.1103/PhysRevD.81.085027 [arXiv:0911.4122 [hep-ph]].
[23] B. Garbrecht and M. J. Ramsey-Musolf, "Cuts, Cancellations and the Closed Time Path: The Soft Leptogenesis Example," Nucl. Phys. B **882** (2014) 145 doi:10.1016/j.nuclphysb.2014.02.012 [arXiv:1307.0524 [hep-ph]].

[24] L. Covi, E. Roulet and F. Vissani, "CP violating decays in leptogenesis scenarios," Phys. Lett. B **384** (1996) 169 doi:10.1016/0370-2693(96)00817-9 [hep-ph/9605319].
[25] M. Flanz, E. A. Paschos, U. Sarkar and J. Weiss, "Baryogenesis through mixing of heavy Majorana neutrinos," Phys. Lett. B **389** (1996) 693 doi:10.1016/S0370-2693(96)01337-8, 10.1016/S0370-2693(96)80011-6 [hep-ph/9607310].
[26] A. Pilaftsis, "Resonant CP violation induced by particle mixing in transition amplitudes," Nucl. Phys. B **504** (1997) 61 doi:10.1016/S0550-3213(97)00469-0 [hep-ph/9702393].
[27] A. Pilaftsis and T. E. J. Underwood, "Resonant leptogenesis," Nucl. Phys. B **692** (2004) 303 doi:10.1016/j.nuclphysb.2004.05.029 [hep-ph/0309342].
[28] E. K. Akhmedov, V. A. Rubakov and A. Y. Smirnov, "Baryogenesis via neutrino oscillations," Phys. Rev. Lett. **81** (1998) 1359 doi:10.1103/PhysRevLett.81.1359 [hep-ph/9803255].
[29] T. Asaka and M. Shaposhnikov, "The nuMSM, dark matter and baryon asymmetry of the universe," Phys. Lett. B **620** (2005) 17 doi:10.1016/j.physletb.2005.06.020 [hep-ph/0505013].
[30] B. Garbrecht, "More Viable Parameter Space for Leptogenesis," Phys. Rev. D **90** (2014) no.6, 063522 doi:10.1103/PhysRevD.90.063522 [arXiv:1401.3278 [hep-ph]].
[31] B. Garbrecht, F. Glowna and P. Schwaller, "Scattering Rates For Leptogenesis: Damping of Lepton Flavour Coherence and Production of Singlet Neutrinos," Nucl. Phys. B **877** (2013) 1 doi:10.1016/j.nuclphysb.2013.08.020 [arXiv:1303.5498 [hep-ph]].
[32] B. Garbrecht and M. Herranen, "Effective Theory of Resonant Leptogenesis in the Closed-Time-Path Approach," Nucl. Phys. B **861** (2012) 17 doi:10.1016/j.nuclphysb.2012.03.009 [arXiv:1112.5954 [hep-ph]].
[33] B. Garbrecht, F. Gautier and J. Klaric, "Strong Washout Approximation to Resonant Leptogenesis," JCAP **1409** (2014) no.09, 033 doi:10.1088/1475-7516/2014/09/033 [arXiv:1406.4190 [hep-ph]].
[34] S. Iso and K. Shimada, "Coherent Flavour Oscillation and CP Violating Parameter in Thermal Resonant Leptogenesis," JHEP **1408** (2014) 043 doi:10.1007/JHEP08(2014)043 [arXiv:1404.4816 [hep-ph]].
[35] M. Drewes, B. Garbrecht, D. Gueter and J. Klaric, "Leptogenesis from Oscillations of Heavy Neutrinos with Large Mixing Angles," JHEP **1612** (2016) 150 doi:10.1007/JHEP12(2016)150 [arXiv:1606.06690 [hep-ph]].
[36] R. N. Mohapatra and J. W. F. Valle, "Neutrino Mass and Baryon Number Nonconservation in Superstring Models," Phys. Rev. D **34** (1986) 1642. doi:10.1103/PhysRevD.34.1642
[37] R. N. Mohapatra, "Mechanism for Understanding Small Neutrino Mass in Superstring Theories," Phys. Rev. Lett. **56** (1986) 561. doi:10.1103/PhysRevLett.56.561
[38] M. C. Gonzalez-Garcia and J. W. F. Valle, "Fast Decaying Neutrinos and Observable Flavor Violation in a New Class of Majoron Models," Phys. Lett. B **216** (1989) 360. doi:10.1016/0370-2693(89)91131-3
[39] G. C. Branco, W. Grimus and L. Lavoura, "The Seesaw Mechanism in the Presence of a Conserved Lepton Number," Nucl. Phys. B **312** (1989) 492. doi:10.1016/0550-3213(89)90304-0
[40] P. S. Bhupal Dev, P. Millington, A. Pilaftsis and D. Teresi, "Flavour Covariant Transport Equations: an Application to Resonant Leptogenesis," Nucl. Phys. B **886** (2014) 569 doi:10.1016/j.nuclphysb.2014.06.020 [arXiv:1404.1003 [hep-ph]].
[41] M. Drewes and B. Garbrecht, "Leptogenesis from a GeV Seesaw without Mass Degeneracy," JHEP **1303** (2013) 096 doi:10.1007/JHEP03(2013)096 [arXiv:1206.5537 [hep-ph]].
[42] P. Hernndez, M. Kekic, J. Lpez-Pavn, J. Racker and N. Rius, "Leptogenesis in GeV scale seesaw models," JHEP **1510** (2015) 067 doi:10.1007/JHEP10(2015)067 [arXiv:1508.03676 [hep-ph]].
[43] P. Hernndez, M. Kekic, J. Lopez-Pavn, J. Racker and J. Salvado, "Testable Baryogenesis in Seesaw Models," JHEP **1608** (2016) 157 doi:10.1007/JHEP08(2016)157 [arXiv:1606.06719 [hep-ph]].
[44] M. Drewes, B. Garbrecht, D. Gueter and J. Klaric, "Testing the low scale seesaw and leptogenesis," arXiv:1609.09069 [hep-ph].
[45] T. Asaka, S. Eijima and A. Watanabe, "Heavy neutrino search in accelerator-based experiments," JHEP **1303** (2013) 125 doi:10.1007/JHEP03(2013)125 [arXiv:1212.1062 [hep-ph]].
[46] A. Blondel *et al.* [FCC-ee study Team], "Search for Heavy Right Handed Neutrinos at the FCC-ee," Nucl. Part. Phys. Proc. **273-275** 1883 doi:10.1016/j.nuclphysbps.2015.09.304 [arXiv:1411.5230 [hep-ex]].
[47] S. Antusch, E. Cazzato and O. Fischer, "Displaced vertex searches for sterile neutrinos at future lepton colliders," JHEP **1612** (2016) 007 doi:10.1007/JHEP12(2016)007 [arXiv:1604.02420 [hep-ph]].
[48] T. Asaka and H. Ishida, "Lepton number violation by heavy Majorana neutrino in B decays," Phys. Lett. B **763** (2016) 393 doi:10.1016/j.physletb.2016.10.070 [arXiv:1609.06113 [hep-ph]].

IOP Conf. Series: Journal of Physics: Conf. Series **873** (2017) 012028 doi:10.1088/1742-6596/873/1/012028

Extension of Standard Model with a Complex Singlet and Iso-Doublet Vector Quarks

Neda Darvishi

Faculty of Physics, University of Warsaw, Pasteura 5, 02-093 Warsaw, Poland

E-mail: neda.darvishi@fuw.edu.pl

Abstract. In this paper the extension of the SM by a neutral complex scalar singlet with a nonzero vacuum expectation value and a heavy vector quark pair is considered. This model provides an extra source of spontaneous CP violation. The focus of this article is to obtain the rate of baryon number generation. We show that the considered model provides a strong enough first-order electroweak phase transition to suppress the baryon-violating sphaleron process.

1. Introduction

Among the various scenarios that try to explain the baryon asymmetry of the Universe (BAU), the electroweak (EW) baryogenesis is one of the most interesting scenarios [1, 2, 3, 4]. All the necessary conditions for a successful baryogenesis according to the Sakharov's criteria[5] (i.e. baryon number violation, C and CP violation and departure from thermal equilibrium) can be found in the Standard Model (SM). It has become apparent that the SM of EW interactions is unable to account for the observed magnitude of the BAU for two reasons: (*i*) The EW phase transition (EWPT) is not strongly first-order and therefore, any baryon asymmetry generated during the EWPT would be washed out by unsuppressed baryon violating processes in the broken phase [6, 7, 8, 9, 10, 11, 12, 13, 14, 15, 16]. (*ii*) There is not enough CP violation from the CKM matrix to generate the baryon asymmetry [17]. In order to address this important issues, various extensions of the SM are considered [16, 18, 19, 20, 21, 22]. In the present work, we obtain the rate of baryon number generation in the model with a neutral complex scalar singlet χ, which accompanies the SM-like Higgs doublet Φ, and a pair of iso-doublet vector quarks $V_L + V_R$. This kind of extension of the SM was discussed in the literature with various motivations, e.g.[23, 24, 25, 26, 27, 28, 29, 30, 31]. We consider the potential with a softly broken global U(1) symmetry, see [32, 33, 34, 35]. This model provides a strong enough first-order EW phase transition via the soft breaking terms in the potential, to suppress the baryon-violating sphaleron process. On the other hand, the mixing of SM quarks with the heavy vector quark pair results in appearance of an additional CP violation. The issue of the CP violation due to a complex singlet with a complex VEV has been previously discussed, see [32, 34, 35].

The Yukawa Lagrangian acquires additional quark mass terms because of the presence of an iso-doublet vector quark and a complex singlet. Diagonalizing the quark mass matrix results in nondiagonal terms coming from the kinetic terms, with appearance of new terms which are functions of the time-dependent phase (complex singlet VEV [36]). These terms lead to the generation of baryon asymmetry.

The content of this paper is as follows. In section (2), we present a general structure of the model and its constrained version. In section (3), the necessary conditions for strong enough first order EWPT in the model will be verified. We obtain the rate of baryon number generation in the section 4. Finally, the section 5 contains our conclusion.

2. Lagrangian of the model

The full Lagrangian of this model is given by

$$\mathcal{L} = \mathcal{L}_{gf}^{SM} + \mathcal{L}_{scalar} + \mathcal{L}_Y(\psi_f, \Phi) + \mathcal{L}_Y(V_q, \chi), \tag{1}$$

where $\mathcal{L}_{gf}^{SM}$ describes the pure gauge boson terms as well the SM boson- SM fermion interaction, $\mathcal{L}_{scalar}$ describes the scalar sector of the model with one SU(2) doublet Φ and a neutral complex scalar (spinless) singlet χ. $\mathcal{L}_Y(\psi_f, \Phi)$ and $\mathcal{L}_Y(V_q, \chi)$ represent the Yukawa interaction of Φ with SM fermions and the Yukawa interaction of singlet scalar with heavy iso-doublet vector quarks and SM quarks.

We consider a pair (left and right handed) of heavy iso-doublet vector quarks, $V_L + V_R$, with V_L and V_R having the same transformation properties under the gauge group of the SM, and where V_L transforms in the same way as a standard model quark doublet Q_L. Essentially, the case of iso-singlet vector quarks have the same result as the case of iso-doublet vector quarks, while their transformation are similar to u_R or d_R with respect to the SM gauge group [36].

The mass terms in the presence of the complex singlet are:

$$\mathcal{L}_Y(V_q, \chi) = \lambda_V \chi \overline{Q}_L V_R + M \overline{V}_L V_R + h.c, \tag{2}$$

where M is the mass of vector quarks (which is considered to be heavy) and λ_V is the coupling between the singlet scalar, vector quarks and the SM quarks. In general the couplings of all three generations of SM quarks to vector quarks should be considered, nevertheless, since only one linear combination of these will actually mix with the vector quarks, we consider only one generations of SM quarks (the heaviest), Q_L.

The SM-like Higgs boson in the model predominantly consists of a neutral CP-even component of the Φ doublet and its mass is $\sim$ 125 GeV. There are two other neutral Higgs like particles, see discussion in [33, 34, 35].

We assume Φ and χ fields have non-zero vacuum expectation values (VEV), v and $we^{i\xi}$, respectively ($v, w, \xi \in \mathbf{R}$). We use the following field decomposition ,

$$\Phi = \begin{pmatrix} \phi^+ \\ \frac{1}{\sqrt{2}}(\phi_1 + i\phi_4) \end{pmatrix}, \chi = \frac{1}{\sqrt{2}}(\phi_2 + i\phi_3), \tag{3}$$

where

$$we^{i\xi} = w\cos\xi + iw\sin\xi = w_1 + iw_2. \tag{4}$$

Masses of the gauge bosons are given by the VEV of the doublet, e.g $M_W^2 = g^2v^2/4$ for the W boson.

The scalar potential of the model can be written as follows [33, 34, 35]

$$V = V_D + V_S + V_{DS}, \tag{5}$$

with the pure doublet and the pure singlet parts (respectively V_D and V_S) and the mixed term V_{DS}. The SM part of the potential is presented by V_D:

$$V_D = -\frac{1}{2}m_{11}^2\Phi^\dagger\Phi + \frac{1}{2}\lambda\left(\Phi^\dagger\Phi\right)^2. \tag{6}$$

The potential for a complex singlet is equal to,

$$\begin{aligned} V_S = \quad & -\frac{1}{2}m_s^2\chi^*\chi - \frac{1}{2}m_4^2(\chi^{*2}+\chi^2) \\ & +\lambda_{s1}(\chi^*\chi)^2 + \lambda_{s2}(\chi^*\chi)(\chi^{*2}+\chi^2) + \lambda_{s3}(\chi^4+\chi^{*4}) \\ & +\kappa_1(\chi+\chi^*) + \kappa_2(\chi^3+\chi^{*3}) + \kappa_3(\chi+\chi^*)(\chi^*\chi). \end{aligned} \tag{7}$$

The doublet-singlet interaction term is,

$$\begin{aligned} V_{DS} = \quad & \Lambda_1(\Phi^\dagger\Phi)(\chi^*\chi) + \Lambda_2(\Phi^\dagger\Phi)(\chi^{*2}+\chi^2) \\ & +\kappa_4(\Phi^\dagger\Phi)(\chi+\chi^*). \end{aligned} \tag{8}$$

There are three quadratic (m_i^2), six dimensionless quartic (λ_i, Λ_i) and four dimensionful parameters κ_i, $i = 1, 2, 3, 4$, describing linear (κ_1) and cubic terms (κ_2, κ_3) and κ_4. The linear term κ_1 can be removed by a translation of the singlet field. To simplify the model, we apply a global U(1) symmetry [33, 34, 35].

$$U(1): \quad \Phi \to \Phi, \; \chi \to e^{i\alpha}\chi. \tag{9}$$

However, a non-zero VEV of χ would lead to a spontaneous breaking of this symmetry and appearing of massless Nambu-Goldstone scalar particles, what is not acceptable. We keep U(1) soft-breaking terms to solve this problem. Therefore, we consider a potential with U(1) soft-breaking terms, where the singlet cubic terms $\kappa_{2,3}$, κ_4 and the singlet quadratic term m_4^2 are kept. Using following notation, $\lambda_s = \lambda_{s1}, \Lambda = \Lambda_1$, we get the potential in the following form

$$\begin{aligned} V = \quad & -\frac{1}{2}m_{11}^2\Phi^\dagger\Phi + \frac{1}{2}\lambda\left(\Phi^\dagger\Phi\right)^2 + \Lambda(\Phi^\dagger\Phi)(\chi^*\chi) \\ & -\frac{1}{2}m_s^2\chi^*\chi + \lambda_s(\chi^*\chi)^2 + \kappa_4(\Phi^\dagger\Phi)(\chi+\chi^*) \\ & -\frac{1}{2}m_4^2(\chi^{*2}+\chi^2) + \kappa_2(\chi^3+\chi^{*3}) + \kappa_3(\chi+\chi^*)(\chi^*\chi). \end{aligned} \tag{10}$$

The potential is symmetric under the $\chi \to \chi^*$ transformation and all parameters are real. We express the complex singlet χ in terms of its real and imaginary parts, $\chi = (\phi_2 + i\phi_3)/\sqrt{2}$, so we have

$$\begin{aligned} V = \quad & -\frac{1}{2}m_{11}^2\Phi^\dagger\Phi + \frac{1}{2}\lambda\left(\Phi^\dagger\Phi\right)^2 - \frac{\mu_1^2}{4}\phi_2^2 \\ & -\frac{\mu_2^2}{4}\phi_3^2 + \frac{1}{2}\Lambda(\Phi^\dagger\Phi)(\phi_2^2+\phi_3^2) \\ & +\frac{1}{4}\lambda_s(\phi_2^2+\phi_3^2)^2 + \frac{1}{\sqrt{2}}\kappa_2(\phi_2^3 - 3\phi_2\phi_3^2) \\ & +\frac{1}{\sqrt{2}}\kappa_3(\phi_2^3+\phi_2\phi_3^2) + \sqrt{2}\kappa_4(\Phi^\dagger\Phi)\phi_2. \end{aligned} \tag{11}$$

The extremum conditions lead to the following constraints,

$$v(-m_{11}^2 + v^2\lambda + 2\sqrt{2}w_1\kappa_4 + \Lambda w^2) = 0, \tag{12}$$

$$w_1(-\mu_1^2 + v^2\Lambda + 2w^2\lambda_s) + \sqrt{2}[3(w_1^2 - w_2^2)\kappa_2 + (3w_1^2 + w_2^2)\kappa_3] + v^2\sqrt{2}\kappa_4 = 0, \tag{13}$$

$$w_2[-\mu_2^2 + v^2\Lambda + 2w^2\lambda_s + 2\sqrt{2}w_1(-3\kappa_2 + \kappa_3)] = 0, \tag{14}$$

where the parameters μ_1^2 and μ_2^2 are defined as

$$\mu_1^2 = m_s^2 + 2m_4^2, \quad \mu_2^2 = m_s^2 - 2m_4^2.$$

We consider the CP violating vacua, i.e. v, w_1 and w_2 different from zero. To have a stable minimum the parameters of the potential need to satisfy the positivity conditions. These conditions read as follows:

$$\lambda,\ \lambda_s > 0, \quad \Lambda > -\sqrt{2\lambda\lambda_s}. \tag{15}$$

In addition, the unitarity and perturbativity conditions limit the quartic terms λ, λ_s and Λ to be below 4π.

3. The electroweak phase transition

The last requirement in Sakharov's criteria, the departure from thermal equilibrium, is an important ingredient to explain the baryon asymmetry within the SM of particle physics. The condition for the first-order EWPT with two degenerate minima at critical temperature, T_c, is [11, 37, 38],

$$\frac{v_{T_c}}{T_c} \geq 1, \tag{16}$$

where, v_{T_c} is VEV at critical temperature, T_c. The one loop thermal corrections to the effective potential at finite temperature T are (see for review, ref. [38]),

$$\Delta V_{thermal} = \sum_i \frac{n_i T^4}{2\pi^2} I_{B,F}\left(\frac{m_i^2}{T^2}\right), \tag{17}$$

where

$$I_{B,F}(y) = \int_0^\infty dx\, x^2\, ln\left[1 \mp e^{-\sqrt{x^2+y}}\right], \tag{18}$$

the minus and plus sign corresponds to the bosons and the fermions, respectively. In Eq.(17) m_i is the field-dependent mass and n_i is the number of degrees of freedom, (see Appendix A and Appendix B). Since the barrier is produced at tree-level, it is sufficient to include the high temperature expansion in the one-loop thermal potential (i.e. only keeping T^2 terms) [39, 40]. Therefore, the one-loop thermal potential using the high temperature approximation is given by,

$$\begin{aligned} V(T) = \quad & \frac{1}{2}\overline{m}_{11}^2\Phi^\dagger\Phi + \frac{1}{2}\lambda(\Phi^\dagger\Phi) + \frac{\overline{\mu}_1^2}{4}\phi_2^2 + \frac{\overline{\mu}_2^2}{4}\phi_3^2 \\ & \frac{1}{2}\Lambda(\Phi^\dagger\Phi)(\phi_2^2 + \phi_3^2) + \frac{1}{4}\lambda_s(\phi_2^2 + \phi_3^2)^2 \\ & +\kappa_2\frac{1}{\sqrt{2}}(\phi_2^3 - 3\phi_2\phi_3^2) + \kappa_3\frac{1}{\sqrt{2}}(\phi_2^3 + \phi_2\phi_3^2) \\ & +\sqrt{2}\kappa_4(\Phi^\dagger\Phi)\phi_2 + \overline{\kappa}_{34}\frac{T^2}{3}\phi_2, \end{aligned} \tag{19}$$

where,

$$\begin{aligned}
\overline{m}_{11}^2 &= -m_{11}^2 + (3\lambda + \Lambda + \frac{2m_W^2 + m_Z^2 + 2m_t^2}{2v^2})\frac{T^2}{3}, \\
\frac{1}{2}\overline{\mu}_1^2 &= -\frac{1}{2}\mu_1^2 + (\Lambda + 2\lambda_s)\frac{T^2}{3}, \\
\frac{1}{2}\overline{\mu}_2^2 &= -\frac{1}{2}\mu_2^2 + (\Lambda + 2\lambda_s)\frac{T^2}{3}, \\
\overline{\kappa}_{34} &= \sqrt{2}(\kappa_3 + \kappa_4).
\end{aligned} \tag{20}$$

The extremum conditions of the effective potential Eq.(17) at temperature T, with respect to the fields ϕ_1, ϕ_2 and ϕ_3 are

$$v(\overline{m}_{11}^2 + \lambda v^2 + \Lambda w^2 + 2\sqrt{2}\kappa_4 w_1) = 0, \tag{21}$$

$$\begin{aligned}
&w_1(\overline{\mu}_1^2 + \Lambda v^2 + 2\lambda_s w^2) + \sqrt{2}\left[3\kappa_2(w_1^2 - w_2^2)\right. \\
&\qquad \left. + \kappa_3(3w_1^2 + w_2^2) + \kappa_4 v^2\right] + \frac{2}{3}\overline{\kappa}_{34}T^2 = 0,
\end{aligned} \tag{22}$$

$$w_2[\overline{\mu}_2^2 + \Lambda v^2 + 2\lambda_s w^2 + 2\sqrt{2}(-3\kappa_2 + \kappa_3)w_1] = 0. \tag{23}$$

The solution of the Eqs.(21),(22) and (23) at extreme temperatures is as follows,

$$v = 0, \; w_1 \approx \frac{-\overline{\kappa}_{34}}{2\lambda_s + \Lambda}, \; w_2 \approx 0. \tag{24}$$

At extreme temperatures the scalar component ϕ_3 and vector quarks decouple from the model because their contribution to the finite temperature effective potential is Boltzmann suppressed and therefore, the potential is similar to the SM plus a real singlet [41].

Now, we will scan over the parameter space of the model under the condition for strong first order EWPT,

- $V_{eff}(v_{T_c}, T_c) = V_{eff}(0, T_c)$ (i.e. two degenerate minima at critical temperature, T_c),
- $v_{T_c}/T_c \geq 1$.

We consider T_c to be smaller than 250 GeV and v_{T_c} below its zero-temperature value $v_0 = 246$ GeV. In our scanning the following regions for parameters space of the model fulfilling positivity and unitarity conditions is considered (see [34]), namely:

$$\begin{aligned}
&\Lambda \in [-0.25, 0.25], \; \lambda_s \in [0, 1], \; \rho_{2,3,4} \in [-1, 1], \; \xi \in [0, \pi], \\
&m_{11}^2, \mu_1^2, \mu_2^2 \in [-90000, 90000]\mathrm{GeV}^2,
\end{aligned} \tag{25}$$

where we used dimensionless parameters $\rho_{2,3,4} = \kappa_{2,3,4}/w$. The mass of the lightest Higgs boson in this model is given by $M_{h_1}^2 \approx m_{11}^2 \approx \lambda v^2$ ($M_{h_1} \approx 125$ GeV), considering the SM-like scenarios at the LHC. Therefore, the coupling λ is taken in the range [34]:

$$\lambda \in [0.2, 0.3]. \tag{26}$$

The model has two additional Higgs neutral scalars M_{h_2} and M_{h_3}, that we take to be [33, 34]

$$M_{h_3} \gtrsim M_{h_2} > 150\,\mathrm{GeV}. \tag{27}$$

It is shown that these ranges of parameters are in agreement with LHC data and measurements of the oblique parameters, in ref [34].

The results of our scanning are shown in Fig.1. In the Fig.1(a), the allowed region of v_{T_c}/T_c as a function of T_c is presented. Within an interval $T_c \in [100, 200]$ the ratio v_{T_c}/T_c can reach 2.5. The Fig.1(b) shows that the ratio $v_{T_c}/T_c \geq 1$ is possible for $|\rho_3| > 10^{-3}$.
Throughout above discussions, we can concludes that a strong enough first-ordered EWPT is possible in this model and hence, a successful BAU can be achieved [38].

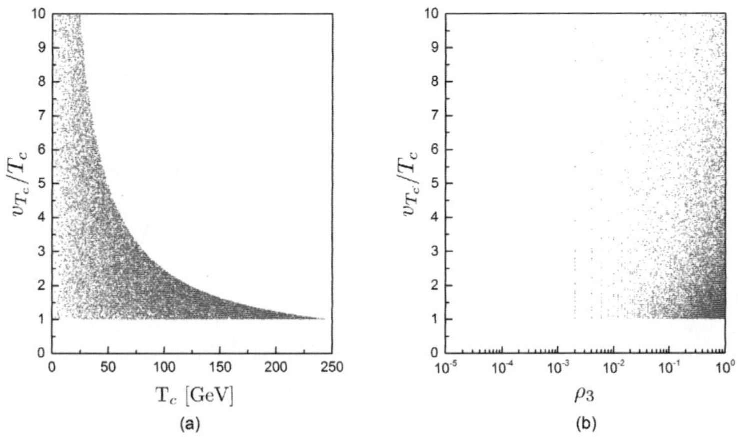

Figure 1. The allowed regions of critical temperature T_c, v_{T_c} and $|\rho_3|$ for strongly first order phase transition. (a) (T_c,v_{T_c}/T_c) and (b) ($|\rho_3|$, v_{T_c}/T_c). The scatter points are selected to satisfy the criterion, $(v_{T_c}/T_c) \geq 1$(see text for details).

4. Baryogenesis

In this section we describe the baryon asymmetry resulting from a mixing of the SM quarks and heavy vector quarks provided by CP violation vacuum [41, 36]. To generate baryon asymmetry, the phase of the singlet VEV should be time-dependent, otherwise, such constant phase can be easily rotated away with the redefinition of the V_L and V_R [36]. Diagonalizing the quark mass matrix from Eq.(2) results in some non-diagonal kinetic terms dependent on mass eigenstate, Q'_L and V'_L. In addition, a couple of time-dependent terms appear in the Lagrangian (see Appendix C), namely

$$\overline{Q'}_L i\gamma^\mu \partial_\mu Q'_L + \overline{V'}_L i\gamma^\mu \partial_\mu V'_L + \Delta\mathcal{L}_k + const. \quad (28)$$

As mentioned above, the CP violation disappears for a constant phase, therefor the following kinetic term, that are not constant needs to be considered,

$$\Delta\mathcal{L}_k = -\frac{\lambda_V^2 w^2}{M^2}\dot{\xi}(\overline{Q'}_L\gamma^0 Q'_L - \overline{V'}_L\gamma^0 V'_L). \quad (29)$$

Such terms result in increasing the baryon density of the universe. The amount of BAU is calculated via the following relation,

$$n_B = -N_f \int \frac{\Gamma_{sph}(T)}{2T}\mu_B dt, \quad (30)$$

where N_f is the number of flavors in the model. The sphaleron rate is defined as $\Gamma_{sph} = K(\alpha_W T)^4$, in the symmetric phase. K is the numerical factor reflecting the uncertainty in the estimate of the transition rate between vacua of different $B+L$ value, that is estimated to be between 0.1 and 1 [42]. The chemical potential, μ_B, is associated with the baryonic charge. The chemical potential for the third generation is given by [36],

$$\mu_B = -\frac{5}{6}\frac{\lambda_V^2 w^2}{M^2}\dot{\xi}. \tag{31}$$

Since by assumption $M \geq T$, the sphaleron fluctuations can not produce V quark pair, therefore we get the number density of baryons n_B at the temperature T as follows

$$n_B = \frac{5K\alpha_W^4}{2}\frac{\lambda_V^2 w^2}{M^2}\delta\xi T^3, \tag{32}$$

where $\delta\xi$ is the total change of the phase ξ. The BAU is determined via the ratio of the baryon number to the entropy [43]. The entropy density is defined as

$$s = \frac{2\pi^2}{45}g^* T^3, \tag{33}$$

therefore, the ratio of the baryon number to the entropy in this model is given by,

$$\frac{n_B}{s} = \frac{225K\alpha_W^4}{4\pi^2 g^*}\frac{\lambda_V^2 w^2}{M^2}\delta\xi, \tag{34}$$

where $\alpha_W = 3.4 \times 10^{-2}$ is the $SU(2)$ gauge coupling and $g^* \sim 100$ is the effective number of degrees of freedom in the thermal equilibrium.

The observed value for n_B/s ratio is [44, 45]

$$\frac{n_B}{s} = 8.7 \pm 0.3 \times 10^{-11}. \tag{35}$$

Comparing Eq.(34)and Eq.(35), one concludes

$$K\frac{\lambda_V^2 w^2}{M^2}\delta\xi = 1.14 \pm 0.3 \times 10^{-3}. \tag{36}$$

The numerical analysis has been performed via scanning over the relevant parameters in the following range,

$$\begin{aligned} &M \in [0.3, 13]\text{TeV}, \\ &\lambda_V \in [0, 1], \\ &w \in [2, 400]\text{GeV}, \\ &\delta\xi \in [0, \pi], \end{aligned} \tag{37}$$

taking $K = 1$. Figure 2 illustrates the parameter space allowing the generation of observed BAU for the n_B/s ratio within 2σ. The results were obtained from scanning in the ranges given by Eq.(37) with the central value of Eq.(35). Also, we have used our results from ref. [34] for the range of w. Fig.2(a) demonstrates the region of parameters $(\delta\xi, w)$, while Fig.2(b) shows the (w, M) points. Since M and w are independent parameters, their correlation is a direct consequence of the constraint (36). Based on these results, we conclude that our model provides successful BAU.

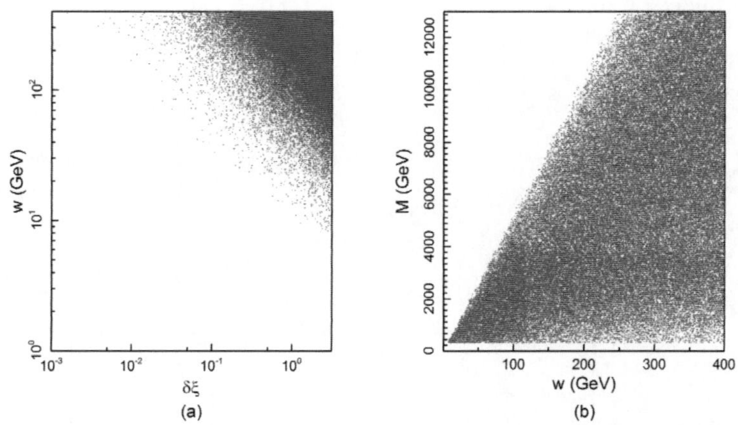

Figure 2. The allowed region for $\delta\xi$, w and M in the Eq.(35) of the given ranges, i.e. Eq.(37), for acceptable BAU within 2σ. (a) is the correlation of $\delta\xi$ and w.(b) is the correlation of w versus M.

5. Conclusion

In the present work the possibility of the first order EWPT for the extended SM with complex singlet and a pair of iso-doublet vector quarks is investigated, showing that such process is strong enough to generate BAU in the presence of non zero value of cubic terms, κ_2, κ_3 and/or κ_4. Afterwards, the parameter space of the model for the valid regions of BAU is scanned, concluding that this model with a heavy iso-doublet vector quark pair could successfully predict an acceptable value for BAU.

Acknowledgements

I am grateful to Prof. Maria Krawczyk for fruitful discussions and for reviewing this work. This work is supported in part by the National Science Centre, Poland, the HARMONIA project under contract UMO-2015/18/M/ST2/00518.

Appendix A. The field-dependent mass m_i

The field-dependent mass m_i of gauge bosons, Goldestone boson, m_{ϕ_1}, m_{ϕ_2} and m_{ϕ_3}, which is used in Eq.(17) are given by,

$$\begin{aligned}
M_W^2 &= \frac{g^2\phi_1^2}{4},\ M_Z^2 = (g^2+g'^2)\frac{\phi_1{}^2}{4}, \\
m_G^2 &= \lambda\phi_1^2 + \Lambda(\phi_2^2+\phi_3^2) + 2\sqrt{2}\kappa_4\phi_2), \\
m_{\phi_1}^2 &= 3\lambda\phi_1^2 + \Lambda(\phi_2^2+\phi_3^2) + 2\sqrt{2}\kappa_4\phi_2), \\
m_{\phi_2}^2 &= 3\lambda_s\phi_2^2 + \lambda_s\phi_3^2 + 3\sqrt{2}(\kappa_2+\kappa_3)\phi_2 + \frac{1}{2}\Lambda\phi_1^2, \\
m_{\phi_3}^2 &= 3\lambda_s\phi_3^2 + \lambda_s\phi_2^2 + \sqrt{2}(-3\kappa_2+\kappa_3)\phi_2 + \frac{1}{2}\Lambda\phi_1^2.
\end{aligned} \tag{A.1}$$

n_i is the number of degrees of freedom in Eq.(17) as,

$$n_W = 6,\ n_Z = 3,\ n_G = 3,\ n_{\phi_1,\phi_2,\phi_3} = 1,\ n_t = 12. \tag{A.2}$$

IOP Conf. Series: Journal of Physics: Conf. Series **873** (2017) 012028 doi:10.1088/1742-6596/873/1/012028

Appendix B. Evaluation of the integral

The integral Eq.(18) is evaluated as follow,

$$\frac{\partial}{\partial y} I_{B,F}(y) = \frac{1}{2}\int_0^\infty dx \frac{x^2}{(x^2+y)^{1/2}} \frac{1}{exp((x^2+y)^{1/2})-1}, \tag{B.1}$$

$$I_{B,F}(y)|_{y=0} = \int_0^\infty dx x^2 ln(1-e^{-x}) = -\frac{\pi^4}{45}, \tag{B.2}$$

$$\frac{\partial}{\partial y} I_{B,F}(y)|_{y=0} = \frac{1}{2}\int_0^\infty dx \frac{x}{e^x-1} = \frac{\pi^2}{12}. \tag{B.3}$$

Appendix C. Rotation matrix

As discussed in sec.4 (see the Eq.(28)) the transformation of Q and V with rotation matrix to Q' and V' reads as follows

$$\begin{aligned}
\begin{bmatrix} Q'_L \\ V'_L \end{bmatrix} &= \begin{bmatrix} a & b \\ -b^* & a^* \end{bmatrix} \begin{bmatrix} Q_L \\ V_L \end{bmatrix} \\
a &= \left[1+\left(\frac{\lambda_V w}{M}\right)^2\right]^{-1/2} \\
b &= \left(\frac{\lambda_V w}{M}\right)\left[1+\left(\frac{\lambda_V w}{M}\right)^2\right]^{-1/2} e^{-i\xi}
\end{aligned} \tag{C.1}$$

References

[1] Cohen A G, Kaplan D B and Nelson A E 1993 *Ann. Rev. Nucl. Part. Sci.* **43** 27–70 (*Preprint* hep-ph/9302210)
[2] Quiros M 1994 *Helv. Phys. Acta* **67** 451–583
[3] Rubakov V A and Shaposhnikov M E 1996 *Usp. Fiz. Nauk* **166** 493–537 [Phys. Usp.39,461(1996)] (*Preprint* hep-ph/9603208)
[4] Carena M and Wagner C E M 1997 (*Preprint* hep-ph/9704347)
[5] Sakharov A D 1967 *Pisma Zh. Eksp. Teor. Fiz.* **5** 32–35 [Usp. Fiz. Nauk161,61(1991)]
[6] Anderson G W and Hall L J 1992 *Phys. Rev.* **D45** 2685–2698
[7] Carrington M E 1992 *Phys. Rev.* **D45** 2933–2944
[8] Dine M, Leigh R G, Huet P, Linde A D and Linde D A 1992 *Phys. Lett.* **B283** 319–325 (*Preprint* hep-ph/9203201)
[9] Dine M, Leigh R G, Huet P Y, Linde A D and Linde D A 1992 *Phys. Rev.* **D46** 550–571 (*Preprint* hep-ph/9203203)
[10] Bochkarev A I, Kuzmin S V and Shaposhnikov M E 1991 *Phys. Rev.* **D43** 369–374
[11] Davies A T, froggatt C D, Jenkins G and Moorhouse R G 1994 *Phys. Lett.* **B336** 464–470
[12] Espinosa J R, Quiros M and Zwirner F 1993 *Phys. Lett.* **B307** 106–115 (*Preprint* hep-ph/9303317)
[13] Shaposhnikov M E 1993 *Phys. Lett.* **B316** 112–120 (*Preprint* hep-ph/9306296)
[14] Nelson A E, Kaplan D B and Cohen A G 1992 *Nucl. Phys.* **B373** 453–478
[15] Cohen A G and Nelson A E 1992 *Phys. Lett.* **B297** 111–117 (*Preprint* hep-ph/9209245)
[16] Comelli D, Pietroni M and Riotto A 1994 *Nucl. Phys.* **B412** 441–458 (*Preprint* hep-ph/9304267)
[17] Shaposhnikov M E 1987 *Nucl. Phys.* **B287** 757–775
[18] Accomando E *et al.* 2006 (*Preprint* hep-ph/0608079)
[19] Haber H E and Surujon Z 2012 *Phys. Rev.* **D86** 075007 (*Preprint* 1201.1730)
[20] Ginzburg I F and Krawczyk M 2005 *Phys. Rev.* **D72** 115013 (*Preprint* hep-ph/0408011)
[21] Gunion J F and Haber H E 2005 *Phys. Rev.* **D72** 095002 (*Preprint* hep-ph/0506227)

[22] Davidson S, González Felipe R, Serôdio H and Silva J P 2013 *JHEP* **11** 100 (*Preprint* 1307.6218)
[23] Branco G C, Parada P A and Rebelo M N 2003 (*Preprint* hep-ph/0307119)
[24] Lebedev O 2011 *Phys. Lett.* **B697** 58–62 (*Preprint* 1011.2630)
[25] Espinosa J R, Gripaios B, Konstandin T and Riva F 2012 *JCAP* **1201** 012 (*Preprint* 1110.2876)
[26] Gabrielli E, Heikinheimo M, Kannike K, Racioppi A, Raidal M and Spethmann C 2014 *Phys. Rev.* **D89** 015017 (*Preprint* 1309.6632)
[27] Costa R, Morais A P, Sampaio M O P and Santos R 2015 *Phys. Rev.* **D92** 025024 (*Preprint* 1411.4048)
[28] Kozaczuk J 2015 *JHEP* **10** 135 (*Preprint* 1506.04741)
[29] Jiang M, Bian L, Huang W and Shu J 2016 *Phys. Rev.* **D93** 065032 (*Preprint* 1502.07574)
[30] Alexander-Nunneley L and Pilaftsis A 2010 *JHEP* **09** 021 (*Preprint* 1006.5916)
[31] Barger V, Langacker P, McCaskey M, Ramsey-Musolf M and Shaughnessy G 2009 *Phys. Rev.* **D79** 015018 (*Preprint* 0811.0393)
[32] Darvishi N 2016 *JHEP* **11** 065 (*Preprint* 1608.02820)
[33] Bonilla C, Sokolowska D, Darvishi N, Diaz-Cruz J L and Krawczyk M 2016 *J. Phys.* **G43** 065001 (*Preprint* 1412.8730)
[34] Darvishi N and Krawczyk M 2016 (*Preprint* 1603.00598)
[35] Krawczyk M, Darvishi N and Sokolowska D 2016 *Acta Phys. Polon.* **B47** 183 (*Preprint* 1512.06437)
[36] McDonald J 1996 *Phys. Rev.* **D53** 645–654
[37] Kane G L Perspectives on Higgs physics II. Singapore: World Scientific
[38] Quiros M 1999 *Proceedings, Summer School in High-energy physics and cosmology: Trieste, Italy, June 29-July 17, 1998* pp 187–259 (*Preprint* hep-ph/9901312) URL http://alice.cern.ch/format/showfull?sysnb=0302087
[39] Dolan L and Jackiw R 1974 *Phys. Rev.* **D9** 3320–3341
[40] Espinosa J R, Konstandin T and Riva F 2012 *Nucl. Phys.* **B854** 592–630 (*Preprint* 1107.5441)
[41] Branco G C, Delepine D, Emmanuel-Costa D and Gonzalez F R 1998 *Phys. Lett.* **B442** 229–237 (*Preprint* hep-ph/9805302)
[42] Ambjorn J, Askgaard T, Porter H and Shaposhnikov M E 1991 *Nucl. Phys.* **B353** 346–378
[43] Cline J M 2006 *Les Houches Summer School - Session 86: Particle Physics and Cosmology: The Fabric of Spacetime Les Houches, France, July 31-August 25, 2006* (*Preprint* hep-ph/0609145)
[44] (WMAP) S T Http://map.gsfc.nasa.gov/site/citations.html
[45] Ade P A R *et al.* (Planck) 2014 *Astron. Astrophys.* **571** A16 (*Preprint* 1303.5076)

IOP Conf. Series: Journal of Physics: Conf. Series **873** (2017) 012029 doi:10.1088/1742-6596/873/1/012029

Right-handed neutrino dark matter in left-right symmetric models

P. S. Bhupal Dev[1], **Rabindra N. Mohapatra**[2], **Yongchao Zhang**[3]

[1] Department of Physics and McDonnell Center for the Space Sciences, Washington University, St. Louis, MO 63130, USA
[2] Maryland Center for Fundamental Physics, Department of Physics, University of Maryland, College Park, MD 20742, USA
[3] Service de Physique Théorique, Université Libre de Bruxelles, Boulevard du Triomphe, CP225, 1050 Brussels, Belgium

E-mail: bdev@physics.wustl.edu, rmohapat@umd.edu, yongchao.zhang@ulb.ac.be

Abstract. We show that in a class of non-supersymmetric left-right extensions of the Standard Model (SM), the lightest right-handed neutrino (RHN) is naturally stable and can therefore play the role of thermal Dark Matter (DM) in the Universe for a wide mass range from TeV to PeV. Our model is based on the gauge group $SU(3)_c \times SU(2)_L \times SU(2)_R \times U(1)_{Y_L} \times U(1)_{Y_R}$ in which a heavy copy of the SM fermions are introduced and the stability of the RHN DM is guaranteed by an automatic Z_2 symmetry present in the leptonic sector. The active neutrino masses in the model arise from the type-II seesaw mechanism. We find a lower bound on the RHN DM mass of order TeV from relic density constraints, as well as an unitarity upper bound in the multi-TeV to PeV scale, depending on the entropy dilution factor. The RHN DM could be made long-lived by soft-breaking of the Z_2 symmetry and provides a concrete example of decaying DM interpretation of the PeV neutrinos observed at IceCube.

1. Introduction

The existence of dark matter (DM) in our Universe has been well established by astrophysical and cosmological observations. To explain this, one needs some new particle(s) and symmetries beyond the Standard Model (SM). Being colorless and electrically neutral, SM-singlet right-handed neutrino (RHN), is a natural DM candidate. Although the right handed neutrino is introduced to implement the seesaw mechanism to understand small neutrino masses, in the conventional seesaw models [1, 2, 3, 4, 5], it has Yukawa couplings to SM lepton and Higgs doublets, making it highly unstable, and therefore, unsuitable for DM. The only exception is when its mass is in the keV range and it has appropriately small Yukawa couplings [6]. In ultraviolet (UV) complete scenarios of seesaw, e.g. those based on the Left-Right (LR) gauge group $SU(2)_L \times SU(2)_R \times U(1)_{B-L}$ [7, 8, 9], even having tiny Yukawa couplings do not help since the gauge interactions of RHNs induce further decay modes, which have to be suppressed/forbidden by imposing additional discrete symmetries to keep the lightest RHN cosmologically stable [10].

In this paper we discuss an alternative class of LR models based on the gauge group $SU(3)_c \times SU(2)_L \times SU(2)_R \times U(1)_{Y_L} \times U(1)_{Y_R}$ [11] which has an accidental Z_2 symmetry, at the renormalizable level that keeps the lightest RHN stable making it a natural DM candidate.

IOP Conf. Series: Journal of Physics: Conf. Series **873** (2017) 012029 doi:10.1088/1742-6596/873/1/012029

In this model there exists a right-handed copy of the SM electroweak sector at the TeV scale (or higher), charged under $SU(2)_R \times U(1)_{Y_R}$. The color $SU(3)$ is common to both sectors. The breaking of the weak gauge symmetry down to $U(1)_{\rm em}$ is similar to the LR models with universal seesaw [13, 14, 15]. But one of the key differences from the universal seesaw models is that the right-handed lepton sector in our model is prevented by gauge symmetry from having any masses connecting the heavy sector to the light ones. Because of this, there appears a remnant automatic Z_2 symmetry in the leptonic sector at the renormalizable level that keeps the lightest RHN stable. If the stabilizing Z_2 symmetry is softly broken by a mass term connecting the heavy and light sector, the RHN DM can be made very long-lived but allowing for its decay into SM leptons. This raises the possibility that for a PeV-scale DM, one can explain the ultra-high energy neutrino events observed at IceCube [16]. We also point out a dilution mechanism which allows us to relax the unitarity bound on the dark matter to the PeV mass level so that it can be applied to the discussion of PeV neutrinos in the context of a thermal annihilating DM.

2. The Model

The model is based on the gauge group $SU(2)_L \times SU(2)_R \times U(1)_{Y_L} \times U(1)_{Y_R}$ [17, 18]. For each generation of SM quark and lepton doublets Q_L, ψ_L and singlets u_R, d_R, e_R, we have the corresponding heavy fermion partners $\mathcal{Q}_R$, Ψ_R and $\mathcal{U}_L$, $\mathcal{D}_L$, $\mathcal{E}_L$, all of which are collected in the upper rows of Table 1. Note that the heavy doublets transform nontrivially under the $SU(2)_R$ gauge symmetry, required by the parity $L \leftrightarrow R$. Analogous to other LR models, the electric charge formula is generalized to

$$Q = I_{3L} + I_{3R} + \frac{1}{2}(Y_L + Y_R)\,. \tag{1}$$

The Higgs sector of the model consists of $SU(2)_{L,R}$ doublets $\chi_{L,R}$ and triplets $\Delta_{L,R}$, which are also presented in Table 1. In the limit of vanishing mixing of the scalars, the doublet χ_L plays the role of the SM Higgs doublet. The gauge symmetries $SU(2)_L$ and $SU(2)_R$ are broken by the vacuum expectation values (VEVs) of the neutral components of these Higgs fields:

$$\langle \chi_L^0 \rangle = v_L \equiv v_{\rm EW}, \qquad \langle \chi_R^0 \rangle = v_R, \qquad \langle \Delta_L^0 \rangle = w_L, \qquad \langle \Delta_R^0 \rangle = w_R\,, \tag{2}$$

which gives also masses to the heavy light gauge bosons. The SM and heavy charged fermion masses are obtained through their Yukawa couplings to the Higgs doublets:

$$\begin{aligned} -\mathcal{L}_Y \supset\ & y_u \overline{Q}_L \widetilde{\chi}_L u_R + y_d \overline{Q}_L \chi_L d_R + y_e \overline{\psi}_L \chi_L e_R \\ & + y'_u \overline{\mathcal{Q}}_R \widetilde{\chi}_R \mathcal{U}_L + y'_d \overline{\mathcal{Q}}_R \chi_R \mathcal{D}_L + y'_e \overline{\Psi}_R \chi_R \mathcal{E}_L + \text{H.c.}\,, \end{aligned} \tag{3}$$

where $\tilde{\chi}_L = i\sigma_2 \chi_L^*$ and similarly for χ_R (σ_2 being the second Pauli matrix). Exact parity symmetry implies that the Yukawa couplings $y_f = y'_f$ (and also $f = f'$ in Eq. (4) below). If the $SU(2)_R$ symmetry breaking is at the TeV scale, there would be new quarks in the few GeV range, which is inconsistent with the current LHC bounds on vector-like fermion masses of order of TeV. To have heavy quarks and leptons at the TeV scale, we adopt the parity violating scenario in which the couplings y_f and y'_f are independent and choose $y'_f \sim \mathcal{O}(1)$.[1] It is worth noting that at this stage the lower bound on M_{W_R} from low-energy flavor changing effects such as $K^0 - \overline{K^0}$ mixing does not apply to our model.

[1] The parity symmetric version could be realized at the TeV scale by doubling the doublet scalars in both the SM and heavy sectors, such that one set of the Yukawa couplings $y_{f,1}$ is responsible for the SM fermion masses and the other one $y_{f,2}$ dominates the heavy fermion masses [11].

IOP Conf. Series: Journal of Physics: Conf. Series **873** (2017) 012029 doi:10.1088/1742-6596/873/1/012029

Table 1. Fermion and scalar contents of the LR model in this paper and their representations under the gauge group $SU(3)_C \times SU(2)_L \times SU(2)_R \times U(1)_{Y_L} \times U(1)_{Y_R}$ [11].

field	representation	field	representation
$Q_L = \begin{pmatrix} u_L \\ d_L \end{pmatrix}$	$(\mathbf{3},\mathbf{2},\mathbf{1},\frac{1}{3},0)$	$\mathcal{Q}_R = \begin{pmatrix} \mathcal{U}_R \\ \mathcal{D}_R \end{pmatrix}$	$(\mathbf{3},\mathbf{1},\mathbf{2},0,\frac{1}{3})$
u_R	$(\mathbf{3},\mathbf{1},\mathbf{1},\frac{4}{3},0)$	$\mathcal{U}_L$	$(\mathbf{3},\mathbf{1},\mathbf{1},0,\frac{4}{3})$
d_R	$(\mathbf{3},\mathbf{1},\mathbf{1},-\frac{2}{3},0)$	$\mathcal{D}_L$	$(\mathbf{3},\mathbf{1},\mathbf{1},0,-\frac{2}{3})$
$\psi_L = \begin{pmatrix} \nu_L \\ e_L \end{pmatrix}$	$(\mathbf{1},\mathbf{2},\mathbf{1},-1,0)$	$\Psi_R = \begin{pmatrix} N \\ \mathcal{E}_R \end{pmatrix}$	$(\mathbf{1},\mathbf{1},\mathbf{2},0,-1)$
e_R	$(\mathbf{1},\mathbf{1},\mathbf{1},-2,0)$	$\mathcal{E}_L$	$(\mathbf{1},\mathbf{1},\mathbf{1},0,-2)$
$\chi_L = \begin{pmatrix} \chi_L^+ \\ \chi_L^0 \end{pmatrix}$	$(\mathbf{1},\mathbf{2},\mathbf{1},1,0)$	$\chi_R = \begin{pmatrix} \chi_R^+ \\ \chi_R^0 \end{pmatrix}$	$(\mathbf{1},\mathbf{1},\mathbf{2},0,1)$
$\Delta_L = \begin{pmatrix} \Delta_L^+/\sqrt{2} & \Delta_L^{++} \\ \Delta_L^0 & -\Delta_L^+/\sqrt{2} \end{pmatrix}$	$(\mathbf{1},\mathbf{3},\mathbf{1},2,0)$	$\Delta_R = \begin{pmatrix} \Delta_R^+/\sqrt{2} & \Delta_R^{++} \\ \Delta_R^0 & -\Delta_R^+/\sqrt{2} \end{pmatrix}$	$(\mathbf{1},\mathbf{1},\mathbf{3},0,2)$
Σ_1	$(\mathbf{1},\mathbf{1},\mathbf{1},-\frac{4}{3},\frac{4}{3})$	Σ_2	$(\mathbf{1},\mathbf{1},\mathbf{1},\frac{2}{3},-\frac{2}{3})$

The masses of light and heavy neutrinos are generated from couplings to the triplet scalars $\Delta_{L,R}$ via the type-II seesaw mechanism [19, 20, 21, 22]:

$$-\mathcal{L}_Y \supset f\overline{\psi_L^C} i\sigma_2 \Delta_L \psi_L + f' \overline{\Psi_R^C} i\sigma_2 \Delta_R \Psi_R + \text{H.c.}\,, \tag{4}$$

which implies that the VEV w_L should be at the eV scale and w_R is at the TeV scale or above. This can be easily obtained by choosing appropriate values of the parameters in the scalar potential.

3. Depleting the heavy quarks

After symmetry breaking at the right-handed scale and the electroweak scale, the model has a large global symmetry: $U(1)_{B,L} \times U(1)_{B,R} \times Z_{2\ell,L} \times Z_{2\ell,R}$, with the two $U(1)_B$ the baryon numbers in the SM and heavy sectors, and the two Z_2's the corresponding lepton numbers.

Protected by the baryon number symmetry, the lightest hadron in the $SU(2)_R$ sector is absolutely stable, which can be one of the heavy baryons of form QQQ, QQq, Qqq, or heavy mesons $\bar{Q}Q$ and $\bar{Q}q$. For the phenomenological purpose of avoiding the heavy lightest baryon also becoming DM and affecting the evolution of the Universe, we add two more Higgs singlets $\Sigma_1(\mathbf{1},\mathbf{1},-\frac{4}{3},+\frac{4}{3})$ and $\Sigma_2(\mathbf{1},\mathbf{1},+\frac{2}{3},-\frac{2}{3})$ which connect the heavy and light singlet quarks via the terms

$$\mathcal{L}_{\text{mix}} = \lambda_U \overline{\mathcal{U}}_L u_R \Sigma_1^* + \lambda_D \overline{\mathcal{D}}_L d_R \Sigma_2^* + \text{H.c.}\,. \tag{5}$$

Once the Σ fields acquire VEVs, the heavy-light quark mass mixing terms $\delta_U = \lambda_U \langle \Sigma_1 \rangle$ and $\delta_D = \lambda_D \langle \Sigma_2 \rangle$ appear in the Lagrangian, and the heavy quark can then decay into the SM fermions, leaving only the lightest RHN stable due to the leptonic Z_2 symmetry. The singlets $\Sigma_{1,2}$ are charged under $U(1)_{Y_L} \times U(1)_{Y_R}$ and break this symmetry down to $U(1)_{\mathcal{Y}}$. At this stage this model is similar to the usual LR models with $U(1)_{B-L}$, however with the quantum number $\mathcal{Y}$ different from $B-L$ and also from the SM hypercharge Y_L (which is a linear combination of $\mathcal{Y}$ and the $U(1)$ subgroup of $SU(2)_R$ after symmetry breaking).

The decay rate of heavy quarks $Q \to q + Z$ can be estimated as

$$\Gamma_Q \simeq \frac{g_L^2}{64\pi \cos^2\theta_w} \frac{\delta_Q^2}{v_R^2} \frac{M_Q^3}{M_Z^2}\,, \tag{6}$$

where θ_w is the SM weak mixing angle and g_L is the gauge coupling of $SU(2)_L$. Requiring that the heavy quarks are depleted before the QCD phase transition epoch $T_{\rm QCD} \approx 200$ MeV imposes a lower bound on the magnitude of $\delta_Q \gtrsim 10^{-6}$ GeV [11]. When this limit is close to being saturated, we expect the heavy quark decay length to be several cm, leading to displaced vertex signatures at hadron colliders.

4. Connecting the SM to DM

The heavy gauge boson Z_R is a linear combination of the $U(1)$ subgroup of $SU(2)_R$, $U(1)_{Y_L}$ and $U(1)_{Y_R}$, thus both the SM and heavy fermions are charged under Z_R. Even in the absence of any scalar and fermion mixings connecting the heavy and light sectors, the RHN N could interact pairwise with the SM particles through the heavy Z_R boson. Up to the leading order of mixings of $h - \Delta_R^0$ and $Z - Z_R$, two N's could also annihilate to the SM fields via the scalar portal (h and Δ_R^0) and gauge portal (Z boson). The SM Higgs h is assumed to be predominantly from the doublet χ_L, and thus the scalar mixing is expected to be of order $\zeta_S \simeq \lambda v_{\rm EW}/w_R$ from the quartic term of form $\lambda(\chi_L^\dagger \chi_L){\rm Tr}(\Delta_R^\dagger \Delta_R)$, where for simplicity we have neglected the mixing with other scalars. The neutral gauge boson mixing is of order $\zeta_Z \simeq \xi_Z (M_Z/M_{Z_R})^2$, where $\zeta_Z = \sin\theta_w(\cot^2\theta_w g_R^2/g_L^2 - 1)^{-1/2}$. Note that when the gauge coupling g_R approaches the theoretical lower limit of $g_L \tan\theta_w$ [23], which is independent of the symmetry breaking pattern [24], the $Z - Z_R$ mixing could be significantly enhanced.

In the minimal version of our model, we do not have the bi-multiplet scalars which transform non-trivially under both $SU(2)_L$ and $SU(2)_R$, thus the charged gauge bosons W and W_R can not mix at the tree level. A tiny mixing arises at one-loop level, induced by the (3rd generation) heavy-light quark mixings, which is estimated to be very small, of order $\zeta_W \sim 10^{-13}$, given the heavy quarks are of TeV mass, the W_R boson with mass of 5 TeV and the parameters $\delta_{U,D} \simeq$ GeV [11].

5. DM annihilation and relic density

In the early universe, all the particles are in equilibrium due to the common $SU(3)_c$ color and $U(1)_\mathcal{Y}$ interactions. As the universe cools down below the DM mass M_N, the DM density goes down and freezes out, as in case of a generic thermal DM candidate. The primary annihilation channels to the SM particles proceed via the scalar and the neutral gauge boson portals:

$$NN \to \left\{ \begin{array}{c} h^{(*)}/\Delta_R^{0\,(*)} \\ Z^{(*)}/Z_R^{(*)} \end{array} \right. \to \text{SM particles}\,. \qquad (7)$$

The h, Δ_R^0 and Z portals are suppressed respectively by the small mixing angles ζ_S and ζ_Z. On the other hand, though the couplings of Z_R to N and the SM fermions are of order one, the Z_R portal is suppressed by the large Z_R mass, except for the Breit-Wigner resonance when $2M_N \simeq M_{Z_R}$. Note that the Δ_R^0 portal is further suppressed by the DM mass when DM is light, as the Yukawa coupling f' in Eq. (4) is proportional to the DM mass via $M_N = 2f'w_R$. Combining all the available channels, we calculate the thermally averaged DM annihilation cross section times velocity $\langle\sigma v\rangle = a + b\langle v^2\rangle + \mathcal{O}(v^4)$, from which we can obtain the current DM relic density as for a standard thermal WIMP particle.

An example is given in Fig. 1, where we set explicitly the quartic coupling $\lambda = 1$ and the heavy scalar mass $M_{\Delta_R^0} = 2$ TeV with width 30 GeV. Three different value of the gauge coupling g_R are chosen with $g_R/g_L = 0.6$, 1 and 1.5. The corresponding Z_R mass and the RH scale $v_R = w_R$ are set to their current experimental lower bounds [25, 26], viz. $M_{Z_R} = 4.6, 3.9, 3.5$ TeV and $w_R = 3.0, 3.2, 2.1$ TeV for $g_R/g_L = 0.6$, 1, 1.5 respectively. The horizontal dashed line shows the observed relic density, as measured by Planck [27]. The various peaks in Fig. 1 are respectively

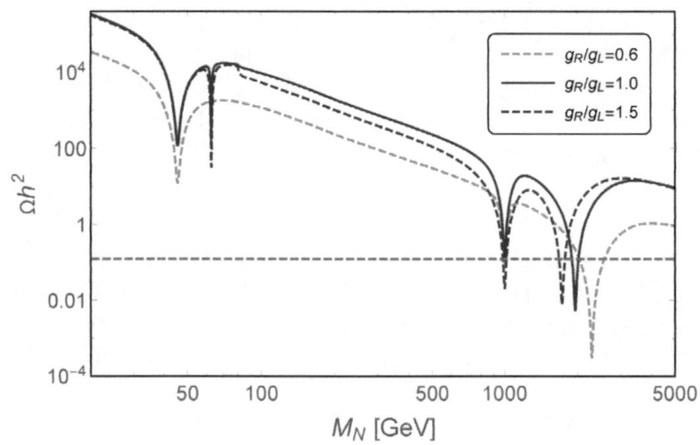

Figure 1. An illustration of the relic density of RHN DM as a function of its mass M_N for different values of g_R/g_L. The horizontal line gives the observed value from Planck data [27].

(from left to right) due to the SM Z and Higgs bosons, Δ_R^0 and Z_R. We find that a TeV scale RHN could accommodate the observed DM relic density, with the annihilation dominated by the heavy scalar and/or Z_R bosons, depending largely on the quartic coupling λ, the gauge coupling g_R and the VEV v_R.

6. Going beyond the unitarity bound

In generic annihilating thermal DM models, the DM mass cannot be arbitrarily large due to the well-known partial wave unitarity limit [28]. For the resonance $R(= \Delta_R^0, Z_R)$ just above the threshold, the annihilation cross section times velocity can be approximated by

$$\langle \sigma v \rangle \simeq 16\pi \frac{\Gamma_{NN}\Gamma_{\rm SM}}{(4M_N^2 - M_R^2)^2 + \Gamma_R^2 M_R^2} \sim \frac{4\pi}{M_N^2} \left(B_{NN} B_{\rm SM} \right) , \tag{8}$$

where Γ_{NN} and $\Gamma_{\rm SM}$ are the partial decay widths of R to the DM pair and SM particles respectively, and B_{NN} and $B_{\rm SM}$ are the corresponding branching ratios. From Eq. (8), we find that the annihilation rate decreases with increasing DM mass, which leads to the unitarity bound of ~ 20 TeV in this model.

It turns out that there is a built-in mechanism for entropy dilution in our model, which makes it possible to go beyond the unitarity limit. The basic idea is that the next-to-lightest heavy neutrino (denoted here by N_2) does not need to be stable and can decay into SM leptons, e.g. $N_2 \to \ell\ell\nu$ via the charged lepton mixing terms in Eq. (10) involving only the heavier-than-N leptons in the $SU(2)_R$ sector. Note that the hadronic channels $N_2 \to \ell q\bar{q}$ are comparatively suppressed by the small quark mixing $\delta_{U,D}$, and the two-body decay $N_2 \to W^\pm \ell^\mp$ is heavily suppressed by the loop-induced $W - W_R$ mixing. Because of higher mass, N_2 freezes out before the DM N but could decay after N freezes out, i.e. $T_{N_2} < T_F$, with appropriate values of leptonic mixing. The decay of N_2 to relativistic SM species will generate entropy which dilutes the relic density of N, and therefore, the initial relic density of N can be significantly larger than the current value, thus allowing for larger M_N values. The dilution factor depends largely on the mixing parameters in the scalar and lepton sector and some of the Yukawa couplings, which could reach up to the factor of 10^6 to accommodate a PeV DM N to explain the IceCube data. More details can be found in [11, 12].

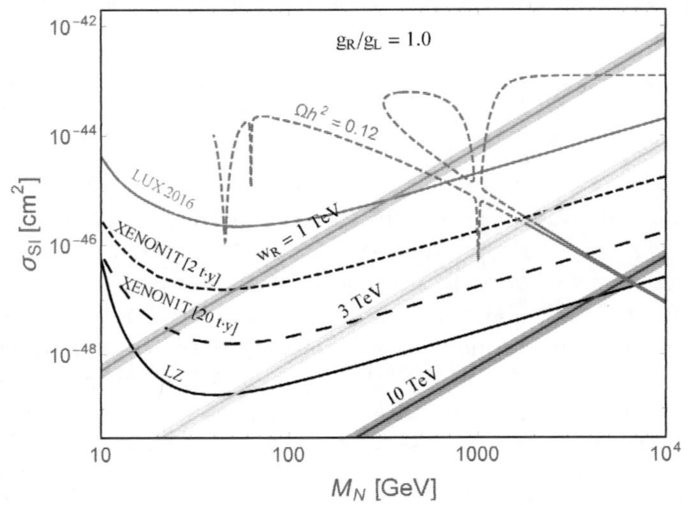

Figure 2. Predictions for the SI DM-nucleon scattering cross sections as functions of DM mass M_N with the gauge coupling $g_R = g_L$ and the RH scale $w_R = 1$, 3 and 10 TeV. The colorful bands are due to the sizable uncertainties of effective DM-nucleon couplings $f_{p,n}$. We show also the current upper limits from LUX [30], as well as the future reaches of XENON1T [32] and LZ [33]. The gray curves correspond to the parameter space producing the observed relic density $\Omega h^2 = 0.12$ (with the dashed part excluded by direct collider searches of Z_R in the dilepton channel).

7. Direct and indirect detections and collider searches

As a Majorana DM candidate, the RHN has both spin-independent (SI) and spin-dependent (SD) interactions with nuclei, which are mediated by the scalars and neutral gauge bosons, respectively.

The SI scattering cross section is dominated by the SM Higgs contribution, with the heavy scalar portal suppressed by its large mass,

$$\sigma_{\rm SI} = \frac{\lambda^2 \mu^2 M_N^2}{2\pi m_h^4 w_R^4} \left[Z f_p + (A-Z) f_n \right]^2 \tag{9}$$

where μ is the reduced DM-nucleus mass, and $f_{p,n}$ the effective Higgs-proton/neutron couplings. A characteristic feature of our model is that $\sigma_{\rm SI} \propto M_N^2$, which originates from the dependence of Yukawa coupling on the DM mass $f' \propto M_N$. As shown in Fig. 2, when the right-handed scale $v_R = w_R$ goes higher, the scalar mixing decreases and the scattering cross section $\sigma_{\rm SI}$ also gets smaller. In this plot we adopt the benchmark values of $v_R = w_R = 1$, 3 and 10 TeV, the gauge coupling $g_R = g_L$ and the quartic coupling $\lambda = 1$. The colorful bands are due to the sizable uncertainties of effective couplings $f_{p,n}$. The most stringent current direct detection limits come from LUX [30] (with comparable limits from PandaX-II [31]), which rule out the lighter RHN DM mass range, if the DM relic density is set to be the observed value (the gray contours). When the collider limits on Z_R mass are also taken into consideration (dashed gray lines), we can set more stringent lower limits of a TeV or so on the DM mass. We show also the projected limits from XENON1T with exposure values of 2 and 20 ton·year [32] and the LZ experiment [33] with orders of magnitude improvement in the SI scattering cross section limits, which can constrain the DM mass M_N and right-handed scale w_R up to the 10 TeV scale in near future experiments.

As for the SD scattering, it is mediated at leading order by the Z and Z_R bosons, both of which are suppressed by the small $Z - Z_R$ mixing angle ζ_Z. When the collider constraints on M_{Z_R} are taken into consideration, the $Z - Z_R$ mixing is so small that even with the future LZ limits it is rather challenging to test the model in the SD channels.

In the early universe, the RHN DM annihilates into the SM particles, e.g. W^+W^-, $\ell^+\ell^-$ and $q\bar{q}$, which could emit energetic gamma-rays. Thus the RHN DM can also be probed by the Fermi-LAT gamma-ray observations. Also DM annihilation injects energy into the thermal bath in the early universe, thus the annihilation cross section could also be constrained by the Planck data of the cosmic microwave background (CMB). However, limited by the small scalar and gauge mixing angles (ζ_S and ζ_Z), these indirect detections do not put additional constraints on the currently allowed model parameter space with the observed relic density.

As in usual thermal DM models, the RHN DM can be pair-produced at high energy hadron colliders, with monojet, mono-h, mono-W/Z and mono-photon signatures. However, suppressed again by the small mixing parameters, none of these channels could effectively probe our model at the LHC. It might be helpful to check some specific parameter regions at future higher energy colliders, such as FCC-hh and SPPC.

8. Decaying RHN DM and IceCube PeV neutrinos

In the model discussed in Section 2, the SM charged leptons e_R do not mix with their heavy partners $\mathcal{E}_L$, thus the lightest RHN is stabilized by the automatic Z_2 symmetry. If we add the following terms to the Lagrangian

$$\mathcal{L}_{\rm mix} = \delta_\ell \bar{\mathcal{E}}_L e_R + {\rm H.c.}\,, \tag{10}$$

to explicitly break the Z_2 symmetry (albeit softly), then the lightest RHN can decay into SM lepton and W boson through the loop mediated $W - W_R$ mixing, i.e. $N \to W^\pm \ell^\mp$. This raises the interesting possibility of explaining the IceCube PeV neutrinos [34] when we set $M_N \sim$ few PeV [16]. However, the two-body decays are disfavored by the IceCube data, as the neutrino spectrum is flatter than required [16].

To fit the IceCube neutrino data, we promote the scalars to be lepton-specific two Higgs doublet model (2HDM) in both the SM and heavy sectors, with the Yukawa couplings given by

$$\begin{aligned} -\mathcal{L}_Y \supset\ & y_u \bar{Q}_L \tilde{\chi}_{L,q} u_R + y_d \bar{Q}_L \chi_{L,q} d_R + y_\ell \bar{\psi}_L \chi_{L,\ell} e_R \\ & + y_u' \bar{\mathcal{Q}}_R \tilde{\chi}_{R,q} \mathcal{U}_L + y_d' \bar{\mathcal{Q}}_R \chi_{R,q} \mathcal{D}_L + y_\ell' \bar{\Psi}_R \chi_{R,\ell} \mathcal{E}_L + {\rm H.c.}\,, \end{aligned} \tag{11}$$

where the electroweak VEV $v_{\rm EW} = \sqrt{v_{L,\ell}^2 + v_{L,q}^2}$ with $v_{L,\ell} = \langle \chi_{L,\ell}^0 \rangle$ and $v_{L,q} = \langle \chi_{L,q}^0 \rangle$. When the VEV ratio $\tan\beta = v_{L,q}/v_{L,\ell} \gg 1$, the singly-charged scalar is predominantly from the lepton doublet, i.e. $\chi_L^\pm \simeq (\chi_{L,\ell}^\pm - \cot\beta \chi_{L,q}^\pm)$, and its hadronic decays are highly suppressed by $\tan\beta$. Mediated by the mixing of $\chi_L^\pm$ with its heavy partner $\chi_R^\pm$ which couples directly to the DM, we can have three-body decay of DM, i.e. $N \to \ell^+\ell^-\nu$, which can successfully fit the IceCube data [16].

9. Conclusion

In summary, we have discussed the phenomenology and cosmology of a new TeV-scale RHN DM candidate in the context of an $SU(3)_c \times SU(2)_L \times SU(2)_R \times U(1)_{Y_L} \times U(1)_{Y_R}$ model, where we introduced an $SU(2)$ analog of the electroweak interaction and matter contents but with a common QCD. After symmetry breaking, there is an automatic Z_2 symmetry in the leptonic sector that guarantees the stability of the lightest RHN, which plays the role of cold DM in the universe. The DM relic density and direct detection constraints imply a lower limit of order 1 TeV for the RHN DM. We have also discussed a dilution mechanism which can push the RHN dark matter mass up to the multi-PeV range (beyond the conventional unitarity limit) so that together with a weak breaking of the Z_2 symmetry, the RHN DM can in principle explain the PeV neutrinos observed at IceCube.

Acknowledgment
This work of B.D. was partly supported by the DFG grant RO 2516/5-1. The work of R. N. M. is supported by National Science Foundation grant no. PHY1620074. Y. Z. would like to thank the IISN and Belgian Science Policy (IAP VII/37) for support.

References

[1] P. Minkowski, Phys. Lett. B **67**, 421 (1977).
[2] R. N. Mohapatra and G. Senjanović, Phys. Rev. Lett. **44**, 912 (1980).
[3] T. Yanagida, Conf. Proc. C **7902131**, 95 (1979).
[4] M. Gell-Mann, P. Ramond and R. Slansky, Conf. Proc. C **790927**, 315 (1979) [arXiv:1306.4669 [hep-th]].
[5] S. L. Glashow, NATO Sci. Ser. B **61**, 687 (1980).
[6] M. Drewes *et al.*, JCAP **1701**, no. 01, 025 (2017) [arXiv:1602.04816 [hep-ph]].
[7] J. C. Pati and A. Salam, Phys. Rev. D **10**, 275 (1974).
[8] R. N. Mohapatra and J. C. Pati, Phys. Rev. D **11** 2558 (1975).
[9] G. Senjanović and R. N. Mohapatra, Phys. Rev. D **12** 1502 (1975).
[10] M. Re Fiorentin, V. Niro and N. Fornengo, JHEP **1611**, 022 (2016) [arXiv:1606.04445 [hep-ph]].
[11] P. S. Bhupal Dev, R. N. Mohapatra and Y. Zhang, JHEP **1611**, 077 (2016) [arXiv:1608.06266 [hep-ph]].
[12] P. S. B. Dev, R. N. Mohapatra and Y. Zhang, arXiv:1610.05738 [hep-ph].
[13] K. S. Babu and R. N. Mohapatra, Phys. Rev. D **41**, 1286 (1990).
[14] R. N. Mohapatra and Y. Zhang, JHEP **1406**, 072 (2014) [arXiv:1401.6701 [hep-ph]].
[15] P. S. B. Dev, R. N. Mohapatra and Y. Zhang, JHEP **1602**, 186 (2016) [arXiv:1512.08507 [hep-ph]].
[16] P. S. B. Dev, D. Kazanas, R. N. Mohapatra, V. L. Teplitz and Y. Zhang, JCAP **1608**, no. 08, 034 (2016) [arXiv:1606.04517 [hep-ph]].
[17] S. M. Barr, D. Chang and G. Senjanovic, Phys. Rev. Lett. **67**, 2765 (1991).
[18] Q. H. Cao, S. L. Chen and P. H. Gu, arXiv:1512.07541 [hep-ph].
[19] J. Schechter and J. W. F. Valle, Phys. Rev. **D 22**, 2227 (1980).
[20] T. P. Cheng and L. F. Li, Phys. Rev. D **22**, 2860 (1980).
[21] G. Lazarides, Q. Shafi and C. Wetterich, Nucl. Phys. **B 181**, 287 (1981).
[22] R. N. Mohapatra and G. Senjanović, Phys. Rev. **D 23**, 165 (1981).
[23] J. Brehmer, J. Hewett, J. Kopp, T. Rizzo and J. Tattersall, JHEP **1510**, 182 (2015) [arXiv:1507.00013 [hep-ph]].
[24] P. S. B. Dev, R. N. Mohapatra and Y. Zhang, JHEP **1605**, 174 (2016) [arXiv:1602.05947 [hep-ph]].
[25] The ATLAS collaboration [ATLAS Collaboration], ATLAS-CONF-2016-045.
[26] CMS Collaboration [CMS Collaboration], CMS-PAS-EXO-16-031.
[27] P. A. R. Ade *et al.* [Planck Collaboration], arXiv:1502.01589 [astro-ph.CO].
[28] K. Griest and M. Kamionkowski, Phys. Rev. Lett. **64**, 615 (1990).
[29] S. Nussinov, arXiv:1408.1157 [hep-ph].
[30] D. S. Akerib *et al.*, arXiv:1608.07648 [astro-ph.CO].
[31] A. Tan *et al.* [PandaX-II Collaboration], Phys. Rev. Lett. **117**, no. 12, 121303 (2016) [arXiv:1607.07400 [hep-ex]].
[32] E. Aprile *et al.* [XENON Collaboration], JCAP **1604**, no. 04, 027 (2016) [arXiv:1512.07501 [physics.ins-det]].
[33] D. S. Akerib *et al.* [LZ Collaboration], arXiv:1509.02910 [physics.ins-det].
[34] M. G. Aartsen *et al.* [IceCube Collaboration], arXiv:1510.05223 [astro-ph.HE].

Dark Matter and CP-violation in the Three-Higgs Doublet Model

Dorota Sokołowska

University of Warsaw, Faculty of Physics, Pasteura 5, 02-093 Warsaw, Poland.

E-mail: dsok@fuw.edu.pl

Abstract. In many scalar Dark Matter models an imposed discrete symmetry will result in CP conservation. We present results for the 3HDM, the Standard Model with two additional inert doublets, where it is possible to have CP-violating effects and a stable Dark Matter candidate. We discuss the new regions of DM relic density opened up by CP-violation and constrain the parameter space of the CP-violating model using recent results from the LHC and DM direct and indirect detection experiments.

1. Introduction

In 2012 a scalar boson with a mass of ≈ 125 GeV was discovered by both ATLAS and CMS experiments at the CERN Large Hadron Collider (LHC) reported [1, 2]. Although the properties of the observed boson are in accordance with those of the Higgs boson of the Standard Model (SM), it may just be one member of an extended scalar sector. So far no signs of detection of physics Beyond SM (BSM) have been reported, but it is well understood that the SM of particle physics is not complete. A good motivation for BSM is the lack of a Dark Matter (DM) candidate in the SM. Although the nature of DM is not yet known, according to the Standard Cosmological Λ-CDM Model it should be a particle which is stable on cosmological time scales, cold, non-baryonic, neutral and weakly interacting. Various candidates for such a state with these characteristics exist in the literature, the most well-studied being Weakly Interacting Massive Particles (WIMPs) [3, 4, 5]. Any such WIMP candidate must be cosmologically stable, usually due to the conservation of a discrete symmetry, and must freeze-out to result in the observed relic density of $\Omega_{\rm DM}h^2 = 0.1199 \pm 0.0027$ [6].

It is clear that the SM scalar sector cannot provide a WIMP candidate. However, it was suggested some time ago that the scalar sector could be extended by the addition of an extra doublet, which may not develop a Vacuum Expectation Value (VEV) while leaving a discrete Z_2 symmetry unbroken [7]. This possibility, which is known as the Inert Doublet Model (IDM), has been studied extensively for the last few years (see, e.g., [8, 9, 10]). Since the IDM involves *1 inert* doublet plus *1 active* Higgs doublet, we shall also refer to it as the I(1+1)HDM.

The I(1+1)HDM remains a viable model for a scalar DM candidate, being in agreement with current experimental constraints, although the allowed parameter space is reduced [11, 12, 13].

In recent papers [14, 15] we studied DM in a model with *2* inert plus *1* active Higgs doublet, which we referred to as the I(2+1)HDM. We showed that the extended scalar sector can provide a viable DM candidate in a region of parameter space which would be excluded in the I(1+1)HDM.

IOP Conf. Series: Journal of Physics: Conf. Series **873** (2017) 012030 doi:10.1088/1742-6596/873/1/012030

In this work we present results for chosen benchmarks for the CP-violating I(2+1)HDM [16]. We found that introduction of complex parameters in the scalar potential changes the annihilation scenarios of the DM candidate, opening new regions of parameter space in agreement with all experimental constraints.

2. The scalar sector of 3HDM

Here we discuss a particular version of a 3HDM with *2* inert and *1* active doublet, which we refer to as the I(2+1)HDM. Let us define a Z_2 symmetry under which the three scalar doublets $\phi_{1,2,3}$ transform, respectively, as $g_{Z_2} = \mathrm{diag}\,(-1,-1,1)$. A Z_2-symmetric 3HDM potential[1] is of the following form [17] :

$$\begin{aligned} V_{3HDM} \;=\; & -\mu_1^2(\phi_1^\dagger\phi_1) - \mu_2^2(\phi_2^\dagger\phi_2) - \mu_3^2(\phi_3^\dagger\phi_3) + \lambda_{11}(\phi_1^\dagger\phi_1)^2 + \lambda_{22}(\phi_2^\dagger\phi_2)^2 + \lambda_{33}(\phi_3^\dagger\phi_3)^2 \\ & +\lambda_{12}(\phi_1^\dagger\phi_1)(\phi_2^\dagger\phi_2) + \lambda_{23}(\phi_2^\dagger\phi_2)(\phi_3^\dagger\phi_3) + \lambda_{31}(\phi_3^\dagger\phi_3)(\phi_1^\dagger\phi_1) \\ & +\lambda'_{12}(\phi_1^\dagger\phi_2)(\phi_2^\dagger\phi_1) + \lambda'_{23}(\phi_2^\dagger\phi_3)(\phi_3^\dagger\phi_2) + \lambda'_{31}(\phi_3^\dagger\phi_1)(\phi_1^\dagger\phi_3), \\ & -\mu_{12}^2(\phi_1^\dagger\phi_2) + \lambda_1(\phi_1^\dagger\phi_2)^2 + \lambda_2(\phi_2^\dagger\phi_3)^2 + \lambda_3(\phi_3^\dagger\phi_1)^2 + h.c. \end{aligned}$$

where CP violation (CPV) is introduced explicitly through complex parameters of the potential: $\mu_{12}^2, \lambda_{1,2,3}$. Here, we study a simplified version of the I(2+1)HDM by imposing the following equalities $\mu_1^2 = \mu_2^2, \lambda_3 = \lambda_2, \lambda_{31} = \lambda_{23}, \lambda'_{31} = \lambda'_{23}$ which we call the "dark democracy" limit. By imposing it, there are only two parameters that remain complex[2]: $\mu_{12}^2 = |\mu_{12}^2|e^{i\theta_{12}}$ and $\lambda_2 = |\lambda_2|e^{i\theta_2}$; θ_{12} and θ_2 as the respective CPV phases. Note that the *inert* sector is protected by a conserved Z_2 symmetry from coupling to the SM particles, therefore, the amount of CPV introduced here is not constrained by SM data.

The doublets are defined as

$$\phi_1 = \begin{pmatrix} H_1^+ \\ \frac{H_1^0+iA_1^0}{\sqrt{2}} \end{pmatrix}, \quad \phi_2 = \begin{pmatrix} H_2^+ \\ \frac{H_2^0+iA_2^0}{\sqrt{2}} \end{pmatrix}, \quad \phi_3 = \begin{pmatrix} G^+ \\ \frac{v+h+iG^0}{\sqrt{2}} \end{pmatrix}, \tag{1}$$

where ϕ_1 and ϕ_2 are the two doublets that are odd under the Z_2 (hence are inert) and ϕ_3 is the active doublet, which is even under the Z_2 and plays the role of the SM-Higgs doublet. The symmetry of the potential is therefore respected by the vacuum alignment.

To make the entire Lagrangian Z_2-symmetric, an even Z_2 parity is assigned to all SM particles, identical to the Z_2 parity of the only doublet that couples to them, i.e., the active doublet ϕ_3 [18] (the Yukawa Lagrangian of the model is identical to the SM Yukawa Lagrangian). With this parity assignment Flavour Changing Neutral Currents are avoided as the extra doublets are forbidden to couple to fermions by Z_2 conservation. Furthermore, the conservation of Z_2-symmetry leads to the lightest particle from doublets $\phi_{1,2}$ to be stable. Among the Z_2-odd particles there are four neutral states, S_i, and two charged states, $S_i^\pm$.

The neutral physical states are composed of all neutral base states from (1), so have a mixed CP-charge. We take S_1 to be the lightest neutral field from the inert doublets:

$$\begin{aligned} S_1 &= \frac{\alpha H_1^0 + \alpha H_2^0 - A_1^0 + A_2^0}{\sqrt{2\alpha^2+2}}, \quad & S_2 &= \frac{-H_1^0 - H_2^0 - \alpha A_1^0 + \alpha A_2^0}{\sqrt{2\alpha^2+2}}, \\ S_3 &= \frac{\beta H_1^0 - \beta H_2^0 + A_1^0 + A_2^0}{\sqrt{2\beta^2+2}}, \quad & S_4 &= \frac{-H_1^0 + H_2^0 + \beta A_1^0 + \beta A_2^0}{\sqrt{2\beta^2+2}}, \end{aligned} \tag{2}$$

[1] Adding extra Z_2-respecting terms e.g. $(\phi_3^\dagger\phi_1)(\phi_2^\dagger\phi_3)$, $(\phi_1^\dagger\phi_2)(\phi_3^\dagger\phi_3)$, $(\phi_1^\dagger\phi_2)(\phi_1^\dagger\phi_1)$ and/or $(\phi_1^\dagger\phi_2)(\phi_2^\dagger\phi_2)$ does not change the phenomenology of the model. The coefficients of these terms, therefore, are set to zero for simplicity.

[2] The remaining parameter, λ_1 governs the self-coupling of inert particles and does not influence DM or LHC phenomenology studied here.

IOP Conf. Series: Journal of Physics: Conf. Series **873** (2017) 012030 doi:10.1088/1742-6596/873/1/012030

where α and β are the rotation angles [16].

Note that the scalar h contained in the doublet ϕ_3 in our model, has exactly the couplings of the SM-Higgs boson. The CP-violation is only introduced in the *inert* sector which is forbidden from mixing with the *active* sector by the Z_2 symmetry. Therefore, the amount of CP-violation is not limited by EDMs and SM-Higgs couplings.

It is useful to write the parameters of the model in terms of the physical observables:

$$|\mu_{12}^2| = \frac{1}{2}(m_{S_2^\pm}^2 - m_{S_1^\pm}^2), \lambda_{23} = \frac{2\mu_2^2}{v^2} + \frac{m_{S_2^\pm}^2 + m_{S_1^\pm}^2}{v^2}, \lambda_{23}' = \frac{1}{v^2}(m_{S_2}^2 + m_{S_1}^2 - m_{S_2^\pm}^2 - m_{S_1^\pm}^2),$$

$$\mu_2^2 = \frac{v^2}{2} g_{S_1 S_1 h} - \frac{v^2|\lambda_2|}{2(1+\alpha^2)}\Big(4\alpha \sin\theta_2 + 2(\alpha^2 - 1)\cos\theta_2\Big) - \frac{m_{S_2}^2 + m_{S_1}^2}{2},$$

$$|\lambda_2| = \frac{1}{v^2}\left[|\mu_{12}^2|\cos(\theta_2 + \theta_{12}) + \sqrt{|\mu_{12}^2|^2\cos^2(\theta_2+\theta_{12}) + \left(\frac{m_{S_2}^2 - m_{S_1}^2}{2}\right)^2 - |\mu_{12}^2|^2}\,\right].$$

We take the masses of $S_{1,2}$, $S_{1,2}^\pm$, the two angles θ_2 and θ_{12} and the Higgs-DM coupling, $g_{S_1 S_1 h}$ (with the Lagrangian term equal to $\frac{v}{2} g_{S_1 S_1 h} h S_1^2$) as the input parameters of the model.

2.1. Constraints on parameters

We take into account various theoretical and experimental constraints that include: the boundedness of the potential, positive-definiteness of the Hessian and S, T, U parameters put on the model, bounds from relic density observations, Gamma-ray searches, DM direct and indirect detection, the contribution of the new scalars to the W and Z gauge boson widths, null searches for charged scalars at LEP and LHC, invisible Higgs decays, Higgs total decay width and the $h \to \gamma\gamma$ signal strength [16].

3. Relevant DM (co)annihilation scenarios

The relic density of the scalar DM candidate, S_1, after freeze-out is given by the solution of the Boltzmann equation:

$$\frac{dn_{S_1}}{dt} = -3Hn_{S_1} - \langle\sigma_{eff} v\rangle(n_{S_1}^2 - n_{S_1}^{eq\,2}), \tag{3}$$

where the thermally averaged effective (co)annihilation cross section contains all relevant annihilation processes of any $S_i S_j$ pair into SM particles:

$$\langle\sigma_{eff} v\rangle = \sum_{ij}\langle\sigma_{ij} v_{ij}\rangle \frac{n_i^{eq}}{n_{S_1}^{eq}}\frac{n_j^{eq}}{n_{S_1}^{eq}}, \quad \frac{n_i^{eq}}{n_{S_1}^{eq}} \sim \exp(-\frac{m_i - m_{S_1}}{T}). \tag{4}$$

Therefore, only processes for which the mass splitting between S_i and the lightest Z_2-odd particle S_1 are comparable to the thermal bath temperature T provide a sizeable contribution to this sum.

The I(2+1)HDM studied here has many features of a Higgs-portal DM model. In a large region of parameter space the crucial DM annihilation channel is $S_1 S_1 \to h \to f\bar{f}$, with the cross-section depending on both the mass of DM and the Higgs-DM coupling. In general, if $m_{\rm DM} < m_h/2$, then a relatively large coupling is needed to produce correct relic density. Processes that produce gauge boson, e.g. $S_1 S_1 \to h \to VV, S_1 S_1 \to VV$, where V is any of the SM gauge bosons, also contribute to the total annihilation cross section. These contributions are suppressed when m_{S_1} is smaller than m_W, however it is known that diagrams with off-shell gauge bosons may be very important.

IOP Conf. Series: Journal of Physics: Conf. Series **873** (2017) 012030 doi:10.1088/1742-6596/873/1/012030

Coannihilation effects play an important role in scenarios with multiple particles that are close in mass. Therefore, processes such as $S_1S_i \to h \to f\bar{f}, S_1S_i \to Z^* \to f\bar{f}, S_1S_j^\pm \to W^{\pm *} \to ff'$ with $i = 2,3,4$ and $j = 1,2$ are included in calculating the effective annihilation cross section. If all inert particles are very close in mass then all channels $S_iS_j \to h \to f\bar{f}, S_iS_j \to VV$ contribute to the final DM relic density.

Below we present the numerical results for the chosen benchmark scenarios for low and medium DM mass. It is convenient to introduce the mass splittings between the DM candidate and other inert scalars as: $\delta_{12} = m_{S_2} - m_{S_1}$, $\delta_{1c} = m_{S_1^\pm} - m_{S_1}$, $\delta_c = m_{S_1^\pm} - m_{S_2^\pm}$, with the CP-violating phases complementing the needed parameter set of $m_{S_{1,2}}, m_{S_{1,2}^\pm}$ θ_2 and θ_{12}. We then define three base benchmarks in low and medium mass region as:

$$A1: \ \delta_{12} = 125\,\text{GeV}, \delta_{1c} = 50\,\text{GeV}, \delta_c = 50\,\text{GeV}, \theta_2 = \theta_{12} = 1.5 \tag{5}$$

$$B1: \ \delta_{12} = 125\,\text{GeV}, \delta_{1c} = 50\,\text{GeV}, \delta_c = 50\,\text{GeV}, \theta_2 = \theta_{12} = 0.82 \tag{6}$$

$$C1: \ \delta_{12} = 12\,\text{GeV}, \delta_{1c} = 100\,\text{GeV}, \delta_c = 1\,\text{GeV}, \theta_2 = \theta_{12} = 1.57 \tag{7}$$

In the case of **Scenario A**, where there are large mass splittings between the DM candidate and all other inert particles ($m_{S_1} \ll m_{S_2}, m_{S_3}, m_{S_4}, m_{S_1^\pm}, m_{S_2^\pm}$), no co-annihilation channels are present. In **Scenario B**, where $m_{S_1} \sim m_{S_3} \ll m_{S_2}, m_{S_4}, m_{S_1^\pm}, m_{S_2^\pm}$, the DM candidate can coannihilate with its only particle close in mass, namely S_3. In **Scenario C**, with all neutral particles close in mass, $m_{S_1} \sim m_{S_3} \sim m_{S_2} \sim m_{S_4} \ll m_{S_1^\pm}, m_{S_2^\pm}$, the DM particle can coannihilate with all other neutral inert particles.

4. Relation between couplings and DM relic density

In the CP-conserving version of the I(2+1)HDM (within the "dark democracy" limit) [14, 15], couplings between inert scalars and gauge bosons are fixed, and given by the rotation angles $\theta_a = \theta_h = \pi/4$. They do not depend on the mass splittings or the value of m_{S_1}. In the CP-violating case the situation is different, as the couplings (normalized to $\frac{ie}{2c_w s_w}$) are given by:

$$\chi_{ZS_1S_3} = \chi_{ZS_2S_4} = \frac{\alpha+\beta}{\sqrt{\alpha^2+1}\sqrt{\beta^2+1}}, \chi_{ZS_1S_4} = \chi_{ZS_2S_3} = \frac{\alpha\beta-1}{\sqrt{\alpha^2+1}\sqrt{\beta^2+1}}, \tag{8}$$

$$\chi^2_{ZS_1S_3} + \chi^2_{ZS_1S_4} = 1, \quad \chi^2_{ZS_2S_3} + \chi^2_{ZS_2S_4} = 1. \tag{9}$$

The strength of gauge-inert interaction depends on parameters α and β, which in turn depend on m_{S_i}. Higgs-inert scalar couplings are also modified with respect to the CP-conserving case [16]. This leads to important differences in DM phenomenology, especially in the region where coannihilation channels are important. Fig. 1 shows the change in values of Z-inert couplings for benchmarks A1, B1 and C1. The introduction of varying values of α and β leads to the following modifications with respect to the (co)annihilation scenarios in the CP-conserving I(2+1)HDM.

4.1. Low DM mass region

For benchmark A1, couplings with the Z boson are modified with respect to the CP-conserving case (Fig. 1), however, as DM does not coannihilate, it does not change the annihilation scenario of S_1. For low DM mass S_1 annihilates mostly through $S_1S_1 \to h \to b\bar{b}$, entering the resonance region with small Higgs-DM coupling for masses close to $m_h/2$. This benchmark resembles both the CP-conserving I(2+1)HDM as well as the IDM.

For benchmark B1, S_1 is close in mass with S_3, opening the coannihilation channel $S_1S_3 \to Z \to f\bar{f}$. Such a scenario in the CP-conserving limit results in too low a relic density for any value of the Higgs-DM coupling due to strong coannihilation. However, in the CP-violating case its strength is reduced. We can therefore change the contribution of this diagram to the

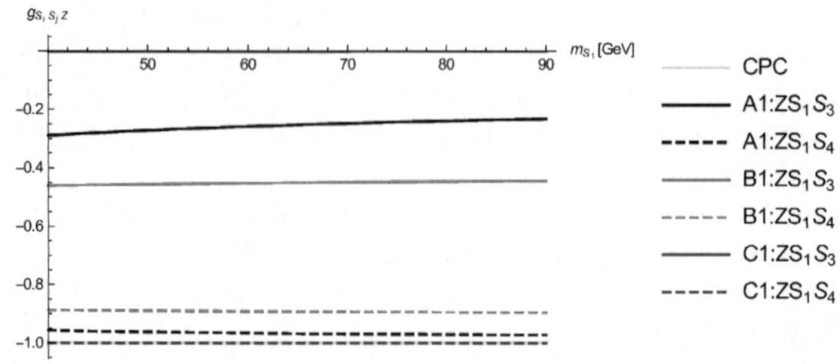

Figure 1. $\chi_{ZS_1S_3} = \chi_{ZS_2S_4}$ and $\chi_{ZS_1S_4} = \chi_{ZS_2S_3}$ for chosen benchmarks. Figure from [16].

relic density calculations not only by introducing the change for the mass splitting, but also by modifying the value of the coupling itself.

For benchmark C1 all particles are close in mass and in principle all coannihilation diagrams could be important. The couplings $g_{S_1S_2h}, g_{S_3S_4h}$ and $g_{ZS_1S_3}$ are suppressed, and the crucial contribution comes from $S_1S_4 \to Z \to q\bar{q}$. In the CP-conserving case, this scenario is only viable in the resonance region, but in the CP-violating case the strength of the coannihilation channels depends on the input parameters and can therefore vary.

Figure 2 shows good relic density regions in the $(m_{S_1}, g_{S_1S_1h})$ plane for benchmarks A1, B1 and C1. Benchmark A1 exhibits the standard behaviour of an $SU(2)$ DM candidate. For benchmark B1 coannihilation channels are important, unlike A1. For large values of $g_{S_1S_1h}$ the dominant channel is $S_1S_1 \to h \to \bar{b}b$ and, as there are also coannihilation channels, $\Omega_{DM}h^2$ is usually too small. For smaller couplings the dominant channel is $S_1S_3 \to Z \to q\bar{q}$, especially strong for low masses. As the mass grows, the coannihilation channel gets weaker, allowing us to obtain the proper relic density. For masses closer to $m_h/2$ the Higgs resonance annihilation dominates. In case of benchmark C1 for small values of $g_{S_1S_1h}$ the dominant channel is $S_1S_4 \to Z \to f\bar{f}$ (light quarks), with a small contribution from $S_2S_3 \to Z \to f\bar{f}$. For larger couplings the process $S_1S_1 \to h \to b\bar{b}$ strongly increases the annihilation cross section. This, combined with the fact that coannihilation channels are generally strong, leaves the region $m_{S_1} > 49$ GeV.

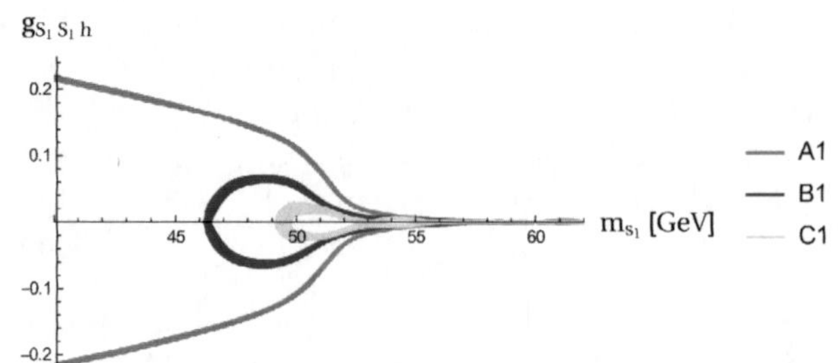

Figure 2. Relic density for low DM mass region in Scenarios A1, B1 and C1. Figure from [16].

4.2. Medium DM mass region

Here, for $m_h/2 < m_{S_1} < m_{W^\pm,Z}$, the crucial channel for all benchmarks is the annihilation of $S_1S_1 \to W^+W^-$ and this vertex does not depend on parameters α and β, but only the

DM mass. This is why all studied benchmarks, as well as the CP-conserving scenarios, follow the similar behaviour, presented in Fig. 3. For larger values of DM mass this annihilation is stronger, and cancellation with $S_1S_1 \to h \to W^+W^-$ is needed to ensure the proper value of relic density. This mechanism is responsible for moving towards negative values of Higgs-DM coupling. In benchmarks B1 and C1 other channels, like $S_1S_4 \to q\bar{q}$ or $S_3S_3 \to W^+W^-$ give small contributions, leading to small deviations from the behaviour of benchmark A1.

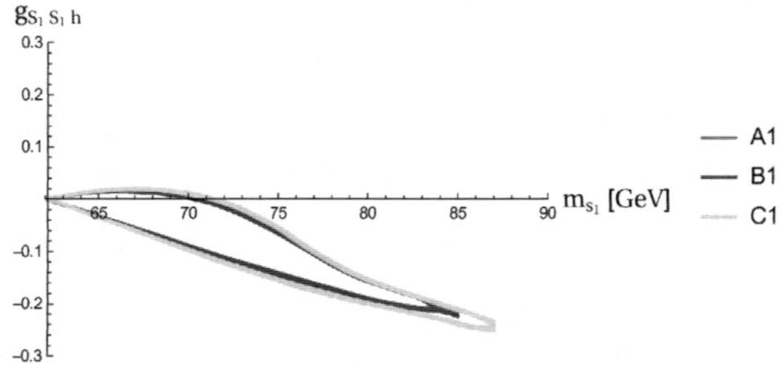

Figure 3. Relic density for medium DM mass region in Scenarios A1, B1 and C1. Note that the medium mass region behaviour of the three scenarios is very similar to each other. Figure from [16].

4.3. Other solutions in low and medium mass region

In the discussion above we have presented results for three sets of parameters in scenarios A1, B1, and C1. By changing the input parameter set we can reach different regions of parameters space. The performed scan shows that by varying the mass splitting and phases θ_2 and θ_{12} we can actually fill the empty regions in plots 2 and 3 within the range given by the CP-conserving scenario with large mass splittings (the standard SU(2) DM candidate, with no coannihilation channels). There is more freedom in the low mass region, as the strength of coannihilation channels depends strongly on α and β. In the medium mass region main annihilation of S_1 into WW pair (both direct and Higgs-mediated) is benchmark independent and the coannihilation processes have smaller impact.

In Fig. 4 results obtained for various additional sets of parameters are presented. We can fill the plot by different B scenarios, where the coannihilation channel $S_1S_3 \to Z \to q\bar{q}$ (with varying $\chi_{ZS_1S_3}$) is crucial. There are also multiple solutions of type C, where all neutral particles have a relatively low mass.

4.4. DM detection experiments

DM direct detection experiments aim to measure the scattering of DM particle off nuclei. This interaction is mediated by the Higgs particle, and therefore results of these experiments constrain the DM mass, as well as its coupling to h. In the low and medium mass region the strongest constraints come from the LUX experiment [19], and they set strong limits on the parameter space of the 3HDM. Results are presented in Fig. 5, where the solid line corresponds to the current LUX limit, while the dashed line shows the projected sensitivity of XENON1T [20]. From the plot we can see that for chosen benchmark points A1, B1 and C1 the only surviving region of this part of parameter space is $50\,\text{GeV} \lesssim m_{S_1} \lesssim 76\,\text{GeV}$, where the Higgs-DM coupling can be small. From Fig. 5 one can see that a large part of parameter space is within the reach of future experiments, however there are still points, which correspond to $g_{S_1S_1h} \to 0$ that will escape detection.

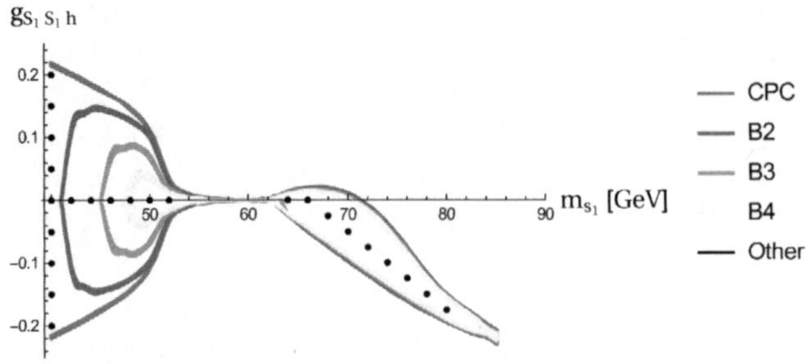

Figure 4. The relic density plots for different B and C scenarios where by changing the angles θ_2 and θ_{12} the whole region not accesible by the CP-conserving limit could be realised in the CP-violating case. Figure from [16].

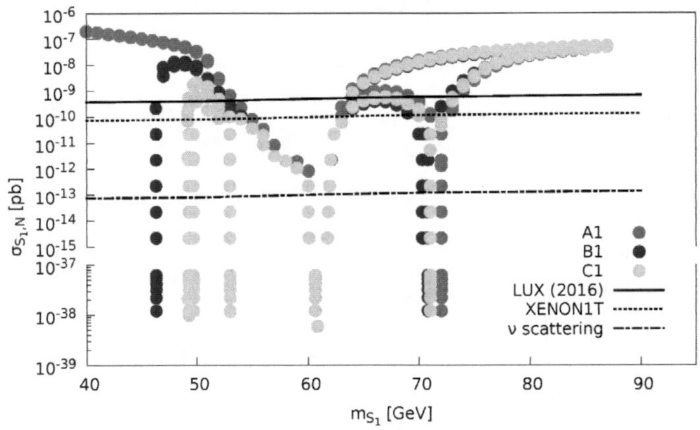

Figure 5. Direct detection limits for low and medium mass regions. Figure from [16].

Recent indirect detection results from Fermi-LAT strongly constrain the DM candidate annihilating into $b\bar{b}$ pair [21], and therefore are crucial for the low DM mass region. The CP-conserving scalar Higgs-portal type of DM with proper relic density and $m_{S_1} \lesssim 53$ GeV is ruled out. The same limit applies to case A1, as the dominant annihilation channel is into $b\bar{b}$ pair (Figure 6). For cases B1 and C1 annihilation channels are different and a good relic density is obtained for smaller values of Higgs-DM coupling. This weakens the annihilation into $b\bar{b}$, leading to most of the parameter space to lie within the allowed region, see Fig. 6.

In both direct and indirect detection plots one can see two branches in the medium mass region correspond to two asymmetrical regions from Fig. 3. They do overlap in the low mass region, where good relic density regions from Fig. 2 are symmetrical. Direct and indirect detection experiments provide a complementary way to constrain the parameter space of the model. It is especially important for masses just above $m_h/2$, where the Higgs-DM coupling is small. This region escapes the possibility of direct detection, however, due to an enhancement from the Breit-Wigner resonance effect it is possible to exclude this region from the results of indirect detection experiments.

4.5. LHC limits

The presence of additional light scalars can modify decays of the Higgs particle. Channel $h \to S_1 S_1$ contributes to the Higgs invisible decay ratio (BR($h \to$ inv.)), as S_1 is stable and will

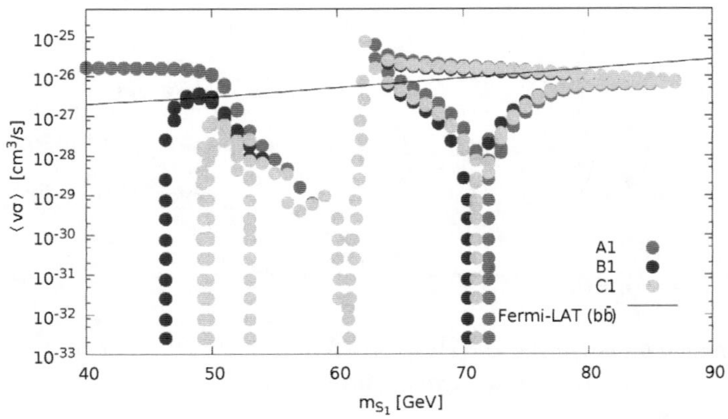

Figure 6. Indirect detection limits for low and medium mass region. Figure from [16].

escape the detector. All other inert particles can contribute to the total decay width of h (μ_{tot}). Results for benchmarks A1, B1, C1, as well as the experimental limits [22, 23, 24], are presented in Fig. 7, 8, 9. In case A1 there is only one particle that contributes to the Higgs decay (the DM candidate S_1). For small values of $g_{S_1S_1h}$ the contribution to the total decay width of the Higgs particle, and its invisible branching ratio, is small enough. There is also a small region fulfilling this constraint for case C1, but not for case B1. This is related to especially large values of $g_{S_iS_ih}$ couplings in case B, that significantly change decay channels of the Higgs particle with respect to the SM case.

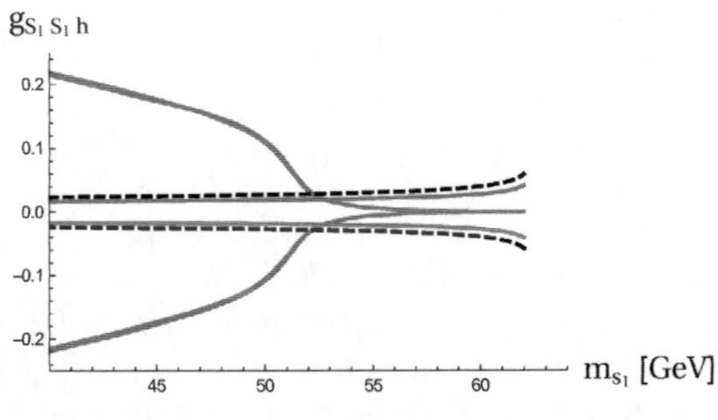

Figure 7. Relic density constraints vs. Higgs invisible branching ratio and Higgs total signal strength bounds for scenario A1. Figure from [16].

We want to stress that the LHC limits can provide stronger constraints for some benchmark points in the low mass region than the dedicated DM detection experiments. It is especially important considering the astrophysical uncertainties that may influence interpretation of results provided by DM detection experiments.

5. Conclusion

In this paper we have studied an extension of the Standard Model (SM) in which two copies of the SM-Higgs doublet which do not acquire a Vacuum Expectation Value (VEV), and hence are *inert*, are added to the scalar sector. In other words, this is a 3HDM with two inert and one active scalar doublet, denoted as the I(2+1)HDM. We have allowed for CP-violation in the

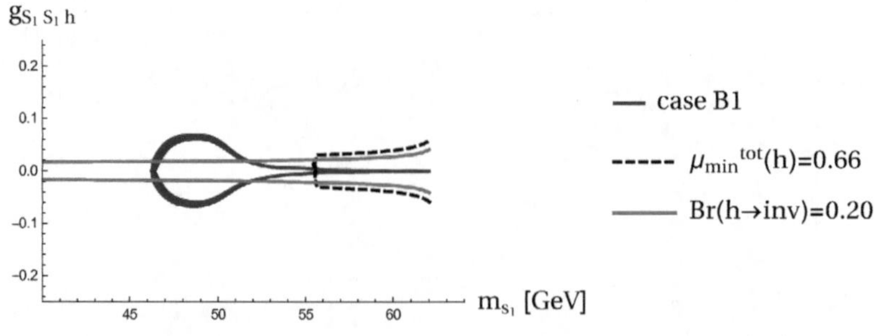

Figure 8. Relic density constraints vs. Higgs invisible branching ratio and Higgs total signal strength bounds for scenario B1. Figure from [16].

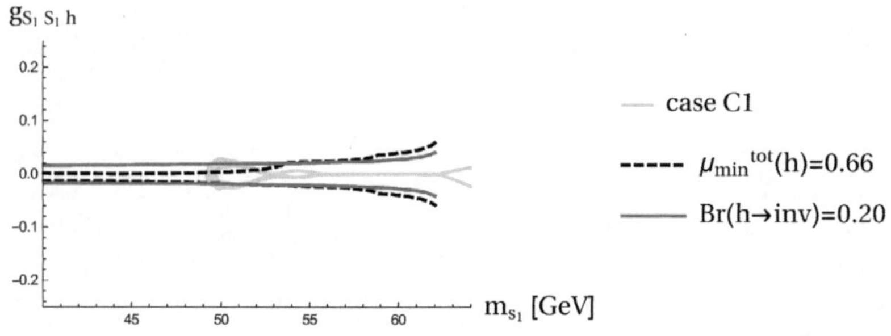

Figure 9. Relic density constraints vs. Higgs invisible branching ratio and Higgs total signal strength bounds for scenario C1. Figure from [16].

inert sector, where the lightest *inert* state is protected from decaying to SM particles through the conservation of a Z_2 symmetry. The lightest neutral particle from the *inert* sector, which has a mixed CP-charge due to CP-violation, is hence a DM candidate in the model.

The extended scalar sector, and the inclusion of complex parameters in the scalar potential, accommodates both the DM candidate and an unbounded amount of possible CPV. We find that with the introduction of CP violation, the strength of the gauge-inert couplings which were fixed in the CP conserving limit, become unconstrained, allowing us for more freedom and giving the access to a previously excluded part of parameter space. In particular, it is possible to have a light DM candidate, with very small DM-Higgs coupling, that is in agreement with all experimental data, unlike in the previously studied the I(1+1)HDM or CP-conserving I(2+1)HDM.

Acknowledgement

DS is partially supported by the HARMONIA project under contract UMO-2015/18/M/ST2/00518 (2016-2019).

References

[1] Aad G *et al.* (ATLAS) 2012 *Phys. Lett.* **B716** 1–29 (*Preprint* 1207.7214)
[2] Chatrchyan S *et al.* (CMS) 2012 *Phys. Lett.* **B716** 30–61 (*Preprint* 1207.7235)
[3] Jungman G, Kamionkowski M and Griest K 1996 *Phys. Rept.* **267** 195–373 (*Preprint* hep-ph/9506380)
[4] Bertone G, Hooper D and Silk J 2005 *Phys. Rept.* **405** 279–390 (*Preprint* hep-ph/0404175)
[5] Bergstrm L 2000 *Rept. Prog. Phys.* **63** 793 (*Preprint* hep-ph/0002126)
[6] Ade P A R *et al.* (Planck) 2016 *Astron. Astrophys.* **594** A13 (*Preprint* 1502.01589)
[7] Deshpande N G and Ma E 1978 *Phys.Rev.* **D18** 2574

[8] Ma E 2006 *Phys. Rev.* **D73** 077301 (*Preprint* hep-ph/0601225)
[9] Barbieri R, Hall L J and Rychkov V S 2006 *Phys. Rev.* **D74** 015007 (*Preprint* hep-ph/0603188)
[10] Lopez Honorez L, Nezri E, Oliver J F and Tytgat M H G 2007 *JCAP* **0702** 028 (*Preprint* hep-ph/0612275)
[11] Krawczyk M, Sokolowska D, Swaczyna P and Swiezewska B 2013 *JHEP* **09** 055 (*Preprint* 1305.6266)
[12] Arhrib A, Tsai Y L S, Yuan Q and Yuan T C 2014 *JCAP* **1406** 030 (*Preprint* 1310.0358)
[13] Ilnicka A, Krawczyk M and Robens T 2016 *Phys. Rev.* **D93** 055026 (*Preprint* 1508.01671)
[14] Keus V, King S F, Moretti S and Sokolowska D 2014 *JHEP* **11** 016 (*Preprint* 1407.7859)
[15] Keus V, King S F, Moretti S and Sokolowska D 2015 *JHEP* **11** 003 (*Preprint* 1507.08433)
[16] Cordero-Cid A, Hernndez-Snchez J, Keus V, King S F, Moretti S, Rojas D and Sokołowska D 2016 *JHEP* **12** 014 (*Preprint* 1608.01673)
[17] Ivanov I P, Keus V and Vdovin E 2012 *J. Phys.* **A45** 215201 (*Preprint* 1112.1660)
[18] Ivanov I P and Keus V 2012 *Phys. Rev.* **D86** 016004 (*Preprint* 1203.3426)
[19] Akerib D S *et al.* (LUX) 2017 *Phys. Rev. Lett.* **118** 021303 (*Preprint* 1608.07648)
[20] Aprile E (XENON1T) 2013 *Springer Proc. Phys.* **148** 93–96 (*Preprint* 1206.6288)
[21] Ackermann M *et al.* (Fermi-LAT) 2015 *Phys. Rev. Lett.* **115** 231301 (*Preprint* 1503.02641)
[22] Aad G *et al.* (ATLAS) 2015 *JHEP* **11** 206 (*Preprint* 1509.00672)
[23] Collaboration C (CMS) 2015
[24] Olive K A *et al.* (Particle Data Group) 2014 *Chin. Phys.* **C38** 090001

IOP Conf. Series: Journal of Physics: Conf. Series **873** (2017) 012031 doi:10.1088/1742-6596/873/1/012031

Latest results from Daya Bay

Vit Vorobel on behalf of Daya Bay Collaboration

Charles University, Faculty of Mathematics and Physics, Prague, Czech Republic

vit.vorobel@mff.cuni.cz

Abstract. The Daya Bay Reactor Neutrino Experiment was designed to measure θ_{13}, the smallest mixing angle in the three-neutrino mixing framework, with unprecedented precision. The experiment consists of eight functionally identical detectors placed underground at different baselines from three pairs of nuclear reactors in South China. Since Dec. 2011, the experiment has been running stably for more than 4 years, and has collected the largest reactor anti-neutrino sample to date. Daya Bay is able to greatly improve the precision on θ_{13} and to make an independent measurement of the effective mass splitting in the electron antineutrino disappearance channel. Daya Bay can also perform a number of other precise measurements, such as a high-statistics determination of the absolute reactor antineutrino flux and spectrum, as well as a search for sterile neutrino mixing, among others. The most recent results from Daya Bay are discussed in this paper, as well as the current status and future prospects of the experiment.

1. Introduction

The big progress in neutrino research during the last two decades brought a lot of experimental data which strongly imply that neutrinos undergo mixing between flavor and mass eigenstates and hence carry non-zero mass. It also represents the first evidence of physics beyond the Standard Model of particle physics. Since its discovery, the neutrino flavor oscillations have been confirmed and precisely measured with data from natural (atmospheric and solar) and man-made (reactor and accelerator) neutrino sources.

The phenomena of neutrino flavor oscillation arise from two remarkable characteristics of neutrinos: small differences between different masses of the three neutrino states, $m_1 \neq m_2 \neq m_3$, and an inequivalence between neutrino flavor and mass eigenstates. Produced in a flavor eigenstate by the weak interaction, a neutrino state evolves as a coherent superposition of mass eigenstates. Interference between the phases of each mass component results in the periodic oscillation of the neutrino flavor. The flavor oscillates with phases given by $\Delta m_{ji}^2 L/4E_\nu$, where L is the propagation distance, E_ν is the neutrino energy, and $\Delta m_{ji}^2 = m_j^2 - m_i^2$ is the difference of the squared masses. The amplitude of flavor oscillation is determined by the amount of mixing between the flavor and mass eigenstates. Using the unitary Pontecorvo-Maki-Nakagawa-Sakata (PMNS) matrix, U_{PMNS}, a neutrino with flavor α can be expressed as a combination of mass states, $|\nu_\alpha\rangle = \sum_i U_{\alpha i}^* |\nu_i\rangle$. In the three-flavor model, U_{PMNS} is commonly parameterized using three mixing angles, θ_{12}, θ_{23}, θ_{13} and an off-diagonal CP-violating phase δ_{CP}. With sensitivity to the small neutrino mass separations, oscillation experiments have provided strong evidence for three distinct neutrino mass states.

2. The Daya Bay Neutrino Experiment

The Daya Bay Reactor Neutrino Experiment was designed to measure θ_{13}, the smallest mixing angle in the three-neutrino mixing framework, with unprecedented precision [1]. The experiment profits from rare constellation of a nuclear power station complex situated near Hong Kong and adjacent mountains. The reactors serve as the neutrinos source while the mountains provide a sufficient overburden suppressing cosmic muons – the strongest background source (see Fig. 1). The Daya Bay and Ling Ao nuclear power plant (NPP) reactors (red circles) were situated on a narrow coastal shelf between the Daya Bay coastline and inland mountains.

At the time of the measurement the facility consisted of six pressurized water reactors (PWRs). The electron antineutrinos are emitted in beta-decay of fission fragments released in the chain reaction. The antineutrino flux and the energy spectrum is determined by the total thermal power of the reactor, the fraction of each fissile isotope in the fuel, the fission rate of each isotope and the energy spectrum of neutrinos from each isotope. All the reactors have the same thermal power 2.9 GW_{th} each and all together produced roughly 3.5×10^{21} $\overline{\nu}_e$/s with energies up to 8 MeV making it one of the most intense $\overline{\nu}_e$ sources on Earth.

Two antineutrino detectors installed in each underground experimental hall near to the reactors (Hall 1 and Hall 2) measured the $\overline{\nu}_e$ flux emitted by the reactors, while four detectors in the far experimental hall (Hall 3) measured a deficit in the $\overline{\nu}_e$ flux due to oscillation in the location where the neutrino oscillation effect is expected to be the strongest. Such configuration allows to suppress reactor related uncertainty in the measured neutrino flux. The disappearance signal is most pronounced at the first oscillation minimum. Based on existing accelerator and atmospheric neutrino oscillation measurements this corresponded to a distance $L_f \approx 1.6$ km for reactor $\overline{\nu}_e$ with a mean energy of ~4 MeV. The detectors were built and initially tested in a surface assembly building (SAB), transported to a Liquid Scintillator hall for filling, and then installed in an experimental hall.

Fig. 1. Layout of the Daya Bay Neutrino Experiment.

IOP Conf. Series: Journal of Physics: Conf. Series **873** (2017) 012031 doi:10.1088/1742-6596/873/1/012031

Detection of antineutrinos is based on the same principle as in the famous experiment of Reines and Cowan [2] who registered reactor antineutrinos in 1956. A sensitive detector part consist of a hydrogen rich liquid scintillator doped with gadolinium (Reines and Cowan used Cd instead as the dopant). Antineutrino interacts via inverse beta-decay (IBD)

$$\overline{\nu}_e + \mathrm{p} \rightarrow e^+ + n$$

with a proton (hydrogen) giving rise to neutron and positron. The positron deposits its kinetic energy to the scintillator then annihilates on electron and generates two gamma quanta 511 keV which together with the deposited positron kinetic energy cause a "prompt signal" within a few nanoseconds. The neutron is first moderated then is captured on Gd typically ~30 ns after the "prompt signal". Consequently, a cascade of gamma quanta with total energy of 8 MeV is emitted and generates a "delayed signal". Appearance of the two signals "prompt" and "delayed" is a signature of the antineutrino registration.

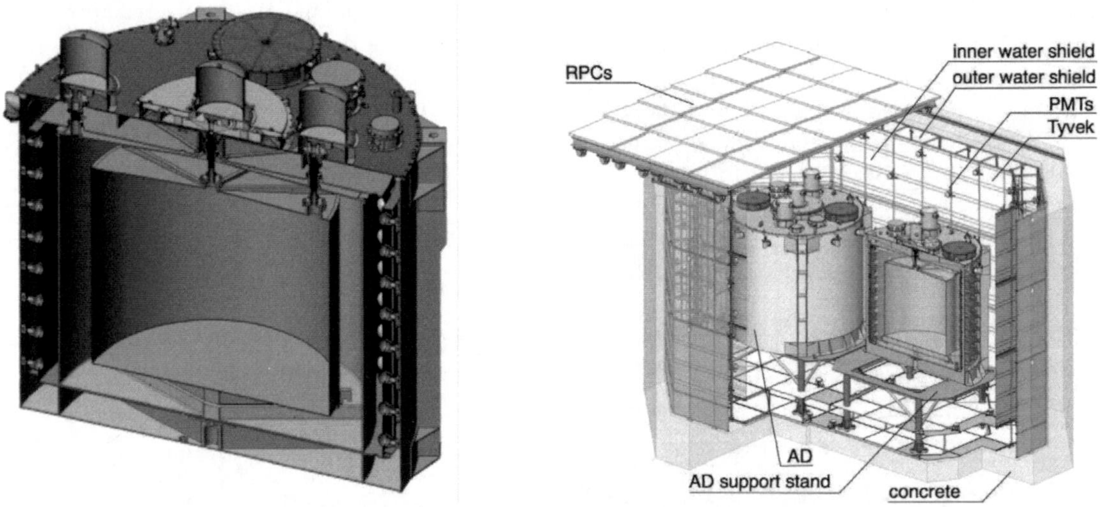

Fig. 2. Scheme of the antineutrino detector (AD) – left, and the near site detection view – right.

The Daya Bay antineutrino detector modules have an onion like structure (see Fig. 2, left). The inner-most volume is filled with 20 tons of the Gd-loaded liquid scintillator (GdLS), serving as the antineutrino target. The second layer – the gamma catcher - is filled with additional 20 tons of normal liquid scintillator (LS) which can register most of the gamma energies from neutron capture or positron annihilation. Neutrino interactions in the gamma catcher will not satisfy the trigger since only the signal of neutron-capture on Gd will trigger a neutrino event. The outer-most layer is normal mineral oil (MO) that shields radiation from the PMT glass from entering the fiducial volume. The two inner vessels are fabricated of PMMA which is transparent for optical photons and chemically resistant against the used liquids, the outer-most 5 m by 5 m tank is made of stainless steel and is equipped with 192 8-inch PMTs. Specular reflectors are located above and below the outer PMMA vessel to improve the light collection uniformity while the vertical wall of the detector is black. Three automated calibration units are used to deploy radioactive sources (^{60}Co, ^{68}Ge, and ^{241}Am-^{13}C) and light-emitting diodes through narrow teflon-bellow penetrations into the GdLS and LS regions.

After filling, the antineutrino detectors were installed in a 10 m deep water pool in each underground experimental hall, as shown in Fig. 2, right. The water shielded the detectors from γ-rays arising from natural radioactivity and muon-induced neutrons, which were primarily emanating from the cavern rock walls. The pool was optically separated into two independent regions, the inner (IWS) and outer water

IOP Conf. Series: Journal of Physics: Conf. Series **873** (2017) 012031 doi:10.1088/1742-6596/873/1/012031

shield (OWS). Both regions were instrumented with PMTs to detect the Cherenkov light produced by cosmogenic muons. A 4-layer resistive plate chamber (RPC) system was installed over the pool, which served in studies of muons and muon-induced backgrounds. Identification of muons which passed through the IWS, OWS, and RPC system enhanced the rejection of background from neutrons generated by muon interactions in the immediate vicinity of the antineutrino detectors. Each detector (ADs, IWS, OWS) operated as an independently triggered system.

3. Detector response

The purpose of calibration was to refine the recorded digital times (TDCs) and charges (ADCs) so that they provided accurate, consistent, and stable estimates of the time, energy, and position of particle interactions within the detectors.

The time at which each detector triggered was recorded with 25 ns precision. A GPS-synchronized trigger time stamp (25 ns resolution) provided a measure of the absolute time for each triggered event. Fast-pulsed calibration LEDs were used to measure the relative time responses of the PMTs within a single detector. The time delays observed in each channel were corrected for LED-to-PMT distances and were fitted as a function of light intensity. The results were recorded to a database and used to correct TDC values during data analysis. The timing calibration was repeated whenever a modification was made to a detector system.

In order to provide proper information about the deposited energy, the registered signal amplitude was corrected for the individual PMT gain, number of active PMTs, spatial non-uniformity of the scintillation light collection and light yield.

The energy resolution of the detectors were adequately modelled using the expression

$$\frac{\sigma_E}{E_{rec}} = \sqrt{a^2 + \frac{b^2}{E_{rec}} + \frac{c^2}{E_{rec}^2}}.$$

Modeling of the detector resolution gave $a = 0.016$, $b = 0.081$ MeV$^{1/2}$, and $c = 0.026$ MeV.

3.1. Nonlinearity

While uncertainty in absolute calibration had negligible impact on the measurement of θ_{13}, it influenced the estimate of the neutrino mass-squared difference. This can be seen as a direct consequence of the oscillation phase dependence on the antineutrino energy. The most significant bias arose from the nonlinear response of the detector. Nonlinearity of the reconstructed positron energy relative to the true reconstruction energy originated from two sources: particle-dependent nonlinear light yield of the scintillator, and charge dependent nonlinearity associated with the electronic readout of the PMT charge signal.

Positron interactions with the scintillator are predominantly identical to electrons, except for their annihilation. Therefore, the visible energy for a positron was effectively modeled as $E_{vis,e^+} = E_{vis,e^-} + 2 \times E_{vis,\gamma}(0.511 \text{ MeV})$.

The scintillation light output for low-energy electrons was suppressed due to ionization quenching. A semi-empirical analytic approach based on Birks' law was used to model this mechanism. The energy-dependent contribution from Cherenkov light, predicted to be at the level of a few percent relative to scintillation light, induced an additional nonlinearity.

The charge nonlinearity induced by the electronics arose from a complex interplay of the time profile of scintillation light and the response of the readout electronics, which could not be easily calibrated at the single PMT channel level. Instead, the resulting nonlinearity was effectively modeled at the detector level.

IOP Conf. Series: Journal of Physics: Conf. Series **873** (2017) 012031 doi:10.1088/1742-6596/873/1/012031

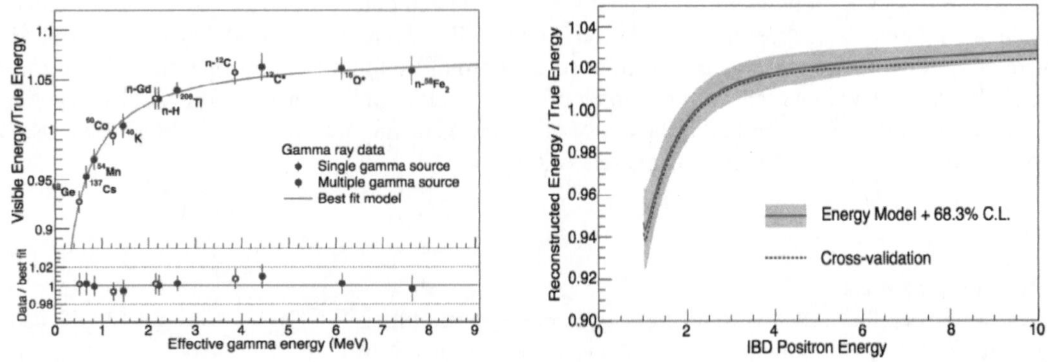

Fig. 3. The ratio of the observed energy to the true energy. Comparison of the model with γ-ray reference data – left, estimate of the detector response to positrons – right.

The complete energy model, relating the reconstructed energy to true particle kinetic energy, contained four free parameters: Birks' constant, the contribution from Cherenkov radiation, and the two parameters of the electronics model. The parameter values were obtained from fit to various calibration data, consisting of twelve gamma lines from both deployed and naturally-occurring sources, as well as the continuous electron spectrum from the decays of ^{12}B produced by cosmogenic muon. Fig. 3 – left, compares the resulting energy response model with the calibration data. The resulting estimate of the detector response to positrons is shown in Fig. 3 – right.

This model of the detector response to positrons was validated using independent calibration reference data. These included the 53 MeV endpoint in the Michel electron spectrum from muon decays, and the continuous β+γ spectra from bismuth and thallium decay.

4. Selection of the antineutrino detection candidates

As discussed in Sec. 2, antineutrino inverse beta decay interactions provide a very characteristic pattern of two time-correlated signals of specific energies. Within the Gd-loaded target region, ~84% of the captures occurred on either ^{157}Gd (~81.5%, 7.95 MeV) or ^{155}Gd (~18.5%, 8.54 MeV). The multiple γ-rays emitted upon capture could be distinguished from natural radioactive backgrounds with energies predominately below 5 MeV.

The remaining ~16% of neutrons captured almost entirely on ^{1}H, releasing a single 2.2 MeV γ-ray. For capture on ^{1}H, it was not possible to clearly discriminate whether the $\overline{\nu}_e$ interacted within the target region or in the gamma-catcher region. Analysis of the n-^{1}H data therefore suffered from larger uncertainties in target volume and detector response, as well as a much more significant background contamination. Despite these obstacles, independent measurements of neutrino oscillation have been obtained using these interactions, with results that were consistent with the analysis of signals identified by n-Gd capture, albeit with less precision.

Fig. 4 – left, shows the distribution of prompt versus delayed energy for signal pairs which satisfied the $\overline{\nu}_e$ inverse beta decay selection criteria. Interactions with prompt energy of 0.7 MeV < E_p < 12 MeV and delayed energy of 6 MeV < E_d < 12 MeV were used for the n-Gd neutrino oscillation analysis (red dotted box). A few-percent contamination from accidental backgrounds (symmetric under interchange of prompt and delayed energy) and ^{9}Li decay and fast neutron backgrounds (high prompt and ~8 MeV delayed energy) are visible within the selected region. Inverse beta decay interactions where the neutron was captured on ^{1}H provided an additional signal region with E_d ~ 2.2 MeV, albeit with much higher background.

IOP Conf. Series: Journal of Physics: Conf. Series **873** (2017) 012031 doi:10.1088/1742-6596/873/1/012031

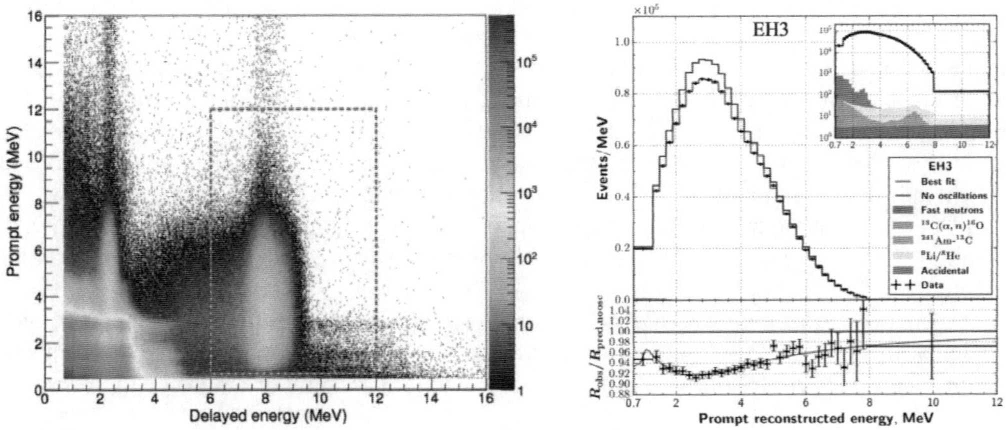

Fig. 4. The distribution of prompt versus delayed energy – left, and the observed reconstructed positron energy spectra observed in the far experimental hall – right.

5. Background

The vast majority of triggered signals in each detector were caused by natural radioactive backgrounds, with less than one in $\sim 10^5$ resulting from reactor $\overline{\nu}_e$interactions. Aside from natural radioactivity, a minor PMT instrumentation-related background was discovered during detector assembly and commissioning. Selecting pairs of prompt plus delayed signals with the proper energies and time separations rejected almost all backgrounds caused by natural radioactivity. Occasionally two such uncorrelated interactions would accidentally satisfy the antineutrino selection criteria. Fig. 4 – right, shows reconstructed positron energy spectrum in far detectors as well as contribution from different background origins.

The majority of prompt-like uncorrelated signals were from natural radioactivity in the detector components and the surrounding environment, and had $E_{rec} < 3$ MeV. Delayed-like uncorrelated signals were primarily from two sources. The first were beta decays of unstable isotopes, an unavoidable by-product of cosmogenic muon interactions with the scintillator. ^{12}B was by far the most prominent of such isotopes, although others such as ^{8}Li, ^{8}B, and ^{9}C are also produced. The second were high-energy γ-rays produced by the capture of neutrons emitted by the ^{241}Am-^{13}C calibration sources located in the calibration unit on the detector lid. The rate, energy spectrum, and other characteristics of these backgrounds were precisely modeled from studies of individual uncorrelated signals. Each day, only a few of the $\sim 10^7$ uncorrelated signals were estimated to form a pair which satisfied the antineutrino selection. Detailed studies of all uncorrelated interactions measured this background contamination to be from 1% to 2% depending on detector, with negligible uncertainty.

All remaining backgrounds are from physical processes that produce a pair of correlated interactions, which potentially mimic inverse beta decay. The vast majority of such correlated background results from cosmogenic muon interactions with the detector or nearby environment. Vetoing signals that occur soon after muon interactions with the detector or muon systems effectively reduced the contamination caused by these backgrounds to less than 0.5%. Lastly, three or more signals would occasionally occur close in time, giving no chance to recognize which pair was most likely the result of an antineutrino interaction. To avoid this ambiguity, sets of signals with multiplicity >2 were rejected. Six potential sources of correlated backgrounds were identified:

• Cosmogenic muons: Cosmogenic muons fragmented stable nuclei in or near the detectors, creating free neutrons and unstable nuclei. The signals from the initial muon interaction, subsequent neutron captures, or beta decays of unstable nuclei could potentially form a pair which satisfied the antineutrino selection. Muons were easily identified by their large scintillation light production or by Cherenkov light in the water shield. Vetoing concurrent or subsequent signals rendered this background negligible, at the cost of 2% of the far and 14%–18% of the near hall live times.

• Fast neutrons: Cosmogenic muon interactions in the environment near the detector generated fast neutrons. A nuclear collision of the neutron within the scintillator could mimic a prompt signal, while the subsequent neutron capture was identical to a true IBD delayed signal.

• β-n decays: Cosmogenic muon interactions occasionally produced the rare unstable isotopes ^{9}Li and ^{8}He, which β-decay with a chance of simultaneously emitting a neutron.

•^{241}Am-^{13}C neutron sources: During detector operation it was found that neutrons from the ^{241}Am-^{13}C calibration sources within the ACUs occasionally introduced several γ-rays, correlated in time, into the detector.

• (α,n) interactions: α's emitted by natural radioactivity within the detector could eject neutrons from stable nuclei, with $^{13}C(\alpha,n)^{16}O$ being the most prevalent interaction. Protons scattered by the neutron or ^{16}O de-excitation γ-rays could mimic a prompt signal, while the eventual neutron capture provided a delayed signal.

• High-multiplicity signals: The pile-up of correlated $\overline{\nu}_e$ signal pairs with uncorrelated radioactive backgrounds resulted in three or more signals within the correlation time for IBD candidates. Such sets of signals with multiplicity ≥3 resulted in ambiguity in determination of the actual prompt and delayed $\overline{\nu}_e$ signals. Complete rejection of these high-multiplicity combinations resolved this ambiguity, but introduced a ~2.5% loss of $\overline{\nu}_e$ efficiency.

6. Results

The presented results are from analysis of data collected in the Daya Bay experiment with 6 detectors in 217 days (Dec/2011 – Jul/2012) and with 8 detectors in 1013 days (Oct/2012 – Jul/2015). During 1230 days of operation, the Daya Bay experiment collected more than million inverse beta decays in near halls and more than 150 thousand IBD have been detected in far hall. Daily rate is ~2500 IBD events in near halls and ~300 IBD in far hall.

6.1. Oscillation analysis based on n-Gd [21]

A much simplified rate only analysis found

$$\sin^2 2\theta_{13} = 0.0850 \pm 0.0030(stat.) \pm 0.0028(syst.)$$

The distortion of the energy spectrum at the far hall relative to near halls was consistent with oscillation, and allowed measurement of $|\Delta m^2_{ee}|$. The parameters of the three-flavor model in best agreement with the observed rate and energy spectra were

$$\begin{aligned} \sin^2 2\theta_{13} &= 0.0841 \pm 0.0027(stat.) \pm 0.0019(syst.), \\ |\Delta m^2_{ee}| &= [2.50 \pm 0.06(stat.) \pm 0.06(syst.)] \times 10^{-3}\, eV^2, \\ \Delta m^2_{32}(NH) &= [2.45 \pm 0.06(stat.) \pm 0.06(syst.)] \times 10^{-3}\, eV^2, \\ \Delta m^2_{32}(IH) &= [-2.55 \pm 0.06(stat.) \pm 0.06(syst.)] \times 10^{-3}\, eV^2. \end{aligned}$$

The Δm^2_{32} values were obtained under the assumptions of normal (NH) and inverted (IH) mass ordering.

Fig. 5 - left, shows the observed electron survival probability as a function of effective baseline L_{eff} divided by the average antineutrino energy $\langle E_\nu \rangle$. Almost one full oscillation disappearance and reappearance cycle was sampled, given the range of L/E_ν values which were measured.

The confidence intervals for Δm^2_{ee} versus $\sin^2 2\theta_{13}$ are shown in Fig. 5 – right. The 1σ, 2σ, and 3σ 2-D confidence intervals are estimated using $\Delta\chi^2$ values of 2.30 (red), 6.18 (green), and 11.83 (blue) relative to the best fit. The upper panel provides the 1-D $\Delta\chi^2$ for $\sin^2 2\theta_{13}$ obtained by profiling $|\Delta m^2_{ee}|$ (blue line), and dash lines mark the corresponding 1σ, 2σ, and 3σ intervals. The right panel is

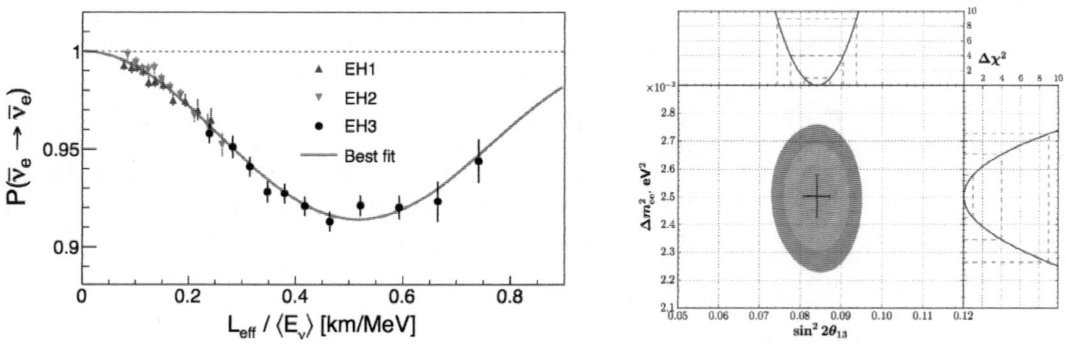

Fig. 5. Oscillation survival probability versus antineutrino proper time – left. Confidence intervals for $\sin^2 2\theta_{13}$ and $|\Delta m^2_{ee}|$ – right.

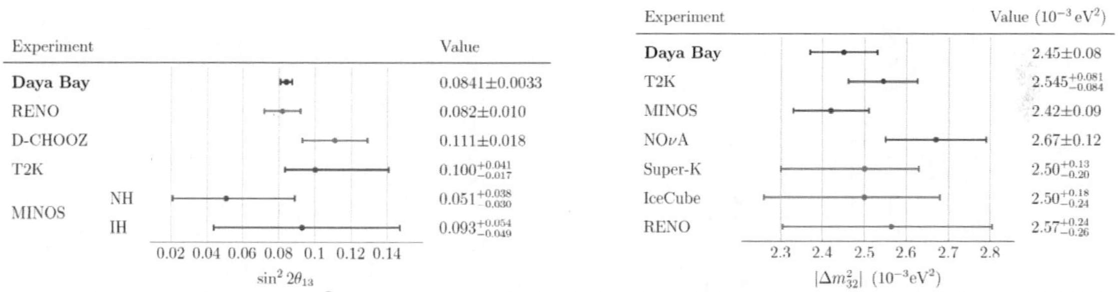

Fig. 6. Comparison of $\sin^2 2\theta_{13}$ provided by RENO [13], D-CHOOZ [14], T2K [15] and MINOS [16], and $|\Delta m^2_{32}|$ values provided by RENO [13], T2K[15], MINOS [17], NOvA [18], Super-K [19] and IceCube [20].

the same, but for $|\Delta m^2_{ee}|$, with $\sin^2 2\theta_{13}$ profiled. The point marks the best estimates, and the error bars display their 1-D 1σ confidence intervals. The Fig. 6 compares the values provided by different experiments.

6.2. Oscillation analysis based on n-H [22]

Alternative data analysis of data taken in 621 days, based on the events in which the neutron from IBD is captured on hydrogen results with

$$\sin^2 2\theta_{13} = 0.071 \pm 0.011.$$

Combination of the n-H and n-Gd results gives 8% improvement in precision

$$\sin^2 2\theta_{13} = 0.082 \pm 0.004.$$

6.3. Search for Light Sterile Neutrino

The large statistics collected with full configuration of eight detectors in the Daya Bay experiment allowed new precise analysis with aim to search for a light sterile neutrino [23]. A relative comparison of the rate and energy spectrum of reactor antineutrinos in the three experimental halls yields no evidence of sterile neutrino mixing in the $2 \times 10^{-4} < |\Delta m^2_{41}| < 0.3$ eV^2 mass range. The resulting limits on $\sin^2 2\theta_{14}$ shown on the Fig. 7 – left, constitute the most stringent constraints to date in the $|\Delta m^2_{41}| < 0.2$ eV^2 region.

Searches for a light sterile neutrino have been independently performed by the MINOS and the Daya Bay experiments using the muon (anti)neutrino and electron antineutrino disappearance channels, respectively. Results from both experiments are combined with those from the Bugey-3 reactor neutrino experiment to constrain oscillations into light sterile neutrinos [24]. The three experiments are sensitive to complementary regions of parameter space, enabling the combined analysis to probe regions allowed by the LSND and MiniBooNE experiments in a minimally extended four-neutrino flavor framework. Stringent limits on $\sin^2 2\theta_{\mu e}$ are set over six orders of magnitude in the sterile mass-squared splitting Δm^2_{41}. The sterile-neutrino mixing phase space allowed by the LSND and MiniBooNE experiments is excluded for $\Delta m^2_{41} < 0.8$ eV^2 at 95% CL$_s$, see the Fig. 7 – right.

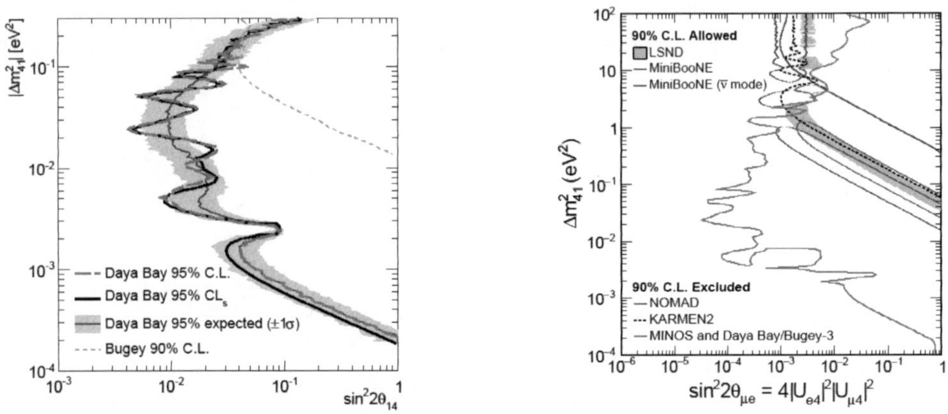

Fig. 7. Constraints for sterile light neutrino provided by Daya Bay [23] – left, and combined analysis of data from MINOS and Daya Bay/Bugey-3 [24] – right.

6.4. *Reactor antineutrino flux and spectrum* [25]

Data collected in 621 days were used to measure the IBD yield in the eight detectors, and the ratio of measured to predicted flux was found to be 0.946 ± 0.020 (0.992 ± 0.021) for the Huber+Mueller (ILL+Vogel) model, see Fig. 8 – left. A 2.9σ deviation was found in the measured IBD positron energy spectrum compared to the predictions. In particular, an excess of events in the region of 4–6 MeV was found in the measured spectrum, with a local significance of 4.4σ, see Fig. 8 – right.

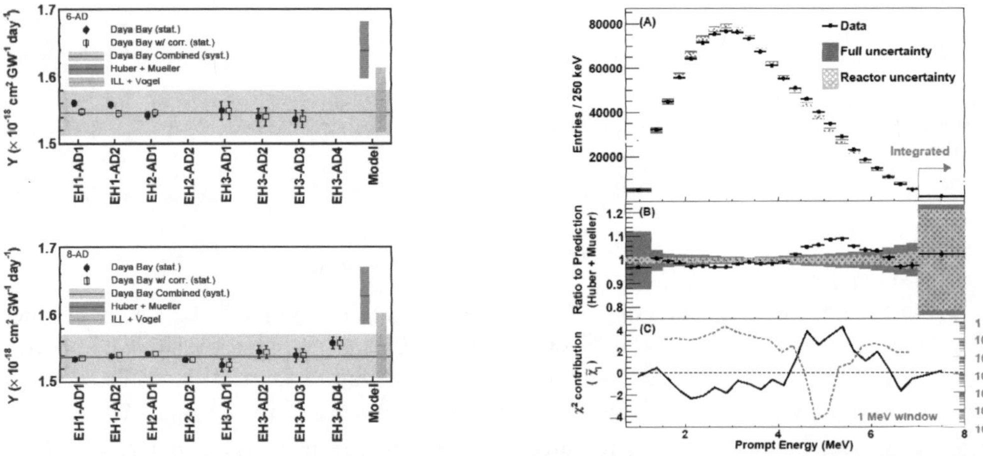

Fig. 8. Yield Y for the IBD events in the 6-AD only (top) and 8-AD only (bottom) – left. Comparison of predicted and measured prompt energy spectra – right.

IOP Conf. Series: Journal of Physics: Conf. Series **873** (2017) 012031 doi:10.1088/1742-6596/873/1/012031

References

[1] An F P et al. (Daya Bay Collaboration) 2012 *Phys. Rev. Lett.* **108** 171803
[2] Cowan C L, Reines F, Harrison F B, Kruse H W, McGuire A D 1956 *Science* **124** 103
[3] Ahn J K et al. (RENO Collaboration) 2012 *Phys. Rev. Lett.* **108** 191802
[4] Abe Y et al. (Double Chooz Collaboration) 2015 *JHEP* **1410** 086 [err: 2015 *JHEP* **1502** 074]
[5] Abe K et al. (T2K Collaboration) 2014 *Phys. Rev. Lett.* **112** 061802
[6] An F P et al. (Daya Bay Collaboration) 2014 *Phys. Rev. Lett.* **112** 061801
[7] An F P et al. (Daya Bay Collaboration) 2015 *Phys. Rev. Lett.* **115** 111802
[8] An F P et al. (Daya Bay Collaboration) *arXiv:1508.04233*
[9] Seo S (RENO Collaboration) 2014 *26th Int. Conf. on Neutrino*
[10] Beringer J et al. (Particle Data Group) 2012 *Phys. Rev. D* **86** 010001 see Sec. 13
[11] Adamson P et al. (MINOS Collaboration) 2014 *Phys. Rev. Lett.* **112** 191801
[12] Abe K et al. (T2K Collaboration) 2014 *Phys. Rev. Lett.* **112** 181801
[13] Choi J H et al. (RENO Collaboration) 2016 *Phys. Rev. Lett.* **116**, 211801 (*arXiv:1511.05849 [hep-ex]*)
[14] Ishitsuka M (Double Chooz Collaboration) 2016 *51st Rencontres de Moriond, Electroweak Interactions and United Theories*
[15] Iwamoto K (T2K Collaboration) 2016 *38th Int. Conf. on High Energy Physics*
[16] Adamson P et al. (MINOS Collaboration) 2013 *Phys. Rev. Lett.* **110** 171801
[17] Evans J (MINOS and MINOS+ Collaborations) 2016 *27th Int. Conf. on Neutrino Physics and Astrophysics*
[18] Vahle P (NOvA Collaboration) 2016 *27th Int. Conf. on Neutrino Physics and Astrophysics*
[19] Moriyama S (Super-Kamiokande Collaboration) 2016 *27th Int. Conf. on Neutrino Physics and Astrophysics*
[20] Koskinen J (Ice Cube Collaboration) 2016 *27th Int. Conf. on Neutrino Physics and Astrophysics*
[21] An F P et al. (Daya Bay Collaboration) 16 Oct 2016 *arXiv:1610.04802v1 [hep-ex]*
[22] An F P et al. (Daya Bay Collaboration) 2016 *Phys. Rev. D* **93** 072011
[23] An F P et al. (Daya Bay Collaboration) 2016 *Phys. Rev. Lett.* **117** no.15 151802
[24] An F P et al. (Daya Bay Collaboration, MINOS Collaboration) 2016 *Phys. Rev. Lett.* **117** no.15 151801
Addendum: 2016 *Phys. Rev. Lett.* **117** no.20 209901
[25] An F P et al. (Daya Bay Collaboration) 2016 *Phys. Rev. Lett.* **116** no.6 061801

IOP Conf. Series: Journal of Physics: Conf. Series **873** (2017) 012032 doi:10.1088/1742-6596/873/1/012032

Latest Results from MINOS and MINOS+

Simon De Rijck, for the MINOS and MINOS+ Collaborations

University of Texas at Austin, Department of Physics, 2515 Speedway, C1600, Austin, TX 78712-1192

E-mail: simonderijck@gmail.com

Abstract. The MINOS and MINOS+ experiments collected accelerator beam neutrino and antineutrino data for 11 years corresponding to peak energies of 3 GeV and 6 GeV, respectively, over a baseline of 735 km. These proceedings report on new limits and constraints set by MINOS and MINOS+ in and beyond the standard three-flavor paradigm. The atmospheric neutrino mass splitting in the three-flavor model is measured to be $(2.42 \pm 0.09) \times 10^{-3}\,\text{eV}^2$ for normal mass ordering and $-(2.48^{+0.09}_{-0.11}) \times 10^{-3}\,\text{eV}^2$ for inverted mass ordering. Constraints are set on sterile neutrinos and antineutrinos in the four-flavor model by looking for sterile-driven ν_μ and $\overline{\nu}_\mu$ disappearance and sterile-driven ν_e and $\overline{\nu}_e$ appearance. A combination of the MINOS four-flavor ν_μ disappearance search and the Daya Bay and Bugey-3 four-flavor $\overline{\nu}_e$ disappearance searches excludes the parameter space allowed by LSND and MiniBooNE for $\Delta m^2_{41} < 0.8\,\text{eV}^2$ at 90% C.L. The size of large extra dimensions is constrained to be smaller than 0.17 μm at 90% C.L. in the limit of a vanishing lightest active neutrino mass. Finally, a search for non-standard ν_e and $\overline{\nu}_e$ appearance in MINOS is presented.

1. Introduction

MINOS and MINOS+ are long baseline neutrino oscillation experiments that ran from 2005 through 2012 and 2013 through 2016, respectively. Charged Current (CC) and Neutral Current (NC) interactions of accelerator beam muon neutrinos from the Neutrinos at the Main Injector (NuMI) beam [1] at Fermilab are measured by two functionally identical detectors located on the NuMI beamline axis. Both detectors are magnetized tracking-sampling calorimeters built of iron plates interleaved with scintillator planes, the latter being read out using wavelength-shifting fibers coupled to multi-anode photomultiplier tubes. The Near Detector (ND) is located 1.04 km downstream of the NuMI target at Fermilab and has a 23.7 t fiducial (980 t total) mass. The Far Detector (FD) is located 735 km downstream of the NuMI target in Soudan, Minnesota and has a 4.2 kt fiducial (5.4 kt total) mass [2].

The magnetic horns in NuMI allow to operate the beam in either ν_μ or $\overline{\nu}_\mu$ mode. During the MINOS era, NuMI provided a beam with a peak neutrino energy of 3 GeV corresponding to an exposure of 10.71×10^{20} protons-on-target (POT) when operating in ν_μ mode and 3.36×10^{20} POT in $\overline{\nu}_\mu$ mode. The peak neutrino energy increased to 6 GeV for MINOS+, with NuMI providing an exposure of 5.80×10^{20} POT in ν_μ mode during the first two years of running. The analysis of the final year of MINOS+ beam data is ongoing.

MINOS was originally designed to perform precision measurements of the atmospheric neutrino oscillation parameters in the standard three-flavor model, Δm^2_{32} and θ_{23}, by observing muon neutrino disappearance occurring along the FD baseline. The shift to higher neutrino energies in MINOS+, away from the FD three-flavor oscillation minimum at 1.5 GeV, motivates searches beyond the standard three-flavor model, including searches for sterile neutrinos in the four-flavor model, large extra dimensions, and non-standard interactions.

2. Standard Three-Flavor Oscillation

In the three-flavor oscillation paradigm, MINOS and MINOS+ employ CC ν_μ beam and atmospheric events to search for ν_μ disappearance in the FD, and CC ν_e beam events to search for ν_e appearance in the FD. An update to the MINOS result [3] was performed by adding the first two years of MINOS+ beam exposure and an extra year of atmospheric events to the MINOS dataset [4]. The FD reconstructed beam ν_μ energy spectrum is shown in Fig. 1a and the 68% and 90% C.L. contours are compared to results from other experiments in Fig. 1b. To date, MINOS and MINOS+ provide one of the best limits on Δm^2_{32}, with $\Delta m^2_{32} = (2.42 \pm 0.09) \times 10^{-3}\,\text{eV}^2$ for normal mass ordering and $\Delta m^2_{32} = -(2.48^{+0.09}_{-0.11}) \times 10^{-3}\,\text{eV}^2$ for inverted mass ordering.

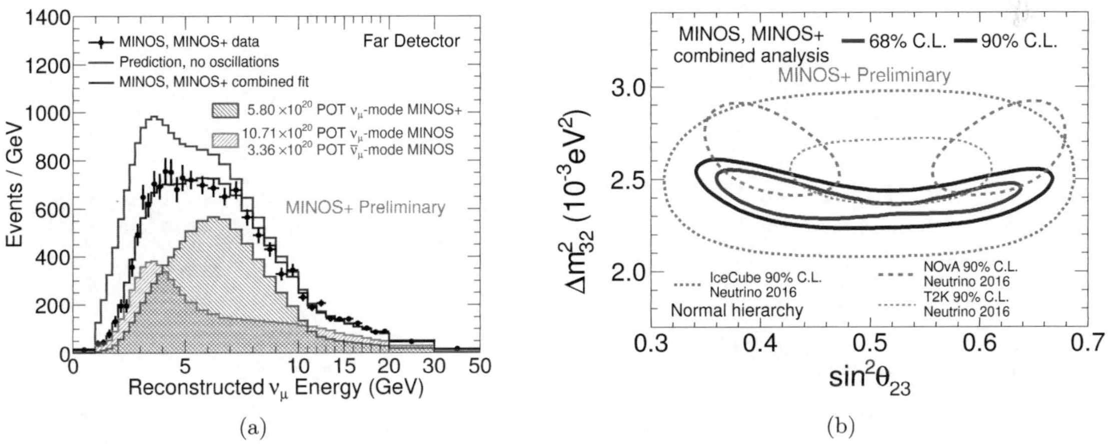

Figure 1: Updated standard oscillation result. (a) The reconstructed beam ν_μ energy spectrum for MINOS and MINOS+ FD selected events. Predictions are shown for no oscillations (red line) and for oscillations at the MINOS and MINOS+ combined best fit (blue line), with the MINOS (hashed magenta) and MINOS+ (hashed blue) contributions shown separately. (b) The 68% (red line) and 90% (blue line) C.L. contours resulting from a fit to 48.67 kt-yrs of atmospheric data combined with disappearance and appearance data from the MINOS beam and disappearance data from the MINOS+ beam during the first two years of operation. The combined MINOS and MINOS+ contours are compared with the 90% C.L. limits of NOνA [5], T2K [6], and IceCube DeepCore [7] presented at Neutrino 2016.

3. Sterile Neutrinos in the Four-Flavor Model

As precision experiments, MINOS and MINOS+ employ their data to look for discrepancies from the three-flavor paradigm that could be accounted for by small modifications to the standard three-flavor model. One such scenario adds one sterile neutrino state that can mix with the three active neutrino states. This minimal extension to the three-flavor model requires three additional mixing angles, θ_{14}, θ_{24}, and θ_{34}, one additional independent mass splitting, Δm^2_{41}, and two additional CP-violating phases, δ_{14} and δ_{24}.

IOP Conf. Series: Journal of Physics: Conf. Series **873** (2017) 012032 doi:10.1088/1742-6596/873/1/012032

3.1. Sterile Neutrino Search through Disappearance

MINOS and MINOS+ are sensitive to θ_{24} by searching for CC ν_μ disappearance and sterile neutrino appearance, the latter showing itself as a deficit of NC ν_i ($i =$ e,μ,τ) events compared to the three-flavor expectation. Unlike in the three-flavor model, significant oscillation along the ND baseline is possible. To account for this, the ratio of the measured FD and ND neutrino energy spectra, the Far-over-Near ratio, is compared to the four-flavor predictions. The Far-over-Near ratio for MINOS and MINOS+ selected CC events is shown in Fig. 2a.

MINOS and MINOS+ put constraints on θ_{24} over many orders of magnitude of Δm^2_{41}. No evidence for a sterile neutrino has been found [4, 8]. The combined MINOS and MINOS+ 90% and 95% C.L. Feldman-Cousins corrected contours are compared to results from other experiments in Fig. 2b.

A combination, employing the $\mathrm{CL_s}$ technique, of the MINOS disappearance result [8], placing limits on θ_{24}, and the $\overline{\nu}_e$ disappearance results of Daya Bay [9] and Bugey-3 [10], both placing limits on θ_{14}, allows a direct comparison with ν_e appearance results in the $(\sin^2 2\theta_{\mu e}, \Delta m^2_{41})$ plane, where

$$\sin^2 2\theta_{\mu e} = \sin^2 2\theta_{14} \sin^2 \theta_{24}. \tag{1}$$

The combined result, shown in Fig. 3a, excludes the LSND and MiniBooNE allowed regions for $\Delta m^2_{41} < 0.8\,\mathrm{eV}^2$ at 90% C.L. [11].

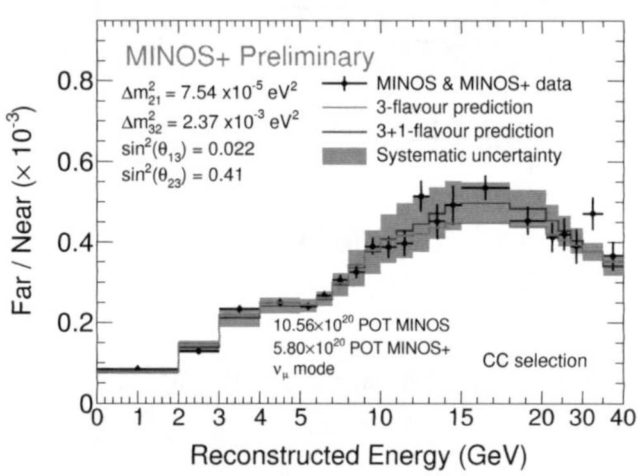

(a)

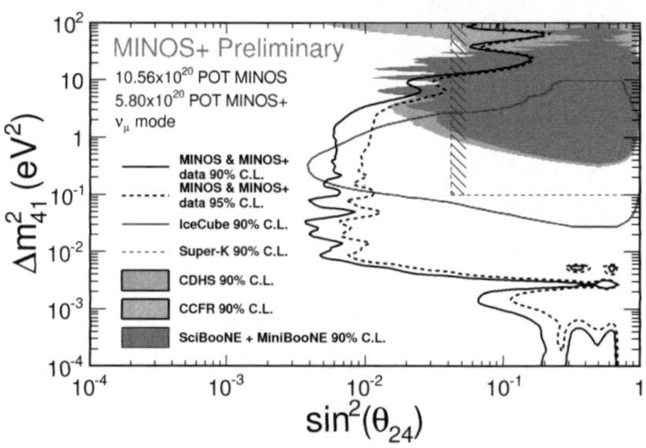

(b)

Figure 2: Constraining the four-flavor model with ν_μ disappearance. (a) The Far-over-Near ratio of observed MINOS and MINOS+ ν_μ CC beam events. The four-flavor best fit result is shown in blue with parameters $\Delta m^2_{41} = 24.01\,\mathrm{eV}^2$, $\Delta m^2_{32} = 2.512 \times 10^{-3}\,\mathrm{eV}^2$, $\theta_{23} = 0.917$, $\theta_{24} = 0.294$, and $\theta_{34} = 2.06 \times 10^{-11}$. (b) The combined MINOS and MINOS+ 90% (black line) and 95% (dashed black line) C.L. Feldman-Cousins corrected contours in the $(\sin^2 \theta_{24}, \Delta m^2_{41})$ plane using CC ν_μ disappearance and searching for a deficit in NC events. The combined MINOS and MINOS+ contours are compared to results from IceCube [12], Super-K [13], CDHS [14], CCFR [15], and SciBooNE + MiniBooNE [16].

3.2. Sterile Neutrino Search through ν_e Appearance

Through CC ν_e appearance in the FD, MINOS+ puts direct constraints on the four-flavor model in the $(\sin^2 2\theta_{\mu e}, \Delta m^2_{41})$ plane. The analysis utilizes an extended version of the framework developed for the MINOS three-flavor CC ν_e analyses [21], selecting only CC ν_e and $\overline{\nu}_e$

events above 6 GeV, away from the standard three-flavor oscillation minimum. The 90% C.L. contour based on the first year of MINOS+ beam data is shown in Fig. 3a.

3.3. Sterile Antineutrino Search through Disappearance

The ND and FD are magnetized, allowing separation of neutrinos and antineutrinos on an event-by-event basis. CC $\overline{\nu}_\mu$ events are selected from the MINOS beam data acquired during both $\overline{\nu}_\mu$ and ν_μ operations. Figure 3b compares the MINOS 90% C.L contour to results from other experiments. MINOS provides the strongest disappearance constraints for $\Delta\overline{m}^2_{41} < 0.5\,\mathrm{eV}^2$.

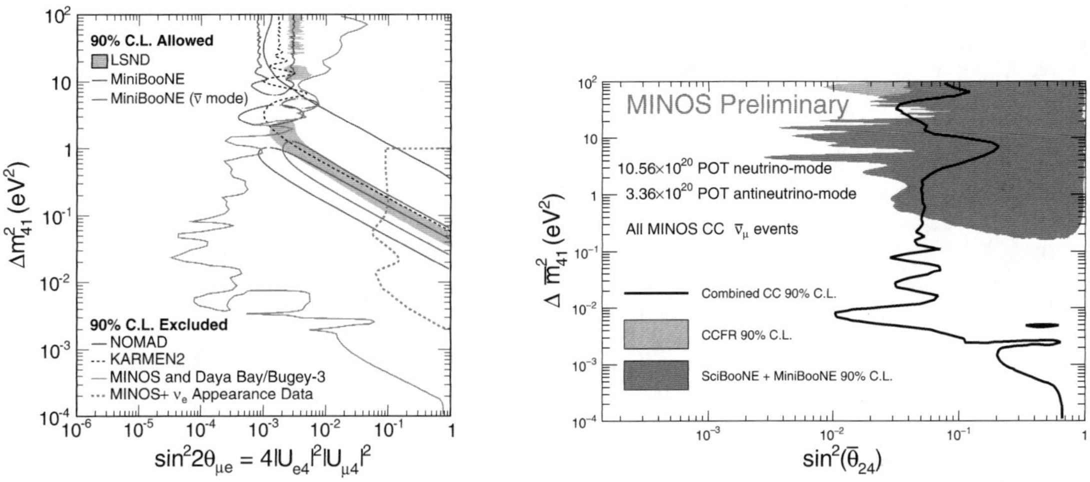

(a) (b)

Figure 3: Constraining the four-flavor model with ν_e appearance and antineutrinos. (a) The MINOS and Daya Bay/Bugey-3 combined 90% CL$_s$ limit (red line) on $\sin^2 2\theta_{\mu e}$ and the 90% C.L. contour (dashed orange line) using ν_e and $\overline{\nu}_e$ appearance events collected during the first year of MINOS+ are compared to the LSND [17] and MiniBooNE [18] 90% C.L. allowed regions. The regions excluded at 90% C.L. by KARMEN2 [19] and NOMAD [20] are also shown. (b) The MINOS 90% C.L. Feldman-Cousins corrected contour in the $(\sin^2\overline{\theta}_{24}, \Delta\overline{m}^2_{41})$ plane using $\overline{\nu}_\mu$ disappearance.

4. Large Extra Dimensions

MINOS and MINOS+ also search for extra dimensions. In the Large Extra Dimension (LED) model of [22, 23, 24, 25], sterile neutrinos arise as Kaluza-Klein states in a large extra dimension compactified on a circle with radius R. These sterile neutrino states mix with the active neutrino states and thereby modify the neutrino oscillation probabilities. The effects of the large extra dimension are governed by R and the lightest active neutrino mass m_0.

Following the Far-over-Near ratio approach of the four-flavor search and in the limit of a vanishing m_0, MINOS beam data constrains R to be smaller than 0.45 μm [26], while adding MINOS+ beam data tightens the constraint to $R < 0.17$ μm at 90% C.L. [4]. Figure 4a shows the 90% C.L. contours based on MINOS beam data and combined MINOS and MINOS+ beam data.

5. Non-Standard Interactions

The oscillation probability observed in the FD would also be modified if neutrinos undergo non-standard interactions with matter when traveling from Fermilab to Minnesota. Such non-standard interactions would induce an additional term in the interaction Hamiltonian which can be parameterized by the coupling coefficients $\epsilon_{\alpha\beta}$ $(\alpha, \beta = \mathrm{e}, \mu, \tau)$,

$$H_{\mathrm{NSI}} = \sqrt{2} G_{\mathrm{F}} N_{\mathrm{e}} \begin{pmatrix} \epsilon_{\mathrm{ee}} & \epsilon^*_{\mathrm{e}\mu} & \epsilon^*_{\mathrm{e}\tau} \\ \epsilon_{\mathrm{e}\mu} & \epsilon_{\mu\mu} & \epsilon^*_{\mu\tau} \\ \epsilon_{\mathrm{e}\tau} & \epsilon_{\mu\tau} & \epsilon_{\tau\tau} \end{pmatrix}, \tag{2}$$

where G_{F} is the Fermi coupling constant and N_{e} is the electron density of the matter through which the neutrinos travel. The oscillation probabilities furthermore depend on three additional phases, $\delta_{\mathrm{e}\mu}$, $\delta_{\mathrm{e}\tau}$, and $\delta_{\mu\tau}$ [27, 28, 29, 30]. Employing ν_{e} and $\overline{\nu}_{\mathrm{e}}$ appearance events in the FD, MINOS has set constraints on $|\epsilon_{\mathrm{e}\tau}|$ as a function of $(\delta_{\mathrm{CP}} + \delta_{\mathrm{e}\tau})$ [31], as shown in Fig. 4b.

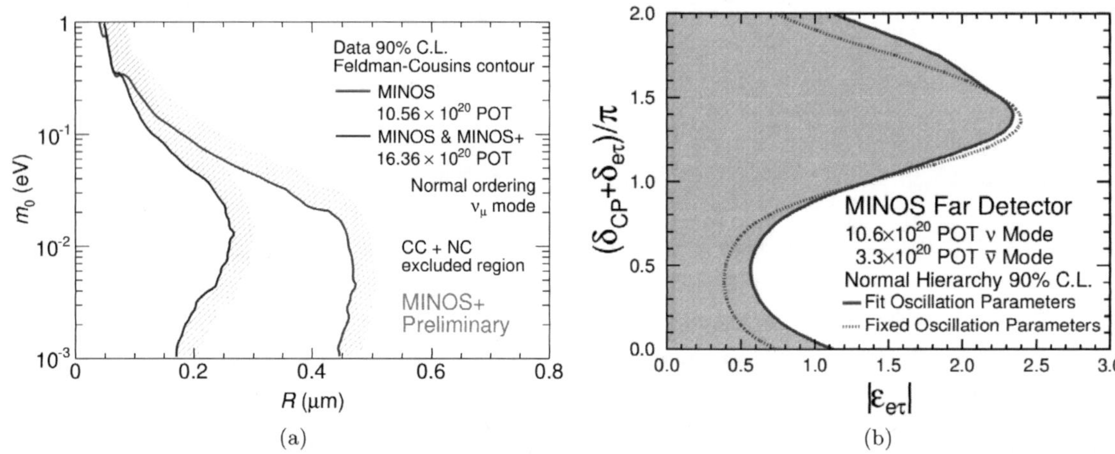

Figure 4: Constraining exotic models beyond the three-flavor paradigm. (a) Comparison of the 90% C.L. contour for the LED model, obtained using the Feldman-Cousins technique, based on 10.56×10^{20} POT MINOS data and the contour obtained when adding 5.80×10^{20} POT MINOS+ data. The large extra dimension size and the smallest neutrino mass are denoted as R and m_0, respectively. (b) The MINOS 90% C.L. contour in the $(|\epsilon_{\mathrm{e}\tau}|, \delta_{\mathrm{CP}} + \delta_{\mathrm{e}\tau})$ plane for normal neutrino mass ordering using ν_{e} and $\overline{\nu}_{\mathrm{e}}$ appearance events in the FD. The shaded areas to the left of the solid contours indicate the MINOS allowed regions where additional oscillation parameters, including the non-standard interaction parameters, were included in the fit. The dotted contours show the limits where the additional oscillation parameters were fixed.

6. Conclusions

Between observing the first atmospheric neutrino event in the MINOS FD on August 1, 2003, and collecting the last beam neutrino event on June 29, 2016, MINOS and MINOS+ collected 2.38×10^{21} POT of accelerator beam neutrino data and 60.9 kt-yrs of atmospheric neutrino data. The addition of the first two years of MINOS+ beam data to the MINOS beam dataset leads to one of the best measurements of the atmospheric mass splitting in the standard three-flavor model, produces leading constraints over many orders of magnitude in the new mass splitting on sterile neutrino and antineutrino searches in the four-flavor model, and results in the strongest constraints on the size of large extra dimensions reported by a neutrino oscillation experiment. Inclusion of the final year of MINOS+ data will further improve our results in and beyond the three-flavor paradigm.

References

[1] Adamson P *et al.* (MINOS Collaboration) 2016 *Nucl. Instrum. Meth. A* **806** 279-306
[2] Michael D G *et al.* (MINOS Collaboration) 2008 *Nucl. Instrum. Meth. A* **596** 190-228
[3] Adamson P *et al.* (MINOS Collaboration) 2014 *Phys. Rev. Lett.* **112** 191801
[4] Evans J *New results from MINOS and MINOS+* Talk at the XXVII International Conference on Neutrino Physics and Astrophysics (Neutrino 2016), London, 2016
[5] Vahle P *New results from NOvA* Talk at the XXVII International Conference on Neutrino Physics and Astrophysics (Neutrino 2016), London, 2016
[6] Tanaka H *Status, recent results and plans for T2K* Talk at the XXVII International Conference on Neutrino Physics and Astrophysics (Neutrino 2016), London, 2016
[7] Koskinen D J *Atmospheric neutrino results from IceCube/DeepCore and plans for PINGU* Talk at the XXVII International Conference on Neutrino Physics and Astrophysics (Neutrino 2016), London, 2016
[8] Adamson P *et al.* (MINOS Collaboration) 2016 *Phys. Rev. Lett.* **117** 151803
[9] An F P *et al.* (Daya Bay Collaboration) 2016 *Phys. Rev. Lett.* **117** 151802
[10] Achkar B *et al.* (Bugey-3 Collaboration) 1995 *Nucl. Phys. B* **434** 503
[11] Adamson P *et al.* (MINOS and Daya Bay Collaborations) 2016 *Phys. Rev. Lett.* **117** 151801
[12] Aartsen M G *et al.* (IceCube Collaboration) 2016 *Phys. Rev. Lett.* **117** 0712801
[13] Abe K *et al.* (Super Kamiokande Collaboration) 2015 *Phys. Rev. D* **91** 052019
[14] Dydak F *et al.* (CDHS Collaboration) 1984 *Phys. Lett. B* **134** 281
[15] Stockdale I E *et al.* (CCFR Collaboration) 1984 *Phys. Rev. Lett.* **52** 1384
[16] Mahn K B M *et al.* (MiniBooNE and SciBooNE Collaborations) 2012 *Phys. Rev. D* **85** 032007
[17] Aguilar A *et al.* (LSND Collaboration) 2001 *Phys. Rev. D* **64** 112007
[18] Aguilar-Arevalo A *et al.* (MiniBooNE Collaboration) 2013 *Phys. Rev. Lett.* **110** 161801
[19] Armbruster B *et al.* (KARMEN Collaboration) 2002 *Phys. Rev. D* **65** 112001
[20] Astier P *et al.* (NOMAD Collaboration) 2003 *Phys. Lett. B* **570** 19
[21] Adamson P *et al.* (MINOS Collaboration) 2013 *Phys. Rev. Lett.* **110** 171801
[22] Dienes K R, Dudas E and Gherghetta T 1999 *Nucl. Phys. B* **557** 25
[23] Arkani-Hamed N, Dimopoulos S, Dvali G R and March-Russell J 2001 *Phys. Rev. D* **65** 024032
[24] Davoudiasl H, Langacker P and Perelstein M 2002 *Phys. Rev. D* **65** 105015
[25] Machado P A N, Nunokawa H and Zukanovich Funchal R 2011 *Phys. Rev. D* **84** 013003
[26] Adamson P *et al.* (MINOS Collaboration) 2016 *Phys. Rev. D* **94** 111101(R)
[27] Ganzalez-Garcia M C, Guzzo M M, Krastev P I, Nunokawa H, Peres O L G, Pleitez V, Valle J W F and Zukanovich Funchal R 1999 *Phys. Rev. Lett.* **82** 3202
[28] Friedland A, Lunardini C and Maltoni M 2004 *Phys. Rev. D* **70** 111301
[29] Friedland A and Lunardini C 2006 *Phys. Rev. D* **74** 033012
[30] Ohlsson T 2013 *Rep. Prog. Phys.* **76** 044201
[31] Adamson P *et al.* (MINOS Collaboration) 2017 *Phys. Rev. D* **95** 012005

IOP Conf. Series: Journal of Physics: Conf. Series **873** (2017) 012033 doi:10.1088/1742-6596/873/1/012033

A model of radiatively induced quark and lepton mass model

Takaaki Nomura

School of Physics, KIAS, Seoul 02455, Korea

E-mail: nomura@kias.re.kr

Abstract. We discuss a radiatively induced quark and lepton mass model in the first and second generation introducing extra $U(1)$ gauge symmetry, discrete Z_2 symmetry, vector-like fermions and exotic scalar fields. Then we analyze the allowed parameter regions which simultaneously satisfy the constraints of FCNCs for the quark sector and of LFVs including $\mu-e$ conversion, observed quark mass and mixing, and the lepton mass and mixing. In addition, the typical value for the $(g-2)_\mu$ in our model is presented. We also show extension of the model in which Majorana type neutrino masses are generated at the two loop level.

1. Introduction

The fermion masses in the standard model(SM) are hierarchical in both quark and lepton sector. In particular, masses of the SM neutrinos are very small compared to the other fermion masses. It is thus challenging to understand the hierarchical structure of fermion masses.

Radiative mass generation models are one of the interesting scenario to explain such a mass hierarchy in which some fermion masses are generated at loop levels and loop suppression factor would be a source of large mass difference. In addition, these models would lead rich phenomenology such as lepton flavor violation(LFV), anomalous muon magnetic dipole moment(muon $g-2$), flavor changing neutral current(FCNC) and existence of dark matter(DM) candidate.

In this proceeding, we review two radiative mass generation models where the second one is an extension of the first one [1, 2]. In our models, the first and second generation SM charged-lepton and quark masses are induced at one-loop level while the masses of third generation fermions are generated via the vacuum expectation value(VEV) of SM Higgs field. To realize our scenario, we add extra local U(1) symmetry to restrict the Yukawa interaction associated with SM Higgs field in anomaly free way, and the vector-like quarks and leptons are introduced to write relevant one-loop diagrams for fermion mass generation. Therefore, the charged-lepton and quark masses in first two generations are generated at the one-loop level induced by the Yukawa interactions among SM fermions, inert scalar fields and vector-like fermions which are invariant under the new U(1). The SM neutrino masses are generated at one-loop level as the other fermion masses in the first model while the second model realize two-loop level mass generation of neutrino [2]. Thus the smallness of the neutrino masses are naturally obtained in the second model. For both models, we search for the parameter range which can fit the observed fermion masses and mixing, taking into account the constraints from flavor physics. Note also that our model includes DM candidate whose relic density can be accommodated with

Table 1. Fermion contents in the model I with their charge assignments under $SU(2)_L \times U(1)_Y \times U(1)_R \times Z_2$, where each of the flavor index is defined as $\alpha \equiv 1-3$ and $i = 1, 2$.

Fermions	Q_L^α	u_R^i	d_R^i	t_R	b_R	$Q'^i_{L(R)}$	L_L^α	e_R^i	ν_R^i	τ_R	$L'^i_{L(R)}$
$SU(3)_C$	**3**	**3**	**3**	**3**	**3**	**3**	**1**	**1**	**1**	**1**	**1**
$SU(2)_L$	**2**	**1**	**1**	**1**	**1**	**2**	**2**	**1**	**1**	**1**	**2**
$U(1)_Y$	$\frac{1}{6}$	$\frac{2}{3}$	$-\frac{1}{3}$	$\frac{2}{3}$	$-\frac{1}{3}$	$\frac{1}{6}$	$-\frac{1}{2}$	-1	0	-1	$-\frac{1}{2}$
$U(1)_R$	0	x	$-x$	0	0	0	0	$-x$	x	0	0
Z_2	$+$	$+$	$+$	$+$	$+$	$-$	$+$	$+$	$+$	$+$	$-$

Table 2. Charge assignment for scalar fields in the model I.

Scalar fileds	Φ	φ	η	S
$SU(2)_L$	**2**	**1**	**2**	**1**
$U(1)_Y$	$\frac{1}{2}$	0	$\frac{1}{2}$	0
$U(1)_R$	0	x	x	0
Z_2	$+$	$+$	$-$	$-$

observed value where we omit the discussion in order to focus on fermion mass sector; detailed discussions can be found in Ref. [2].

2. Model I

In this section, we introduce a model which is based on $G_{SM} \times \mathrm{U}(1)_R \times Z_2$ symmetry; G_{SM} is the SM gauge symmetry and $\mathrm{U}(1)_R$ is extra U(1) gauge symmetry. Then we discuss fermion mass matrices, FCNC in quark sector and LFVs [1]. In our analysis, we assume Z' associated with $\mathrm{U}(1)_R$ is sufficiently heavy avoiding experimental constraints and discussion of it is omitted.

2.1. Particle contents

The fermion contents are summarized in Table 1 where we introduce $SU(2)_L$ doublet exotic vector-like quarks $Q' \equiv [u', d']^T$ and leptons $L' \equiv [N', E']^T$ with two flavors. In addtion we introduce two right-handed neutrinos ν_R^i $(i = 1, 2)$, which constitute Dirac fields after the spontaneous electroweak symmetry breaking as the other fermion sectors in the SM. Then a gauged $U(1)_R$ symmetry is imposed where only the first and second family of right-handed SM fermions and ν_R^i have none-zero charge x; the $\mathrm{U}(1)_R$ is anomaly free by the charge assignment. Notice here that we require the third generation couple to the SM-like Higgs directly and masses of third generation fermions are generated at tree level.

The contents of scalar fields are summarized in Table 2 where we add two $SU(2)_L$ singlets φ and S, and one $SU(2)_L$ doublet scalar η to the SM Higgs-like field Φ; here Φ and φ have the VEVs, symbolized by $\langle\Phi\rangle \equiv v/\sqrt{2}$ and $\langle\varphi\rangle \equiv v'/\sqrt{2}$, that spontaneously break the electroweak and $U(1)_R$ symmetry respectively, while S and η do not have VEVs that are assured by the Z_2 symmetry. Note also that S is assumed to be real scalar field.

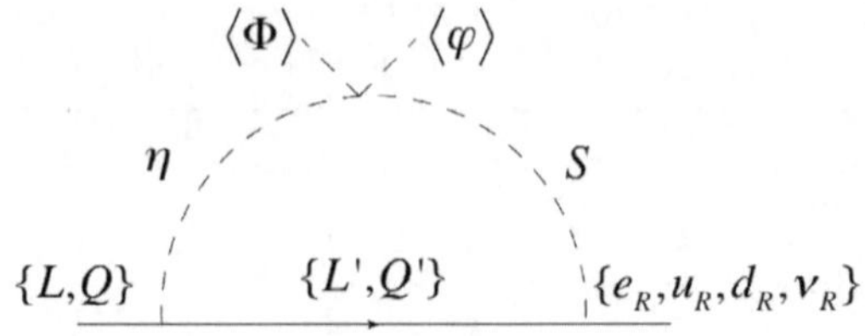

Figure 1. The one loop diagram for generating fermion masses.

Under symmetries in the model, the vector-like fermion mass term and Yukawa interactions are given by

$$-\mathcal{L}_Q = (y_u)_{ij}\bar{Q}'_i u_{R_j}(i\sigma_2)\eta^* + (y_d)_{ij}\bar{Q}'_{ij}\eta d_{R_j} + (y_Q)_{\alpha j}\bar{Q}_\alpha Q'_j S$$
$$+ (y_t)_\alpha \bar{Q}_\alpha t_{R_j}(i\sigma_2)\Phi^* + (y_b)_\alpha \bar{Q}_\alpha \Phi b_R + m_{Q'_k}\bar{Q}'_k Q'_k + \text{c.c.}, \quad (1)$$
$$-\mathcal{L}_L = (y_\nu)_{ij}\bar{L}'_i \nu_{R_j}(i\sigma_2)\eta^* + (y_\ell)_{ij}\bar{L}'_{ij}\eta e_{R_j} + (y_L)_{\alpha j}\bar{L}_\alpha L'_j S$$
$$+ (y_\tau)_\alpha \bar{L}_\alpha \tau_{R_j}(i\sigma_2)\Phi + (y_{\nu_\tau})_\alpha \bar{L}_\alpha \Phi \nu_{\tau R} + m_{L'_k}\bar{L}'_k L'_k + \text{c.c.}, \quad (2)$$

where σ_2 is the second Pauli matrix.

The scalar potential in the model is given by

$$V = m_\varphi^2|\varphi|^2 + m_S^2 S^2 + m_\Phi^2|\Phi|^2 + m_\eta^2|\eta|^2 + \lambda_0\left[(\Phi^\dagger\eta)S\varphi^* + \text{c.c.}\right] \quad (3)$$
$$+ \lambda_\varphi|\varphi|^4 + \lambda_S S^4 + \lambda_\Phi|\Phi|^4 + \lambda_\eta|\eta|^4 + \lambda_{\varphi S}|\varphi|^2 S^2 + \lambda_{\varphi\Phi}|\varphi|^2|\Phi|^2 + \lambda_{\varphi\eta}|\varphi|^2|\eta|^2$$
$$+ \lambda_{S\Phi}S^2|\Phi|^2 + \lambda_{S\eta}S^2|\eta|^2 + \lambda_{\Phi\eta}|\Phi|^2|\eta|^2 + \lambda'_{\Phi\eta}|\Phi^\dagger\eta|^2. \quad (4)$$

We write scalar fields by components as

$$\Phi = \begin{bmatrix} w^+ \\ \frac{v+\phi+iz}{\sqrt{2}} \end{bmatrix}, \quad \eta = \begin{bmatrix} \eta^+ \\ \frac{\eta_R+i\eta_I}{\sqrt{2}} \end{bmatrix}, \quad \varphi = \frac{v' + \varphi_R + i\varphi_I}{\sqrt{2}}, \quad (5)$$

where $w^\pm$, z, and φ_I are Nambu-Goldstone(NG) bosons absorbed by the longitudinal degrees of freedom of charged SM gauged boson $W^\pm$, neutral SM gauge boson Z, and neutral $U(1)_R$ gauge boson Z'. After the spontaneous symmetry breaking, neutral scalar bosons mix each other as follows:

$$\begin{bmatrix} S \\ \eta_R \end{bmatrix} = O_R \begin{bmatrix} H_1 \\ H_2 \end{bmatrix}, \quad \begin{bmatrix} \varphi_R \\ \phi \end{bmatrix} = O_\beta \begin{bmatrix} h_1 \\ h_2 \end{bmatrix}, \quad O_a \equiv \begin{bmatrix} c_a & s_a \\ -s_a & c_a \end{bmatrix}, \quad (6)$$

where $c_a \equiv \cos a$, $s_a \equiv \sin a$, $H_i(i = 1, 2)$ is the mass eigenstate of the inert neutral scalar boson, and $h_i(i = 1, 2)$ is the mass eigenstate of the Z_2 even neutral scalar boson.

By these interactions, first and second generation fermions masses are generated at one loop level as shown in Fig. 1. We discuss formula for the mass matrices below.

2.2. Quark sector

In this subsection, we analyze the quark sector including mass matrices and FCNC. The mass matrix in our form is written in terms of tree level mass matrix and one-loop one as

$$(M_{u(d)})_{\alpha\beta} = (M_{u(d)}^{tree})_{\alpha 3} + (M_{u(d)}^{one-loop})_{\alpha j}, \quad (7)$$

with

$$(M^{tree}_{u(d)})_{\alpha 3} = \frac{(y_{t(b)})_\alpha}{\sqrt{2}}, \tag{8}$$

$$(M^{one-loop}_{u(d)})_{\alpha j} = s_R c_R \sum_{k=1,2} \frac{(y_Q)_{\alpha k} m_{Q'_k} (y_{u(d)})_{kj}}{\sqrt{2}(4\pi)^2} \int_0^1 dx_1 \frac{x_1 m^2_{Q'_k} + (1-x_1) m^2_{H_1}}{x_1 m^2_{Q'_k} + (1-x_1) m^2_{H_2}}. \tag{9}$$

As in the SM, up- and down-type quark mass matrices are diagonalized by $M^{diag.}_u = V_{u_L} M_u V_{u_R}$, and $M^{diag.}_d = V_{d_L} M_d V_{d_R}$, where $V's$ are unitary matrix. Then CKM matrix is defined by $V_{CKM} \equiv V^\dagger_{u_L} V_{d_L}$ which can be parametrized by three mixing angles and one phase as follows:

$$V_{CKM} \equiv V^\dagger_{u_L} V_{d_L} \equiv \begin{bmatrix} 1 & 0 & 0 \\ 0 & c_{23} & s_{23} \\ 0 & -s_{23} & c_{23} \end{bmatrix} \begin{bmatrix} c_{13} & 0 & s_{13} e^{-i\delta} \\ 0 & 1 & 0 \\ -s_{13} e^{i\delta} & 0 & c_{13} \end{bmatrix} \begin{bmatrix} c_{12} & s_{12} e^{-i\delta} & 0 \\ -s_{12} & c_{12} & 0 \\ 0 & 0 & 1 \end{bmatrix}. \tag{10}$$

In our numerical analysis, we assume to take the following forms to evade the stringent constraint of $B_0 - \bar{B}_0$ mixing in the numerical analysis:

$$V^\dagger_{u_L} = \begin{bmatrix} 1 & 0 & 0 \\ 0 & c_{23} & s_{23} \\ 0 & -s_{23} & c_{23} \end{bmatrix} \begin{bmatrix} c_{13} & 0 & s_{13} e^{-i\delta} \\ 0 & 1 & 0 \\ -s_{13} e^{i\delta} & 0 & c_{13} \end{bmatrix}, \quad V_{d_L} = \begin{bmatrix} c_{12} & s_{12} e^{-i\delta} & 0 \\ -s_{12} & c_{12} & 0 \\ 0 & 0 & 1 \end{bmatrix}. \tag{11}$$

The stringent constraints come from FCNCs where we consider $Q - \bar{Q}$ mixing inducing the mass difference between a meson and an anti-meson. Here we symbolize these observables by Δm_M where $M = D, K, B$ indicates the type of mesons. Then our formulas are given by calculating box-type one-loop diagrams by [3]

$$\Delta m_M \approx \frac{1}{2(4\pi)^2} \sum_{\rho,\sigma}^{1,2} \left(\mathcal{F}^-_{\rho\sigma} C^M_{Q_2} (\mathcal{M}^M_{Q_2})^2_{\rho\sigma} + \mathcal{F}^+_{\rho\sigma} C^M_{Q_4} (\mathcal{M}^M_{Q_4})^2_{\rho\sigma} \right), \tag{12}$$

with

$$(\mathcal{M}^D_{Q_{2[4]}})^2_{\rho\sigma} \equiv (y'_u)_{\rho 2} m_{Q'_\rho} (y'_Q)_{1\rho} (y'_{Q[u]})^{[\dagger]}_{1\sigma} m_{Q'_\sigma} (y'_{u[Q]})^{[\dagger]}_{\sigma 2}, \tag{13}$$

$$(\mathcal{M}^K_{Q_{2[4]}})^2_{\rho\sigma} \equiv (y'_d)_{\rho 2} m_{Q'_\rho} (y'_Q)_{1\rho} (y'_{Q[d]})^{[\dagger]}_{1\sigma} m_{Q'_\sigma} (y'_{d[Q]})^{[\dagger]}_{\sigma 2}, \tag{14}$$

$$(\mathcal{M}^B_{Q_{2[4]}})^2_{\rho\sigma} \equiv (y'_d)_{\rho 3} m_{Q'_\rho} (y'_Q)_{1\rho} (y'_{Q[d]})^{[\dagger]}_{1\sigma} m_{Q'_\sigma} (y'_d[Q]))^{[\dagger]}_{\sigma 3}, \tag{15}$$

$$C^M_{Q_2} = -\frac{5}{24} \left(\frac{m_M}{m_c + m_u} \right)^2 m_M f^2_M, \quad C^M_{Q_4} = \left[\frac{1}{24} + \frac{1}{4} \left(\frac{m_M}{m_c + m_u} \right)^2 \right] m_M f^2_M, \tag{16}$$

and

$$\mathcal{F}^\pm_{\rho\sigma} \equiv \left[\frac{s^2_{2R}}{2} (F_1[H_1] + F_1[H_2]) + c^2_{2R} (F_2[H_1, H_2] + F_1[H_2, H_1]) \pm c^2_R F_2[\eta_I, H_1] \pm s^2_R F_2[\eta_I, H_2] \right]_{\rho\sigma}, \tag{17}$$

$$F_1[m_a] \equiv \int_0^1 da \int_0^{1-a} db \frac{1-a-c}{\left[a m^2_{Q'_\rho} + b m^2_{Q'_\sigma} + (1-a-b) m^2_a \right]^2}, \tag{18}$$

$$F_2[m_a, m_b] \equiv \int_0^1 da \int_0^{1-a} db \int_0^{1-a-b} dc \frac{1}{\left[a m^2_{Q'_\rho} + b m^2_{Q'_\sigma} + c m^2_a + (1-a-b-c) m^2_b \right]^2}, \tag{19}$$

Table 3. Summary of constrains from LFV processes $\ell_j \to \ell_\alpha \gamma$ and $\mu \to e$ conversion.

Process	(j, α)	Experimental bounds (90% CL)	References
$\mu^- \to e^- \gamma$	$(2,1)$	$BR(\mu \to e\gamma) < 4.2 \times 10^{-13}$	[5]
$\tau^- \to e^- \gamma$	$(3,1)$	$Br(\tau \to e\gamma) < 3.3 \times 10^{-8}$	[6]
$\tau^- \to \mu^- \gamma$	$(3,2)$	$BR(\tau \to \mu\gamma) < 4.4 \times 10^{-8}$	[6]
$\mu \to e$ conversion	$(2,1)$	$R(Ti) < 4.3 \times 10^{-12} \to \mathcal{O}(10^{-18})$(future bound)	[7]→ [8]

where the Yukawa couplings are defined to be $y'_{u(d)} \equiv y_{u(d)} V^\dagger_{u(d)_R}$, $y'_Q \equiv V_{u_L} y_Q$, and we assume $V_L = V_R$ in our analytical convenience. Here experimental and input values [4] are given as follows:

$$m_u \approx 2.3[\text{MeV}],\ m_c \approx 1275[\text{MeV}],\ m_t \approx 173.2[\text{GeV}], \tag{20}$$

$$m_d \approx 4.8[\text{MeV}],\ m_s \approx 95[\text{MeV}],\ m_b \approx 4.18[\text{GeV}], \tag{21}$$

$$m_D \approx 1864.84[\text{MeV}],\ m_K \approx 497.614[\text{MeV}],\ m_B \approx 5279.50[\text{MeV}], \tag{22}$$

$$f_D \approx 212[\text{MeV}],\ f_K \approx 159.8[\text{MeV}],\ f_B \approx 200[\text{MeV}]. \tag{23}$$

Then we apply phenomenological bounds requiring that new physics contribution to Δm_M should be less than observed values [4]

$$\Delta m_D \lesssim 6.25 \times 10^{-12}[\text{MeV}],\ \Delta m_K \lesssim 3.484 \times 10^{-12}[\text{MeV}],\ \Delta m_B \lesssim 3.356 \times 10^{-10}[\text{MeV}]. \tag{24}$$

2.3. Lepton sector

In this subsection, we discuss the lepton sector including mass matrix and LFVs, where neutrinos are supposed to be Dirac neutrino. As in the quark sector, mass matrices can be given by

$$(M_{\nu(\ell)})_{\alpha\beta} = (M_\ell^{tree})_{\alpha 3} + (M_{\nu(\ell)}^{one-loop})_{\alpha j}, \tag{25}$$

with

$$(M_\ell^{tree})_{\alpha 3} = \frac{(y_T)_\alpha}{\sqrt{2}}, \tag{26}$$

$$(M_{\nu(\ell)}^{one-loop})_{\alpha j} = s_R c_R \sum_{k=1,2} \frac{(y_{\nu(L)})_{\alpha k} m_{L'_k} (y_S)_{kj}}{\sqrt{2}(4\pi)^2} \int_0^1 dx_2 \frac{x_2 m^2_{L'_k} + (1-x_2) m^2_{H_1}}{x_2 m^2_{L'_k} + (1-x_2) m^2_{H_2}}. \tag{27}$$

The LFV processes give the constraints on our parameters. We firstly consider $\ell_j \to \ell_\alpha \gamma$ process, and its branching ratio is given by

$$BR(\ell_j \to \ell_\alpha \gamma) \approx \frac{48\pi^3 \alpha_{em} C_j}{G_F^2 m^2_{\ell_j}} |a_{j\alpha}|^2 \tag{28}$$

where $C_j = (1, 1/5)$ for $(i = \mu, \tau)$, $\alpha_{em} \approx 1/137$ is the fine-structure constant, $G_F \approx 1.17 \times 10^{-5}$ GeV^{-2} is the Fermi constant, $a_{j\alpha}$ is computed as

$$a_{j\alpha} = -\frac{s_R c_R}{2\sqrt{2}(4\pi)^2} \sum_{k=1}^{2} (y'_\ell)_{\alpha k} m_{L'_k} (y_{L'})_{k,j} \left[F_{\ell_j \to \ell_\alpha \gamma}(m_{H_1}, m_{L'_k}) - F_{\ell_j \to \ell_\alpha \gamma}(m_{H_2}, m_{L'_k}) \right], \tag{29}$$

$$F_{\ell_j \to \ell_\alpha \gamma}(m_1, m_2) = \frac{2m_1^4 - 4m_1^2 m_2^2 + m_2^4 + 4m_1^4 \ln\left[\frac{m_2}{m_1}\right]}{2(m_1^2 - m_2^2)^3}. \tag{30}$$

Through the same diagram for the above LFVs, with $j = \alpha = 2$, there exists the contribution to $(g-2)_\mu$ denoted by Δa_μ which is simply given by

$$\Delta a_\mu \approx -\frac{m_\mu a_{22}}{2}. \tag{31}$$

This value can be tested by current and future experiments [9, 10, 11].

Secondly we consider the $\mu - e$ conversion process which can be found in the same diagram as the process of $\ell_j \to \ell_\alpha \gamma$ with γ line being attached to nucleons and additional contribution is taken into account by replacing γ with Z boson. Then the $\mu - e$ conversion rate R is obtained as [12]

$$R \equiv \frac{\Gamma(\mu \to e)}{\Gamma_{\text{capt}}} = \frac{C_{\mu e}}{\Gamma_{\text{capt}}} \left| Z \left(b_{21}^\gamma - \frac{a_{21}}{m_\mu} \right) - b_{21}^Z \frac{(2Z+N)A_u + (Z+2N)A_d}{2(s_{tw}c_{tw})^2} \right|^2, \tag{32}$$

$$b_{21}^V = \frac{s_R c_R}{2\sqrt{2}(4\pi)^2} \sum_{k=1}^{2} (y'_\ell)_{\alpha k} m_{L'_k} (y_{L'})_{k,j} \left[F_{\mu e}(m_{H_1}, m_{L'_k}, m_V) - F_{\mu e}(m_{H_2}, m_{L'_k}, m_V) \right], \tag{33}$$

$$F_{\mu e}(m_1, m_2, m_3) = \int_0^1 dx_3 \int_0^{1-x_3} dx_4 \frac{x_4(1-x_4)}{x_3 m_1^2 + (1-x_3)m_2^2 + x_4(x_3+x_4-1)m_3^2}, \tag{34}$$

where $C_{\mu e} \equiv 4\alpha_{\text{em}}^5 \frac{Z_{\text{eff}}^4 |F(q)|^2 m_\mu^5}{Z}$, $A_u \equiv -\frac{1}{2} - \frac{4}{3}s_{tw}^2$, $A_d \equiv -\frac{1}{2} + \frac{2}{3}s_{tw}^2$, $\sin^2\theta_w \equiv s_{tw}^2 \approx 0.23$, $V \equiv (\gamma, Z)$, $m_\gamma = 0$, and $m_Z \approx 91.19$ GeV. The values of Γ_{capt}, Z, N, Z_{eff}, and $F(q)$ depend on the type of nuclei. Here we consider Titanium case, because its sensitivity will be improved by several orders of magnitude [8] in near future compared to the current bound [7], as can be seen in the Table 3. In this case, we apply $\Gamma_{\text{capt}} = 2.59 \times 10^6$ sec^{-1}, $Z = 22$, $N = 26$, $Z_{\text{eff}} = 17.6$, and $|F(-m_\mu^2)| = 0.54$ [13].

2.4. Numerical analysis

Now that we have all the formulae for the quark and lepton sector, we carry out numerical analysis to search for the preferred parameter range. Then we randomly select values of the fifteen parameters within the corresponding ranges

$$\begin{aligned}
&m_{\eta_I} \in [250\text{ GeV}, 500\text{ GeV}], \quad m_{H_1} \in [600\text{ GeV}, 800\text{ GeV}], \quad m_{H_2} \in [4\text{ TeV}, 6\text{ TeV}], \\
&m_{Q'_1} \in [4.5\text{ TeV}, 5\text{ TeV}], \quad m_{Q'_2} \in [1.7\text{ TeV}, 2.2\text{ TeV}], && (35)\\
&m_{L'_1} \in [7\text{ TeV}, 7.5\text{ TeV}], \quad m_{L'_2} \in [9\text{ TeV}, 10\text{ TeV}], && (36)\\
&\{(y_u)_{12}, (y_u)_{22}\} \in [-1, 1], \quad (y_u)_{11} \in [-0.1, 0.1], \quad (y_u)_{21} \in [-0.002, 0.002], \\
&\{(y_\nu)_{21}, (y_\nu)_{22}\} \in [-7 \times 10^{-12}, 7 \times 10^{-12}], \quad (y_\nu)_{11} \in [-3 \times 10^{-13}, 3 \times 10^{-13}], \\
&(y_\nu)_{12} \in [-2 \times 10^{-13}, 2 \times 10^{-12}], && (37)
\end{aligned}$$

to reproduce quark and charged lepton masses, CKM matrix, and to fit neutrino oscillation data while satisfying the constraints of LFVs and FCNC. Note that the other Yukawa couplings such as y_Q, y_d, y_L, y_ℓ are numerically solved by using the best fit values of the measurements for quark [14] and lepton sector [15]. Generating randomly 10^7 sample points, we obtain 734 allowed points where the most stringent constraint comes from the process of $\mu \to e\gamma$. We also show some indication of the model in Fig. 2; the left figure is the scattering plot for Δm_D and Δm_K normalized by [MeV]$\times 10^{12}$, and the black solid lines represents the experimental upper bound while the right figure is the scattering plot for $\Delta a_\mu \times 10^{12}$ and $R_{Ti} \times 10^{17}$. Our Δm_K values are the same order as our phenomenological constraint and some parameter sets are excluded, and Δm_D values can be comparable to the constraint, which can be tested in the near future.

IOP Conf. Series: Journal of Physics: Conf. Series **873** (2017) 012033 doi:10.1088/1742-6596/873/1/012033

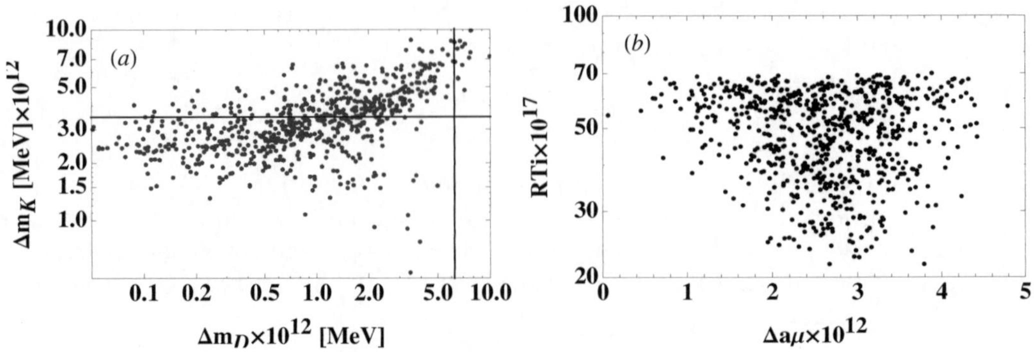

Figure 2. (a): The scattering plot in terms of Δm_D and Δm_K normalized by [MeV]$\times 10^{12}$, and the black solid lines represents the observed values. (b): The scattering plot in terms of $\Delta a_\mu \times 10^{12}$ and $R_{Ti} \times 10^{17}$.

Table 4. The charge assignments of lepton and scalar field sector in model II.

Fieldes	L_L^α	e_R^i	τ_R	$L'^i_{L,R}$	N_R^i	N_L^i	Φ	φ	η	Φ_2	S	S^-
$SU(2)_L$	**2**	**1**	**1**	**2**	**1**	**1**	**2**	**1**	**2**	**2**	**1**	**1**
$U(1)_Y$	$-\frac{1}{2}$	-1	-1	$-\frac{1}{2}$	0	0	$\frac{1}{2}$	0	$\frac{1}{2}$	$\frac{1}{2}$	0	-1
$U(1)_R$	0	$-x$	0	0	x	0	0	x	x	$-x$	0	0
Z_2	$+$	$+$	$+$	$-$	$-$	$-$	$+$	$+$	$-$	$+$	$-$	$+$

We also find that the maximal value for $(g-2)_\mu$ is around 5×10^{-12}, which is lower than the current bound by three order of magnitude, and that R_{Ti} is also much lower than the current bound, however, it will be tested in the future experiment such as COMET [8], which will reach $R \approx 10^{-18}$.

3. Model II

In this section, we extend the model in previous section and discuss neutrino mass matrix which is generated at two-loop level [2].

3.1. Structure of the model

The structure of the model is similar to that in previous section where the applied symmetry is the same. The lepton sector and scalar filed sectors are extended as in the Table. 4; we assigned Z_2 odd parity to right-handed neutrino to forbid neutrino mass at tree and one-loop level, and introduced new gauge singlet fermion N_L, Higgs doublet with U(1)$_R$ charge Φ_2 and charged singlet scalar S^-. Note that we do not have Dirac mass term of SM neutrino due to Z_2 parity while $N_{R,L}$ has Dirac mass via VEV of φ. In our extension, we have two physical charged scalar components and the mass eigensates are given by

$$\begin{bmatrix} S^\pm \\ \phi_2^\pm \end{bmatrix} \equiv \begin{bmatrix} \cos C & \sin C \\ -\sin C & \cos C \end{bmatrix} \begin{bmatrix} H_1^\pm \\ H_2^\pm \end{bmatrix}. \tag{38}$$

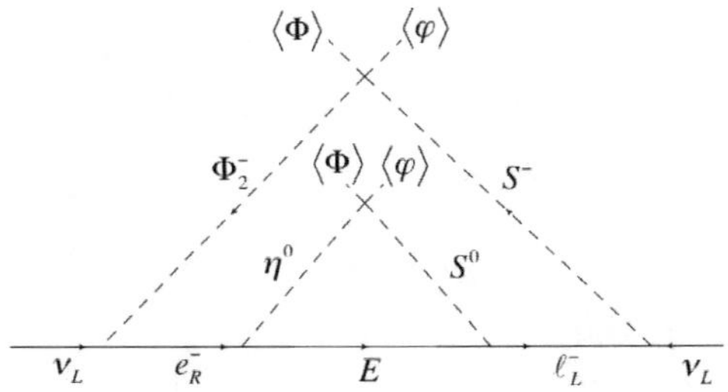

Figure 3. The two loop diagram for generating Majorana mass term for active neutrinos.

The new terms in Lagrangian of the model II are

$$-\mathcal{L} \supset (y'_\ell)_{\alpha j}\bar{L}_\alpha \Phi_2 e_{R_j} + (y'_S)_{\alpha\beta}\bar{L}^C_\alpha (i\sigma_2) L_\beta S^+ - \lambda_1[\Phi^T(i\sigma_2)\Phi_2]\varphi S^- + \text{c.c.}, \tag{39}$$

and Yukawa interactions with y_ν coupling in Eq. (2) are now forbidden by Z_2 symmetry.

3.2. Neutrino mass matrix

As a consequence of the extending the model, the dominant contribution to the active neutrino mass matrix m_ν is now given at two-loop level where the corresponding diagram is shown in Fig. 3. Calculating the diagram, we then obtain the analytic form of neutrino mass matrix:

$$(m_\nu)_{ab} = -s_R c_R s_C c_C \frac{(y_\ell)_{ai}(y'^\dagger_\ell)_{ij}\mathcal{M}_{E_j}(y^\dagger_S)_{jk}(y'^\dagger_S)_{kb} + [(y_\ell)_{ai}(y'^\dagger_\ell)_{ij}\mathcal{M}_{E_j}(y^\dagger_S)_{jk}(y'^\dagger_S)_{kb}]^T}{2(4\pi)^4}, \tag{40}$$

$$\begin{aligned}
&\mathcal{M}_{E_j} \equiv m_{L'_j}\int[dx]\int[dX]\frac{y}{(x_4-1)^2}\ln\left(\frac{\Delta_3[,E_j,H_1^\pm,H_2]\Delta_3[E_j,H_2^\pm,H_1]}{\Delta_3[E_j,H_1^\pm,H_1]\Delta_3[E_j,H_2^\pm,H_2]}\right),\\
&\Delta_3[a,b,c] \equiv x m_a^2 - \frac{y(x_3 m_b^2 + x_4 m_c^2)}{x_4^2 - x_4}, \quad \int[dx] \equiv \int_0^1 dx \int_0^{1-x} dy\,\delta(1-x-y),\\
&\int[dX] \equiv \int_0^1 dx_1 \int_0^{1-x_1} dx_2 \int_0^{1-x_1-x_2} dx_3 \int_0^{1-x_1-x_2-x_3} x_4\delta(1-x_1-x_2-x_3-x_4).
\end{aligned} \tag{41}$$

The neutrino mass matrix $(m_\nu)_{ab}$ can generally be diagonalized by use of the mixing matrix $V_{\rm MNS}$ (PMNS) and can be written in terms of experimental values depending on the normal hierarchy (NH) and inverted hierarchy (IH) as follows:

$$(m_\nu)^{exp.}_{ab} = (V_{\rm MNS} D_\nu V^T_{\rm MNS})_{ab}, \quad D_\nu \equiv (m_{\nu_1}, m_{\nu_2}, m_{\nu_3})$$

$$(\text{NH}):\ |m^{exp.}_\nu| \approx \begin{bmatrix} 0.0845-0.475 & 0.0629-0.971 & 0.0411-0.964 \\ * & 1.44-3.49 & 1.94-2.85 \\ * & * & 1.22-3.33 \end{bmatrix} \times 10^{-11}\ \text{GeV}, \tag{42}$$

$$(\text{IH}):\ |m^{exp.}_\nu| \approx \begin{bmatrix} 0.993-4.96 & 0.00261-3.83 & 0.00280-3.95 \\ * & 0.00380-3.08 & 0.345-2.61 \\ * & * & 0.000647-3.30 \end{bmatrix} \times 10^{-11}\ \text{GeV}, \tag{43}$$

$$V_{\rm MNS} = \begin{bmatrix} c_{13}c_{12} & c_{13}s_{12} & s_{13}e^{-i\delta} \\ -c_{23}s_{12} - s_{23}s_{13}c_{12}e^{i\delta} & c_{23}c_{12} - s_{23}s_{13}s_{12}e^{i\delta} & s_{23}c_{13} \\ s_{23}s_{12} - c_{23}s_{13}c_{12}e^{i\delta} & -s_{23}c_{12} - c_{23}s_{13}s_{12}e^{i\delta} & c_{23}c_{13} \end{bmatrix} \begin{bmatrix} e^{i\alpha_1/2} & 0 & 0 \\ 0 & e^{i\alpha_2/2} & 0 \\ 0 & 0 & 1 \end{bmatrix}, \tag{44}$$

IOP Conf. Series: Journal of Physics: Conf. Series **873** (2017) 012033 doi:10.1088/1742-6596/873/1/012033

where we have used the following neutrino oscillation data at 3σ [15]:

$$\begin{aligned}
&(\mathrm{NH}): 0.278 \le s_{12}^2 \le 0.375,\ 0.392 \le s_{23}^2 \le 0.643,\ 0.0177 \le s_{13}^2 \le 0.0294,\ \delta \in [-\pi,\pi], \\
&\qquad m_{\nu_3} = (\sqrt{23.0} - \sqrt{26.5}) \times 10^{-11}\ \mathrm{GeV},\quad m_{\nu_2} = (\sqrt{0.711} - \sqrt{0.818}) \times 10^{-11}\ \mathrm{GeV}, \quad (45)\\
&(\mathrm{IH}): 0.278 \le s_{12}^2 \le 0.375,\ 0.403 \le s_{23}^2 \le 0.640,\ 0.0183 \le s_{13}^2 \le 0.0297,\ \delta \in [-\pi,\pi], \\
&\qquad m_{\nu_1} = (\sqrt{22} - \sqrt{25.4}) \times 10^{-11}\ \mathrm{GeV},\quad m_{\nu_2} = (\sqrt{22.711} - \sqrt{26.218}) \times 10^{-11}\ \mathrm{GeV}, \quad (46)
\end{aligned}$$

and the two Majorana phases $\alpha_{1,2}$ are taken to be $\alpha_{1,2} \in [-\pi, \pi]$ in both cases. Notice that we assume one of three neutrino masses is zero, which is predicted by the theoretical aspect that y'_S is anti-symmetric matrix, i.e. the rank of neutrino mass matrix is reduced to two. Applying the above properties, we can rewrite two components of y'_S by experimental values and one of the component of y'_S as follows [1]:

$$(\mathrm{NH}):\ (y'_S)_{e\tau} = \left(\frac{s_{12}c_{23}}{c_{12}c_{13}} + \frac{s_{13}s_{23}}{c_{13}} e^{-i\delta}\right)(y'_S)_{\mu\tau},\quad (y'_S)_{e\mu} = \left(\frac{s_{12}c_{23}}{c_{12}c_{13}} - \frac{s_{13}s_{23}}{c_{13}} e^{-i\delta}\right)(y'_S)_{\mu\tau}, \tag{47}$$

$$(\mathrm{IH}):\ (y'_S)_{e\tau} = -\frac{s_{23}}{t_{13}}(y'_S)_{\mu\tau},\quad (y'_S)_{e\mu} = \frac{c_{23}}{t_{13}}(y'_S)_{\mu\tau}, \tag{48}$$

where the explicit form of y'_S is given as

$$(y'_S) \equiv \begin{bmatrix} 0 & (y'_S)_{e\mu} & (y'_S)_{e\tau} \\ -(y'_S)_{e\mu} & 0 & (y'_S)_{\mu\mu} \\ -(y'_S)_{e\tau} & -(y'_S)_{\mu\tau} & 0 \end{bmatrix}. \tag{49}$$

In our numerical analysis, only $(y'_S)_{\mu\tau}$ is an input parameter, and we will search for the allowed parameter region by comparing the experimental values in Eqs. (42) and (43).

3.3. Numerical analysis

Finally, we carry out numerical analysis to search for the parameter region that can fit the experimental data. To reduce the number of free parameters, we first fix some parameters such that; $s_\beta = 0.3$, $s_R = 0.1$, $s_C = 1$, $m_{h_1} = 4$ TeV. Then we randomly select values of masses as $M_X \equiv m_{H_1} \in [0, 10]$ TeV, $m_{H_2^\pm} \in [11M_X/10, 12M_X/10]$ TeV, $m_{L'_1} \in [11M_X/10, 9]$ TeV, $m_{L'_2} \in [m_{L'_1}, 9.5]$ TeV, $m_{L'_3} \in [m_{L'_2}, 10]$ TeV, and $m_{\eta^\pm} = m_{H_2} \in [11M_X/10, 12(11.1)M_X/10]$. Then, we randomly select values of the dimensionless parameters within $[-1, 1]$ to search for allowed parameter region. As a result we find the allowed ranges for our Yukawa couplings in both NH and IH cases:

$$(y'_S)_{\mu\tau} \in [-0.027, 0.027], \tag{50}$$

$$\begin{aligned}
y_L &= \begin{bmatrix} 0.01 - 0.135 & -(0.025 - 0.02) & -(0.0265 - 0.02) \\ 0.03 - 0.036 & 0.01 - 0.015 & -(0.045 - 0.04) \\ -(0.027 - 0.02) & -(0.0208 - 0.02) & -(0.028 - 0.02) \end{bmatrix}, \\
y'_\ell &= \begin{bmatrix} -(0.016 - 0.01) & 0.04 - 0.0453 & -(0.02 - 0.01) \\ -(0.0063 - 0.001) & 0.04 - 0.05 & -(0.031 - 0.01) \\ 0.001 - 0.0012 & -(0.0174 - 0.01) & 0.03 - 0.036 \end{bmatrix}, \\
y_\ell &= \begin{bmatrix} 0.03 - 0.037 & -(0.0445 - 0.04) & 0.001 - 0.0088 \\ 0.01 - 0.0123 & 0.02 - 0.0275 & -(0.032 - 0.01) \\ 0.01 - 0.03 & 0.001 - 0.0094) & 0.001 - 0.005 \end{bmatrix},
\end{aligned} \tag{51}$$

[1] The detailed analysis for both hierarchies can be found in Ref. [16].

IOP Conf. Series: Journal of Physics: Conf. Series **873** (2017) 012033 doi:10.1088/1742-6596/873/1/012033

which can reproduce neutrino oscillation data and satisfies the constraints from LFVs Therefore we find that the hierarchy in the Yukawa couplings becomes smaller compared to the SM case. Note that the allowed parameter region can accommodate observed relic density of DM assuming H_1 to be our DM where detailed discussion can be found in Ref. [2].

4. Summary and discussion

We have reviewed two models in which light fermion masses are generated at loop levels. The Yukawa couplings with SM Higgs are restricted by introducing extra U(1) gauge symmetry and discrete Z_2 symmetry. Then first and second generation charged-lepton and quark masses are generated at one-loop level by introducing vector-like fermions and exotic scalars. In the first model, the SM neutrino masses are generated at one-loop level as the same way as other fermion sectors, which require tiny Yukawa coupling constants. On the other hand, the neutrino masses are generated at two-loop level in the second model which is the extension of first one.

We presented the formulas for fermion mass matrices, lepton flavor violations and flavor changing neutral current effect on meson anti-meson (Q-$\bar{Q}$) mixing. Then numerical calculation is carried out to search for the parameter range to fit the observed data. We find that our Yukawa couplings can be less hierarchical compared to SM case due to loop induced mass for first two generations. In particular, in the second model, we can generate tiny neutrino mass with $\mathcal{O}(10^{-3})$ to $\mathcal{O}(1)$ Yukawa coupling constants through two-loop level effect. We also show some implication of the model in μ-e conversion and Q-$\bar{Q}$ mixing which can be tested in future experiment. In addition, collider physics of our exotic particles will be interesting and it is left as future works.

References

[1] Nomura T, Okada H 2016 *Phys. Lett.* B **761** 190
[2] Nomura T, Okada H 2016 *Phys. Rev.* D **94** no. 9 093006
[3] Gabbiani F, Gabrielli E, Masiero A, Silvestrini L 1996 *Nucl. Phys.* B **477** 321
[4] Okada H, Okada N, Orikasa Y, Yagyu K 2016 *Phys. Rev.* D **94** no. 1 015002
[5] Baldini A M *et al.* [MEG Collaboration] 2016 *Eur. Phys. J.* C **76** no. 8 434
[6] Adam J, *et al.* [MEG Collaboration] 2013 *Phys. Rev. Lett.* **110** 201801
[7] Dohmen C, *et al.* [SINDRUM II Collaboration] 1993 *Phys. Lett.* B **317** 631
[8] Hungerford E V [COMET Collaboration] 2009 *AIP Conf. Proc.* **1182** 694
[9] Bennett G W, *et al.* [Muon G-2 Collaboration] 2006 *Phys. Rev.* D **73** 072003
[10] Jegerlehner F, Nyffeler A 2009 *Phys. Rept.* **477** 1
[11] Benayoun M, David P, Delbuono L, Jegerlehner F 2012 *Eur. Phys. J.* C **72** 1848
[12] Hisano J, Moroi T, Tobe K, Yamaguchi M 1996 *Phys. Rev.* D **53** 2442
[13] Alonso R, Dhen M, Gavela M B, Hambye T 2013 *JHEP* **1301** 118
[14] Olive K A, *et al* (Particle Data Group) 2014 *Chin. Phys.* C **38** 090001
[15] Forero D V, Tortola M, Valle J W F 2014 *Phys. Rev.* D **90** no. 9 093006
[16] Herrero-Garcia J, Nebot M, Rius N, Santamaria A 2014 *Nucl. Phys.* B **885** 542

IOP Conf. Series: Journal of Physics: Conf. Series **873** (2017) 012034 doi:10.1088/1742-6596/873/1/012034

A simple criterium for CP conservation in the most general 2HDM

I. F. Ginzburg,
Sobolev Institute of Mathematics, Novosibirsk, 630090, Russia;
Novosibirsk State University, Novosibirsk, 630090, Russia,
M. Krawczyk,
Faculty of Physics, University of Warsaw,
ul. Pasteura 5, 02-093 Warsaw, Poland

Abstract. We find a set of necessary and sufficient conditions for the CP conservation in the most general 2HDM in terms of observable quantities. This set contains two simple and relatively easily testable conditions instead of usually discussed more complex conditions.

1. Introduction

CP violation is one of the important yet not well understood aspect of the fundamental physics. Modern LHC data [1] allow to conclude that the observed particle $h(125)$ is Higgs boson with spin-CP parity 0^{++} only under assumption that this particle has a definite parity. Generally it can have no definite parity, what happens in many models. In this case mentioned data give no information about $h(125)$ parity [2].

In the Standard Model the CP violation is described by the CKM matrix, but its origin remains unclear. The extension of SM with two Higgs doublets, called Two Higgs Doublet Model (2HDM), has been introduced in 1974 with the main aim for providing an extra source of CP violation [3]. Later, many variants of 2HDM were considered with different features, providing different physical realizations (see for review [6, 7]). Each of these variants is described by Lagrangian with many parameters. The choice of the set of these parameters, describing the same physical reality, (choice of basis in the parameter space) is ambiguous. In the non-minimal models, like 2HDM, the observed situation with properties of Higgs boson similar to those in the SM (*SM-like scenario* [4] or *alignment limit* [5]) can be described by different non-equivalent sets of parameters.

We discuss two type of problems which appear in the study of CP violation in these models.

(A) Let we have a variant of 2HDM, constructed as a model for description of some set of phenomena.
How to know whether CP symmetry is violated or not in this model without detailed calculations of each particular effect? Here the basis independent recipe is desirable.

(B) Consider 2HDM as an approximation in the description of Nature. How do we know whether CP violation is described by this approximation or the observed CP violation is an effect of some weaker interactions.
How to check CP violation in the experiments with particles, which appear in 2HDM?

2. 2HDM

The 2HDM describes a system of two spinless isospinor fields ϕ_1, ϕ_2 with hypercharge $Y = 1$. The most general form of the 2HDM potential is as follows

$$\begin{aligned} V = & -\frac{m_{11}^2}{2}\left(\phi_1^\dagger\phi_1\right) - \frac{m_{22}^2}{2}\left(\phi_2^\dagger\phi_2\right) - \left[\frac{m_{12}^2}{2}\left(\phi_1^\dagger\phi_2\right) + \text{h.c.}\right] \\ & + \frac{\lambda_1}{2}\left(\phi_1^\dagger\phi_1\right)^2 + \frac{\lambda_2}{2}\left(\phi_2^\dagger\phi_2\right)^2 + \lambda_3\left(\phi_1^\dagger\phi_1\right)\left(\phi_2^\dagger\phi_2\right) + \lambda_4\left(\phi_1^\dagger\phi_2\right)\left(\phi_2^\dagger\phi_1\right) \\ & + \left[\frac{\lambda_5}{2}\left(\phi_1^\dagger\phi_2\right)^2 + \lambda_6\left(\phi_1^\dagger\phi_1\right)\left(\phi_1^\dagger\phi_2\right) + \lambda_7\left(\phi_2^\dagger\phi_2\right)\left(\phi_1^\dagger\phi_2\right) + \text{h.c.}\right]. \end{aligned} \tag{1}$$

The potential parameters are restricted by the requirement that the potential should be positive at large quasiclassical values of ϕ_i (*positivity constraints*). We assume also that these coefficients are not too big so that one can use estimates based on the lowest non-trivial approximation of the perturbation theory.

After EWSB the 2HDM contains 3 neutral Higgs bosons $h_a \equiv h_{1,2,3}$ (in general with indefinite CP parity) and charged Higgs bosons $H^\pm$ with masses M_a and $M_\pm$, respectively (*the numbering of h_a is independent on an order of masses M_a*).

2.1. Reparametrization freedom

2HDM describes system of two fields with identical quantum numbers. Therefore, its description in terms of original fields ϕ_i or in terms of their linear superpositions ϕ_i' are equivalent; this statement verbalizes a *reparameterization* (RPa) freedom of the model. The RPa group consists of RPa transformations $\hat{\mathcal{F}}$ of the form:

$$\begin{pmatrix}\phi_1'\\ \phi_2'\end{pmatrix} = \hat{\mathcal{F}}_{gen}(\theta,\tau,\rho)\begin{pmatrix}\phi_1\\ \phi_2\end{pmatrix}, \quad \hat{\mathcal{F}}_{gen} = e^{i\rho_0}\begin{pmatrix}\cos\theta\, e^{i\rho/2} & \sin\theta\, e^{i(\tau+\rho/2)}\\ -\sin\theta\, e^{-i(\tau+\rho/2)} & \cos\theta\, e^{-i\rho/2}\end{pmatrix}. \tag{2}$$

This transformation induces a transformation of the parameters of the Lagrangian in such a way that the new Lagrangian, written in terms of fields ϕ_i', describes the same physical content. We refer to these different choices as - the different RPa bases. A subgroup of the RPa group – the *rephasing* group RPh – describes a freedom in choice of the relative phase of fields ϕ_i.

Transformation (2) is parameterized by angles θ, ρ, τ and ρ_0. The parameter ρ_0 describes an overall phase transformation of the fields and can be ignored since it does not affect the parameters of the potential. The parameter ρ describes the RPh symmetry transformation of system.

The $U(1)_{EM}$ symmetry preserving ground state of this system is given by a global minimum of the potential and reads

$$\langle\phi_1\rangle = \begin{pmatrix}0\\ v_1/\sqrt{2}\end{pmatrix}, \quad \langle\phi_2\rangle = \begin{pmatrix}0\\ v_2 e^{i\xi}/\sqrt{2}\end{pmatrix}, \tag{3}$$

with a standard parameterization $\tan\beta = v_2/v_1$, or $v_1 = v\cos\beta$, $v_2 = v\sin\beta$.

2.2. The Higgs basis

We use below the RPa basis with $v_2 = 0$ (the Higgs, or Georgi, basis [8]), in which the 2HDM potential can be written in the form [9]

$$\begin{aligned} V_{HB} = & M_\pm^2\left(\Phi_2^\dagger\Phi_2\right) + \frac{\Lambda_1}{2}\left(\Phi_1^\dagger\Phi_1 - \frac{v^2}{2}\right)^2 + \frac{\Lambda_2}{2}\left(\Phi_2^\dagger\Phi_2\right)^2 + \Lambda_3\left(\Phi_1^\dagger\Phi_1 - \frac{v^2}{2}\right)\left(\Phi_2^\dagger\Phi_2\right) \\ & + \Lambda_4\left(\Phi_1^\dagger\Phi_2\right)\left(\Phi_2^\dagger\Phi_1\right) + \left[\frac{\Lambda_5}{2}\left(\Phi_1^\dagger\Phi_2\right)^2 + \Lambda_6\left(\Phi_1^\dagger\Phi_1 - \frac{v^2}{2}\right)\left(\Phi_1^\dagger\Phi_2\right) + \Lambda_7\left(\Phi_2^\dagger\Phi_2\right)\left(\Phi_1^\dagger\Phi_2\right) + \text{h.c.}\right]. \end{aligned} \tag{4}$$

IOP Conf. Series: Journal of Physics: Conf. Series **873** (2017) 012034 doi:10.1088/1742-6596/873/1/012034

For this basis we use capital letters to denote fields and parameters of potential, Φ_i and Λ_j respectively.

3. Relative couplings

In the discussion below we use the relative couplings for each neutral Higgs boson[1] h_a (in 2HDM $a = 1, 2, 3$):

$$\chi_a^P = \frac{g_a^P}{g_{\text{SM}}^P}, \quad \chi_a^\pm = \frac{g(H^+H^-h_a)}{2M_\pm^2/v}, \quad \chi_a^{H^+W^-} = \frac{g(H^+W^-h_a)}{M_W/v}. \tag{5}$$

The quantities χ_a^P (where $P = V\,(W, Z)$, $q = t, b, ...$, $\ell = \tau, ...$) are the ratios of the couplings of h_a with the fundamental particles P to the corresponding couplings for the would be SM Higgs boson with mass M_a. The other relative couplings describe interaction of h_a with the charged Higgs boson $H^\pm$. Couplings χ_a^V and $\chi_a^\pm$ are real due to Hermiticity of Lagrangian, and are directly measurable. Couplings $\chi_a^{H^+W^-}$, χ_a^q and χ_a^ℓ are generally complex.

There are useful sum rules among these couplings, namely

$$(a)\ \sum_a (\chi_a^V)^2 = 1, \quad (b)\ (\chi_a^V)^2 + |\chi_a^{H^+W^-}|^2 = 1. \tag{6}$$

Both real and imaginary parts of Yukawa couplings χ_a^q and χ_a^ℓ can be measured in principle, using distributions of Higgs bosons decay products in $h_a \to \bar{q}q$, $h_a \to \bar{\ell}\ell$. The absolute value of the coupling $\chi_a^{H^+W^-}$ is fixed by the sum rule (6 b) and is well measurable.

The unitarity of the rotation matrix describing transition from components of fields ϕ_i to the physical Higgs fields h_a, allows to obtain the following relations for couplings $\chi_a^{H^+W^-}$ (the factor $e^{i\rho}$ represents the rephasing freedom in the Higgs basis):

$$\begin{aligned}
\chi_1^{H^+W^-} &\equiv \left(\chi_1^{H^-W^+}\right)^* &&= -e^{i\rho}\,\frac{\chi_1^V\chi_2^V - i\chi_3^V}{\sqrt{1-(\chi_2^V)^2}},\\
\chi_2^{H^+W^-} &\equiv \left(\chi_2^{H^-W^+}\right)^* &&= e^{i\rho}\,\sqrt{1-(\chi_2^V)^2},\\
\chi_3^{H^+W^-} &\equiv \left(\chi_3^{H^-W^+}\right)^* &&= -e^{i\rho}\,\frac{\chi_2^V\chi_3^V + i\chi_1^V}{\sqrt{1-(\chi_2^V)^2}}.
\end{aligned} \tag{7}$$

Note, that here we discuss couplings which appear in the Lagrangian. Radiative corrections (RC) change these couplings, however in most cases these corrections are small and therefore the corresponding observable differ weakly from those presented in the Lagranian. In this sense we treat the latter ones as being measurable. Therefore the identities (6), (7) are valid with accuracy to RC.

For the interaction of Higgs bosons with the gauge bosons $V = W, Z$ one should distinguish interactions of different structure, that are vectorial $g_a^V h_a V_\mu V^\mu$, tensor $g_{aT} h_a V_{\mu\nu} V^{\mu\nu}$ and axial-tensor $\tilde{g}_{aT} h_a \tilde{V}_{\mu\nu} V^{\mu\nu}$ with corresponding coupling constants g_a^V, g_{aT}^V and $\tilde{g}_{aT}^V$, respectively (where $V_{\mu\nu} = \partial_\mu V_\nu - \partial_\nu V_\mu$, $\tilde{V}_{\mu\nu} = \epsilon_{\mu\nu\alpha\beta} V^{\alpha\beta}$). These interactions can be separated in the experiment from each other by the study of angular correlations in the decays like $h_a \to ZZ \to e^+e^-\mu^+\mu^-$ [1]. **In the paper we concentrate on vectorial couplings appearing in Lagrangian** (and modified weakly by RC).

In the 2HDM tensor and axial-tensor interactions appear only due to RC (mainly from t-loops). In the most of cases they give very small contributions to the observed decays rates of h_a and hardly observable.

[1] We omit the adjective "relative" below.

The axial-tensor interaction can imitate CP violation. This interaction was studied in ref. [11] for CP conserved variant of 2HDM, in which one of neutral Higgs bosons is pseudoscalar A. It was found that the corresponding decay $A \to e^+e^-\mu^+\mu^-$, etc. can be observable in very narrow range of parameters only ($M_A \approx 2m_t$, low $\tan\beta$). If it will be realized, the study of angular correlation of leptons allows to distinguish CP odd nature of initial state (as it is done currently at LHC in the study of CP properties of the Higgs boson [1]) without violation of CP symmetry.

4. A minimal complete set of observables in 2HDM

In ref. [9] a *minimal complete* set of *directly measurable quantities (observables)* defining the 2HDM was found ($a = 1, 2, 3$), namely :

$$\begin{array}{l} (1)\ \textit{v.e.v. of Higgs field}\ v = 246\ GeV,\ \ (2\text{-}5)\ \textit{masses of Higgs bosons}\ M_a,\ M_\pm,\\ (6\text{-}7)\ \textit{2 (out of 3) couplings}\ \chi_a^V,\ \ (8\text{-}10)\ \textit{3 couplings}\ \chi_a^\pm,\\ (11)\ \textit{quartic coupling}\ g(H^+H^-H^+H^-). \end{array} \tag{8}$$

In the most general 2HDM, these 11 observables are independent of each other. In particular variants of 2HDM additional relations between these parameters may appear.

The parameters of potential in the Higgs basis are expressed through these observables and free parameter ρ (it appears in $\Lambda_{5,6,7}$ via couplings $\chi_a^{H^+W^-}$ given in eq. (7)):

$$\begin{array}{l} \Lambda_1 = \sum\limits_a (\chi_a^V)^2 M_a^2/v^2\,; \quad \Lambda_5 = \sum\limits_a (\chi_a^{H^-W^+})^2 M_a^2/v^2;\\ \Lambda_4 = (\sum\limits_a M_a^2 - M_\pm^2)/v^2 - \Lambda_1; \quad \Lambda_3 = 2\left(M_\pm^2/v^2\right)\sum\limits_a \chi_a^V \chi_a^\pm;\\ \Lambda_6 = \sum\limits_a \chi_a^V \chi_a^{H^-W^+} M_a^2/v^2; \quad \Lambda_7 = 2\left(M_\pm^2/v^2\right)\sum\limits_a \chi_a^{H^-W^+}\chi_a^\pm\,;\\ \Lambda_2 = 2g(H^+H^-H^+H^-)\,. \end{array} \tag{9}$$

The parameters of the potential (1) with the defined values of $\tan\beta$ and ξ (3) are obtained from parameters (9) with the aid of transformation (2) in the following form

$$\begin{pmatrix}\phi_1\\ \phi_2\end{pmatrix} = \hat{\mathcal{F}}_{HB}\begin{pmatrix}\Phi_1\\ \Phi_2\end{pmatrix}, \quad \hat{\mathcal{F}}_{HB} = \hat{\mathcal{F}}_{gen}(\theta = -\beta, \tau = \xi, -\rho). \tag{10}$$

5. Conditions for a CP conservation

In the models like 2HDM neutral scalar particles coincide with their antiparticles. Therefore, in such models one can discuss the P-parity violation, but not C-parity. When in addition we consider fermions, this P violation is transformed to the CP violation (see e.g. [6, 7]).

5.1. Basic facts

The CP symmetry is conserved in the 2HDM and similar models containing spinless bosons if [2]

- *Each observable physical neutral spinless boson has definite P-parity* (11a)
- *There are no P-violating interactions between these bosons.* (11b)

In the 2HDM we denote neutral spinless P-even bosons as h_1, h_2 (they are called often as h and H) and P-odd boson as h_3 (it is called often as A). The condition (11b) means an absence of the interactions[3] ($i, j, k = 1, 2$)

$$h_i h_j h_3, \quad h_3 h_3 h_3, \quad h_i h_j h_k h_3, \quad h_i h_3 h_3 h_3. \tag{12}$$

[2] Note that CP violation, obliged by fermions and (or) vector bosons, appears in the interactions of spinless particles via radiative correction.

[3] Similar conditions were discussed in [12].

It is well known that in the 2HDM CP conservation holds if

There exists the RPa basis in which:
- *all parameters of potential are real,* (13a)
- *relative phase* (3) $\xi = 0$. (13b)

In is worth mentioning, that the condition (13a) forbids an explicit CP violation while conditions (13a) and (13b) together forbid a spontaneous CP violation.

The statement (11) only describes CP-conservation, but does not provide a criterion for CP conservation in the considered model. The description (13) is RPa basis-dependent. Below we discuss both (A) and (B) set-ups of our problem, presented in the Introduction.

Many authors consider solution of the problem (A) as a necessary step in solving the problem (B). Our approach is different – we look for a solution of the problem (B), with a solution of problem (A) appearing as a by-product.

5.2. Method of the CP-odd basis-independent invariants

Many author found the basis independent criterium for CP violation conservation in terms of parameters of the Higgs potential – this corresponds to the problem (A) stated in the Introduction. For this goal they constructed the RPa-invariant CP-odd combinations of parameters of the potential and as a condition for CP conservation they demand vanishing of all these invariants. For 2HDM three such invariants $Im\mathcal{J}_{1,2,3}$ were found in ref. [10], the procedure for construction of such invariants for multi-Higgs models is developed in ref. [13].

To solve the corresponding problem (B) the invariants [10] for 2HDM were expressed via measurable quantities in refs. [14, 15]. In the terms of quantities (8), the corresponding conditions for the CP conservation read as

$$\begin{aligned} Im\mathcal{J}_1 &= \sum_{i,j,k} \varepsilon_{ijk} \frac{2M_i^2 M_\pm^2}{v^4} \chi_i^V \chi_k^V \chi_j^\pm = 0\,, \quad & Im\mathcal{J}_2 &= 2\chi_1^V \chi_2^V \chi_3^V \sum_{i,j,k} \varepsilon_{ijk} \frac{M_i^4 M_k^2}{v^6} = 0\,, \\ Im\mathcal{J}_{30} &= 4 \sum_{i,j,k} \varepsilon_{ijk} \mathcal{T}_i \chi_j^V \chi_k^\pm = 0\,, & \text{where } \mathcal{T}_i &= \frac{(M_\pm^2 \chi_i^\pm + M_i^2 \chi_i^V) M_i^2 M_\pm^2}{v^6}\,. \end{aligned} \tag{14}$$

Note that in this approach there are four CP-odd invariants but one should check vanishing only of two of them (see e.g. [7]). Since the choice of these two invariants is not fixed from beginning, the presented set contains three conditions, instead of necessary two. In our opinion the equations (14) are too complicated and their experimental verification, discussed in [14, 15, 16], requires very complex procedure.

5.3. A direct criterium for CP conservation

In the direct method we approach the mentioned above problem (B) directly. We start with a description of CP conservation (11) and use only observables, which by definition are basis independent.

In Refs. [17], [18] we proposed conditions for CP conservation, based only on one condition[4] (11a), without checking up of condition (11b), in the form

$$\prod_a \chi_a^V = 0\,, \quad \prod_a \chi_a^\pm = 0\,, \quad \left|\prod_a \chi_a^f\right| = \prod_a \left|\chi_a^f\right|\,. \tag{15}$$

[4] Particular version of such approach was used in [19].

IOP Conf. Series: Journal of Physics: Conf. Series **873** (2017) 012034 doi:10.1088/1742-6596/873/1/012034

Below we simplify these conditions to prove that the set of new conditions is *a necessary and a sufficient* one.

- **The direct criterium**

In the 2HDM for the CP conserved case all h_a should have definite P-parity. In particular, one of them is P-odd, while two others are P-even (11a). Therefore, the **necessary condition** for a CP conservation is an existence of one neutral Higgs boson (we denote it h_3), which doesn't couple to the CP-even states VV and H^+H^-. So it reads:

$$\boxed{\begin{array}{c}\text{There exists a neutral Higgs boson } h_3 \text{ for which vectorial coupling}\\ g_3^V \equiv g(h_3VV) = 0 \text{ and } g_3^{\pm} \equiv g(h_3H^+H^-) = 0\,.\end{array}} \qquad (16)$$

Now, one has to check condition (11b). In order to do this, we substitute Eq-s. (16), (7) into (9), choosing $\rho = 0$. Choice $\rho = 0$ together with constraint $\chi_3^V = 0$ in (7) makes both $\chi_{1,2}^{H^+W^-}$ real while $\chi_3^{H^+W^-}$ is imaginary. Next, we insert these $\chi_a^{H^+W^-}$ into (9). The parameter Λ_5 contains real negative quantity $(\chi_3^{H^+W^-})^2$ while in Λ_6 and Λ_7 this imaginary $\chi_3^{H^+W^-}$ is multiplied by χ_3^V or $\chi_3^{\pm}$, which are equal to 0 (16). Hence, with this choice all parameters of potential Λ_a in the Higgs basis are real. Therefore, in view of a statement (13), the CP-symmetry of model is not violated. In particular, the CP violating interactions (12) don't appear. (Besides, it is easy to check that conditions (16) ensure compliance of conditions (14) and first two conditions (15).) Therefore we conclude that **the conditions** (16) **are necessary and sufficient** for establishing CP conservation in the 2HDM (problem B).

Note, that the eq-s (9) after substitution equation (16) together with (10) can be treated as a solution of the problem (A).

- **A discussion of the direct criterium**

Certainly, for the experimental verification one can use different conditions instead of one or both conditions (16). In particular, the condition $g(h_3VV) = 0$ means that the scalar h_3 has no P-even admixture. This leads immediately to the identities $g(h_3h_kZ) = 0$ $(k = 1, 2)$. This conclusion allows to replace checking up of the condition $g(h_3VV) = 0$ by checking of one of conditions $g(h_3h_kZ) = 0$, i. e. non-observation of corresponding decays, as it was proposed[5] in [19]. Besides, the conditions for CP conservation, written via the CP-odd invariants (14) are fulfilled if some masses M_a are degenerated and some special relations among observable couplings χ_a^V and $\chi_a^{\pm}$ took place. This combination can be treated as an alternative form of our necessary and sufficient condition in this particular case.

- **Consequence for the Yukawa interaction.**

The Yukawa interaction can violate CP symmetry of the model only in the case of a P non-conservation for the scalars. The conditions (16) guarantee that h_a are pure state of CP parity (h_1 and h_2 are CP even, h_3 is CP odd). In this case Yukawa interactions cannot generate CP violation, and the third condition from the equation (15) is fulfilled automatically.

6. Possibilities for a verification

The verification of the CP conservation requires an observation of all scalars of the model. In the realized in Nature SM-like scenario this looks difficult (see e.g. [18]). Moreover, one should

[5] Let us remind that such condition (including all its forms discussed in [19]) is insufficient to establish CP conservation in the considered model in the general case, since does not guarantee the fulfillment of the condition (11b), i. e. non-appearance of CP odd vertices (12).

IOP Conf. Series: Journal of Physics: Conf. Series **873** (2017) 012034 doi:10.1088/1742-6596/873/1/012034

check if some measurable quantities are equal to zero. In any case, these measurements cannot be performed with a high accuracy. From this point of view the proposal to change a direct criterium to a condition for a non-observation of decay $h_3 \to h_1 Z$, etc. given in [19] looks attractive. Nevertheless, one can not hope for a high accuracy in testing CP conservation in the 2HDM. Let us remind that the possible observation of weak enough decay $h_3 \to ZZ \to \ell_1^+ \ell_1^- \ell_2^+ \ell_2^-$ doesn't contradict CP conservation. If this decay will be observed the correlations between momenta of leptons should be studied. The CP conservation for h_3 will be confirmed if these correlations show axial-tensor nature of this interaction, i.e. pseudoscalar nature of h_3.

If in future the experiments will show, within the definite experimental uncertainties, the agreement with criteria (16) the important problem arises for the further studies, namely *whether one can expect that the more accurate measurements will show violation of CP in our 2HDM, i. e. violation of criteria* (16), *or the observed CP violation is in fact an effect beyond approximation given by 2HDM, and some new weaker interactions should be implemented in the description of Nature.*

This situation is similar to that in atomic physics. Atomic interaction (QED) conserves P-parity. Parity non-conservation appears at the smallest distances due to next level weak interaction. It is observed as a small effect in rare atomic transitions.

7. Conclusions

We present here a compact set of necessary and sufficient conditions for CP conservation in 2HDM (16), which are common for all mechanisms of CP violation. We prove that the verification of CP conservation in 2HDM requires to measure two simple and relatively easily testable observables instead of more complex conditions with CP-odd invariants (14) discussed by many authors [10], [14, 15, 16].

Appendix. Necessary condition for CP conservation in Multi Higgs Doublet Model

The criterion of CP conservation in the Multi Higgs Doublet Models – nHDM are also of interest. Here, the method of CP-odd invariants allows to construct many equations, which can be used for obtaining conditions for CP conservation. Both complete set of these equations and their expressions via measurable quantities are absent up to now (see e. g. [13]).

The direct method used above allows to formulate for the nHDM simple necessary conditions for the CP conservation.

After EWSB the nHDM contains $2n-1$ neutral Higgs bosons h_a, generally with indefinite CP parity, and $n-1$ charged Higgs boson $H_b^{\pm}$ with masses M_a, $M_{b\pm}$, respectively. The couplings $h_a VV$ obey first sum rule (6). In the case of CP conservation one can split spinless neutral particles h_a into two groups: P-even $h_1, ..., h_n$ and P-odd $h_{n+1}, ..., h_{2n-1}$. Similarly to (16), the condition for a CP conservation in the nHDM can be written as

$$\boxed{\begin{array}{c}\text{There exist } n-1 \text{ neutral Higgs bosons } h_c\text{, forwhich } g(h_c VV)=0\,,\\ g(h_c H_b^+ H_b^-)=0 \;\; (\text{with } \; n+1\leqslant c\leqslant 2n-1, \;\; 1\leqslant b\leqslant n-1)\,.\end{array}} \tag{17}$$

These $n(n-1)$ conditions are *necessary* for CP conservation. We don't know now whether these conditions are *sufficient* or not.

Acknowledgments

We are thankful to G.C. Branco, H. Haber, I.P. Ivanov, P. Osland, A. Pilaftsis, M. Rebelo, R. Santos for discussions. This work was supported in part by grants RFBR 15-02-05868 and HARMONIA project under contract UMO-2015/18/M/ST2/00518 (2016-2019).

IOP Conf. Series: Journal of Physics: Conf. Series **873** (2017) 012034 doi:10.1088/1742-6596/873/1/012034

References

[1] The CMS Collaboration. *Phys. Rev. Lett.* **110**, 081803 (2013); arXiv:1212.6639v2 [hep-ex]; *Phys. Rev.* **D 92**, 012004 (2015); arXiv:1411.3441 [hep-ex]; The ATLAS Collaboration. *Eur. Phys. J.* **C75**, 231 (2015); arXiv:1503.03643; *Eur. Phys. J.* **C**; arXiv:1506.05669 [hep-ex]; The ATLAS and CMS Collaborations, arXiv:1606.02266 [hep-ex]

[2] I.F. Ginzburg, *What tell us LHC data about Higgs boson parity.* arXiv:1610.08196 [hep-ph]

[3] T.D. Lee, *Phys. Rev.* **D8** (1973) 1226.

[4] I.F. Ginzburg, M. Krawczyk, P. Osland, 2nd ECFA/DESY Study 1998-2001, 997-1001; hep-ph/9909455;// Nucl. Instrum. Meth. **A 472** (2001) 149-154; hep-ph/0101229; AIP Conf. Proc. Series 578 (2001) 304, IFT-2001-2, 304; hep-ph/0101331;// Proc. SUSY02 Conference 703-706; hep-ph/0211371

[5] See e.g. J. Bernon, J. F. Gunion, H.E. Haber, Y. Jiang, S. Kraml, *Phys. Rev.* **D 92**, 075004 (2015); **D 93**, 035027 (2016); A. Pilaftsis. *Phys. Rev.* **D 93**, 075012 (2016)

[6] J.F. Gunion, H.E. Haber, G. Kane, S. Dawson, *The Higgs Hunter's Guide* (Addison-Wesley, Reading, 1990)

[7] G. C. Branco, L. Lavoura and J. P. Silva, *CP Violation*, Int. Ser. Monogr. Phys. **103** (1999) 1; G. C. Branco, P. M. Ferreira, L. Lavoura, M. N. Rebelo, M. Sher and J. P. Silva, Phys. Rept. **516**, 1 (2012) [arXiv:1106.0034 [hep-ph]].

[8] H. Georgi and D. V. Nanopoulos, Phys. Lett. B **82**, 95 (1979).

[9] I.F. Ginzburg, K.A. Kanishev, *2HDM in terms of observable quantities*, *Phys. Rev.* **D 92** (2015) 015024; arXiv:1502.06346 [hep-ph].

[10] J. F. Gunion and H. E. Haber, *Conditions for CP-violation in the general two-Higgs-doublet model*, *Phys. Rev.* **D 72** (2005) 095002 [hep-ph/0506227]; S. Davidson, H. E. Haber, *Phys. Rev.* **D72** (2005) 035004 [hep-ph/0504050]; G. C. Branco, M. N. Rebelo, J. I. Silva-Marcos, *CP-odd invariants in models with several Higgs doublets*, *Phys. Lett.* **B 614** (2005) 187 [hep-ph/0502118]; H. E. Haber and D. O'Neil, *Basis-independent methods for the two-Higgs-doublet model. II. The Significance of tanbeta*, *Phys. Rev.* **D 74**, 015018 (2006) [E: *Phys. Rev.* **D 74**, 059905 (2006)]; *Basis-independent methods for the two-Higgs-doublet model III: The CP-conserving limit, custodial symmetry, and the oblique parameters S, T, U*, *Phys. Rev.* **D 83**, 055017 (2011); D. O'Neil, *Phenomenology of the Basis-Independent CP-Violating Two-Higgs Doublet Model* arXiv:0908.1363 [hep-ph]

[11] J.F.Gunion , H.E. Haber and C. Kao, *Searching for the CP-Odd Higgs Boson of the Minimal Supersymmetric Model at Hadron Supercolliders*, *Phys. Rev.* **D46** (1992) 2907-2917

[12] G. Cvetic, M. Nowakowski, A.Pilaftsis, *CP violation with bosons*, *Phys. Lett.* **B 301** (1993) 77; [hep-ph/9301272].

[13] I. de Medeiros Varzielas, S. F. King, C. Luhn, T. Neder, *CP-odd invariants for multi-Higgs models: applications with discrete symmetry*, arXiv:1603.06942 [hep-ph]

[14] L. Lavoura, J. P. Silva, *Fundamental CP violating quantities in a SU(2) x U(1) model with many Higgs doublets*, *Phys. Rev.* **D 50** (1994) 4619 [hep-ph/9404276]; F. J. Botella, J. P. Silva, *Jarlskog-like invariants for theories with scalars and fermions*, *Phys. Rev.* **D 51** (1995) 3870 [hep-ph/9411288].

[15] B. Grzadkowski, O. M. Ogreid, P. Osland, *Diagnosing CP properties of the 2HDM*, *JHEP* **1401** (2014) 105 [arXiv:1309.6229 [hep-ph]; *Measuring CP violation in Two-Higgs-Doublet models in light of the LHC Higgs data*, arXiv:1409.7265 [hep-ph]; *Testing the presence of CP violation in the 2HDM*, arXiv:1504.06076 [hep-ph]

[16] B.Grzadkowski, M. Ogreid, P. Osland, *Spontaneous CP violation in the 2HDM: physical conditions and the alignment limit*, arXiv:1609.04764 [hep-ph]

[17] I. F. Ginzburg. *Nonminimal Higgs Models, Dark Matter, and Evolution of the Universe*, *JETP Lett.* **99** # 12 (2014) 742-751; arXiv: 1410.0873 [hep-ph]

[18] I.F. Ginzburg, *Multi-Higgs models. Perspectives for identification of wide set of models in future experiments at colliders in the SM-like situation*, arXiv:1502.07197 [hep-ph]

[19] D. Fontes, J. C. Romao, R. Santos, J. P. Silva, *Undoubtable signs of CP-violation in Higgs decays at the LHC run 2*, arXiv:1506.06755 [hep-ph]

IOP Conf. Series: Journal of Physics: Conf. Series **873** (2017) 012035 doi:10.1088/1742-6596/873/1/012035

CP-odd invariants for multi-Higgs models and applications with discrete symmetry

Ivo de Medeiros Varzielas

CFTP, Departamento de Física, Instituto Superior Técnico,
Universidade de Lisboa, Avenida Rovisco Pais 1, 1049 Lisboa, Portugal

E-mail: ivo.de@udo.edu

Abstract.
CP-odd invariants are useful for studying the CP properties of Lagrangians in any basis. We explain how to build basis invariants for the scalar sector, and how to distinguish CP-odd invariants from CP-even invariants. Up to a certain order, we use these methods to systematically build all the CP-odd invariants. The CP-odd invariants signal either explicit or spontaneous violation of CP. Making use of the CP-odd invariants, we determine the CP properties of potentials with 3 and with 6 Higgs fields arranged as triplets of specific discrete symmetries in the $\Delta(3n^2)$ or $\Delta(6n^2)$ series (inlcuding A_4, S_4, $\Delta(27)$ and $\Delta(54)$ as well as the cases for $n > 3$).

1. Introduction

This contribution to the proceedings of DISCRETE 2016 is based on [1].

CP violation (CPV) is present in the Standard Model (SM) in the Yukawa sector, due to the replication of fermion in 3 generations. Physics does not depend on choice of basis, and indeed the amount of CPV in the SM can be calculated in a weak basis invariant which can be expressed in terms of the quark mass matrices [2, 3]. In the SM scalar sector, with a single Higgs doublet, the possibility of CPV does not exist. CPV is enabled in the scalar sector of the 2 (or more) Higgs Doublet Model (HDM) [4]. The extension of the scalar content of the SM is well motivated, as additional CPV sources are required to account for the baryon asymmetry of the Universe [5, 6]. In the scalar sector, CPV can appear explicitly or through the vacuum expectation values (VEVs), the latter case consisting in Spontaneous CPV (SCPV).

There is often some subtletly to the CP properties of theories (e.g. it is possible to have complex parameters in a Lagrangian and for it to conserve CP), which increases the appeal of an invariant approach, and the generalisation of the powerful invariant approach to CP [3] to theories with discrete symmetries recently lead to other relevant CP-odd invariants (CPIs) for the respective Yukawa sector [7, 8]. The invariant approach to CP as applied to scalar potentials in [9, 10, 11, 12, 13, 14] was further developed in [1].

In order to find CPIs for arbitrary potentials, we use either the kind of diagrams introduced for the 2HDM in [13] or a new method based on "contraction matrices", and catalogued several new CPIs. We then apply them to several potentials with 3 or 6 Higgs fields with a discrete symmetry from the $\Delta(3n^2)$ and $\Delta(6n^2)$ series. These symmetries are considered frequently in the context of flavour and CP models [15, 16, 17, 18, 19, 20, 21, 22, 23, 24, 25, 26, 27, 28, 29, 30, 31, 32, 33, 34, 35, 36, 37, 38].

IOP Conf. Series: Journal of Physics: Conf. Series **873** (2017) 012035 doi:10.1088/1742-6596/873/1/012035

In particular, the 3HDM [39, 40, 41, 42, 43, 44, 45, 46, 47, 48, 49, 50, 51, 52, 53, 54, 55] and the 6HDM [56, 57, 58, 59, 60, 61, 62] where the 3 or 6 Higgses are related by a discrete symmetry as one or two (flavour) triplets have been extensively studied. Nevertheless, our systematic study through the new CPIs revealed several novel results.

2. CP-odd invariants for potentials

In our notation, we write scalar potentials in a standard form. This is similar to the notation in [12, 13] but with some important differences. An even potential of N scalar fields φ_i, where $\phi = (\varphi_1, \ldots, \varphi_N)$ and $\phi^* = (\varphi_1^*, \ldots, \varphi_N^*)$, can be rewritten as

$$V = \phi^{*a}\, Y_a^b\, \phi_b + \phi^{*a}\phi^{*c}\, Z_{ac}^{bd}\, \phi_b\phi_d \, . \tag{1}$$

The lower indices on Y and Z are contracted with ϕ^* and upper indices with ϕ. The tensors Y and Z contain the couplings from the potential (the formalism can be extended to account also for tri-linear terms). Z_{ac}^{bd} is by definition unchanged if $a \leftrightarrow c$ or $b \leftrightarrow d$.

In particular, potentials of Higgs doublets are rewritten in the standard form with their components: n Higgs doublets $H_{i\alpha} = (h_{i,1}, h_{i,2})$, with $\alpha = 1, 2$ for $SU(2)_L$ index and i identifies the doublet from 1 to n, are recast as

$$\phi = (\varphi_1, \varphi_2, \ldots, \varphi_{2n-1}, \varphi_{2n}) = (h_{1,1}, h_{1,2}, \ldots, h_{n,1}, h_{n,2}) \, , \tag{2}$$

The invariance of the potential under whatever symmetries it is invariant under (e.g. $SU(2)_L \times U(1)_Y$) will be manifest in the structure of the tensors Y and Z: if under G, ϕ transforms in some representation $\rho(g)$, $g \in G$,

$$\phi_a \;\mapsto\; [\rho(g)]_a^{a'}\, \phi_{a'} \, , \tag{3}$$

$$\phi^{*a} \;\mapsto\; \phi^{*a'} [\rho^\dagger(g)]_{a'}^a \, , \tag{4}$$

then

$$Y_a^b = \rho_a^{a'}\, Y_{a'}^{b'}\, \rho^{\dagger}{}_{b'}^{b} \, , \tag{5}$$

$$Z_{ac}^{bd} = \rho_a^{a'}\, \rho_c^{c'}\, Z_{a'c'}^{b'd'}\, \rho^{\dagger}{}_{b'}^{b}\, \rho^{\dagger}{}_{d'}^{d} \, , \tag{6}$$

where we denote $\rho_a^{a'} = [\rho(g)]_a^{a'}$.

An arbitrary basis transformation acts on the fields by a unitary $N \times N$ matrix

$$\phi_a \;\mapsto\; V_a^{a'}\, \phi_{a'} \, , \tag{7}$$

$$\phi^{*a} \;\mapsto\; \phi^{*a'} V^{\dagger}{}_{a'}^{a} \, , \tag{8}$$

which changes tensor components accordingly:

$$Y_a^b \;\mapsto\; V_a^{a'}\, Y_{a'}^{b'}\, V^{\dagger}{}_{b'}^{b} \, , \tag{9}$$

$$Z_{ac}^{bd} \;\mapsto\; V_a^{a'}\, V_c^{c'}\, Z_{a'c'}^{b'd'}\, V^{\dagger}{}_{b'}^{b}\, V^{\dagger}{}_{d'}^{d} \, . \tag{10}$$

In our notation, complex conjugation changes indices of a field:

$$\phi_a \;\mapsto\; (\phi_a)^* \equiv \phi^{*a} \, , \tag{11}$$

$$\phi^{*a} \;\mapsto\; (\phi^{*a})^* \equiv \phi_a \, , \tag{12}$$

therefore,

$$\phi^{*a}\, Y_a^b\, \phi_b \;\mapsto\; \phi_a\, (Y_a^b)^*\, \phi^{*b} = \phi^{*b}\, (Y_a^b)^*\, \phi_a = \phi^{*a}\, (Y_b^a)^*\, \phi_b \, , \tag{13}$$

IOP Conf. Series: Journal of Physics: Conf. Series **873** (2017) 012035 doi:10.1088/1742-6596/873/1/012035

and we obtain (because $V^* = V$):

$$(Y_b^a)^* = Y_a^b \,. \tag{14}$$

Similarly, for the Z tensor we have

$$(Z_{bd}^{ac})^* = Z_{ac}^{bd} \,. \tag{15}$$

A general CP transformation acts with a unitary matrix U:

$$\phi_a \mapsto \phi^{*a'} U_{a'}^a \,, \tag{16}$$

$$\phi^{*a} \mapsto U^{\dagger a'}_{\ a} \phi_{a'} \,, \tag{17}$$

and thus

$$\phi^{*a} Y_a^b \phi_b \mapsto U^{\dagger a'}_{\ a} \phi_{a'} Y_a^b \phi^{*b'} U_{b'}^b = U^{\dagger a'}_{\ a} \phi_{a'} (Y_b^a)^* \phi^{*b'} U_{b'}^b \tag{18}$$

$$= \phi^{*b'} U_{b'}^b (Y_b^a)^* U^{\dagger a'}_{\ a} \phi_{a'} \tag{19}$$

$$= \phi^{*a} U_a^{a'} (Y_{a'}^{b'})^* U^{\dagger b}_{\ b'} \phi_b \,. \tag{20}$$

A comparison of this with the original term, and the respective exercise for Z, reveals how the general CP transformation acts on the tensors:

$$Y_a^b \mapsto U_a^{a'} (Y_{a'}^{b'})^* U^{\dagger b}_{\ b'} \,, \tag{21}$$

$$Z_{ac}^{bd} \mapsto U_a^{a'} U_c^{c'} (Z_{a'c'}^{b'd'})^* U^{\dagger b}_{\ b'} U^{\dagger d}_{\ d'} \,. \tag{22}$$

This allows us to recast the condition for the potential to be CP invariant: if there is any U such that the left- and right-hand sides of Eqs. (21) and (22) are identical. The trivial CP:

$$U_{a'}^a = \delta_{a'}^a \,. \tag{23}$$

obeys the condition for real Y and Z tensors. We refer to this trivial CP throughout as CP_0.

When considering SCPV, we build spontaneous CPIs (SCPIs) by using also VEVs to form basis invariants. It is therefore useful to note they transform under basis change as:

$$v_a \mapsto V_a^{a'} v_{a'} \,, \tag{24}$$

$$v^{*a} \mapsto v^{*a'} V^{\dagger a}_{\ a'} \,, \tag{25}$$

where $v \equiv (v_1, \ldots)$, $v_i = \langle \varphi_i \rangle$. Under general CP transformations:

$$v_a \mapsto v^{*a'} U_{a'}^a \,, \tag{26}$$

$$v^{*a} \mapsto U^{\dagger a'}_{\ a} v_{a'} \,. \tag{27}$$

2.1. Simplest basis invariants

If one has a combination of Y and Z tensors with all indices contracted, it forms a basis invariant. Thus, the simplest invariant is the Trace of Y:

$$Y_a^a. \tag{28}$$

The possibilities with two Y tensors are:

$$Y_a^a Y_b^b \,, \quad Y_b^a Y_a^b. \tag{29}$$

IOP Conf. Series: Journal of Physics: Conf. Series **873** (2017) 012035 doi:10.1088/1742-6596/873/1/012035

We note that these contractions can be mapped to elements of the permutation group by identifying the permutation that takes the ordering in the upper indices to the ordering in the lower indicies:

$$Y^a_a Y^b_b \Leftrightarrow a \mapsto a\,, \quad b \mapsto b\,, \tag{30}$$

and

$$Y^a_b Y^b_a \Leftrightarrow a \mapsto b\,, \quad b \mapsto a\,. \tag{31}$$

We write thus:

$$Y^a_{\sigma(a)} Y^b_{\sigma(b)}\,, \quad \sigma \in S_2\,, \tag{32}$$

where σ is an element of the group S_2. Out of the two basis invariants with two Y tensors, one is simply the square of the Trace of Y, i.e. it can be expressed in terms of smaller basis invariants. We refer to basis invariants that can't be written in terms of smaller ones as irreducible.

Similarly for Z tensors, the simplest invariants are:

$$Z^{ab}_{\sigma(a)\sigma(b)}\,, \quad \sigma \in S_2\,, \tag{33}$$

or explicitly:

$$Z^{ab}_{ab}\,, \quad Z^{ab}_{ba}\,. \tag{34}$$

Due to the inherent symmetry Z tensors have under exchange of upper (or lower) indices, these two basis invariants are the same. With two Z tensors, the possible contractions are 24:

$$Z^{ab}_{\sigma(a)\sigma(b)} Z^{cd}_{\sigma(c)\sigma(d)}\,, \quad \sigma \in S_4\,, \tag{35}$$

but there are only two new irreducible basis invariants, which can be taken to be

$$Z^{ab}_{bd} Z^{cd}_{ac}\,, \quad Z^{ab}_{cd} Z^{cd}_{ab}. \tag{36}$$

In general, a basis invariant $I^{(n_Z,m_Y)}_\sigma$ with m_Y Y tensors and n_Z Z tensors is

$$I^{(n_Z,m_Y)}_\sigma \equiv Y^{a_1}_{\sigma(a_1)} \dots Y^{a_{m_Y}}_{\sigma(a_{m_Y})} Z^{b_1 b_2}_{\sigma(b_1)\sigma(b_2)} \dots Z^{b_{2n_Z-1} b_{2n_Z}}_{\sigma(b_{2n_Z-1})\sigma(b_{2n_Z})}\,, \quad \sigma \in S_{m_Y+2n_Z}. \tag{37}$$

Many basis invariants are not CP-odd. Under a general CP transformation, a coupling tensor goes to its complex conjugate and is acted on by unitary transformations U. Being basis invariant means the U matrices drop out, so within a basis invariant, a general CP transformation reduces to converting tensors into their complex conjugates (by exchanging upper and lower indices). For example, the Trace of Y:

$$Y^a_a \to CP (Y^{a'}_{a''})^* U^{\dagger a}_{\ a'} U^{a''}_a = (Y^{a'}_{a''})^* \delta^{a''}_{a'} = (Y^a_a)^* = Y^a_a. \tag{38}$$

Likewise:

$$\begin{aligned} I^{(n_Z,m_Y)}_\sigma &\equiv Y^{a_1}_{\sigma(a_1)} \dots Y^{a_{m_Y}}_{\sigma(a_{m_Y})} Z^{b_1 b_2}_{\sigma(b_1)\sigma(b_2)} \dots Z^{b_{2n_Z-1} b_{2n_Z}}_{\sigma(b_{2n_Z-1})\sigma(b_{2n_Z})} \\ \to CP Y^{\sigma(a_1)}_{a_1} &\dots Y^{\sigma(a_{m_Y})}_{a_{m_Y}} Z^{\sigma(b_1)\sigma(b_2)}_{b_1 b_2} \dots Z^{\sigma(b_{2n_Z-1})\sigma(b_{2n_Z})}_{b_{2n_Z-1} b_{2n_Z}} = [I^{(n_Z,m_Y)}_\sigma]^*\,. \end{aligned} \tag{39}$$

From a basis invariant I differing from its CP conjugate I^*, we can build a CPI $\mathcal{I}$:

$$\mathcal{I} = I - I^*. \tag{40}$$

In order to better find useful CPIs we employ diagrams and in particular contraction matrices that reveal which basis invariants can form CPIs.

IOP Conf. Series: Journal of Physics: Conf. Series **873** (2017) 012035 doi:10.1088/1742-6596/873/1/012035

For the diagrams, the rules are as follows

$$X^{a\cdot}_{\cdot\cdot}X^{\cdot\cdot}_{a\cdot} = \quad (41)$$

Lines don't need to be distinguished:

$$Z^{ab}_{\cdot\cdot}Z^{\cdot\cdot}_{ab} = \quad (42)$$

Contracting indices on same tensor:

$$X^{a\cdot}_{a\cdot} = \quad (43)$$
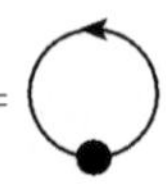

For the contraction matrices, we list all the Y tensors, then all the Z tensors as rows and columns and have $0, 1, 2...$ if there are $0, 1, 2, ...$ lines connecting from upper indices of that row's tensor to the lower index of that column's tensor. Some examples are:

$$Y^a_a =$$

$$= (1) \quad (44)$$

$$Y^a_a Y^b_b =$$
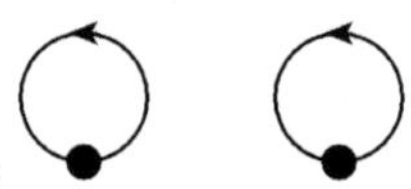
$$= \begin{pmatrix} 1 & 0 \\ 0 & 1 \end{pmatrix} \quad (45)$$

$$Y^a_b Y^b_a =$$
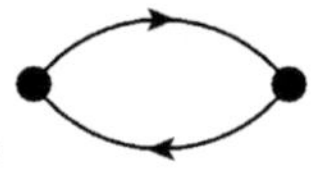
$$= \begin{pmatrix} 0 & 1 \\ 1 & 0 \end{pmatrix} \quad (46)$$

$$Z^{ab}_{ab} =$$
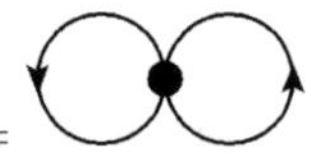
$$= (2) \quad (47)$$

$$Z^{ac}_{bc} Y^b_a =$$
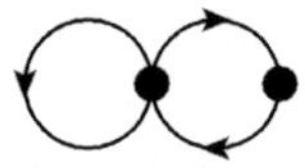
$$= \begin{pmatrix} 0 & 1 \\ 1 & 1 \end{pmatrix} \quad (48)$$

$$Z^{ab}_{cd} Z^{cd}_{ab} =$$
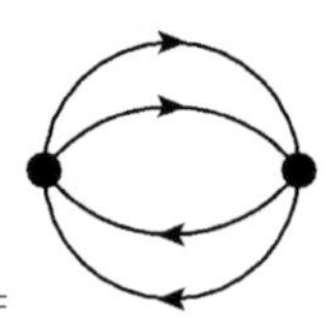
$$= \begin{pmatrix} 0 & 2 \\ 2 & 0 \end{pmatrix} \quad (49)$$

$$Z^{ab}_{ac} Z^{cd}_{bd} =$$
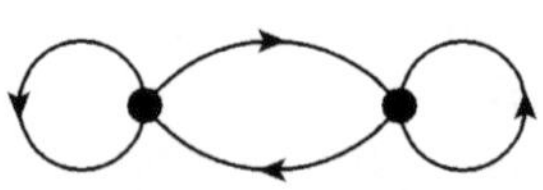
$$= \begin{pmatrix} 1 & 1 \\ 1 & 1 \end{pmatrix} \quad (50)$$

A CPI can be built from a basis invariant that changes if all upper indices are exchanged with the lower indices, which, up to rearranging the tensors, corresponds to diagrams that are distinct when the arrows are reversed and to contraction matrices that not symmetric (i.e. are

IOP Conf. Series: Journal of Physics: Conf. Series **873** (2017) 012035 doi:10.1088/1742-6596/873/1/012035

distinct by exchanging rows and columns). The smallest CPI can be built from the following basis invariant:

$$I_1 \equiv Z^{ab}_{ae} Z^{cd}_{bf} Y^e_c Y^f_d =$$

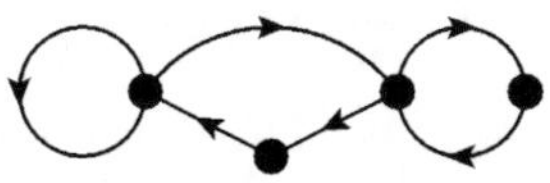

$$= \begin{pmatrix} 0 & 0 & 1 & 0 \\ 0 & 0 & 0 & 1 \\ 0 & 0 & 1 & 1 \\ 1 & 1 & 0 & 0 \end{pmatrix} \qquad (51)$$

and its distinct CP conjugate

$$I_1^* \equiv Z_{ab}^{ae} Z_{cd}^{bf} Y_e^c Y_f^d = \qquad (52)$$

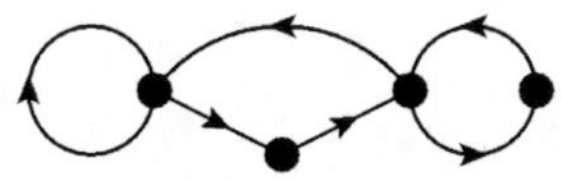

where in the matrix the Y tensors are the first rows and columns (even though they are ordered differently in the expressions and diagrams).

For the purpose of the potentials considered, the useful CPIs are order 6 in Z tensors:

$$I_2^{(6)} = Z^{a_1a_2}_{a_7a_{10}} Z^{a_3a_4}_{a_{11}a_6} Z^{a_5a_6}_{a_9a_8} Z^{a_7a_8}_{a_3a_{12}} Z^{a_9a_{10}}_{a_5a_4} Z^{a_{11}a_{12}}_{a_1a_2} =$$

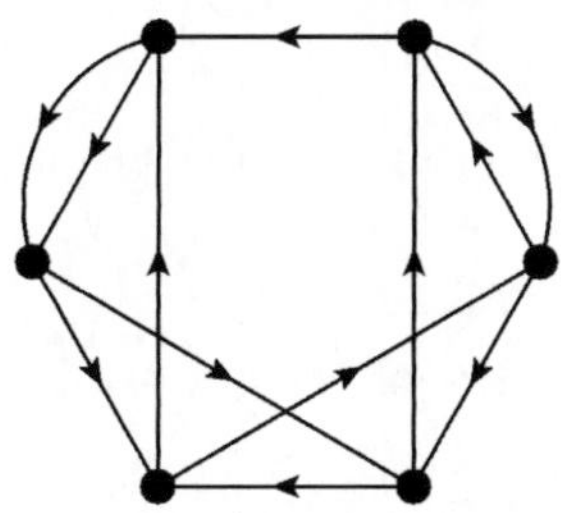

$$= \begin{pmatrix} 0 & 0 & 0 & 0 & 0 & 2 \\ 0 & 0 & 0 & 1 & 1 & 0 \\ 0 & 1 & 0 & 0 & 1 & 0 \\ 1 & 0 & 1 & 0 & 0 & 0 \\ 1 & 0 & 1 & 0 & 0 & 0 \\ 0 & 1 & 0 & 1 & 0 & 0 \end{pmatrix} \qquad (53)$$

$$I_4^{(6)} = Z^{a_1a_2}_{a_{11}a_{10}} Z^{a_3a_4}_{a_5a_8} Z^{a_5a_6}_{a_7a_{12}} Z^{a_7a_8}_{a_9a_6} Z^{a_9a_{10}}_{a_1a_4} Z^{a_{11}a_{12}}_{a_3a_2} =$$

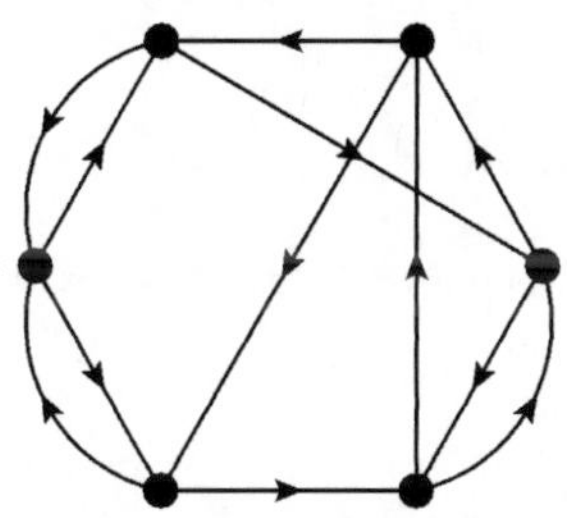

$$= \begin{pmatrix} 0 & 0 & 0 & 0 & 1 & 1 \\ 0 & 0 & 0 & 0 & 1 & 1 \\ 0 & 1 & 0 & 1 & 0 & 0 \\ 0 & 1 & 1 & 0 & 0 & 0 \\ 1 & 0 & 0 & 1 & 0 & 0 \\ 1 & 0 & 1 & 0 & 0 & 0 \end{pmatrix} \qquad (54)$$

A full list of CPIs up to order 6 in Z tensors can be found in the appendix of [1].

In addtion, SCPIs can be built by contracting in a similar fashion VEVs (which transform under basis changes and general CP transformations as the respective field does). The SCPI that is relevant for the potentials considered is:

$$J_1^{(3,2)} \equiv Z^{a_1a_2}_{a_4a_5} Z^{a_3a_4}_{a_2a_6} Z^{a_5a_6}_{a_7a_8} v_{a_1} v_{a_3} v^{*a_7} v^{*a_8} = \qquad (55)$$

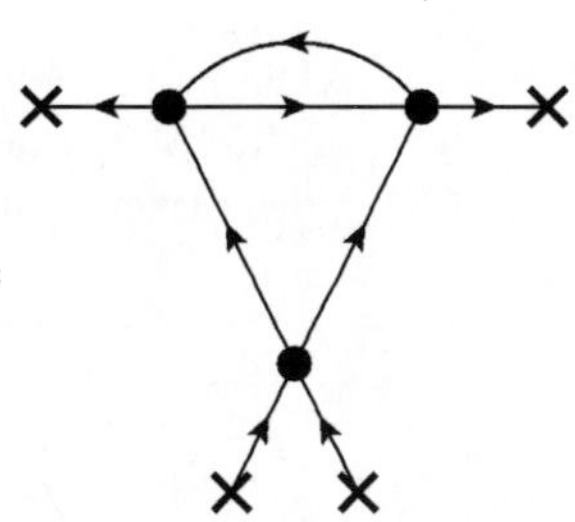

3. Explicit CP violation

The potential of one triplet of $\Delta(27)$ has interesting CP properties. We write

$$V_0(\varphi) = -\, m_\varphi^2 \sum_i \varphi_i \varphi^{*i} + r \left(\sum_i \varphi_i \varphi^{*i} \right)^2 + s \sum_i (\varphi_i \varphi^{*i})^2 \qquad (56)$$

IOP Conf. Series: Journal of Physics: Conf. Series **873** (2017) 012035 doi:10.1088/1742-6596/873/1/012035

$$V_{\Delta(27)}(\varphi) = V_0(\varphi) + \left[d\left(\varphi_1\varphi_1\varphi^{*2}\varphi^{*3} + \text{cycl.}\right) + \text{h.c.}\right] \tag{57}$$

for $SU(2)_L$ singlet scalars, where we consider an additional $U(1)$ flavour symmetry charging the triplet to exclude tri-linear terms, making the potential similar to the respective potentials for $SU(2)_L$ doublets

$$\begin{aligned} V_0(H) &= -m_h^2 \sum_{i,\alpha} h_{i\alpha}h^{*i\alpha} + \sum_{i,j,\alpha,\beta}\left[r_1(h_{i\alpha}h^{*i\alpha})(h_{j\beta}h^{*j\beta}) + r_2(h_{i\alpha}h^{*i\beta})(h_{j\beta}h^{*j\alpha})\right] \\ &\quad + s\sum_{i,\alpha,\beta}(h_{i\alpha}h^{*i\alpha})(h_{i\beta}h^{*i\beta}) \end{aligned} \tag{58}$$

$$V_{\Delta(27)}(H) = V_0(H) + \sum_{\alpha,\beta}\left[d\left(h_{1\alpha}h_{1\beta}h^{*2\alpha}h^{*3\beta} + \text{cycl.}\right) + \text{h.c.}\right] \tag{59}$$

We consider also the potential with two triplets of $\Delta(3n^2)$ for $n > 3$. We write:

$$\begin{aligned} V_1(\varphi,\varphi') &= +\tilde{r}_1\left(\sum_i \varphi_i\varphi^{*i}\right)\left(\sum_j \varphi'_j\varphi'^{*j}\right) + \tilde{r}_2\left(\sum_i \varphi_i\varphi'^{*i}\right)\left(\sum_j \varphi'_j\varphi^{*j}\right) \\ &\quad + \tilde{s}_1\sum_i\left(\varphi_i\varphi^{*i}\varphi'_i\varphi'^{*i}\right) \\ &\quad + \tilde{s}_2\left(\varphi_1\varphi^{*1}\varphi'_2\varphi'^{*2} + \varphi_2\varphi^{*2}\varphi'_3\varphi'^{*3} + \varphi_3\varphi^{*3}\varphi'_1\varphi'^{*1}\right) \\ &\quad + i\,\tilde{s}_3\left[(\varphi_1\varphi'^{*1}\varphi'_2\varphi^{*2} + \text{cycl.}) - (\varphi^{*1}\varphi'_1\varphi'^{*2}\varphi_2 + \text{cycl.})\right]. \end{aligned} \tag{60}$$

$$V_{\Delta(3n^2)}(\varphi,\varphi') = V_0(\varphi) + V'_0(\varphi') + V_1(\varphi,\varphi')\ . \tag{61}$$

Explicit CPV is confirmed for these potentials as at least one CPI is non-zero. For $SU(2)_L$ singlets and $\Delta(27)$:

$$\mathcal{I}_4^{(6)} = -\frac{3}{32}\left(d^3 - d^{*3}\right)\left(d^3 + 6dd^*s + d^{*3} - 8s^3\right) \tag{62}$$

and for $SU(2)_L$ doublets the expression is the same apart from a different fraction. The 6 solutions of $\mathcal{I}_4^{(6)} = 0$ in terms of d and s (parameters of the $\Delta(27)$ potential) correspond to imposing one of the CP symmetries identified in [63], which naturally eliminates the possibility of explicit CPV.

For $\Delta(3n^2)$ we calculate:

$$\mathcal{I}_2^{(6)} = \frac{3}{512}i\tilde{s}_2\tilde{s}_3(-3\tilde{r}_2^2 + \tilde{s}_3^2)[-\tilde{s}_1^2 + \tilde{s}_1\tilde{s}_2 + \tilde{r}_2(-2\tilde{s}_1 + \tilde{s}_2) + \tilde{s}_3^2] \tag{63}$$

The solutions for $\mathcal{I}_2^{(6)} = 0$ match CP symmetries that can be imposed on the potential: e.g. trivial CP_0 forces $\tilde{s}_3 = 0$; another CP symmetry forces $\tilde{s}_2 = 0$ and after finding $(-3\tilde{r}_2^2 + \tilde{s}_3^2)$ in other CPIs, we suspected there was a CP symmetry CP forcing that combination of parameters to vanish and indeed found it.

A summary table with our results for explicit CPV can be found in [1].

IOP Conf. Series: Journal of Physics: Conf. Series **873** (2017) 012035 doi:10.1088/1742-6596/873/1/012035

4. Spontaneous CP violation

The $\Delta(27)$ invariant potentials has interesting SCPV properties, as its SCPV can be geometric [39, 42, 43, 44, 47, 48, 53]. We analysed it with a SCPI and obtained

$$\mathcal{J}_1^{(3,2)} = \frac{1}{4}(d^{*3} - d^3)Q(|v_i|) \tag{64}$$

$$+\frac{1}{2}(dd^{*2} - 2d^* s^2 + d^2 s)(v_2 v_3 v_1^{*2} + v_1 v_3 v_2^{*2} + v_1 v_2 v_3^{*2})$$

$$-\frac{1}{2}(d^2 d^* - 2ds^2 + d^{*2} s)(v_2^* v_3^* v_1^2 + v_1^* v_3^* v_2^2 + v_1^* v_2^* v_3^2)\ . \tag{65}$$

As the potential has explicit CPV, SCPV only applies after imposing a CP symmetry, which simplifies the expression. E.g., for CP_0 (forces $\mathrm{Arg}(d) = 0$):

$$\mathcal{J}_1^{(3,2)} = \tfrac{1}{2}(d^3 - 2ds^2 + d^2 s) \tag{66}$$

$$\left[(v_2 v_3 v_1^{*2} + v_1 v_3 v_2^{*2} + v_1 v_2 v_3^{*2}) - (v_2^* v_3^* v_1^2 + v_1^* v_3^* v_2^2 + v_1^* v_2^* v_3^2)\right] \tag{67}$$

The results obtained through the SCPI matched the known results. For example, for CP_0, it can be checked from the dependence on the VEVs that $\langle\varphi\rangle = (1, \omega, \omega^2)$ makes the SCPI vanish (and indeed this VEV preserves a CP symmetry present in the potential when CP_0 is imposed). Conversely, it confirms the VEV $\langle\varphi\rangle = (\omega, 1, 1)$ has SCPV, as the SCPI is non-vanishing.

5. Conclusion

In summary, we developed the formalism for finding CP-odd basis invariants for scalar potentials, and performed a systematic search finding several new such invariants (up to 6 Z tensors). These methods for studying explicit and spontaneous CP violation are valid for any potential when brought to standard form, for any $SU(2)_L$ assignments of the scalars, and are therefore very useful in studying the CP properties of potentials in multi-Higgs models. We exemplified the use of the CP-odd invariants by considering 3 and 6 Higgs doublet models symmetric under $\Delta(3n^2)$ and $\Delta(6n^2)$ groups and identifying their CP properties.

Acknowledgments

IdMV has received funding from Fundação para a Ciência e a Tecnologia (FCT) through the contract IF/00816/2015. This project has received funding from the European Union's Seventh Framework Programme for research, technological development and demonstration under grant agreement no PIEF-GA-2012-327195 SIFT.

References

[1] de Medeiros Varzielas I, King S F, Luhn C and Neder T 2016 *Phys. Rev.* **D94** 056007 (*Preprint* 1603.06942)
[2] Jarlskog C 1985 *Phys. Rev. Lett.* **55** 1039
[3] Bernabeu J, Branco G C and Gronau M 1986 *Phys. Lett.* **169B** 243–247
[4] Lee T D 1973 *Phys. Rev.* **D8** 1226–1239 [,516(1973)]
[5] Sakharov A D 1967 *Pisma Zh. Eksp. Teor. Fiz.* **5** 32–35 [Usp. Fiz. Nauk161,61(1991)]
[6] Kuzmin V A, Rubakov V A and Shaposhnikov M E 1985 *Phys. Lett.* **B155** 36
[7] Branco G C, de Medeiros Varzielas I and King S F 2015 *Phys. Rev.* **D92** 036007 (*Preprint* 1502.03105)
[8] Branco G C, de Medeiros Varzielas I and King S F 2015 *Nucl. Phys.* **B899** 14–36 (*Preprint* 1505.06165)
[9] Mendez A and Pomarol A 1991 *Phys. Lett.* **B272** 313–318
[10] Lavoura L and Silva J P 1994 *Phys. Rev.* **D50** 4619–4624 (*Preprint* hep-ph/9404276)
[11] Botella F J and Silva J P 1995 *Phys. Rev.* **D51** 3870–3875 (*Preprint* hep-ph/9411288)
[12] Branco G C, Rebelo M N and Silva-Marcos J I 2005 *Phys. Lett.* **B614** 187–194 (*Preprint* hep-ph/0502118)
[13] Davidson S and Haber H E 2005 *Phys. Rev.* **D72** 035004 [Erratum: Phys. Rev.D72,099902(2005)] (*Preprint* hep-ph/0504050)

IOP Conf. Series: Journal of Physics: Conf. Series **873** (2017) 012035 doi:10.1088/1742-6596/873/1/012035

[14] Gunion J F and Haber H E 2005 *Phys. Rev.* **D72** 095002 (*Preprint* hep-ph/0506227)
[15] Feruglio F, Hagedorn C and Ziegler R 2013 *JHEP* **07** 027 (*Preprint* 1211.5560)
[16] Ding G J, King S F, Luhn C and Stuart A J 2013 *JHEP* **05** 084 (*Preprint* 1303.6180)
[17] Feruglio F, Hagedorn C and Ziegler R 2014 *Eur. Phys. J.* **C74** 2753 (*Preprint* 1303.7178)
[18] King S F, Neder T and Stuart A J 2013 *Phys. Lett.* **B726** 312–315 (*Preprint* 1305.3200)
[19] Luhn C 2013 *Nucl. Phys.* **B875** 80–100 (*Preprint* 1306.2358)
[20] Ding G J, King S F and Stuart A J 2013 *JHEP* **12** 006 (*Preprint* 1307.4212)
[21] Li C C and Ding G J 2014 *Nucl. Phys.* **B881** 206–232 (*Preprint* 1312.4401)
[22] Ding G J and Zhou Y L 2015 *Chin. Phys.* **C39** 021001 (*Preprint* 1312.5222)
[23] King S F and Neder T 2014 *Phys. Lett.* **B736** 308–316 (*Preprint* 1403.1758)
[24] Ding G J and King S F 2014 *Phys. Rev.* **D89** 093020 (*Preprint* 1403.5846)
[25] Ding G J and Zhou Y L 2014 *JHEP* **06** 023 (*Preprint* 1404.0592)
[26] Neder T 2015 *J. Phys. Conf. Ser.* **598** 012025 (*Preprint* 1405.0019)
[27] Li C C and Ding G J 2015 *JHEP* **08** 017 (*Preprint* 1408.0785)
[28] Hagedorn C, Meroni A and Molinaro E 2015 *Nucl. Phys.* **B891** 499–557 (*Preprint* 1408.7118)
[29] Ding G J, King S F and Neder T 2014 *JHEP* **12** 007 (*Preprint* 1409.8005)
[30] Bjrkeroth F, de Anda F J, de Medeiros Varzielas I and King S F 2015 *JHEP* **06** 141 (*Preprint* 1503.03306)
[31] Li C C and Ding G J 2015 *JHEP* **05** 100 (*Preprint* 1503.03711)
[32] Di Iura A, Hagedorn C and Meloni D 2015 *JHEP* **08** 037 (*Preprint* 1503.04140)
[33] Ballett P, Pascoli S and Turner J 2015 *Phys. Rev.* **D92** 093008 (*Preprint* 1503.07543)
[34] Neder T 2015 *J. Phys. Conf. Ser.* **631** 012019 (*Preprint* 1503.09041)
[35] Turner J 2015 *Phys. Rev.* **D92** 116007 (*Preprint* 1507.06224)
[36] Ding G J and King S F 2016 *Phys. Rev.* **D93** 025013 (*Preprint* 1510.03188)
[37] Bjrkeroth F, de Anda F J, de Medeiros Varzielas I and King S F 2016 *Phys. Rev.* **D94** 016006 (*Preprint* 1512.00850)
[38] Li C C, Yao C Y and Ding G J 2016 *JHEP* **05** 007 (*Preprint* 1601.06393)
[39] Branco G C, Gerard J M and Grimus W 1984 *Phys. Lett.* **B136** 383–386
[40] de Adelhart Toorop R, Bazzocchi F, Merlo L and Paris A 2011 *JHEP* **03** 035 [Erratum: JHEP01,098(2013)] (*Preprint* 1012.1791)
[41] de Adelhart Toorop R, Bazzocchi F, Merlo L and Paris A 2011 *JHEP* **03** 040 (*Preprint* 1012.2091)
[42] de Medeiros Varzielas I and Emmanuel-Costa D 2011 *Phys. Rev.* **D84** 117901 (*Preprint* 1106.5477)
[43] de Medeiros Varzielas I, Emmanuel-Costa D and Leser P 2012 *Phys. Lett.* **B716** 193–196 (*Preprint* 1204.3633)
[44] Bhattacharyya G, de Medeiros Varzielas I and Leser P 2012 *Phys. Rev. Lett.* **109** 241603 (*Preprint* 1210.0545)
[45] Ivanov I P and Vdovin E 2013 *Eur. Phys. J.* **C73** 2309 (*Preprint* 1210.6553)
[46] Ivanov I P and Vdovin E 2012 *Phys. Rev.* **D86** 095030 (*Preprint* 1206.7108)
[47] de Medeiros Varzielas I and Pidt D 2014 *J. Phys.* **G41** 025004 (*Preprint* 1307.0711)
[48] de Medeiros Varzielas I and Pidt D 2013 *JHEP* **11** 206 (*Preprint* 1307.6545)
[49] Keus V, King S F and Moretti S 2014 *JHEP* **01** 052 (*Preprint* 1310.8253)
[50] Keus V, King S F, Moretti S and Sokolowska D 2014 *JHEP* **11** 016 (*Preprint* 1407.7859)
[51] Ivanov I P and Nishi C C 2015 *JHEP* **01** 021 (*Preprint* 1410.6139)
[52] Keus V, King S F, Moretti S and Sokolowska D 2015 *JHEP* **11** 003 (*Preprint* 1507.08433)
[53] Fallbacher M and Trautner A 2015 *Nucl. Phys.* **B894** 136–160 (*Preprint* 1502.01829)
[54] Ivanov I P and Silva J P 2016 *Phys. Rev.* **D93** 095014 (*Preprint* 1512.09276)
[55] Emmanuel-Costa D, Ogreid O M, Osland P and Rebelo M N 2016 *JHEP* **02** 154 [Erratum: JHEP08,169(2016)] (*Preprint* 1601.04654)
[56] Ivanov I P and Lavoura L 2013 *Eur. Phys. J.* **C73** 2416 (*Preprint* 1302.3656)
[57] Ivanov I P and Nishi C C 2010 *Phys. Rev.* **D82** 015014 (*Preprint* 1004.1799)
[58] Ivanov I P 2010 *JHEP* **07** 020 (*Preprint* 1004.1802)
[59] Carcamo Hernandez A E, de Medeiros Varzielas I, Kovalenko S G, Ps H and Schmidt I 2013 *Phys. Rev.* **D88** 076014 (*Preprint* 1307.6499)
[60] Keus V, King S F and Moretti S 2014 *Phys. Rev.* **D90** 075015 (*Preprint* 1408.0796)
[61] Nishi C C 2015 *JHEP* **03** 034 (*Preprint* 1411.4909)
[62] de Medeiros Varzielas I, Fischer O and Maurer V 2015 *JHEP* **08** 080 (*Preprint* 1504.03955)
[63] Nishi C C 2013 *Phys. Rev.* **D88** 033010 (*Preprint* 1306.0877)

IOP Conf. Series: Journal of Physics: Conf. Series **873** (2017) 012036 doi:10.1088/1742-6596/873/1/012036

CP-symmetry of order 4 and its consequences

Igor P. Ivanov

CFTP, Instituto Superior Técnico, Universidade de Lisboa, Avenida Rovisco Pais 1, 1049–001 Lisboa, Portugal

E-mail: igor.ivanov@tecnico.ulisboa.pt

Abstract. Extended Higgs sectors offer rich opportunities for various forms of CP-violation. Here, we describe a new form of *CP-conservation* and discuss its consequences. We give a concrete example of a three-Higgs-doublet model dubbed CP4-3HDM with a CP-symmetry of order 4 and no other other accidental symmetries. If the vacuum conserves this symmetry, the model is CP-conserving with pairwise mass-degenerate extra neutral Higgs bosons. These fields cannot be classified as CP-even or CP-odd but they can be combined into complex physical fields which are CP-half-odd, that is, they pick up the i factor upon CP transformation. These CP-half-odd scalars can be Yukawa-coupled to the fermion bilinears in a CP-conserving way. We discuss fundamental and phenomenological features of the model, and stress a peculiar clash between the CP-symmetry and any convention for the particle-antiparticle assignment.

1. Introduction

In spite of the long and extensive investigation of the phenomenon of CP-violation, its fundamental origin remains enigmatic [1]. In the Standard Model (SM), CP-violation is introduced by hand in the form of complex quark Yukawa couplings. In models with extended Higgs sectors, — a mainstream direction of theoretical exploration of opportunities beyond the SM (bSM), — it can originate from the scalar sector. In rich Higgs sectors, several options are available, see the recent review [2]. CP-violation (CPV) can be explicit, when the Higgs potential itself is not invariant under any definition of CP-symmetry, or spontaneous, as a side effect of the Higgs phenomenon taking place in an explicitly CP-conserving model. Even in the two-Higgs-doublet model (2HDM) [3, 4], CPV comes in various forms, some of which were found in the last few years [5, 6, 7, 8]. Understanding how CP-violation actually happens may additionally shed some light on the flavor sector hierarchy, which may be intimately intertwined with it, and on generation of the baryon asymmetry of the Universe. In short, any novel form of CP-violation deserves a closer theoretical study as it may tell us something new and lead to testable predictions.

When building a model with desired CP properties, one must keep in mind that quantum field theory does not uniquely specify how discrete symmetry transformations act on fields [9, 10, 11, 1]. This action must be assigned, and it can happen that this assignment is not unique. One often mentions that discrete transformations of complex fields can involve unconstrained phase factors [9]. In models with several fields with identical quantum numbers, this freedom becomes much larger. Consider for definiteness N complex scalar fields ϕ_i, $i = 1, \dots, N$. The standard convention for the CP-transformation,

$$\phi_i(\mathbf{x}, t) \xrightarrow{CP} (CP)\phi_i(\mathbf{x}, t)(CP)^{-1} = \phi_i^*(-\mathbf{x}, t)\,, \quad (1)$$

is basis-dependent: the same transformation viewed in another basis will contain an extra unitary transformation accompanying the conjugation. If one wants to know whether a given N-Higgs-doublet model (NHDM) is CP-conserving, one needs to check if the potential is unvariant under

$$\phi_i(\mathbf{x},t) \xrightarrow{CP} X_{ij}\phi_j^*(-\mathbf{x},t)\,, \quad (2)$$

with any unitary $X \in U(N)$. These are often called generalized CP-transformations (GCP) [12, 11] as if contrasted to the "standard" CP. Upon a basis change $\phi'_i = U_{ij}\phi_j$, the new fields ϕ' transform under the same transformation (2) with the matrix $X' = UXU^T$. What looks standard in one basis becomes generalized in another, and a generalized one can become standard. In short, if it happens that the model is invariant under (2) with *any* whatever fancy unitary X, the model is called CP-conserving and this transformation plays the role of "the CP-symmetry". Indeed, all experimentally observable quantities which could signal the presence of CP-violation can be related to the CP-violating basis invariant combinations of input parameters and, — being basis-invariant, — they do not feel the effect of X [1].

The fact that this rule involves U^T rather than $U^\dagger$ has an important consequence: not every unitary X in (2) can be diagonalized by a basis change. The simplest form of X one can achieve is the block-diagonal form [12, 13, 7], with the blocks being either phases or 2×2 matrices

$$\begin{pmatrix} \cos\alpha & \sin\alpha \\ -\sin\alpha & \cos\alpha \end{pmatrix} \text{ as in [12], or } \begin{pmatrix} 0 & e^{i\alpha} \\ e^{-i\alpha} & 0 \end{pmatrix} \text{ as in appendix 2C of [13].} \quad (3)$$

Each block contains its own parameter α which can be arbitrary. Notice that applying GCP twice, one gets a usual family transformation, $\phi_i \to (XX^*)_{ij}\phi_j$, with XX^* not necessarily being identify. This opens up the possibility of CP-transformations of higher order [7]: if $\alpha = \pi/p$ with integer p, then one needs to apply it $2p$ times to obtain identity. By CPT-invariance of the standard interaction terms, this implies the T-symmetry of higher order. Although such a possibility sounds exotic, it is well consistent with all requirements of the local causal quantum field theory. Conversely, if one assumes that a GCP symmetry is of order two, then there exists a basis change which makes it diagonal, as in the left form in Eq. (3) with $\alpha = 0$ (the case of $\alpha = \pi$ reduces to it via an additional rephasing). Thus, order-2 GCP implies the possibility to bring the GCP transformation to the standard form (1), which in turn means that the Higgs potential is purely real in this basis. This link between the explicit CP-conservation and the existence of a real basis was formulated as Theorem 1 in [14]. Although it was formulated there in the context of 2HDM, the statement is valid for NHDM but it hinges on the assumption that the CP-symmetry is of order-2. For higher-order GCP, the link breaks down as demonstrated in [15].

Until recently, the possibility of having higher-order CP symmetry did not raise much phenomenological interest. In all concrete examples considered so far, imposing such a symmetry led to models with other accidental symmetries, including CP symmetries of order two, see 2HDM examples in [5, 7, 8]. It was viewed just as a compact way of defining such models, rather than a path towards *new* models beyond the usual "order-2 CP + family symmetry" combination. A rare exception is [16] where the higher-order CP symmetries were classified as distinct opportunities for model building.

The recent works [15, 17] gave the first concrete example of a multi-Higgs model, dubbed CP4-3HDM, in which the lagrangian was symmetric only under one specific CP-symmetry (2) of order 4 (which is labeled CP4) and its powers, without any other accidental symmetry [18]. This is the first example of a CP-conserving model without the usual CP-symmetry (1) in any basis. The unusual features of this model include:

- there exists no basis change which could make all coefficients real, thus barring the extention of Theorem 1 of [14] beyond 2HDM;

- in the CP-conserving vacuum, the four extra neutral Higgs bosons are pairwise mass-degenerate, and this degeneracy arises from the discrete symmetry CP4, without any continuous symmetry. It serves an example of the "state degeneracy beyond Kramers doubling", which was described in appendix 2C of [13] but no concrete example of which was known;
- despite CP-conservation, the extra physical neutral Higgses cannot be classified as CP-even or CP-odd. However, they can be combined into complex neutral fields Φ and φ which are CP-half-odd, that is, they transform under CP as

$$\Phi(\mathbf{x},t) \xrightarrow{CP} i\Phi(-\mathbf{x},t)\,, \quad \varphi(\mathbf{x},t) \xrightarrow{CP} i\varphi(-\mathbf{x},t)\,; \tag{4}$$

- one can couple these CP-half-odd scalars to the fermion bilinears via the Yukawa interactions; this coupling can respect CP4 provided the CP-transformation also acts in a generalized way on fermion generations. This technical possibility leads to a clash between the definitions of CP-symmetry and particle-antiparticle assignment for the fermions, which arguably cannot be resolved within this CP-conserving toy model.

In this contribution, we will describe the CP4-3HDM model itself, derive the physical Higgs bosons with the exotic CP-properties, and explore their couplings with the fermions.

2. CP4-3HDM and CP-half-odd scalars

2.1. The potential and physical Higgs spectrum

The model CP4-3HDM [15] is based on three Higgs doublets ϕ_i, $i = 1, 2, 3$, whose self-interaction potential is $V = V_0 + V_1$ contains the phase-insensitive part

$$\begin{aligned} V_0 &= -m_{11}^2(\phi_1^\dagger\phi_1) - m_{22}^2(\phi_2^\dagger\phi_2 + \phi_3^\dagger\phi_3) + \lambda_1(\phi_1^\dagger\phi_1)^2 + \lambda_2\left[(\phi_2^\dagger\phi_2)^2 + (\phi_3^\dagger\phi_3)^2\right] \\ &+ \lambda_3(\phi_1^\dagger\phi_1)(\phi_2^\dagger\phi_2 + \phi_3^\dagger\phi_3) + \lambda_3'(\phi_2^\dagger\phi_2)(\phi_3^\dagger\phi_3) + \lambda_4(|\phi_1^\dagger\phi_2|^2 + |\phi_1^\dagger\phi_3|^2) + \lambda_4'|\phi_2^\dagger\phi_3|^2\,, \end{aligned} \tag{5}$$

with all parameters being real, and the phase-sensitive part

$$V_1 = \lambda_5(\phi_3^\dagger\phi_1)(\phi_2^\dagger\phi_1) + \frac{\lambda_6}{2}\left[(\phi_2^\dagger\phi_1)^2 - (\phi_1^\dagger\phi_3)^2\right] + \lambda_8(\phi_2^\dagger\phi_3)^2 + \lambda_9(\phi_2^\dagger\phi_3)(\phi_2^\dagger\phi_2 - \phi_3^\dagger\phi_3) + h.c. \tag{6}$$

with real λ_5, λ_6, and complex λ_8, λ_9. This potential is invariant under the generalized CP transformation (2) with

$$X = \begin{pmatrix} 1 & 0 & 0 \\ 0 & 0 & i \\ 0 & -i & 0 \end{pmatrix}, \tag{7}$$

which corresponds to the right form of Eq. 3 with $\alpha = \pi/2$. Since $XX^* = \mathrm{diag}(1, -1, -1)$, one needs to apply it four times to get the identity transformation. This is the order-4 GCP dubbed CP4. For generic values of the coefficients, this potential has no other Higgs-family or CP-symmetries apart from powers of CP4 [18]. We note that this potential can be further simplified: with an appropriate $SO(2)$ rotation between ϕ_2 and ϕ_3 one can set either λ_5 or λ_6 to zero. However, in order to keep the same notation as in [15, 17], we will not apply this simplification.

We select the CP-conserving vacuum alignment: $\langle\phi_1^0\rangle = v/\sqrt{2}$, $\langle\phi_2\rangle = \langle\phi_3\rangle = 0$. For physical scalars, we get the SM-like Higgs with mass $m_{h_{SM}}^2 = 2\lambda_1 v^2 = 2m_{11}^2$, and a pair of degenerate charged Higgses with $m_{H^\pm}^2 = \lambda_3 v^2/2 - m_{22}^2$. In the neutral scalar sector, the mass matrices for real $h_{2,3}$ and imaginary $a_{2,3}$ components of $\phi^0_{2,3}$ split. These matrices are not identical but they have the same eigenvalues

$$M^2, m^2 = m^2_{H^\pm} + \frac{1}{2}v^2\left(\lambda_4 \pm \sqrt{\lambda_5^2 + \lambda_6^2}\right) \tag{8}$$

and are diagonalized by the same rotation with the angle α defined as $\tan 2\alpha = \lambda_5/\lambda_6$ but in the opposite directions for h's and a's. Denoting the two heavier scalars as H, A and the two ligher scalars as h, a, we find that the diagonalizing rotation brings ϕ^0_2 and ϕ^0_3 to

$$c_\alpha\phi^0_2 + s_\alpha\phi^0_3 = \frac{1}{\sqrt{2}}(H + ia)\,, \quad -s_\alpha\phi^0_2 + c_\alpha\phi^0_3 = \frac{1}{\sqrt{2}}(h + iA)\,. \tag{9}$$

The real neutral fields H, A, h, a are not CP-eigenstates:

$$H \xrightarrow{CP} A\,, \quad A \xrightarrow{CP} -H\,, \quad h \xrightarrow{CP} -a\,, \quad a \xrightarrow{CP} h\,. \tag{10}$$

One can combine them into neutral fields, $\Phi = (H - iA)/\sqrt{2}$, $\varphi = (h + ia)/\sqrt{2}$, which *are* CP and mass eigenstates:

$$\Phi(\mathbf{x}, t) \xrightarrow{CP} i\Phi(-\mathbf{x}, t)\,, \quad \varphi(\mathbf{x}, t) \xrightarrow{CP} i\varphi(-\mathbf{x}, t)\,. \tag{11}$$

One can then quantify the CP properties with the global quantum number q defined modulo 4, and assign $q = +1$ to Φ, φ, and $q = -1$ to their conjugate fields. All other neutral fields are either CP-odd, $q = +2$, or CP-even, $q = 0$. Since CP is a good symmetry of the lagrangian and of the vacuum, it commutes with the hamiltonian. Therefore, in any transition between initial and final states with definite q, this quantum number is conserved.

We stress that during these manipulations with the fields, we did not change the definition of CP-transformation itself. The CP-transformation in (11) is the same $\phi_i \to X_{ij}\phi_j^*$ with X given in (7). We just selected another basis to represent the scalars degrees of freedom.

2.2. How did the conjugation disappear?

The striking feature of the law (11) is that these are complex fields, yet they do not get complex-conjugated upon the CP-transformation. In fact, this law looks as a generalized P-transformation rather than CP. One can question whether identifying it with CP is legitimate at all. Even if it is, how did it happen that the conjugation disappeared upon a mere basis change?

An intuitive answer to these doubts is as follows. The C-transformation is expected, physically, to exchange one-particle states $a^\dagger|0\rangle$ with the corresponding antiparticles. If particles can be clearly distinguished from antiparticles via their gauge couplings, no ambiguity arises. However if we deal with several mass-degenerate *gauge-blind* scalars, then the particle vs. antiparticle distinction is blurred. The freedom of particle-antiparticle assignment and, consequently, of the basis changes allowed becomes larger. It is this enlarged basis change freedom that erases the complex conjugation upon CP.

To see how it occurs, let us step aside from the CP4-3HDM and consider an even simpler situation with two complex scalar fields ϕ_1 and ϕ_2, which are mass-degenerate and gauge-blind (that is, these fields do not participate in conserved gauge interactions). Suppose the model is invariant under the CP-symmetry $\phi_1 \xrightarrow{CP} \phi_2^*$ and $\phi_2 \xrightarrow{CP} \phi_1^*$. Let us perform a usual basis rotation and define new complex fields $\eta = (\phi_1 + \phi_2)/\sqrt{2}$ and $\xi = (-\phi_1 + \phi_2)/\sqrt{2}$. Then, η and ξ have "normal" CP properties: $\eta \xrightarrow{CP} \eta^*$, $\xi \xrightarrow{CP} -\xi^*$. In total, we have four real

IOP Conf. Series: Journal of Physics: Conf. Series **873** (2017) 012036 doi:10.1088/1742-6596/873/1/012036

degrees of freedom, two of them being CP-even ($\mathrm{Re}\,\eta$ and $\mathrm{Im}\,\xi$) and two are CP-odd ($\mathrm{Re}\,\xi$ and $\mathrm{Im}\,\eta$). Since all four of them are mass-degenerate, we are allowed to recombine them differently: $\Phi = \mathrm{Re}\,\eta - i\mathrm{Im}\,\xi$ and $\tilde{\Phi} = \mathrm{Re}\,\xi - i\mathrm{Im}\,\eta$. Then, we immediately see that upon *the same* CP-transformation, $\Phi \xrightarrow{CP} \Phi$ and $\tilde{\Phi} \xrightarrow{CP} -\tilde{\Phi}$. The conjugation disppeared.

To get further insight, we can express the two-step passage $(\phi_1, \phi_2) \to (\eta, \xi) \to (\Phi, \tilde{\Phi})$ as a single transformation:

$$\begin{pmatrix} \Phi \\ \tilde{\Phi} \end{pmatrix} = \frac{1}{\sqrt{2}} \begin{pmatrix} 1 & 1 \\ -1 & 1 \end{pmatrix} \begin{pmatrix} \phi_1 \\ \phi_2^* \end{pmatrix} . \tag{12}$$

We are allowed to perform this non-holomorphic basis change in the two-complex-dimensional space because there is no physical quantity which distinguishes ϕ's from ϕ^*'s. This enhanced basis change belongs to the group $O(4)$ which is larger than the complex-structure-preserving group $U(2)$ usually associated with basis changes. We have more freedom to reshuffle CP-even and CP-odd degrees of freedom, and it is this freedom that undoes the complex conjugation.

In short, when defining the general CP transformation in models with mass-degenerate gauge-blind scalars, the question of conjugating the fields or not becomes a basis-dependent choice.

In this example we used for the purpose of illustration the CP-transformation of order 2. A very similar phenomenon happens in the 3HDM with CP4. We start with a usual GCP which involves conjugation (2) and observe that the physical neutral Higgses with the same mass reside in two different complex fields (9). Since after the spontaneous symmetry breaking the neutral Higgses do not participate in the electromagnetic interactions, one can combine h with a and H with A to form new complex fields. This combination is made possible by the enhanced freedom of basis changes.

We mention that this is not the first example of CP-transformation acting on complex fields without conjugation. A similar situation happens in models with a complex scalar singlet $S = h_1 + ih_2$, see for example chapter 23.6 of the book [1]. The general argument is that when *building* a model gauge-blind singlets, we are free to assign CP properties of h_1 and h_2. This assignment defined the CP-transformation of the complex field S itself. If both h_1 and h_2 happen to be CP-even, then $S \xrightarrow{CP} S$. Of course, CP acts on the SM fields in the regular way; it is only in the singlet sector that it acts trivially. In the absence of fermion and gauge interactions, this transformation, if it leaves the lagrangian invariant, can be identified with the CP symmetry of the model. Thus, in models with one Higgs doublet and one complex singlet, there is no CP-violation in the scalar sector even if the S self interaction potential has complex coefficients. One would need to enlarge the particle content of the model, for example, by adding vector-like quarks, in order for CP-violation to take place and CP-violating effects to become observable [19, 20, 21].

In the case of CP4-3HDM, we do not even need to *assume* this choice. The pairwise degeneracy between components of the two inert doublets happens automatically as a result of the discrete symmetry imposed. Thus, it provides an example where this construction with non-congugated gauge-singlets appears naturally.

2.3. Coupling CP-half-odd scalars Yukawa

The simplest phenomenologically viable version of CP4-3HDM assumes that the inert doublets ϕ_2 and ϕ_3 do not couple to fermions, just like in the usual Inert doublet model (IDM) [22, 23, 24, 25]. However if one extends CP4 to the fermion sector and, in particular, assumes that it can also mix the fermion generations, then it is possible to couple CP-half-odd scalars with fermions via the usual Yukawa-type interactions.

Consider the quark Yukawa sector

$$-\mathcal{L}_Y = \bar{Q}_{Li}\Gamma_{a,ij} d_{Rj}\,\phi_a + \bar{Q}_{Li}\Delta_{a,ij} u_{Rj}\,\tilde{\phi}_a + h.c. \tag{13}$$

The scalar doublets transform under the CP as $\phi_a \to X_{ab}\phi_b^*$ with X given by (7), while the $n_f = 3$ generations of fermions are transformed

$$\psi_i \xrightarrow{CP} Y_{ij}^* \cdot \gamma^0 \mathcal{C} \bar{\psi}_j^T\,, \quad Y \in U(n_f)\,. \tag{14}$$

As usual, one can find a basis in which

$$Y = \begin{pmatrix} 1 & 0 & 0 \\ 0 & 0 & e^{i\alpha} \\ 0 & e^{-i\alpha} & 0 \end{pmatrix}. \tag{15}$$

Q_L, d_R, and u_R can in principle transform differently under the CP-transformation, with the three matrices Y_L, Y_{dR}, and Y_{uR}, but for the sake of demonstration we assume $Y_{dR} = Y_{uR} = Y_L = Y$. The requirement that the Yukawa sector is CP4-invariant translates into

$$\begin{aligned} &Y^\dagger \Gamma_1^* Y = \Gamma_1\,, \quad -iY^\dagger \Gamma_2^* Y = \Gamma_3\,, \quad iY^\dagger \Gamma_3^* Y = \Gamma_2\,, \\ &Y^\dagger \Delta_1^* Y = \Delta_1\,, \quad iY^\dagger \Delta_2^* Y = \Delta_3\,, \quad -iY^\dagger \Delta_3^* Y = \Delta_2\,. \end{aligned} \tag{16}$$

These requirements can be simultaneously satisfied in two cases:

- **case 1**: $\alpha = \pm\pi/4 + \pi k$:

$$\Gamma_1 = \begin{pmatrix} g_1 & 0 & 0 \\ 0 & g_2 & 0 \\ 0 & 0 & g_2^* \end{pmatrix}, \quad \Gamma_2 = \begin{pmatrix} 0 & 0 & 0 \\ 0 & 0 & g_{23} \\ 0 & g_{32} & 0 \end{pmatrix}, \quad \Gamma_3 = \begin{pmatrix} 0 & 0 & 0 \\ 0 & 0 & \pm g_{32}^* \\ 0 & \mp g_{23}^* & 0 \end{pmatrix}. \tag{17}$$

- **case 2**: $\alpha = \pm\pi/2$:

$$\Gamma_1 = \begin{pmatrix} g_1 & 0 & 0 \\ 0 & g_2 & g_3 \\ 0 & -g_3^* & g_2^* \end{pmatrix}, \quad \Gamma_2 = \begin{pmatrix} 0 & g_{12} & g_{13} \\ g_{21} & 0 & 0 \\ g_{31} & 0 & 0 \end{pmatrix}, \quad \Gamma_3 = \pm \begin{pmatrix} 0 & -g_{13}^* & g_{12}^* \\ g_{31}^* & 0 & 0 \\ -g_{21}^* & 0 & 0 \end{pmatrix}. \tag{18}$$

In both cases g_1 is real and all other entries are complex and independent. The expressions for Δ_a are of the same form, with parameters d_i instead of g_i and with the exchange of index $2 \leftrightarrow 3$. For leptons, one gets the similar construction as for the down quarks. Notice that in case 1, the GCP transformation is in fact of order 8 within the fermion sector. Thus, we have constructed the desired CP-conserving Yukawa sector in CP4-3HDM. The extra doublets do not have to be inert after all. We also mention that a similar problem of extended higher-order GCP to Yukawa sector was studied in [8] for 2HDM.

After spontaneous symmetry breaking, generation 2 and 3 fermions are mass-degenerate. For example, in the lepton sector, with the notation $\ell_i = (e, \mu, \tau)$ we have $m_\mu = m_\tau$ as the result of the conserved CP4. Expressing the Yukawa interactions via physical scalar bosons Φ, φ, and the physical fermions, we obtain for case 1 the following interaction pattern [17]:

$$-\mathcal{L}_Y = (\bar{\mu}\tau)(g\Phi - \tilde{g}\varphi) + (\bar{\tau}\gamma_5\mu)(\tilde{g}^*\Phi + g^*\varphi) + h.c., \tag{19}$$

where complex parameters g and $\tilde{g}$ are expressed via the entries of the matrices (17). Each interaction term here is separately CP4-invariant. For example, the μ-τ mixing GCP transformation with matrix Y given by (15) renders the bilinear $\bar{\mu}\tau$ CP-half-odd: $\bar{\mu}\tau \xrightarrow{CP} -i\bar{\mu}\tau$, and its $-i$ factor compensates the i factor from Φ. Also, notice that insertion of γ_5 changes q-charge by two units: $\bar{\mu}\gamma_5\tau \xrightarrow{CP} i\bar{\mu}\gamma_5\tau$. This is equivalent of an extra CP-oddness introduced with γ^5 , just like it happens in the usual case. A slightly longer expression is obtained for case 2:

$$\begin{aligned} -\mathcal{L}_Y &= (\bar{e}\mu + \bar{\tau}e)(g_+\Phi - \tilde{g}_+\varphi) - (\bar{\mu}e - \bar{e}\tau)(\tilde{g}_-^*\Phi + g_-^*\varphi) \\ &+ (\bar{e}\gamma_5\mu - \bar{\tau}\gamma_5 e)(g_-\Phi - \tilde{g}_-\varphi) - (\bar{\mu}\gamma_5 e + \bar{e}\gamma_5\tau)(\tilde{g}_+^*\Phi + g_+^*\varphi) + h.c., \end{aligned} \tag{20}$$

where the coupling constants are again expressed via the Yukawa matrices entries (18).

3. The clash of definitions

The resulting Yukawa interactions (19) and (20) exhibit a peculiar asymmetric pattern of couplings of the CP-half-odd scalars and their conjugates to fermion pairs. It is tempting to interpret interaction terms such as $\bar{\mu}\tau\Phi$ as a source of lepton flavour violation. However when reading physical processes off such interactions, one must not forget that particle-antiparticle assignments become tricky in the case of mass-degenerate fields.

Indeed, let us first consider scalars. After quantization, a complex scalar field is written in terms of creation and annihilation operators that satisfy the standard commutation relations, and it reads

$$\phi(\mathbf{x}, t) = \int \tilde{dp} \left[a(\mathbf{p})e^{-ipx} + b^\dagger(\mathbf{p})e^{ipx}\right], \tag{21}$$

where $px \equiv Et - \mathbf{px}$ and $\tilde{dp} \equiv d^3p/[2E(2\pi)^3]$. The standard assignment is that the one-particle states $a^\dagger(\mathbf{p})|0\rangle$ and $b^\dagger(\mathbf{p})|0\rangle$ correspond to a particle and its antiparticle. As a natural consequence of this convention, one postulates that the C-transformation acts on operators by exchanging a and b. In our case, we have CP-half-odd fields Φ and φ with CP-properties $\Phi \xrightarrow{CP} i\Phi$. Then it unavoidably follows that the corresponding operators a_Φ and b_Φ are *not* exchanged under CP4:

$$a_\Phi(\mathbf{p}) \xrightarrow{CP} ia_\Phi(-\mathbf{p}), \quad b_\Phi(\mathbf{p}) \xrightarrow{CP} -ib_\Phi(-\mathbf{p}). \tag{22}$$

This means that $a_\Phi^\dagger|0\rangle$ is a one-particle state labelled Φ, which is *its own antiparticle*. The other one-particle state $b_\Phi^\dagger|0\rangle$ is labelled as Φ^* and it is *another* particle, not antiparticle of Φ. Therefore, the field Φ in the interaction lagrangian can lead either to production of a particle Φ^* or to annihilation of a different particle Φ. Diagrammatically, when drawing a line ending in such a vertex, one must label it differently for the incoming and outgoing lines.

A similar convention must be used for fermions. The interaction $\bar{\mu}\tau\Phi$ in case 1 describes the Φ decay to a $\mu^+\mu^-$ pair (or $\tau^+\tau^-$ transition into Φ^*), while $\bar{\tau}\mu\Phi^*$ describes the Φ^* decay to a $\tau^+\tau^-$ pair. As a result, Φ and Φ^* have different decay preferences, but since they are not antiparticles of each other, these results are hardly surprising. The situation is less trivial in case 2, where at least the lepton-flavor-violating coupling between e and μ/τ exists.

However, in contrast with scalars, the fermions are charged and participate in the electromagnetic interactions via $(\bar{\mu}\gamma^\mu\mu + \bar{\tau}\gamma^\mu\tau)A_\mu$. This interaction is apparently diagonal in the fermion flavor. However expressing it via creation and annihilation operators, one sees that it leads not only to (virtual) subdiagrams $\mu^- \to \mu^-\gamma$ but also to $\mu^-\tau^+ \to \gamma$. One arrives at the counter-intuitive conclusion that, although a fermion can emit a photon without changing its flavor, it must pick up a *different* antifermion to annihilate into a photon.

One might find this conclusion sufficiently disturbing to revert the fermion-antifermion convention back to the usual one. That is, one assumes that a given fermion field contains the creation operator of a particle and the annihilation operator of *its antiparticle*. With this physically appealing definition, fermion annihilates together with its antifermion. But then the Yukawa interactions (19) and (20) will be manifestly CP-violating: each interaction term gives

IOP Conf. Series: Journal of Physics: Conf. Series **873** (2017) 012036 doi:10.1088/1742-6596/873/1/012036

preference to a fermion over its antifermion. However the model is still CP-conserving by all accepted standards.

To summarize this discussion, our model reveals a clash between two different conventions for particle-antiparticle assignments for charged fermions. One is "technical", it is consistent with the conserved CP-symmetry, but it leads to counter-intuitive transitions like $\mu^-\tau^+ \to \gamma$. The other is "physical"; it requires that, at tree-level, particles can only annihilate with their antiparticles into a photon. But in this case one must accept that a CP-conserving model leads to manifest CP-violation.

There is a third way: to simply avoid assigning who is antiparticle of whom. In this case, there is no such transformation as C-parity, and CP4 is just a peculiar symmetry linking different fields, not the CP symmetry. However it is not clear how one should phrase the physical phenomenon of CP violation and baryogenesis within this "C-agnostic" point of view.

Yet another possibility is that it is premature to draw any phenomenological conclusion from the above observations because this is a toy model with mass-degenerate μ and τ. It will be interesting to see whether in a phenomenologically relevant version of CP4-3HDM with spontaneously broken CP4 any interaction of this type persists and leads to observable signals. To this end, we notice that our model resembles the 2HDM with "maximal CP-symmetry" suggested and explored in [5, 6]. That model is also based on a GCP transformation of essentially the same family-mixing kind but only with two doublets. This has two key consequences in which it differs from CP4-3HDM. First, when applied twice on two doublets, it leads to an overall sign flip on both doublets, which is proportional to the unit matrix and can be removed by an overall sign flip. Thus, the transformation is effectively of order two, not four. In our case, an extra doublet stays invariant and it prevents removing the overall sign. Second, with two doublets, this initial symmetry must be broken by the vevs. As a result, it leads to a bunch of remarkable flavor violating signatures. In our unbroken-CP4 case, it is the first doublet which acquires the vev, and the symmetry persists. If broken, it may lead to similar flavor-violating signals, but since they arise now in addition to the first Yukawa structure, the strength of these effects can be controlled.

4. Towards phenomenology of CP4-3HDM

We have not yet explored the phenomenology of CP4-3HDM. Since it realizes the novel form of CP-conservation, it is definitely worth investigating its phenomenological manifestations, either in its simplest version with two inert doublets, or with CP4 symmetry extended to the Yukawa sector and then spontaneous broken. Below, we list several issues to be studied in future.

- In CP4-3HDM with exact CP4 and inert doublets, there are two mass-degenerate real scalar DM candidates. Unlike simpler scalar DM models with two DM candidates, such as IDM with $\lambda_5 = 0$, these two DM candidades do not coannihilate via Z-boson. This strongly affects the relic density calculations for light DM masses.
- The model we constructed is based on a novel form of CP-conservation, CP4. However it is not clear whether there can exist, even within the toy model, any phenomenological manifestation which can distinguish this and the usual CP. For that, one would need to find a process which cannot be mimicked by any CP-conserving model with the usual definition of CP. The only suggestions we have is to look into pairs of CP-half-odd scalars Φ. Since
$$(\mathcal{CP})a^\dagger_\Phi a^\dagger_\Phi(\mathcal{CP})^{-1} = -a^\dagger_\Phi a^\dagger_\Phi\,, \tag{23}$$
we find that such a pair must be only in even-partial-wave states, which are CP-odd. This peculiar assignment is impossible for any pair of CP-even/CP-odd scalar fields. However, we have not yet found any specific process that would recevie this peculiar feature.

- For a model with CP4 extended to the Yukawa sector, one would need to break CP4 spontaneously to produce a physical fermion sector. Although the CP4-symmetric Yukawa structures (18) contain many free parameters, it is not *a priori* clear if one can use it to fit all quark masses and mixing and, simulnatenously, avoid too large FCNCs. Fitting flavor sectors within this models and looking for its characteristic features is the next step along these lines. Including the neutrino sector in this description is also highly desirable.

Acknowledgments

This work was supported by the Portuguese *Fundação para a Ciência e a Tecnologia* (FCT) through the Investigator contract IF/00989/2014/CP1214/CT0004 under the IF2014 Program and in part by the contracts UID/FIS/00777/2013 and CERN/FIS-NUC/0010/2015, which are partially funded through POCTI, COMPETE, QREN, and the European Union.

References

[1] Branco G C, Lavoura L and Silva J P 1999 *CP Violation* (*The International Series of Monographs on Physics* no 103) (Oxford: Clarendon Press)
[2] Ivanov I P 2017 Building and testing models with extended Higgs sectors (*Preprint* 1702.03776)
[3] Lee T D 1973 *Phys. Rev.* **D8** 1226–1239 [,516(1973)]
[4] Branco G C, Ferreira P M, Lavoura L, Rebelo M N, Sher M and Silva J P 2012 *Phys. Rept.* **516** 1–102 (*Preprint* 1106.0034)
[5] Maniatis M, von Manteuffel A and Nachtmann O 2008 *Eur. Phys. J.* **C57** 739–762 (*Preprint* 0711.3760)
[6] Maniatis M and Nachtmann O 2009 *JHEP* **05** 028 (*Preprint* 0901.4341)
[7] Ferreira P M, Haber H E and Silva J P 2009 *Phys. Rev.* **D79** 116004 (*Preprint* 0902.1537)
[8] Ferreira P M and Silva J P 2010 *Eur. Phys. J.* **C69** 45–52 (*Preprint* 1001.0574)
[9] Feinberg G and Weinberg S 1959 *Il Nuovo Cimento* **14** 571–592
[10] Lee T D and Wick G C 1966 *Phys. Rev.* **148** 1385–1404 [,445(1966)]
[11] Grimus W and Rebelo M N 1997 *Phys. Rept.* **281** 239–308 (*Preprint* hep-ph/9506272)
[12] Ecker G, Grimus W and Neufeld H 1987 *J. Phys.* **A20** L807
[13] Weinberg S 2005 *The Quantum theory of fields. Vol. 1: Foundations* (Cambridge University Press) ISBN 9780521670531, 9780511252044
[14] Gunion J F and Haber H E 2005 *Phys. Rev.* **D72** 095002 (*Preprint* hep-ph/0506227)
[15] Ivanov I P and Silva J P 2016 *Phys. Rev.* **D93** 095014 (*Preprint* 1512.09276)
[16] Chen M C, Fallbacher M, Mahanthappa K T, Ratz M and Trautner A 2014 *Nucl. Phys.* **B883** 267–305 (*Preprint* 1402.0507)
[17] Aranda A, Ivanov I P and Jiménez E 2016 When the C in CP does not matter: anatomy of CP-half-odd scalars and their Yukawa interactions (*Preprint* 1608.08922)
[18] Ivanov I P, Keus V and Vdovin E 2012 *J. Phys.* **A45** 215201 (*Preprint* 1112.1660)
[19] Bento L and Branco G C 1990 *Phys. Lett.* **B245** 599–604
[20] Branco G C, Parada P A and Rebelo M N 2003 A Common origin for all CP violations (*Preprint* hep-ph/0307119)
[21] Darvishi N 2016 *JHEP* **11** 065 (*Preprint* 1608.02820)
[22] Deshpande N G and Ma E 1978 *Phys. Rev.* **D18** 2574
[23] Ma E 2006 *Phys. Rev.* **D73** 077301 (*Preprint* hep-ph/0601225)
[24] Barbieri R, Hall L J and Rychkov V S 2006 *Phys. Rev.* **D74** 015007 (*Preprint* hep-ph/0603188)
[25] Lopez Honorez L, Nezri E, Oliver J F and Tytgat M H G 2007 *JCAP* **0702** 028 (*Preprint* hep-ph/0612275)

IOP Conf. Series: Journal of Physics: Conf. Series **873** (2017) 012037 doi:10.1088/1742-6596/873/1/012037

CP as a Symmetry of Symmetries

Andreas Trautner

Bethe Center for Theoretical Physics und Physikalisches Institut der Universität Bonn, Nussallee 12, 53115 Bonn, Germany

E-mail: atrautner@uni-bonn.de

Abstract. It is explained that the Standard Model combined charge conjugation and parity transformation (CP) is a simultaneous complex conjugation outer automorphism transformation of gauge and space-time symmetries. Simple examples are given for the general concept of outer automorphisms ("symmetries of symmetries"), as well as for their possible actions on physical theories. It is highlighted that complex conjugation outer automorphisms do not, in general, exist for all symmetries. Examples are given for cases in which the physical CP transformation is violated as a consequence of requiring another symmetry. A toy model is illustrated in which CP is spontaneously violated in the broken phase of a continuous gauge symmetry, while an unbroken outer automorphism protects the topological vacuum angle at $\theta = 0$.

1. Introduction

One of the main questions raised by the Standard Model of particle physics (SM) is the origin of the SM flavor structure. By observation, the violation of flavor and the violation of matter-anti matter symmetry are highly intertwined in the SM. In particular, CP violation has never been observed without flavor violation, implying that also the strong CP problem can be regarded as inherent part of the flavor puzzle. Ultimately, a complete theory of flavor should also be a theory of CP violation.

Among many other approaches, the flavor puzzle has been addressed with discrete symmetries. In this context, the interplay of CP transformations and discrete flavor symmetries has been investigated [1, 2]. It has been found that CP transformations are described by particular outer automorphisms of discrete symmetries [2, 3]. In review, this is not surprising as it has already been understood much earlier that outer automorphisms also describe CP transformations of gauge [4] and space-time symmetries [5].

A crucial difference between discrete and continuous groups is that many discrete groups are inconsistent with outer automorphisms which correspond to physical CP transformations in generic models [3]. For such groups it has been found that CP violating phases are calculable and take quantized ("geometrical") values. This has been observed in models of explicit [3, 6, 7] and spontaneous CP violation [8, 3] alike. Moreover, very recently a model has been found in which geometrical CP violation is predicted in the dynamical breaking of a continuous gauge symmetry to a discrete subgroup, while an unbroken outer automorphism enforces $\theta = 0$ for the topological vacuum angle of the gauge group [9].

The aim of this talk is to give a brief overview of the topic of predictive CP violation and outer automorphisms. For this, the concept of an outer automorphism (a "symmetry of a symmetry") is discussed based on simple examples. Then it is explained how the well established

CP transformation of the SM can be understood as a simultaneous outer automorphism transformation of gauge and space-time symmetries. Extrapolating from the SM, a general definition of a physical CP transformation is given. Based on this we discuss the existence of model independent CP transformations for the most interesting continuous groups, as well as the absence of such transformations for a certain class of discrete groups. Finally, we illustrate a toy model of spontaneous geometrical CP violation in which the outer automorphism is unbroken and enforces $\theta = 0$. For the ease of the presentation we will skip many mathematical details.

2. What is an outer automorphism?

Outer automorphisms are transformations that map a group to itself, while not being part of the group themselves. Unfortunately, outer automorphisms are rarely treated in the standard introductory literature on group theory. Some details for the case of Lie groups can be found in [10] while a pedagogical introduction for finite groups has been given in [11]. Here, we will only discuss a very simple example of an outer automorphism, which suffices to understand the basic idea.

Consider the symmetry group $\mathbb{Z}_3$ which is generated by a single element a and the relation $\mathsf{a}^3 = \mathsf{id}$. A complete list of group elements is $\left\{\mathsf{id}, \mathsf{a}, \mathsf{a}^2\right\}$. *Inner* automorphisms are conjugation maps $\mathsf{g} \mapsto \mathsf{g}' = \mathsf{h}\mathsf{g}\mathsf{h}^{-1}$ (with $\mathsf{g}, \mathsf{g}', \mathsf{h} \in G$) and they partition the group G into conjugacy classes. Skipping over mathematical rigor, *outer* automorphisms (Outs) can be thought of as conjugation maps $\mathsf{g} \mapsto \mathsf{g}' = \mathsf{u}\mathsf{g}\mathsf{u}^{-1}$ with elements u that are *not* part of the group. For the case of $\mathbb{Z}_3$ there is a single Out which can be generated by the transformation

$$\mathrm{Out}(\mathbb{Z}_3) : \quad \mathsf{a} \mapsto u(\mathsf{a}) = \mathsf{a}^2 . \tag{1}$$

We note that this map induces a permutation of group elements within the list $\left\{\mathsf{id}, \mathsf{a}, \mathsf{a}^2\right\}$. In general, Outs act as permutation of group elements and conjugacy classes while leaving the unsorted list of group elements invariant. Also, Outs do not change the structure (i.e. the multiplication algebra) of the group. This justifies to call outer automorphisms the non-trivial "symmetries of a symmetry".

In a concrete sense, Outs act as a permutation of representations of a group, Out : $\boldsymbol{r}_i \mapsto \boldsymbol{r}_j$ where i, j label the irreducible representations (irreps) of the group. In order to be consistent with the group structure, the corresponding representation matrix U of an Out has to fulfill [12, 11]

$$U\, \rho_{\boldsymbol{r}'}(\mathsf{g})\, U^{-1} = \rho_{\boldsymbol{r}}(u(\mathsf{g})) , \qquad \forall \mathsf{g} \in G , \tag{2}$$

where $\rho_{\boldsymbol{r}}(\mathsf{g})$ denotes the matrix representation of g in the representation $\boldsymbol{r}$, and $u(\mathsf{g})$ is the abstract Out transformation on the group elements which maps $\boldsymbol{r} \mapsto \boldsymbol{r}'$. In particular, this can include the special case $\boldsymbol{r}' = \boldsymbol{r}^*$ [2]. After having made a basis choice for $\boldsymbol{r}$ and $\boldsymbol{r}'$, U is fixed by this relation up to a central element of the group and a global phase.

For the case of $\mathbb{Z}_3$, the simultaneous action of $\mathrm{Out}(\mathbb{Z}_3) \cong \mathbb{Z}_2$ on group elements and representations is depicted in table 1. The transformation (1) turns out to be a complex conjugation Out in the sense that $\boldsymbol{r}_i \mapsto \boldsymbol{r}_j = \boldsymbol{r}_i{}^*$ for all irreps of $\mathbb{Z}_3$.

2.1. Outer automorphisms of groups

For any group, the possible Outs are dictated by the structure of the group itself. For the specific cases of simple Lie groups and for finite groups some general statements can be made. The Outs of simple Lie groups are given by the symmetries of the Dynkin diagram of the corresponding Lie algebra. In the case of finite groups, Outs are symmetries of the character table under permutation of rows and columns (cf. table 1). However, this relation is not 1 : 1, meaning that there can be non-trivial Outs which leave the character table invariant. Vice versa, there can also be symmetries of the character table which do not correspond to an Out. Especially for

Table 1. Character table of $\mathbb{Z}_3$. The arrows illustrate the action of the unique $\mathbb{Z}_2$ outer automorphism on the conjugacy classes and representations. We use the definition $\omega := \mathrm{e}^{2\pi\mathrm{i}/3}$.

$\mathbb{Z}_3$	id	a ⟷	a^2
1	1	1	1
1′ ⟲	1	ω	ω^2
1″ ⟲	1	ω^2	ω

Table 2. Outer automorphisms of all simple Lie groups as well as for some randomly chosen examples of finite discrete groups. The action on representations denoted by $\boldsymbol{r}_i \to \boldsymbol{r}_j$ denotes some unspecified permutation of representations other than a complex conjugation outer automorphism.

Simple Lie Group	Algebra	Out	Action on reps
SU(N)	$A_{n>1}$	$\mathbb{Z}_2$	$\boldsymbol{r} \to \boldsymbol{r}^*$
SO(8)	$D_{n=4}$	S_3	$\boldsymbol{r}_i \to \boldsymbol{r}_j$
SO($2N$)	$D_{n>4}$	$\mathbb{Z}_2$	$\boldsymbol{r} \to \boldsymbol{r}^*$
E_6	E_6	$\mathbb{Z}_2$	$\boldsymbol{r} \to \boldsymbol{r}^*$
all others		/	/

Group	Out	Action on reps
$\mathbb{Z}_3$	$\mathbb{Z}_2$	$\boldsymbol{r} \to \boldsymbol{r}^*$
$\mathrm{A}_{n\neq 6}$	$\mathbb{Z}_2$	$\boldsymbol{r} \to \boldsymbol{r}^*$
$\mathrm{S}_{n\neq 6}$	/	/
T_7	$\mathbb{Z}_2$	$\boldsymbol{r}_i \to \boldsymbol{r}_j$
$\Delta(27)$	$\mathrm{GL}(2,3)$	$\boldsymbol{r}_i \to \boldsymbol{r}_j$
$\Delta(54)$	S_4	$\boldsymbol{r}_i \to \boldsymbol{r}_j$

discrete groups, the number of distinct Outs can easily surpass the order of the group. A useful tool in finding Outs of small finite groups is `GAP` [13]. Some examples for groups and their Outs are given in table 2.

2.2. Physics of outer automorphisms

A generic physical model (possibly a QFT) can be specified by stating the underlying symmetry group as well as the field content in terms of representations of the group. The symmetry group may have a number of Outs, and a central question is how these are related to the physics described by the model.

As described above, the general action of an Out on the irreps of a group is a permutation map $\boldsymbol{r}_i \mapsto \boldsymbol{r}_j$. For a model which includes an irrep $\boldsymbol{r}_i$ there are, in general, two logical possibilities for the action of an Out:

(i) Either, both $\boldsymbol{r}_i$ and $\boldsymbol{r}_j$ are included in the model,

(ii) or, $\boldsymbol{r}_j$ is not part of the model.

IOP Conf. Series: Journal of Physics: Conf. Series **873** (2017) 012037 doi:10.1088/1742-6596/873/1/012037

Let us briefly discuss these two cases, and give two well-known examples.

In the first case (i), the Out transformation would, in principle, be a possible symmetry transformation of the model in the sense that it permutes already present (symmetry invariant) operators. However, as it is not a symmetry of the model *by assumption*, it will generally be broken by the values of some couplings. Turning this around it is clear that the Out will join the symmetries of the model *iff* the couplings are tuned to very specific values. Altogether, the Out behaves very much like an explicitly (softly) broken symmetry transformation. Regions in the parameter space with enhanced symmetry, thus, can be identified by Outs. The existence of such regions has interesting implications in general. For example, parameter values with enhanced symmetry will be boundaries of the RGE flow.

In the second case (ii), the Out is in no way a possible symmetry of the model as the target representation $\boldsymbol{r}_j$ is not part of the model. Performing the Out mapping nevertheless, would map the model to a physically distinct model. We see that the Out is broken by the representation content of the model. Transformations of this kind cannot be conserved by tuning of parameters and they are called explicitly and *maximally* broken.

Let us give two examples for the possible action of Outs based on the SM.

(i) CP transformation in the SM. Anticipating the results of the next section, the CP transformation of the SM maps $\boldsymbol{r} \mapsto \boldsymbol{r}^*$ for all representations (details will be given below). For example, it maps

$$Q_{\mathrm{L}} \equiv (\mathbf{3},\mathbf{2})^{\mathrm{L}}_{1/6} \mapsto (\overline{\mathbf{3}},\overline{\mathbf{2}})^{\mathrm{R}}_{-1/6} \sim Q^*_{\mathrm{L}} \,. \tag{3}$$

Performed on the action, this transformation is equivalent to a mapping in the parameter space $V_{\mathrm{CKM}} \mapsto (V_{\mathrm{CKM}})^*$. Is this a symmetry of the model? Only experimental input can decide this question. Experimentally it is well established that $\delta_{\mathrm{CKM}} \neq \{0,\pi\}$, for what reason this Out, which corresponds to a physical CP transformation, is explicitly broken by the values of couplings.

(ii) P transformation in the SM. Pictorially, parity maps $\boldsymbol{r}_{\mathrm{L}} \mapsto \boldsymbol{r}_{\mathrm{R}}$ where L and R denote chiral representations of the Lorentz-group. In the SM, for example

$$Q_{\mathrm{L}} \equiv (\mathbf{3},\mathbf{2})^{\mathrm{L}}_{1/6} \mapsto (\mathbf{3},\mathbf{2})^{\mathrm{R}}_{1/6} \equiv Q_{\mathrm{R}} \,. \tag{4}$$

However, $\mathrm{Q_R}$ is not part of the model to begin with, meaning that this transformation would map the SM to some other model. We see that this transformation is broken explicitly and *maximally*, i.e. not by the values of couplings but by the representation content of the model.

Even though these examples may appear trivial, they nicely illustrate the possible action of Outs in general, also for other models and also for transformations other than the well-known P or CP. For a first application of this to models with Outs other than CP see [12, 11]. As the concept of symmetries of symmetries is as general as the concept of symmetry, we expect that many more applications of such transformations will come to light in the future.

With this formulation at hand, it is very easy to argue that CP cannot be broken maximally, i.e. by leaving out or adding representations. For a real valued action, the Lagrangian must always involve both, $\boldsymbol{r}$ and $\boldsymbol{r}^*$ in mutually hermitean conjugate operators. Therefore, a simultaneous mapping $\boldsymbol{r} \mapsto \boldsymbol{r}^*$ of all representations cannot be broken maximally.

3. CP as a special (outer) automorphism

Let us show that the usual CP transformation of the SM is a simultaneous complex conjugation Out of all present symmetries.

IOP Conf. Series: Journal of Physics: Conf. Series **873** (2017) 012037 doi:10.1088/1742-6596/873/1/012037

The most general possible CP transformation for (SM) gauge and one generation of (chiral) fermion fields is given by (cf. e.g. [4])

$$\begin{aligned} W^a_\mu(x) &\mapsto R^{ab}\,\mathcal{P}_\mu^{\ \nu}\,W^b_\nu(\mathcal{P}\,x)\;, \\ \Psi^i_\alpha(x) &\mapsto \eta_{\mathsf{CP}}\,U^{ij}\,\mathcal{C}_{\alpha\beta}\,\Psi^{*j}_{\ \beta}(\mathcal{P}\,x)\;. \end{aligned} \tag{5}$$

Here W^a_μ and Ψ^i_α denote the gauge and fermion fields, $\mathcal{P} = \mathrm{diag}(1,-1,-1,-1)$ is the usual spatial reflection, and η_{CP} denotes an arbitrary phase. Greek lowercase letters run within the corresponding Lorentz group representation, while lowercase Roman letters run within the respective gauge group representation and we implicitly sum over repeated indices. $\mathcal{C}$ and U here are general unitary matrices, allowing the fermions to rotate in the Dirac representation space of the Lorentz group and in the representation space of the gauge group, respectively. R is orthogonal and can be chosen to be real due to the reality of the gauge fields.

The CP transformation (5) is a conserved symmetry of the gauge-kinetic action *iff*

$$\begin{aligned} &\text{(i)}: & R_{aa'}\,R_{bb'}\,f_{a'b'c} &= f_{abc'}\,R_{c'c}\;, \\ &\text{(ii)}: & U\,(-\mathsf{T}_a^{\mathrm{T}})\,U^{-1} &= R_{ab}\,\mathsf{T}_b\;, \\ &\text{(iii)}: & \mathcal{C}\,(-\gamma^{\mu\,\mathrm{T}})\,\mathcal{C}^{-1} &= \gamma^\mu\;. \end{aligned} \tag{6}$$

Here, T_a and f_{abc} are the generators and structure constants of the gauge group fulfilling the usual relation $[\mathsf{T}_a, \mathsf{T}_b] = \mathrm{i} f_{abc}\,\mathsf{T}_c$, while γ^μ denote the usual Dirac γ-matrices which form the generators $S^{\mu\nu} = \mathrm{i}/4 \times [\gamma^\mu, \gamma^\nu]$ of the Dirac spinor representation.

The equations (6) are special cases of the general consistency condition (2) for the gauge and space-time symmetries. This shows that the most general physical CP transformation is a special automorphism of both, the gauge and space-time symmetries of a model. Namely, the physical CP transformation in the SM corresponds to an automorphism which simultaneously maps all present (gauge and space-time) symmetry representations to their own complex conjugate representations, $\boldsymbol{r}_i \mapsto \boldsymbol{r}_i^{\,*}\ \forall\, i$. Such an automorphism must be outer for a group that has complex representations. To prevent confusion we emphasize that neither C nor P alone are complex conjugation automorphisms of the Lorentz group. Only the combined CP transformation is a complex conjugation Out of the Lorentz group (cf. e.g. [5]).

If the field Ψ would, in addition, also transform under some additional global symmetry, it is imperative that also the corresponding representation of the global symmetry is mapped onto its own complex conjugate representation. That is, also for all additional symmetry groups beyond the SM the CP transformation should map representations to their own complex conjugate representations.[1] Extrapolating from the SM, we can use these insights in order to formulate a group theoretical definition of a physical CP transformation. In words:

A physical CP transformation is an (outer) automorphism transformation which simultaneously maps all present representations of all symmetries (global, local, space-time) to their respective complex conjugate representations.

This definition generalizes previous definitions of CP given for simple Lie groups [4] or the Poincaré group [5]. Of course, it is also consistent with the standard textbook treatments of CP.

We note that a physical CP transformation does not have to be unique (i.e. there can be multiple, equally valid CP transformations) nor is it guaranteed to exist at all for a given symmetry.

[1] Strictly speaking this is true only for fields which are charged under the SM gauge group or which couple to SM neutral operators that are charged under the new symmetry, implying that completely neutral fields or completely decoupled sectors could behave differently.

IOP Conf. Series: Journal of Physics: Conf. Series **873** (2017) 012037 doi:10.1088/1742-6596/873/1/012037

As an important remark, we note that it is, of course, always possible to impose *any* complex conjugation (i.e. physical CP) transformation on a given model – even if there exists no corresponding complex conjugation automorphism of the symmetry group G of the model. However, this will generally enhance the linear symmetry of the model, possibly up to a decoupling of the model (as it happens for our example in section 5). It would then be unreasonable to speak of a G-symmetric setting for what reason this case is not considered here.

Let us also remark that there are, in general, Outs which have nothing to do with physical CP whatsoever. In general, it is model dependent whether or not a given automorphism constitutes a physical CP transformation. For example, an automorphism which corresponds to a physical CP transformation for one model may not be a physical CP transformation once additional representations are added to the model. This will be the case for discrete groups of type I below. In contrast, if a symmetry allows for a complex conjugation automorphism, this always corresponds to a model independent physical CP transformation.

Finally, we stress that our above definition does not require the CP transformation to be of order two, i.e. it equally allows for CP transformations of higher order.

4. Do CP transformations exist for all models/symmetry groups?

Following the previous section we understand model independent physical CP transformations as complex conjugation Outs of a group, and the general answer to the question in the subheading is no.

4.1. CP transformations for Lie groups

For the particular case of (semi-)simple Lie groups it has been shown that complex conjugation Outs always exist [4]. This does, of course, not imply that CP is automatically conserved in a model based on such a group. In contrast, models which allow for CP transformations are generally not predictive with respect to the CP violating phase, meaning that it would need an experiment to find out whether or not CP is violated explicitly. There are, however, situations in which all potentially CP violating parameters are unphysical and CP is automatically conserved (cf. [14] for a very detailed treatment of when this is the case).

For other groups, it has to be checked on a case-by-case basis whether they allow for a complex conjugation automorphism transformation or not. The physically interesting case of the Poincaré or Lorentz group has been investigated in [5] and it has been found that there exists a complex conjugation Out which serves as a model independent physical CP transformation.

4.2. CP and discrete groups

A prominent set of groups for which complex conjugation Outs do not, in general, exist are (finite) discrete groups [3]. A simple indicator to decide whether or not a complex conjugation Out exists for a given discrete group has been presented in [3] (see also [15]). In particular, all (finite) discrete groups can be classified into two categories depending on whether or not a complex conjugation Out exists. We will briefly discuss the categories in the following. Examples for each category of finite groups are shown in table 3.

Groups of the so-called "type I" do not allow for a complex conjugation Out. Depending on the representation content of a specific model based on a type I group, it can happen that the physical CP transformation $\boldsymbol{r}_i \mapsto \boldsymbol{r}_i^* \ \forall\ i$ is inconsistent with the symmetry of the model. If this is the case, CP is violated *as a consequence* of requiring the type I symmetry. Explicitly it has been found that the necessarily complex Clebsch–Gordan coefficients of type I groups then enter CP violating observables such as asymmetries in oscillations and decays [3]. These phases can be calculated and they take discrete "geometrical" values (fractions of 2π). Models based

IOP Conf. Series: Journal of Physics: Conf. Series **873** (2017) 012037 doi:10.1088/1742-6596/873/1/012037

Table 3. Some examples for discrete groups of type I, II A, and II B with their common names and SmallGroups library ID of GAP, see [3] for details.

(a) Examples for groups of type I.

G	$\mathbb{Z}_5 \rtimes \mathbb{Z}_4$	T_7	$\Delta(27)$	$\mathbb{Z}_9 \rtimes \mathbb{Z}_3$
SG id	$(20,3)$	$(21,1)$	$(27,3)$	$(27,4)$

(b) Examples for groups of type II A.

G	S_3	A_4	T'	S_4	A_5
SG id	$(6,1)$	$(12,3)$	$(24,3)$	$(24,12)$	$(60,5)$

(c) Examples for groups of type II B.

G	$\Sigma(72)$	$((\mathbb{Z}_3 \times \mathbb{Z}_3) \rtimes \mathbb{Z}_4) \rtimes \mathbb{Z}_4$
SG id	$(72,41)$	$(144,120)$

on type I groups, hence, can be truly predictive with respect to CP violation. This way of CP violation has been termed explicit geometrical CP violation [16, 6, 7].

We stress that not all models based on type I groups automatically violate CP. In non-generic settings CP can be conserved even though the model features a type I symmetry. This can happen only if the group has an automorphism that maps $\boldsymbol{r}_j \mapsto \boldsymbol{r}_j{}^*$ for a subset of representations j (including a faithful representation), and if the model contains only representations out of this subset.

While CP then can be conserved, it could also be violated spontaneously. In this case, the discrete phases originating from the complex Clebsch-Gordan coefficients of the type I group can become relative physical phases of VEVs [12]. This is known as spontaneous geometrical T violation [8].

The second category of finite groups are groups of "type II". These groups allow for an Out that simultaneously maps all irreps to their own complex conjugate irreps. This allows for a model independent definition of a physical CP transformation. CP violation in type II groups then works completely analogous to the case of simple Lie groups. Namely, it is always possible to perform a CP transformation in consistency with the symmetry. The CP transformation could be violated by the value of certain couplings, but CPV cannot be predicted.

A special subset of type II groups are the groups of the so-called type II B. For them, a complex conjugation Out exists, but it generates a CP transformation of order larger than two. In this case, CP can be defined but conserving it typically requires an additional global symmetry [3].

4.3. Higher order CP transformations

As an aside we note that the presence of complex couplings is in general *not* sufficient to conclude that CP is violated. More specifically, it has been found that there are models with higher-order CP transformations in which couplings are necessarily complex even though CP is manifestly conserved [17]. Exactly the same situation arises for groups of the type II B, which also have only higher-order CP transformations and necessarily complex Clebsch-Gordan coefficients entering couplings [3]. A generic consequence of conserved higher-order CP transformations is the presence of additional linear symmetries [18] as well as the presence of exotic CP "half–odd" states with potentially very interesting phenomenological consequences [19].

5. CP violation in the breaking of continuous to discrete

An interesting observation is that type I groups can arise as subgroups of simple Lie groups. While simple Lie groups allow for a complex conjugation Out which serves to define a model independent physical CP transformation, a type I subgroup does not have a complex conjugation Out and, hence, also no possible physical CP transformation in a generic setting. As argued above, this implies that CP is violated in generic settings based on a type I group. An immediate question, therefore, is what happens to the CP transformation if a simple Lie group is broken to a type I subgroup.

This has been investigated in a concrete model [9] based on a gauged SU(3) symmetry which is dynamically broken to a T_7 subgroup by the vacuum expectation value (VEV) of a complex scalar ϕ in the **15**-plet representation of SU(3) (Dynkin indices $\{2,1\}$). The model is given by

$$\mathscr{L} \;=\; (D_\mu\,\phi)^\dagger\,(D^\mu\,\phi) - \frac{1}{4}\,G^a_{\mu\nu}\,G^{\mu\nu,a} - \mathscr{V}(\phi)\;, \tag{7}$$

where $D_\mu = \partial_\mu - \mathrm{i}g\,A_\mu$ is the gauge covariant derivative and $G^a_{\mu\nu}$ the field strength tensor. The scalar potential is given by

$$\mathscr{V}(\phi) \;=\; -\,\mu^2\,\phi^\dagger\phi + \sum_{i=1}^{5}\lambda_i\,\mathcal{I}_i^{(4)}(\phi)\;, \tag{8}$$

with $\mu^2 > 0$, and $\mathcal{I}_i^{(4)}$ denote the five possible quartic SU(3) invariants in the self-contraction of the **15**-plet. These invariants are real and hence $\lambda_i \in \mathbb{R}$ without loss of generality. For some choice of parameters, this potential has a global minimum $\langle\phi\rangle$ that spontaneously breaks $\mathrm{SU}(3) \to \mathrm{T}_7$ [20, 21]. Crucially, the VEV does *not* break the Out transformation of SU(3). Thus, if the Out was imposed to be an exact symmetry initially, it continues to be an exact symmetry also in the broken phase.

Requiring CP conservation at the level of SU(3) amounts to require the complex conjugation automorphism to be conserved. The actual symmetry breaking chain in the model is then given by

$$\mathrm{SU}(3) \rtimes \mathbb{Z}_2 \xrightarrow{\langle\phi\rangle} \mathrm{T}_7 \rtimes \mathbb{Z}_2\;. \tag{9}$$

However, at the level of T_7, the $\mathbb{Z}_2$ Out transformation does not correspond to a physical CP transformation. This can be understood because some of the physical (T_7) states in the broken phase are complex linear combinations of SU(3) states, in such a way that the Out does not map them to their own complex conjugate.

At the level of T_7, physical CP is thus violated by calculable phases, irrespective of the presence of an unbroken Out [9]. This has been explicitly cross-checked by computing a non-vanishing CP-odd basis invariant in the broken phase of the model [9].

Note that there is also no other possible CP transformation at the level of T_7. This can easily be concluded from the fact that T_7 is of type I, and that the branching of the SU(3) fields to

T_7 is such that all T_7 irreps are populated. If one would plainly enforce a CP transformation on the model by simply requiring the transformation $\boldsymbol{r}_{T_7} \mapsto \boldsymbol{r}^*_{T_7}$ for all T_7 representations to be a symmetry, all couplings vanish and one ends up with a free theory.

The conserved Out acts on the SU(3) gauge fields like a CP transformation, i.e. it prohibits a possible topological term

$$\mathscr{L}_\theta = \theta \frac{g^2}{32\pi^2} G^a_{\mu\nu} \widetilde{G}^{\mu\nu,a} . \tag{10}$$

The unbroken Out continues to prohibit the θ-term in the broken phase, where the Out, however, cannot be any longer interpreted as a physical CP transformation of the T_7 states. This provides a symmetry reason for $\theta = 0$ also in the broken phase, despite the fact that CP is violated by other T_7 invariant couplings [9].

Whether or not a similar construction can serve as a solution to the strong CP problem of the SM remains to be seen. In any case it is interesting to see that one and the same outer automorphism can have different physical interpretations depending on the linearly realized symmetries of the ground state of a theory.

6. Conclusion

It has been shown how outer automorphisms can be understood as non-trivial symmetries of a symmetry. For physics purpose these transformations are best thought of as (certain) permutations among the irreps of a group. We have then shown that CP transformations in the SM are complex conjugation outer automorphisms of gauge and space-time symmetries. Extrapolating from the SM, we have given a general definition of a physical CP transformation as a simultaneous mapping of all symmetry representations to their own respective complex conjugate representations. It has been pointed out that physical CP transformations are clashing with certain discrete symmetries, in which case CP violating phases with quantized values can be predicted. Finally, we have presented a toy model in which a continuous gauge symmetry is dynamically broken to a discrete subgroup that does not allow for a physical CP transformation. In this case geometrical CP violation arises while an unbroken outer automorphism transformation warrants $\theta = 0$.

Acknowledgments

This work has been supported by the German Science Foundation (DFG) within the SFB-Transregio TR33 "The Dark Universe".

References

[1] Feruglio F, Hagedorn C and Ziegler R 2013 *JHEP* **1307** 027 (*Preprint* 1211.5560)
[2] Holthausen M, Lindner M and Schmidt M A 2013 *JHEP* **1304** 122 (*Preprint* 1211.6953)
[3] Chen M C, Fallbacher M, Mahanthappa K, Ratz M and Trautner A 2014 *Nucl.Phys.* **B883** 267 (*Preprint* 1402.0507)
[4] Grimus W and Rebelo M 1997 *Phys.Rept.* **281** 239–308 (*Preprint* hep-ph/9506272)
[5] Buchbinder I L, Gitman D M and Shelepin A L 2002 *Int. J. Theor. Phys.* **41** 753–790 (*Preprint* hep-th/0010035)
[6] Branco G C, de Medeiros Varzielas I and King S F 2015 *Phys. Rev.* **D92** 036007 (*Preprint* 1502.03105)
[7] Varzielas I d M 2015 *J. Phys. Conf. Ser.* **631** 012020 (*Preprint* 1503.02633)
[8] Branco G, Gerard J and Grimus W 1984 *Phys.Lett.* **B136** 383
[9] Ratz M and Trautner A 2017 *JHEP* **02** 103 (*Preprint* 1612.08984)
[10] Fuchs J and Schweigert C 1997 *Symmetries, Lie algebras and representations: A graduate course for physicists* (Cambridge, UK: University Press) 438 p.
[11] Trautner A 2016 *CP and other Symmetries of Symmetries* Ph.D. thesis Munich, Tech. U., Universe (*Preprint* 1608.05240) URL http://inspirehep.net/record/1482355/files/arXiv:1608.05240.pdf
[12] Fallbacher M and Trautner A 2015 *Nucl. Phys.* **B894** 136–160 (*Preprint* 1502.01829)
[13] The GAP Group 2012 *GAP – Groups, Algorithms, and Programming, Version 4.5.5*

IOP Conf. Series: Journal of Physics: Conf. Series **873** (2017) 012037 doi:10.1088/1742-6596/873/1/012037

[14] Haber H E and Surujon Z 2012 *Phys.Rev.* **D86** 075007 (*Preprint* 1201.1730)
[15] Bickerstaff R and Damhus T 1985 *International Journal of Quantum Chemistry* **XXVII** 381–391
[16] Chen M C and Mahanthappa K 2009 *Phys.Lett.* **B681** 444–447 (*Preprint* 0904.1721)
[17] Ivanov I P and Silva J P 2015 (*Preprint* 1512.09276)
[18] Grimus W 1988 *Fortsch. Phys.* **36** 201
[19] Aranda A, Ivanov I P and Jiménez E 2016 (*Preprint* 1608.08922)
[20] Luhn C 2011 *JHEP* **1103** 108 (*Preprint* 1101.2417)
[21] Merle A and Zwicky R 2012 *JHEP* **1202** 128 (*Preprint* 1110.4891)

IOP Conf. Series: Journal of Physics: Conf. Series **873** (2017) 012038 doi:10.1088/1742-6596/873/1/012038

Phenomenological prospects for two Higgs doublet models with controlled FCNC

Miguel Nebot[1]

[1] Centro de Física Teórica de Partículas (CFTP), Instituto Superior Técnico (IST), U. de Lisboa (UL), Av. Rovisco Pais, P-1049-001 Lisboa, Portugal.

E-mail: miguel.r.nebot.gomez@tecnico.ulisboa.pt

Abstract. Flavour changing processes like $t \to hu, hc$, $h \to \tau e, \tau\mu$ as well as hadronic decays $h \to bs, bd$, are analysed within a class of two Higgs doublet models where flavour changing neutral scalar currents at tree level are present. The most remarkable characteristic of this class of models is that the flavour-violating couplings are entirely determined by the fermion masses, the ratio of vacuum expectation values $\tan\beta$ and the CKM and PMNS matrices. The flavour structure of the scalar currents results from a symmetry of the Lagrangian and consequently it is natural and stable under the renormalization group evolution. We show that, in some of the models, the rates of the mentioned flavour changing processes can reach values at the discovery level for the LHC operating at 13 TeV, even taking into account stringent bounds on low energy processes.

1. Introduction

The second run of the Large Hadron Collider (LHC), with a center of mass energy $\sqrt{s} = 13$ TeV, is starting to probe the flavour changing couplings of the 125 GeV scalar boson h [1, 2]. These couplings can contribute to rare top decays like $t \to hq$ ($q = u, c$) and may lead as well to (a) flavour changing leptonic decays such as $h \to \tau^{\pm}\ell^{\mp}$, ($\ell = \mu, e$), and (b) flavour changing hadronic decays like $h \to bs$ and $h \to bd$. In the Standard Model (SM) these decays are highly suppressed since the required couplings vanish at tree level. However, Higgs Flavour Violating Neutral Couplings (HFVNC) can arise in many extensions of the SM, including in particular the Two Higgs Doublet Models (2HDM) [3–5]. Any extension of the SM featuring HFVNC has to comply with strict experimental limits on processes mediated by flavour changing neutral currents (FCNC) as well as with the constraints on CP violating transitions producing, for example, electric dipole moments of quarks and leptons [6].

We analyse the allowed size of HFVNC in a class of 2HDM, denoted BGL models, which was first proposed for the quark sector [7], and then generalised [8] and extended to the leptonic sector [9]. BGL models have the remarkable property of having HFVNC with a flavour structure entirely determined by the Cabibbo-Kobayashi-Maskawa (CKM) and the Pontecorvo-Maki-Nakagawa-Sakata (PMNS) matrices, denoted V and U respectively, in addition to fermion masses and the ratio of vacuum expectation values $\tan\beta$ of the two scalar doublets. HFVNC have been widely addressed in the literature [10–28]. The distinctive feature in BGL models is the natural suppression of many of the most dangerous HFVNC through combinations of small mixing matrix elements and/or light fermion masses. This is a consequence of the introduction of

IOP Conf. Series: Journal of Physics: Conf. Series **873** (2017) 012038 doi:10.1088/1742-6596/873/1/012038

a symmetry in the Lagrangian and hence the suppression is entirely natural. Another important characteristic of BGL models is that, depending on the specific type model model within this class, HFVNC exist either in the up or in the down sector, but not in both sectors simultaneously; analogous considerations apply to the leptons. A generalisation of BGL models which includes FCNC in both the up and the down sectors at the cost of few extra parameters has been recently proposed in [29]. In the general 2HDM there are three neutral scalars in the so-called Higgs basis [30–32], H^0, R^0 and A. The couplings of H^0 to fermions in the fermion mass eigenstate basis are flavour diagonal. On the other hand R^0 and A have HFVNC with an arbitrary structure. BGL models are remarkable because the flavour structure of HFVNC is limited to depend on V and U. The neutral scalar mass eigenstates are then linear combinations of H^0 and R^0 together with the pseudoscalar neutral field A. In these models, the symmetry constrains the scalar potential not to violate CP either explicitly or spontaneously, therefore rotating (with angle β) the fields from the symmetry basis to the Higgs basis, makes the charged fields and the pseudoscalar field A already physical fields. It also results in the two other neutral physical fields being related to H^0 and R^0 through a single angle rotation.

Previous work [33] gave a detailed analysis of the allowed mass ranges for the new scalars under the assumption that the discovered Higgs h could be identified with H^0. In some of the BGL models these masses can be in the range of a few hundred GeV and thus within reach of searches at the LHC running at 13 TeV. The present analysis [34] addresses the general case where h is a mixture of H^0 and R^0, which is determined by an angle denoted $\beta-\alpha$. The intensity of the HFVNC of h depends crucially on $\tan\beta \equiv v_2/v_1$, with v_i the vacuum expectation values of the scalar doublets, and $\cos(\beta-\alpha)$. BGL models share some features with implementations of the minimal flavour violation hypothesis (MFV) [16, 35–37]. Nevertheless it is important to stress that BGL models have the unique property of coming from a symmetry, which produces a reduced number of free parameters, allowing for definite predictions once constraints on these parameters given by present experimental bounds are considered.

One can summarise the challenge with the following question: can one have regions in the $\tan\beta$ versus $\alpha-\beta$ plane, where, in some of the BGL models, the HFVNC of h can give rates for the rare processes $t \to hq$, $h \to \mu\tau$ consistent with discovery at LHC-13 TeV? Of course, these scenarios have to be consistent with the stringent constraints on all Standard Model (SM) processes associated to the Higgs production through different mechanisms (e.g. gluon-gluon fusion, vector boson fusion, Higgs-strahlung) and its subsequent decays into ZZ, WW, $\gamma\gamma$, $b\bar{b}$ and $\tau\bar{\tau}$. In addition, constraints from low energy phenomenology have to be considered: those obtained in [33] and also the new ones due to the presence of a mixing H^0-R^0. Although processes such as $h \to bs$ and $h \to bd$ are probably out of reach of the LHC, they may become important for the physics of the future Linear Collider, and thus they are also included in the analysis.

The discussion is organised in the following manner. In the next section we review BGL models and set the notation. In Section 3 we analyse top flavour changing decays $t \to hq$ ($q = u, c$) in BGL models with HFVNC in the up quark sector. Flavour changing decays of the Higgs are addressed in Section 4. We consider in particular neutrino models with HFVNC in the charged lepton sector giving rise to $h \to \ell\tau$ decays. Up type BGL models, with HFVNC in the down quark sector, are also considered since they give rise to $h \to bs$, $h \to bd$ decays. In Section 5 we discuss some aspects of the analysis and investigate the discovery regions and the existing correlations among the decays of interest.

2. BGL Models

The quark Yukawa interactions in the context of 2HDM can be written as:

$$\begin{aligned}\mathcal{L}_Y = &- \bar{Q}_L^0\left[\Gamma_1\,\Phi_1 + \Gamma_2\,\Phi_2\right] d_R^0 - \bar{Q}_L^0\left[\Delta_1\,\tilde{\Phi}_1 + \Delta_2\,\tilde{\Phi}_2\right] u_R^0 \\ &- \bar{L}_L^0\left[\Pi_1\,\Phi_1 + \Pi_2\,\Phi_2\right] l_R^0 - \bar{L}_L^0\left[\Sigma_1\,\tilde{\Phi}_1 + \Sigma_2\,\tilde{\Phi}_2\right] \nu_R^0 + \text{h.c.},\end{aligned} \tag{1}$$

IOP Conf. Series: Journal of Physics: Conf. Series **873** (2017) 012038 doi:10.1088/1742-6596/873/1/012038

where Γ_i, Δ_i Π_i and Σ_i are matrices in flavour space. The requirement that Γ_i, Δ_i lead to tree level FCNC with strength completely controlled by the Cabibbo-Kobayashi-Maskawa matrix V, was achieved by Branco, Grimus and Lavoura (BGL) [7] through a symmetry of the Lagrangian of the form:

$$Q^0_{Lj} \mapsto \exp(i\tau)\, Q^0_{Lj}\,, \qquad u^0_{Rj} \mapsto \exp(i2\tau)\, u^0_{Rj}\,, \qquad \Phi_2 \mapsto \exp(i\tau)\, \Phi_2\,, \tag{2}$$

where $\tau \neq 0, \pi$, with all other quark fields transforming trivially. The index j can be fixed as either 1, 2 or 3. Alternatively, the symmetry can be chosen as:

$$Q^0_{Lj} \mapsto \exp(i\tau)\, Q^0_{Lj}\,, \qquad d^0_{Rj} \mapsto \exp(i2\tau)\, d^0_{Rj}\,, \quad \Phi_2 \mapsto \exp(-i\tau)\, \Phi_2\,. \tag{3}$$

Equation (2) leads to Higgs FCNC in the down sector, whereas the symmetry of eq. (3) leads to Higgs FCNC in the up sector. These two alternative choices combined with the three possible elections of the index j give rise to six different realizations of 2HDM with the flavour structure, in the quark sector, controlled by the V matrix. Up-type BGL models are those with HFVNC in the down sector, defined bythe symmetry imposed through eq. (2); each one of the three implementations is labelled u-type, c-type or t-type depending on the value of the index j, respectively 1, 2 or 3. Likewise for the down-type models. In the leptonic sector, with Dirac neutrinos, there is perfect analogy with the quark sector, and the corresponding symmetry applied to the leptonic fields leads to six different realizations with the strength of Higgs mediated flavour changing neutral currents now controlled by the Pontecorvo-Maki-Nakagawa-Sakata matrix, U. This results in thirty six different implementations of BGL models. However, as shown in [9], in the case of Majorana neutrinos, there are only 18 models corresponding to the neutrino types, and therefore with HFVNC in the charged lepton sector.
The discrete symmetry of Eqs. (2) or (3) constrains the Higgs potential to be of the form:

$$\begin{aligned} V = {}& \mu_1 \Phi_1^\dagger \Phi_1 + \mu_2 \Phi_2^\dagger \Phi_2 - m_{12}\left(\Phi_1^\dagger \Phi_2 + \Phi_2^\dagger \Phi_1\right) + 2\lambda_3 \left(\Phi_1^\dagger \Phi_1\right)\left(\Phi_2^\dagger \Phi_2\right) \\ & + 2\lambda_4 \left(\Phi_1^\dagger \Phi_2\right)\left(\Phi_2^\dagger \Phi_1\right) + \lambda_1 \left(\Phi_1^\dagger \Phi_1\right)^2 + \lambda_2 \left(\Phi_2^\dagger \Phi_2\right)^2 . \end{aligned} \tag{4}$$

The term in m_{12} is a soft symmetry breaking term; its introduction prevents the appearance of an would-be Goldstone boson due to the continuous global symmetry of the potential which arises when the BGL symmetry is exact. Since such a potential cannot violate CP either explicitly or spontaneously, the scalar and pseudoscalar neutral fields do not mix among themselves and there are only two important rotation angles, β and α. The angle β parametrises the rotation to the Higgs basis, singling out the three neutral fields: H^0, with couplings to the quarks proportional to mass matrices, R^0 which is a neutral scalar and A which is a neutral pseudoscalar (in addition the physical charged Higgs fields $H^\pm$ and the pseudo-Goldstone bosons). In BGL models, A and $H^\pm$ are already physical fields, while H^0 and R^0 may still mix. In the limit in which H^0 does not mix with R^0, H^0 is identified with the Higgs field h discovered by ATLAS [1] and CMS [2]. In this limit, H^0 does not mediate tree level flavour changes; α is defined in such a way that the mixing angle between these fields, $(\beta - \alpha)$, acquires the value $\pi/2$ in that case. Expanding around the vacuum expectation values of the neutral fields [38], $\phi^0_j = \frac{1}{\sqrt{2}}(v_j + \rho_j + i\eta_j)$, we can write:

$$\begin{pmatrix} \mathrm{H}^0 \\ \mathrm{R}^0 \end{pmatrix} \equiv \begin{pmatrix} \cos\beta & \sin\beta \\ -\sin\beta & \cos\beta \end{pmatrix} \begin{pmatrix} \rho_1 \\ \rho_2 \end{pmatrix}, \quad \begin{pmatrix} H \\ h \end{pmatrix} \equiv \begin{pmatrix} \cos\alpha & \sin\alpha \\ -\sin\alpha & \cos\alpha \end{pmatrix} \begin{pmatrix} \rho_1 \\ \rho_2 \end{pmatrix}. \tag{5}$$

The angle β is given by $\tan\beta \equiv v_2/v_1$; we use in the following the shorthand notation $\tan\beta \equiv t_\beta$, $\cos(\beta - \alpha) \equiv c_{\beta\alpha}$ and $\sin(\beta - \alpha) \equiv s_{\beta\alpha}$.

IOP Conf. Series: Journal of Physics: Conf. Series **873** (2017) 012038 doi:10.1088/1742-6596/873/1/012038

In terms of quark mass eigenstates and the scalar fields in the Higgs basis, the Yukawa couplings are:

$$\begin{aligned}\mathcal{L}_{\text{Yuk}} = & -\frac{\sqrt{2}\,H^+}{v}\,\bar{u}\left[V N_d \gamma_R - N_u^\dagger V \gamma_L\right] d + \text{h.c.} \\ & -\frac{\text{H}^0}{v}\left[\bar{u}\,D_u\,u + \bar{d}\,D_d\,d\right] \\ & -\frac{\text{R}^0}{v}\left[\bar{u}\,(N_u\gamma_R + N_u^\dagger\gamma_L)\,u + \bar{d}\,(N_d\gamma_R + N_d^\dagger\gamma_L)\,d\right] \\ & + i\frac{A}{v}\left[\bar{u}\,(N_u\gamma_R - N_u^\dagger\gamma_L)\,u - \bar{d}\,(N_d\gamma_R - N_d^\dagger\gamma_L)\,d\right]\end{aligned} \tag{6}$$

with γ_L and γ_R the left-handed and right-handed chirality projectors, and D_d and D_u are the diagonal mass matrices for down and up quarks, respectively. Equation (6) defines N_d and N_u, the matrices which give the flavour structure and strength of FCNC. In general 2HDM, N_d and N_u are entirely arbitrary. On the contrary, BGL models have the remarkable feature of having N_d and N_u entirely determined by fermion masses, V and the angle β, with no other free parameters. In BGL up-type models the matrices N_d has the simple form:

$$\left(N_d^{(u_j)}\right)_{rs} = \left[t_\beta \delta_{rs} - (t_\beta + t_\beta^{-1}) V_{jr}^* V_{js}\right] (D_d)_{ss}\,, \tag{7}$$

with no sum in j implied. The upper index (u_j) indicates that labels a symmetry of the form in eq. (2), i.e. an up-type model with index j leading to FCNC in the down-sector. All FCNC are proportional to the factor $(t_\beta + t_\beta^{-1})$, multiplying products of entries involving one single row of V. The corresponding N_u matrix is:

$$\left(N_u^{(u_j)}\right)_{rs} = \left[t_\beta - (t_\beta + t_\beta^{-1})\delta_{rj}\right] (D_u)_{ss}\,\delta_{rs}\,. \tag{8}$$

N_u is a diagonal matrix and the t_β dependence is not the same for each diagonal entry: it is proportional to $-t_\beta^{-1}$ for the (jj) element and to t_β for all other elements. The index j fixes the row of the V matrix which suppresses the flavour changing neutral currents in eq. (7). For down-type models, which correspond to the symmetry in eq. (3), the rôles of N_d and N_u are exchanged:

$$\left(N_u^{(d_j)}\right)_{rs} = \left[t_\beta \delta_{rs} - (t_\beta + t_\beta^{-1}) V_{rj} V_{sj}^*\right] (D_u)_{ss}\,, \tag{9}$$

$$\left(N_d^{(d_j)}\right)_{rs} = \left[t_\beta - (t_\beta + t_\beta^{-1})\delta_{rj}\right] (D_d)_{ss}\,\delta_{rs}\,. \tag{10}$$

The flavour changing neutral currents are suppressed, in down-type models, by the columns of the V matrix.

The next ingredient is to allow for the possibility of h being a linear combination of H^0 and R^0; this is parametrised by the angle $(\beta - \alpha)$:

$$h = s_{\beta\alpha}\,\text{H}^0 + c_{\beta\alpha}\,\text{R}^0\,, \quad H = c_{\beta\alpha}\,\text{H}^0 - s_{\beta\alpha}\,\text{R}^0\,. \tag{11}$$

This mixing will be constrained by data from the LHC observables concerning “the Higgs” (see section 5). The quark Yukawa couplings of h can be written as

$$\mathcal{L}_{hq\bar{q}} = -Y_{ij}^D\,\bar{d}_{Li}\,d_{Rj}\,h - Y_{ij}^U\,\bar{u}_{Li}\,u_{Rj}\,h + \text{h.c.}\,, \tag{12}$$

and similarly for the leptonic sector with the coefficients denoted by Y^{ℓ}_{ij} and Y^{ν}_{ij}. From Eqs. (6) and (11), one can read:

$$
\begin{aligned}
Y^D_{ij} &= \frac{1}{v}\left[s_{\beta\alpha}\,(D_d)_{ij} + c_{\beta\alpha}\,(N_d)_{ij}\right] \\
Y^U_{ij} &= \frac{1}{v}\left[s_{\beta\alpha}\,(D_u)_{ij} + c_{\beta\alpha}\,(N_u)_{ij}\right]
\end{aligned} \tag{13}
$$

Explicitely, for $i \neq j$, we have the following flavour violating Yukawa couplings in the different types of BGL models[1]:

(i) up-type u_k model, k fixed as 1 (u) or 2 (c) or 3 (t), with HFVNC in the down quark sector:

$$Y^D_{ij}(u_k) = -V^*_{ki}V_{kj}\,\frac{m_{d_j}}{v}\,c_{\beta\alpha}(t_\beta + t_\beta^{-1})\,,\ i \neq j,\ \text{no sum in } k\,, \tag{14}$$

(ii) down-type d_k model, k fixed as 1 (d) or 2 (s) or 3 (b), with HFVNC in the up quark sector:

$$Y^U_{ij}(d_k) = -V_{ik}V^*_{jk}\,\frac{m_{u_j}}{v}\,c_{\beta\alpha}(t_\beta + t_\beta^{-1})\,,\ i \neq j,\ \text{no sum in } k\,, \tag{15}$$

(iii) leptonic sector, neutrino-type, ν_k model, k fixed as 1 (ν_1) or 2 (ν_2) or 3 (ν_3), with HFVNC in the charged lepton sector:

$$Y^{\ell}_{ij}(\nu_k) = -U_{ik}U^*_{jk}\,\frac{m_{l_j}}{v}\,c_{\beta\alpha}(t_\beta + t_\beta^{-1})\,,\ i \neq j,\ \text{no sum in } k\,. \tag{16}$$

For Dirac neutrinos one can write similar expressions for charged lepton type models, but in this case FCNC appear in the neutrino sector and are suppressed by the extremely small neutrino masses.

3. Flavour changing decays of top quarks

We analyse now flavour changing decays of top quarks $t \to hq$. According to eqs. (13) and (15), the couplings of h with a top t and a u or c quark, in a model of type d_ρ, are

$$Y^U_{qt}(d_\rho) = -V_{q\rho}V^*_{t\rho}\,\frac{m_t}{v}\,c_{\beta\alpha}(t_\beta + t_\beta^{-1})\,, \quad q = u, c\,. \tag{17}$$

The corresponding $t \to hq$ decay rate is

$$\Gamma_{(d_\rho)}(t \to hq) = \frac{m_t^3}{32\pi v^2}\left(1 - \frac{m_h^2}{m_t^2}\right)^2 |V_{q\rho}|^2|V_{t\rho}|^2 c^2_{\beta\alpha}(t_\beta + t_\beta^{-1})^2\,. \tag{18}$$

Apart from the global factor $c^2_{\beta\alpha}(t_\beta + t_\beta^{-1})^2$, every other factor in eq. (18) is fixed once the specific down-type model d_ρ and the decay channel $t \to hc$ or $t \to hu$ are chosen. Consequently, for a given model, $t \to hq$ processes constrain the factor $c^2_{\beta\alpha}(t_\beta + t_\beta^{-1})^2$. The branching ratio for $t \to hq$ in the d_ρ type model is

$$\mathrm{Br}_{(d_\rho)}(t \to hq) = \frac{\Gamma_{(d_\rho)}(t \to hq)}{\Gamma(t \to Wb)} = f(x_h, y_W)\,\frac{|V_{q\rho}V_{t\rho}|^2}{|V_{tb}|^2}\,c^2_{\beta\alpha}(t_\beta + t_\beta^{-1})^2\,, \tag{19}$$

where

$$f(x_h, y_W) = \frac{1}{2}\,(1 - x_h)^2\left(1 - 3y_W^2 + 2y_W^3\right)^{-1}\,, \text{ with } x_h = \frac{m_h^2}{m_t^2}\,,\ y_W = \frac{M_W^2}{m_t^2}\,. \tag{20}$$

IOP Conf. Series: Journal of Physics: Conf. Series **873** (2017) 012038 doi:10.1088/1742-6596/873/1/012038

Table 1. V factors in eq. (18), $\lambda \simeq 0.22$ [39, 40].

Model	$t \to hu$	$t \to hc$
d	$\lvert V_{ud}V_{td}\rvert^2\,(\sim \lambda^6) = 7.51 \cdot 10^{-5}$	$\lvert V_{cd}V_{td}\rvert^2\,(\sim \lambda^8) = 4.01 \cdot 10^{-6}$
s	$\lvert V_{us}V_{ts}\rvert^2\,(\sim \lambda^6) = 8.20 \cdot 10^{-5}$	$\lvert V_{cs}V_{ts}\rvert^2\,(\sim \lambda^4) = 1.53 \cdot 10^{-3}$
b	$\lvert V_{ub}V_{tb}\rvert^2\,(\sim \lambda^6) = 1.40 \cdot 10^{-5}$	$\lvert V_{cb}V_{tb}\rvert^2\,(\sim \lambda^4) = 1.68 \cdot 10^{-3}$

Using the top quark pole mass $m_t = 173.3$ GeV [39], $m_h = 125.0$ GeV and $M_W = 80.385$ GeV, one obtains $f(x_h, y_W) = 0.1306$. Table 1 shows the V factors involved in different decay channels and models. The most interesting models for $t \to hc$ are the s and b models, where the suppression is only at the λ^4 level, compared to the d model which has a strong suppression, at the λ^8 level, for the same decay. The d model has the curiosity that the suppression is higher for $t \to hc$ than for $t \to hu$, unlike in the s and b models. Now, considering the upper bounds 0.79% from the ATLAS [41] and 0.56% from the CMS [42, 43] collaborations, we have, for b and s-type models,

$$|c_{\beta\alpha}(t_\beta + t_\beta^{-1})| \lesssim 4.9\,. \tag{21}$$

Notice that, for this value, perturbative unitarity constraints from the scalar sector have to be considered (see [34]).

4. Flavour changing Higgs decays

4.1. The decays $h \to \ell\tau$ ($\ell = \mu, e$)

For leptonic HFVNC, the most interesting BGL models are the ν models. As with quarks, there are three neutrino-type BGL models, depending on the column of the U matrix involved in leptonic FCNC. With notation analogous to the one of the quark sector and considering eq. (16) for the coupling of h to μ and τ,

$$Y^{\ell}_{\mu\tau}(\nu_\rho) = \frac{1}{v}\, c_{\beta\alpha}\, (N_\ell^{(\nu_\sigma)})_{\mu\tau} = -c_{\beta\alpha}(t_\beta + t_\beta^{-1})\, U_{\mu\sigma} U^*_{\tau\sigma}\, \frac{m_\tau}{v}\,, \tag{22}$$

and the decay rate is

$$\Gamma_{(\nu_\sigma)}(h \to \mu\bar{\tau}) + \Gamma_{(\nu_\sigma)}(h \to \bar{\mu}\tau) = c^2_{\beta\alpha}(t_\beta + t_\beta^{-1})^2 |U_{\mu\sigma}U_{\tau\sigma}|^2\, \Gamma_{\rm SM}(h \to \tau\bar{\tau})\,, \tag{23}$$

with $\Gamma_{\rm SM}(h \to \tau\bar{\tau}) = \frac{m_h}{8\pi}\frac{m_\tau^2}{v^2}$. Notice, again, the appearance of the same factor $c^2_{\beta\alpha}(t_\beta + t_\beta^{-1})^2$. Table 2 lists the PMNS mixing matrix factors for the different ν - type models. The first direct search for lepton-flavour-violating decays of the observed Higgs boson performed by the CMS collaboration [44], led to the observation of a slight excess of signal events, with a significance of 2.4 standard deviations. The best fit value is:

$$\text{Br}(h \to \mu\bar{\tau} + \tau\bar{\mu}) = \left(0.84\,^{+0.39}_{-0.37}\right)\%\,, \tag{24}$$

which sets a constraint on the branching fraction $\text{Br}(h \to \mu\bar{\tau} + \tau\bar{\mu}) < 1.51\%$ at the 95% confidence level. The ATLAS collaboration presented a result based on hadronic τ decays [45], giving $\text{Br}(h \to \mu\bar{\tau} + \tau\bar{\mu}) = (0.77 \pm 0.62)\%$. While analyses at 8 TeV have been extended to

[1] These couplings, as well as the diagonal ones, can be extracted from equations (7) – (10) making, where necessary, the corresponding changes of quarks by leptons.

Table 2. U factors in eq. (23); estimates, e.g. 1/9, 1/36, correspond to a tri-bimaximal U (except for $|U_{e3}|$, of course); analogous information for $h \to e\mu$ and $h \to e\tau$ decays is included.

Model	$h \to e\mu$	$h \to e\tau$	$h \to \mu\tau$
ν_1	$\|U_{e1}U_{\mu 1}\|^2(\sim \frac{1}{9}) = 0.105$	$\|U_{e1}U_{\tau 1}\|^2(\sim \frac{1}{9}) = 0.118$	$\|U_{\mu 1}U_{\tau 1}\|^2(\sim \frac{1}{36}) = 0.028$
ν_2	$\|U_{e2}U_{\mu 2}\|^2(\sim \frac{1}{9}) = 0.089$	$\|U_{e2}U_{\tau 2}\|^2(\sim \frac{1}{9}) = 0.126$	$\|U_{\mu 2}U_{\tau 2}\|^2(\sim \frac{1}{9}) = 0.115$
ν_3	$\|U_{e3}U_{\mu 3}\|^2 = 0.0128$	$\|U_{e3}U_{\tau 3}\|^2 = 0.0097$	$\|U_{\mu 3}U_{\tau 3}\|^2(\sim \frac{1}{4}) = 0.234$

additional channels [46, 47], results at 13 TeV are yet unclear [48]. Assuming the h width to be $\Gamma_h \simeq \Gamma_h^{[\mathrm{SM}]}(= 4.03$ MeV), one can use the SM branching ratio $\mathrm{Br}_{\mathrm{SM}}(h \to \tau\bar{\tau}) = 0.0637$ in eq. (23), and obtain the estimate

$$|c_{\beta\alpha}(t_\beta + t_\beta^{-1})| \sim 1\,, \tag{25}$$

necessary to produce $\mathrm{Br}(h \to \mu\bar{\tau} + \tau\bar{\mu})$ of order 10^{-2}.

4.2. The flavour changing decays $h \to bq$ ($q = s, d$)

We now address FCNC in the down sector. The most promising experimental signatures correspond to $h \to bq$ decays, with $q = s, d$. The relevant h couplings to down quarks in eq. (12) are, according to eq. (14):

$$Y^D_{qb}(u_k) = -c_{\beta\alpha}(t_\beta + t_\beta^{-1})\, V^*_{kq} V_{kb}\, \frac{m_b}{v}\,, \ q \neq b, \text{ no sum in } k\,. \tag{26}$$

Once again, it should be emphasised that once the up-type model u_k is chosen, the strength of the flavour changing couplings only depends on the combination $c_{\beta\alpha}(t_\beta + t_\beta^{-1})$ together with the down quark masses and V factors which are already known. The decay rate of h to pairs of quarks $q_i q_j$ ($i \neq j$) is

$$\Gamma_{(u_k)}(h \to \bar{q}_i q_j + q_i \bar{q}_j) = \frac{3\, m_h}{8\pi}\left[\frac{1}{2}|Y_{ij}|^2 + \frac{1}{2}|Y_{ji}|^2\right]\,, \tag{27}$$

and thus

$$\Gamma_{(u_k)}(h \to \bar{b}q + b\bar{q}) = c^2_{\beta\alpha}(t_\beta + t_\beta^{-1})^2\, |V_{kq}|^2 |V_{kb}|^2\, \Gamma_{\mathrm{SM}}(h \to b\bar{b})\,. \tag{28}$$

With $\Gamma_h \simeq \Gamma_h^{[\mathrm{SM}]}$, one can make the following estimate

$$\mathrm{Br}_{(u_k)}(h \to \bar{b}q + b\bar{q}) = c^2_{\beta\alpha}(t_\beta + t_\beta^{-1})^2\, |V_{kq}|^2 |V_{kb}|^2\, \mathrm{Br}_{\mathrm{SM}}(h \to b\bar{b})\,, \tag{29}$$

where $\mathrm{Br}_{\mathrm{SM}}(h \to b\bar{b}) = 0.578$. The relevant CKM factors for $h \to bs$ and $h \to bd$ in u_k BGL models are given in Table 3.

To a good approximation we have

- in models c and t, $\mathrm{Br}(h \to \bar{b}s + b\bar{s}) \sim c^2_{\beta\alpha}(t_\beta + t_\beta^{-1})^2\, \lambda^4 \sim 10^{-3}\, c^2_{\beta\alpha}(t_\beta + t_\beta^{-1})^2$,
- in model u, $\mathrm{Br}(h \to \bar{b}s + b\bar{s}) \sim c^2_{\beta\alpha}(t_\beta + t_\beta^{-1})^2\, \lambda^8 \sim 10^{-7}\, c^2_{\beta\alpha}(t_\beta + t_\beta^{-1})^2$,
- in all u, c and t models, $\mathrm{Br}(h \to \bar{b}d + b\bar{d}) \sim c^2_{\beta\alpha}(t_\beta + t_\beta^{-1})^2\, \lambda^6 \sim 10^{-5}\, c^2_{\beta\alpha}(t_\beta + t_\beta^{-1})^2$.

We stress that, a priori, in models without the $h \to \mu\tau$ constraint, one can reach values for $\mathrm{Br}(h \to b\bar{s} + s\bar{b})$ not far from 10^{-1}. This can happen in charged lepton models of the charm and top types with $c_{\beta\alpha}(t_\beta + t_\beta^{-1})$ ranging from 5 to 10 (again, perturbative unitarity constraints from the scalar sector can be relevant for these values of $c_{\beta\alpha}(t_\beta + t_\beta^{-1})$).

Table 3. V factors entering eq. (28).

Model	$h \to bd$	$h \to bs$
u	$\lvert V_{ud}V_{ub}\rvert^2\,(\sim \lambda^6) = 1.33 \cdot 10^{-5}$	$\lvert V_{us}V_{ub}\rvert^2\,(\sim \lambda^8) = 7.14 \cdot 10^{-7}$
c	$\lvert V_{cd}V_{cb}\rvert^2\,(\sim \lambda^6) = 8.52 \cdot 10^{-5}$	$\lvert V_{cs}V_{cb}\rvert^2\,(\sim \lambda^4) = 1.59 \cdot 10^{-3}$
t	$\lvert V_{td}V_{tb}\rvert^2\,(\sim \lambda^6) = 7.90 \cdot 10^{-5}$	$\lvert V_{ts}V_{tb}\rvert^2\,(\sim \lambda^4) = 1.61 \cdot 10^{-3}$

5. Analysis and results

An interesting aspects of flavour violation in BGL models is the possibility to establish clear correlations between various flavour violating processes owing to the reduced number of parameters which are involved: apart from the CKM and/or PMNS mixing matrices and fermion masses, FCNC only depend on the values of $\tan\beta$ and $\cos(\beta - \alpha)$. Moreover, the suppression depends on the specific BGL model, and thus the correlations differ from model to model. Although the focus of this presentation is in tree level flavour violating processes involving the Higgs boson already discovered at the LHC, the analysis has to take into consideration the flavour conserving Higgs constraints already obtained from Run 1 of the LHC. In particular one has to comply with the measured signal strengths for the different combinations of production mechanism and decay channel. For a detailed account and the latest combined results of the ATLAS and CMS collaborations, see [49]. They involve the flavour conserving couplings of the Higgs and put constraints in the t_β vs. $\alpha - \beta$ available space (we refer the reader to [34] for details). Figure 1 illustrates the effect of these constraints alone in a few models.

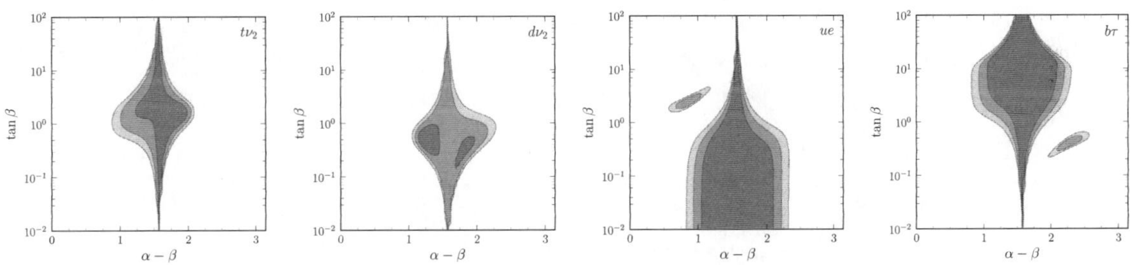

Figure 1. Effect of flavour conserving Higgs constraints alone on t_β vs. $\alpha - \beta$ in a sample of models; darker to lighter colouring represents 68%, 95% and 99% CL regions.

Furthermore, although an extended analysis of the phenomenology of the models under consideration, of the type presented in [33], is beyond the present scope, some attention has been devoted to potential bounds on the combination $c_{\beta\alpha}^2 (t_\beta + t_\beta^{-1})^2$ from low energy processes like $B_d - \bar{B}_d$, $B_s - \bar{B}_s$, $K^0 - \bar{K}^0$ and $D^0 - \bar{D}^0$ mixings (see, appendix B.2 of [34] in particular and also [50]).

As pointed out before, given the nature of BGL models, we can analyse different kinds of correlations among the different HFVNC observables considered in the previous sections:

- within the same quark sector:
  - $t \to hc$ vs. $t \to hu$, in down-type models, where there are tree level FCNC in the up quark sector,
  - $h \to bs$ vs. $h \to bd$, in up-type models, where there are tree level FCNC in the down quark sector,

IOP Conf. Series: Journal of Physics: Conf. Series **873** (2017) 012038 doi:10.1088/1742-6596/873/1/012038

- within the quark and the lepton sector in neutrino-type models (where there are tree level FCNC in the charged lepton sector),
  - $t \to hq$ vs. $h \to \mu\tau$, in down-neutrino-type models,
  - $h \to bq$ vs. $h \to \mu\tau$, in up-neutrino-type models.

For an exhaustive analysis of the different correlations of interest, on the constraints operating in each case and additional relevant aspects for the interpretation of them, we refer again to reference [34], and content ourselves with an illustrative sample in terms of the plots presented in figure 2.

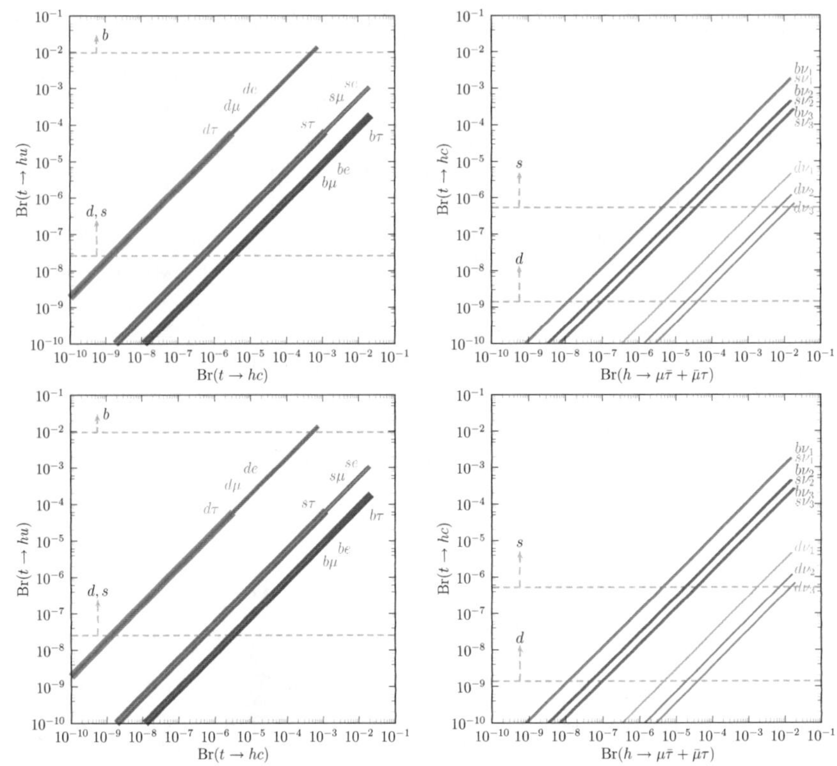

Figure 2. Correlations of different observables in BGL models.

Conclusions

We have discussed prospects for the observability of several flavour changing rare processes involving the Higgs in the context of a class of two Higgs doublet models which feature flavour changing neutral currents which have a controlled intensity and which are shaped by a symmetry to depend on the CKM and PMNS mixing matrices.

Acknowledgments

The author thanks the organization of the Discrete 2016 Symposium for the excellent development of the conference. MN also acknowledges support from *Fundação para a Ciência e a Tecnologia* (FCT, Portugal) through postdoctoral grant SFRH/BPD/112999/2015 and through the project CFTP-FCT Unit 777 (UID/FIS/00777/2013) which are partially funded through POCTI (FEDER), COMPETE, QREN and EU.

Bibliography

[1] Aad G *et al.* (ATLAS) 2012 *Phys. Lett.* **B716** 1–29 (*Preprint* 1207.7214)
[2] Chatrchyan S *et al.* (CMS) 2012 *Phys. Lett.* **B716** 30–61 (*Preprint* 1207.7235)
[3] Gunion J F, Haber H E, Kane G L and Dawson S 2000 *Front. Phys.* **80** 1–448
[4] Branco G, Ferreira P, Lavoura L, Rebelo M, Sher M *et al.* 2012 *Phys.Rept.* **516** 1–102 (*Preprint* 1106.0034)
[5] Ivanov I P 2017 (*Preprint* 1702.03776)
[6] Raidal M *et al.* 2008 *Eur. Phys. J.* **C57** 13–182 (*Preprint* 0801.1826)
[7] Branco G, Grimus W and Lavoura L 1996 *Phys.Lett.* **B380** 119–126 (*Preprint* hep-ph/9601383)
[8] Botella F, Branco G and Rebelo M 2010 *Phys.Lett.* **B687** 194–200 (*Preprint* 0911.1753)
[9] Botella F, Branco G, Nebot M and Rebelo M 2011 *JHEP* **1110** 037 (*Preprint* 1102.0520)
[10] Hall L J and Wise M B 1981 *Nucl. Phys.* **B187** 397
[11] Han T and Marfatia D 2001 *Phys. Rev. Lett.* **86** 1442–1445 (*Preprint* hep-ph/0008141)
[12] Kanemura S, Ota T and Tsumura K 2006 *Phys. Rev.* **D73** 016006 (*Preprint* hep-ph/0505191)
[13] Carcamo Hernandez A E, Martinez R and Rodriguez J A 2007 *Eur. Phys. J.* **C50** 935–948 (*Preprint* hep-ph/0606190)
[14] Davidson S and Grenier G J 2010 *Phys. Rev.* **D81** 095016 (*Preprint* 1001.0434)
[15] Harnik R, Kopp J and Zupan J 2013 *JHEP* **03** 026 (*Preprint* 1209.1397)
[16] Dery A, Efrati A, Hiller G, Hochberg Y and Nir Y 2013 *JHEP* **1308** 006 (*Preprint* 1304.6727)
[17] Bhattacharyya G, Das D and Kundu A 2014 *Phys.Rev.* **D89** 095029 (*Preprint* 1402.0364)
[18] Celis A, Cirigliano V and Passemar E 2014 *Phys. Rev.* **D89** 095014 (*Preprint* 1403.5781)
[19] Lee C J and Tandean J 2015 *JHEP* **04** 174 (*Preprint* 1410.6803)
[20] Heeck J, Holthausen M, Rodejohann W and Shimizu Y 2015 *Nucl. Phys.* **B896** 281–310 (*Preprint* 1412.3671)
[21] Crivellin A, D'Ambrosio G and Heeck J 2015 *Phys. Rev. Lett.* **114** 151801 (*Preprint* 1501.00993)
[22] Dorner I, Fajfer S, Greljo A, Kamenik J F, Konik N and Niandic I 2015 *JHEP* **06** 108 (*Preprint* 1502.07784)
[23] de Medeiros Varzielas I and Hiller G 2015 *JHEP* **06** 072 (*Preprint* 1503.01084)
[24] Das D and Kundu A 2015 *Phys. Rev.* **D92** 015009 (*Preprint* 1504.01125)
[25] Bhattacherjee B, Chakraborty S and Mukherjee S 2016 *Mod. Phys. Lett.* **A31** 1650174 (*Preprint* 1505.02688)
[26] Altunkaynak B, Hou W S, Kao C, Kohda M and McCoy B 2015 *Phys. Lett.* **B751** 135–142 (*Preprint* 1506.00651)
[27] He X G, Tandean J and Zheng Y J 2015 *JHEP* **09** 093 (*Preprint* 1507.02673)
[28] Chiang C W, Fukuda H, Takeuchi M and Yanagida T T 2015 (*Preprint* 1507.04354)
[29] Alves J M, Botella F J, Branco G C, Cornet-Gomez F and Nebot M 2017 (*Preprint* 1703.03796)
[30] Georgi H and Nanopoulos D V 1979 *Phys. Lett.* **B82** 95
[31] Lavoura L and Silva J P 1994 *Phys. Rev.* **D50** 4619–4624 (*Preprint* hep-ph/9404276)
[32] Botella F J and Silva J P 1995 *Phys. Rev.* **D51** 3870–3875 (*Preprint* hep-ph/9411288)
[33] Botella F, Branco G, Carmona A, Nebot M, Pedro L and Rebelo M 2014 *JHEP* **1407** 078 (*Preprint* 1401.6147)
[34] Botella F J, Branco G C, Nebot M and Rebelo M N 2016 *Eur. Phys. J.* **C76** 161 (*Preprint* 1508.05101)
[35] Buras A, Gambino P, Gorbahn M, Jager S and Silvestrini L 2001 *Phys.Lett.* **B500** 161–167 (*Preprint* hep-ph/0007085)
[36] D'Ambrosio G, Giudice G, Isidori G and Strumia A 2002 *Nucl.Phys.* **B645** 155–187 (*Preprint* hep-ph/0207036)
[37] Bobeth C, Bona M, Buras A J, Ewerth T, Pierini M, Silvestrini L and Weiler A 2005 *Nucl. Phys.* **B726** 252–274 (*Preprint* hep-ph/0505110)
[38] Lee T 1973 *Phys.Rev.* **D8** 1226–1239
[39] Olive K A *et al.* (Particle Data Group) 2014 *Chin. Phys.* **C38** 090001
[40] Wolfenstein L 1983 *Phys.Rev.Lett.* **51** 1945
[41] Aad G *et al.* (ATLAS) 2014 *JHEP* **06** 008 (*Preprint* 1403.6293)
[42] Chatrchyan S *et al.* (CMS) 2014 *Phys. Rev.* **D90** 032006 (*Preprint* 1404.5801)
[43] Khachatryan V *et al.* (CMS) 2014 *Phys. Rev.* **D90** 112013 (*Preprint* 1410.2751)
[44] Khachatryan V *et al.* (CMS) 2015 *Phys. Lett.* **B749** 337–362 (*Preprint* 1502.07400)
[45] Aad G *et al.* (ATLAS) 2015 *JHEP* **11** 211 (*Preprint* 1508.03372)
[46] Khachatryan V *et al.* (CMS) 2016 *Phys. Lett.* **B763** 472–500 (*Preprint* 1607.03561)
[47] Aad G *et al.* (ATLAS) 2017 *Eur. Phys. J.* **C77** 70 (*Preprint* 1604.07730)
[48] CMS Collaboration, report CMS-PAS-HIG-16-005
[49] Aad G *et al.* (ATLAS, CMS) 2016 *JHEP* **08** 045 (*Preprint* 1606.02266)
[50] Nebot M and Silva J P 2015 *Phys. Rev.* **D92** 085010 (*Preprint* 1507.07941)

IOP Conf. Series: Journal of Physics: Conf. Series **873** (2017) 012039 doi:10.1088/1742-6596/873/1/012039

Assessing lepton-flavour non-universality from $B \to K^* \ell\ell$ angular analyses

B Capdevila

Universitat Autonoma de Barcelona and IFAE, 08193 Bellaterra, Spain

E-mail: bcapdevila@ifae.es

Abstract. The recent years of activity at the LHC have brought to light several anomalies in exclusive semileptonic $b \to s\ell^+\ell^-$ decays. Even though the latest model independent analyses are pointing to sizable New Physics contributions to different Wilson coefficients (especially C_9), the possibility of interpreting these results unambiguously in terms of New Physics is spoilt by non-perturbative QCD corrections entering at amplitude level. In this proceeding, we present a brief summary of the latest results of our global fits. Also, the origin of the above mentioned QCD corrections, plus their interplay with possible New Physics contributions, is shortly reviewed. A discussion about the properties of the observable R_K leads to the central part of this work: the proposal of a new set of observables able to test for lepton-flavour universality violation, where hadronic uncertainties cancel (in the Standard Model). To conclude, the properties of these observables are analysed within the Standard Model as well as within several New Physics benchmark scenarios.

1. Introduction

After the Higgs search era at the LHC, flavour physics has become one of the most promising windows for studying possible Beyond the Standard Model (BSM) effects. In particular, processes driven by the flavour-changing neutral current (FCNC) transition $b \to s\ell^+\ell^-$ have been providing the framework for the raise of several deviations from the Standard Model (SM) in various observables. In 2015, the LHCb analysis [1] of the 3 fb^{-1} data on $B \to K^*\mu^+\mu^-$ confirmed a $\sim 3\sigma$ anomaly in two large K^*-recoil bins of the angular observable P_5' [2, 3] that was already present in the 2013 results with 1 fb^{-1} [4]. The same experiment measured the observable $R_K = \mathcal{B}(B \to K\mu^+\mu^-)/\mathcal{B}(B \to Ke^+e^-)$ [5] in the dilepton mass range from 1 to 6 GeV^2 and found a 2.6σ tension with its SM value, predicted to be equal to 1 (to a high accuracy). Further deviations were observed by LHCb in the two large-recoil bins of the branching ratio of $B_s \to \phi\mu^+\mu^-$ [6]. And finally, a few months ago, the Belle experiment performed an independent measurement of P_5' [7] which confirmed the deviation w.r.t. SM observed by LHCb.

The relevance of the mentioned tensions (P_5', R_K, $\mathcal{B}(B_s \to \phi\mu^+\mu^-)$ and several low-recoil bins of $B \to V\ell^+\ell^-$ and $B \to P\ell^+\ell^-$ decays) is that all of then are sensitive to the same effective couplings, the Wilson coefficients $C^{(\prime)}_{7,9,10}$ induced by the four-fermion operators in the effective Hamiltonian approach (see Fig. 1):

$$\mathcal{O}_7^{(\prime)} = \frac{\alpha}{4\pi} m_b \left[\bar{s}\sigma_{\mu\nu}P_{R(L)}b\right] F^{\mu\nu}, \quad \mathcal{O}_9^{(\prime)} = \frac{\alpha}{4\pi}\left[\bar{s}\gamma^\mu P_{L(R)}b\right]\left[\bar{\ell}\gamma_\mu\ell\right], \quad \mathcal{O}_{10}^{(\prime)} = \frac{\alpha}{4\pi}\left[\bar{s}\gamma^\mu P_{L(R)}b\right]\left[\bar{\ell}\gamma_\mu\gamma_5\ell\right],$$
$$C_7^{\rm SM}(\mu_b) = -0.29, \qquad C_9^{\rm SM}(\mu_b) = 4.07, \qquad C_{10}^{\rm SM}(\mu_b) = -4.31, \tag{1}$$

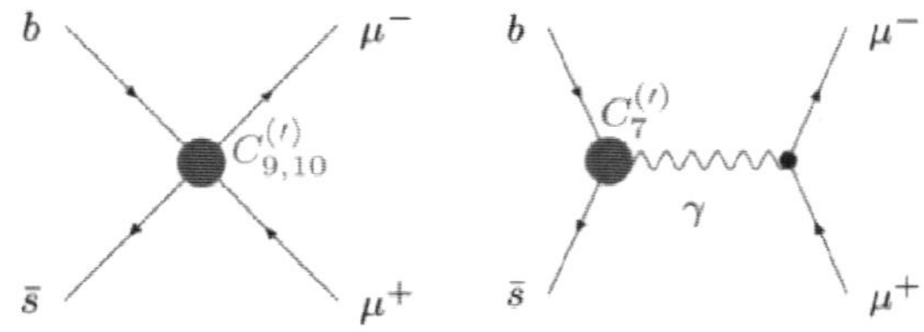

processes	$\mathscr{C}_7^{(\prime)}$	$\mathscr{C}_9^{(\prime)}$	$\mathscr{C}_{10}^{(\prime)}$
$B \to X_s\gamma, B \to K^*\gamma$	✓		
$B \to X_s\mu^+\mu^-$	✓	✓	✓
$B_s \to \mu^+\mu^-$			✓
$B_{(s)} \to (K^{(*)}, \phi)\mu^+\mu^-$	✓	✓	✓

Figure 1. Effective couplings $C^{(\prime)}_{7,9,10}$ contributing to $b \to s\ell^+\ell^-$ transitions and sensitivity of the various radiative and (semi-)leptonic $B_{(s)}$ decays modes to them.

where $P_{L,R} = (1 \mp \gamma_5)/2$, m_b stands for the mass of the b quark and $\mu_b = 4.8$ GeV denotes the energy scale. The right-handed (primed) Wilson coefficients are not given since they either vanish or can be neglected in the SM.

The next point that needs to be addressed is whether a New Physics (NP) contribution to the coefficients $C^{(\prime)}_{7,9,10}$ could simultaneously account for the various tensions in data. Plausible BSM scenarios with the required characteristics to generate this type of contributions include SM extensions with Z' bosons or lepto-quarks. The couplings $C^{(\prime)}_{7,9,10}$ can be constrained through various observables in radiative and (semi-)leptonic $B_{(s)}$ decays, each of them sensitive to a different subset of coefficients (see Fig. 1). Therefore, in order to assess the extend of potential NP effects it is required to perform a combined study of these observables, including their correlations. Several analysis of this fashion can be found in the literature, see Refs. [8, 9, 10, 11, 12, 13, 14]. Differences among them include the use of different data sets, different form factor inputs and different/additional observables.

In order to be able to interpret the observed deviations in the (semi-)leptonic $B_{(s)}$ decays in terms of NP effects a detailed analysis of all possible hadronic uncertainties has to be done (see, for instance, [15]). The theoretical predictions are plagued by perturbative and non-perturbative QCD effects, and it can be argued that some of the non-perturbative effects may mimic a NP signal. In a very recent paper [16], some of the authors carefully discuss the different sources of hadronic uncertainties and provide robust arguments disfavouring the possibility that these non-perturbative effects are the origin of the observed anomalies.

Another tentative path for assessing possible NP effects goes along the line pointed by the deviation observed in R_K. The tension in this observable can be understood if NP would affect predominantly $b \to s\mu^+\mu^-$ compared to $b \to se^+e^-$ [5] and thus violate lepton-flavour universality. Also the fact that no deviation was observed in $B \to K^*e^+e^-$ data at very large K^* recoil [17] supports the hints suggested by R_K. In Ref. [18] we assume lepton-flavour universality is maximally violated between the electronic and muonic modes and exploit the resulting new paradigm to build observables where the presence of NP in $b \to s\mu^+\mu^-$ can be proved in a clean way, via comparison of $b \to s\mu\mu$ and $b \to see$ observables.

Last December, Belle presented a separate measurement [19] of P_5' in the muon and electron channels, and hence of the observable $Q_5 = P_5^{\mu\prime} - P_5^{e\prime}$ proposed in Ref. [18]. Though not yet statistically significant, the result points to a violation of lepton-flavour universality in P_5' in compliance with R_K.

The structure of this proceeding is the following: in Sec. 2 we report the features and most important results of the analysis in Ref. [11]. The origin of the hadronic uncertainties and how they can interfere with a NP signal is sketched in Sec. 3. Finally, in Sec. 4 we present the observables postulated in [18] and review their characteristics.

Coefficient	Best fit	1σ	3σ	Pull_{SM}	p-value (%)
C_7^{NP}	-0.02	$[-0.04, -0.00]$	$[-0.07, 0.04]$	1.1	13.0
C_9^{NP}	-1.05	$[-1.25, -0.85]$	$[-1.62, -0.40]$	**4.7**	61.0
C_{10}^{NP}	0.55	$[0.34, 0.77]$	$[-0.05, 1.24]$	2.8	24.0
$C_{7'}^{\text{NP}}$	0.02	$[-0.00, 0.04]$	$[-0.05, 0.09]$	0.9	13.0
$C_{9'}^{\text{NP}}$	0.06	$[-0.18, 0.30]$	$[-0.65, 0.78]$	0.3	12.0
$C_{10'}^{\text{NP}}$	-0.03	$[-0.20, 0.14]$	$[-0.55, 0.48]$	0.2	12.0
$C_9^{\text{NP}} = C_{10}^{\text{NP}}$	-0.18	$[-0.36, 0.02]$	$[-0.68, 0.50]$	0.9	13.0
$C_9^{\text{NP}} = -C_{10}^{\text{NP}}$	-0.59	$[-0.74, -0.44]$	$[-1.06, -0.17]$	**4.3**	51.0
$C_{9'}^{\text{NP}} = C_{10'}^{\text{NP}}$	0.00	$[-0.24, 0.24]$	$[-0.74, 0.71]$	0.0	12.0
$C_{9'}^{\text{NP}} = -C_{10'}^{\text{NP}}$	0.03	$[-0.08, 0.13]$	$[-0.30, 0.35]$	0.2	12.0
$C_9^{\text{NP}} = -C_{9'}^{\text{NP}}$	-1.00	$[-1.20, -0.78]$	$[-1.55, -0.32]$	**4.4**	54.0
$C_9^{\text{NP}} = -C_{10}^{\text{NP}} = -C_{9'}^{\text{NP}} = -C_{10'}^{\text{NP}}$	-0.61	$[-0.45, -0.45]$	$[-1.17, -0.17]$	**4.3**	50.0
$C_9^{\text{NP}} = -C_{10}^{\text{NP}} = C_{9'}^{\text{NP}} = -C_{10'}^{\text{NP}}$	-0.24	$[-0.33, -0.15]$	$[-0.53, 0.03]$	2.7	23.0

Table 1. Best-fit point, confidence intervals, pulls for the SM hypothesis and p-value for different 1D NP scenarios, including $b \to see$ data but assuming NP only in $b \to s\mu\mu$.

2. A short review of the global fits

The reference fits in [11] are obtained using the branching ratios and angular observables for $B \to K^*\mu^+\mu^-$ and $B_s \to \phi\mu^+\mu^-$, the branching ratios of the charged and neutral modes $B \to K\mu^+\mu^-$, the branching ratios of $B \to X_s\mu^+\mu^-$, $B_s \to \mu^+\mu^-$ and $B \to X_s\gamma$, as well as the isospin asymmetry A_I and the time-dependent CP asymmetry $S_{K^*\gamma}$ of $B \to K^*\gamma$. For the theoretical predictions, lattice form factors from Refs. [20, 21] are used in the low-recoil region, while we resort to light-cone sum rule (LCSR) form factors from Ref. [22], with their correlations assessed from the large-recoil symmetries, in the low-q^2 region. The only exception being $B_s \to \phi$, which requires the use of the form factors in Ref. [23].

The hypothesis tested in our analysis is modeled by treating the NP contributions to the Wilson coefficients $\{C_i^{\text{NP}}\}$ as parameters allowed to vary freely. We estimate the value of these parameters by performing a frequentist fit including experimental and theoretical correlation matrices. In Tab. **??** we present our most updated 1D fit results, where the new $B_s \to \phi$ form factors [23] and the new experimental results on $\mathcal{B}(B^0 \to K^*(892)^0\mu^+\mu^-)$ [24] are included. We report no significant changes in the fit results from the ones in Ref. [11]. In the last two columns we provide information about the goodness-of-fit by displaying the SM-pull, i.e. the number of standard deviations by how much the best fit point is preferred over the SM point $\{C_i^{\text{NP}}\} = 0$, and the p-value for each scenario. A scenario with a large SM-pull, or equivalently a large p-value, indicates a parameter selection that allows for a better description of the data. Therefore, results in Tab. **??** establish the hypothesis of having a contribution to the C_9 coefficient of $\sim$

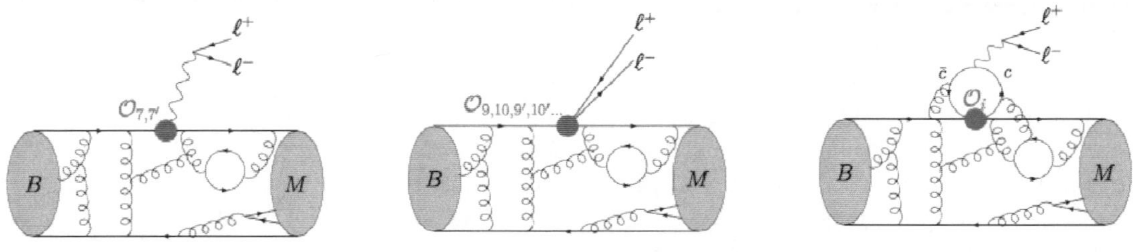

Figure 2. Illustration of factorisable (first two diagrams) and non-factorisable (third diagram) QCD corrections to exclusive $B \to M\ell^+\ell^-$ matrix elements.

-25 % w.r.t. its SM value as the most favored one.

3. Some comments about hadronic uncertainties

Theoretical computations in the framework of semi-leptonic B decays necessarily require accounting for contributions coming from QCD effects both of perturbative and non-perturbative nature. From the amplitude level perspective, predictions involve tree-level diagrams with insertions of the operators $\mathcal{O}_{7,9,10}$ (generated at loop level in the SM), as well as one-loop diagrams with an insertion of the charged-current operator

$$\mathcal{O}_2 = [\bar{s}\gamma^\mu P_L c]\,[\bar{c}\gamma_\mu P_L b] \tag{2}$$

(generated at tree level in the SM). In contributions of the first type, the leptonic and the hadronic currents factorise, and QCD corrections are restricted to the hadronic $B \to M$ current (first two diagrams in Fig. 2). This class of *factorisable QCD corrections* thus forms part of the hadronic form factors parametrizing the $B \to M$ transition. On the other hand, contributions of the second type receive *non-factorisable QCD corrections* (third diagram in Fig. 2) that cannot be absorbed into form factors.

3.1. Form factor uncertainties and power corrections

In order to be protected from uncertainties stemming from factorisable QCD corrections enclosed in the form factors, one can exploit the large-recoil symmetries of QCD to build observables such that their form factor sensitivity is minimised [2, 3, 25, 26, 27, 28]. Following this line of thought, a set of observables that only exhibit a mild form factor dependence (suppressed by powers of α_s and Λ/m_b) was proposed, the so-called optimised observables P_i.

It is essential for the analysis of the observables to control the correlations of the errors of the different form factors. For that purpose, our approach relies on the assessment of the correlations by means of the large-recoil symmetry relations. This method guarantees a model independent determination of the correlations from first principles but, as a drawback, the result is only valid up to order Λ/m_b corrections that need to be estimated. To estimate these *factorisable power corrections* we follow the technique first introduced in Ref. [29], which was improved shortly after in Ref. [15]. Numerically, we allow for a generic size of 10% power corrections to the form factor, as suggested by fitting to the particular LCSR form factors from Refs. [22, 23].

The size of the factorisable power corrections has been under intense debate recently due to attemps to explain the LHCb anomalies because of their impact. This possibility has been dismissed by the arguments provided in Ref. [16].

3.2. Impact of long-distance $c\bar{c}$ loops

Contributions to the amplitude coming from insertions of the $\mathcal{O}_2$ effective operator are of non-factorisable type, meaning that cannot be encoded into form factors. These contributions, commonly referred as long-distance charm-loop effects, can mimic a shift in the Wilson coefficient C_9 and thus have been suggested as a solution of the anomaly in $B \to K^*\mu^+\mu^-$ [30, 31]. Unlike a NP contribution, due to the non-local structure of the mentioned corrections, they are expected to show a non-constant q^2-dependence (q^2 stands for the dilepton invariant mass). This contribution always accompanies the perturbative SM part of C_9, and enter in the structure of the effective Wilson coefficient like

$$C_9^{\text{eff}\,i}(q^2) = C_{9\,\text{pert}}^{\text{eff SM}}(q^2) + C_9^{\text{NP}} + C_9^{c\bar{c}\,i}(q^2) \tag{3}$$

Charm-loop effects are also expected to depend on the helicity of the final state. For this reason we allow for three different $C_9^{c\bar{c}\,i}$ contributions, and thus three different $C_9^{\text{eff}\,i}$, depending on the transversity amplitude under consideration $i = 0, \|, \perp$.

The theoretical computation of this-long distance contribution is technically very challenging. At the moment, there is only a partial computation in the literature [22], which resorts on light-cone sum rules combined with a dispersion relation.

4. Observables testing lepton-flavour universality violation

Looking only to $B \to K^*\mu^+\mu^-$ data, one might be tempted to explain the anomaly in terms of long-distance charm, even despite all the arguments that support the contrary [16]. However, this temptation could just be the result of a excessively narrow point view. Actually, just by broadening the frame to include $B \to K^*e^+e^-$ data, the hypothesis of long-distance charm effects as the responsible for the observed tensions starts to loose weight, because of R_K. The deviation observed in this observable cannot be explained by long-distance charm, while it adds coherently with the pattern of deviations observed in $B \to K^*\mu^+\mu^-$, being possible to explain both tensions by introducing a signal of lepton-flavour universality violation (LFUV) that couples only to the muonic channel and not the the electronic one. In other words, both the tensions in $B \to K^*\mu^+\mu^-$ and in R_K are alleviated by adding a constant contribution to $C_{9\mu}$ but leaving C_{9e} SM-like.

R_K alone, though, is not enough, since it has very little discrimination power among different generic NP scenarios. It is thus necessary to complement R_K with more observables able to test for LFUV, we proposed in [18] several of them: Q_i, B_i and M.

4.1. A new basis of observables: Q_i

A particularly interesting set of observables with the desired properties comes from the direct comparison of $P_i^{(\prime)}(B \to K^*\mu\mu)$ and $P_i^{(\prime)}(B \to K^*ee)$. These observables are the so-called $Q_i = P_i^{(\prime)\,\mu} - P_i^{(\prime)\,e}$. Being defined in terms of optimised observables, Q_i inherit their properties and show a reduced sensitivity to hadronic uncertainties. In particular, these observables are very protected against long-distance charm-loop contributions in the SM, since the effective operator $\mathcal{O}_2$ couples identically to muons and electrons. A measurement of Q_i different from zero would point to NP in an unambigous way, confirming the violation of lepton flavour universality observed in R_K. Obviously, in presence of NP the problem of hadronic uncertainties reemerges, but then we enter in a completely different field, that of a NP Discovery. To ilustrate this crucial property of Q_i, in Fig. 3 we portray $P_5^{\prime\,\mu}$ and Q_5 both in the SM and in a NP scenario with $C_{9\mu}^{\text{NP}} = -1.1$ (taking $C_{9e}^{\text{NP}} = 0$). Notice how tiny are the uncertainties in the SM predictions (grey boxes) for Q_5 compared with $P_5^{\prime\,\mu}$. On the other hand, when we allow for a NP contribution, the theoretical uncertainties in Q_5 grow, as expected, but are more limited than the ones of $P_5^\prime$.

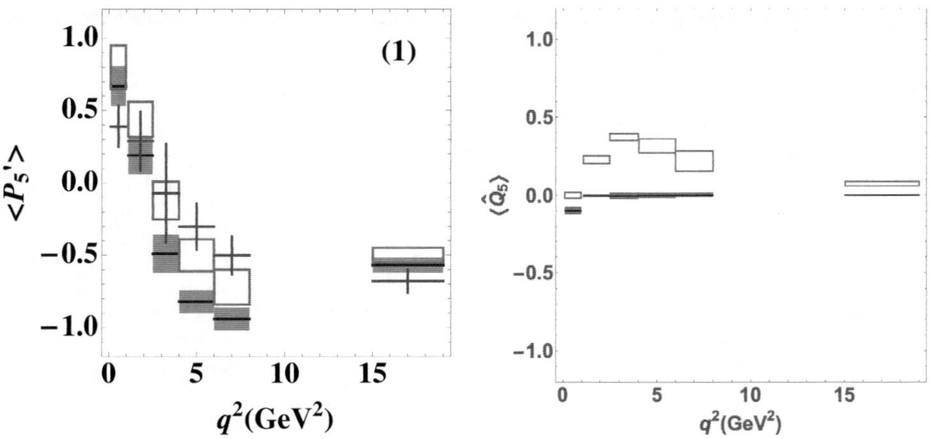

Figure 3. Predictions for P_5' and Q_5 in the SM (grey boxes) and in presence of NP (red boxes) in $C_{9\mu}^{\rm NP} = -1.1$ (with $C_{9e}^{\rm NP} = 0$).

Under the assumption of maximal LFUV between the muonic and electronic modes, the chronic dichotomy of NP or charm is traded by the scenario of a Discovery, where charm-loop uncertainties only enter into the picture when discussing the type of NP. Precisely, for the purpose of distinguishing between potential NP scenarios, the angular analysis of a subset of Q_i observables ($i = 1, 2, 4, 5$) provides precious information. The observables[1] $\hat{Q}_1$ and $\hat{Q}_4$ offer excellent tests for the presence of right-handed currents in $C'_{9\mu}$ and $C'_{10\mu}$, as it can be seen by the very distinctive signature of $\hat{Q}_1$ and the position of the bins in each end of the large-recoil region of $\hat{Q}_4$ (Fig. 4). Additional tests for distinguishing scenarios with NP contributions only in $C_{9\mu}$ from scenarios that allow for NP both in $C_{9\mu}$ and $C_{10\mu}$ can be found in the last two bins of the large recoil region of $\hat{Q}_2$ and $\hat{Q}_5$ and the first two bins of $\hat{Q}_4$ (see Fig. 5).

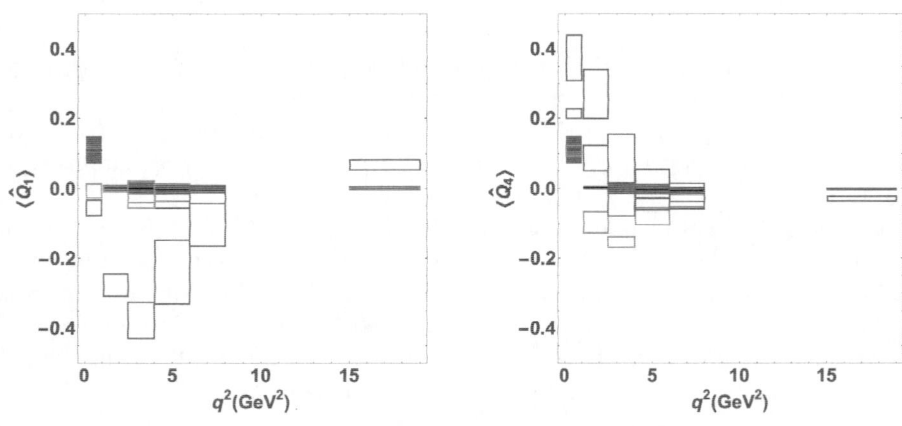

Figure 4. Predictions for $\hat{Q}_1$ and $\hat{Q}_4$ in the SM (grey boxes) and in presence of NP (red boxes) in $C_{9\mu}^{\rm NP} = -1.1$, $C_{9\mu}^{\rm NP} = -C_{10\mu}^{\rm NP} = -0.65$ and $C_{9\mu}^{\rm NP} = -C'_{9\mu} = -1.18$ & $C_{10\mu}^{\rm NP} = C'_{10\mu} = 0.38$ (with $C_{ie}^{(\prime)\rm NP} = 0$, $i = 9, 10$).

Belle has been the first experiment to measure the observable Q_5 [19], finding a 1.2σ tension

[1] The hat notation specifies that F_L is assumed to be the one measured by LHCb (see [18] for definitions).

w.r.t. the Standard Model prediction in the relevant bin [4, 8] GeV^2 (this tension is reduced to 0.6σ in the presence of a NP contribution $C_{9\mu}^{\rm NP} = -1.1$). The low statistical significance of this result makes impossible to draw any conclusion, although it is interesting to notice its connection with the deviation in R_K.

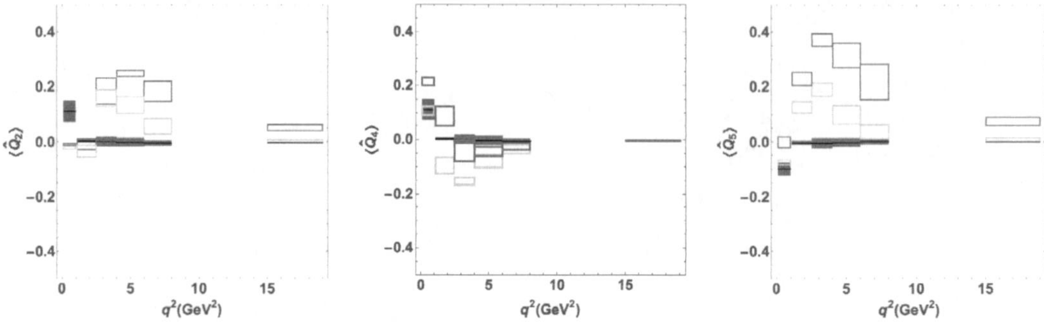

Figure 5. Predictions for $\hat{Q}_2$, $\hat{Q}_4$ and $\hat{Q}_5$ in the SM (grey boxes) and in presence of NP (red boxes) in $C_{9\mu}^{\rm NP} = -1.1$ and $C_{9\mu}^{\rm NP} = -C_{10\mu}^{\rm NP} = -0.65$ (with $C_{ie}^{(\prime)\rm NP} = 0$, $i = 9, 10$).

4.2. Additional LFUV observables

One can think of exploiting the angular coefficients in electron and muon modes, J_i^e and J_i^μ, in order to build observables sensitive to certain Wilson coefficients, and in some cases, insensitive to long-distance charm contributions.

Following this idea, in Ref. [18] we proposed

$$B_5 = \frac{J_5^\mu}{J_5^e} - 1 \qquad B_{6s} = \frac{J_{6s}^\mu}{J_{6s}^e} - 1 \qquad M = \frac{B_5 B_{6s}}{B_{6s} - B_5} \tag{4}$$

The observables B_5 and B_{6s} are form factor independent at all orders (up to corrections because of the different lepton masses) in the SM and share the charm-loop protection properties of the Q_i observables. In addition, B_5 and B_{6s}, assuming no right-handed currents, are proportional to the difference $(C_{10\mu} - C_{10e})$ which gives them unique capabilities for testing NP effects in C_{10}.

Finally, the last LFUV observable proposed in Ref. [18] is the M observable. This observable has the very singular property of being insensitive to contributions coming from long-distance charm-loops not only in the SM but also in presence of NP only in C_9, assuming *transversity-independent charm*[2]. If we consider *transversity-dependent charm* contributions or allow for the presence of NP in C_{10}, charm effects reemerge in M. Even though, its strong shielding from hadronic uncertainties makes M very sensitive to NP at low-q^2.

5. Conclusions

Recent experimental results provided by the LHCb and Belle experiments are showing a pattern of tensions with SM predictions in several $b \to s\ell^+\ell^-$ decay modes. Global analyses including all the available data show that scenarios with a large negative $C_9^{\rm NP}$ are preferred over the SM by typically more than 4σ. Explanations for the anomalies in terms of non-factorisable power corrections, although not fully excluded, are clearly disfavored by the theoretical arguments

[2] Charm-loop contributions that do not depend on the helicity of the final state, $C_9^{c\bar{c}} = C_9^{c\bar{c}\,\perp} = C_9^{c\bar{c}\,\parallel} = C_9^{c\bar{c}\,0}$.

provided in [16] and also by the appearance of tensions in the LFUV observables R_K and Q_5, measured by LHCb and Belle respectively. If confirmed, this tensions would unambiguously point to a NP scenario involving different couplings for muons and electrons. The current situation urges for the measurement of the newly proposed lepton flavour universality tests.

Acknowledgments
I would like to thank J. Matias, S. Descotes-Genon, J. Virto and L. Hofer; this proceeding is based on work made in collaboration with them. This work has received financial support from the grant FPA2014-61478-EXP and the Centro de Excelencia Severo Ochoa SEV-2012-0234.

References

[1] R. Aaij *et al.* [LHCb Collaboration], JHEP **1602** (2016) 104 doi:10.1007/JHEP02(2016)104 [arXiv:1512.04442 [hep-ex]].
[2] S. Descotes-Genon, J. Matias, M. Ramon and J. Virto, JHEP **1301** (2013) 048 doi:10.1007/JHEP01(2013)048 [arXiv:1207.2753 [hep-ph]].
[3] S. Descotes-Genon, T. Hurth, J. Matias and J. Virto, JHEP **1305** (2013) 137 doi:10.1007/JHEP05(2013)137 [arXiv:1303.5794 [hep-ph]].
[4] R. Aaij *et al.* [LHCb Collaboration], Phys. Rev. Lett. **111** (2013) 191801 doi:10.1103/PhysRevLett.111.191801 [arXiv:1308.1707 [hep-ex]].
[5] R. Aaij *et al.* [LHCb Collaboration], Phys. Rev. Lett. **113** (2014) 151601 doi:10.1103/PhysRevLett.113.151601 [arXiv:1406.6482 [hep-ex]].
[6] R. Aaij *et al.* [LHCb Collaboration], JHEP **1307** (2013) 084 doi:10.1007/JHEP07(2013)084 [arXiv:1305.2168 [hep-ex]].
[7] A. Abdesselam *et al.* [Belle Collaboration], arXiv:1604.04042 [hep-ex].
[8] S. Descotes-Genon, J. Matias and J. Virto, Phys. Rev. D **88** (2013) 074002 doi:10.1103/PhysRevD.88.074002 [arXiv:1307.5683 [hep-ph]].
[9] W. Altmannshofer and D. M. Straub, Eur. Phys. J. C **73** (2013) 2646 doi:10.1140/epjc/s10052-013-2646-9 [arXiv:1308.1501 [hep-ph]].
[10] F. Beaujean, C. Bobeth and D. van Dyk, Eur. Phys. J. C **74** (2014) 2897 Erratum: [Eur. Phys. J. C **74** (2014) 3179] doi:10.1140/epjc/s10052-014-2897-0, 10.1140/epjc/s10052-014-3179-6 [arXiv:1310.2478 [hep-ph]].
[11] S. Descotes-Genon, L. Hofer, J. Matias and J. Virto, JHEP **1606** (2016) 092 doi:10.1007/JHEP06(2016)092 [arXiv:1510.04239 [hep-ph]].
[12] W. Altmannshofer and D. M. Straub, Eur. Phys. J. C **75** (2015) no.8, 382 doi:10.1140/epjc/s10052-015-3602-7 [arXiv:1411.3161 [hep-ph]].
[13] W. Altmannshofer and D. M. Straub, arXiv:1503.06199 [hep-ph].
[14] T. Hurth, F. Mahmoudi and S. Neshatpour, Nucl. Phys. B **909** (2016) 737 doi:10.1016/j.nuclphysb.2016.05.022 [arXiv:1603.00865 [hep-ph]].
[15] S. Descotes-Genon, L. Hofer, J. Matias and J. Virto, JHEP **1412** (2014) 125 doi:10.1007/JHEP12(2014)125 [arXiv:1407.8526 [hep-ph]].
[16] B. Capdevila, S. Descotes-Genon, L. Hofer and J. Matias, arXiv:1701.08672 [hep-ph].
[17] R. Aaij *et al.* [LHCb Collaboration], JHEP **1504** (2015) 064 doi:10.1007/JHEP04(2015)064 [arXiv:1501.03038 [hep-ex]].
[18] B. Capdevila, S. Descotes-Genon, J. Matias and J. Virto, JHEP **1610** (2016) 075 doi:10.1007/JHEP10(2016)075 [arXiv:1605.03156 [hep-ph]].
[19] S. Wehle *et al.* [Belle Collaboration], arXiv:1612.05014 [hep-ex].
[20] R. R. Horgan, Z. Liu, S. Meinel and M. Wingate, Phys. Rev. D **89** (2014) no.9, 094501 doi:10.1103/PhysRevD.89.094501 [arXiv:1310.3722 [hep-lat]].
[21] C. Bouchard *et al.* [HPQCD Collaboration], Phys. Rev. D **88** (2013) no.5, 054509 Erratum: [Phys. Rev. D **88** (2013) no.7, 079901] doi:10.1103/PhysRevD.88.079901, 10.1103/PhysRevD.88.054509 [arXiv:1306.2384 [hep-lat]].
[22] A. Khodjamirian, T. Mannel, A. A. Pivovarov and Y.-M. Wang, JHEP **1009** (2010) 089 doi:10.1007/JHEP09(2010)089 [arXiv:1006.4945 [hep-ph]].
[23] A. Bharucha, D. M. Straub and R. Zwicky, JHEP **1608** (2016) 098 doi:10.1007/JHEP08(2016)098 [arXiv:1503.05534 [hep-ph]].
[24] R. Aaij *et al.* [LHCb Collaboration], JHEP **1611** (2016) 047 doi:10.1007/JHEP11(2016)047 [arXiv:1606.04731 [hep-ex]].
[25] F. Kruger and J. Matias, Phys. Rev. D **71** (2005) 094009 doi:10.1103/PhysRevD.71.094009 [hep-ph/0502060].

[26] B. Grinstein and D. Pirjol, Phys. Rev. D **70** (2004) 114005 doi:10.1103/PhysRevD.70.114005 [hep-ph/0404250].
[27] J. Matias, F. Mescia, M. Ramon and J. Virto, JHEP **1204** (2012) 104 doi:10.1007/JHEP04(2012)104 [arXiv:1202.4266 [hep-ph]].
[28] C. Bobeth, G. Hiller and D. van Dyk, JHEP **1007** (2010) 098 doi:10.1007/JHEP07(2010)098 [arXiv:1006.5013 [hep-ph]].
[29] S. Jager and J. Martin Camalich, JHEP **1305** (2013) 043 doi:10.1007/JHEP05(2013)043 [arXiv:1212.2263 [hep-ph]].
[30] J. Lyon and R. Zwicky, arXiv:1406.0566 [hep-ph].
[31] M. Ciuchini, M. Fedele, E. Franco, S. Mishima, A. Paul, L. Silvestrini and M. Valli, JHEP **1606** (2016) 116 doi:10.1007/JHEP06(2016)116 [arXiv:1512.07157 [hep-ph]].

Flavon-induced lepton flavour violation

Venus Keus

Department of Physics and Helsinki Institute of Physics, Gustaf Hällströmin katu 2, FIN-00014 University of Helsinki, Finland

E-mail: venus.keus@helsinki.fi

Abstract. ATLAS and CMS have observed a flavor violating decay of the Higgs to muon and tau. The fact that flavour violating couplings of the Higgs boson are exactly zero in the Standard Model suggests the mixing of the Higgs with another scalar with flavour violating couplings. We use the flavon field from the Froggatt-Nielsen mechanism, responsible for generating the lepton Yukawa matrices, for this purpose. The parameter space is constrained from experimental bounds on charged lepton flavor violation in other processes, however, we show that a substantial region of parameter space survives these bounds while producing a large enough $Br(h \to \mu\tau)$.

1. Introduction

Fermion Yukawa couplings in the Standard Model (SM) are free parameters and there is no explanation for the hierarchy among the fermion masses within the SM. A popular Beyond Standard Model (BSM) framework that naturally generates the SM fermion Yukawa couplings was suggested by Froggatt and Nielsen [1].

Lepton Flavour Violating (LFV) couplings of the Higgs boson are zero in the SM, however, ATLAS and CMS experiments have reported on a LFV decay of the observed boson [2, 3][1] at the LHC with the combined branching ratio

$$Br(h \to \mu\tau) = 0.82^{+0.33}_{-0.32}. \tag{1}$$

Recent results from the LHC [4] has reduced this value to $0.55^{+0.28}_{-0.28}$.

In a recent paper [5], we aimed at explaining LFV decays of the Higgs by employing the Froggatt-Nielsen mechanism. This mechanism requires the addition of a scalar field, *flavon*, singlet under the SM gauge group, charged under an extra $U(1)$-symmetry which breaks spontaneously due to the flavon field acquiring a Vacuum Expectation Value (VEV) which then generates the SM Yukawa matrices through higher order operators. The Higgs boson then acquires LFV couplings through mixing with the flavon.

The resulting mass eigenstates could then mediate LFV processes which are constrained by experimental data and exclude regions of the parameter space. We show that it is possible to acquire a large enough $Br(h \to \mu\tau)$ in the surviving regions of the parameter space of the model.

2. The Froggatt-Nielsen framework

In the Froggatt-Nielsen mechanism the SM Yukawa interaction are generated through higher order operators which are obtained by integrating out heavy states at some scale Λ. Such

[1] The process $h \to l_i l_j$ is assumed to be the sum of $h \to l_i^+ l_j^-$ and $h \to l_i^- l_j^+$.

Particle	$f^c_{L,i}$	$f_{R,i}$	H	Φ
$U(1)$ charge	$q_{\bar{L},i}$	$q_{R,i}$	q_h	q_ϕ

Table 1: The $U(1)$ charges of SM fermions $f_{R,L}$, SM Higgs field H and the flavon Φ.

operators which are invariant under a $U(1)$ symmetry are of the form

$$c_{ij}\,\frac{\Phi^{n_{ij}}}{\Lambda^{n_{ij}}}\bar{f}_{L,i}f_{R,j}\;H + \text{h.c.}\ , \tag{2}$$

where c_{ij} are order one coefficients, $f_{L,R}$ are SM fermions and Φ is the flavon which acquires a VEV,

$$\Phi = \frac{1}{\sqrt{2}}(v_\phi + \phi) \tag{3}$$

with $\phi = \text{Re}\phi + i\text{Im}\phi$. The Yukawa couplings are then given by

$$Y_{ij} = c_{ij}\,\left(\frac{v_\phi}{\sqrt{2}\Lambda}\right)^{n_{ij}} \equiv c_{ij}\;\epsilon^{n_{ij}}\ , \tag{4}$$

where ϵ is a small parameter. The $U(1)$ symmetry requires

$$n_{ij} = -\frac{1}{q_\phi}(q_{\bar{L},i} + q_{R,j} + q_h)\ , \tag{5}$$

where q_i are the charges identified in Table 1.

In principle, one can introduce several flavons each responsible for describing the hierarchy of the masses in the quark, charged and neutral lepton sectors. Here, we focus on the leptonic sector and allow for only the flavon and the leptons to transform under the $U(1)$ symmetry. As a result, after the Higgs doublet acquires a VEV,

$$H = \begin{pmatrix} 0 \\ \frac{v+h}{\sqrt{2}} \end{pmatrix} \tag{6}$$

the effective Lagrangian takes the form

$$\mathcal{L}_{eff}\ \supset\ \frac{v}{\sqrt{2}}\,Y_{ij}\,\left(1 + \frac{h}{v} + n_{ij}\,\frac{\phi}{v_\phi}\right)\bar{l}'_{L,i}l'_{R,j}\ . \tag{7}$$

Here, l' stands for the lepton fields in the weak basis and l for the mass eigenstate basis. Rotating the lepton fields from gauge eigenstates to the mass eigenstates requires diagonalising the lepton Yukawa matrix which is done using two unitary matrices:

$$Y_{\text{diag}} = U_L Y\,U_R^\dagger. \tag{8}$$

The effective Lagrangian then becomes

$$\mathcal{L}_{eff}\ \supset\ \frac{v}{\sqrt{2}}\bar{l}_L\,Y_{\text{diag}}\,l_R + \frac{h}{\sqrt{2}}\,\bar{l}_L\,Y_{\text{diag}}\,l_R + \frac{\phi}{\sqrt{2}}\frac{v}{v_\phi}\,\bar{l}_L\,\kappa\,l_R + \text{h.c.}, \tag{9}$$

IOP Conf. Series: Journal of Physics: Conf. Series **873** (2017) 012040 doi:10.1088/1742-6596/873/1/012040

where the flavon vertex involves the matrix

$$\kappa = U_L \left(Y_{ij} n_{ij}\right) U_R^\dagger \,, \tag{10}$$

with $n_{ij} = a_i + b_j$. Setting $q_\Phi = -1$ and $q_h = 0$ throughout this paper, we have

$$\kappa_{ij} = y_j \sum_{k=1}^{3} q_{\bar{L},k} (U_L)_{ik} (U_L)^*_{jk} + y_i \sum_{k=1}^{3} q_{R,k} (U_R)_{ik} (U_R)^*_{jk} \,, \tag{11}$$

where y_i are the Yukawa matrix eigenvalues. This is the source of lepton flavor violation in our model. Upon the Higgs–flavon mixing, such flavor changing couplings also appear in the interactions of the physical Higgs–like boson.

2.1. The scalar potential

The $U(1)$-symmetric scalar potential is of the form

$$V(H,\Phi) = -\mu_h^2 (H^\dagger H) + \lambda_h (H^\dagger H)^2 - \mu_\phi^2 (\Phi^\dagger \Phi) + \lambda_\phi (\Phi^\dagger \Phi)^2 + \lambda_{h\phi} (H^\dagger H)(\Phi^\dagger \Phi) \,. \tag{12}$$

The minimisation conditions lead to the following mass eigenstates H_1 and H_2 which are rotated by the following matrix

$$\begin{pmatrix} H_1 \\ H_2 \end{pmatrix} \equiv \begin{pmatrix} \cos\theta & \sin\theta \\ -\sin\theta & \cos\theta \end{pmatrix} \begin{pmatrix} h \\ \mathrm{Re}\phi \end{pmatrix} \tag{13}$$

where the mixing angle is defined as

$$\tan 2\theta = \frac{\lambda_{h\phi} v_h v_\phi}{\lambda_h v_h^2 - \lambda_\phi v_\phi^2} \tag{14}$$

and is constrained by theoretical and experimental data, reviewed and summarised for Higgs-portal models in [6] which are taken into account in our calculations. The masses of H_1 and H_2 are as follows

$$m_{H_{1,2}}^2 \;=\; \lambda_h v_h^2 + \lambda_\phi v_\phi^2 \pm \sqrt{\left(\lambda_h v_h^2 - \lambda_\phi v_\phi^2\right)^2 + \lambda_{h\phi}^2 v_h^2 v_\phi^2} \tag{15}$$

where we take the lighter boson, H_1, to be the SM-like Higgs boson, the observed scalar at the LHC with mass 125 GeV.

In the case of a global $U(1)$ symmetry which is broken by the flavon VEV, the model predicts a massless Goldstone boson which is phenomenologically unacceptable. In the case of a gauged $U(1)$ symmetry, we find that due to tight constraints on a flavor non-universal Z', the model does not lead to a large enough $Br(h \to \mu\tau)$. Here, we assume a discretized $U(1) \to Z_N$ by introducing a soft explicit breaking term $\tilde{m}^2 \Phi^2 + \text{h.c.}$ which induces $m_{\mathrm{Im}\phi} \sim \tilde{m}$. Note that our approach is one of an effective field theory and while tree level processes are well under control, the loop contributions are only indicative in nature and depend on the details of the UV completion.

	Observable	Present limit
1	$\mathrm{BR}(\mu \to eee)$	1.0×10^{-12} [7]
2	$\mathrm{BR}(\tau \to eee)$	3.0×10^{-8} [8]
3	$\mathrm{BR}(\tau \to \mu\mu\mu)$	2.0×10^{-8} [8]
4	$\mathrm{BR}(\tau^- \to \mu^- e^+ e^-)$	1.7×10^{-8} [9]
5	$\mathrm{BR}(\tau^- \to e^- \mu^+ \mu^-)$	2.7×10^{-8} [9]
6	$\mathrm{BR}(\tau^- \to e^+ \mu^- \mu^-)$	1.7×10^{-8} [9]
7	$\mathrm{BR}(\tau^- \to \mu^+ e^- e^-)$	1.5×10^{-8} [9]
8	$\mathrm{BR}(\mu \to e\gamma)$	5.7×10^{-13} [10]
9	$\mathrm{BR}(\tau \to \mu\gamma)$	4.4×10^{-8} [8]
10	$\mathrm{BR}(\tau \to e\gamma)$	3.3×10^{-8} [8]
11	$\mathrm{CR}(\mu\text{-}e, Au)$	7.0×10^{-13} [11]

Table 2: Current experimental bounds on the branching ratios of three–body LFV decays, magnetic transitions and the conversion rate of $\mu \to e$.

Particle	e_L^c	e_R	μ_L^c	μ_R	τ_L^c	τ_R	H	ϕ
Charge	6	0	4	0	2	1	0	-1

Table 3: $\mathrm{U}(1)/\mathrm{Z}_N$ charge assignment.

3. Bounds on lepton flavor violating processes

The flavon interaction in equation (9) induces LFV processes mediated by the three scalar mass eigenstates which are strongly constrained by experiment. We impose the current bounds from the three–body decay $l_i \to l_j l_k l_l$, magnetic transition $l_i \to l_j \gamma$ and $\mu \to e$ conversion processes presented in Table 2 which put limits on the flavon couplings parametrized by

$$\tilde{\kappa}_{ij} = \frac{1}{\sqrt{2}} \frac{v}{v_\phi} \kappa_{ij} \,, \tag{16}$$

where the flavon–lepton coupling is $\tilde{\kappa}_{ij}\, \bar{l}_{L,i}\, l_{R,j} \phi +$ h.c.

We assign charges listed in Table 3 to produce the following Yukawa texture and $\tilde{\kappa}$ matrix

$$Y = \begin{pmatrix} 3.4\,\epsilon^6 & -0.6\,\epsilon^6 & 3.5\,\epsilon^7 \\ 5.4\,\epsilon^4 & 6.1\,\epsilon^4 & -3.1\,\epsilon^5 \\ 0.5\,\epsilon^2 & 0.5\,\epsilon^2 & 7.3\,\epsilon^3 \end{pmatrix}, \quad \tilde{\kappa} = \frac{v}{v_\phi} \begin{pmatrix} 1\times 10^{-5} & -1\times 10^{-6} & -3\times 10^{-6} \\ -2\times 10^{-5} & 2\times 10^{-3} & 6\times 10^{-4} \\ 3\times 10^{-4} & -4\times 10^{-3} & 2\times 10^{-2} \end{pmatrix},$$

which reproduces the correct lepton masses for $\epsilon = 0.1$ and the shown order one coefficients (their precise values are given in [5]).

3.1. Negligible Higgs–flavon mixing

It is instructive to consider a scenario where the Higgs-flavon mixing is close to zero in which case only $\mathrm{Re}\phi$ and $\mathrm{Im}\phi$ mediate the LFV processes. To make the discussion clearer, we also decouple the Imϕ in this sub-section.

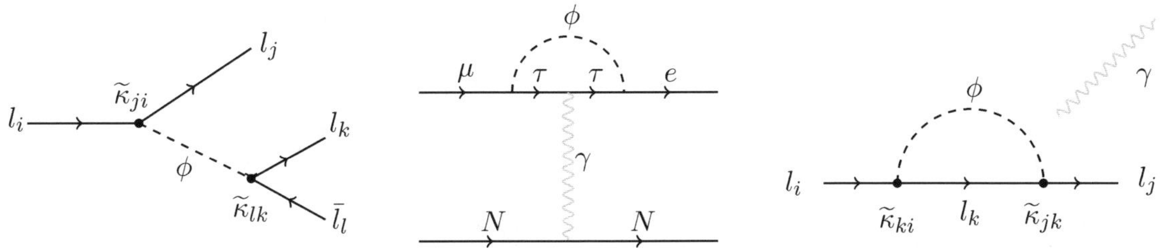

Figure 1: The $l_i \to l_j l_k l_l$ (left), $\mu \leftrightarrow e$-conversion (center) and $l_i \to l_j \gamma$ (right) processes mediated by the flavon ϕ. The decay $l_i \to l_j l_k l_l$ also receives important contributions at one loop.

We have taken into account all processes represented in Fig. 1 at tree and loop level. Here, we only show the details of the $\mu \to e\gamma$ process which puts the strongest bounds on the model. Neglecting the light lepton contributions, we find

$$\Gamma(\mu \to e\gamma) = \frac{\alpha m_\mu^3 m_\tau^2}{1024\pi^4} \frac{|\widetilde{\kappa}_{e\tau}|^2|\widetilde{\kappa}_{\tau\mu}|^2 + |\widetilde{\kappa}_{\tau e}|^2|\widetilde{\kappa}_{\mu\tau}|^2}{m_{\mathrm{Re}\phi}^4} \left[\frac{3}{2} - \log\left(\frac{m_\phi^2}{m_\tau^2}\right)\right]^2, \tag{17}$$

The invariant amplitude $A^L_{\mu\to e\gamma}$ is

$$A^L_{\mu\to e\gamma} = -\frac{ie}{32\pi^2}\widetilde{\kappa}^*_{\tau e}\widetilde{\kappa}^*_{\mu\tau}\left[\frac{3}{2} - \log\left(\frac{m_\phi^2}{m_\tau^2}\right)\right]\frac{m_\tau}{m_\phi^2} \tag{18}$$

and the corresponding equation for A^R is obtained by replacing $\widetilde{\kappa}_{ij}$ with $\widetilde{\kappa}^*_{ji}$.

Figure 2 shows the allowed values of v_ϕ vs m_ϕ after imposing the $\mu \to e\gamma$ bounds.

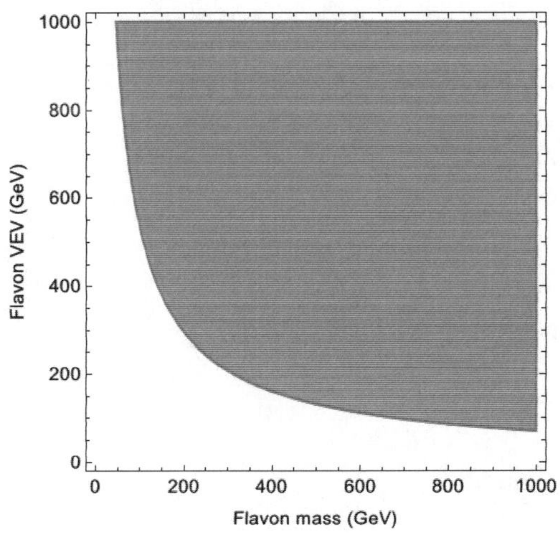

Figure 2: Allowed parameter space (shaded) for the texture at hand (Eq.17) with negligible Higgs-flavon mixing. The strongest constraint is imposed by $Br(\mu \to e\gamma)$.

IOP Conf. Series: Journal of Physics: Conf. Series **873** (2017) 012040 doi:10.1088/1742-6596/873/1/012040

3.2. Substantial Higgs–flavon mixing

In the case of non-zero Higgs-flavon mixing, all three scalar mass eigenstates, H_1, H_2 and $\mathrm{Im}\phi$, mediate LFV processes. The relevant interaction terms are

$$\begin{aligned}\mathcal{L} \;\supset\; & \left[\cos\theta\,\frac{Y_{ij}^{\rm diag}}{\sqrt{2}} + \sin\theta\,\widetilde{\kappa}_{ij}\right]\bar{l}_i P_R l_j\; H_1 + \left[-\sin\theta\,\frac{Y_{ij}^{\rm diag}}{\sqrt{2}} + \cos\theta\,\widetilde{\kappa}_{ij}\right]\bar{l}_i P_R l_j\; H_2 \\ & +\; i\widetilde{\kappa}_{ij}\;\bar{l}_i P_R l_j\;\mathrm{Im}\phi + \mathrm{h.c.}\end{aligned} \tag{19}$$

The couplings of H_1, H_2 to quarks are flavor–diagonal and are rescaled compared to the corresponding SM couplings by $\cos\theta$ and $-\sin\theta$, respectively.

Note that in this case due to the couplings of H_1 and H_2 to quarks, the $\mu \leftrightarrow e$ conversion process is possible at tree level and the $\mu \to e\gamma$ processes gets significant contributions from 2–loop Barr-Zee diagrams with the top quark and the W in the loop.

Again, the most limiting constraints comes from the $\mu \to e\gamma$ process with the 1– and 2–loop contributions to the amplitude

$$A^L_{\mu\to e\gamma} = A^L_{\mu\to e\gamma}(1-\mathrm{loop}) + A^L_{\mu\to e\gamma}(2-\mathrm{loop})\;. \tag{20}$$

At one loop we have

$$\begin{aligned}A^L_{\mu\to e\gamma}(1-\mathrm{loop}) \;=\; & -\frac{iem_\tau}{32\pi^2}\widetilde{\kappa}^*_{\tau e}\widetilde{\kappa}^*_{\mu\tau}\left\{\frac{\sin^2\theta}{m^2_{H_1}}\left[\frac{3}{2} - \log\left(\frac{m^2_{H_1}}{m^2_\tau}\right)\right]\right. \\ & \left. + \frac{\cos^2\theta}{m^2_{H_2}}\left[\frac{3}{2} - \log\left(\frac{m^2_{H_2}}{m^2_\tau}\right)\right] - \frac{1}{m^2_{\mathrm{Im}\phi}}\left[\frac{3}{2} - \log\left(\frac{m^2_{\mathrm{Im}\phi}}{m^2_\tau}\right)\right]\right\}.\end{aligned} \tag{21}$$

The 2–loop amplitude receives contributions from the top quark and the W boson [12],

$$A^L_{\mu\to e\gamma}(2-\mathrm{loop}) = A^L_t + A^L_W, \tag{22}$$

with

$$A^L_t = -i\frac{e\alpha v G_F}{6\sqrt{2}\pi^3}\sin\theta\cos\theta\;\widetilde{\kappa}^*_{\mu e}\left[f(z_{tH_1}) - f(z_{tH_2})\right]\;, \tag{23}$$

and

$$\begin{aligned}A^L_W \;=\; & i\frac{e\alpha v G_F}{16\sqrt{2}\pi^3}\sin\theta\cos\theta\;\widetilde{\kappa}^*_{\mu e} \\ & \times\left\{\left[3f(z_{WH_1}) + 5g(z_{WH_1}) + \frac{3}{4}g(z_{WH_1}) + \frac{3}{4}h(z_{WH_1}) + \frac{f(z_{WH_1}) - g(z_{WH_1})}{2z_{WH_1}}\right]\right. \\ & \left. - \left[3f(z_{WH_2}) + 5g(z_{WH_2}) + \frac{3}{4}g(z_{WH_2}) + \frac{3}{4}h(z_{WH_2}) + \frac{f(z_{WH_2}) - g(z_{WH_2})}{2z_{WH_2}}\right]\right\}.\end{aligned} \tag{24}$$

Here the loop functions are:

$$f(z) = \frac{1}{2}z\int_0^1 dx\frac{1-2x(1-x)}{x(1-x)-z}\log\left(\frac{x(1-x)}{z}\right), \tag{25}$$

$$g(z) = \frac{1}{2}z\int_0^1 dx\frac{1}{x(1-x)-z}\log\left(\frac{x(1-x)}{z}\right), \tag{26}$$

$$h(z) = z^2\frac{\partial}{\partial z}\left(\frac{g(z)}{z}\right) = \frac{z}{2}\int_0^1\frac{dx}{z-x(1-x)}\left[1 + \frac{z}{z-x(1-x)}\log\left(\frac{x(1-x)}{z}\right)\right]. \tag{27}$$

The arguments of these functions are defined by $z_{tH_i} = m_t^2/m_{H_i}^2$ and $z_{WH_i} = m_W^2/m_{H_i}^2$, with $i = 1, 2$. The $A^R_{\mu\to e\gamma}(2-\text{loop})$ amplitude is obtained by replacing $\widetilde{\kappa}^*_{ji}$ with $\widetilde{\kappa}_{ij}$. The $\mu \to e\gamma$ decay width is then calculated to be

$$\Gamma(\mu \to e\gamma) = \frac{m_\mu^3}{4\pi}\left(|A^L|^2 + |A^R|^2\right) . \tag{28}$$

In Fig. 3 (left) we show the surviving region in the $(v_\phi, \sin\theta)$ plane after imposing these LFV bounds for and exemplary value of $m_{H_2} = 500$ GeV.

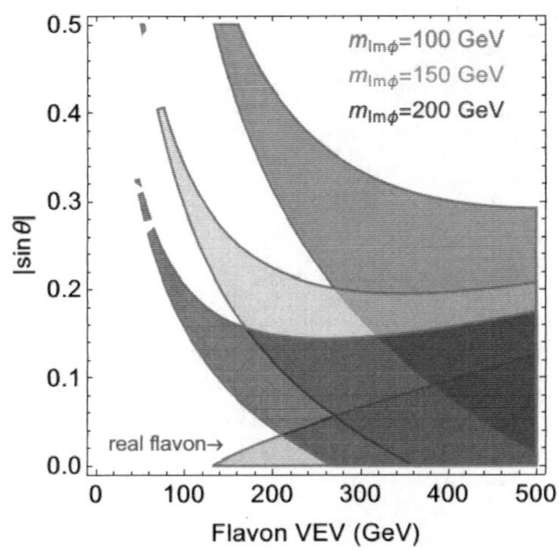

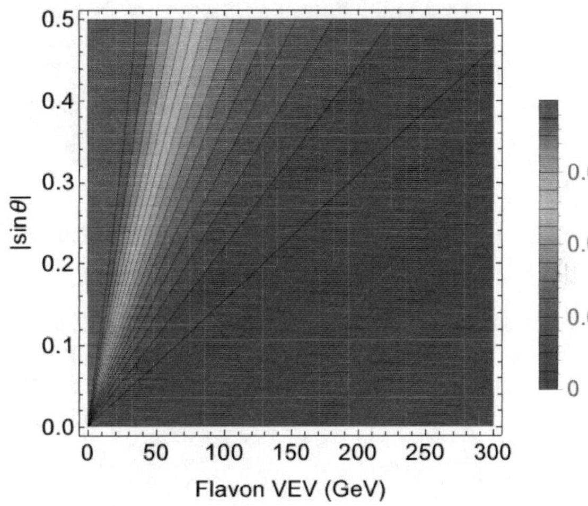

Figure 3: Left: parameter space allowed by the LFV constraints for $m_{\text{Im}\phi} = 100,\ 150,\ 200$ GeV. We have set $m_{H_2} = 500$ GeV. (The discontinuities appear for technical reasons.) Right: $\text{BR}_{\text{eff}}(H_1 \to \mu\tau)$ as a function of v_ϕ and $|\sin\theta|$.

We see that for $|\sin\theta| \simeq 0.3$ and $v_\phi \sim 100$ GeV the model survives the LFV bounds for a range of $m_{\text{Im}\phi}$ around 150-200 GeV. Note that direct collider constraints on Imϕ are very loose due to its small couplings to leptons. Fig. 3 (right) then shows that one expects a substantial decay rate $H_1 \to \mu\tau$ for $v_\phi \sim 100$ GeV.

It is important to note that such a scenario also leads to the enhancement of the lepton diagonal couplings of Higgs which are constrained by the LHC data. We choose $\sin\theta < 0$ which leads to some cancellations in $H_1 \to l_i l_i$ for our Yukawa texture. We have

$$\Gamma(H_1 \to \mu\tau) = \frac{m_{H_1}}{8\pi}\sin^2\theta\left(|\widetilde{\kappa}_{\mu\tau}|^2 + |\widetilde{\kappa}_{\tau\mu}|^2\right) \tag{29}$$

$$\Gamma(H_1 \to \tau\tau) = \frac{m_{H_1}}{8\pi}\left[\cos\theta\,\frac{Y_\tau^{\text{diag}}}{\sqrt{2}} + \sin\theta\,\widetilde{\kappa}_{\tau\tau}\right]^2 , \tag{30}$$

and analogously for $H_1 \to \mu\mu$. In our convention, a negative θ reduces $\Gamma(H_1 \to \tau\tau)$ without affecting the LFV rates.

Lastlly, to incorporate the difference between the H_1 and h production cross sections, we introduce the effective branching ratio $\text{Br}_{\text{eff}}(H_1 \to l_i l_j)$ through

$$\sigma(H_1)\,\text{Br}(H_1 \to l_i l_j) = \sigma(h)\,\frac{\Gamma(H_1 \to l_i l_j)}{\Gamma^{\text{total}}_{\text{SM}}(h)} \equiv \sigma(h)\,\text{Br}_{\text{eff}}(H_1 \to l_i l_j) , \tag{31}$$

and apply the LHC result in equation (1) to this effective branching ratio. We show our results in Fig. 4 where all of the constraints are satisfied and the observed $\text{BR}_{\text{eff}}(H_1 \to \mu\tau)$ for $\sin\theta = -0.3$ is presented.

IOP Conf. Series: Journal of Physics: Conf. Series **873** (2017) 012040 doi:10.1088/1742-6596/873/1/012040

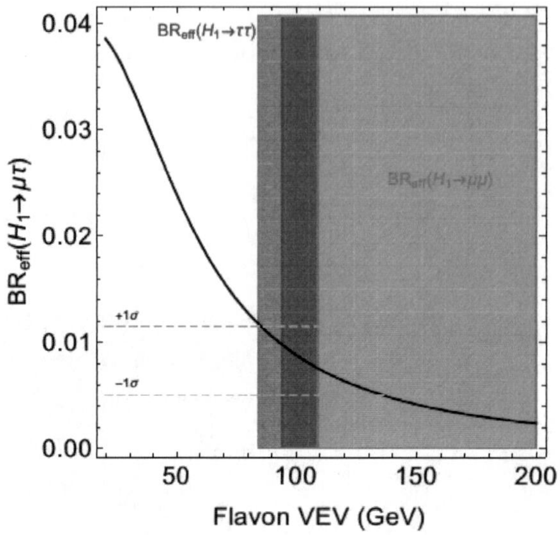

Figure 4: $\mathrm{Br}_{\mathrm{eff}}(H_1 \to \mu\tau)$ vs v_ϕ (black curve) for $\sin\theta = -0.3$. The red region is allowed by $\mathrm{Br}_{\mathrm{eff}}(H_1 \to \mu\mu)$ at 95% CL, the blue region is allowed by $\mathrm{Br}_{\mathrm{eff}}(H_1 \to \tau\tau)$, while their overlap (purple) is consistent with both. The dashed lines show the $\pm 1\sigma$ limits on the observed $\mathrm{Br}_{\mathrm{eff}}(H_1 \to \mu\tau)$.

4. Conclusion

We have employed a lepton–specific Froggatt–Nielsen framework which naturally generates the lepton Yukawa couplings and leads to lepton flavor violation at the observable level. The flavon predicted by this mechanism mixes with the SM Higgs boson and introduces a $h \to \mu\tau$ decay channel with the branching ratio at the percent level with a flavon VEV at the electroweak scale. We have found this scenario to be consistent with the LFV and Higgs data constraints.

Acknowledgments

VK's research is partially supported by the Academy of Finland project "The Higgs Boson and the Cosmos", project no. 267842. She also acknowledges the H2020-MSCA-RISE-2014 grant no. 645722 (NonMinimalHiggs) and would like to thank the organizers of DISCRETE 2016 for the kind hospitality and the lively scientific atmosphere provided by the conference.

References

[1] Froggatt C. D. and Nielsen H. B. 1979 *Nucl. Phys.* B **147** 277
[2] Khachatryan V. *et al.* [CMS Collaboration] 2015 *Phys. Lett.* B **749** 337
[3] Aad G. *et al.* [ATLAS Collaboration] 2015 *JHEP* **1511** 211
[4] Aad G. *et al.* [ATLAS Collaboration] 2017 *Eur. Phys. J.* C **77**, no. 2, 70
[5] Huitu K., Keus V., Koivunen N. and Lebedev O. 2016 *JHEP* **1605** 026
[6] Falkowski A., Gross C. and Lebedev O. 2015 *JHEP* **1505** 057
[7] Bellgardt U. *et al.* [SINDRUM Collaboration] 1988 *Nucl. Phys.* B **299** 1
[8] Amhis Y. *et al.* [Heavy Flavor Averaging Group Collaboration] 2012 *arXiv:1207.1158*
[9] Hayasaka K. *et al.* 2010 *Phys. Lett.* B **687** 139
[10] Adam J. *et al.* [MEG Collaboration] 2013 *Phys. Rev. Lett.* **110** 201801
[11] Bertl W. H. *et al.* [SINDRUM II Collaboration] 2006 *Eur. Phys. J.* C **47** 337
[12] Harnik R., Kopp J. and Zupan J. 2013 *JHEP* **1303** 026

IOP Conf. Series: Journal of Physics: Conf. Series **873** (2017) 012041 doi:10.1088/1742-6596/873/1/012041

Self-consistent radiative corrections to false vacuum decay

B Garbrecht[1] **and P Millington**[2]

[1] Physik Department T70, Technische Universität München
James-Franck-Straße, 85748 Garching, Germany
[2]School of Physics and Astronomy, University of Nottingham,
Nottingham NG7 2RD, United Kingdom

E-mail: garbrecht@tum.de, p.millington@nottingham.ac.uk

Abstract. With the Higgs mass now measured at the sub-percent level, the potential metastability of the electroweak vacuum of the Standard Model (SM) motivates renewed study of false vacuum decay in quantum field theory. In this note, we describe an approach to calculating quantum corrections to the decay rate of false vacua that is able to account fully and self-consistently for the underlying inhomogeneity of the solitonic tunneling configuration. We show that this method can be applied both to theories in which the instability arises already at the level of the classical potential and those in which the instability arises entirely through radiative effects, as is the case for the SM Higgs vacuum. We analyse two simple models in the thin-wall regime, and we show that the modifications of the one-loop corrections from accounting fully for the inhomogeneity can compete at the same level as the two-loop homogeneous corrections.

1. Introduction

The dynamics of first-order phase transitions have relevance across high-energy physics, astroparticle physics and cosmology. Given a ~ 125 GeV Higgs boson and a ~ 174 GeV top quark, state-of-the-art perturbative calculations of the Standard Model (SM) effective potential indicate that the electroweak vacuum may suffer an instability [1, 2, 3, 4, 5, 6] at a scale of around 10^{11} GeV, leading to a catastrophic transition towards a high-scale global minimum that lies far above the Planck scale. In the cosmology of the early Universe, first-order thermal phase transitions may play a role in the generation of the Baryon Asymmetry of the Universe [7, 8] and have the potential to produce stochastic gravitational-wave backgrounds [9, 10, 11, 12]. Further study of the latter is certainly warranted given the recent discovery of gravitational waves by LIGO [13] and the potential for such a stochastic background to be observed by the forthcoming LISA satellite array or pulsar-timing arrays. Moreover, in analysing vacuum decay, we must study non-perturbative solutions of non-linear field equations, and the necessary techniques are also of relevance to the formation and evolution of (non-)topological defects.

In the case of the SM electroweak vacuum, the predicted lifetime is subject to a number of uncertainties. The main experimental uncertainty is due to the determination of the top-quark mass [14, 15]. More significantly, the exit point for the tunneling can be close to the Planck scale, and one therefore cannot neglect the impact of new physics and higher-dimension (non-renormalisable) operators [16, 17, 18, 19, 20], which can significantly impact the tunneling rate. Moreover, one should consider the effect of gravitational corrections [21, 22, 23, 24], as well as

IOP Conf. Series: Journal of Physics: Conf. Series **873** (2017) 012041 doi:10.1088/1742-6596/873/1/012041

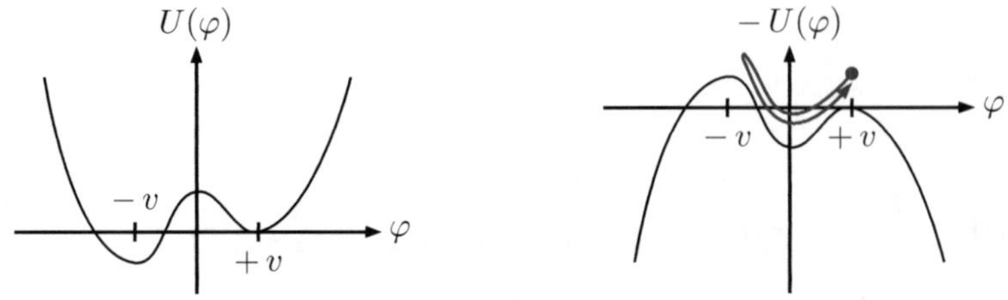

Figure 1: Sketches of the tree-level potential (left) and the upturned potential (right) for the model in Eq. (1). The blue line (right) indicates the trajectory of the bounce.

the role played by "seeded" decay, which can result from impurities [25] due, for instance, to black holes [26, 27, 28]. On the theoretical side, we must be careful about the interpretation of non-convex regions in the perturbatively-calculated effective potential [29, 30, 31], as well as the implementation of the renormalisation-group (RG) improvement of the effective potential [32, 33]. In this note, we will concentrate on the additional uncertainty introduced by neglecting the spatial inhomogeneity of the underlying solitonic field configuration. This inhomogeneity has not been captured fully in existing analyses of the SM vacuum stability, where the tunneling rate is calculated using the effective potential of a homogeneous field configuration. To this end, we will outline a Green's function method, based on Refs. [34, 35, 36], that allows quantum-corrected tunneling rates to be calculated self-consistently within the space-time dependent background of the tunneling configuration.

We begin by reviewing the semi-classical calculation of the tunneling rate, before discussing the calculation of the first quantum corrections. We then apply the Green's function method to two toy models. In the first model (reported from Ref. [34]), the vacuum instability is present at tree level, and we will calculate the quantum corrections analytically in the inhomogeneous background field configuration. In the second model (reported from Ref. [36]), the vacuum instability emerges only at the one-loop level, and, therein, we will apply a self-consistent numerical procedure to calculate the quantum-corrected tunneling rate. We will find that the modifications to the one-loop results from fully accounting for the background inhomogeneity can compete at the same level as the two-loop (homogeneous) corrections.

2. Semi-classical tunneling rate

We begin by reviewing the calculation of the semi-classical tunneling rate for the archetypal quartic theory with Euclidean Lagrangian

$$\mathcal{L} \;=\; \frac{1}{2!}\left(\partial_\alpha \Phi\right)^2 \;+\; U(\Phi)\,, \tag{1a}$$

$$U(\Phi) \;=\; -\,\frac{1}{2!}\,\mu^2\,\Phi^2 \;+\; \frac{1}{3!}\,g\,\Phi^3 \;+\; \frac{1}{4!}\,\lambda\,\Phi^4 \;+\; U_0\,. \tag{1b}$$

Herein, $\Phi \equiv \Phi(x)$ is a real scalar field, ∂_α indicates the partial derivative with respect to the Euclidean four-coordinate $x_\alpha \equiv (\mathbf{x}, x_4)$, $g, \lambda > 0$ are dimensionful and dimensionless couplings, respectively, and U_0 is a constant. For $\mu^2 > 0$, the system undergoes spontaneous symmetry breaking, and the field obtains a vacuum expectation value $\varphi \equiv \langle \Phi \rangle = v_\pm \simeq \pm v - 3g/(2\lambda)$, where $v = \sqrt{6}\mu/\sqrt{\lambda}$. The cubic coupling g, which we take to be small ($3g/(2\lambda) \ll v$), breaks the $\mathbb{Z}_2$ symmetry, such that the minima at $v_\pm$ are not fully degenerate. The calculation that follows was first presented by Coleman in Ref. [37], based on the work of Langer [38] on the statistical description of the decay of metastable states.

An initially homogeneous state of the false vacuum ($\varphi = +v$) will spontaneously fragment via the nucleation of bubbles, which then expand and coalesce to fill the space with the true vacuum ($\varphi = -v$). The rate of nucleation of such bubbles depends upon the difference in the free energies of the two vacuum states, and the tunneling field configuration corresponds to the lowest intervening saddle point of the free energy. Since the free energy is related to the Euclidean action, we are interested in the classical Euclidean equation of motion

$$-\Delta^{(4)}\varphi + U'(\varphi) = 0\,, \tag{2}$$

where $\Delta^{(4)}$ is the four-dimensional Laplace-Beltrami operator and $'$ denotes differentiation with respect to the field φ. The Euclidean equation of motion is equivalent to that of a classical particle moving in an upturned potential $-U(\varphi)$ (see Fig. 1), and the tunneling corresponds to the trajectory that starts at the false vacuum in the distant Euclidean 'past,' reaches close to the true vacuum at the origin of Euclidean 'time,' and rolls back to the false vacuum in the distant 'future.' Coleman called this solution *the bounce*, and its boundary conditions are $\varphi(x)|_{x_4\to\pm\infty} = +v$ and $\partial\varphi(x)/\partial x_4|_{x_4=0} = 0$. In order to obtain a finite action, we set $U_0 = (\mu v/2)^2 - gv^3/6$, so that the false vacuum has zero energy density, and require $\varphi(x)|_{|\mathbf{x}|\to\infty} = +v$, i.e. that the system far from the bubble is unaffected by its nucleation. After transforming to four-dimensional hyperspherical coordinates, the boundary conditions become $\varphi(r)|_{r\to\infty} = +v$ and $\partial\varphi(r)/\partial r|_{r=0} = 0$, where $r = \sqrt{x\cdot x}$. Since the turning point at $r = 0$ occurs close to $\varphi = -v$, we see that bounce corresponds to a four-dimensional spherical bubble, separating false vacuum on the outside from true vacuum on the inside.

In the thin-wall regime, where the diameter of the bubble is much larger than the width of the bubble wall, the gradients of the field are small everywhere except in a narrow region centred on the bubble wall. Since this occurs at some large radius $R \propto 1/g$, we can therefore neglect the damping term $3/r\ \mathrm{d}\varphi/\mathrm{d}r$ in the hyperspherical equation of motion

$$-\frac{\mathrm{d}^2\varphi}{\mathrm{d}r^2} - \frac{3}{r}\frac{\mathrm{d}\varphi}{\mathrm{d}r} + U'(\varphi) = 0\,. \tag{3}$$

The relevant solution is the well-known kink (see Fig. 2)

$$\varphi(r) = v\tanh\big[\gamma(r-R)\big]\,, \tag{4}$$

where $\gamma \equiv \mu/\sqrt{2}$. To fix the radius of the critical bubble R, we extremise the bounce action

$$B = \int \mathrm{d}^4x\left[\frac{1}{2}\left(\frac{\partial\varphi}{\partial x_\mu}\right)^2 + U(\varphi)\right] \tag{5}$$

with respect to R, giving $R = 12\gamma/(gv)$. Wick rotating back to Minkowski spacetime, $x_4 = ix_0$, we find that the bubble expands along a hyperbolic trajectory $R^2 = |\mathbf{x}|^2 - (ct)^2$, nucleating at $t = 0$ with three-dimensional radius R and quickly accelerating to the speed of light (c).

The critical bubble actually corresponds to a maximum of the free energy, and the spectrum of the Klein-Gordon operator therefore has a negative eigenvalue

$$\lambda_0 = \frac{1}{B}\frac{\delta^2 B}{\delta R^2} = -\frac{3}{R^2}\,, \tag{6}$$

associated with dilatations of the bounce. As a result, the Euclidean path integral

$$Z[0] = \int \mathcal{D}\Phi\, e^{-S[\Phi]/\hbar} \tag{7}$$

is ill-defined. Moreover, by virtue of the four-dimensional translational invariance of the action, the spectrum also contains four zero eigenvalues. The integral over these zero modes can be

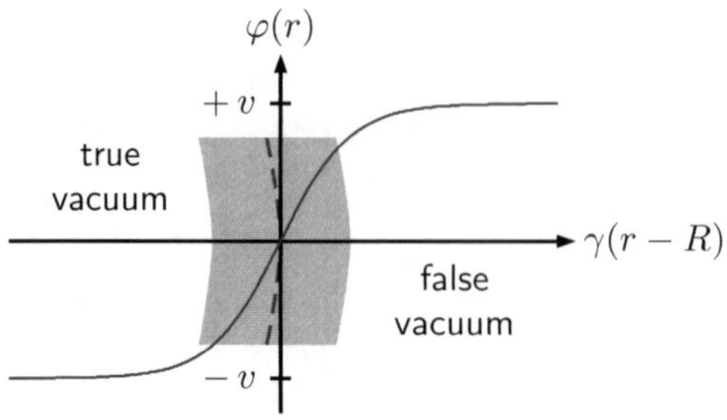

Figure 2: Sketch of the classical kink solution [Eq. (4)]. The grey band indicates the position of the bubble wall.

traded for an integral over the collective coordinates of the bounce, and this accounts for the fact that the nucleation of the critical bubble can occur anywhere in the four-volume VT. The integral over the negative mode can be performed by the method of steepest decent by deforming the contour of functional integration into the complex plane. As a result, the path integral obtains an imaginary part, which is related to the tunneling rate per unit volume as

$$\Gamma/V \;=\; 2|\mathrm{Im}\,Z[0]|/(VT) \;\sim\; e^{-B/\hbar}\,. \tag{8}$$

Hereafter, we will denote the solution to the classical equation of motion in Eq. (4) by $\varphi^{(0)}$ and the classical bounce action by $B^{(0)} \equiv S[\varphi^{(0)}]$.

3. Quantum corrections

In order to find the first quantum corrections (see Refs. [39] and [40]) and to fix the proportionality in Eq. (8), we can perform a saddle-point evaluation of the path integral by expanding around the classical solution, writing $\Phi = \varphi^{(0)} + \hbar^{1/2}\phi$ and truncating at first order in $\hbar$. In this way, we find

$$Z[0] \;=\; -\,\frac{i}{2}\,e^{-B^{(0)}/\hbar}\left|\frac{\lambda_0\det^{(5)}G^{-1}(\varphi^{(0)})}{(VT)^2\big(\frac{B^{(0)}}{2\pi\hbar}\big)^4(4\gamma^2)^5\det^{(5)}G^{-1}(v)}\right|^{-1/2}, \tag{9}$$

where the factors of $(VT)^2\big(\frac{B^{(0)}}{2\pi\hbar}\big)^4$ have arisen from the integral over the translational zero modes, the superscript "(5)" indicates that the determinant is over only the positive-definite eigenvalues and $G^{-1}(\varphi^{(0)})$ is the Klein-Gordon or *fluctuation operator*

$$G^{-1}(\varphi^{(0)};x,y) \;\equiv\; \frac{\delta^2 S[\Phi]}{\delta\Phi(x)\delta\Phi(y)}\bigg|_{\Phi\,=\,\varphi^{(0)}} \;=\; \delta^4(x-y)\Big(-\;\Delta^{(4)}\;+\;U''(\varphi^{(0)})\Big)\,. \tag{10}$$

The factor $(4\gamma^2)^5\det^{(5)}G^{-1}(v)$ is introduced for normalisation. Substituting the above result into Eq. (8), the tunneling rate per unit volume is found to be

$$\Gamma/V \;=\; \left(\frac{B^{(0)}}{2\pi\hbar}\right)^2(2\gamma)^5|\lambda_0|^{-1/2}\exp\left[-\,\frac{1}{\hbar}\Big(B^{(0)}\;+\;\hbar\,B^{(1)}\Big)\right], \tag{11}$$

where the one-loop (order $\hbar$) corrections are captured by the *fluctuation determinant*

$$B^{(1)} \;=\; \frac{1}{2}\Big(\ln\,\det^{(5)}G^{-1}(\varphi^{(0)})-\ln\,\det^{(5)}G^{-1}(v)\Big) \;=\; \frac{1}{2}\,\mathrm{tr}^{(5)}\Big(\ln\,G^{-1}(\varphi^{(0)})-\ln\,G^{-1}(v)\Big)\,. \tag{12}$$

IOP Conf. Series: Journal of Physics: Conf. Series **873** (2017) 012041 doi:10.1088/1742-6596/873/1/012041

This can be evaluated by a number of approaches, including, for instance, the heat kernel method (see, e.g., Ref. [41]), the Gel'fand-Yaglom theorem [42] and the direct integration method due to Baacke and Junker [43, 44, 45]. The latter is particularly well suited to numerical applications.

In the model considered so far [Eq. (1)], the instability is already present at the level of the classical potential. However, in the case of the SM, the instability emerges only at the quantum level, and the relevant quantum path, viz. the tunneling trajectory, is therefore non-perturbatively far away from the classical path. To see the significance of this observation, we should consider the spectrum of the fluctuation operator. In the tree-level case, an expansion around the solution to the classical equations of motion leads to a fluctuation operator with a non-positive-definite spectrum. On the other hand, when the instability results from radiative effects, the tree-level fluctuation operator will have a positive-definite spectrum, whilst the negative and zero eigenvalues will appear only at the level of the quantum fluctuation operator. In order to capture the instability consistently, we must therefore perform the saddle-point evaluation of the path integral by expanding around the solutions to the quantum equations of motion, both for the one- and two-point functions. Such a procedure, which we refer to as the *method of external sources* [35], can be formulated rigorously within the nPI effective action formalism. Consider, for illustration, the 2PI effective action [46]

$$\Gamma[\phi,\Delta] \;=\; -\,\hbar\,\ln\,Z[\mathcal{J},\mathcal{K}] \,+\, \mathcal{J}_x[\phi,\Delta]\phi_x \,+\, \frac{1}{2}\mathcal{K}_{xy}[\phi,\Delta]\big(\phi_x\phi_y \,+\, \hbar\,\Delta_{xy}\big)\,. \tag{13}$$

By choosing the extremal sources $\mathcal{J}$ and $\mathcal{K}$ such that

$$\frac{\delta S[\Phi]}{\delta\Phi}\bigg|_{\Phi\,=\,\varphi} \,-\, \mathcal{J}_x[\phi,\Delta] \,-\, \mathcal{K}_{xy}[\phi,\Delta]\varphi_y \;=\; \frac{\delta\Gamma[\phi,\Delta]}{\delta\phi_x}\bigg|_{\phi\,=\,\varphi} \;=\; 0 \tag{14}$$

and $\mathcal{G}^{-1}=G^{-1}-\mathcal{K}$ is the dressed inverse two-point function, we can drive the system along the quantum saddle point, as desired.

3.1. *Tree-level instability*

The tunneling rate per unit volume can be written in terms of the effective action as

$$\mathit{\Gamma}/V \;=\; 2|\mathrm{Im}\,e^{-\Gamma/\hbar}|/(VT)\,, \tag{15}$$

and, for the model in Eq. (1), it is convenient to work at the level of the 1PI effective action:

$$\Gamma[\varphi^{(1)}] \;=\; S[\varphi^{(1)}] \,+\, \frac{i\pi\hbar}{2} \,+\, \frac{\hbar}{2}\,\ln\left|\frac{\lambda_0\det^{(5)}G^{-1}(\varphi^{(1)})}{\frac{1}{4}(VT)^2\big(\frac{\mathcal{N}^{-2}}{2\pi\hbar}\big)(4\gamma^2)^5\mathrm{det}^{(5)}G^{-1}(v)}\right| \,+\, \mathcal{O}(\hbar^2)\,. \tag{16}$$

For further discussion of the normalisation of the zero modes $\mathcal{N}$, see Ref. [36]. Writing the quantum-corrected bounce as $\varphi^{(1)}=\varphi^{(0)}+\hbar\,\delta\varphi$, we can expand the 1PI effective action around the classical kink solution $\varphi^{(0)}$ to obtain

$$\mathit{\Gamma}/V \;=\; \left(\frac{\mathcal{N}^{-2}}{2\pi\hbar}\right)^2(2\gamma)^5|\lambda_0|^{-1/2}\exp\left[-\,\frac{1}{\hbar}\Big(B^{(0)}+\hbar\,B^{(1)}\,+\,\hbar^2\,B^{(2)}\,+\,\hbar^2\,B^{(2)\prime}\Big)\right]. \tag{17}$$

The two-loop 1PR corrections (appearing at order $\hbar^2$) are due to expanding the action of the corrected bounce $S[\varphi^{(1)}]$ and the fluctuation determinant $B^{(1)}[\varphi^{(1)}]$ around $\varphi^{(0)}$:

$$B^{(2)} \;=\; -\,\frac{1}{2}\int \mathrm{d}^4x\;\varphi^{(0)}(x)\Pi(\varphi^{(0)};x)\delta\varphi(x) \;=\; -\,\frac{1}{2}\,B^{(2)\prime}\,. \tag{18}$$

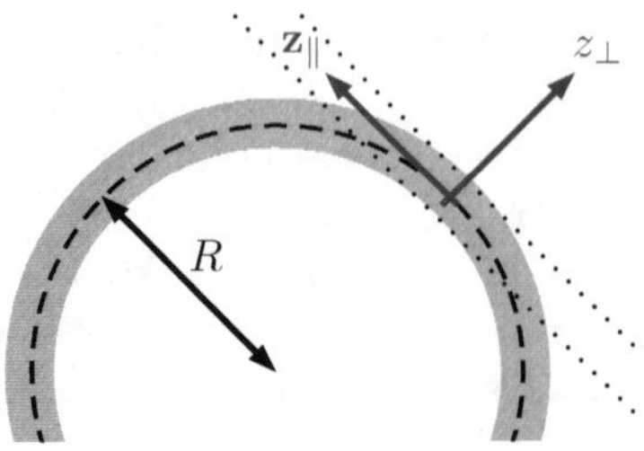

Figure 3: Sketch of the coordinate system $(z_\perp, \mathbf{z}_\parallel)$ applied in the planar-wall limit.

Returning to the model in Eq. (1), we now wish to calculate the quantum corrections to the tunneling rate, and the starting point will be the Green's function determined within the background of the kink.[1] In the thin-wall regime, this calculation can be performed analytically (see Ref. [34]). Specifically, we can treat the bubble wall as approximately planar, introducing a set of coordinates aligned parallel and perpendicular to the bubble wall (see Fig. 3). In the limit $\mu R \gg 1$, we can replace the discrete summation over angular momentum modes by an integral over a continuous momentum conjugate to the parallel coordinates $\mathbf{z}_\parallel$:

$$\frac{j(j+1)\hbar}{\mu^2 R^2} \longrightarrow \frac{k^2}{\mu^2}\,. \tag{19}$$

Making the change of variables $u^{(\prime)} \equiv \varphi(r^{(\prime)})/v$, the Green's function has the form

$$\begin{aligned} G(u,u',m) \;=\; & \frac{1}{2\gamma m}\left[\vartheta(u-u')\left(\frac{1-u}{1+u}\right)^{m/2}\left(\frac{1+u'}{1-u'}\right)^{m/2}\right.\\ & \times\left(1-3\,\frac{(1-u)(1+m+u)}{(1+m)(2+m)}\right)\left(1-3\,\frac{(1-u')(1-m+u')}{(1-m)(2-m)}\right) + \left.(u\longleftrightarrow u')\right]\,, \end{aligned} \tag{20}$$

where we have defined

$$m \;\equiv\; 2\left(1+\frac{k^2}{4\gamma^2}\right)^{1/2}\,. \tag{21}$$

So as to make a meaningful perturbative truncation and to enhance the radiative effects, we supplement the model in Eq. (1) with N additional scalar fields:

$$\mathcal{L} \;\supset\; \sum_{i=1}^{N}\left[\frac{1}{2}(\partial_\alpha X_i)^2 \,+\, \frac{1}{2}m_\chi^2 X_i^2 \,+\, \frac{1}{4}\lambda\,\Phi^2\,X_i^2\right]\,. \tag{22}$$

The dominant corrections in the $1/N$ expansion [52] of the tunneling rate are shown in Fig. 4. In the limit $m_\chi^2 \gg \gamma^2$, the renormalised X tadpole is given by

$$\Sigma^R(u) \;=\; \frac{\lambda\gamma^2}{8\pi^2}\frac{\gamma^2}{m_\chi^2}\left[72+(1-u^2)(40-3u^2)\right]\,. \tag{23}$$

We can also calculate the quantum corrections to the bounce, which are given explicitly by

$$\delta\varphi(u) \;=\; -\,\frac{v}{\gamma}\int_{-1}^{1}\mathrm{d}u'\,\frac{u'G(u,u',m)|_{k=0}}{1-u'^2}\left(\Pi^R(u')+N\,\Sigma^R(u')\right)\,, \tag{24}$$

[1] Green's function methods have been applied previously to self-consistent calculations in the Hartree-Fock approximation of the pure $\lambda\Phi^4$ theory in Refs. [47, 48, 49, 50] and the $\lambda\Phi^4 + g\Phi^6$ theory in Ref. [51].

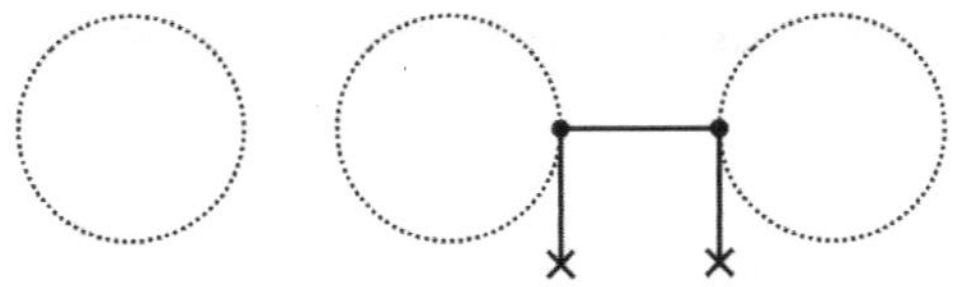

Figure 4: The dominant corrections to the tunneling rate at order $\hbar$ (left) and order $\hbar^2$ (right). Dotted lines are X Green's functions, solid lines are Φ Green's functions and crosses correspond to insertions of the classical background field $\varphi^{(0)}$.

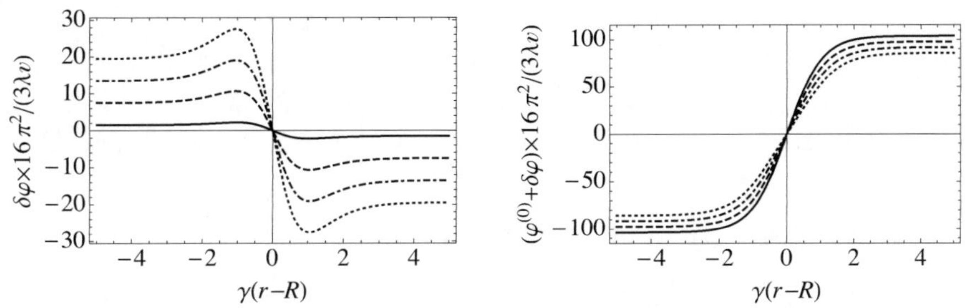

Figure 5: The quantum correction $\delta\varphi$ (left) and quantum-corrected bounce $\varphi^{(1)} = \varphi^{(0)} + \delta\varphi$ (right) for $N\gamma^2/m_X^2$ equal to 0 (solid), 0.5 (dashed), 1 (dash-dotted) and 1.5 (dotted).

where we have included the (subdominant) renormalised Φ tadpole (Π^R) for completeness:

$$\Pi^R(u) \;=\; \frac{3\lambda\gamma^2}{16\pi^2}\left[6 + (1-u^2)\left(5 - \frac{\pi}{\sqrt{3}}\,u^2\right)\right]. \tag{25}$$

Figure 5 shows the quantum correction and quantum-corrected bounce for a range of values of $N\gamma^2/m_X^2$. The broadening of the bubble wall reduces the bounce action, leading to an increase in the tunneling rate. This is in qualitative agreement with the analysis of Ref. [47].

4. Radiatively-generated instability

We now turn our attention to a model where the instability instead arises at the quantum level. We start with a classically scale-invariant model (for $g = 0$) with Lagrangian

$$\mathcal{L} \;=\; \frac{1}{2!}\left(\partial_\alpha\Phi\right)^2 + \frac{1}{2!}\sum_{i=1}^{N}\left(\partial_\alpha X_i\right)^2 + \frac{1}{4}\lambda\,\Phi^2\sum_{i=1}^{N}X_i^2 + \frac{1}{4}\kappa\sum_{i,\,j=1}^{N}X_i^2X_j^2 + \frac{1}{3!}\,g\,\Phi^3 + U_0\,. \tag{26}$$

This model suffers logarithmic infra-red divergences at the one-loop level, which are regularised by introducing a scale m. This regulator mass is translated into a symmetry breaking scale v via dimensional transmutation through the well-known Coleman-Weinberg mechanism [53]. Thus, at the one-loop level, the potential contains two minima, the degeneracy of which is broken by hand through the small cubic interaction. Choosing renormalisation conditions as in Ref. [36], the renormalised Coleman-Weinberg (CW) one-loop effective potential is

$$U^R_{\rm eff}(\varphi) \;=\; \left(\frac{\lambda}{16\pi}\right)^2\varphi^4\left[N\left(\ln\frac{3\varphi^2}{\rho M^2} - \frac{3}{2}\right) + F(\rho) + \frac{1}{6}\,g\,\varphi^3 + U_0\right] + \mathcal{O}(g^2)\,, \tag{27}$$

where $\rho \equiv 6\kappa/\lambda$ and

$$F(\rho) \;\equiv\; \ln 3 + \frac{8}{(1-\rho^2)}\left(3 + \rho + \frac{1+3\rho}{1-\rho}\,\ln\rho\right). \tag{28}$$

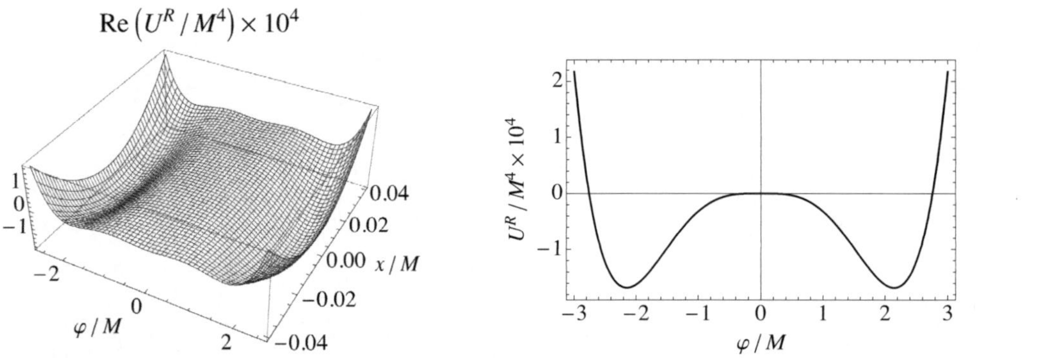

Figure 6: Three-dimensional plot of the renormalised effective potential in the unitary gauge for the model in Eq. (26) (left), as well as the cross-section in the plane $\chi = 0$ (right).

For $g \ll 32\pi^2 v/(\lambda^2 N)$, the global minima of the potential lie at

$$\varphi \approx \pm v = \pm\sqrt{\frac{\rho M^2}{3}}\exp\left[\frac{1}{2}+\frac{F}{2N}\right]. \tag{29}$$

By using auxiliary fields to induce the symmetry breaking, we remain in the regime of validity of the one-loop approximation. The effective potential is plotted in Fig. 6.

For large N, the $1/N$ power counting is such that (i) we must treat the equation of motion for φ at the one-loop level:

$$-\Delta^{(4)}\varphi(x) + \Sigma^R(\varphi;x)\varphi(x) = 0\,, \qquad \Sigma^R(\varphi;x) = \frac{\lambda N}{2}S(\varphi;x,x) + \delta\Sigma\,, \tag{30}$$

where $S(\varphi;x,y)$ is the X Green's function and $\delta\Sigma$ contains the relevant counterterms; (ii) we can treat the equation of motion for the X field Green's function at tree-level:

$$\left(-\Delta^{(4)} + \frac{\lambda}{2}\varphi^2\right)S(\varphi;x,y) = \delta^4(x-y)\,; \tag{31}$$

and (iii) the Φ Green's function decouples entirely. Working again in the thin- and planar-wall limits, we employ an iterative numerical procedure to determine the self-consistent bounce:

1. Calculate a first approximation to the bounce by promoting the homogeneous field configuration in the CW effective potential to a space-time dependent one:
$$-\Delta^{(4)}\varphi + U_{\rm eff}^{R\prime}(\varphi) = 0\,. \tag{32}$$
2. Solve for the X field Green's function.
3. Calculate the tadpole correction, renormalising in the homogeneous false vacuum.
4. Insert the tadpole into the quantum equation of motion and solve for the bounce.
5. Iterate over steps 2 to 5 until the solution has converged sufficiently.

This procedure essentially undoes a gradient expansion, putting back the full dependence on the inhomogeneity of the solitonic configuration. The resulting self-consistent bounce and one-loop tadpole, as obtained in Ref. [36], are shown in Fig. 7.

The dominant dependence on the gradients occurs in the one-loop fluctuation determinant $B^{(1)}$. The results are shown in Fig. 8. We see from the right-hand plot that the deviation between the self-consistent and (homogeneous) CW results scales like λN relative to the CW result. Scaling as λN times the homogeneous one-loop terms, the gradient effects can therefore compete with or even be more important than the two-loop effects, which scale as $\lambda\kappa N^2$ or $\lambda^2 N$. This numerically confirms arguments presented in Ref. [54].

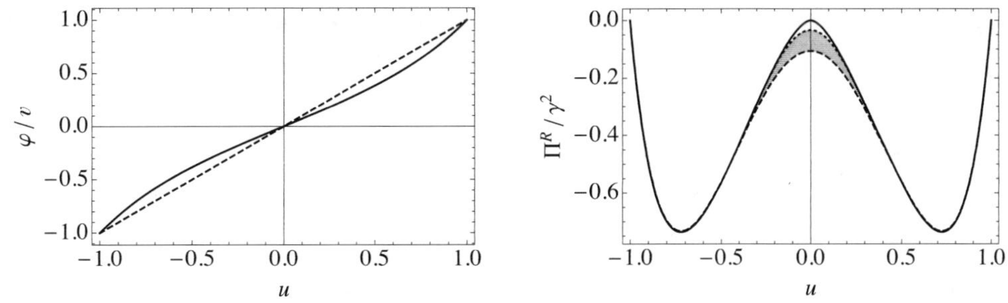

Figure 7: Left: The self-consistent bounce compared with the kink in Eq. (4). Right: The renormalised tadpole correction. The solid line is the CW (homogeneous) result, and the dotted and dashed lines are the self-consistent results for the first and last parameter points in the sampling range: $M = 1$ with $0.04 \leq \lambda^2 N \leq 0.4$.

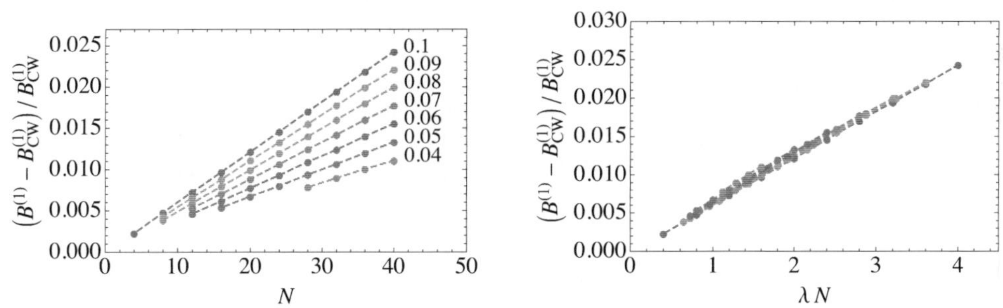

Figure 8: The fractional difference in the self-consistent fluctuation determinant $B^{(1)}$ relative to the Coleman-Weinberg result $B^{(1)}_{\rm CW}$ as a function of N (left) and λN (right). The labels in the left-hand plot indicate values of λ.

5. Conclusions and future directions

We have described a Green's function method for calculating the quantum corrections to the decay rate of false vacua that fully accounts for the background inhomogeneity of the solitonic, tunneling configuation. We have applied this technique to calculate the quantum-corrected decay rate analytically in the archetypal Φ^4 theory and outlined a numerical procedure for the self-consistent calculation of the tunneling rate when the instability arises only at the quantum level. There, we have shown that the modification of the one-loop terms due to accounting fully for gradient effects can compete with the (homogeneous) two-loop terms. In the thin-wall regime, the gradient effects are suppressed due to the symmetry of the bounce solution about the mid-point of the bubble wall, and such a suppression is not anticipated in the thick-wall regime. A study of this regime is of particular relevance to the instability of the SM electroweak vacuum and will be presented elsewhere.

Acknowledgments

The work of PM is supported by STFC grant ST/L000393/1. BG acknowledges the support of the Deutsche Forschungsgemeinschaft (DFG) through the cluster of excellence 'Origin and Structure of the Universe'. PM would like to thank Wen-Yuan Ai for on-going collaboration, the organisers of DISCRETE2016 for their hospitality and the opportunity to present this work, and the participants for their questions, comments and suggestions.

IOP Conf. Series: Journal of Physics: Conf. Series **873** (2017) 012041 doi:10.1088/1742-6596/873/1/012041

References

[1] Elias-Miró J, Espinosa J R, Giudice G F, Isidori G, Riotto A and Strumia A 2012 *Phys. Lett.* B **709** 222–8
[2] Degrassi G, Di Vita S, Elias-Miró J, Espinosa J R, Giudice G F, Isidori G and Strumia A 2012 *J. High Energy Phys.* JHEP08(2012)098
[3] Alekhin S, Djouadi A and Moch S 2012 *Phys. Lett.* B **716** 214–9
[4] Buttazzo D, Degrassi G, Giardino P P, Giudice G F, Sala F, Salvio A and Strumia A 2013 *J. High Energy Phys.* JHEP12(2013)089
[5] Bednyakov A V, Kniehl B A, Pikelner A F and Veretin O L 2015 *Phys. Rev. Lett.* **115** 201802
[6] Di Luzio L, Isidori G and Ridolfi G 2016 *Phys. Lett.* B **753** 150–60
[7] Morrissey D E and Ramsey-Musolf M J 2012 *New J. Phys.* **14** 125003
[8] Chung D J H, Long A J and Wang L-T 2013 *Phys. Rev.* D **87** 023509
[9] Witten E 1984 *Phys. Rev.* D **30** 272–85
[10] Kosowsky A, Turner M S and Watkins R 1992 *Phys. Rev.* D **45** 4514–35
[11] Kosowsky A, Turner M S and Watkins R 1992 *Phys. Rev. Lett.* **69** 2026–9
[12] Caprini C, Durrer R, Konstandin T and Servant G 2009 *Phys. Rev.* D **79** 083519
[13] Abbott B P et al. [LIGO Scientific and Virgo Collaborations] 2016 *Phys. Rev. Lett.* **116** 061102
[14] Bezrukov F, Kalmykov M Y, Kniehl B A and Shaposhnikov M 2012 *J. High Energy Phys.* JHEP10(2012)140
[15] Masina I 2013 *Phys. Rev.* D **87** 053001
[16] Branchina V and Messina E 2013 *Phys. Rev. Lett.* **111** 241801
[17] Branchina V, Messina E and Platania A 2014 *J. High Energy Phys.* JHEP09(2014)182
[18] Lalak Z, Lewicki M and Olszewski P 2014 *J. High Energy Phys.* JHEP05(2014)119
[19] Branchina V, Messina E and Sher M 2015 *Phys. Rev.* D **91** 013003
[20] Eichhorn A, Gies H, Jaeckel J, Plehn T, Scherer M M and Sondenheimer R 2015 *J. High Energy Phys.* JHEP04(2015)022
[21] Branchina V, Messina E and Zappala D 2016 *Europhys. Lett.* **116** 21001
[22] Czerwińska O, Lalak Z, Lewicki M and Olszewski P 2016 *J. High Energy Phys.* JHEP10(2016)004
[23] Burda P, Gregory R and Moss I G 2015 *Phys. Rev. Lett.* **115** 071303
[24] Rajantie A and Stopyra S 2017 *Phys. Rev.* D **95** 025008
[25] Grinstein B and Murphy C W 2015 *J. High Energy Phys* JHEP12(2015)063
[26] Gregory R, Moss I G and Withers B 2014 *J. High Energy Phys.* JHEP03(2014)081
[27] Burda P, Gregory R and Moss I G 2015 *J. High Energy Phys.* JHEP08(2015)114
[28] Burda P, Gregory R and Moss I G 2016 *J. High Energy Phys.* JHEP06(2016)025
[29] Weinberg E J and Wu A 1987 *Phys. Rev.* D **36** 2474-80
[30] Alexandre J and Farakos K 2008 *J. Phys.* A: *Math. Theor.* **41** 015401
[31] Branchina V, Faivre H and Pangon V 2009 *J. Phys.* G **36** 015006
[32] Einhorn M B and Jones D R T 2007 *J. High Energy Phys.* JHEP04(2007)051
[33] Gies H and Sondenheimer R 2015 *Eur. Phys. J.* C **75** 68
[34] Garbrecht B and Millington P 2015 *Phys. Rev.* D **91** 105021
[35] Garbrecht B and Millington P 2016 *Nucl. Phys.* B **906** 105–32
[36] Garbrecht B and Millington P 2015 *Phys. Rev.* D **92** 125022
[37] Coleman S R 1977 *Phys. Rev.* D **15** 2929–36 [Erratum: *Phys. Rev.* D **16** 1248]
[38] Langer J S 1969 *Annals Phys.* **54** 258–75
[39] Callan C G and Coleman S R 1977 *Phys. Rev.* D **16** 1762–8
[40] Konoplich R V 1987 *Theor. Math. Phys.* **73** 1286–95 [*Teor. Mat. Fiz.* **73** 379–92]
[41] Vassilevich D V 2003 *Phys. Rept.* **388** 279–360
[42] Gel'fand I M and Yaglom A M 1960 *J. Math. Phys.* **1** 48
[43] Baacke J and Junker S 1993 *Mod. Phys. Lett.* A **8** 2869–74
[44] Baacke J and Junker S 1994 *Phys. Rev.* D **49** 2055–73
[45] Baacke J and Junker S 1994 *Phys. Rev.* D **50** 4227–28
[46] Cornwall J M, Jackiw R and Tomboulis E 1974 *Phys. Rev.* D **10** 2428–45
[47] Bergner Y and Bettencourt L M A 2004 *Phys. Rev.* D **69** 045002
[48] Bergner Y and Bettencourt L M A 2004 *Phys. Rev.* D **69** 045012
[49] Baacke J and Kevlishvili N 2005 *Phys. Rev.* D **71** 025008
[50] Baacke J and Kevlishvili N 2007 *Phys. Rev.* D **75** 045001 [Erratum: *Phys. Rev.* D **76** 029903]
[51] Bergner Y and Bettencourt L M A 2003 *Phys. Rev.* D **68** 025014
[52] 't Hooft G 1974 *Nucl. Phys.* B **72** 461–73
[53] Coleman S R and Weinberg E J 1973 *Phys. Rev.* D **7** 1888–910
[54] Weinberg E J 1993 *Phys. Rev.* D **47** 4614–27

IOP Conf. Series: Journal of Physics: Conf. Series **873** (2017) 012042 doi:10.1088/1742-6596/873/1/012042

Scale symmetry and quantum corrections

P Olszewski[1]

[1]Institute of Theoretical Physics, Faculty of Physics, University of Warsaw, ul. Pasteura 5, Warsaw, Poland

E-mail: Pawel.Olszewski@fuw.edu.pl

Abstract. We use the scale invariant regularization method to study loop corrections in a scale symmetric model of two interacting scalar fields. The method resembles the usual procedure of dimensional regularization, but the renormalization scale μ is a function of dynamical fields. The model is free of the trace anomaly and accommodates spontaneous breaking of the symmetry. It is also necessarily nonrenormalizable. We show how the nonstandard method results in beta functions and corrections to the effective scalar potential different than in the usual regularization, while at the same time it does not spoil the Callan-Symanzik equation.

1. Introduction

Scale symmetry is broken in nature. In models of particle interactions this fact is often accounted for by means of a quantum anomaly of the scale symmetry. An alternative possibility is to construct a theory, where this symmetry is not anomalous but broken spontaneously [1]. The idea that a fundamental particle theory posseses the global dilatation invariance is an exciting one as it could recast the Higgs naturalness problem in a new light, requiring a dynamical explanation for the hierarchy of masses. This prospect is enabled by using a modification of dimensional regularization, the scale invariant (SI) method, based on the idea that the renormalization scale μ is a function of dynamical fields rather that a parameter, and that it acquires a nonzero value through spontaneous breaking of the scale symmetry [2, 3, 4, 5, 6, 7].

Using this method we perform perturbative regularization of a simple model with two scalar fields. The model is classically scale invariant and we arrange it to allow for the spontaneous breaking by requiring that the effective potential possesses a flat direction. The usage of scale invariant regularization constitutes a major qualitative change as it enables the Golstone boson of the broken scale symmetry to remain massless to all orders. Our goal is to study this regularization in some detail. We show why it may be used only in the broken phase of the symmetry and that it requires us to introduce infinitely many nonrermalizable interactions to the Lagrangian. Further, we calculate renormalization group functions and corrections to the effective potential. The calculation is performed up to the level of two-loop Fynanman diagrams, in order to obtain modifications introduced by the *new* regularization as compared to the *old* procedure with $\mu = const.$

IOP Conf. Series: Journal of Physics: Conf. Series **873** (2017) 012042 doi:10.1088/1742-6596/873/1/012042

2. Scale transformations

We will focus in this presentation on toy models that involve only scalar fields. To begin, let us consider the Lagrangian for a single field with a quartic interaction

$$\mathcal{L} = \frac{1}{2}(\partial\varphi)^2 - \lambda\varphi^4 \,, \tag{1}$$

We wish to extend the algebra of Poincare symmetry by adding the generator of scale transformations (dilatations), D. The only new nonzero commutator in the algebra is

$$[D, P_\mu] = iP_\mu \,, \tag{2}$$

We will assume that any field, Φ, transforms (scales) under a nontrivial representation of this group,

$$\begin{aligned} U(a) &= e^{iaD} && (3)\\ \Phi(x) \to U(a)\Phi(x)U^{-1}(a) &= e^{ad_\Phi}\Phi(e^a x) && (4)\\ \delta\Phi(x) = i[D,\Phi] &= (d_\Phi + x^\nu\partial_\nu)\,\Phi(x)\,. && (5)\end{aligned}$$

Denote the number of space-time dimensions by $n = 4 - 2\epsilon$. d_Φ is the engineering dimension of the field, ($d_\varphi = \frac{n}{2} - 1 = 1 - \epsilon$ for scalars). It is fixed by the requirement that the kinetic part of its action is invariant under dilatations,

$$\begin{aligned} \frac{1}{2}(\partial\varphi(x))^2\,\mathrm{d}^n x &\to \frac{1}{2}e^{2(d_\varphi+1)a}\left(\widetilde{\partial}\varphi(\widetilde{x})\right)^2 e^{-na}\mathrm{d}^n\widetilde{x}\,, \quad \text{where} \quad \widetilde{x} = e^a x\,,\ \widetilde{\partial}_\mu = \frac{\partial}{\partial\widetilde{x}^\mu} && (6)\\ \delta\,\frac{1}{2}(\partial\phi)^2 &= [2(d_\varphi+1) + x^\nu\partial_\nu]\,\frac{1}{2}(\partial\varphi)^2 = [n + x^\nu\partial_\nu]\,\frac{1}{2}(\partial\varphi)^2 && (7)\end{aligned}$$

By assumption, dilatation transforms dynamical fields but not dimensional parameters, like masses,

$$\delta\,\mu = i[D,\mu] = 0\,. \tag{8}$$

This applies in particular to the renormalization scale μ, introduced to the Lagrangian as part of the procedure of dimensional regularization,

$$\mathcal{L} = \frac{1}{2}\,(\partial\varphi)^2 - \lambda\mu^{2\epsilon}\varphi^4\,. \tag{9}$$

As a consequence, the action is no longer invariant wrt dilatations when $n \neq 4$,

$$\begin{aligned} \varphi^4(x)\,\mathrm{d}^n x &\to e^{-2\epsilon\,a}\,\varphi^4(\widetilde{x})\,\mathrm{d}^n\widetilde{x} && (10)\\ \delta\,\varphi^4 &= (4 + x^\nu\partial_\nu)\varphi^4 \neq (n + x^\nu\partial_\nu)\varphi^4\,. && (11)\end{aligned}$$

One can also observe that the Noether current for dilatations,

$$D^\mu = \frac{\delta\,\mathcal{L}}{\delta\partial_\mu\varphi}\delta\varphi - x^\mu\mathcal{L}\,, \tag{12}$$

when continued to $n \neq 4$ dimensions, is no longer conserved. Upon using the equation of motion $\partial_\mu\frac{\delta\mathcal{L}}{\delta\partial_\mu\varphi} = \frac{\delta\mathcal{L}}{\delta\varphi}$, and $\delta\partial_\mu\varphi = (d_\varphi+1)\partial_\mu\varphi + x^\nu\partial_\nu\partial_\mu\varphi$ we have

$$\partial_\mu D^\mu = d_\varphi\frac{\delta\mathcal{L}}{\delta\varphi}\varphi + (d_\varphi+1)\frac{\delta\,\mathcal{L}}{\delta\partial_\mu\varphi}\partial_\mu\varphi - n\mathcal{L} = 2\epsilon\lambda\mu^{2\epsilon}\varphi^4 \neq 0\,. \tag{13}$$

On the other hand, $\partial_\mu D^\mu$ is equal to the trace of improved energy-momentum tensor $T_{\mu\nu}$, see [8],

$$\partial_\mu D^\mu = T^\mu_\mu \quad , \quad T_{\mu\nu} = \Theta_{\mu\nu} - \frac{1}{6}\left(\partial_\mu\partial_\nu - g_{\mu\nu}\Box\right)\varphi^2 \quad , \quad \Theta_{\mu\nu} = \frac{\delta\,\mathcal{L}}{\delta\partial_\mu\varphi}\partial_\nu\varphi - g_{\mu\nu}\mathcal{L}\,. \tag{14}$$

We observe that the scale invariance, present in four dimensions, is no longer preserved when the theory is continued to $n \neq 4$ dimensions in course of the usual dimensional regularization.

IOP Conf. Series: Journal of Physics: Conf. Series **873** (2017) 012042 doi:10.1088/1742-6596/873/1/012042

3. Scale invariant regularization

Introduction of the parameter μ, that is dimensionful but not charged under the transformation generated by D, results in a Lagrangian that is not a homogeneous function of dimension n under dilatations:

$$\delta\mathcal{L} \neq (n + x^\nu \partial_\nu)\mathcal{L} = \partial_\nu(x^\nu \mathcal{L}) \,. \tag{15}$$

This has been recognized as the source of the trace anomaly, that is the fact that $T^\mu_\mu \neq 0$ persists in the limit $n \to 4$ after including loop corrections [9, 10]. A solution, known as scale-invariant (SI) regularization, originally proposed in [2, 3], and more recently studied and used in [4, 5, 6, 7], is to reinterpret μ as a function of fields, such that it does transform under dilatations with dimension 1,

$$\mu = \mu(\Phi) \quad , \quad i[D,\mu] = \frac{\partial\,\mu}{\partial\,\Phi_i}\, i[D,\Phi_i] = (1 + x^\nu\partial_\nu)\,\mu \,. \tag{16}$$

As a simple example we will use

$$\mu(\phi_2) \;=\; t\,\phi_2^{\frac{1}{1-\epsilon}} \quad , \quad t \in \mathbb{R}^+ \tag{17}$$

$$\mathcal{L} \;=\; \frac{1}{2}\,(\partial\phi_1)^2 + \frac{1}{2}\,(\partial\phi_2)^2 - \mu(\phi_2)^{2\epsilon}\, V(\phi_1,\phi_2) \,. \tag{18}$$

The parameter t is a positive number. Its presence exemplifies the freedom in choosing the function $\mu(\Phi)$. Regularized quantities should be independent of what $\mu(\Phi)$ was used. This leads to the hypothesis that coupling constants functionally depend on μ, $\lambda = \lambda[\mu(\Phi)]$. That dependence reduces to the familiar *running*, $\lambda = \lambda(t)$, upon a choice like the one in (17).

To use an interaction term like $\phi_2^{\frac{2\epsilon}{1-\epsilon}}\phi_1^4$ in perturbative loop calculations, one has to expand ϕ_2 around a nonzero background value $\phi_2 = \langle\phi_2\rangle + \phi_2'$. This results in an infinite number of interactions involving the dynamical fluctuation ϕ_2',

$$\phi_2^{\frac{2\epsilon}{1-\epsilon}}\phi_1^4 = \langle\phi_2\rangle^{\frac{2\epsilon}{1-\epsilon}}\left[\langle\phi_1\rangle + \phi_1'\right]^4 \left[1 + \underbrace{2\epsilon\frac{\phi_2'}{\langle\phi_2\rangle} - \epsilon\frac{\phi_2'^2}{\langle\phi_2\rangle^2} + \frac{2}{3}\epsilon\frac{\phi_2'^3}{\langle\phi_2\rangle^3} - \frac{1}{2}\epsilon\frac{\phi_2'^4}{\langle\phi_2\rangle^4} + \ldots}_{new}\right] \,, \tag{19}$$

where the ellipsis stands for terms with higher powers of ϵ or $\frac{\phi_2'}{\langle\phi_2\rangle}$. As marked in (19), we will refer to all the interactions of ϕ_2' stemming from the expansion of $\mu(\phi_2)$ around $\langle\phi_2\rangle$ as *new*. We also recover the familiar term $\mu_0^{2\epsilon}\phi_1^4$ with $\mu_0 = t\,\langle\phi_2\rangle^{\frac{1}{1-\epsilon}} = const$, identical to the interaction in (9). One needs only a finite number of *new* interactions, as long as a finite number of external ϕ_2 legs and loops is considered.

After having calculated loop corrections one is interested only in finite terms, $\mathcal{O}(\epsilon^0)$ and first order poles $\mathcal{O}(\epsilon^{-1})$. Since the *new* terms contain positive powers of ϵ, through one-loop diagrams they can contribute only to finite corrections. For the *new* interactions to produce poles, and consequently modifications in renormalization group functions, one has to include them in two-loop diagrams. Notably, among poles produced this way one finds

$$\frac{1}{\epsilon}\frac{\langle\phi_1\rangle^{4+2m}}{\langle\phi_2\rangle^{2m}} \,, \quad m = 1, 2, \ldots \tag{20}$$

Hence, corresponding interaction terms need to be included in the Lagrangian from the start, in order to regularize the model,

$$\mathcal{L} \;\supset\; \frac{\lambda^B_{(4+2m)}}{4+2m}\,\phi_2^{\frac{2\epsilon}{1-\epsilon}}\,\frac{\phi_1^{4+2m}}{\phi_2^{2m}} \quad , \quad m = 1, 2, \ldots \tag{21}$$

$$\lambda^B_{(4+2m)} \;=\; t^{2\epsilon}\, Z_{(4+2m)}\,\lambda_{(4+2m)}(t) \,, \tag{22}$$

IOP Conf. Series: Journal of Physics: Conf. Series **873** (2017) 012042 doi:10.1088/1742-6596/873/1/012042

where the $Z_{(4+2m)}$ factor contains appropriate counterterms.

The expansion in (19) is justified if ϕ_2 has a nonzero vacuum expectation value that breaks the scale symmetry spontaneously. Similarly, each term in (21) has a clear interpretation only upon the expansion of ϕ_2, when it is seen to produce an infinite series of nonrenormalizable interactions suppressed by powers of $\langle\phi_2\rangle$. Consequently, the model can be analyzed perturbatively only when $\langle\phi_2\rangle \gg \langle\phi_1\rangle$.

The assumption that the scale symmetry is broken is of course otherwise required, if one hopes to account for the observed nature of elementary particles. Existence of massive states would be excluded by an unbroken scale symmetry. Nontrivial (other than power-like) scaling of Green functions with momenta still arises as a consequence of the symmetry being spontaneously broken rather than anomalous [1].

4. The model with two scalar fields and a flat direction

A simple model that accommodates the spontaneous scale symmetry breaking and uses the SI regularization is

$$\begin{aligned}
\mathcal{L} &= \frac{1}{2}(\partial\phi_1)^2 + \frac{1}{2}(\partial\phi_2)^2 \; - \; \mu^{2\epsilon}\, V^{(0)}(\phi_1,\phi_2) \; + \; \mathcal{L}_{ct} && (23)\\
V^{(0)} &= \frac{\overline{\lambda_0}}{4!}\phi_2^4 + \frac{\lambda_2}{4}\phi_1^2\phi_2^2 + \frac{\lambda_4}{4!}\phi_1^4 + \sum_{m=1}^{\infty} \frac{\lambda_{(4+2m)}}{4+2m}\frac{\phi_1^{(4+2m)}}{\phi_2^{2m}} && (24)\\
\mathcal{L}_{ct} &= \frac{1}{2}(Z_{\phi_1}-1)\,(\partial\phi_1)^2 + (Z_{\phi_2}-1)\frac{1}{2}\,(\partial\phi_2)^2 \; - \; \mu^{2\epsilon}\left[(Z_0-1)\frac{\overline{\lambda_0}}{4!}\phi^4 \; + \; etc.\right] && (25)\\
\mu &= t\left(\sqrt{Z_{\phi_2}}\phi_2\right)^{\frac{1}{1-\epsilon}} \; . && (26)
\end{aligned}$$

The effective potential, V, is a homogeneous function of the fields,

$$\begin{aligned}
V &= V^{(0)} + V^{(1)} + \ldots && (27)\\
V(\phi_1,\phi_2) &= R^4\, W(\theta) \quad , \quad R^2 = \phi_1^2 + \phi_2^2 \quad , \quad \tan\theta = \frac{\phi_1}{\phi_2} \; . && (28)
\end{aligned}$$

The bar over λ_0 in (24) signifies the fact that we treat it as a function of all the other couplings $\lambda_0 = \overline{\lambda_0}(\lambda_2, \lambda_4, \ldots)$, defined by the condition that there exists the angle $\theta = \overline{\theta}(\lambda_2, \lambda_4, \ldots)$ such that

$$W(\overline{\theta}) = 0 \quad \text{and} \quad W'(\overline{\theta}) = 0 \; . \tag{29}$$

In other words, we demand that the potential has a flat direction, so that the field values

$$(\langle\phi_1\rangle, \langle\phi_2\rangle) \; = \; R\left(\tan\overline{\theta}, 1\right) \; , \; R \in \mathbb{R} \tag{30}$$

are the degenerate vacua of the model. Two scalar modes (σ, h),

$$\begin{bmatrix}\sigma\\ h\end{bmatrix} = \begin{bmatrix}\cos\overline{\theta} & \sin\overline{\theta}\\ -\sin\overline{\theta} & \cos\overline{\theta}\end{bmatrix}\begin{bmatrix}\phi_2\\ \phi_1\end{bmatrix} \; , \tag{31}$$

diagonalize the mass matrix to the form $M^2 = diag(0, m_h^2)$.

$\overline{\lambda_0}$ and $\overline{\theta}$ receive corrections at each loop level. Tuning the quartic coupling of the ϕ_2 field to be equal to $\overline{\lambda_0}$ has the interpretation of a regularization condition. It is analogous to fixing the cosmological constant in a theory without the scale symmetry, in order to obtain a desired value for the vacuum energy.

IOP Conf. Series: Journal of Physics: Conf. Series **873** (2017) 012042 doi:10.1088/1742-6596/873/1/012042

In what follows we will choose values of the couplings in nonpolynomial interactions to be zero,

$$\lambda_{(4+2m)} = 0 \quad , \quad m = 1, 2, \ldots \tag{32}$$

Nonetheless, one should note that, as explained earlier, due to the presence of *new* interactions, setting these values to zero does not place them in a fixed point of the renormalization group evolution,

$$\beta_{(4+2m)}\Big|_{\{\lambda_{(4+2m)}=0\}} \neq 0 \,. \tag{33}$$

After that choice, we have at the tree level

$$\overline{\lambda_0}^{(0)} = 9\frac{\lambda_2^2}{\lambda_4} \tag{34}$$

$$\tan^2\overline{\theta}^{(0)} = -3\frac{\lambda_2}{\lambda_4}\,, \quad \lambda_2 < 0 \tag{35}$$

$$m_h^{2\,(0)} = -\lambda_2 \left\langle \phi_2 \right\rangle^2 \left(1 - 3\frac{\lambda_2}{\lambda_4}\right) \,. \tag{36}$$

This model (or one closely similar) was studied previously in [6, 7, 1, 11, 12]. But quantum corrections were considered there at most up to the level of one loop. Computing corrections at higher loop orders is of interest since only there one obtains modifications of the renormalization group caused by the *new* interactions in (19), when compared to the case of using the regular dimensional regularisation with $\mu = const.$ Two loop corections to renormalization group functions and to the effective scalar potential were obtained in [13]. We will now present the main results of that work.

5. Two loop corrections

To regularize the model described in the previous section we use a minimal subtraction scheme. This means that the counterterms contain only poles in ϵ, and no finite contributions - in particular those arising from the presence of *new* interactions - are canceled. (We will denote the loop order of an expression by a superscript in parenthesis.)

$$Z_N = 1 + \sum_{k=1}^{\infty} \frac{a_{N,k}}{\epsilon^k} \quad , \quad a = a^{(1)} + a^{(2)} + \ldots \quad , \quad a^{(n)} = \mathcal{O}(\lambda^n) \quad , \quad N = \phi_1, \phi_2, 0, 2, 4, \ldots \tag{37}$$

Consider obtaining loop corrections using two different Lagrangians, both of which are given by (23) and (32), but differ in the meaning of μ. The first Lagrangian realizes the SI regularization method and uses (26), while in the second one we use (in the *old* way) $\mu = \langle \phi_2 \rangle = const.$ We will present the differences between renormalization group functions and scalar potentials obtained for these two Lagrangians. We will refer to these differences as *new* corrections:

$$
\begin{aligned}
V &= V^{old} + V^{new} \quad , \quad \beta = \beta^{old} + \beta^{new} \quad , \quad \gamma_\phi = \gamma_\phi^{old} + \gamma_\phi^{new} \\
\beta^{(1)\,new} &= 0\,, \quad \text{for all couplings} \\
\beta_4^{(1)} &= \frac{3}{(4\pi)^2}(\lambda_4^2 + \lambda_2^2) \\
\beta_2^{(1)} &= \frac{1}{(4\pi)^2}(\lambda_4 + 4\lambda_2 + \lambda_0)\lambda_2 \\
\beta_0^{(1)} &= \frac{3}{(4\pi)^2}(\lambda_2^2 + \lambda_0^2) \\
\gamma_\phi^{(1)} &= 0\,, \quad \gamma_\phi^{(2)\,new} = 0\,, \quad \text{for both fields} \\
\beta_0^{(2)\,new} &= -\frac{1}{2(4\pi)^4}\left(48\lambda_2^3 + 4\lambda_2^2\lambda_0 + 21\lambda_0^3\right) \\
\beta_2^{(2)\,new} &= -\frac{\lambda_2}{6(4\pi)^4}\left(48\lambda_4\lambda_2 + 6\lambda_4\lambda_0 + 123\lambda_2^2 + 86\lambda_2\lambda_0 + 3\lambda_0^2\right) \\
\beta_4^{(2)\,new} &= \frac{1}{2(4\pi)^4}\left[\lambda_2^2(24\lambda_2 - 7\lambda_0) + \lambda_4(-14\lambda_2^2 + 16\lambda_2\lambda_0 - 3\lambda_0^2) + \lambda_4^2(-80\lambda_2 + 6\lambda_0)\right] \\
\beta_6^{(2)\,new} &= \frac{1}{4(4\pi)^4}\,\lambda_4\lambda_2(7\lambda_4 - 14\lambda_2 + \lambda_0) \\
\beta_8^{(2)\,new} &= \frac{1}{2(4\pi)^4}\,\lambda_4\lambda_2^2 \\
\beta_{(8+2m)}^{(2)\,new} &= 0\,, \quad m = 1, 2, \ldots
\end{aligned}
\tag{38}
$$

The potential V is obtained in the background field method, which means that it is computed as a sum of vacuum 1PI diagrams for the propagating fields ϕ_1', ϕ_2', using $\langle\phi_1\rangle$- and $\langle\phi_2\rangle$-dependent Feynman rules. Thus, it is formally a function of two variables denoted by $\langle\phi_1\rangle$ and $\langle\phi_2\rangle$, but we will skip the angle brackets in the following presentation, $V = V(\phi_1, \phi_2)$.

$$
\begin{aligned}
V^{(1)\,new} &= \frac{1}{48(4\pi)^2}\left[\lambda_4\lambda_4\frac{\phi_1^6}{\phi_2^2} - (16\lambda_4\lambda_2 + 18\lambda_2^2 - \lambda_4\lambda_0)\phi_1^4 - (48\lambda_2 + 25\lambda_0)\lambda_2\,\phi_1^2\phi_2^2 - 7\lambda_0^2\phi_2^4\right] \\
V^{(2)\,new} &= \frac{1}{192(4\pi)^4}\left[\overline{\ln} V_{(+)} + \overline{\ln} V_{(-)}\right]\Big[\big(-144\lambda_4^2\lambda_2 - 111\lambda_4\lambda_2^2 - 168\lambda_2^3 + 9\lambda_4^2\lambda_0 - 40\lambda_2^2\lambda_0\big)\,\phi_1^4 \\
&- \big(192\lambda_4\lambda_2^2 + 705\lambda_2^3 + 37\lambda_4\lambda_2\lambda_0 + 368\lambda_2^2\lambda_0 + 106\lambda_2\lambda_0^2\big)\,\phi_1^2\phi_2^2 \\
&- \big(48\lambda_2^3 + 46\lambda_2^2\lambda_0 + 63\lambda_0^3\big)\,\phi_2^4 + \big(18\lambda_4^2\lambda_2 - 24\lambda_4\lambda_2^2 + 3\lambda_2^3 + 3\lambda_4\lambda_2\lambda_0\big)\,\frac{\phi^6}{\phi_2^2} \\
&+ 3\lambda_4\lambda_2^2\,\frac{\phi_1^8}{\phi_2^4}\Big] + t\text{-independent terms}\,, \quad \text{where} \quad \overline{\ln} A \equiv \ln\frac{A}{(t\,\phi_2)^2 4\pi e^{-\gamma_E}} - 1\,, \\
&V_{(\pm)} = 1/2\Big[V_{11} + V_{22} \pm \big[(V_{11} - V_{22})^2 + 4V_{12}^2\big]^{1/2}\Big], \quad V_{ij} = \frac{\partial^2 V^{(0)}}{\partial\phi_i\partial\phi_j}\,.
\end{aligned}
\tag{39}
$$

Consider now the Callan-Symanzik equation,

$$
\frac{\mathrm{d}V}{\mathrm{d}\ln t} = \Big(t\frac{\partial}{\partial t} + \sum_{m=0}\beta_{2m}\frac{\partial}{\partial\lambda_{2m}} - \phi_1\gamma_{\phi_1}\frac{\partial}{\partial\phi_1} - \phi_2\gamma_{\phi_2}\frac{\partial}{\partial\phi_2}\Big)V = 0\,. \tag{40}
$$

It too can be split into the *old*, established result that we can cross out and the *new* part (that contains expressions in (38) and (39) marked as *new*). The latter reads

$$\frac{\partial V^{(2)\,new}}{\partial \ln t} + \sum_{2m=0}^{4} \beta^{(2)\,new}_{\lambda_{2m}} \frac{\partial V^{(0)}}{\partial \lambda_{2m}} + \sum_{m=0}^{2} \beta^{(1)\,old}_{\lambda_{2m}} \frac{\partial V^{(1)\,new}}{\partial \lambda_{2m}} = 0\,, \tag{41}$$

where one should remember that $V^{(0)}$ contains $\frac{\lambda_6}{6}\frac{\phi_1^6}{\phi_2^2}$ and $\frac{\lambda_8}{8}\frac{\phi_1^8}{\phi_2^4}$.

Using the explicit expressions in (38) and (39) one readily checks that (41) is satisfied. We observe that at the level of two-loop corrections, the SI regularization method results in nontrivial modifications both in the running of coupling constants and in the corrections to the scalar effective potential. Yet these changes do not invalidate the Callan-Symanzik equation.

6. Summary

We have described the scale invariant regularization method and used it to renormalize a model with two interacting scalar fields. It was shown that the method requires presence of nonrenormalizable interactions. Lagrangian of the model is classically scale invariant and this symmetry is not broken by quantum anomaly, but it is broken spontaneously. By performing a perturbative calculation in the broken phase of the symmetry, we obtained renormalization group functions for the coupling constants and corrections to the effective potential at the order of two loops. This enabled us to explicitly check validity of the Callan-Symanzik equation in the framework of scale invariant regularization.

Acknowledgments

This work has been supported by National Science Centre, Poland, under research grant DEC-2016/21/N/ST2/03312.

References

[1] Tamarit C 2013 *JHEP* **12** 098 (*Preprint* 1309.0913)
[2] Englert F, Truffin C and Gastmans R 1976 *Nucl. Phys.* **B117** 407–432
[3] Fradkin E S and Vilkovisky G A 1978 *Phys. Lett.* **B73** 209–213
[4] Shaposhnikov M and Zenhausern D 2009 *Phys. Lett.* **B671** 162–166 (*Preprint* 0809.3406)
[5] Shaposhnikov M and Zenhausern D 2009 *Phys. Lett.* **B671** 187–192 (*Preprint* 0809.3395)
[6] Shaposhnikov M E and Tkachov F V 2009 (*Preprint* 0905.4857)
[7] Armillis R, Monin A and Shaposhnikov M 2013 *JHEP* **10** 030 (*Preprint* 1302.5619)
[8] Pokorski S 2005 *GAUGE FIELD THEORIES* (Cambridge University Press) ISBN 9780511037801, 9780521478168 URL http://www.cambridge.org/uk/catalogue/catalogue.asp?isbn=0521265371
[9] Akyeampong D A and Delbourgo R 1973 *Nuovo Cim.* **A18** 94–104
[10] Akyeampong D A and Delbourgo R 1974 *Nuovo Cim.* **A19** 219–224
[11] Ghilencea D M 2016 *Phys. Rev.* **D93** 105006 (*Preprint* 1508.00595)
[12] Ghilencea D M 2016 *PoS* **CORFU2015** 040 (*Preprint* 1605.05632)
[13] Ghilencea D M, Lalak Z and Olszewski P 2016 *Eur. Phys. J.* **C76** 656 (*Preprint* 1608.05336)

IOP Conf. Series: Journal of Physics: Conf. Series **873** (2017) 012043 doi:10.1088/1742-6596/873/1/012043

Scalar color octets and triplets in the SUSY with R-symmetry

Wojciech Kotlarski

Faculty of Physics, University of Warsaw, Pasteura 5, 02-093 Warsaw, Poland
Institute for Nuclear and Particle Physics, TU Dresden, Zellescher Weg 19, 01069 Dresden, Germany

E-mail: wojciech.kotlarski@fuw.edu.pl

Abstract. In this note we report on the recent progress in the study of the strongly interacting sector of the Minimal R-symmetric Supersymmetric Standard Model (MRSSM). First, we discuss the limits originating from the search of the sgluon pair production in the same-sign lepton final state at the LHC. Next, we present the first available in literature calculation of the NLO SQCD corrections to the same-sign squark pair production in the MRSSM. We discuss the relative size of k-factors compared to the MSSM and some technical details that led to this result.

1. Introduction

The Minimal Supersymmetric Standard Model (MSSM), with its simple structure and rich phenomenology, has been for many years the primary realization of supersymmetry considered in the literature. But as the Large Hadron Collider continues its second run with no MSSM in sight, the question is being raised whether low-scale supersymmetry exists at all. While it is possible that this is indeed not the case, it is also possible that supersymmetry is just not realized in the minimal, MSSM way. This is the reason behind increased attention in the non-minimal SUSY models observed in recent years.

One of the models which proved to be very promising in this context is the Minimal R-symmetric Supersymmetric Standard Model (MRSSM) [1]. The model ameliorates the flavor problem of the MSSM [1, 2] while reducing fine-tuning [3] and exhibiting distinct, interesting phenomenology. Though very promising, it is also much less explored - with the MSSM having a 30-year head start. Until now, the main effort has been devoted to the study of its electroweak sector, especially in the context of the discovery of the Higgs boson [3–5] and studies of dark matter [6]. The phenomenology of the strongly interacting sector, which is the most important from the point of view of direct detection at the LHC, was much less investigated. The focus there has so far been mainly on the sgluon production and LO analyses [7–13].[1] As the super-QCD (SQCD) k-factors tend to be large, a question arises of how the latter studies change in the presence of quantum corrections. To fill this gap, in this work, after the recap of the recent

[1] Strictly speaking, some of these works do not deal with the MRSSM but with a model in which $\mathcal{N} = 1$ SUSY is extended in the strongly interacting sector by $\mathcal{N} = 2$ gauge companions, which does have a very similar SQCD sector.

Table 1: Strongly interacting superfield content of the MRSSM together with R-charges of the component fields. The superfield in the last line is absent in the MSSM. It comprises of the right-handed component of the Dirac gluino and two real scalar gluons.

superfield		boson		fermion	
left-handed (s)quark	$\hat{Q}_L$	$\tilde{q}_L$	1	q_L	0
right-handed (s)quark	$\hat{Q}_R$	$\tilde{q}_R^\dagger$	1	$\bar{q}_R$	0
gluon vector superfield	$\hat{g}$	g	0	$\tilde{g}$	$+1$
adjoint chiral superfield	$\hat{O}$	O	0	$\tilde{O}$	-1

results of sgluon analysis, we report on the progress in calculation of the next-to-leading order (NLO) SQCD corrections to the same-sign squark pair production.

The paper is structured as follows. In section 2, we briefly recap the main features of the MRSSM. In section 3, we discuss the phenomenology of the color octet scalars (sgluons) production at the 13 TeV LHC. Finally, in section 4, we report on the calculation of the NLO SQCD corrections to the same-sign squark pair production. To that end, we discuss the calculation of virtual corrections together with their renormalization and derivation of the SUSY-restoring counterterms. We also describe the treatment of soft and collinear divergences and resonances appearing in real corrections. We end with a comparison of cross sections between the MSSM and the MRSSM for 2 benchmark points with different squark-gluino mass hierarchies.

2. The Model

The guiding principle behind the construction of the MRSSM is the presence of the unbroken R-symmetry at the low scale. However, with the R-charge assignment mimicking R-parity, the construction of phenomenologically viable model does require the extension of the MSSM field content [1]. In the strongly interacting sector this amounts to adding a color-octet (electroweak singlet) chiral superfield $\hat{O}$, whose fermionic component allows one to construct a Dirac mass term for the Dirac gluino $\tilde{g}_D$,

$$\tilde{g}_D \equiv \begin{pmatrix} \tilde{g} \\ \bar{\tilde{O}} \end{pmatrix}, \tag{1}$$

where, as shown in table 1, $\tilde{g}$ and $\tilde{O}$ are the gluino and the octino Weyl fermions, respectively.

The addition of the superfield $\hat{O}$ has important phenomenological consequences. For one, its scalar, CP-odd component O_A can be arbitrarily light as its tree-level mass is controlled purely by the soft-breaking parameter. This makes it ideal to search for at the LHC, also in the framework of simplified models. Also, heavier squark pair production is dominated by same-sign squarks, a process mediated by the exchange of a gluino in the t-channel. In the MSSM, due to gluino's Majorana nature, it is possible to produce $\tilde{q}_L\tilde{q}_L$, $\tilde{q}_R\tilde{q}_R$ and $\tilde{q}_L\tilde{q}_R$ final states while in the MRSSM only the $\tilde{q}_L\tilde{q}_R$ combination is allowed. This lowers the inclusive squark production cross section, weakening the LHC exclusion limits. This comes at the expense of the gluino mass limits, which are more stringent for Dirac gluinos.

To illustrate this discussion, in figure 1 we compare the MRSSM and MSSM cross sections for the pair production of various strongly interacting SUSY particles present in these models. In the plot we use the `MMHT2014lo68cl` [14] PDFs interfaced through `LHAPDF6` [15] with factorization and renormalization scales set equal to arithmetic mean of the masses of final-state particles. As can be seen, assuming the gluino mass of 2 TeV, for masses of squarks of around 1 TeV the

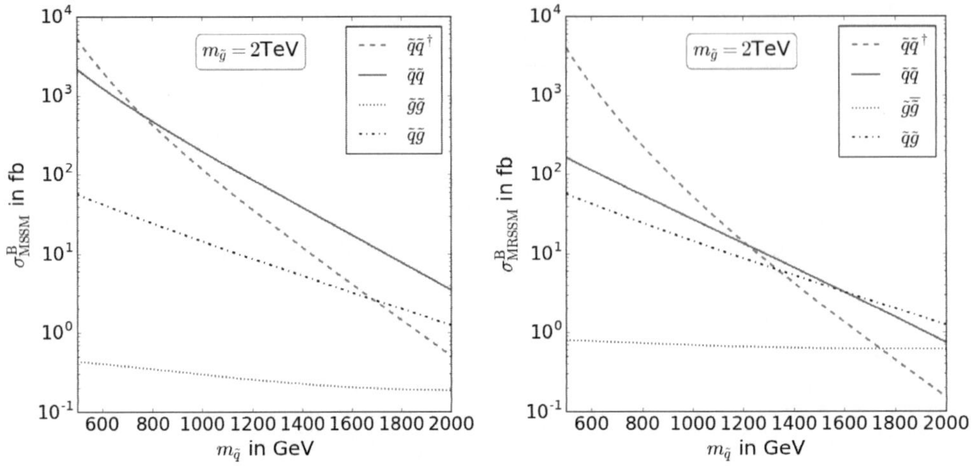

Figure 1: Comparison of the selected cross sections between the MSSM and the MRSSM in function of the squark mass. Different quark flavors (excluding stops) and "chiralities" are added together.

MSSM cross section is dominated by the (inclusive) same-sign squark pair production. As stated in the previous paragraph, the analogous cross section in the MRSSM is much smaller while the opposite-sign squark production cross section is quite similar to the MSSM. The sgluon pair production cross section, which is an exclusive feature of the MRSSM, is also of the same order.

Since we are primarily interested in the differences between the MSSM and the MRSSM, we will focus our discussion in the next two sections on production of sgluon pairs and same-sign squarks.

3. Sgluon pair production at the 13 TeV LHC

The partonic cross sections for the pair production of scalar or pseudoscalar sgluons are given by

$$\hat\sigma(q\bar q \to OO) = \frac{2\pi\alpha_s^2}{9\hat s}\beta_O^3, \tag{2}$$

$$\hat\sigma(gg \to OO) = \frac{3\pi\alpha_s^2}{32\hat s}\left(27\beta_O - 17\beta_O^3 + 6(-3 + 2\beta_O^2 + \beta_O^4)\,\mathrm{arctanh}\,\beta_O\right), \tag{3}$$

where $\hat s$ is the standard partonic Mandelstam variable and $\beta_O \equiv \sqrt{1 - 4m_O^2/\hat s}$ is the sgluon's velocity in the center-of-mass system of colliding partons. In the MRSSM, the masses of the scalar (S) and the pseudoscalar (A) components of the field O from table 1, where $O \equiv \frac{1}{2}(O_S + \imath O_A)$, are related through the gluino mass M_O^D (in the normalization used in Ref. [4]) as

$$m_{O_S}^2 = m_{O_A}^2 + 4(M_O^D)^2. \tag{4}$$

Since LHC constraints require gluinos to be $\gtrsim 1.5$ TeV, light scalar sgluon would require a very light pseudoscalar, which is highly constrained. For phenomenological analysis, we therefore focus on the scenario where $m_{O_A} \simeq 1$ TeV.

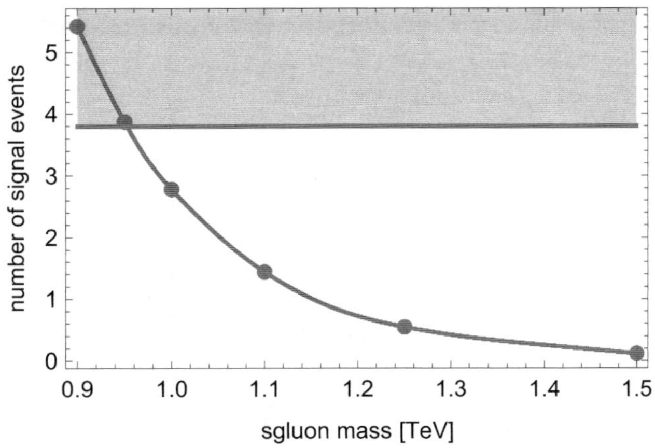

Figure 2: Predicted number of observed signal events as a function of the pseudoscalar sgluon mass (blue points). Solid line shows interpolation between these points. Red region is excluded by ATLAS for SR3b of [16] at 95% CL. Interpreted in the context of sgluon production, it corresponds to a lower limit on the sgluon mass $m_{O_A} \gtrsim 0.95$ TeV.

Once produced, pseudoscalar sgluons will decay almost exclusively to top quark pairs (see Appendix E of [17]). This produces a 4-top quark final state, a signature that is rare in the SM. Using the search of the same-sign lepton production performed by ATLAS [16], we have constrained the sgluons' contribution to this signature based on a `aMC@NLO` [18] simulation of the sgluon pair production and decay [11]. This, as shown in figure 2, directly translates to the lower limit on the pseudoscalar sgluon mass of 0.95 TeV.

Possibility of the existence of a relatively light color-charged particle that decays directly to standard model particles is one of the distinguishing features of models with Dirac gluinos. If ever discovered, it will be a clear falsification of the MSSM.

4. Same-sign squark pair production at the NLO

The second important channel is the same-sign squark pair production. The partonic cross section for the pair production of left and right squarks of flavor i and equal masses in the MSSM and in the MRSSM is given by

$$\hat{\sigma}(q_i q_i \to \tilde{q}_{i,L}\tilde{q}_{i,R}) = \frac{\pi\alpha_s^2}{\hat{s}}\left[-\frac{8}{9}\sqrt{1-\frac{4m_{\tilde{q}_i}^2}{\hat{s}}} + \left(-\frac{4}{9} - \frac{8(m_{\tilde{g}}^2 - m_{\tilde{q}_i}^2)}{9\hat{s}}\right) \right.$$
$$\left. \times \ln \frac{\hat{s}\left(1-\sqrt{1-\frac{4m_{\tilde{q}_i}^2}{\hat{s}}}\right) + 2(m_{\tilde{g}}^2 - m_{\tilde{q}_i}^2)}{\hat{s}\left(1+\sqrt{1-\frac{4m_{\tilde{q}_i}^2}{\hat{s}}}\right) + 2(m_{\tilde{g}}^2 - m_{\tilde{q}_i}^2)}\right], \tag{5}$$

where $m_{\tilde{g}}$ is (Dirac or Majorana) gluino mass and $\hat{s}$ is as in eqs 2 and 3. As noted in section 2, in the MSSM one also has additional channels for left-left and right-right squark pair production. While these tree-level results are enough for qualitative discussions as done in figure 1, a detailed comparison of the MSSM and the MRSSM does require going beyond that. We therefore present here the calculation of NLO SQCD corrections to the left-right same-sign squark pair production in the MRSSM.

The setup of the calculation is as follows. We created a custom `FeynArts` [19] MRSSM model file containing UV and DREG (dimensional regularization) to DRED (dimensional reduction) [20, 21] transition countertems. Both the real and virtual amplitudes were generated using `FeynArts` and evaluated in `FormCalc` [22]. The amplitudes were then processed in *mathematica* and exported to a standalone `C++` code.

4.1. Virtual corrections

The evaluation of the virtual amplitude follows standard procedures, so here we only outline the most important details. The amplitude is regularized using the so-called 't Hooft-Veltman scheme (HV in the notation of Ref. [23]). In the HV, internal gluons are kept D-dimensional, which breaks supersymmetry due to the mismatch between gluon and gluino number of degrees of freedom. This is corrected by supplementing the result with a set of counterterms that translate the UV-poles from DREG to DRED, as DRED was proven to not break SUSY at the one-loop level (for reviews of checks see e.g. [24, 25]). For the process at hand, only the renormalisation constant of the squark-quark-gluino vertex needs to be changed. Denoting the renormalization constant of this vertex by $\delta\hat{g}_s$, we find that it has to satisfy

$$\delta\hat{g}_s = \delta g_s + \frac{g_s^3}{16\pi^2}\left(\frac{2C_A}{3} - \frac{C_F}{2}\right), \tag{6}$$

where δg_s is the strong coupling constant counterterm in DREG and zero momentum subtraction scheme (more precisely, we use zero momentum subtraction for contributions from particles heavier than the bottom quark and $\overline{\text{MS}}$ for contributions from bottom and light quarks). The remaining counterterms are fixed by the on-shell renormalization conditions.

After this procedure the amplitude is UV finite and supersymmetric but it still contains the infrared (IR) singularities.

4.2. Real corrections

The IR singularities of the virtual amplitude cancel after adding real corrections - soft ones cancel according to the Kinoshita-Lee-Nauenberg theorem [26, 27] while collinear ones are removed through mass factorization [28, 29]. To extract soft and collinear divergences, we employed the two-cut phase space slicing method as documented in [30]. In short, this amounts to splitting the real emission contribution phase space according to emitted gluon energy into soft region defined as

$$E_g \leq \delta_s \frac{\sqrt{\hat{s}}}{2}, \tag{7}$$

where $\delta_s \ll 1$, and its complement. This region of phase space does contain both soft and/or collinear divergences. Using the eikonal approximation, the divergences can be extracted in D-dimensions as poles in 4-D.

However, the complement of the phase space defined by eq. 7 still contains collinear divergences. These are extracted by slicing the phase space according to collinearity condition. We use the collinear condition from [31] as it allows, contrary to the one used in [30], to decouple the soft and collinear limits.

The IR finiteness of the result was checked by evaluating squared matrix elements for a single phase space point, showing cancellation of divergences at the level of a double numerical precision.

4.3. Treatment of on-shell resonances in real emission diagrams

The calculation of the real emission corrections poses one more difficulty. Figure 3 shows the diagrams for $gu \to \tilde{u}_L\tilde{u}_R\bar{u}$ process. This process contains a contribution from the s-channel production of a gluino. These kinds of effects, observed already in the SM for single-top

IOP Conf. Series: Journal of Physics: Conf. Series **873** (2017) 012043 doi:10.1088/1742-6596/873/1/012043

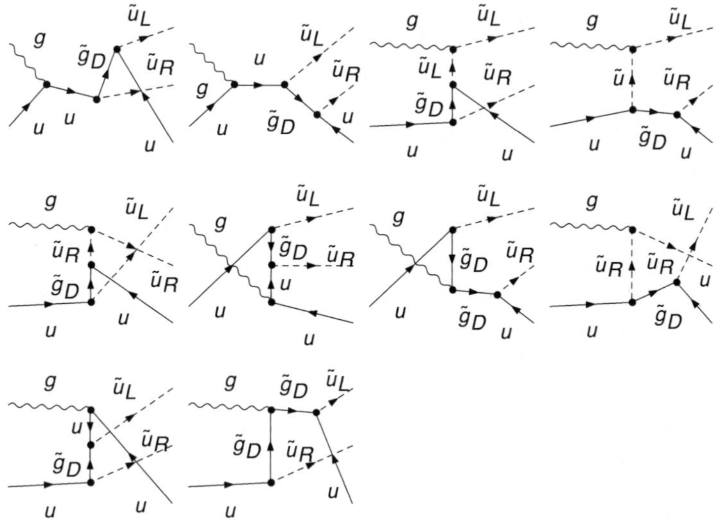

Figure 3: Feynman diagrams for $gu \to \tilde{u}_L\tilde{u}_R\bar{u}$ subprocess. For $m_{\tilde{g}_D} > m_{\tilde{u}_{L/R}}$, amplitude exhibit an on-shell singularity.

production, require a careful definition of the process of interest. The amplitude for diagrams in figure 3 can be split into resonant and non-resonant parts as

$$M_{tot} = M_{nr} + M_r. \tag{8}$$

For $m_{\tilde{g}_D} > m_{\tilde{u}_{L/R}}$, the M_r becomes infinite when $(p_{\tilde{g}_D} + p_{\tilde{u}_{L/R}})^2 = m^2_{\tilde{g}_D}$. We therefore employ the diagram removal (DR) procedure of Ref. [32], where resonant diagrams are removed at the amplitude level. As this procedure breaks gauge invariance, it requires a careful choice of the gauge for the external gluon. We choose a light-cone gauge, with the gauge vector η set equal to the momentum of the second incoming parton.

4.4. Results

For numerical studies, we consider 2 MRSSM scenarios with a different gluino-squark mass hierarchy. Both points have common squark masses $m_{\tilde{q}} = 1.5$ TeV, $m_{O_A} = 5$ TeV, $m_t = 172$ GeV and differ only in the value of the gluino mass, with 1 TeV mass in the BMP1 and 2 TeV one in the BMP2. The scalar sgluon mass m_{O_S} is fixed by the eq. 4. We perform all the calculations for the LHC running at center-of-mass energy of 13 TeV using `MMHT2014nlo68cl` PDFs interfaced through `LHAPDF6`.

The total, NLO cross section in the BMP1 is

$$\sigma^{\text{BMP1}}_{\text{MRSSM}}(pp \to \tilde{u}_L\tilde{u}_R) = 9.94312 \pm 0.00058 \text{ fb}. \tag{9}$$

This can be compared with the MSSM result obtained using `MadGraph5_aMC@NLO` [18] and an NLO-capable UFO [33] model `SUSYQCD_UFO` [34], which is

$$\sigma^{\text{BMP1}}_{\text{MSSM}}(pp \to \tilde{u}_L\tilde{u}_R) = 9.431 \pm 0.003 \text{ fb}. \tag{10}$$

The difference is of the order of 5%, and results exclusively from the virtual part.

Similarly, for the BMP2 and after employing the diagram removal procedure, we get

$$\sigma^{\text{BMP2}}_{\text{MRSSM}}(pp \to \tilde{u}_L \tilde{u}_R) = 3.01357 \pm 0.00020 \text{ fb} \tag{11}$$

versus the MSSM's

$$\sigma^{\text{BMP2}}_{\text{MSSM}}(pp \to \tilde{u}_L \tilde{u}_R) = 2.78114 \pm 0.00055 \text{ fb}, \tag{12}$$

giving an 8% difference.

Since the Born and the real matrix elements are identical for those processes in both models, it is not surprising that differences, though not negligible, do not exceed a few percent.

The results for the MRSSM are in full agreement with our second calculation based on `MadGraph5_aMC@NLO` and `GoSam` [35, 36], interfaced using `BLHA2` interface [37]. However, we postpone the description of this method to further publication.

In the appendix, we give the decomposition of numbers in eqs 9 and 11 into tree, virtual, soft and/or collinear and hard non-collinear contributions. For phase space slicing parameters δ_s and δ_c we chose values 10^{-5} and 10^{-6}, respectively. We checked extensively that the final result is independent of precise numerical values of those (unphysical) parameters. We also verified results for separate partonic channels between our `C++` and `MadGraph5_aMC@NLO` implementations.

The above-mentioned `C++` code, containing $pp \to \tilde{u}_L \tilde{u}_R$ and $pp \to \tilde{u}_L \tilde{u}_L^*$ processes, will become public soon together with a more detailed description of both calculations.

5. Conclusions

The Minimal R-symmetric Supersymmetric Standard Model is a well-motivated and viable alternative to the MSSM. Recent studies have shown that within the MRSSM framework one can accommodate the Higgs boson discovered at the LHC while staying in agreement with constraints from electroweak precision observables. Also, the first very promising studies of the dark matter sector were performed.

With respect to this progress, the study of the SQCD sector is lagging behind. With that in mind, we have presented in this note our recent progress in that matter. This included both the discussion of the recently derived limits on the sgluon mass from the LHC Run 2 data, as well as the calculation of the NLO SQCD corrections to the same-sign squark pair production. To cross-check the latter results, we did two separate calculations, one based on analytic results encapsulated into a standalone `C++` code and the other using a well-established framework of `MadGraph5_aMC@NLO`, which are in full agreement.

Our calculations showed a moderate difference of 5 - 10% between MSSM and MRSSM for the considered benchmark points, which can be traced back to the difference in the virtual matrix element.

The developed `C++` code is already also capable of calculating the $pp \to \tilde{u}_L \tilde{u}_L^*$ process. We plan to release the code to the public together with a description of the theoretical setup of the calculation. This will be followed shorty by a phenomenological analysis.

Acknowledgments

I would like to thank Philip Diessner, Jan Kalinowski, Sebastian Liebschner and Dominik Stoeckinger, with whom the work on which this note is based was done. Work supported in part by the German DFG grant STO 876/4-1, the Polish National Science Centre HARMONIA project under contract UMO-2015/18/M/ST2/00518 (2016-2019) and the PL-Grid Infrastructure.

IOP Conf. Series: Journal of Physics: Conf. Series **873** (2017) 012043 doi:10.1088/1742-6596/873/1/012043

Appendix

In this appendix we present the decomposition of the total cross sections, given in eqs 9 and 11, into separate partonic channels. Each partonic channel consists of the tree level, virtual and real emission soft/collinear and hard non-collinear contributions. In the description of each channel, squarks are denoted as `suL` (left up squarks) and `suR` (right up squarks), and the `+X` denotes that the processes should be understood inclusively with respect to the presence of an additional parton in the final state. Hence, for a given channel, the tree level and virtual contributions might be 0 if a given channel does not have a Born configuration. The p-values quoted along the statistical errors describe the probabilities, that the quoted errors are not reliable.

Output comes directly from the created by us, and advertised in the text, `C++` code.

Numerical results for the BMP1

```
Results for subprocess uu > suLsuR(+X)
--------------------------------------------------------------------------------
                        tree:  5.67655e+00 +/- 5.7e-06 fb ( p-value =  0.0e+00 )
                     virtual:  3.53574e+00 +/- 3.5e-05 fb ( p-value =  9.9e-17 )
real (soft and collinear): -5.71007e+01 +/- 4.4e-05 fb ( p-value =  0.0e+00 )
real (hard non-collinear):  5.78352e+01 +/- 5.8e-04 fb ( p-value =  0.0e+00 )
--------------------------------------------------------------------------------
                         sum:  9.94681e+00 +/- 5.8e-04 fb

Results for subprocess gu > suLsuR(+X)
--------------------------------------------------------------------------------
                        tree:  0.00000e+00 +/- 0.0e+00 fb ( p-value =  0.0e+00 )
                     virtual:  0.00000e+00 +/- 0.0e+00 fb ( p-value =  0.0e+00 )
real (soft and collinear): -5.03323e-02 +/- 5.0e-09 fb ( p-value =  0.0e+00 )
real (hard non-collinear):  4.66389e-02 +/- 4.7e-07 fb ( p-value =  0.0e+00 )
--------------------------------------------------------------------------------
                         sum: -3.69334e-03 +/- 4.7e-07 fb

Results for subprocess sum
-------------------------------------------------------
                        tree:  5.67655e+00 +/- 5.7e-06 fb
                     virtual:  3.53574e+00 +/- 3.5e-05 fb
real (soft and collinear): -5.71511e+01 +/- 4.4e-05 fb
real (hard non-collinear):  5.78819e+01 +/- 5.8e-04 fb
-------------------------------------------------------
                         sum:  9.94312e+00 +/- 5.8e-04 fb
```

Numerical results for the BMP2

```
Results for subprocess uu > suLsuR(+X)
--------------------------------------------------------------------------------
                        tree:  2.01180e+00 +/- 2.0e-06 fb ( p-value =  0.0e+00 )
                     virtual:  6.97868e-01 +/- 7.0e-06 fb ( p-value =  0.0e+00 )
real (soft and collinear): -1.99882e+01 +/- 1.6e-05 fb ( p-value =  0.0e+00 )
real (hard non-collinear):  2.02943e+01 +/- 2.0e-04 fb ( p-value =  0.0e+00 )
--------------------------------------------------------------------------------
                         sum:  3.01582e+00 +/- 2.0e-04 fb
```

IOP Conf. Series: Journal of Physics: Conf. Series **873** (2017) 012043 doi:10.1088/1742-6596/873/1/012043

```
Results for subprocess gu > suLsuR(+X)
------------------------------------------------------------------------------
                      tree:  0.00000e+00 +/- 0.0e+00 fb ( p-value =  0.0e+00 )
                   virtual:  0.00000e+00 +/- 0.0e+00 fb ( p-value =  0.0e+00 )
real (soft and collinear): -1.69119e-02 +/- 1.7e-09 fb ( p-value =  0.0e+00 )
real (hard non-collinear):  1.46572e-02 +/- 1.5e-07 fb ( p-value =  0.0e+00 )
------------------------------------------------------------------------------
                       sum: -2.25477e-03 +/- 1.5e-07 fb

Results for subprocess sum
-------------------------------------------------------
                      tree:  2.01180e+00 +/- 2.0e-06 fb
                   virtual:  6.97868e-01 +/- 7.0e-06 fb
real (soft and collinear): -2.00051e+01 +/- 1.6e-05 fb
real (hard non-collinear):  2.03090e+01 +/- 2.0e-04 fb
-------------------------------------------------------
                       sum:  3.01357e+00 +/- 2.0e-04 fb
```

References

[1] Kribs G D, Poppitz E and Weiner N 2008 *Phys. Rev.* **D78** 055010 (*Preprint* 0712.2039)

[2] Dudas E, Goodsell M, Heurtier L and Tziveloglou P 2014 *Nucl. Phys.* **B884** 632–671 (*Preprint* 1312.2011)

[3] Bertuzzo E, Frugiuele C, Gregoire T and Ponton E 2015 *JHEP* **04** 089 (*Preprint* 1402.5432)

[4] Dießner P, Kalinowski J, Kotlarski W and Stoeckinger D 2014 *JHEP* **12** 124 (*Preprint* 1410.4791)

[5] Diessner P, Kalinowski J, Kotlarski W and Stoeckinger D 2015 *Adv. High Energy Phys.* **2015** 760729 (*Preprint* 1504.05386)

[6] Diessner P, Kalinowski J, Kotlarski W and Stoeckinger D 2016 *JHEP* **03** 007 (*Preprint* 1511.09334)

[7] Choi S Y, Drees M, Kalinowski J, Kim J M, Popenda E and Zerwas P M 2009 *Phys. Lett.* **B672** 246–252 (*Preprint* 0812.3586)

[8] Plehn T and Tait T M P 2009 *J. Phys.* **G36** 075001 (*Preprint* 0810.3919)

[9] Calvet S, Fuks B, Gris P and Valery L 2013 *JHEP* **04** 043 (*Preprint* 1212.3360)

[10] Kotlarski W, Kalinowski A and Kalinowski J 2013 *Acta Phys. Polon.* **B44** 2149–2154

[11] Kotlarski W 2017 *JHEP* **02** 027 (*Preprint* 1608.00915)

[12] Choi S Y, Drees M, Freitas A and Zerwas P M 2008 *Phys. Rev.* **D78** 095007 (*Preprint* 0808.2410)

[13] Kribs G D and Martin A 2013 (*Preprint* 1308.3468)

[14] Harland-Lang L A, Martin A D, Motylinski P and Thorne R S 2015 *Eur. Phys. J.* **C75** 204 (*Preprint* 1412.3989)

[15] Buckley A, Ferrando J, Lloyd S, Nordström K, Page B, Rüfenacht M, Schönherr M and Watt G 2015 *Eur. Phys. J.* **C75** 132 (*Preprint* 1412.7420)

[16] Aad G *et al.* (ATLAS) 2016 *Eur. Phys. J.* **C76** 259 (*Preprint* 1602.09058)

[17] Kotlarski W 2016 *Analysis of the R-symmetric supersymmetric models including quantum corrections* Ph.D. thesis U. of Warsaw (*Preprint* 1611.06622) URL http://inspirehep.net/record/1499505/files/arXiv:1611.06622.pdf

IOP Conf. Series: Journal of Physics: Conf. Series **873** (2017) 012043 doi:10.1088/1742-6596/873/1/012043

[18] Alwall J, Frederix R, Frixione S, Hirschi V, Maltoni F, Mattelaer O, Shao H S, Stelzer T, Torrielli P and Zaro M 2014 *JHEP* **07** 079 (*Preprint* 1405.0301)

[19] Hahn T 2001 *Comput. Phys. Commun.* **140** 418–431 (*Preprint* hep-ph/0012260)

[20] Siegel W 1979 *Phys.Lett.* **B84** 193

[21] Capper D, Jones D and van Nieuwenhuizen P 1980 *Nucl.Phys.* **B167** 479

[22] Hahn T and Perez-Victoria M 1999 *Comput. Phys. Commun.* **118** 153–165 (*Preprint* hep-ph/9807565)

[23] Signer A and Stoeckinger D 2009 *Nucl. Phys.* **B808** 88–120 (*Preprint* 0807.4424)

[24] Jack I and Jones D R T 1997 [Adv. Ser. Direct. High Energy Phys.21,494(2010)] (*Preprint* hep-ph/9707278)

[25] Stoeckinger D 2005 *JHEP* **03** 076 (*Preprint* hep-ph/0503129)

[26] Kinoshita T 1962 *J. Math. Phys.* **3** 650–677

[27] Lee T D and Nauenberg M 1964 *Phys. Rev.* **133**(6B) B1549–B1562

[28] Collins J C, Soper D E and Sterman G F 1985 *Nucl. Phys.* **B261** 104–142

[29] Bodwin G T 1985 *Phys. Rev.* **D31** 2616 [Erratum: Phys. Rev.D34,3932(1986)]

[30] Harris B W and Owens J F 2002 *Phys. Rev.* **D65** 094032 (*Preprint* hep-ph/0102128)

[31] Dawson S, Kao C, Wang Y and Williams P 2007 *Phys. Rev.* **D75** 013007 (*Preprint* hep-ph/0610284)

[32] Frixione S, Laenen E, Motylinski P, Webber B R and White C D 2008 *JHEP* **07** 029 (*Preprint* 0805.3067)

[33] Degrande C, Duhr C, Fuks B, Grellscheid D, Mattelaer O and Reiter T 2012 *Comput. Phys. Commun.* **183** 1201–1214 (*Preprint* 1108.2040)

[34] Degrande C, Fuks B, Hirschi V, Proudom J and Shao H S 2016 *Phys. Lett.* **B755** 82–87 (*Preprint* 1510.00391)

[35] Cullen G, Greiner N, Heinrich G, Luisoni G, Mastrolia P, Ossola G, Reiter T and Tramontano F 2012 *Eur. Phys. J.* **C72** 1889 (*Preprint* 1111.2034)

[36] Cullen G *et al.* 2014 *Eur. Phys. J.* **C74** 3001 (*Preprint* 1404.7096)

[37] Alioli S *et al.* 2014 *Comput. Phys. Commun.* **185** 560–571 (*Preprint* 1308.3462)

IOP Conf. Series: Journal of Physics: Conf. Series **873** (2017) 012044 doi:10.1088/1742-6596/873/1/012044

Gravitational waves from domain walls in the Standard Model

T Krajewski[1], Z Lalak[1], M Lewicki[1,2] and P Olszewski[1]

[1]Institute of Theoretical Physics, Faculty of Physics, University of Warsaw, ul. Pasteura 5, Warsaw, Poland
[2] ARC Centre of Excellence for Particle Physics at the Terascale (CoEPP) & CSSM, Department of Physics, University of Adelaide, South Australia 5005, Australia

E-mail: Tomasz.Krajewski@fuw.edu.pl, Zygmunt.Lalak@fuw.edu.pl, Marek.Lewicki@adelaide.edu.au, Pawel.Olszewski@fuw.edu.pl

Abstract. We study domain walls interpolating between the physical electroweak vacuum and the global minimum of the Standard Model scalar potential appearing at very high field strengths. Such domain walls could be created in the early Universe under the assumption of validity of the Standard Model up to very high energy scales. The creation of the network of domain walls which ends up in the electroweak vacuum percolating through the Universe is not as difficult to obtain as one may expect, although it requires certain tuning of initial conditions. Our numerical simulations confirm that such domain walls would swiftly decay. Moreover we have found that for the standard cosmology the energy density of gravitational waves emitted from domain walls is too small to be observed in present and planned detectors.

1. Introduction

After the discovery of the Higgs boson at the Large Hadron Collider [1, 2] all of the parameters of the Standard Model (SM) are known with accuracy high enough for the quantitative study of the RG improved effective SM potential. For the present experimental values the potential drops below the electroweak vacuum values at field strengths of the order of 10^{11} GeV and leads to a new minimum at superplanckian field strengths.

Early in the history of the Universe, when it was hotter and denser of when it rapidly expanded during during inflation [3, 4], the Higgs field fluctuations might have been large enough to overcome the potential barrier between the two minima. As a result a network of domain walls interpolating between patches of the Universe occupied by the Higgs field laying in different minima was created. Our main interest is evolution of this network. To remain independent from a specific model of the early Universe we investigate dependence of our results on initial conditions: a mean value of the field strength and an amplitude of fluctuations.

If minima of the potential are not degenerate, domain walls interpolating between these minima are unstable. We have found using numerical simulations based on Press, Ryden, Spergel (PRS) algorithm [5] that the conformal time needed for networks of SM domain walls to decay is relatively short and of the order of $10^{-9}\ \frac{\hbar}{\mathrm{GeV}}$.

Very short lifetime of the SM domain walls makes them consistent with the observational data protecting them from numerous problems which long-lived domain walls can cause. The effective equation of state of a network of cosmological domain walls predicts the negative pressure. The

IOP Conf. Series: Journal of Physics: Conf. Series **873** (2017) 012044 doi:10.1088/1742-6596/873/1/012044

energy density of the network of stable domain walls decreases (with the expansion) slower then the energy density of both radiation and dust, so long lived domain walls tend to dominate the Universe. Moreover domain walls which pose a significant fraction of the energy density of the Universe at recombination would produce unacceptably large fluctuations of the Cosmic Microwave Background Radiation (CMBR).

During the process of the decay of a network of domain walls the energy of the field is transferred to other degrees of freedom, including production of gravitational waves (GWs). The recent observation of GWs at the LIGO and the Virgo experiments [6] promoted spectrum of GWs to one of the most promising observables which could in principle probe domain walls in the early Universe.

We estimate the spectrum of GWs produced during the decay of SM domain walls, in the case of the standard cosmology, using a direct calculation utilizing a lattice simulation. Our prediction for present energy density of produced GWs is orders of magnitude smaller than the sensitivity of present and planned detectors of GWs. However the present energy density could be greater if the evolution of the Universe before formation of domain walls was different then in the standard scenario.

2. RG improved SM effective potential

Previous studies of the evolution of domain walls focused on the dynamics driven by simple polynomial potentials. Even though the SM potential at the tree level is of this form, quantum corrections introduce significant modifications for large field strengths and generate a new minimum at superplanckian field strengths.

Our work is based on the general statement that the (space-time dependent) expectation value of the field strength $\Psi_{cl}(x)$ satisfies the following equation of motion (eom):

$$\frac{\delta}{\delta\Psi_{cl}(x)}\Gamma_{\text{eff}}[\Psi_{cl}] = 0, \tag{1}$$

where Γ_{eff} is the 1PI effective action.

We have used the following approximation for the 1PI effective action of the SM Γ_{SM} in our studies:

$$\Gamma_{\text{SM}}[\phi] \approx \int d^4x \sqrt{|\det g|}\left[\frac{1}{2}g^{\mu\nu}\partial_\mu\phi\partial_\nu\phi - V_{\text{SM}}(\phi)\right], \tag{2}$$

where ϕ is a real scalar field which models the Higgs field in our simulations and V_{SM} is the effective potential of the SM in zero temperature. The eom for the approximated effective action proposed in eq. (2) is of the form:

$$\frac{\partial^2\phi}{\partial\eta^2} + \frac{2}{a}\left(\frac{da}{d\eta}\right)\frac{\partial\phi}{\partial\eta} - \Delta\phi + a^2\frac{\partial V_{\text{eff}}}{\partial\phi} = 0, \tag{3}$$

assuming the Friedman-Robertson-Walker metric background:

$$g = dt^2 - a^2(t)\delta_{ij}dx^i dx^j = a^2(\eta)\left(d\eta^2 - \delta_{ij}dx^i dx^j\right), \tag{4}$$

where Latin indices correspond to spatial coordinates, t is cosmic time and η denotes conformal time (such that $d\eta = \frac{1}{a(t)}dt$). The contribution to the Friedman equations from the Higgs field energy-momentum tensor was neglected in our simulations.

Let us consider the following generalization of eq. (3)

$$\frac{\partial^2\phi}{\partial\eta^2} + \alpha\left(\frac{d\log a}{d\log\eta}\right)\frac{1}{\eta}\frac{\partial\phi}{\partial\eta} - \Delta\phi + a^\beta\frac{\partial V}{\partial\phi} = 0, \tag{5}$$

proposed in [5]. We used the Crank-Nicholson scheme for discretization of space-like derivatives and the "staggered leapfrog" for time derivatives which gives us the second order discretization scheme in both time and space. Eq. (5) with values $\alpha = 3$ and $\beta = 0$ combined with used discretization scheme are known as PRS algorithm [5]. This algorithm was widely used in the past for numerical simulations of the dynamics of domain walls for example in [7, 8].

As an approximation of $V_{\rm SM}$ we have used RG improved perturbative effective potential $\widetilde{V}_{\rm SM}$ presented in fig. 1. The loop level up to which we work when computing $\widetilde{V}_{\rm SM}$ numerically is two for effective quartic coupling $\lambda_{\rm eff}$ and three for RGEs. We have used expressions for RGEs, coupling constants and their values renormalised at the scale $\mu = m_{top}$ from [9].

The potential $\widetilde{V}_{\rm SM}$ resembles a mexican hat only at scales $\mathcal{O}(10^2$ GeV). At larger scales it is determined by the behaviour of $\lambda_{\rm eff}$. The potential reaches a maximum around the value $|h|_{\rm max} \sim 10^{10}$ GeV, $V_{\rm max} \sim 10^{34}$ GeV^4. Slope of the potential is much steeper for field values $|h| > |h|_{\rm max}$ than it is for $|h| < |h|_{\rm max}$, and potential quickly decreases below value h_{EW} at the electroweak (EW) minimum for field strengths $|h| \gtrsim 10^{10}$GeV.

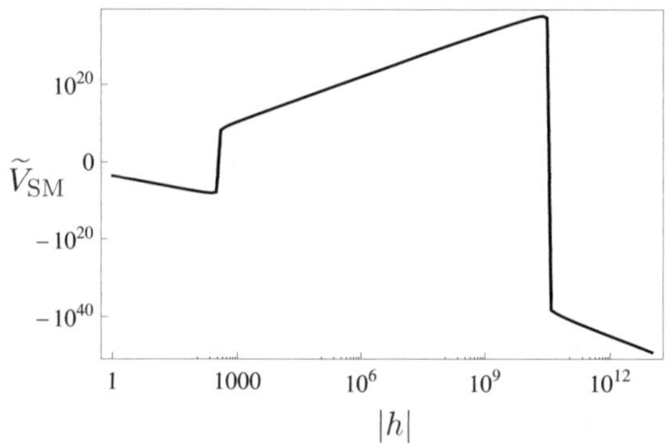

Figure 1: The RG improved effective SM potential of $\widetilde{V}_{\rm SM}$.

Assuming validity of the SM up to such high scales, one witnesses an intricate vacuum structure, which in principle allows for formation and non-trivial evolution of domain walls in the early Universe.

3. The width of domain walls

The estimation of the physical width of domain walls is critical for numerical simulations of their dynamics. The width must be at least a few times larger than the lattice spacing (i.e. the physical distance between neighboring points) used in the simulation in order to assure sufficient accuracy to model profiles of walls. On the other hand, statistical properties of the evolution of the network of domain walls will be reproduced poorly in simulations with too small lattice spacing, because only few walls (spreading over many lattice points) will fit into finite lattice. Simulations with the physical width of walls varying from 2 to 100 lattice spacing were used in the past [5, 10, 11, 12, 7, 13, 14, 15].

We estimated the width of domain walls in the SM using the approach based on the first integral of the eom (see [16] for derivation). Firstly we calculated the value h_2 of the Higgs field which gives the same value of the effective potential as the value $h_{\rm EW}$ and bigger than the local maximum. The planar wall solution of the eom in the Minkowski background is given in the implicit form:

$$X(\varphi_2) - X(\varphi_1) = \int_{\varphi_1}^{\varphi_2} \frac{d\varphi}{\sqrt{2\left(V_{\rm SM}(\varphi) - V_{\rm SM}(h_{\rm EW})\right)}}, \tag{6}$$

IOP Conf. Series: Journal of Physics: Conf. Series **873** (2017) 012044 doi:10.1088/1742-6596/873/1/012044

and the wall's tension by the following integral formula:

$$\Sigma_{\rm SM}(\varphi_1,\varphi_2) := \int_{\varphi_1}^{\varphi_2} \frac{V_{\rm SM}(\varphi)d\varphi}{\sqrt{2\left(V_{\rm SM}\left(\varphi\right) - V_{\rm SM}(h_{\rm EW})\right)}}. \tag{7}$$

We have found values $\tilde{\varphi_1}$ and $\tilde{\varphi_2}$ such that $V_{\rm SM}\left(\tilde{\varphi_1}\right) = V_{\rm SM}\left(\tilde{\varphi_2}\right)$ and $\frac{\Sigma_{\rm SM}(\tilde{\varphi_1},\tilde{\varphi_2})}{\Sigma_{\rm SM}(h_{\rm EW},h_2)} \approx 97\%$, obtaining our estimation of the width of domain walls in the SM (using the convention $\hbar = 1 = c$, so the lengths and times are expressed in units of GeV^{-1}):

$$w_{\rm SM} := X(\tilde{\varphi_2}) - X(\tilde{\varphi_1}) \approx 4 \times 10^{-9}\ \text{GeV}^{-1}. \tag{8}$$

The potential energy density as a function of the distance in the direction perpendicular to the domain wall surface for the solution (6) is presented in fig. 2. The value $x = 0$ on the horizontal axis corresponds to $\varphi(0) = \tilde{\varphi_1}$.

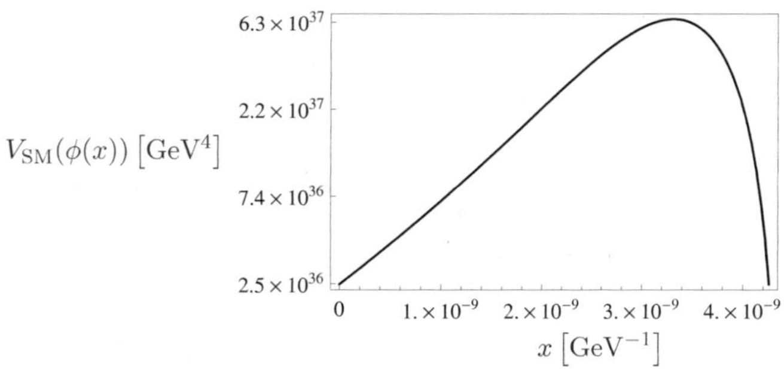

Figure 2: The potential energy density $V_{\rm SM}(\phi(x))$ of the planar SM domain wall as a function of the distance x in the direction perpendicular to the wall surface.

We have chosen the physical lattice spacing l to be equal to 10^{-10} GeV^{-1}, so the width of the walls is of the order of 40 l. We use consistently the units of 10^{10} GeV through the paper.

4. The Decay of domain walls

Cosmological domain walls are subject to many experimental constraints. The equation of state of the network of domain walls is generally predicted to $-2/3 < w_X < -1/3$, which is ruled out by the present observational data for a single component Dark Energy. Furthermore the energy density of a network of domain walls decreases slower then the energy density of both: the radiation and the dust, so long lived domain walls would dominate the Universe. Moreover domain walls present during the recombination would produce unacceptably large fluctuations of CMBR. Finally the final product of the decay of SM domain walls have to be the EW vacuum. For these reasons we concentrated on the determination of the time and the final state of the process of the decay of SM domain walls.

Our simulations were initialized with three initial values:

- initial conformal time η_{start},
- initial mean value of the field strength θ,
- initial standard deviation σ.

Initial conformal time η_{start} is determined by the time at which domain walls are formed in the early Universe. However the initial time of the simulation must be earlier, in order to smooth out the initial numerical fluctuations by the field evolution. The range for initial conformal time in our simulations is from $10^{-4}\ l = 10^{-14}$ GeV^{-1} to $10\ l = 10^{-9}$ GeV^{-1}.

IOP Conf. Series: Journal of Physics: Conf. Series **873** (2017) 012044 doi:10.1088/1742-6596/873/1/012044

Following [10], we assumed that an initial distribution of the field strength is given by the gaussian probability distribution:

$$P(\phi) = \frac{1}{\sqrt{2\pi}\sigma} e^{-\frac{(\phi-\theta)^2}{2\sigma^2}}. \tag{9}$$

We studied the evolution of networks of domain walls initialized with different values of θ and σ in order to accommodate variety of processes leading to the formation of the walls.

Initial conditions cannot be deduced from the dynamics of domain walls and must be derived from a model of the evolution of the early Universe. Our results can be thought as a constraint on the space of models of the early Universe.

4.1. Dependence on the initial time η_{start}

We run simulations for different $\sigma = \mathcal{O}\left(10^{10}\text{ GeV}\right)$ and $\theta = 0$ with five different initialization conformal times η_{start}. Even though observed decay times are of the same order for all considered initialization times, we observe a differences in the evolution of early formed (i.e. $\eta_{start} < l = 10^{-10}\text{ GeV}^{-1}$) and late formed (i.e. $\eta_{start} > l$) domain walls. For initial conditions providing the configuration of the field where the electroweak vacuum strongly dominates, late domain walls decay faster then early ones. This issue is presented in fig. 3a in which we plot the evolution of the ratio of the number of lattice sites occupied by the field strength belonging to the basin of attraction of the electroweak minimum to the size of the lattice $\frac{V_{EW}}{V}$. For more moderate contribution of EW vacuum in initial conditions the decay time of late domain walls can be longer than for early ones. The example of this situation is illustrated in fig. 3b.

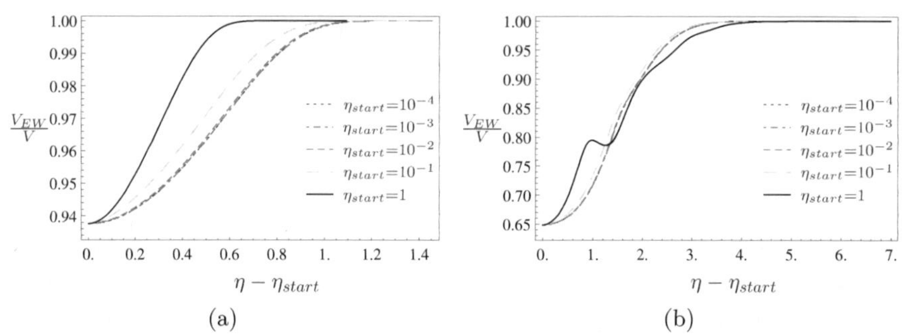

Figure 3: The time dependence of the fraction of lattice sites occupied by field on the electroweak side of the barrier $\frac{V_{EW}}{V}$ for small (a) and moderate (b) contribution of EW vacuum at initialization.

For nearly equal contribution of both vacua at initialization time late domain walls decay longer than early ones, as illustrated by the plot in fig. 4a. It is worth stressing that for simulations presented in fig. 4a V_{EW} decreases initially, however finally the EW vacuum dominates the lattice. The decay of the domain walls leading to the final state without the EW vacuum is possible even for the initial configuration with slight dominance of the EW vacuum. These scenario is realized by the example shown in fig. 4b.

We observe decay times of networks of domain walls ranging from $0.7\, l = 7 \times 10^{-11}\text{ GeV}^{-1}$ for initial conditions with the contribution of the high energy minimum of the order of a few percents, up to $1\, l = 1.3 \times 10^{-9}\text{ GeV}^{-1}$ in the case of nearly equal contributions of both vacua at initialization. In all cases the decay time displays the weak dependence on the initialization time η_{start}.

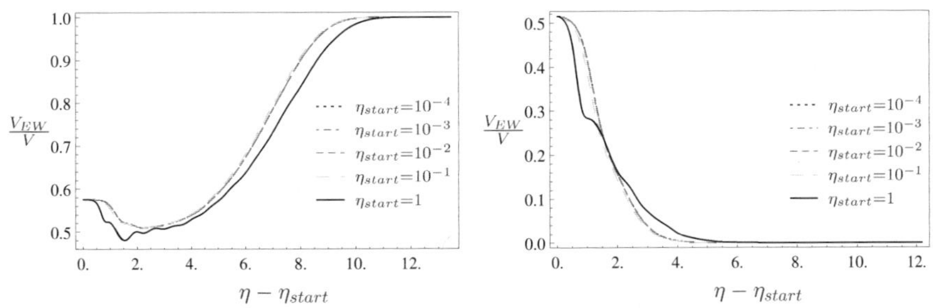

(a) (b)

Figure 4: The time dependence of the fraction of lattice sites occupied by field on the electroweak side of the barrier $\frac{V_{EW}}{V}$ for nearly equal contributions of both vacua at initialization time.

4.2. Dependence on mean value and variation of the field strength

In [16] we considered cases when the sum of the mean field value θ and some fraction $\frac{1}{f}$ of the standard deviation σ is equal to the position v_{max} of the local maximum of the potential at the initialization time η_{start}. We restricted the initial standard deviation to $10^{-3}\theta \leq \sigma \leq \theta$.

For $f = 1$ simulations start with ratio $\frac{V_{EW}}{V}$ equal to 84 %. In this case the evolution of networks displays the weak dependence on the value of σ and for all simulations the final state is the EW vacuum. Decay times in simulations with initialization times $10^{-4}\ l \leq \eta_{start} \leq 1\ l$ are all of the order of $2\ l = 2 \times 10^{-10}$ GeV^{-1}. The time dependence of $\frac{V_{EW}}{V}$ for the network of domain walls initialized at the time $\eta_{start} = 10^{-4}\ l$ is presented in fig. 5a. Late domain walls decay in the time of the order of $14\ l = 1.4 \times 10^{-9}$ GeV^{-1}. The time dependence of $\frac{V_{EW}}{V}$ for the network of domain walls initialized at the time $\eta_{start} = 10\ l$ is presented in fig. 5b.

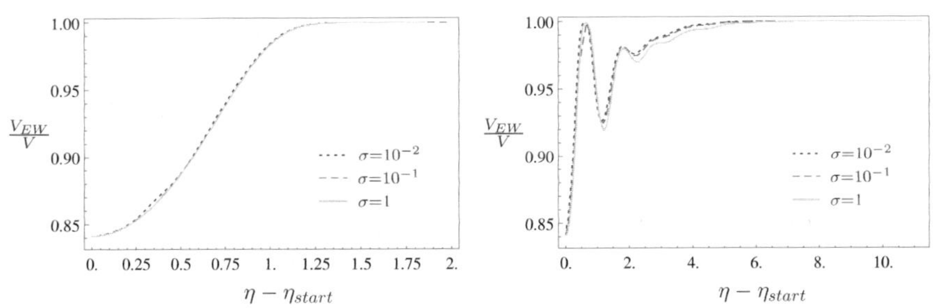

(a) (b)

Figure 5: The time dependence of the fraction of lattice sites occupied by field on the electroweak side of the barrier $\frac{V_{EW}}{V}$ for different σ for two initialization times: $\eta_{start} = 10^{-4}\ l$ (a) and $\eta_{start} = 10\ l$ (b).

For $f = 5$ simulations start with ratio $\frac{V_{EW}}{V}$ equal to 58 %. For $f = 5$ decay times for both early and late domain walls are longer then for $f = 1$. For early domain walls these times are of the order of $6\ l = 6 \times 10^{-10}$ GeV^{-1} and for late domain walls are of the order of $20\ l = 2 \times 10^{-9}$ GeV^{-1}. The time dependence of $\frac{V_{EW}}{V}$ for networks of domain walls initialized at the time $\eta_{start} = 10^{-4}\ l$ is presented in fig. 6a and for ones initialized at the time $\eta_{start} = 10\ l$ is presented in fig. 6b.

IOP Conf. Series: Journal of Physics: Conf. Series **873** (2017) 012044 doi:10.1088/1742-6596/873/1/012044

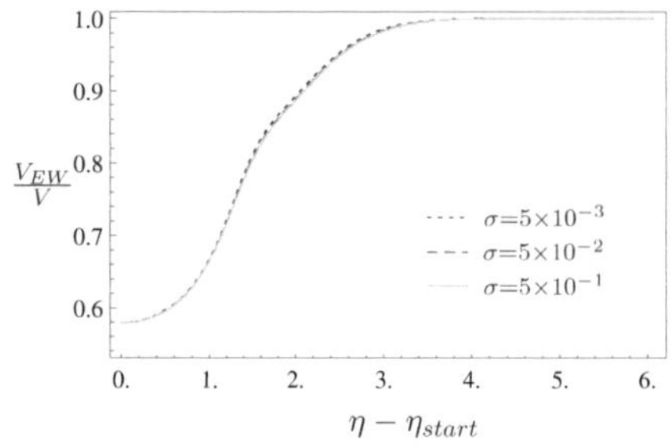

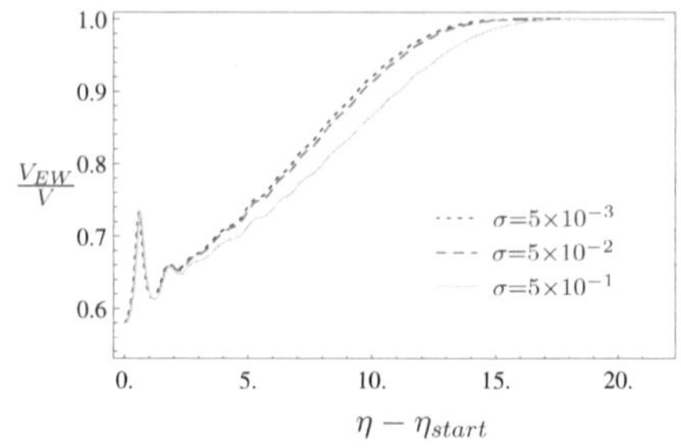

Figure 6: Time dependence of the fraction of lattice sites occupied by field on the electroweak side of the barrier $\frac{V_{EW}}{V}$ for different σ for two initialization times: $\eta_{start} = 10^{-4}\ l$ (a) and $\eta_{start} = 10\ l$ (b).

5. Computing the spectrum of gravitational waves

In our simulations we computed numerically spectrum of gravitational waves' energy density per unit logarithmic frequency interval at conformal time at which the decay of domain walls has ended η_{dec}. The algorithm which we used for calculation was proposed in [17] for calculation of the spectrum of primordial GWs and used in the case of domain walls in [15, 16].

During the evolution of the Universe gravitational waves produced by the network of domain walls were stretched by the expansion. The spectrum at the conformal time η_{EQ} at which the energy densities of matter and radiation are equal can be computed, according to [17], as:

$$\frac{d\rho_{gw}}{d\log|k|}(\eta_{EQ}, k) = \frac{a(\eta_{dec})^4}{a(\eta_{EQ})^4}\frac{d\rho_{gw}}{d\log|k|}(\eta_{dec}, k). \tag{10}$$

The factor $\frac{a(\eta_{dec})^4}{a(\eta_{EQ})^4}$ is determined by the linear growth of the scale factor in the radiation domination epoch:

$$\frac{da}{d\eta} = a(\eta)^2 H(\eta) = const =: \dot{a}_{in}, \qquad a(\eta) = \dot{a}_{in}\eta + a_{in}, \tag{11}$$

where $H(\eta)$ is the value of the Hubble constant at the conformal time η and a_{in} is a constant which sets the value of the scale factor a at the time when radiation starts to dominate the Universe. The value of the Hubble constant at the time of equality of matter and radiation energy densities can be calculated from present values of the Hubble constant H_0 and fractions of the critical density Ω_M, Ω_R assuming simple scaling of energy densities:

$$H_{EQ}{}^2 = \frac{2H_0{}^2\Omega_M{}^4}{\Omega_R{}^3} = 2\times 10^{-55}\frac{\mathrm{eV}^2}{\hbar^2}. \tag{12}$$

The value of the Hubble constant at conformal time after the decay of domain walls can be computed from parameters of the simulation:

$$H_{dec} = \frac{1}{a_{dec}\eta_{dec}}\left(1 - \frac{a_{in}}{a_{dec}}\right) = 10^{19}\left(1 - \frac{a_{in}}{a_{dec}}\right)\left(\frac{10^{-10}\ \hbar\mathrm{GeV}^{-1}}{a_{dec}\eta_{dec}}\right)\frac{\mathrm{eV}}{\hbar}. \tag{13}$$

The ratio $\frac{a_{in}}{a_{dec}}$ is relevant for the theories in which radiation domination epoch begins at very low energy scales. In most inflationary scenarios this value is negligible. Substituting (12) and

IOP Conf. Series: Journal of Physics: Conf. Series **873** (2017) 012044 doi:10.1088/1742-6596/873/1/012044

(13) we obtain:

$$\frac{a(\eta_{dec})}{a(\eta_{EQ})} = \sqrt{\frac{H_{EQ}}{H_{dec}}} = 7.1 \times 10^{-24} \left(\frac{10^{19} \frac{\text{eV}}{\hbar}}{H_{dec}} \right)^{\frac{1}{2}}. \tag{14}$$

Furthermore assuming that the energy density of GWs scales in the matter domination epoch as a^{-4} we can express the present spectrum as:

$$\frac{d\rho_{gw}}{d\log|k|}(\eta_0, k) = (1 + z_{EQ})^{-4} \frac{a(\eta_{dec})^4}{a(\eta_{EQ})^4} \frac{d\rho_{gw}}{d\log|k|}(\eta_{dec}, k), \tag{15}$$

where η_0 is the present time and z_{EQ} is the red-shift to the epoch of matter-radiation equality. The energy density of gravitational waves $\frac{d\rho_{gw}}{d\log|k|}(\eta)$ is usually presented as a fraction $\Omega_{gw}(\eta)$ of the critical density $\rho_{cr}(\eta) := {M_{Pl}}^2 H^2(\eta)$.

Finally we estimated the red-shift of the wave frequency to be equal to:

$$f_0 = \frac{a(\eta_{dec})}{a(\eta_0)} \frac{k}{2\pi} = 5.07 \times 10^6 \left(\frac{10^{19} \frac{\text{eV}}{\hbar}}{H_{dec}} \right)^{\frac{1}{2}} \left(\frac{k}{10^{10} \frac{\text{GeV}}{\hbar c}} \right) \text{ Hz}. \tag{16}$$

6. Spectrum of gravitational waves

Computing the spectrum of gravitational waves in lattice simulations encounters many complications. The modified eom (5) cannot be used in this calculation, because the modification of eom disturbs the dynamics of the short wavelength fluctuations. For the unmodified eom the width of the domain walls decreases as $\propto a^{-2}$, significantly restricting the dynamical range of simulations. As a result only late domain walls ($\eta_{start} = \mathcal{O}(10^{-9}\ \text{GeV}^{-1})$) can be investigated in our simulations. Moreover as noticed in [15], used algorithm produces a spectrum that diverges as k^3 for random initialization of the field strength. Following [15] we introduced a cut-off scale in the Fourier transform of the initialization distribution of the order of the width of domain walls.

Fig. 7a presents the spectrum of GWs Ω_{gw} at the end of the simulation as a function of the frequency f in units of Hz. Ω_{gw} was computed as an average over 5 simulations which were initialized with condition $\theta = v_{max} + \sigma$ for $\sigma = 10^8$ GeV and $\sigma = 10^9$ GeV at the initial conformal time $\eta_{start} = 10\ l$ with $a_{start} := a(\eta_{start}) = \frac{1}{5}$. Simulations ended at the conformal time equal to $\eta_{end} = 20\ l$ with $a_{end} := a(\eta_{end}) = 4$ with the width of domain walls at this time of the order of $10\ l$. We assumed that $a_{in} = 0$ and the value of Hubble constant at the end of simulations is equal to $\frac{1}{a_{end}\eta_{end}}$. The obtained spectrum is peaked at wave vectors of the order of $k_{peak} \sim 0.05\ l^{-1}$, corresponding to the frequency $f_{peak} \sim 3 \times 10^8$ GeV. In fig. 7b we plotted the spectrum of GWs at present time.

7. Summary

In this paper we investigated the possibility of formation of a network of SM domain walls in the early Universe. We concentrated on evolution and decay of such a network and left open the question of an underlying cosmological model which leads to initial conditions facilitating its creation. We modeled these processes using numerical simulations performed on the lattice using PRS algorithm. Main observables in our simulations were the conformal time at which the network of domain walls decays and the spectrum of emitted GWs.

We estimated the width of SM domain walls to be of the order of $4 \times 10^{-9}\ \frac{\hbar c}{\text{GeV}}$. The width of domain walls in our simulations must be a few times bigger than the lattice spacing. We used lattice spacing equal to $l = 10^{-10}\ \frac{c\hbar}{\text{GeV}}$ which results in units for (conformal) time $10^{-10}\ \frac{\hbar}{\text{GeV}}$ and energy 10^{10} GeV.

IOP Conf. Series: Journal of Physics: Conf. Series **873** (2017) 012044 doi:10.1088/1742-6596/873/1/012044

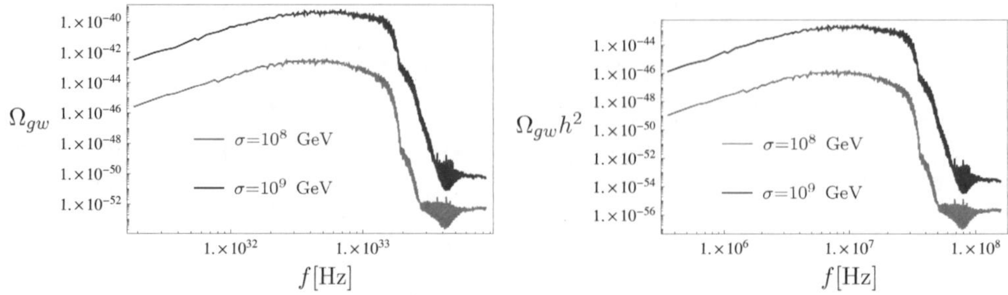

(a) (b)

Figure 7: The spectrum of gravitational waves Ω_{gw} emitted from SM domain walls for $a_{in} = 0$ (standard cosmology) at the time of the decay (a) and at present time (b).

Our simulations show that domain walls, interpolating between minima of the effective potential of the SM, decay shortly after their formation. Decay times range from 8×10^{-11} $\hbar\text{GeV}^{-1}$ to 3×10^{-9} $\hbar\text{GeV}^{-1}$. Such short lifetimes exclude a scenario in which SM domain walls dominate the Universe leading to large distortion of the CMBR. An interesting conclusion from our studies is, that the creation of the network which ends up in the electroweak vacuum percolating through the Universe is not as difficult to obtain as one may think, although it requires certain tuning of initial conditions.

The second signature of the existence of the network of SM domain walls are gravitational waves emitted from decaying domain walls. Computed in our numerical simulations spectra of GWs (solid) are compared with predicted sensitivities (dashed) of future detectors: aLIGO [18], ET [19], LISA, LISA:TNG [20], DECIGO and BBO [21] in fig. 8.

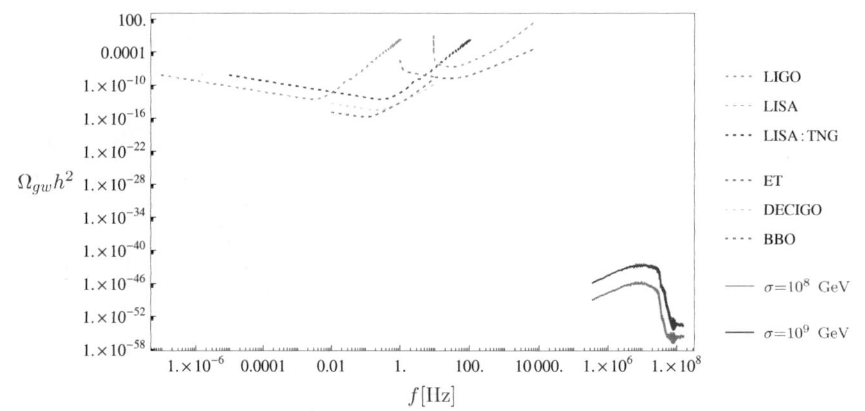

Figure 8: Predicted sensitivities (dashed) for future GWs detectors: aLIGO, ET, LISA, LISA:TNG, DECIGO and BBO compared with the spectrum of GWs (solid) calculated in lattice simulations for the initial values of $\sigma = 10^8,\ 10^9$ GeV and the standard cosmology.

Fig. 8 shows that the detection in upcoming years of GWs emitted by SM domain walls is excluded if one assume the standard cosmological evolution and validity of SM up to very high scales. However, both of these assumptions can be weakened. The present energy density of GWs produced by domain walls could be greater if the evolution of the Universe before the formation of domain walls was different then in the standard scenario. Models predicting very low scale of inflation [22] or including new components of energy density which shorten the radiation domination period [23, 24] would result in larger $\frac{a_{in}}{a_{dec}}$ (so lower H_{dec}). Moreover recent experimental bounds on the size of nonrenormalizable interactions of SM particles are very weak. The recent studies [25] have shown that the inclusion of the nonrenormalizable operators in the Lagrangian density modify the behaviour of the effective potential $V_{\rm eff}$ at high scales and can even make EW vacuum stable. The scenario of nearly degenerate minima of Higgs potential due to inclusion of $|H|^6$ operator has been recently investigated in [26]. The authors of [26] argue that in this case the decay time of domain walls is longer, leading to the increased energy density of emitted GWs, even up to the sensitivity range of future interferometer detectors.

The future detection of GWs emitted from decaying Higgs domain walls, although not likely, would point toward non-standard cosmological scenarios and its significance for both particle physics and cosmology is hard to overestimate.

Acknowledgments

This work has been supported by the Polish NCN grants DEC-2012/04/A/ST2/00099 and 2014/13/N/ST2/02712 and MNiSW grant IP2015 043174 and by the ARC Centre of Excellence for Particle Physics at the Terascale (CoEPP) (CE110001104) and the Centre for the Subatomic Structure of Matter (CSSM).

References

[1] Aad G *et al.* (ATLAS Collaboration) 2012 *Phys.Lett.* **B716** 1–29 (*Preprint* 1207.7214)
[2] Chatrchyan S *et al.* (CMS Collaboration) 2012 *Phys.Lett.* **B716** 30–61 (*Preprint* 1207.7235)
[3] Hook A, Kearney J, Shakya B and Zurek K M 2015 *JHEP* **01** 061 (*Preprint* 1404.5953)
[4] Kobakhidze A and Spencer-Smith A 2013 *Phys. Lett.* **B722** 130–134 (*Preprint* 1301.2846)
[5] Press W H, Ryden B S and Spergel D N 1989 *Astrophys. J.* **347** 590–604
[6] Abbott B P *et al.* (Virgo, LIGO Scientific) 2016 *Phys. Rev. Lett.* **116** 061102 (*Preprint* 1602.03837)
[7] Lalak Z, Lola S and Magnowski P 2008 *Phys. Rev.* **D78** 085020 (*Preprint* 0710.1233)
[8] Lazanu A, Martins C J A P and Shellard E P S 2015 *Phys. Lett.* **B747** 426–432 (*Preprint* 1505.03673)
[9] Buttazzo D, Degrassi G, Giardino P P, Giudice G F, Sala F, Salvio A and Strumia A 2013 *JHEP* **12** 089 (*Preprint* 1307.3536)
[10] Coulson D, Lalak Z and Ovrut B A 1995 *Particles, strings and cosmology. Proceedings, 19th Johns Hopkins Workshop and 5th PASCOS Interdisciplinary Symposium, Baltimore, USA, March 22-25, 1995* pp 429–444 (*Preprint* hep-ph/9508226)
[11] Lalak Z 1996 *High energy physics* (*Preprint* hep-ph/9702405)
[12] Oliveira J C R E, Martins C J A P and Avelino P P 2005 *Phys. Rev.* **D71** 083509 (*Preprint* hep-ph/0410356)
[13] Kawasaki M and Saikawa K 2011 *JCAP* **1109** 008 (*Preprint* 1102.5628)
[14] Leite A M M and Martins C J A P 2011 *Phys. Rev.* **D84** 103523 (*Preprint* 1110.3486)
[15] Hiramatsu T, Kawasaki M and Saikawa K 2014 *JCAP* **1402** 031 (*Preprint* 1309.5001)
[16] Krajewski T, Lalak Z, Lewicki M and Olszewski P 2016 *JCAP* **1612** 036 (*Preprint* 1608.05719)
[17] Dufaux J F, Bergman A, Felder G N, Kofman L and Uzan J P 2007 *Phys. Rev.* **D76** 123517 (*Preprint* 0707.0875)
[18] Aasi J *et al.* (LIGO Scientific) 2015 *Class. Quant. Grav.* **32** 074001 (*Preprint* 1411.4547)
[19] Hild S *et al.* 2011 *Class. Quant. Grav.* **28** 094013 (*Preprint* 1012.0908)
[20] Larson S L, Hiscock W A and Hellings R W 2000 *Phys. Rev.* **D62** 062001 (*Preprint* gr-qc/9909080)
[21] Yagi K and Seto N 2011 *Phys. Rev.* **D83** 044011 (*Preprint* 1101.3940)
[22] Artymowski M, Lalak Z and Lewicki M 2016 (*Preprint* 1607.01803)
[23] Lewicki M, Rindler-Daller T and Wells J D 2016 *JHEP* **06** 055 (*Preprint* 1601.01681)
[24] Huang F P, Wan Y, Wang D G, Cai Y F and Zhang X 2016 *Phys. Rev.* **D94** 041702 (*Preprint* 1601.01640)
[25] Lalak Z, Lewicki M and Olszewski P 2014 *JHEP* **05** 119 (*Preprint* 1402.3826)
[26] Kitajima N and Takahashi F 2015 *Phys. Lett.* **B745** 112–117 (*Preprint* 1502.03725)

IOP Conf. Series: Journal of Physics: Conf. Series **873** (2017) 012045 doi:10.1088/1742-6596/873/1/012045

The Sym2Int program: going from symmetries to interactions

Renato M. Fonseca

AHEP Group, Instituto de Física Corpuscular, C.S.I.C./Universitat de València
Parque Científico, Calle Catedrático José Beltrán, 2 E-46980 Paterna (Valencia) – Spain

E-mail: renato.fonseca@ific.uv.es

Abstract. Model builders often need to find the most general Lagrangian which can be constructed from a given list of fields. These fields are actually representations of the Lorentz and gauge groups (and maybe of some discrete symmetry group as well). I will describe a simple program (Sym2Int) which helps to automate this task by listing all possible interactions between Lorentz/gauge group representations.

1. Symmetries of the fields and their interactions

Symmetries are an ever present feature in beyond-the-Standard-Model models: there is the Lorentz symmetry of flat space-time, there is the gauge symmetry containing $SU(3)_c \times SU(2)_L \times U(1)_Y$, and perhaps there are even more symmetries (for example discrete ones associated to flavor). Therefore, to build a model one first needs to find the most general Lagrangian which can be formed with a given list of fields, each transforming under some irreducible representation of the full symmetry group.

This is very often (or perhaps always) done manually, taking time and potentially leading either to the accidental inclusion of forbidden interactions, or the omission of some of the allowed ones. Furthermore, it is worth emphasizing the following two points. The first one is that the list of interactions in the Standard Model (SM) is small, but with the addition of just a few extra fields their number can increase considerably. The second thing worth keeping in mind is that all SM fields transform under small representations of the gauge group, which simplifies the task of manually computing the allowed interactions.

The aim of the Sym2Int package for Mathematica is to automatize this task: with minimal effort one can quickly obtain a full picture of a model's interactions and at the same time reduce the probability of making mistakes. It also introduces the possibility of cross-checking results obtained by hand. The program can be downloaded from the webpage

renatofonseca.net/sym2int.php

There are already many tools which automate various of the steps required to confronted theory with data (see figure 1). However, to my knowledge the very first step described previously (i.e, finding a model's interactions) has not been addressed fully. In fact, with the notable exception of Susyno [1] and SARAH [2], most tools assume as a starting point the Lagrangian (at the very least). In this sense, the Sym2Int program (which uses Susyno functions) can be seen

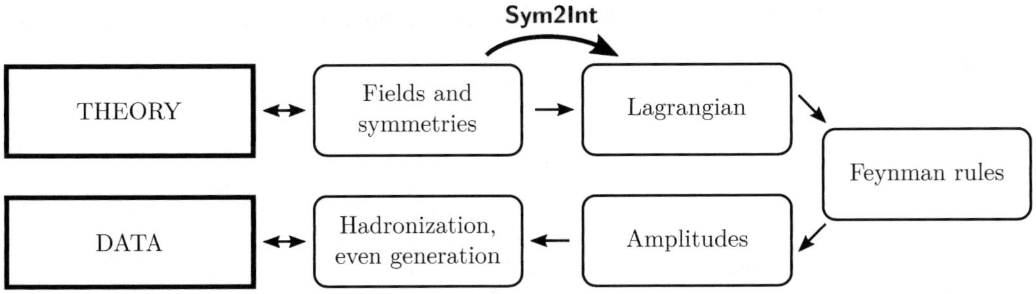

Figure 1. Chain of steps needed in order to explore the phenomenology of a model. The Sym2Int package helps with the first one, by automatizing the task of finding the interactions allowed by symmetries.

as a missing link in this long chain of programs which are needed to automatize the study of the phenomenology of a model. As we will see latter on, the code counts and lists the allowed interactions of a model, but it can also provide the explicit contractions of the field components (for example, the color indices, the $SU(2)_L$ indices, and so on) which are needed to write down the Lagrangian.

Going from fields and symmetries to a Lagrangian requires group theory code, which is available in Susyno. In fact, this program already takes as input symmetries and fields/representations, building from there a Lagrangian. However, it does so for supersymmetric models and furthermore there is an emphasis on the calculation of renormalization group equations. The Sym2Int package uses extensively the group theory code of Susyno but gets rid of the emphasis on theses two aspects (supersymmetry and renormalization group equations): instead, for a generic model, Sym2Int simply focuses on providing a list of allowed couplings and, if the user so wishes, it also shows how to contract the indices of the fields which participate in each interaction.

The program can deal with any representation of the Lorentz and gauge groups, and it will also work with abelian discrete symmetries. There is currently no support for non-abelian discrete symmetries; this will hopefully be added in the future. As for the type of interactions, Sym2Int will compute all allowed terms up to an arbitrary mass dimension specified by the user (it is therefore not limited to renormalizable operators). Unfortunately, due to their peculiar transformation property under gauge transformations, non-renormalizable terms with derivatives ∂_μ and/or gauge bosons are not yet handled by the program in a fully satisfactory way.[1]

Having introduced the Sym2Int program, in the rest of this document I will describe how to use it. This information and some more details are available on the program's webpage [4]; furthermore, the webpage [5] may be of some use since it explains how to use Susyno's group theory code in a stand-alone way.

2. Installing and running the program

The *sym2Int.m* file which can be obtained from the program's webpage [4] should be placed in a folder where Mathematica can find it. This could be for example *(Mathematica base dir.)/AddOns/Applications* on Linux and Mac OS or *(Mathematica base dir.)\AddOns\Applications* on Windows. Since the program relies on Susyno code, this program must also be installed (it is available from the webpage [6]).

[1] This shortcoming will probably also be addressed in a latter version of the code. Regarding this matter, I would like to thank Andreas Trautner for bringing to my attention reference [3].

IOP Conf. Series: Journal of Physics: Conf. Series **873** (2017) 012045 doi:10.1088/1742-6596/873/1/012045

Once this is done, Sym2Int can be loaded with the command

```
<<Sym2Int`
```

3. Defining a model

The program assumes that the defining elements of a model are (i) a **gauge group** and (ii) a **list of fields** which are irreducible representations of the gauge and Lorentz groups. For bookkeeping reasons, a name for the model is also required:

```
gaugeGroup[<model name>] ^= <gauge group>;
fields[<model name>] ^= {<field 1>, <field 2>, ...};
GenerateListOfCouplings[<model name>];
```

In turn, each field consists of (i) a **name**, (ii) a **gauge representation**, (iii) a **Lorentz representation** [such as scalar ("S"), right/left-handed Weyl fermion ("R"/"L"), vector ("V"), etc.], (iv) an indication of whether the **field is real** ("R") **or complex** ("C"), and (v) the **number of copies/flavors**.

```
<field> = {<name>, <gauge rep>, <Lorentz rep: "S", "R", "L", "V", etc.>,
           <"R" or "C">, <#copies>}
```

For example, the SM can be specified as follows:

```
gaugeGroup[SM] ^= {SU3, SU2, U1};
fld1 = {"u", {3, 1, 2/3}, "R", "C", 3};
fld2 = {"d", {3, 1, -1/3}, "R", "C", 3};
fld3 = {"Q", {3, 2, 1/6}, "L", "C", 3};
fld4 = {"e", {1, 1, -1}, "R", "C", 3};
fld5 = {"L", {1, 2, -1/2}, "L", "C", 3};
fld6 = {"H", {1, 2, 1/2}, "S", "C", 1};
fields[SM] ^= {fld1, fld2, fld3, fld4, fld5, fld6};
GenerateListOfCouplings[SM];
```

With this input, the program prints the following table with a list of all possible renormalizable interactions among the given fields (table 1).

#	Operator	Dim.	Self conj.?	Repeated fields	Symmetry and number of parameters
1	H[C] H[R]	2	True		{1, 1}
2	u[C] Q[R] H[R]	4	False		{9, 1}
3	d[C] Q[R] H[C]	4	False		{9, 1}
4	e[R] L[C] H[R]	4	False		{9, 1}
5	H[C] H[C] H[R] H[R]	4	True	{H[C], H[R]}	{{S, S}, 1, 1}

Table 1. Standard Model renormalizable interactions (those with gauge bosons are not shown).

By default, the list stops at mass dimension 4. To go further, for example up to dimension 5, one can use the `MaxOrder` command:

```
GenerateListOfCouplings[SM, MaxOrder -> 5];
```

The output is the one shown in table 2, which contains an extra row corresponding to the Weinberg operator.

#	Operator	Dim.	Self conj.?	Repeated fields	Symmetry and number of parameters
1	H[C] H[R]	2	True		{1, 1}
2	u[C] Q[R] H[R]	4	False		{9, 1}
3	d[C] Q[R] H[C]	4	False		{9, 1}
4	e[R] L[C] H[R]	4	False		{9, 1}
5	H[C] H[C] H[R] H[R]	4	True	{H[C], H[R]}	{{S, S}, 1, 1}
6	L[R] L[R] H[R] H[R]	5	False	{L[R], H[R]}	{{S, S}, 6, 1}

Table 2. Standard Model interactions up to dimension 5 (those with gauge bosons or derivatives are not shown).

Let us now take a look at the information contained in these tables which are printed by the program. For that discussion, it will be convenient to use the word *operator* in a rather liberal way: it stands for all gauge and Lorentz invariant contractions of a given combination of fields. As such, even if there is more than one independent gauge and Lorentz invariant contraction of the fields (say) A, B, C and D, I will still consider them to be all part of a single ABCD operator. So, with this clarification, the output tables contains the following information:

- The **first column** simply numbers each operator.
- The **second column** indicates what are the fields which participate in the interaction; a "`[C]`" after the field's name means that it is conjugated, while a "`[R]`" means that it is not.
- The **third column** shows, for convenience, the mass dimension of the operator: each boson(fermion) contributes 1(3/2) unit(s).
- The **forth column** tells us whether or nor the operator is self-conjugated.
- The **fifth column** lists the fields which appear more than once in the operator (taking into account that conjugated fields with a "`[C]`" are not the same as the unconjugated ones with a "`[R]`").
- The **sixth column** contains information on the permutation symmetry and number of parameters associated to the operator. In general is has the structure "`{<symmetry (if any)>, <nFlav>, <nContr>}`".
  - If there is any repeated field(s) in the operator, the **first entry** indicates what is the permutation symmetry associated to them. Otherwise, this entry is omitted.
  - Some fields have flavors, i.e. they come with a multiplicity. So, the **next entry** indicates the number of couplings/numbers needed to encode each independent contraction of the gauge and Lorentz indices of the fields.
  - The **last entry** indicates the number of independent contractions of the gauge and Lorentz indices of the fields.[2]

[2] This is true in most cases. However, in cases with complex permutation symmetries, this description ceases to be accurate. For the correct/full description of this number, see the discussion in section 4 concerning table 5.

IOP Conf. Series: Journal of Physics: Conf. Series **873** (2017) 012045 doi:10.1088/1742-6596/873/1/012045

All this should be clear, with perhaps the exception of the last column, which deals with parameter counting and also with symmetries under permutations of fields. This is a subtle yet very important issue which we will now discuss in detail.

4. Counting parameters

The number of parameters needed to encode a model is not the same as the number of rows in the output table for various simple reasons. Let us go through them:

(i) The number of copies of each field must be taken into account. For example, in the above SM example, there is a u^*QH interaction[3] on row #2, but since both u and Q have 3 flavors we need a tensor Y_{ij} with dimensions 3×3 to encode all the couplings of this operator. There is then a need for 9 parameters, and the program presents this counting on the the second(first) entry of the last column is there are(there aren't) repeated fields in the operator.

(ii) The coupling multiplying a combination of fields $\mathcal{O}$ is real if $\mathcal{O} = \mathcal{O}^*$ and complex otherwise, meaning that operators which are self-conjugate (forth column of the output matrix) require half as many real couplings as those which are not self-conjugate.[4] For example, the 9 parameters needed for the u^*QH operator in the SM are complex, while H^*H requires a single real parameter.

(iii) Sometimes it is possible to contract the gauge and Lorentz indices of the fields in more than one way: for example, there are two terms of the form $\Delta^*\Delta^*\Delta\Delta$ for an $SU(2)$ complex scalar triplet Δ. This number is shown in the last entry of the sixth column.

(iv) Finally, consider the well-known non-renormalizable operator L_iL_jHH responsible for neutrino mass generation in the SM (row #6 of table 2). The flavor indices i and j go from 1 to 3, so a 3×3 coupling tensor κ_{ij} is required. Therefore, naively one would think from the previous considerations that there are 9 complex degrees of freedom association to this operator. However, it is well known that the leptons L contract symmetrically in this particular case, hence $\kappa_{ij} = \kappa_{ji}$, so there are only 6 complex parameters and indeed that is the number displayed on the last column (second entry). In fact, the two Higgs fields also contract symmetrically (otherwise, since there is only one copy of this field, there would be no operator!). That is what the "{"S", "S"}" in the first entry of the last column means: there are two fields which appear more than once in the operator ("L[R]" and "H[R]" as indicated in the fifth column) and both of them contract symmetrically (an "A" would indicate an anti-symmetry).

In practice, this discussion implies that the number of couplings associated to each operator is obtained by making the product of the two numbers in the sixth column. However, note that sometimes an operator might have different contractions of the gauge/Lorentz components of the fields with different symmetries; in that case the sixth column will be of the form "{<symmetry1>, <nFlav1>, <nContr1>} | {<symmetry2>, <nFlav2>, <nContr2>} | ..." and the number of degrees of freedom is given by the sum of the products <nFlavI><nContrI>.

This last point is better illustrated with another example: the Two Higgs Doublet Model (THDM). There are two ways of inserting a second Higgs in the SM: one of them is to say that

[3] The star represents just field conjugation. I omit the exact contraction of Lorentz and gauge indices.

[4] Once we introduce flavors things may become a bit more complicated. For example, in a self-conjugate term $X_{ij}\phi_i^*\phi_j$ where the indices $i,j = 1,\cdots,n$ stand for different flavors of a field ϕ, it is not true that X_{ij} equals X_{ij}^*. Instead, on must make a permutation of the (ij) indices jointly with the conjugation operation: $X_{ij} = X_{ji}^*$. However, note that it still holds that X_{ij} contains n^2 real degrees of freedom (and not $2n^2$ real $= n^2$ complex degrees of freedom).

IOP Conf. Series: Journal of Physics: Conf. Series **873** (2017) 012045 doi:10.1088/1742-6596/873/1/012045

the fields are still "{fld1, fld2, fld3, fld4, fld5, fld6}" but now there are two copies of the Higgs:

```
fld6 = {"Hi", {1, 2, 1/2}, "S", "C", 2};
```

Another possibility is to keep "fld6" as it was before, and add a "fld7" with the same quantum numbers:

```
fld6 = {"H1", {1, 2, 1/2}, "S", "C", 1};
fld7 = {"H2", {1, 2, 1/2}, "S", "C", 1};
fields[THDM] ^= {fld1, fld2, fld3, fld4, fld5, fld6, fld7};
```

In the first case, we get table 3, while in the second one we get table 4. At first sight, these two results seem very different, but it is easy to check that they represent the same list of interactions. Table 3 is more compact because the two scalar doublets are treated as two copies/flavours of a single field "Hi"; for example, the quartic couplings are all encoded in row #5 of table 3 while in table 4 they are shown on the last six rows.[5]

#	Operator	Dim.	Self conj.?	Repeated fields	Symmetry and number of parameters
1	H[C] H[R]	2	True		{4, 1}
2	u[C] Q[R] H[R]	4	False		{18, 1}
3	d[C] Q[R] H[C]	4	False		{18, 1}
4	e[R] L[C] H[R]	4	False		{18, 1}
5	H[C] H[C] H[R] H[R]	4	True	{H[C], H[R]}	{{S, S}, 9, 1} | {{A, A}, 1, 1}

Table 3. List of interactions in the Two Higgs Doublet Model (those with gauge bosons are not shown).

Let us now focus on the first perspective, namely that there is a single scalar field with two flavors: H_i with $i = 1, 2$. Row #5 of table 3 is saying that one can write the quartic interactions as

$$V^{(4)} = \lambda_{ijkl}^{SS}\left(H_i^* H_j^* H_k H_l\right)_{SS} + \lambda_{ijkl}^{AA}\left(H_i^* H_j^* H_k H_l\right)_{AA}, \quad (1)$$

where $(\cdots)_{SS}$ and $(\cdots)_{AA}$ stand for two different ways of contracting the $SU(2)_L$ indices of the doublets: one is symmetric under the exchange of both the (ij) and (kl) indices, while the other contraction is anti-symmetric under both these exchanges.[6] These symmetries are transmitted to the tensors of couplings λ_{ijkl}^{SS} and λ_{ijkl}^{AA}, so that $\lambda_{ijkl}^{SS} = \lambda_{jikl}^{SS} = \lambda_{ijlk}^{SS}$ while $\lambda_{ijkl}^{AA} = -\lambda_{jikl}^{AA} = -\lambda_{ijlk}^{AA}$. It is then straightforward to show that λ_{ijkl}^{SS} is described by 9 independent real numbers, while λ_{ijkl}^{AA} requires a single one.

The origin of the two terms in equation (1) can be traced back to the fact that the product $\overline{\mathbf{2}} \times \overline{\mathbf{2}} \times \mathbf{2} \times \mathbf{2}$ in $SU(2)$ contains two singlets. But, even so, it is worth mentioning that one

[5] The counting of quartic parameters in the two cases goes as follows. The operator associated to row #5 of table 3 requires $9 \times 1 + 1 \times 1 = 10$ real couplings. As for table 4, rows #10, #13 and #15 correspond to self-conjugated operators, so they need a total of $1 \times 1 + 1 \times 2 + 1 \times 1 = 4$ real couplings, while rows #11, #12 and #14 need a total of $1 \times 1 + 1 \times 1 + 1 \times 1 = 3$ complex couplings, hence there is a total of 10 degrees of freedom in the quartic sector — just as in table 3.

[6] The explicit form of these contractions is quite simple: $\left(H_i^* H_j^* H_k H_l\right)_{SS} = \left(H_i^\dagger H_k\right)\left(H_j^\dagger H_l\right) + \left(H_j^\dagger H_k\right)\left(H_i^\dagger H_l\right)$ and $\left(H_i^* H_j^* H_k H_l\right)_{AA} = \left(H_i^\dagger H_k\right)\left(H_j^\dagger H_l\right) - \left(H_j^\dagger H_k\right)\left(H_i^\dagger H_l\right)$ (up to some arbitrary normalization factors). Still, we will not be discussing explicit contraction of indices since that is not necessary for counting parameters/degrees of freedom of a model.

IOP Conf. Series: Journal of Physics: Conf. Series **873** (2017) 012045 doi:10.1088/1742-6596/873/1/012045

#	Operator	Dim.	Self conj.?	Repeated fields	Symmetry and number of parameters
1	H1[C] H1[R]	2	True		{1, 1}
2	H1[C] H2[R]	2	False		{1, 1}
3	H2[C] H2[R]	2	True		{1, 1}
4	u[C] Q[R] H1[R]	4	False		{9, 1}
5	u[C] Q[R] H2[R]	4	False		{9, 1}
6	d[C] Q[R] H1[C]	4	False		{9, 1}
7	d[C] Q[R] H2[C]	4	False		{9, 1}
8	e[R] L[C] H1[R]	4	False		{9, 1}
9	e[R] L[C] H2[R]	4	False		{9, 1}
10	H1[C] H1[C] H1[R] H1[R]	4	True	{H1[C], H1[R]}	{{S, S}, 1, 1}
11	H1[C] H1[C] H1[R] H2[R]	4	False	H1[C]	{S, 1, 1}
12	H1[C] H1[C] H2[R] H2[R]	4	False	{H1[C], H2[R]}	{{S, S}, 1, 1}
13	H1[C] H1[R] H2[C] H2[R]	4	True		{1, 2}
14	H1[C] H2[C] H2[R] H2[R]	4	False	H2[R]	{S, 1, 1}
15	H2[C] H2[C] H2[R] H2[R]	4	True	{H2[C], H2[R]}	{{S, S}, 1, 1}

Table 4. List of interactions in the Two Higgs Doublet Model (those with gauge bosons are not shown). This result may appear different from the one in table 3, but in reality it is the same: both tables list the same interactions.

does not need to split $V^{(4)}$ in two terms as in equation (1). For example, consider the sum $(\cdots)_{SS} + (\cdots)_{AA} \equiv (\cdots)_{mixed}$. Then one could just write

$$V^{(4)} = \lambda_{ijkl}^{mixed} \left(H_i^* H_j^* H_k H_l \right)_{mixed} \tag{2}$$

since this is the same as $\lambda_{(ij)(kl)}^{mixed} \left(H_i^* H_j^* H_k H_l \right)_{SS} + \lambda_{[ij][kl]}^{mixed} \left(H_i^* H_j^* H_k H_l \right)_{AA}$ where round(square) brackets in the indices denotes symmetrization(anti-symmetrization). So there is a trade-off: the notation gets more compact if we replace equation (1) by equation (2) but, in the latter case, the counting of independent parameters is not as obvious. This observation also highlights the fact that, if we hide different field copies in a flavor index, then there might very well be an ambiguity in the counting of interaction terms (as in equations (1) and (2), which actually stand for the same quartic potential). The approach of the `Sym2Int` program is to always split the interaction terms according to their properties under field permutation symmetries, which makes the counting of parameters easier.

Field permutation symmetries can be more complicated than what has been seen so far, and for completeness we should look at a more general case. Consider the quartic operator of table 3: both H_i^* and H_i appear twice, hence the relevant permutation group is $S_2 \times S_2$. Crucially, the S_2 group only has two 1-dimensional (irreducible) representations: the trivial/symmetric (S) and the alternating/anti-symmetric (A) one. What happens if there are more than $n > 2$ repeated fields participating in an interaction? Take, for example, a model with m real scalar doublets $D_{i=1,\cdots,m}$ with no hypercharge.[7] Is it possible to write down quartic interactions of the form $D_i D_j D_k D_l$? If yes, how many couplings are needed as a function of m? To answer these

[7] This would lead to fractional electric charges, but we do not care about such things here.

questions one needs to understand how do four $SU(2)$ doublets contract. It is well known that $\mathbf{2} \times \mathbf{2} = \mathbf{1}_A + \mathbf{3}_S$, so it is tempting to think that $\mathbf{2} \times \mathbf{2} \times \mathbf{2} \times \mathbf{2} = (\mathbf{1}_A + \mathbf{3}_S) \times (\mathbf{1}_A + \mathbf{3}_S) = \mathbf{1}_A + \mathbf{1}_S + \cdots$. However, this last statement is not correct: the two $SU(2)$ singlets in the product of four doublets are neither completely symmetric under permutations nor completely anti-symmetric. Instead, they form an irreducible doublet of the S_4 group, which is the relevant permutation group. In practical terms, this means that we could write the $D_iD_jD_kD_l$ quartic interactions as

$$V^{(4)} = \lambda^{(1)}_{ijkl} \left(D_iD_jD_kD_l\right)_{(1)} + \lambda^{(2)}_{ijkl} \left(D_iD_jD_kD_l\right)_{(2)} \,, \tag{3}$$

where $(\cdots)_{(1),(2)}$ are the two different contractions of four doublets and, under a $\pi \in S_4$ permutation of the flavour indices,

$$\begin{pmatrix} \lambda^{(1)}_{\pi(ijkl)} \\ \lambda^{(2)}_{\pi(ijkl)} \end{pmatrix} = R\left(\pi\right) \begin{pmatrix} \lambda^{(1)}_{ijkl} \\ \lambda^{(2)}_{ijkl} \end{pmatrix} \,, \tag{4}$$

where crucially the twenty-four $R\left(\pi\right)$ matrices cannot be simultaneously diagonalized. In other words, these matrices form a 2-dimensional irreducible representation of S_4. Knowing this, how many independent couplings are there between $\lambda^{(1)}_{ijkl}$ and $\lambda^{(2)}_{ijkl}$? The answer is $\frac{1}{12}m^2(m-1)(m+1)$ and `Sym2Int` computes this expression for the user:[8]

```
gaugeGroup[modelX] ^= {SU2};
fields[modelX] ^= {{"D", {2}, "S", "R", m}};
GenerateListOfCouplings[modelX];
```

produces table 5. Note that in the second row there is neither an "`S`" nor an "`A`" symmetry: instead, there is a `{2,2}`. The reason is that irreducible representations of the S_n group can be labeled/associated with the partitions of the integer n. So, for S_4 the irreducible representations are {4}, {3,1}, {2,2}, {2,1,1}, {1,1,1,1} with dimensions 1, 3, 2, 3 and 1.[9] The program prints for the user the relevant partitions, except in the cases of $\{n\}$ and {1,1, ...,1} which are associated to the completely symmetry ("`S`") and the completely anti-symmetric ("`A`") representations. Finally, note that while there are two independent contractions of four doublets, in the sixth column of row #2 of table 5, the last entry is a `1`. In reality, this number stands for the number of gauge and Lorentz contractions with a given permutation symmetry. In the present case, since {2,2} is a 2-dimensional representation of S_4,there is (last entry =1) × (dimension of {2,2} =2) independent contractions of the gauge (and Lorentz) indices.

#	Operator	Dim.	Self conj.?	Repeated fields	Symmetry and number of parameters
1	D[R] D[R]	2	True	D[R]	$\{A, \frac{1}{2}(-1+m)\,m, 1\}$
2	D[R] D[R] D[R] D[R]	4	True	D[R]	$\{\{2, 2\}, \frac{1}{12}(-1+m)\,m^2\,(1+m), 1\}$

Table 5. List of interactions in a model with m real scalar doublets D.

[8] For this purpose, `Sym2Int` uses code from `Susyno` which in turn is based on the plethysm function described in the manual of the `LiE` package [7].

[9] The function `SnIrrepDim` from `Susyno` can be used to find these dimensions [5].

IOP Conf. Series: Journal of Physics: Conf. Series **873** (2017) 012045 doi:10.1088/1742-6596/873/1/012045

5. Other groups, representations, and program options

The program is meant to work with any group and gauge/Lorentz representations (obviously, practical constraints associated to computational time and/or memory space do apply). To indicate a gauge group the user only has to write down its name (`SU2`, `SU5`, `SO10`, `E6`, ...). As for gauge representations of dimension D of some group, in the case of $\boldsymbol{D}$ and $\overline{\boldsymbol{D}}$, it is enough to provide the numbers D and $-D$ (respectively). However, there might exist more than one representation with size D, and so primes are added to their names ($\boldsymbol{D}'$, $\boldsymbol{D}''$, ...). For those representations it is necessary to use the notation of Dynkin coefficients (see [5, 1] for details). For example, one can indicate the representations $\mathbf{3}$, $\overline{\mathbf{3}}$, $\mathbf{15}$ and $\overline{\mathbf{15}}$ of $SU(3)$ either with the numbers `3`, `-3`, `15`, `-15` or with the Dynkin coefficients `{1,0}`, `{0,1}`, `{2,1}`, `{1,2}`, while the $\mathbf{15}'$ can only be specified by its Dynkin coefficients, which are `{4,0}`.

As for Lorentz representations, in most cases fields are either scalars (`"S"`), left/right-handed Weyl spinors (`"L"`/`"R"`) or perhaps vectors (`"V"`), in which case it is enough to provide one of these four strings as input. More generally, irreducible representations of the Lorentz can be labeled with with two spins (i.e., two non-negative half-integers) j_L and j_R. `Sym2Int` will accept such an input if it is really necessary: for instance `"S"`, `"L"`, `"R"`, `"V"` is equivalent to `{0,0}`, `{1/2,0}`, `{0,1/2}`, `{1/2,1/2}`.

Furthermore, the function `GenerateListOfCouplings` has several options.

- There is support for abelian discrete symmetries through the option `DiscreteSym`. Multiplicative conservation of the field charges will then be enforced by the program:

```
GenerateListOfCouplings[<model>, DiscreteSym->{<charge of field #1>,
<charge of field #2>, ...}];
```

- The function `GenerateListOfCouplings` not only prints the list of interactions but it also returns this data in a particular format (see section 6). If the table with a model's interactions is not necessary/desired, it can be suppress with the `Verbose->False` option.
- Hermitian conjugated operators are not shown by default. This can be changed with the option `HCTerms->True`.
- Computing the permutation symmetries which were discussed in section 4 is time consuming hence, if this information is not needed, one can use the option `CalculateSnSymmetries->False` to speed up the calculation of the list of interactions.
- As mentioned previously, beyond listing the allowed interactions, `Sym2Int` can also calculate how the field components are contracted in each case. To do that, the option `CalculateInvariants -> True` must be used. Note that, depending on the number and size of the fields, this can be very time consuming. For more details on this part of the program, the user is referred to the online documentation [4].

6. Saving the program's results for further processing

The program prints by default a human-readable table with all the allowed operators. However, one might want to further process `Sym2Int`'s results in Mathematica, and in that case one should save and use instead the output returned by the function `GenerateListOfCouplings`. This output consists of a list with the following 8 pieces of information for each operator:

(i) The number associated to the operator.

(ii) The combination of fields which are interacting. Each field is identified by its position in the list provided by the user as input (if the field does not appear conjugated in the operator) or minus its position in the list provided by the user as input (if the field does appear conjugated in the operator).

(iii) The mass dimension of the operator.

(iv) Is the combination of fields self-conjugated?

(v) The list of fields which appear more than once in the operator (the identification of each field is done as in 2.).

(vi) Permutation symmetry and number of parameters associated to the operator.

(vii) Information on the explicit expression(s) of the gauge and Lorentz invariant contraction (or contractions if there are several) of the fields which appear in the operator. In other words, it provides the information needed to write down the Lagragian. (Requires the use of the option `CalculateInvariants -> True`; otherwise this slot in the list will contain the '`Null`' value.)

(viii) The exact same data/format/style as the relevant row of the table printed by the `GenerateListOfCouplings` function.

To get a quick view of what this means in practice, consider the Standard Model as defined previously in section 3. The code

```
operatorsSM=GenerateListOfCouplings[SM]; Grid[operatorsSM, Frame -> All]
```

yields the result

1	{-6, 6}	2	True	{}	{{{}, 1, 1}}		{1, H[C] H[R], 2, True, , {1, 1}}
2	{-1, 3, 6}	4	False	{}	{{{}, 9, 1}}		{2, u[C] Q[R] H[R], 4, False, , {9, 1}}
3	{-2, 3, -6}	4	False	{}	{{{}, 9, 1}}		{3, d[C] Q[R] H[C], 4, False, , {9, 1}}
4	{4, -5, 6}	4	False	{}	{{{}, 9, 1}}		{4, e[R] L[C] H[R], 4, False, , {9, 1}}
5	{-6, -6, 6, 6}	4	True	{-6, 6}	{{{{2}, {2}}, 1, 1}}		{5, H[C] H[C] H[R] H[R], 4, True, {H[C], H[R]}, {{S, S}, 1, 1}}

.

Acknowledgements
I would like to thank the organizers of the DISCRETE 2016 Symposium in Warsaw, where this computer code was first presented. I am also grateful to the other participants which provided interesting suggestions concerning the program. Lastly, this work was financially supported by the Spanish state through the grants Juan de la Cierva-formación FJCI-2014-21651, FPA2014-58183-P, Multidark CSD2009-00064 and SEV-2014-0398 (from the *Ministerio de Economía, Industria y Competitividad*), as well as the grant PROMETEOII/2014/084 (from the *Generalitat Valenciana*).

References

[1] R. M. Fonseca, *Calculating the renormalisation group equations of a SUSY model with Susyno*, Comput. Phys. Commun. **183** (2012) 2298–2306, arXiv:1106.5016 [hep-ph].

[2] F. Staub, *SARAH 4: A tool for (not only SUSY) model builders*, Comput. Phys. Commun. **185** (2014) 1773–1790, arXiv:1309.7223 [hep-ph].

[3] B. Henning, X. Lu, T. Melia and H. Murayama, *Hilbert series and operator bases with derivatives in effective field theories*, Commun. Math. Phys. **347** (2016) 2 363–388, arXiv:1507.07240 [hep-th].

[4] R. M. Fonseca, *Sym2Int: Listing all interactions of a model*, http://renatofonseca.net/sym2int.php, retrieved March 14, 2017.

[5] R. M. Fonseca, *Tutorial: Group theory with Susyno*, http://renatofonseca.net/susyno/group_theory_tutorial.php, retrieved March 14, 2017.

[6] R. M. Fonseca, *Susyno*, http://renatofonseca.net/susyno.php, retrieved March 14, 2017.

[7] M. A. A. van Leeuwen, A. M. Cohen and B. Lisser, *LiE, a package for Lie group computations*, Computer Algebra Nederland, Amsterdam, ISBN 90-74116-02-7, 1992, http://www-math.univ-poitiers.fr/~maavl/LiE, retrieved March 14, 2017.

IOP Conf. Series: Journal of Physics: Conf. Series **873** (2017) 012046 doi:10.1088/1742-6596/873/1/012046

Discrete and continuous description of physical phenomena

I F Ginzburg

Sobolev Institute of Mathematics, Novosibirsk, 630090, Russia;
Novosibirsk State University, Novosibirsk, 630090, Russia

E-mail: ginzburg@math.nsc.ru

Abstract.
The values z of many observable quantities in the Nature are determined at some discrete set of times t_n, separated by a small interval Δt (which may also represent a coordinate, etc.). Let the z value in neighbour point $t_{n+1} = t_n + \Delta t$ be expressed by the *evolution equation* as as $z(t_{n+1}) \equiv z(t_n + \Delta t) = f(z(t_n))$. This equation gives *a discrete description* of phenomenon. Considering phenomena at $t \gg \Delta t$ this equation is transformed often into the differential equation allowing to determine $z(t)$ leading to *continuous description*. It is usually assumed that the continuous description describes correctly the main features of a phenomenon at values $t \gg \Delta t$.

In this paper I show that the true behavior of some physical systems can differ strongly from that given by the continuous description. The observation of such effects may lead to the desire to supplement the original evolutionary model by additional mechanisms, the origin of which require special explanation. We will show that such construction may not be necessary – simple evolution model can describe properly different observable effects.

This text contains no new calculations. Most of the discussed facts are well known. New is the treatment of the results.

Let the phenomenon described by quantity $z_n \equiv z(t_n)$ be determined at some discrete set of times t_n with steps Δt and $z(t_{n+1}) = z(t_n + \Delta t)$ can be expressed by *the evolution equation*

$$z(t_{n+1}) \equiv z(t_n + \Delta t) = f(z(t_n)). \tag{1}$$

This is *a discrete description* of the phenomenon. This equation is transformed often into the differential equation for $z(t) \leftarrow z_n$, giving *the continuous description* of the phenomenon [1]:

$$\frac{z(t_{n+1}) - z(t_n)}{\Delta t} \Rightarrow \frac{dz(t)}{dt} = \frac{f(z(t_n)) - z_n}{\Delta t} \equiv f_1(z(t)). \tag{2}$$

It is usually assumed that the latter equation correctly describes the main features of phenomenon at time $t \gg \Delta t$.

In this report I show that in many cases the transition to the continuous description results in an incorrect description of reality, described by evolution equation (1). *Vise versa*, the interpretation of some observations, based on the "standard" continuous description, can result in an incorrect description of elementary mechanisms.

IOP Conf. Series: Journal of Physics: Conf. Series **873** (2017) 012046 doi:10.1088/1742-6596/873/1/012046

1. Model

For definiteness, we consider the livestock of carps z_n in a pond without pikes (n is the number of year). In this example time spacing is limited from below, $\Delta t = 1$ year (time between spawning). The simplest model consider only natural growth, $z_{n+1} = kz_n$ (k is called *reproduction factor*). At $k > 1$ this equation describes unlimited growth $z_n = k^n z_0$.

In reality this growth is stopped by competition for food and other effects (*pressure of population*). We describe this pressure by adding in the right hand of the simplest model the term $-\ell z_n^2$. As a result, our physical model takes the form of the evolution equation $z_{n+1} = kz_n - \ell z_n^2$. The change of variables $x_n = (\ell/k)z_n$ transforms this equation in the more convenient form

$$x_{n+1} = f_1(x_n)\,, \qquad f_1(x) = kx(1-x)\,, \qquad x_n \geq 0\,. \tag{3}$$

We are interested situation when the lifetime of population is not limited, the evolution starts from initial condition $x_0 > 0$ and $k > 1$.

Note that our problem has two equilibrium solutions,

$$x_\infty = f(x_\infty) \Rightarrow \begin{cases} x_{(\infty,1)} = 0 & (\textit{stable at} \quad k < 1)\,, \\ x_{(\infty,2)} = \dfrac{k-1}{k} & (\textit{stable at} \quad k > 1) \end{cases}. \tag{4}$$

2. Continuous description. Differential equation

Ordinarily to describe phenomena at the scales larger than Δt (in our case Δt is 1 year) the above difference equation is conventionally reduced to the differential equation [1]. To this end, the continuous variable – *time* t is introduced (in the initial evolution equation the variable t has only integer values n). Following (2), at $t \gg 1$ our equation (3) is transformed to continuous limit with the aid of standard sequence of relations:

$$\frac{x_{n+1} - x_n}{\Delta t} = x_n(k - 1 - kx_n) \Rightarrow \frac{dx}{dt} = x(k-1-kx)\,, \qquad x(0) = x_0\,. \tag{5}$$

This differential equation is solved easily,

$$x = \frac{(k-1)x_0}{kx_0 + (k-1-kx_0)e^{-(k-1)\,t}}\,. \tag{6}$$

This equation describes the evolution up to infinite time for an **arbitrary value of k** and for **on arbitrary initial value $x(0) > 0$**. At $t \to \infty$ and $k > 1$ the quantity x tends monotonically to the limiting value $x_{(\infty,2)} = (k-1)/k$ (4), determined for the incident (discrete) evolution equation.

3. Real evolution. Discrete description

Below we reproduce some known results of the study of evolution model (3) (see e.g. [2]). Our task is to present some interpretations, which are not discussed usually.

In contrast to the continuous description the equation (3) describes evolution of system during infinite time only for $0 < x_n < 1$, in particular for

$$0 < x_0 < 1\,. \tag{7}$$

Indeed, if $x_n > 1$, the process is stopped since eq. (3) gives senseless $x_{n+1} < 0$.

The details of picture are different for different values of the reproduction factor k.

3.1. The region $k < 4$.

Below we consider the asymptotical solutions of eq. (3) at large time n and their stability.

• **The case $k < 3$.**

For the study of the asymptotical solution of eq. (3) we substitute in this equation the expression $x_n = x_\infty + \delta_n \equiv (k-1)/k + \delta_n$ and, considering δ_n to be small at large n, linearize this equation (neglecting terms $\propto \delta_n^2$)). It results in the equation for x_∞ and the recurrence relation for δ_n,

$$x_\infty = kx_\infty(1-x_\infty), \qquad \delta_{n+1} = (2-k)\delta_n . \tag{8}$$

The equation for x_∞ has two solutions (4).

A. $k < 1$.

At $k < 1$ the trivial solution $x_\infty \equiv x_{(\infty,1)} = 0$ is realized (by the population going extinct).

B. $k > 1$.

At $k > 1$ nontrivial solution $x_\infty \equiv x_{(\infty,2)} = (k-1)/k$ is realized. The relation for δ_n shows that at $|2-k| < 1$ (i. e. at $1 < k < 3$) the quantity δ_n tends to 0 at $n \to \infty$, and the solution converges to $x_\infty \equiv x_{(\infty,2)}$.

◊ At $1 < k < 2$ the quantity x_n tends to grow monotonically to x_∞ for larger n just as the solution of differential equation (6). Continuous description works for initial values (7).

◊ At $2 < k < 3$ the quantity x_n also tends to x_∞ but in contrast with the solution of the differential equation, this increase is not monotonic but (damped) oscillating. The continuous description describes the main features of evolution (limiting value) but not important details.

• **The case $3 < k < 3.45$. Doubling of period.**

According to eq. (8) at $k > 3$ we have $|\delta_{n+1}| > |\delta_n|$. Therefore, the asymptotic solution (4) becomes unstable. The point $k = 3$ is a branching point. What next?

Let us check "the simplest" hypothesis that at $n \to \infty$ the number of carps is reproduced through two years (not yearly), with different populations in even and odd years. In other words, we suggest that $k = 3$ is the point of the doubling of the period.

The iteration of eq. (3) gives the equation

$$x_{n+2} = f_2(x_n), \quad f_2(y) = f_1(f_1(y)) \equiv k^2y(1-y)[1-ky(1-y)] . \tag{9}$$

The possible stationary states are given by equation of 4-th order $x_\infty - f_2(x_\infty) = 0$. Two solutions of this equation are also known solutions (4) of equation $x_\infty - f_1(x_\infty) = 0$. To find other solutions we consider equation $[x_\infty - f_2(x_\infty)]/[x_\infty - f_1(x_\infty)] = 0$ which has the form $k^2y^2 - k(k+1)y + k + 1 = 0$. Its solution exists only at $k > 3$ (we introduced new asymptotical values $x_{(\infty,3)}$ and $x_{(\infty,4)}$ so that $x_{(\infty,3)} > x_{(\infty,4)}$):

$$y = x_{(\infty,3,4)} \equiv \frac{k+1 \pm \sqrt{(k+1)(k-3)}}{2k} . \tag{10}$$

It is easy to check that

$$f_1(x_{(\infty,3,4)}) = x_{(\infty,4,3)}, \qquad x_{(\infty,3)} > x_{(\infty,2)} = (k-1)/k > x_{(\infty,4)} .$$

Therefore, at large n the quantity x_n oscillates between the values $x_{(\infty,3)}$ and $x_{(\infty,4)}$ with a period of 2 years. This phenomenon is called the *period doubling*.

Now we find the range of validity of this solution. For this goal we repeat above calculations with substitution $x_n = x_{(\infty,3,4)} + \delta_n$ in Eq. (9). In this case the linearized recurrence relation for δ_n has the form

$$\delta_{n+2} = (4 + 2k - k^2)\,\delta_n . \tag{11}$$

Just as above we find that the solutions $x_{(\infty,3,4)}$ (10) are stable only at $|4 + 2k - k^2| < 1$, i. e. for $3 < k < 1 + \sqrt{6} \approx 3.45$.

• **The case 3.45 < k < 4 and general picture at k < 4.**

With the growth of k the analytical study of the problem becomes complex. The Fig. 1 (obtained with "Mathematica" package) represents limiting values x_∞ at different k. Here in the abscissa the reproduction coefficient k is shown and in the ordinate – values x_∞). At $1 < k < 3$ population tends to the single limiting value given by eq. (4). At $3 < k < 3.45$ the population oscillates (with period 2 year –the period doubling). At $3.45 < k < 3.54$ we have a subsequent doubling of the period; the population varies periodically between four values. With the growth of k the number of limiting values grows (new period doublings replace each other). At $k > 3.569$ zones of chaotic behavior appears. With the growth to $k = 4$ these zones spread to entire interval (0, 1).

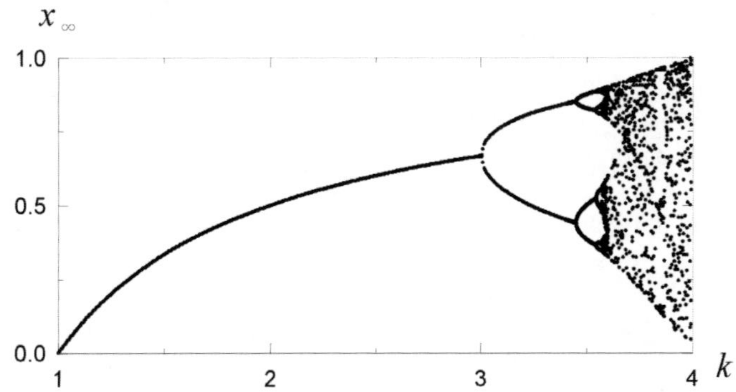

Figure 1. *Limiting values x_∞ at different k.*

3.2. The case $k = 4$. Chaotic behaviour.

At $k = 4$ evolution equation (3) can be solved exactly. We seek a solution in the form

$$x_n = \frac{1 - \cos 2\pi\alpha_n}{2}. \tag{12}$$

Now eq. (3) gives a simple sequence of identities:

$$x_{n+1} = 4\,\frac{1 + \cos 2\pi\alpha_n}{2}\,\frac{1 - \cos 2\pi\alpha_n}{2} = \sin^2(2\pi\alpha_n) = \frac{1 - \cos 4\pi\alpha_n}{2}.$$

Comparing with (12), we find $\alpha_{n+1} = 2\alpha_n \;\Rightarrow\; \alpha_n = 2^n\,\alpha_0$. The adding of any integer number to α_n does not change x_n. Therefore, in this solution only $\{\alpha_n\}$ – fractional part of α_n – makes sense, and

$$\alpha_n = \{2^n\,\alpha_0\}\,. \tag{13}$$

Now we discuss the meaning of this solution from different perspectives.

- Let us consider two initial values $x_0(\alpha_0)$ and $x_0'(\alpha_0')$ with $|\alpha_0 - \alpha_0'| = \varepsilon \ll 1$ and $2^{-r} > \varepsilon > 2^{-(r+1)}$. "The respectable" solution should be stable, it means that it is naturally to expect that the solutions with initial values x_0 and x_0' are close to each other even after long time. On the contrary, in our case at $n > r$ (after r years) the difference between x_n and x_n' becomes unpredictably large.
- The variable z_n is the number of carps averaged over some period. It varies weakly due to natural and accidental deathes. Therefore, the initial value z_0 (or x_0 or α_0) is determined with some uncertainty ε with $2^{-r} > \varepsilon > 2^{-(r+1)}$. Let us present this initial state in the binary-decimal form like $\alpha_0 = 0.1101000100...$. Here the r-th term is known while the $r+1$-th term is unknown. Solution (13) means that in each next year α_{n+1} is obtained from α_n by shift of point from one digit to the right with elimination of signs before the point.

Therefore, the population of carps is predicted via an initial state (by means of solution (13)) during first r years (with downward accuracy) but the population in $r+1$-th year is unpredictable.

- The typical initial value α_0 is irrational. By definition of irrational, the numbers – consecutive values $\{\alpha_n\}$ are non correlated. In other words, for almost all randomly chosen x_0 (or α_0) the range of produced values $\{\alpha_n\}$ is distributed uniformly in the segment [0, 1].

Therefore, at $k = 4$ the limiting value $x_{n\to\infty}$ does not exist. The values x_n vary with time unpredictably. Let us underline that even the averaged value $\langle x_n\rangle$ differs strongly from the value 3/4 predicted by eq. (5)[1] .

3.3. The case $k > 4$.
At $0 < x_n < 1$ the maximal value of the quantity $x_n(1 - x_n)$ is 1/4. Therefore, to prevent the restraining of process at some stage, it should be $k \le 4$. We expected that the mean time of life of population at $k > 4$ is small, but at small values of $k - 4$ it is not very small. The numerical experiment shows infinitesimal time of life even for $k - 4 = 0.01$. (Certainly, for some rational values $x(0)$ time of life can be large, but the rounding errors kill this opportunity).

3.4. **General picture**
In two tables below we compare some features of discrete and continuous description of our system in the regions of parameters allowing evolution during infinite time.

Table 1. Range of parameters and initial values

	discrete description	*continuous description*
range of parameter k of equation	restricted, $k \le 4$	unlimited
range of initial values $x(0)$	restricted, $x(0) < 1$	infinite

Table 2. Behaviour at large time, $t \gg 1$

k	*discrete description*	*continuous description*
$k < 1$	monotonic approaching to $x_{(\infty,1)} = 0$	monotonic
$1 < k < 2$	monotonic approaching to $x_{(\infty,2)} = 0$	monotonic
$2 < k < 3$	non-monotonic approaching to $x_{(\infty,2)} = 0$	monotonic
$3 < k < 3.45$	2 year oscillations between 2 values	monotonic
$3.45 < k < 3.569$	2^n year oscillations between 2^n values	monotonic
$3.569 < k < 4$	unpredictable, chaotic behaviour appear	regular evolution
$k = 4$	unpredictable, chaotic behaviour	regular evolution
$k > 4$	population is dying out quickly	regular evolution

4. Discussion

Most of presented facts are well known [2]. We discuss lessons from this picture for two fields of studies[2].

[1] At $k = 4$ the probability $p(y)dy$ that the quantity x_n is within interval y, $y + dy$ is calculated with the aid of eq. (12). For almost all initial values x_0 we have $p(y) = [\pi y(1-y)]^{-1/2}$. For these x_0 mean value $\langle x_n\rangle = 1/2$ and fluctuation $\langle \Delta x_n\rangle \sim \langle x_n\rangle$.

[2] The discussed discrete approach results in many new physical phenomena. Besides, the considered phenomena produce some difficulties in the computational algorithms at the transition from continuous description to the discrete

4.1. Lessons for using continuous approach for description of a real system

The standard approach in the study of physical phenomena is to change discrete evolution equation to the continuous one and subsequent use of differential equation for the description of long-time evolution of system. Our discussion demonstrates that this approach is valid only for some relatively narrow region of parameters.

- The approximation of discrete variation of some physical quantity $x_{n+1} = f(x_n)$ by continuous one is justified only at relatively low temp of variation of this quantity $k = (df/dx)/_{x=0}$ in the basic difference equation. With growth of this temp k the behavior of such system can be changed strong.
- There are the threshold values k_0 of parameter k. The differential equation describes evolution of physical quantity x at $k < k_0$ while at $k > k_0$ such description become incorrect. The quantity k_0 is different for different physical properties. In our example at $k < k_{01} = 2$ continuous approach describes real picture well. At $k_{01} < k < k_{02} = 3$ continuous approach gives good description of main picture with incorrect description of important details (it skips non-monotonic variation of $x(t)$).
- At $k > k_{02} = 3$ the continuous approach absolutely inapplicable for description of the system during long time.

Certainly, these k_0 are different in different physical problems.

For the validity of continuous approach in the multi-dimensional problems the eigenvalues of reproduction coefficient matrix $\{k\}$ should be small enough (this limitation can be broken even at small values of separate elements of matrix $\{k\}$ due to large number of its component.)

4.2. Lessons for construction of model for physical phenomenon.

Next problem is the construction of model for physical phenomenon, based on the preliminary understanding of the process and observations. Typical procedure is construction of evolution equation, its description in the continuous approach, and obtaining of coefficients from observation. If observed behaviour differs from that predicted in this approach not very strongly, we conclude that our model is correct in general, but some additional mechanisms should be taken into account for description of details. If we observe new phenomena, which don't appear in our continuous approach, we conclude that our incident understanding of problem was incorrect, and new mechanisms are responsible for this phenomenon.

Our analysis shows that such approach can be often incorrect. Simple model describes phenomenon, and only continuous approach can appear wrong.

4.3. Examples

.

- Let population dies out in finite time. In the traditional approach it would be natural to seek an explanation in the existence of some new force which kills population. We understand that in the discrete approach it will be due to inappropriate initial conditions or parameters (e. g. $x_n > 1$ or $k > 4$) with the same simple evolution equation.
- Let we observe periodic behavior. In the traditional approach it would be natural to seek an explanation in the existence of *additional* mechanism providing periodicity (periodic external force or resilient restoring force). Our example for $3 < k < 3,569$ shows that even complex periodic behaviour can be explained without these additional mechanisms.

one. These problems have huge literature, which is beyond our scope here.

- Let we observe chaotic behavior. In the traditional approach it would be natural to seek an explanation in the existence of *additional* mechanisms, like random external force (open system). Our example for $4 > k > 3,569$ shows that even complex chaotic behaviour can be explained without these additional mechanisms.

Acknowledgments
I am thankful to I.P. Ivanov, G.L. Kotkin, M. Krawczyk A. Pogosov and V.G. Serbo for discussions. This work was supported in part by grants RFBR 15-02-05868 and HARMONIA project under contract UMO-2015/18/M/ST2/00518 (2016-2019).

References
[1] See e.g. Tricomi F G 1961 *Differential equations* (BLACKIE & SON LIMITED)
[2] See e.g. Feigenbaum M. J. 1978 *J. Stat. Phys.* (1978) **19**, p. 25; 1979 *J. Stat. Phys.* (1979) **21**, p. 669; 1979 *Lecture Notes in Physics* (1979) **93**, p. 163; 1980 *Comm. Math. Phys.* (1980) **77**, p. 65; 1980 *Los Alamos Science* (1980) **1** No. 1, pp. 4-27; Chirikov B V 1979 *Phys. Rept.* (1979) **52**, p. 263

IOP Conf. Series: Journal of Physics: Conf. Series **873** (2017) 012047 doi:10.1088/1742-6596/873/1/012047

Light neutrino masses from a non-Hermitian Yukawa theory

J Alexandre[1], C M Bender[2] and P Millington[3]

[1]Department of Physics, King's College London,
London WC2R 2LS, United Kingdom
[2]Department of Physics, Washington University,
St. Louis, MO 63130, U.S.A.
[3]School of Physics and Astronomy, University of Nottingham,
Nottingham NG7 2RD, United Kingdom

E-mail: jean.alexandre@kcl.ac.uk, cmb@wustl.edu, p.millington@nottingham.ac.uk

Abstract. Working within the context of $\mathcal{PT}$-symmetric quantum mechanics, we begin by describing a non-Hermitian extension of QED that is both Lorentz invariant and consistent with unitarity. We show that the non-Hermitian Dirac mass matrix of this theory exhibits an exceptional point, corresponding to an effectively massless theory whose conserved current is either right- or left-chiral dominated. With this inspiration, we are able to construct a non-Hermitian model of light Dirac neutrino masses from Hermitian and anti-Hermitian Yukawa couplings that are both of order unity. We finish by highlighting potential phenomenological implications of this model.

1. Introduction

The observation of neutrino flavour oscillations has provided compelling empirical evidence that (at least two of) the neutrinos of the Standard Model (SM) of particle physics have nonzero masses [1]. Observations of the Cosmic Microwave Background (CMB) by ESA's Planck satellite, when combined with data from supernovae light curves and Baryon Acoustic Oscillations (BAO), put an upper limit on the sum of the SM neutrino masses of [2]

$$\sum_j m_j \ < \ 0.23 \text{ eV} \tag{1}$$

at the 95% confidence level. The masses of the SM neutrinos therefore appear to be "unnaturally" small compared with those of the other SM fermions.

Neutrino masses provide strong evidence for the presence of physics beyond the SM, and it is now widely accepted that the SM should be understood as a "low-energy" effective field theory (EFT), albeit an incredibly successful and highly predictive one. If we impose conservation of lepton number (L) on the SM EFT, neutrinos must be of Dirac type and acquire their masses through a coupling to the SM Higgs field in the same way as the other SM fermions. In this case, we must stomach around 12 orders of magnitude difference between the top-quark and neutrino Yukawa couplings. Instead, if we maintain that conservation of lepton number is only accidental at low energies, the SM EFT contains a dimension-5 operator [3] — the Weinberg

IOP Conf. Series: Journal of Physics: Conf. Series **873** (2017) 012047 doi:10.1088/1742-6596/873/1/012047

operator — which violates lepton number by two units. After spontaneous symmetry breaking, this operator, schematically of the form $(L\phi)^2/\Lambda$, where L is the charged-lepton doublet and ϕ is the Higgs doublet, gives rise to a Majorana mass term that is suppressed by the scale of new physics Λ. At tree-level, the Majorana mass term may be mediated by the exchange of heavy right-handed neutrinos (type I see-saw [4, 5, 6, 7, 8]), by a scalar triplet (type II see-saw [8, 9, 10, 11, 12]) or a fermion $SU(2)_L$ triplet (type III see-saw [13]). See-saw mechanisms are compatible with supersymmetric ultra-violet completions of the SM, and they can be realised naturally within grand unified theories (see, e.g., [14] and references therein). Gauge-invariant fermion masses may also be generated radiatively at the three-loop level in $U(1)$ gauge theories by means of global anomalies mediated by scalar or pseudoscalar fields [15]. The presence of such pseudoscalar fields on manifolds with quantum torsion can lead to chirality-violating, right-handed Majorana mass terms generated gravitationally at the two-loop level by analogous anomalous operators [16].

In addition to explaining the smallness of the light neutrino masses, the out-of-equilibrium decay of the heavy right-handed neutrinos in the early Universe can play a pivotal role in explaining the observed matter-anti-matter asymmetry in scenarios of leptogenesis [17, 18] (for reviews, see Refs. [19, 20, 21, 22]). Therein, lepton number violating decays produce an initial excess in lepton number that is converted to a baryon number excess (B) via the B + L-violating sphaleron processes of the standard electroweak theory.

However, as theoretically appealing as the see-saw mechanism may be, the origin of the light neutrino masses is yet to be verified experimentally, and it is therefore prudent to consider alternatives. To this end, we recall that quantum field theories are often constructed with the following three properties in mind:

(i) locality — ensuring microcausality, i.e. that the position-space canonical commutation relations (CCRs) vanish for space-like separations.

(ii) Lorentz invariance in vacuum.

(iii) Hermiticity — guaranteeing real eigenvalues and unitary evolution.

A local, Lorentz-invariant theory with an Hermitian Hamiltonian is necessarily $\mathcal{CPT}$-invariant; this is the celebrated $\mathcal{CPT}$ theorem.

By sacrificing locality, it has been shown that neutrino oscillations can be accommodated through $\mathcal{CPT}$-violating Dirac mass terms without enlarging the neutrino sector of the SM [23]. Fermion masses can also arise dynamically in Lorentz-Invariance-Violating (LIV) $U(1)$ gauge theories, involving higher-order spatial derivatives of the gauge field [24, 25]. Such models can emerge in the low-energy limit of quantum gravity theories [26], and flavour-mixing interactions between the fermion and gauge fields may provide an alternative explanation for flavour oscillations [27, 28]. The impact of non-Hermitian terms on neutrino oscillations in matter have also been considered [29]. In the context of open quantum systems, it has been shown that dissipative effects can give rise to effective neutrino masses, leading to flavour oscillations for initially massless or mass-degenerate neutrinos [30]. Since dissipative quantum systems can be described by non-Hermitian effective Hamiltonians, this suggests to consider possible modifications of the neutrino sector in which the constraint of Hermiticity is relaxed.

Our starting point is the following observation: *All Hermitian matrices have real eigenvalues, but not all matrices with real eigenvalues are Hermitian.* Consider, for example, the complex (but non-Hermitian) 2×2 matrix

$$\mathbf{A} \;=\; \begin{bmatrix} a+ib & c \\ c & a-ib \end{bmatrix}\,, \tag{2}$$

where $a, b, c \in \mathbb{R}$. Its spectrum is

$$\lambda(\mathbf{A}) \;=\; \left\{a \,\pm\, \sqrt{c^2 \,-\, b^2}\right\}\,. \tag{3}$$

IOP Conf. Series: Journal of Physics: Conf. Series **873** (2017) 012047 doi:10.1088/1742-6596/873/1/012047

There are three regimes:

(i) When $|c| > |b|$, we are in the *unbroken phase*, and the spectrum of $\mathbf{A}$ contains two distinct, real eigenvalues $\lambda(\mathbf{A}) = \{a \pm \sqrt{c^2 - b^2}\} \subset \mathbb{R}$.

(ii) When $|c| < |b|$, we are in the *broken phase*, and the spectrum of $\mathbf{A}$ comprises a complex-conjugate pair of eigenvalues $\lambda(\mathbf{A}) = \{a \pm i\sqrt{b^2 - c^2}\} \subset \mathbb{C}$.

(iii) When $|c| = |b| \neq 0$, the eigenvalues merge, and we lose an eigenvector. We have one real eigenvalue $\lambda(\mathbf{A}) = \{a\} \subset \mathbb{R}$, and the matrix $\mathbf{A}$ is defective, having the Jordan normal form

$$\mathbf{A}_J = \begin{bmatrix} a & 1 \\ 0 & a \end{bmatrix} . \tag{4}$$

At $c = \pm b$, the eigenvalues exhibit a square-root singularity, and this is known as an *exceptional point*.

In the context of Hamiltonian systems, Hermiticity is a sufficient but not necessary condition for obtaining real eigenvalues and unitary evolution. The latter can instead be guaranteed by the less restrictive and more physical constraint of $\mathcal{PT}$ symmetry, i.e. symmetry of the Hamiltonian under the discrete space-time symmetries of parity ($\mathcal{P}$) and time-reversal ($\mathcal{T}$). It can then be shown that the evolution is unitary with respect to the positive semi-definite inner product $\langle A|B\rangle = A^{\mathcal{CPT}} \cdot B$ rather than the usual inner product $\langle A|B\rangle = A^{\dagger} \cdot B$ [31]. Thus, for non-Hermitian $\mathcal{PT}$-symmetric quantum theories, Hermitian conjugation is superseded by $\mathcal{CPT}$ conjugation. A comprehensive introduction to $\mathcal{PT}$-symmetric quantum mechanics can be found in Ref. [32].

Consider again the matrix $\mathbf{A}$. Its form is equivalent to the Hamiltonian of a coupled system with gain and loss. The matrix $\mathbf{A}$ is not Hermitian, but it is $\mathcal{PT}$-symmetric: the parity transformation corresponds to interchanging the two systems, i.e.

$$\mathcal{P} : \mathbf{A} \longrightarrow \mathbf{A}^{\mathcal{P}} = \mathbf{P}\mathbf{A}\mathbf{P}^{\mathsf{T}} , \qquad \mathbf{P} = \begin{pmatrix} 0 & 1 \\ 1 & 0 \end{pmatrix} , \tag{5}$$

and time-reversal is enacted by complex conjugation, interchanging gain and loss, i.e.

$$\mathcal{T} : \mathbf{A} \longrightarrow \mathbf{A}^{\mathcal{T}} = \mathbf{A}^* . \tag{6}$$

When the coupling between the systems is strong ($|c| > |b|$), we are in a region of unbroken $\mathcal{PT}$ symmetry (the unbroken phase), and the eigenvectors of $\mathbf{A}$ are simultaneous eigenvectors of $\mathcal{PT}$. On the other hand, when the coupling is weak ($|c| < |b|$), the eigenvectors of $\mathbf{A}$ are no longer eigenvectors of $\mathcal{PT}$, and the $\mathcal{PT}$ symmetry is broken (the broken phase). The $\mathcal{PT}$ phase transition occurs precisely at the exceptional point. Hermitian quantum field theories do not exhibit exceptional points. However, the radius of convergence of their perturbation series in powers of the coupling constant is precisely the distance in the complex coupling-constant plane to the nearest exceptional point [33].

A non-Hermitian extension of quantum electrodynamics (QED) was first studied in Ref. [34] (see also Ref. [35]), wherein the four-vector potential was taken to transform as an axial vector with an imaginary bare coupling to the $U(1)$ current. The $\mathcal{C}$ operator of this theory was constructed in Ref. [36], and it was shown that its S matrix describes unitary evolution. The theory of a free Dirac fermion with an anti-Hermitian, parity-violating mass term was first introduced in Ref. [37] and studied further in Ref. [38]. The authors of Ref. [39] recognised that such an extension of the Dirac equation can maintain a nonzero mass matrix whilst simultaneously having massless dispersion relations, potentially allowing massless neutrinos to undergo flavour oscillations. In what follows, we report the results of Ref. [40], expanding upon the suggestions made in Refs. [38] and [39] that such non-Hermitian mass matrices may have important implications for neutrino physics.

IOP Conf. Series: Journal of Physics: Conf. Series **873** (2017) 012047 doi:10.1088/1742-6596/873/1/012047

2. Non-Hermitian extension of QED

Following Ref. [40], we begin with a non-Hermitian extension of QED of the form[1]

$$\mathcal{L} \;=\; -\frac{1}{4}F_{\mu\nu}F^{\mu\nu} \;+\; \bar{\psi}(x)\big(i\gamma^{\mu}D_{\mu} \;-\; m \;-\; \mu\gamma^5\big)\psi(x) \;+\; \mathcal{L}_{\text{gf}}\;, \tag{7}$$

where $\psi(x)$ is a four-component Dirac fermion, $F_{\mu\nu} = \partial_\mu A_\nu - \partial_\nu A_\mu$ is the field-strength tensor, $\mathcal{L}_{\text{gf}}$ is the gauge fixing term, and we include both vector and axial vector couplings to the $U(1)$ gauge field A_μ:

$$D_\mu \;=\; \partial_\mu \;+\; i\big(g_V \;+\; g_A\gamma^5\big)A_\mu\;, \tag{8}$$

where $g_V, g_A \in \mathbb{R}$. The Lagrangian contains both an Hermitian mass term $m\bar{\psi}\psi$ and an anti-Hermitian mass term $\mu\bar{\psi}\gamma^5\psi$. The latter changes sign on Hermitian conjugation by virtue of the anti-commuting nature of the gamma matrices, i.e. $\{\gamma^\mu, \gamma^5\} = 0\ \forall\, \mu = 0,1,2,3$. By the usual definitions of the discrete symmetry transformations on Fock space, that is the definitions appropriate for a theory with an Hermitian Hamiltonian, the anti-Hermitian mass term is $\mathcal{C}$ even, $\mathcal{P}$ odd, and $\mathcal{T}$ even. Thus, it appears $\mathcal{CP}$ and $\mathcal{CPT}$ odd. We recall that the usual *Hermitian* term $\bar{\psi}i\gamma^5\psi$ is $\mathcal{C}$ even, $\mathcal{P}$ odd, and $\mathcal{T}$ odd.

In the massless limit $m = \mu = 0$, the Lagrangian is invariant under the combined vector and axial vector gauge transformation

$$A_\mu \;\longrightarrow\; A_\mu \;-\; \partial_\mu\phi\;, \tag{9a}$$

$$\psi \;\longrightarrow\; \exp\big[i\big(g_V \;+\; g_A\gamma^5\big)\phi\big]\psi\;, \qquad \bar{\psi} \;\longrightarrow\; \bar{\psi}\exp\big[-i\big(g_V \;-\; g_A\gamma^5\big)\phi\big]\;. \tag{9b}$$

It can be shown [38] that the Dirac equation is the one obtained by varying the Lagrangian with respect to $\bar{\psi}$ for fixed ψ. This fact is not obvious since one would obtain a different equation of motion by varying with respect to ψ and taking the Dirac conjugate. Following the former procedure, we have

$$\big(i\gamma^\mu D_\mu \;-\; m \;-\; \mu\gamma^5\big)\psi(x) \;=\; 0\;. \tag{10}$$

In order to obtain the dispersion relations, we can use the usual trick of acting again with the Dirac operator, yielding

$$\big(D^2 \;-\; m^2 \;+\; \mu^2\big)\psi(x) \;=\; 0\;. \tag{11}$$

The energies are therefore given by

$$\omega^2 \;=\; \mathbf{p}^2 \;+\; M^2\;, \qquad M^2 \;=\; m^2 \;-\; \mu^2\;, \tag{12}$$

which are real when $m^2 \geq \mu^2$. We see that the dispersion relation is the same for particles and anti-particles, implying that there is no $\mathcal{CPT}$ violation.

The fact that the non-Hermitian extension of QED yields identical dispersion relations for particles and anti-particles is indicating that we should construct alternative definitions of the $\mathcal{C}$, $\mathcal{P}$ and $\mathcal{T}$ transformations that are appropriate to a non-Hermitian theory and under which this theory is $\mathcal{C}$ even, $\mathcal{P}$ odd and $\mathcal{T}$ odd [37]. Under these transformations, the theory remains $\mathcal{CP}$ odd, but it is now $\mathcal{CPT}$ and $\mathcal{PT}$ even, thereby falling within the category of $\mathcal{PT}$-symmetric quantum field theories. The dynamical degrees of freedom for a non-Hermitian theory are not related by Hermitian (Dirac) conjugation, and it is for this reason that the correct equation of motion is obtained by varying with respect to $\bar{\psi}$ and not that obtained by taking the Dirac conjugate of the equation of motion found by varying with respect to ψ.

For comparison, in curved space-time, we can generate a similar Hermitian and locally $\mathcal{CPT}$-violating operator through coupling to the background curvature connection. This can

[1] We do not suggest that this non-Hermitian extension of QED is realised in nature. Instead, it will act as a playground in which to study the behaviour of non-Hermitian gauge theories.

be achieved, for example, in homogeneous anisotropic Bianchi Type II, VIII, and IX spacetimes for a locally rotating system [41] (see also Ref. [42]). One has

$$\mathcal{L} \;=\; \sqrt{-g}\,\bar{\psi}(x)\big(i\gamma^a D_a \;-\; m\big)\psi(x)\;, \tag{13}$$

where $g = \det g_{\mu\nu}$ is the determinant of the curved space-time metric. The covariant derivative is given by

$$D_a \;\equiv\; \partial_a \;-\; \frac{i}{4}\,\omega_{bca}\,\sigma^{bc}\;, \tag{14}$$

where

$$\omega_{bca} \;=\; e_{b\lambda}\big(\partial_a e_c^\lambda \;+\; \Gamma^\lambda_{\mu\nu}e_c^\mu e_a^\nu\big)\;, \tag{15}$$

is the spin connection, $\sigma^{ab} = \frac{i}{2}\big[\gamma^a, \gamma^b\big]$ is the generator of tangent-space Lorentz transformations, e_a^μ is a vielbein and $\Gamma^\lambda_{\mu\nu}$ is the curved-space Christoffel symbol. We use lower-case Latin and Greek indices to identify the coordinates of the tangent and curved spaces, respectively. After some algebra, this Lagrangian can be rewritten in the form

$$\mathcal{L} \;=\; \sqrt{-g}\bar{\psi}(x)\big(i\gamma^a\partial_a \;-\; m \;-\; \gamma^5\gamma^a B_a\big)\psi(x)\;, \tag{16}$$

where

$$B^d \;=\; \epsilon^{abcd}e_{b\lambda}\big(\partial_a e_c^\lambda \;+\; \Gamma^\lambda_{\nu\mu}e_c^\nu e_a^\mu\big)\;. \tag{17}$$

If we consider B^a to be a fixed external field, we have a Hermitian and $\mathcal{CPT}$-odd mass term $\gamma^5\gamma^a B_a$. In this case, the dispersion relations are

$$\big(\omega \;\mp\; B_0\big)^2 \;=\; \big(\mathbf{p} \mp \mathbf{B}\big)^2 \;+\; m^2\;, \tag{18}$$

differing between particles and anti-particles.

The presence of the γ^5 in the anti-Hermitian mass term in Eq. (7) indicates that we are treating the left and right chiralities differently. In fact, it is the left- and right-chiral sectors of this theory to which the analogy of a coupled system of gain and loss applies. Transforming to an explicit chiral basis, we have

$$\mathcal{L} \;=\; \begin{pmatrix}\psi_L^\dagger & \psi_R^\dagger\end{pmatrix}\begin{pmatrix} i\bar{\sigma}\cdot D_- & -\,m_+ \\ -\,m_- & i\sigma\cdot D_+\end{pmatrix}\begin{pmatrix}\psi_L \\ \psi_R\end{pmatrix}\;, \tag{19}$$

where $\sigma^\mu = (\sigma^0, \sigma^i)$ and $\bar{\sigma}^\mu = (\sigma^0, -\,\sigma^i)$, σ^i are the Pauli matrices, and

$$m_\pm \;=\; m \;\pm\; \mu\;, \qquad D_\pm^\mu \;=\; \partial^\mu \;+\; ig_\pm A^\mu \qquad g_\pm \;=\; g_V \;\pm\; g_A\;. \tag{20}$$

The mass matrix has the form

$$\mathbf{m} \;=\; \begin{pmatrix} 0 & m_+ \\ m_- & 0\end{pmatrix}\;. \tag{21}$$

We see immediately that this matrix becomes defective in the limit $\mu \to \pm\, m$. For $\mu = +\,m$, we obtain a massless theory with chirality flips biased from left to right; for $\mu = -\,m$, we obtain a massless theory with chirality flips biased from right to left. The conserved current for this theory is

$$j^\mu \;=\; \bar{\psi}\gamma^\mu\bigg(1 \;+\; \frac{\mu}{m}\gamma^5\bigg)\psi \;=\; \psi_L^\dagger\bar{\sigma}^\mu\psi_L\bigg(1 \;-\; \frac{\mu}{m}\bigg) \;+\; \psi_R^\dagger\sigma^\mu\psi_R\bigg(1 \;+\; \frac{\mu}{m}\bigg)\;. \tag{22}$$

Thus, for $\mu = +\,m$, the left-chiral current decouples, and for $\mu = -\,m$, the right-chiral current decouples.

We can look at this another way. For $\mu = +m$, the Lagrangian has the form

$$\mathcal{L} \;=\; \psi_L^\dagger i\bar{\sigma}\cdot D_-\psi_L \;+\; \psi_R^\dagger i\sigma\cdot D_+\psi_R \;-\; 2m\psi_L^\dagger\psi_R \,, \tag{23}$$

and we have the Weyl equations

$$i\sigma\cdot D_+\psi_R \;=\; 0\,, \qquad i\bar{\sigma}\cdot D_-\psi_L = 2m\psi_R\,. \tag{24}$$

We can integrate out the left chirality, giving the on-shell Lagrangian

$$\mathcal{L}_{\text{on-shell}} \;=\; \psi_R^\dagger i\sigma\cdot D_+\psi_R\,, \tag{25}$$

that of a massless, right-chiral Weyl fermion. Trivially, one can show that the full vector plus axial vector gauge invariance is recovered in this limit:

$$A_\mu \;\longrightarrow\; A_\mu \;-\; \partial_\mu\phi\,, \qquad \psi_R \;\longrightarrow\; \exp(ig_+\phi)\psi_R\,. \tag{26}$$

Proceeding similarly for $\mu = -m$, we may integrate out the right chirality, giving the on-shell Lagrangian of a massless, left-chiral Weyl fermion, invariant under the gauge transformation

$$A_\mu \;\longrightarrow\; A_\mu \;-\; \partial_\mu\phi\,, \qquad \psi_L \;\longrightarrow\; \exp(ig_-\phi)\psi_L\,. \tag{27}$$

It can also be shown (see Ref. [40]) that this behaviour persists to all loop orders. For instance, in Feynman gauge, the one-loop photon polarisation tensor is given by

$$\Pi^{\mu\nu}(p) \;=\; -\,\frac{g_V^2\,+\,g_A^2}{2\pi^2}\big(p^\mu p^\nu - \eta^{\mu\nu}p^2\big)\big(B_{21}+B_1\big) \;+\; \frac{g_A^2}{\pi^2}\,\eta^{\mu\nu}M^2B_0\,, \tag{28}$$

where $B_{0;1;21} \equiv B_{0;1;21}(p,M,M)$ are the Passarino-Veltman scalar form factors [43]. Thus,

$$p_\mu\Pi^{\mu\nu}(p) \;=\; \frac{g_A^2}{\pi^2}\,p^\nu M^2B_0 \underset{\mu\,\to\,\pm m}{\longrightarrow} 0\,, \tag{29}$$

and the Ward-Takahashi identity is recovered. Moreover, the right-left and left-right components of the one-loop vertex correction take the forms

$$\Lambda^\mu_{RL} \;=\; \frac{g_+g_-}{4\pi^2}\,m_-\,\Big[(g_++g_-)\big(p^\mu C_{11}+q^\mu C_{12}\big) \;+\; g_-(p^\mu+q^\mu)C_0\Big]\,, \tag{30a}$$

$$\Lambda^\mu_{LR} \;=\; \frac{g_+g_-}{4\pi^2}\,m_+\,\Big[(g_++g_-)\big(p^\mu C_{11}+q^\mu C_{12}\big) \;+\; g_+(p^\mu+q^\mu)C_0\Big]\,, \tag{30b}$$

again in terms of scalar form factors $C_{0;11;12} \equiv C_{0;11;12}(p,q,M,0,M)$ [43]. We see that the vertex corrections vanish, respectively, in the limits $m = +\mu$ and $m = -\mu$, such that the structure of the mass matrix is preserved at the loop level.

The behaviour of this theory is summarised graphically in Fig. 1. By varying the anti-Hermitian mass term μ, we can move smoothly from a massless, left-chiral Weyl theory (at $\mu = -m$) to a massless, right-chiral Weyl theory (at $\mu = +m$), recovering a massive Dirac fermion at $\mu = 0$. For $-m \le \mu < 0$, the $U(1)$ current is dominated by the left chirality; for $0 < \mu \le +m$, the right chirality dominates. It is tempting to entertain the possibility that the light neutrinos of the SM might be described by a generalisation of the former case to the Higgs-Yukawa theory, and we will do so in the next section.

IOP Conf. Series: Journal of Physics: Conf. Series **873** (2017) 012047 doi:10.1088/1742-6596/873/1/012047

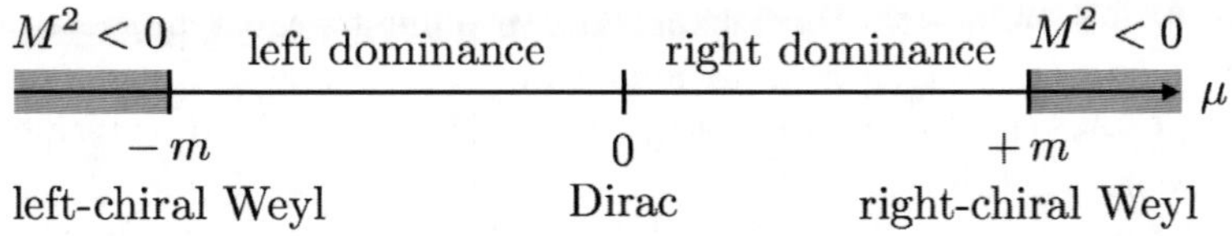

Figure 1: Schematic representation of the phases of the non-Hermitian models considered, indicating the continuum of theories that range from a massless, left-chiral Weyl theory (at $\mu = -m$) to a massless, right-chiral Weyl theory (at $\mu = +m$). The shaded regions indicate the tachyonic regimes, where $|\mu| > |m|$, giving $M^2 < 0$.

3. Non-Hermitian Higgs-Yukawa theory

Inspired by the behaviour of the non-Hermitian $U(1)$ theory described in the previous section, we now turn our attention to a one-generation, chiral Higgs-Yukawa theory with non-Hermitian Yukawa couplings:

$$\mathcal{L} = \bar{L}_L i \not{D} L_L + \bar{\nu}_R i \not{\partial} \nu_R - h_- \bar{L}_L \tilde{\phi} \nu_R - h_+ \bar{\nu}_R \tilde{\phi}^\dagger L_L . \tag{31}$$

Here, $L_L = (\nu_L \; e_L)^\mathsf{T}$ is the charged-lepton doublet, ν_R is a right-handed singlet, $\tilde{\phi} = i\sigma_2 \phi^*$ is the isospin conjugate of the Higgs doublet and D_μ is the covariant derivative of the electroweak gauge groups of the SM.

We can emulate the behaviour of the non-Hermitian $U(1)$ theory of the preceding section by taking the non-Hermitian Yukawa couplings h_+ and h_- to have the structure

$$h_\pm \equiv h \pm \eta \in \mathbb{R} . \tag{32}$$

After spontaneous symmetry breaking and working in unitary gauge, the Higgs doublet and its isospin conjugate take the forms

$$\phi = \frac{1}{\sqrt{2}} \begin{pmatrix} 0 \\ v + H \end{pmatrix} , \qquad \tilde{\phi} = \frac{1}{\sqrt{2}} \begin{pmatrix} v + H \\ 0 \end{pmatrix} , \tag{33}$$

where $v \sim 246$ GeV is the vacuum expectation value of the Higgs field. Neglecting the Z- and Higgs-boson (H) couplings, the neutrino sector of the Lagrangian in Eq. (31) becomes

$$\mathcal{L} \supset \bar{\nu}_L i \not{\partial} \nu_L + \bar{\nu}_R i \not{\partial} \nu_R - m_- \bar{\nu}_L \nu_R - m_+ \bar{\nu}_R \nu_L , \tag{34}$$

where

$$m_\pm \equiv m \pm \mu = \frac{v}{\sqrt{2}} (h \pm \eta) . \tag{35}$$

The non-Hermitian Yukawa couplings have given rise to Hermitian ($m = vh/\sqrt{2}$) and anti-Hermitian ($\mu = v\eta/\sqrt{2}$) masses that are analogous to those of the model in Eq. (7). For aesthetic reasons, we have changed $\mu \to -\mu$ relative to the preceding section. The resulting neutrino mass matrix has eigenvalues

$$\pm M = \pm \frac{v}{\sqrt{2}} \sqrt{h^2 - \eta^2} , \tag{36}$$

and we see that the neutrino becomes massless in the limit $h \to \pm\eta$. In fact, for $h \to \eta$, we obtain the original SM (with one generation): a theory of massless, left-handed neutrinos. On the other hand, for $\eta \lesssim h$ ($\eta > 0$),[2] we can obtain an arbitrarily small neutrino mass with the

[2] We correct an erroneous inequality appearing below Eq. (4.11) in Ref. [40], which should instead read $0 < \eta < h$.

neutrino current still dominated by the left-chiral component. Most interestingly, these small masses can be generated for Hermitian and non-Hermitian Yukawa couplings that are both of order unity. We emphasise that the masses of the left- and right-handed neutrinos are degenerate at tree level in this construction.

For a single generation, we are restricted to having real-valued Yukawa couplings. To see this, let us suppose that

$$-\mathcal{L} \supset h_-\bar{L}_L\tilde{\phi}\nu_R + h_+^*\bar{\nu}_R\tilde{\phi}^\dagger L_L , \qquad h_\pm \in \mathbb{C} . \tag{37}$$

The mass eigenvalues are the roots of

$$M^2 = \frac{v^2}{2}\Big(|h|^2 - |\eta|^2 - 2i\mathrm{Im}\, h^*\eta\Big) , \tag{38}$$

which can be real and nonzero iff h and η are real, distinct and nonzero.[3] This is no longer the case if we extend to multiple flavours.

In the presence of more than one generation, the Yukawa couplings in Eq. (31) are promoted to matrices in flavour space, and the relevant part of the Lagrangian becomes

$$\mathcal{L} \supset \bar{L}_{L,k}i\not{D}L_{L,k} + \bar{\nu}_{R,\alpha}i\not{\partial}\nu_{R,\alpha} - [h_-]_{k\alpha}\bar{L}_{L,k}\tilde{\phi}\nu_{R,\alpha} - [h_+^*]_{k\alpha}\bar{\nu}_{R,\alpha}\tilde{\phi}^\dagger L_{L,k} . \tag{39}$$

The left-handed and right-handed flavours are indexed by the lower-case Roman character k and lower-case Greek character α, respectively. If we take N left- and N right-handed flavours, the left-handed doublet $L_{L,k}$ and the right-handed singlet $\nu_{R,\alpha}$ transform in the fundamental representations of two flavour groups $U_L(N)$ and $U_R(N)$.

For the case of two flavours ($N = 2$) and after spontaneous symmetry breaking, the mass eigenvalues are given by the roots of

$$M^2_{1(2)} = \frac{v^2}{4}\bigg[\mathrm{tr}\,\boldsymbol{h}_+^\dagger\boldsymbol{h}_- - (+)\Big(2\,\mathrm{tr}\big(\boldsymbol{h}^\dagger\boldsymbol{h}_-\big)^2 - \big(\mathrm{tr}\,\boldsymbol{h}_+^\dagger\boldsymbol{h}_-\big)^2\Big)^{1/2}\bigg] . \tag{40}$$

It is clear that if $\boldsymbol{h} = \pm\boldsymbol{\eta}$, we obtain a massless spectrum. Instead, if $\det\boldsymbol{h}_+^\dagger\boldsymbol{h}_- = 0$, we obtain one massless ($M_1^2 = 0$) and one massive state:

$$M_2^2 = \frac{v^2}{2}\,\mathrm{tr}\,\boldsymbol{h}_+^\dagger\boldsymbol{h}_- = \frac{v^2}{2}\Big[\mathrm{tr}\,\boldsymbol{h}^\dagger\boldsymbol{h} - \mathrm{tr}\,\boldsymbol{\eta}^\dagger\boldsymbol{\eta} - 2i\,\mathrm{Im}\,\mathrm{tr}\,\boldsymbol{h}^\dagger\boldsymbol{\eta}\Big] . \tag{41}$$

The mass is real and non-tachyonic so long as $\mathrm{Im}\,\mathrm{tr}\,\boldsymbol{h}^\dagger\boldsymbol{\eta} = 0$ and $\mathrm{tr}\,\boldsymbol{h}^\dagger\boldsymbol{h} > \mathrm{tr}\,\boldsymbol{\eta}^\dagger\boldsymbol{\eta}$. In this two-flavour case and by tuning the Yukawa couplings so that $\mathrm{tr}\,\boldsymbol{h}^\dagger\boldsymbol{h} \gtrsim \mathrm{tr}\,\boldsymbol{\eta}^\dagger\boldsymbol{\eta}$, the mass-splitting $\Delta M^2 = M_2^2 - M_1^2$ can be made arbitrarily small, whilst at the same time maintaining Hermitian and non-Hermitian Yukawa couplings of order unity.

We have so far considered only the Dirac masses. However, we are not precluded from adding to the theory additional Majorana mass terms of the form

$$-\mathcal{L} \supset \frac{1}{2}\bar{\nu}^{\mathcal{C}}_{R,\alpha}m_{R,\alpha\beta}\nu_{R,\beta} + \mathrm{H.c.} . \tag{42}$$

By including such terms and block diagonalising the full mass matrix, we obtain the non-Hermitian generalisation of the see-saw formula

$$\boldsymbol{m}_L = -\boldsymbol{m}_-\boldsymbol{m}_R^{-1}\boldsymbol{m}_+^{\mathsf{T}} . \tag{43}$$

[3] We correct an erroneous factor of 1/2 in Eq. (4.15) of Ref. [40].

For $N = 2$, the mass eigenvalues are

$$M_{1(2)} = -\frac{v^2}{4}\Big[\operatorname{tr}\boldsymbol{h}_-\boldsymbol{m}_R^{-1}\boldsymbol{h}_+^{\mathsf{T}} - (+)\Big(2\operatorname{tr}\big(\boldsymbol{h}_-\boldsymbol{m}_R^{-1}\boldsymbol{h}_+^{\mathsf{T}}\big)^2 - \big(\operatorname{tr}\boldsymbol{h}_-\boldsymbol{m}_R^{-1}\boldsymbol{h}_+^{\mathsf{T}}\big)^2\Big)^{1/2}\Big], \tag{44}$$

and we trivially obtain a massless spectrum for $\boldsymbol{h} = \pm\boldsymbol{\eta}$. Instead, if

$$\det\boldsymbol{h}_-\boldsymbol{m}_R^{-1}\boldsymbol{h}_+^{\mathsf{T}} = 0, \tag{45}$$

we obtain the spectrum

$$M_1 = 0, \qquad M_2 = -\frac{v^2}{2}\operatorname{tr}\boldsymbol{h}_-\boldsymbol{m}_R^{-1}\boldsymbol{h}_+^{\mathsf{T}}, \tag{46}$$

which is real so long as

$$\operatorname{Im}\operatorname{tr}\boldsymbol{h}_-\boldsymbol{m}_R^{-1}\boldsymbol{h}_+^{\mathsf{T}} = 0. \tag{47}$$

We can now obtain an arbitrarily small but finite mass splitting ΔM^2 by arranging for

$$\operatorname{Re}\operatorname{tr}\boldsymbol{h}\boldsymbol{m}_R^{-1}\boldsymbol{h}^{\mathsf{T}} \gtrsim \operatorname{Re}\operatorname{tr}\Big(\boldsymbol{\eta}\boldsymbol{m}_R^{-1}\boldsymbol{\eta}^{\mathsf{T}} + \boldsymbol{h}\boldsymbol{m}_R^{-1}\boldsymbol{\eta}^{\mathsf{T}} - \boldsymbol{\eta}\boldsymbol{m}_R^{-1}\boldsymbol{h}^{\mathsf{T}}\Big), \tag{48}$$

and this mass splitting may be suppressed independent of the scale of the Majorana masses.

The anti-Hermitian terms are $\mathcal{CP}$ odd, and one might naively anticipate additional sources of $\mathcal{CP}$ violation. With the inclusion of the lepton number violating Majorana mass terms, it would therefore be interesting to consider this model in the context of leptogenesis and the generation of the Baryon Asymmetry of the Universe. Furthermore, the ability to suppress the neutrino masses independent of the Majorana mass scale may allow the see-saw scale to be lowered, and a full analysis of the constraints on the non-Hermitian Yukawa couplings from neutrino oscillation data and the current limits on lepton-flavour and lepton-number violating observables, including neutrinoless double beta decay, is warranted (for a review, see Ref. [14]). It would also be of interest to consider analogous non-Hermitian extensions of left-right symmetric models [44, 45, 46]. Finally, we remark that the consequences of the anti-Hermitian Yukawa couplings for the behaviour of the Higgs sector in the presence of fermion condensates remains to be studied.

4. Concluding remarks

We have described a $U(1)$ gauge theory coupled to a fermion with an anti-Hermitian, parity-violating mass term. We have shown that this model actually encompasses a continuum of theories, ranging from a massless left-Weyl theory to a massless right-Weyl theory, and we recover a massive Dirac fermion in the Hermitian limit. Moreover, we have illustrated that the behaviour of this theory is stable under loop corrections.

By constructing an analogous Higgs-Yukawa model, we have described how the presence of Hermitian and non-Hermitian Yukawa couplings of order unity can give rise to small neutrino masses, whilst at the same time leading to a suppression of the right-chiral component of the neutrino current. In and of itself, this model would not alleviate the need for fine-tuning in the SM, since we must arrange for the Hermitian and anti-Hermitian Yukawa couplings to differ only by one part in 10^{12}. However, the small mass splitting needed to fit the observed neutrino spectrum could reasonably arise, for instance, through some radiative breaking of a high-scale degeneracy of the Hermitian and non-Hermitian Yukawa couplings. This point may be addressed in further work.[4] With this goal in mind, it would be interesting to consider ways of generating the anti-Hermitian mass term, for example, through non-perturbative effects (see, e.g., Refs. [27, 28, 25]).

[4] Particular thanks go to Andreas Trautner for emphasising the need to explain this relative fine-tuning of the Hermitian and non-Hermitian Yukawa couplings, as well as to Nikolaos Mavromatos for his constructive and encouraging remarks on this point.

Acknowledgments
The work of PM is supported by STFC grant ST/L000393/1. PM would like to thank the organisers of DISCRETE2016 for their hospitality and the opportunity to present this work, and the participants for their questions, comments and suggestions.

References
[1] Nakamura K and Petcov S T 2016 In *Review of Particle Physics* [Particle Data Group] *Chin. Phys.* C **40** 100001 pp 246–78
[2] Ade P A R *et al.* [Planck Collaboration] 2016 *Astron. Astrophys.* **594** A13
[3] Weinberg S 1979 *Phys. Rev. Lett.* **43** 1566–70
[4] Minkowski P 1977 *Phys. Lett.* B **67** 421–8
[5] Gell-Mann M, Ramond P and Slansky R 1979 *Supergravity* ed van Nieuwenhuizen P and Freedman D Z (Amsterdam, Netherlands: North-Holland) pp 315–21
[6] Yanagida T 1979 *Proc of the Workshop on Unified Theories and the Baryon Number of the Universe* ed Sawada O and Sugamoto A (Tsukuba, Japan: KEK) pp 95–9
[7] Mohapatra R N and Senjanović G 1980 *Phys. Rev. Lett.* **44** 912–5
[8] Schechter J and Valle J W F 1980 *Phys. Rev.* D **22** 2227–35
[9] Magg M and Wetterich C 1980 *Phys. Lett.* B **94** 61–4
[10] Cheng T P and Li L-F 1980 *Phys. Rev.* D **22** 2860–8
[11] Lazarides G, Shafi Q and Wetterich C 1981 *Nucl. Phys.* B **181** 287–300
[12] Mohapatra R N and Senjanović G 1981 *Phys. Rev.* D **23** 165–80
[13] Foot R, Lew H, He X-G and Joshi G C 1989 *Z. Phys.* C **44** 441–4
[14] Deppisch F F, Dev P S B and Pilaftsis A 2015 *New J. Phys.* **17** 075019
[15] Pilaftsis A 2013 *Mod. Phys. Lett.* A **28** 1350083
[16] Mavromatos N E and Pilaftsis A 2012 *Phys. Rev.* D **86** 124038
[17] Fukugita M and Yanagida T 1986 *Phys. Lett.* B **174** (1986) 45–7
[18] Kuzmin V A, Rubakov V A and Shaposhnikov M E 1985 *Phys. Lett.* **155B** 36–42
[19] Pilaftsis A 1999 *Int. J. Mod. Phys.* A **14** 1811–58
[20] Buchmüller W, Peccei R D and Yanagida T 2005 *Ann. Rev. Nucl. Part. Sci.* **55** 311–355
[21] Davidson S, Nardi E and Nir Y 2008 *Phys. Rept.* **466** 105–77
[22] Blanchet S and Di Bari P 2012 *New J. Phys.* **14** 125012
[23] Barenboim G, Borissov L, Lykken J D and Smirnov A Y 2002 *J. High Energy Phys.* JHEP10(2002)001
[24] Alexandre J and Vergou A 2011 *Phys. Rev.* D **83** 125008
[25] Alexandre J, Brister J and Houston N 2012 *Phys. Rev.* D **86** 025030
[26] Mavromatos N E 2011 *Phys. Rev.* D **83** 025018
[27] Alexandre J, Leite J and Mavromatos N E 2013 *Phys. Rev.* D **87** 125029
[28] Alexandre J, Leite J and Mavromatos N E 2014 *Phys. Rev.* D **90** 045026
[29] Ohlsson T 2016 *Europhys. Lett.* **113** 61001
[30] Benatti F and Floreanini R 2001 *Phys. Rev.* D **64** 085015
[31] Bender C M, Brody D C and Jones H F 2002 *Phys. Rev. Lett.* **89** 270401 [Erratum: 2004 *Phys. Rev. Lett.* **92** 119902]
[32] Bender C M 2005 *Contemp. Phys.* **46** 277–92
[33] Bender C M and Wu T T 1969 *Phys. Rev.* **184** 1231–60
[34] Bender C M and Milton K A 1999 *J. Phys.* A: *Math. Gen.* **32** L87–92
[35] Milton K A 2004 *Czech. J. Phys.* **54** 85–91
[36] Bender C M, Cavero-Pelaez I, Milton K A and Shajesh K V 2005 *Phys. Lett.* B **613** 97–104
[37] Bender C M, Jones H F and Rivers R J 2005 *Phys. Lett.* B **625** 333–40
[38] Alexandre J and Bender C M 2015 *J. Phys.* A: *Math. Theor.* **48** 185403
[39] Jones-Smith K and Mathur H 2014 *Phys. Rev.* D **89** 125014
[40] Alexandre J, Bender C M and Millington P 2015 *J. High Energy Phys.* JHEP11(2015)111
[41] Debnath U, Mukhopadhyay B and Dadhich N 2006 *Mod. Phys. Lett.* A **21** 399–408
[42] de Cesare M, Mavromatos N E and Sarkar S 2015 *Eur. Phys. J.* C **75** 514
[43] Passarino G and Veltman M J G 1979 *Nucl. Phys.* B **160** 151–207
[44] Mohapatra R N and Pati J C 1975 *Phys. Rev.* D **11** 566–71
[45] Mohapatra R N and Pati J C 1975 *Phys. Rev.* D **11** 2558–61
[46] Senjanović G and Mohapatra R N 1975 *Phys. Rev.* D **12** 1502–5

IOP Conf. Series: Journal of Physics: Conf. Series **873** (2017) 012048 doi:10.1088/1742-6596/873/1/012048

Prospects for 2HDM charged Higgs searches

M Krawczyk[1], **S Moretti**[2], **P Osland**[3], **GM Pruna**[4], **R Santos**[5,6]

[1] Faculty of Physics, University of Warsaw, Pasteura 5, 02-093 Warsaw, Poland
[2] School of Physics and Astronomy, University of Southampton, Highfield, Southampton SO17 1BJ, United Kingdom
[3] Department of Physics and Technology, University of Bergen, Postboks 7803, N-5020 Bergen, Norway
[4] Paul Scherrer Institute, CH-5232 Villigen PSI, Switzerland
[5] Centro de Física Teórica e Computacional, Faculdade de Ciências, Universidade de Lisboa, Campo Grande, Edifício C8 1749-016 Lisboa, Portugal,
[6] Instituto Superior de Engenharia de Lisboa - ISEL, 1959-007 Lisboa, Portugal

E-mail: Maria.Krawczyk@fuw.edu.pl, S.Moretti@soton.ac.uk, Per.Osland@uib.no, giovanni-marco.pruna@psi.ch, rasantos@fc.ul.pt

Abstract. We discuss the prospects for charged Higgs boson searches at the LHC, within the two-Higgs-doublet models (2HDM). The 2HDM is generally less constrained than the corresponding sector of the MSSM, but there are still severe theoretical and experimental constraints that already exclude significant regions of the naive parameter space. Explicit searches in the $H^+ \to \tau^+\nu$ and $H^+ \to t\bar{b}$ channels are further restricting parts of the 2HDM parameter space.

1. Introduction

The discovery of the Higgs particle [1, 2] raises the obvious question of whether there are more Higgs-like particles. While the simplicity of the minimal Standard Model is very attractive, there are well-known issues that remain to be understood. So far, searches for additional neutral Higgs bosons have found nothing up to a mass of the order of 1 TeV (see, for example, Ref. [3]).

The charged-Higgs sector of the 2HDM has many features in common with the MSSM, but it is a priori less constrained. The mass is essentially unknown, but constrained relative to other parameters of the model. Compared to the MSSM, there is more room for a wide range of masses. In particular, the charged Higgs boson mass could a priori be rather different from the neutral Higgs masses.

In the next section we shall introduce some basic notation, and review the most important constraints. Then, in the following sections we will separately discuss three mass ranges. In section 3 we discuss the light-mass case, where a top quark may decay to a charged Higgs boson, $t \to H^+b$. In section 4 we discuss the intermediate mass region, where the charged Higgs boson is too heavy to be produced via top decay, but where the Model II is excluded by the $b \to X_s\gamma$ constraint, up to of the order of 480 GeV [4][1]. In section 5 we consider still higher masses, where the $b \to X_s\gamma$ constraint does not apply. For a complementary review, see Ref. [6].

[1] A recent study concludes that this limit is even higher, at 570 GeV [5].

IOP Conf. Series: Journal of Physics: Conf. Series **873** (2017) 012048 doi:10.1088/1742-6596/873/1/012048

2. Model review

The 2HDM is a simple extension of the SM, described in detail in Refs. [7, 8].

2.1. Notations

We decompose the SU(2) doublets as follows:

$$\Phi_a = \begin{pmatrix} \varphi_a^+ \\ (v_a + \eta_a + i\chi_a)/\sqrt{2} \end{pmatrix}, \quad a = 1,2 \tag{2.1}$$

with the vacuum expectation value (vev) v_a real. The ratio between these is denoted $\tan\beta = v_2/v_1$, and plays an important role in the parametrization of the model.

The potential is written as

$$\begin{aligned} V(\Phi_1,\Phi_2) &= -\frac{1}{2}\left\{m_{11}^2\Phi_1^\dagger\Phi_1 + m_{22}^2\Phi_2^\dagger\Phi_2 + \left[m_{12}^2\Phi_1^\dagger\Phi_2 + \text{H.c.}\right]\right\} \\ &+ \frac{\lambda_1}{2}(\Phi_1^\dagger\Phi_1)^2 + \frac{\lambda_2}{2}(\Phi_2^\dagger\Phi_2)^2 + \lambda_3(\Phi_1^\dagger\Phi_1)(\Phi_2^\dagger\Phi_2) + \lambda_4(\Phi_1^\dagger\Phi_2)(\Phi_2^\dagger\Phi_1) \\ &+ \frac{1}{2}\left[\lambda_5(\Phi_1^\dagger\Phi_2)^2 + \text{H.c.}\right] + \left\{\left[\lambda_6(\Phi_1^\dagger\Phi_1) + \lambda_7(\Phi_2^\dagger\Phi_2)\right](\Phi_1^\dagger\Phi_2) + \text{H.c.}\right\}. \end{aligned} \tag{2.2}$$

In many studies, the terms with λ_6 and λ_7 are left out, in order to control flavour-changing neutral interactions.

The spectrum consists of three neutral Higgs boson (the lightest one is here assumed to be the discovered one, at 125 GeV), and a charged pair, $H^\pm$. If CP is conserved, there are two CP-even ones (h and H, arising from η_1 and η_2 in Eq. (2.1)), and an odd one (A, arising, together with the neutral Goldstone boson, from χ_1 and χ_2). If, on the other hand CP is not conserved, they will mix to mass eigenstates H_1, H_2 and H_3, with H_1 being the lightest one.

The Yukawa interactions can be written as

$$-\mathcal{L}_{\text{Yukawa}} = \overline{Q}_L\Phi_a F_a^D D_R + \overline{Q}_L\widetilde{\Phi}_a F_a^U U_R + \overline{L}_L\Phi_a F_a^L L_R + \text{H.c.}, \quad a=1,2 \tag{2.3}$$

where $\widetilde{\Phi}_a = i\sigma_2\Phi^*$ and Q_L and L_L are left-handed quark and lepton doublets, whereas D_R, U_R and L_R are the corresponding right-handed fields.

Different choices for the F_a define different "Models": Model I is defined by Φ_2 (and only Φ_2) coupling to all fermions, whereas in Model II the d-type quarks and charged leptons couple exclusively to Φ_1, whereas u-type quarks couple exclusively to Φ_2. This is similar to the structure of the MSSM [9]. These coefficients F_a are proportional to the quark (and lepton) masses, as required by the Higgs mechanism. The implications for the charged-Higgs coupling are that they will consist of two parts, one with a left-handed and the other with a right-handed projector, one proportional to the down-type mass, the other to the up-type mass. The different models give different combinations of projectors and mass factors. For example, for Model II and third-generation quarks we have (with all fields incoming)

$$\begin{aligned} H^+b\bar{t}: &\quad \frac{ig}{2\sqrt{2}\,m_W}V_{tb}[m_b(1+\gamma_5)\tan\beta + m_t(1-\gamma_5)\cot\beta], \\ H^-t\bar{b}: &\quad \frac{ig}{2\sqrt{2}\,m_W}V_{tb}^*[m_b(1-\gamma_5)\tan\beta + m_t(1+\gamma_5)\cot\beta], \end{aligned} \tag{2.4}$$

where g is the SU(2) coupling and V the CKM matrix.

There are also Models X and Y, which are basically variants of Models I and II, with different leptonic couplings.

IOP Conf. Series: Journal of Physics: Conf. Series **873** (2017) 012048 doi:10.1088/1742-6596/873/1/012048

2.2. Important constraints

A recent summary of theoretical and experimental constraints can be found in Ref. [10]. The theoretical constraints include positivity, unitarity and perturbativity, whereas the experimental ones arise from low-energy phenomena, such as various B-meson decays where the existence of a charged Higgs boson would contribute additional effects, as well as precision measurements at LEP constraining radiative corrections related to the electroweak bosons W and Z.

Among the constraints coming from B-meson decay, the $B \to X_s\gamma$ decay, which is experimentally well measured and in the SM dominated by a one-loop diagram involving the W^+ will in the 2HDM Model II also get a contribution from H^+ exchange. Careful analyses of this effect have led to a mass bound $M_{H^\pm} \gtrsim 480$ GeV [4] (recently updated to $M_{H^\pm} \gtrsim 570$ GeV [5]), largely independent of $\tan\beta$.

The relevant LEP precision data, derived from the vacuum polarization, or radiative corrections to the W and Z self-energies are usually quantified in terms of the parameters S, T and U [11, 12, 13]. If the charged Higgs is heavy, the smallness of T [14] imposes strong constraints on the allowed splitting between $M_{H^\pm}$, M_H and M_A [15, 16].

3. The low-mass region, $M_{H^\pm} \lesssim m_t$

In the low-mass region, where the H^+ can be produced via t decay, Model II and Model Y are excluded by the $B \to X_s\gamma$ constraint. For Model I, all Yukawa couplings are proportional to $\cot\beta$. Thus, the sensitivity is larger at low $\tan\beta$.

The LHC experiments have already set very strong bounds on the branching ratio for $t \to H^+b$, assuming the decay $H^+ \to \tau^+\nu$ [17, 18]. These bounds effectively exclude some range of low values of $\tan\beta$, as illustrated in Fig. 1, where the region *above* the dashed bars is excluded by the two experiments at the 95% CL.

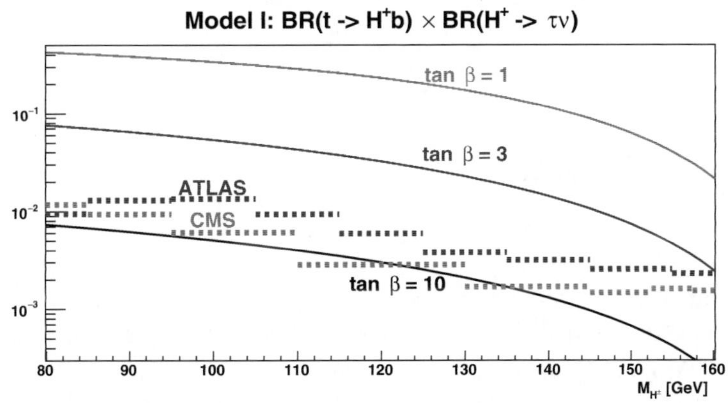

Figure 1. Product of branching ratios, $\text{BR}(t \to H^+b) \times \text{BR}(H^+ \to \tau^+\nu)$, for Model I, and three values of $\tan\beta$, as indicated (solid). The figure is adapted from Ref. [10], showing also 8 TeV exclusion limits from ATLAS [17] (dashed, blue) and CMS [18] (dashed, red).

It is seen that in the low-mass region, only Model I with high values of $\tan\beta$, $\tan\beta \gtrsim \mathcal{O}(5)$ is still allowed.

Model X is like Model I in its couplings to quarks, and would thus have the same branching ratio for $t \to H^+b$. However, the leptonic coupling is proportional to $\tan\beta$ rather than $\cot\beta$. Thus, at low values of $\tan\beta$ the exclusion in Model X is somewhat less severe than in Model I.

In this mass region, assuming the $H^+ \to W^+A/H$ and $H^+ \to W^+h$ rates to be negligible (due to phase space and couplings, respectively), we can approximate the $H^+ \to \tau^+\nu$ branching ratio as

$$\mathrm{BR}(H^+ \to \tau^+\nu) \simeq \frac{\Gamma(H^+ \to \tau^+\nu)}{\Gamma(H^+ \to \tau^+\nu) + \Gamma(H^+ \to c\bar{s})}. \tag{3.1}$$

In Model I, with both rates proportional to $\cot^2\beta$, the branching ratio is independent of $\tan\beta$ and of the order of 0.7 [10], whereas in Model X the rate $\Gamma(H^+ \to \tau^+\nu)$ is instead proportional to $\tan^2\beta$ and the branching ratio quickly approaches unity. Thus, the deterioration in exclusion reach from Model I to Model X is rather modest.

In the decay of $t \to H^+b$, one has also searched for $H^+ \to c\bar{s}$ [19], but the Model I branching ratio is lower by a factor $\sim 3/7$ and this channel is not compatible with the $\tau\nu$ channel.

4. The intermediate-mass region, $m_t \lesssim M_{H^\pm} \lesssim 480$ GeV

Also in the intermediate-mass region, Model II and Model Y are excluded by the $B \to X_s\gamma$ constraint. But in contrast to the low-mass region, the charged Higgs boson can not be produced via top decay. A variety of production mechanisms are involved. In the 5-flavour scheme[2], where b quarks are considered parts of the proton and have non-zero distribution functions, the relevant partonic processes are

$$g\bar{b} \to H^+\bar{t}, \quad gg \to H_j \to H^+W^-. \tag{4.1}$$

Two comments are here in order:

- As indicated, the production can proceed via a neutral Higgs boson in the s-channel. We recall the relevant couplings (all fields incoming)

$$\begin{aligned} H^\mp W^\pm h: &\quad \frac{\mp ig}{2}\cos(\beta-\alpha)(p_\mu - p_\mu^\mp), \\ H^\mp W^\pm H: &\quad \frac{\pm ig}{2}\sin(\beta-\alpha)(p_\mu - p_\mu^\mp), \\ H^\mp W^\pm A: &\quad \frac{g}{2}(p_\mu - p_\mu^\mp), \end{aligned} \tag{4.2}$$

  where p_μ and $p_\mu^\mp$ refer to the momenta of the neutral and charged scalars. Experimental data [3] favour the alignment limit, where

$$\cos(\beta-\alpha) \to 0. \tag{4.3}$$

  Thus, in this limit only the heavier Higgs bosons (H and A) can play a role here. Allowing for CP non-conservation (considering H_1 rather than h) does little to modify this conclusion.
- In Model I (and Model X) all couplings of the H^+ to quarks are proportional to $\cot\beta$. Thus, the production cross section decreases with increasing values of $\tan\beta$.
  This is different from the MSSM, which has a Model II structure, and where the cross section has a minimum for intermediate values of $\tan\beta$, of the order of $\sqrt{m_t/m_b}$.

As suggested by Eq. (4.1), the kinematics may allow for s-channel resonant effects in the production [21]. However, the smallness of T does not allow for a large mass splitting between $H^\pm$, H and A. This restricts the possible resonant enhancement.

Likewise, there are various possible decay modes:

$$H^+ \to \tau^+\nu, \quad H^+ \to W^+h, \quad (\text{or } H^+ \to W^+H_1), \quad H^+ \to t\bar{b} \tag{4.4}$$

[2] For a complete discussion on the flavour scheme choice in inclusive charged Higgs production associated with fermions see IV.3.2 of [20] and references therein.

As mentioned above, the $H^+ \to W^+h$ vanishes in the alignment limit. For the more general case, allowing for CP violation, the relevant coupling is given by

$$H^{\mp}W^{\pm}H_1 : \qquad \frac{g}{2}[\pm i\cos\alpha_2\sin(\beta-\alpha_1)+\sin\alpha_2](p_\mu - p_\mu^{\mp}). \tag{4.5}$$

Here, $\alpha_1 = \alpha + \pi/2$ and α_2 are two of the rotation angles describing mixing of the neutral Higgs fields [22]. A non-zero value of α_2 would indicate CP violation.

The decay $H^+ \to W^+h$ (or $H^+ \to W^+H_1$) requires some deviation from the alignment limit, whereas $H^+ \to W^+A$ and $H^+ \to W^+A$ are limited by phase space. Thus, in practice, the channels $H^+ \to \tau^+\nu$ and $H^+ \to t\bar{b}$ are the most relevant ones, with the latter dominating by a $\tan\beta$-independent factor of the order of $(m_t/m_\tau)^2$.

4.1. The $\tau\nu$ channel

The tau channel benefits from being rather "clean", and interesting limits have been obtained. Considering as an example, the case $M_{H^\pm} = 300$ GeV, we note that both ATLAS and CMS obtain a 95% CL bound $\sigma\cdot\mathrm{BR}(H^+ \to \tau^+\nu) \leq (0.16-0.17)$ pb at 8 TeV [17, 18]. A recent ATLAS result at 13 TeV is 0.41 pb.[3] This is to be compared with the model expectation for the cross section and the branching ratio. Approximate 14 TeV cross sections are quoted in Table 1 for a few $\tan\beta$ values. Accurate values depend on the masses of the heavier neutral states. Barring any significant decay to W^+h, the relevant branching ratio is given by $(m_\tau/m_t)^2 \sim 10^{-4}$. It is seen that the theoretical expectation is far below the current experimental limit.

Table 1. Experimental limits on $\sigma\cdot\mathrm{BR}(H^+ \to \tau^+\nu)$ [pb] at 8 and 13 TeV together with Model I cross sections [10] at 14 TeV, multiplied by a branching ratio taken to be 10^{-4}. Three intermediate values of $M_{H^\pm}$ are considered, and for the theory cross sections we take $(M_2, M_3) = (500, 600)$ GeV.

	200 GeV	300 GeV	400 GeV
ATLAS [17], 8 TeV	0.47	0.17	0.061
CMS [18], 8 TeV	0.36	0.16	0.054
ATLAS [23], 13 TeV	1.83	0.41	0.20
$\tan\beta = 1$, 14 TeV	1.18×10^{-3}	0.646×10^{-3}	0.287×10^{-3}
$\tan\beta = 3$, 14 TeV	1.41×10^{-4}	0.882×10^{-4}	0.514×10^{-4}
$\tan\beta = 30$, 14 TeV	1.42×10^{-6}	0.91×10^{-6}	0.72×10^{-6}

Furthermore, as long as the relevant search region is around $\tan\beta = 1$, there is no significant difference between Model I and Model X, so Model X is no more and no less affected by the experimental bounds on this channel than Model I.

4.2. The $t\bar{b}$ channel

While the $H^+ \to t\bar{b}$ channel suffers from high QCD background rates, sophisticated analyses have started to constrain these models. 8 TeV exclusion limits from CMS are shown in Fig 2 and reproduced in Table 2, together with preliminary ATLAS results at 13 TeV and theoretical expectations for three values of $\tan\beta$. For this channel, assuming $\mathrm{BR}(H^+ \to t\bar{b}) = 100\%$, we find that low values of $\tan\beta$ are excluded. On the other hand, $\sigma(\tan\beta = 30) \sim 10^{-2}$ pb, which is well below the exclusion limit.

[3] The charged-Higgs production cross section roughly scales with the glue-glue luminosity, which in the mass range 200-400 GeV increases by factors of 4–5 from 8 to 13 TeV.

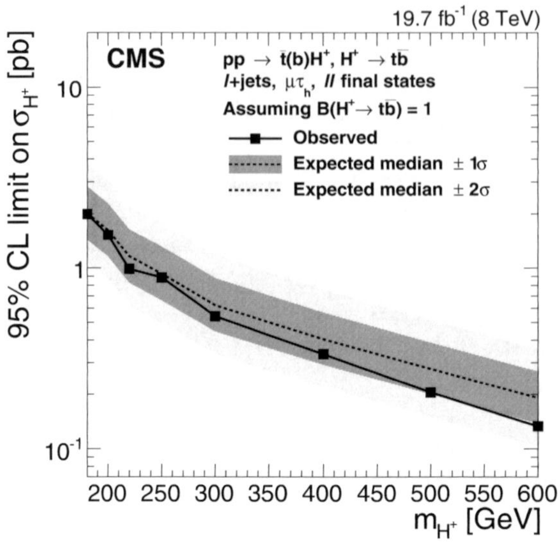

Figure 2. CMS upper limit on $\sigma(pp \to t(b)H^+)$ for the combination of the $\mu\tau_h$, $\ell+$jets and $\ell\ell'$ final states assuming $\mathrm{BR}(H^+ \to t\bar{b}) = 100\%$. [Reprinted with kind permission from JHEP and the authors, Fig. 10 of [18].]

Table 2. Experimental limits on $\sigma \cdot \mathrm{BR}(H^+ \to t\bar{b})$ [pb] together with Model I cross sections [10] multiplied by a branching ratio taken to be 1. Three intermediate values of $M_{H^\pm}$ are considered, and for the theory cross sections we take $(M_2, M_3) = (500, 600)$ GeV.

	200 GeV	300 GeV	400 GeV
CMS [18], 8 TeV	1.53	0.53	0.33
ATLAS [24], 13 TeV		1.06	1.17
$\tan\beta = 1$, 14 TeV	11.8	6.46	2.87
$\tan\beta = 3$, 14 TeV	1.41	0.882	0.514
$\tan\beta = 30$, 14 TeV	1.42×10^{-2}	0.91×10^{-2}	0.72×10^{-2}

We conclude that in this intermediate mass region the search in the $H^+ \to t\bar{b}$ is starting to constrain the model. Since a hadronic final state is considered in this search, there is no difference between Model I and Model X.

5. The high-mass region, 480 GeV $\lesssim M_{H^\pm}$

The high-mass region is distinguished by all Yukawa models being allowed, and also all decay modes listed in Eq. (4.4).

In distinction from the low- and intermediate-mass regions discussed above, we now need to consider also the Model II cross sections. Cross sections are quoted for both models in Table 3 [10]. Note that the cross sections for the four models are pairwise the same,

$$\sigma(\text{Model X}) = \sigma(\text{Model I}), \quad \text{and} \quad \sigma(\text{Model Y}) = \sigma(\text{Model II}), \tag{5.1}$$

but the decay rates would be different if lepton modes were considered. Because of the dependence on the H and A (or H_2, H_3) masses, there is some uncertainty associated with the cross sections. The main features are:

- The cross sections fall off at high masses, mainly due to falling pdf's and phase space reduction, and
- The Model I cross section falls with $\tan\beta$ approximately as $1/\tan^2\beta$, whereas the Model II cross section has a minimum around $\tan\beta = \mathcal{O}(\sqrt{m_t/m_b})$ due to the competition among the two terms in the coupling (2.4).

Table 3. Theoretical cross sections [10] for $\sqrt{s} = 14$ TeV and a range of mass values $M_{H^\pm} = 500 - 1000$ GeV, for $(M_2, M_3) = (500, 600)$ GeV, quoted in the format "Model I/Model II" [pb]. Overall factors 10^{-1} and 10^{-2} are extracted where indicated.

	500 GeV	600 GeV	700 GeV	800 GeV	900 GeV	1000 GeV
$\tan\beta = 1$	1.10/1.27	0.54/0.64	0.30/0.37	0.18/0.22	0.11/0.14	0.072/0.090
$\tan\beta = 3$ $[10^{-1}]$	1.61/1.81	0.60/0.72	0.34/0.41	0.20/0.25	0.13/0.16	0.080/0.10
$\tan\beta = 30$ $[10^{-2}]$	0.29/24.24	0.059/12.95	0.033/7.31	0.020/4.30	0.013/2.61	0.0080/1.63

5.1. The $\tau\nu$ channel

Search results from the $\tau\nu$ channel are given in Table 4. For comparison, we quote the theoretical expectation for $\tan\beta = 1$, for both Model I and Model II. It is seen that the expectation, at $\tan\beta = 1$ is lower by factors ranging from 200 to 600. For higher values of $\tan\beta$, the discrepancy would be even larger. Thus, searches in the $\tau\nu$ channel are not yet constraining the 2HDM in this mass range.

Table 4. Experimental limits on $\sigma \cdot \mathrm{BR}(H^+ \to \tau^+\nu)$ [pb] together with the theoretical expectation for $\tan\beta = 1$, in the format "Model I/Model II", approximating $\mathrm{BR}(H^+ \to \tau\nu) \simeq (m_\tau/m_t)^2 \simeq 10^{-4}$. Overall factors 10^{-2} and 10^{-4} are extracted where indicated.

	500 GeV	600 GeV	700 GeV	800 GeV	900 GeV	1000 GeV
ATLAS [17] $[10^{-2}]$, 8 TeV	2.46	1.11				0.427
ATLAS [23] $[10^{-2}]$, 13 TeV	9.25	5.15	3.35	2.36	2.10	1.80
$\tan\beta = 1$ $[10^{-4}]$, 14 TeV	1.10/1.27	0.54/0.64	0.30/0.37	0.18/0.22	0.11/0.14	0.07/0.09

5.2. The $t\bar{b}$ channel

While the $H^+ \to t\bar{b}$ channel suffers from large QCD backgrounds, the constraints obtained are more interesting. Experimental exclusion limits are given in Table 5, and can be directly compared with the theoretical expectations quoted in Table 3, approximating $\mathrm{BR}(H^+ \to t\bar{b}) = 1$. It is seen that in the range $M_{H^\pm} = 500 - 600$ GeV low values of $\tan\beta$, namely $\tan\beta = \mathcal{O}(1)$ are on the verge of being excluded for both Model I and Model II. Since no leptonic couplings are involved, this conclusion applies equally to Models X and Y.

Table 5. Experimental limits on $\sigma \cdot \mathrm{BR}(H^+ \to t\bar{b})$ [pb].

	500 GeV	600 GeV	700 GeV	800 GeV	900 GeV	1000 GeV
ATLAS [25], 8 TeV	0.68	0.24				
ATLAS [25], 8 TeV	2.89	1.42	0.70	0.37	0.24	0.15
CMS [18], 8 TeV	0.20	0.13				
ATLAS [24], 13 TeV	1.32	1.01	0.53	0.34	0.31	0.18

6. Summary

Within the 2HDM, the search for charged Higgs bosons has mainly proceeded in two channels, $H^+ \to \tau^+\nu$ and $H^+ \to t\bar{b}$. In the $\tau\nu$ channel, the searches have already excluded low values of $\tan\beta$ for Yukawa Models I and X and low masses, $M_{H^\pm} < m_t$. At higher masses, this channel is not yet competitive.

In the intermediate-mass region, $m_t < M_{H^\pm} < 480$ GeV, searches in the $t\bar{b}$ channel exclude the 2HDM at low values of $\tan\beta$, $\tan\beta \lesssim \mathcal{O}(3)$. In the high-mass region, 480 GeV $< M_{H^\pm}$, $\tan\beta$ values of order unity are also on the verge of being excluded up to about 600 GeV.

The bosonic decay modes, $H^+ \to W^+A$, $H^+ \to W^+H$ and $H^+ \to W^+h$ should also be kept in mind, though the former suffer from limited phase space, and the latter from the vanishing of the coupling in the alignment limit.

References

[1] Aad G *et al.* [ATLAS Collaboration], 2012 Observation of a new particle in the search for the Standard Model Higgs boson with the ATLAS detector at the LHC, Phys. Lett. B **716** 1 [arXiv:1207.7214 [hep-ex]].

[2] Chatrchyan S *et al.* [CMS Collaboration], 2012 Observation of a new boson at a mass of 125 GeV with the CMS experiment at the LHC, Phys. Lett. B **716** 30 [arXiv:1207.7235 [hep-ex]].

[3] Canelli F, 2016 in Proceedings of the 38th International Conference on High Energy Physics, PoS(ICHEP2016)014.

[4] Misiak M *et al.*, 2015 Updated NNLO QCD predictions for the weak radiative B-meson decays, Phys. Rev. Lett. **114** no.22, 221801 [arXiv:1503.01789 [hep-ph]].

[5] Misiak M and Steinhauser, M, 2017 Weak Radiative Decays of the B Meson and Bounds on $M_{H^\pm}$ in the Two-Higgs-Doublet Model, arXiv:1702.04571 [hep-ph].

[6] S. Moretti, arXiv:1612.02063 [hep-ph].

[7] Gunion J F, Haber H E, Kane G L and Dawson S, 2000 The Higgs Hunter's Guide, Front. Phys. **80** 1.

[8] Branco G C, Ferreira P M, Lavoura L, Rebelo M N, Sher M and Silva J P, 2012 Theory and phenomenology of two-Higgs-doublet models, Phys. Rept. **516** 1 [arXiv:1106.0034 [hep-ph]].

[9] Djouadi A, 2008 The Anatomy of electro-weak symmetry breaking. II. The Higgs bosons in the minimal supersymmetric model, Phys. Rept. **459** 1 [hep-ph/0503173].

[10] Akeroyd A G *et al.*, 2016 Prospects for charged Higgs searches at the LHC, arXiv:1607.01320 [hep-ph].

[11] Kennedy D C and Lynn B W, 1989 Electroweak Radiative Corrections with an Effective Lagrangian: Four Fermion Processes, Nucl. Phys. B **322** 1.

[12] Peskin M E and Takeuchi T, 1990 A New constraint on a strongly interacting Higgs sector, Phys. Rev. Lett. **65** 964.

[13] Altarelli G and Barbieri R, 1991 Vacuum polarization effects of new physics on electroweak processes, Phys. Lett. B **253** 161.

[14] Olive K A *et al.* [Particle Data Group], 2014 Chin. Phys. C **38** 090001.

[15] Grimus W, Lavoura L, Ogreid O M and Osland P, 2008 A Precision constraint on multi-Higgs-doublet models, J. Phys. G **35** 075001 [arXiv:0711.4022 [hep-ph]].

[16] Grimus W, Lavoura L, Ogreid O M and Osland P, 2008 The Oblique parameters in multi-Higgs-doublet models, Nucl. Phys. B **801** 81 [arXiv:0802.4353 [hep-ph]].

[17] Aad G *et al.* [ATLAS Collaboration], 2015 Search for charged Higgs bosons decaying via $H^\pm \to \tau^\pm\nu$ in fully hadronic final states using pp collision data at $\sqrt{s} = 8$ TeV with the ATLAS detector, JHEP **1503** 088 [arXiv:1412.6663 [hep-ex]].

[18] Khachatryan V *et al.* [CMS Collaboration], 2015 Search for a charged Higgs boson in pp collisions at $\sqrt{s} = 8$ TeV, JHEP **1511** 018 [arXiv:1508.07774 [hep-ex]].

[19] Khachatryan V *et al.* [CMS Collaboration], 2015 Search for a light charged Higgs boson decaying to $c\bar{s}$ in pp collisions at $\sqrt{s} = 8$ TeV, JHEP **1512** 178 doi:10.1007/JHEP12(2015)178 [arXiv:1510.04252 [hep-ex]].

[20] De Florian D *et al.* [LHC Higgs Cross Section Working Group], Handbook of LHC Higgs Cross Sections: 4. Deciphering the Nature of the Higgs Sector, arXiv:1610.07922 [hep-ph].

[21] Asakawa E, Brein O and Kanemura S, 2005 Enhancement of $W^{\pm}H^{\mp}$ production at hadron colliders in the two Higgs doublet model, Phys. Rev. D **72** 055017 [hep-ph/0506249].

[22] El Kaffas A W, Khater W, Ogreid O M and Osland P, Consistency of the two Higgs doublet model and CP violation in top production at the LHC, Nucl. Phys. B **775** (2007) 45 [hep-ph/0605142].

[23] Aaboud M *et al.* [ATLAS Collaboration], Search for charged Higgs bosons produced in association with a top quark and decaying via $H^{\pm} \to \tau\nu$ using pp collision data recorded at $\sqrt{s} = 13$ TeV by the ATLAS detector, Phys. Lett. B **759** (2016) 555 [arXiv:1603.09203 [hep-ex]].

[24] The ATLAS collaboration [ATLAS Collaboration], Search for charged Higgs bosons in the $H^{\pm} \to tb$ decay channel in pp collisions at $\sqrt{s} = 13$ TeV using the ATLAS detector, ATLAS-CONF-2016-089.

[25] Aad G *et al.* [ATLAS Collaboration], 2016 Search for charged Higgs bosons in the $H^{\pm} \to tb$ decay channel in pp collisions at $\sqrt{s} = 8$ TeV using the ATLAS detector, JHEP **1603** 127 [arXiv:1512.03704 [hep-ex]].

IOP Conf. Series: Journal of Physics: Conf. Series **873** (2017) 012049 doi:10.1088/1742-6596/873/1/012049

Sensitivity of CLIC at 380 GeV to the top FCNC decay $t \to cH$

Aleksander Filip Żarnecki
on behalf of the CLICdp collaboration
Faculty of Physics, University of Warsaw, Pasteura 5, 02-093 Warszawa, Poland
E-mail: zarnecki@fuw.edu.pl

Abstract. In the Standard Model (SM), flavour changing neutral current (FCNC) top decays, possible at loop level only, are very strongly suppressed. Observation of any such decay would be a direct signature of physics beyond the SM. Large enhancements are possible in many "new physics" scenarios and the largest enhancement is in most cases expected for the $t \to cH$ decay. A full study for CLIC was based on the WHIZARD simulation of FCNC top decays within the 2HDM(III) model. Beam polarization and beam-induced background were taken into account. Top pair production events with the FCNC decay $t \to cH$ can be identified based on kinematic constrains and flavour tagging information. Due to a large overlap in the kinematic space with standard top pair events, the final signal selection-efficiency is small, at the 10% level. Expected limits on $BR(t \to cH) \times BR(H \to b\bar{b})$ are compared with earlier results based on parton level simulation.

1. Introduction

Top physics, together with Higgs boson studies and searches for Beyond the Standard Model (BSM) phenomena, is one of the three pillars of the research program for future high energy e^+e^- colliders. As the top quark is the heaviest known elementary particle, with an expected value of the Yukawa coupling of the order of one, the precise determination of its properties is a key to the understanding of electroweak symmetry breaking. Determination of top properties is also essential for many "new physics" searches, as the top quark gives large loop contributions to many precision measurements sensitive to BSM effects. Stringent constraints on the "new physics" scenarios are also expected from direct searches for rare top decays. Both future linear colliders, the International Linear Collider (ILC) and the Compact Linear Collider (CLIC), provide an opportunity to study the top quark with unprecedented precision via direct production of $t\bar{t}$ pairs in e^+e^- collisions. Presented in this contribution are prospects of constraining the branching ratio for the flavour changing top decay $t \to cH$ with CLIC running at $\sqrt{s} = 380$ GeV.

2. Experimental conditions

The Conceptual Design Report (CDR) for CLIC was presented in 2012 [1]. CLIC is based on a two-beam acceleration scheme, required to generate a high RF gradient of about 100 MV/m. In the recently updated implementation plan for CLIC [2], a construction in three stages is proposed, with 5 to 7 years of data taking at each stage. The first stage with a footprint of

11 km will allow one to reach a center-of-mass energy of 380 GeV, giving access to most Higgs boson and top quark measurements. The plan assumes collecting 500 fb^{-1} at 380 GeV with additional 100 fb^{-1} collected at the $t\bar{t}$ threshold. The CLIC baseline design includes polarisation for the electron beam, while positron polarisation is considered as a possible upgrade.

The detector concepts proposed for CLIC are based on jet reconstruction and jet energy measurements with the "Particle Flow" approach [3]. Single particle reconstruction and identification exploits high calorimeter granularity, and the best possible jet energy estimate is obtained by combining calorimetric measurements of neutral particles with precise track momentum measurements for the charged ones. Very efficient flavour tagging is possible with a high precision pixel vertex detector placed very close to the interaction point. The background to processes with missing energy can be strongly suppressed thanks to very good detector hermeticity, with instrumentation extending down to a minimum angle of $\theta_{\text{min}} \sim 10$ mrad. Presented in this contribution are detailed simulation results based on the ILD detector concept adopted for CLIC [4].

3. Theoretical expectations

3.1. Standard Model

Flavour-Changing Neutral Current (FCNC) top quark decays, $t \to cX$ ($X = \gamma,\ g,\ Z,\ H$), are strongly suppressed in the Standard Model. Only charged current decays are allowed at the tree level and the loop level contributions are suppressed by the Glashow-Iliopoulos-Maiani (GIM) mechanism [5]. The cancellation is not perfect because of the non-zero down-quark masses and the dominant contribution to the FCNC decays comes from diagrams including a b quark in the loop. The corresponding partial widths are proportional to the square of the element V_{cb} of the Cabibbo-Kobayashi-Maskawa (CKM) quark-mixing matrix [6, 7] and to the fourth power of the b quark to the W boson mass ratio. These two suppression factors result in extremely small branching ratios expected [8] for the Standard Model:

$$\begin{aligned} BR(t \to cg) &\sim 5 \cdot 10^{-12}, \\ BR(t \to c\gamma) &\sim 5 \cdot 10^{-14}, \\ BR(t \to cZ) &\sim 1 \cdot 10^{-14}, \\ BR(t \to cH) &\sim 3 \cdot 10^{-15}. \end{aligned}$$

Observation of any such decay would be a direct signature for "new physics".

One should note that for one of the diagrams contributing to the $t \to cH$ channel the GIM mechanism is not strictly applicable due to the Higgs coupling being proportional to the quark mass. But the contribution of this diagram is suppressed by the b quark to W boson mass ratio, resulting from the Higgs coupling. So, in spite of the GIM mechanism violation, the expected FCNC branching ratio for this channel is smallest in the Standard Model.

3.2. Beyond the Standard Model

Many extensions of the Standard Model predict significant enhancement of the FCNC top decays [8]. The enhancement can be due to the direct tree level FCNC coupling, but in most models it is observed at the loop level and results from contributions of new particles or from the modified particle couplings. The decay $t \to cH$ seems to be the most promising channel, as most BSM scenarios predict significant deviation in the (light) Higgs boson couplings or contributions from the additional Higgs bosons to the loop diagrams. For the Two Higgs Doublet Model (2HDM), which is one of the simplest extensions of the Standard Model, loop contributions can be enhanced up to the level of $BR \sim 10^{-4}$ [9]. For the "non standard" scenarios, 2HDM(III) or "Top 2HDM", where one of the Higgs doublets couples to the top quark only, tree level FCNC couplings are also allowed and an enhancement up to 10^{-2} is possible [10]. The $t \to cH$ decay is

IOP Conf. Series: Journal of Physics: Conf. Series **873** (2017) 012049 doi:10.1088/1742-6596/873/1/012049

also difficult to constrain at the LHC, because of the high QCD background to the dominating Higgs boson decay channels. The expected limit from HL-LHC is $BR(t \to cH) < 2 \cdot 10^{-4}$ [8, 11].

3.3. Parton level study

Before a detailed analysis based on the full detector simulation results could be started, a feasibility study was performed at the parton level, with very coarse modeling of detector effects. Signal and background event samples were generated with WHIZARD [12, 13] using an internal CIRCE1 option for modeling the beam energy spectra. Signal events were generated using the dedicated implementation of 2HDM(III) in SARAH [14]. Model parameters were tuned to obtain $BR(t \to cH) = 10^{-3}$ for the assumed Higgs boson mass $m_H = 125$ GeV. The analysis was based on the 6-fermion final state generated by WHIZARD, with only the $H \to b\bar{b}$ decay channel considered for the signal. As a $t\bar{t}$ sample can be selected at e^+e^- colliders with high purity, it was also assumed that the dominant background to FCNC events comes from the standard top quark decay channels.

Top pair production events with the FCNC decay $t \to cH$ can be identified based on the kinematic constrains and flavour tagging information. To model the detector performance, a very simplified parametrisation was used:

- detector acceptance for leptons: $|\cos\theta_l| < 0.995$;
- detector acceptance for jets: $|\cos\theta_j| < 0.975$;
- jet energy smearing: $\sigma_E/E = S/\sqrt{E[\text{GeV}]}$ (or $S/10$ for jet energy $E > 100$ GeV);
- no energy smearing for reconstructed final state leptons (electrons and muons);
- fixed b quark tagging efficiency ε_b, c and light quark misstagging efficiencies, ε_c and ε_{uds}.

For the jet energy resolution, three scenarios were considered: $S = 30\%$ (optimistic scenario), 50% (realistic scenario) and 80% (pessimistic scenario). For the tagging efficiencies, two scenarios, based on LCFI+ [15] results, were eventually used: $\varepsilon_b = 80\%$, $\varepsilon_c = 8\%$ and $\varepsilon_{uds} = 0.8\%$ (looser b quark selection) or $\varepsilon_b = 70\%$, $\varepsilon_c = 2\%$ and $\varepsilon_{uds} = 0.2\%$ (tighter selection).

Both fully-hadronic and semi-leptonic decay channels of the produced top pairs were considered. After imposing the b-tagging criteria (three tagged jets required: two b from the Higgs boson decay and one from the "spectator" top quark, the one with charged current decay), a kinematic fit was performed for each event, for both signal and background hypothesis. The χ^2 formula for both cases included constraints on the masses of two top candidates, W mass constraint and the Higgs boson or second W mass constraint. The final selection was based on the ratio of the χ^2 values for best signal and best background hypothesis. The selection was optimised so as to obtain the highest sensitivity to the FCNC decays. For jet energy resolution given by $S = 30\%$ or 50%, loose b-tagging settings were chosen while for the worst resolution ($S = 80\%$), tighter flavour selection is required. Expected limits on $BR(t \to cH) \times BR(h \to b\bar{b})$, for e^+e^- collisions at $\sqrt{s} = 380$ GeV, as a function of the integrated luminosity are shown in figure 1. Although the signal selection efficiency is small, limited by a large overlap in the kinematic space with standard top pair events, results indicate that with high integrated luminosity this decay can be probed down to $BR \sim 10^{-5}$ [16].

4. Event simulation and reconstruction

Detailed detector level analysis was performed for e^+e^- collisions at CLIC, for $\sqrt{s} = 380$ GeV. As for the parton level study, signal event samples were generated using the 2HDM(III) model implemented in SARAH. However, to improve the description of the energy distribution, a dedicated CLIC beam spectra file was used. An electron beam polarisation of 80% was also taken into account. Events generated with WHIZARD 2.2.8 were passed to PYTHIA 6.4 for hadronisation with quark masses and other settings adjusted to the configuration used previously

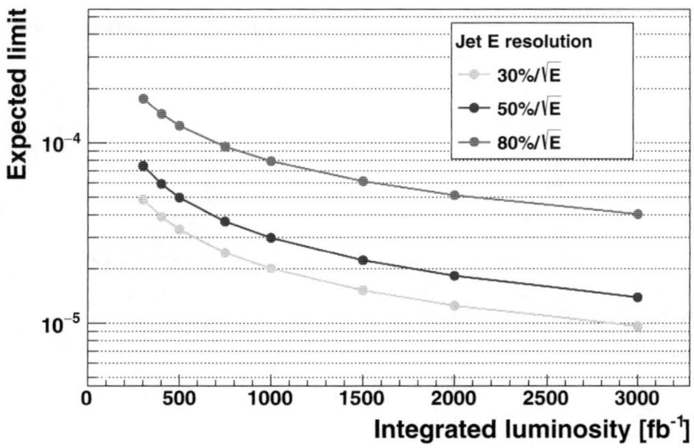

Figure 1. Expected 95% C.L. limits on the $BR(t \to cH) \times BR(h \to b\bar{b})$ as a function of the integrated luminosity for e^+e^- collisions at $\sqrt{s} = 380$ GeV. Results are based on a parton level study assuming different jet energy resolution parameters, as indicated in the plot. Both hadronic and semi-leptonic decay channels are considered.

in CLIC CDR studies [4]. The generated signal sample was then processed with a standard event simulation and reconstruction chain of the CLICdp collaboration using the CLIC_ILD_CDR500 detector configuration. The background sample considered in the analysis included a full set of six-fermion event samples produced previously for CLICdp studies of top pair production at $\sqrt{s} = 380$ GeV. All sub-samples corresponded to an integrated luminosity of at least 500 fb^{-1}. The procedure used for signal event generation was also used to generate an additional "test" sample of Standard Model top pair-production events, which was then used to validate the consistency of simulation settings between old (background) and new (signal and test) samples.

In the final step, a dedicated event analysis procedure was applied to all signal and background events. Because of the large event sample considered, iLCDirac [17] was used as an interface to grid resources, and the processing was based on the MARLIN [18] framework with the ilcsoft version v01-17-09. The reconstructed object collection resulting from loose background rejection cuts (LooseSelectedPandoraPFANewPFOs) was used as an input for jet reconstruction with the Valencia algorithm [19] as well as for primary and secondary vertex (re)reconstruction, and flavour tagging with LCFI+.

5. Selection of signal events

Results presented in this contribution were based on the analysis of hadronic top decays only, i.e. events with all W bosons decaying to two quarks (six jet final state). Hadronic events can be selected by looking at the correlation of the measured transverse momentum, p_T, and the total energy of the event, E, as shown for the considered background sample in figure 2 (left). Shown in figure 2 (right) is the probability that the considered p_T and E values resulted from the hadronic top pair decay. It shows that an efficient selection of hadronic events in this plane can be made with a single cut on the $E - 2p_T$ value. The selection efficiency can be still improved slightly by considering also the total longitudinal momentum of an event, p_z. We defined the effective variable describing the energy balance in the event as

$$E_{\text{balance}} = \sqrt{(E - 2p_T - \sqrt{s})^2 + 4p_Z^2}\ ,$$

and for hadronic events, $E_{\text{balance}} < 100$ GeV was required.

IOP Conf. Series: Journal of Physics: Conf. Series **873** (2017) 012049 doi:10.1088/1742-6596/873/1/012049

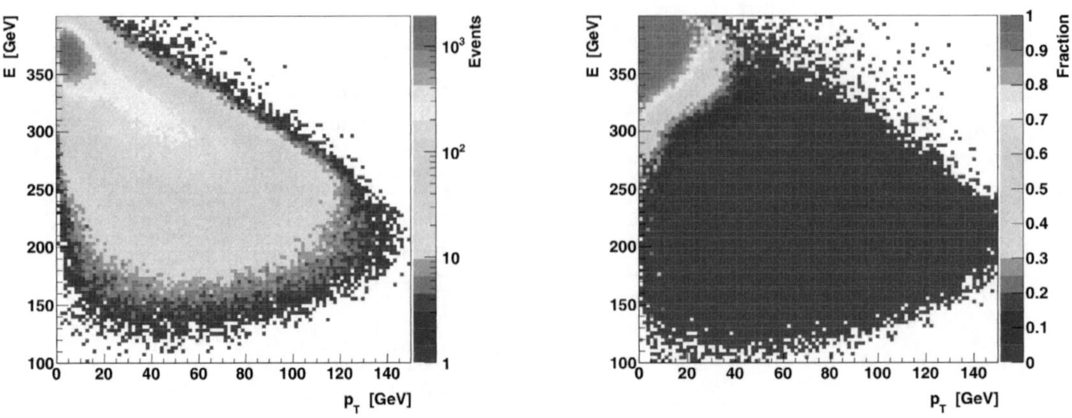

Figure 2. Left: correlation between the measured transverse momentum, p_T, and the total energy of the event, E, for the background events at $\sqrt{s} = 380$ GeV. Right: fraction of hadronic events as a function of the p_T and E.

Before the kinematic fit was used to test the signal vs. background hypothesis, additional preselection cuts were applied to select candidate events. We required that at least three out of six jets reconstructed with the Valencia algorithm were tagged as b-quark jets (with tag probability of at least 0.4) and the fourth jet was tagged as a c or b quark (corresponding to c quark from $t \to cH \to cb\bar{b}$ decay). The estimated efficiency of the described preselection cuts is about 34% for signal events and about 2.4% for background events (including the hadronic branching fraction).

The best signal hypothesis was then selected by comparing χ^2 values for different jet assignments. As significant correlations were observed between the reconstructed Higgs boson mass and the "signal" top quark mass (the one decaying to cH), and between the reconstructed W boson mass and the "spectator" top quark mass, these measurements should not be treated as independent. Therefore the χ^2 function is defined based on the measured values of the following parameters:

- "signal" and "spectator" top quark masses;
- "signal" and "spectator" top quark boosts (energy to mass ratios);
- the ratio of the Higgs boson mass to the "signal" top quark mass;
- the ratio of the W boson mass to the "spectator" top quark mass.

A similar approach was used to select the best background hypothesis (replacing the Higgs boson by another W and requiring a b-tag only for two jets).

After the kinematic fit, additional selection criteria were imposed to optimise signal event selection:

- signal hypothesis fit resulting in $\chi^2_{\text{sig}} < 14$;
- difference of two reconstructed top masses $\Delta M_t < 45$ GeV;
- product of b-tag probabilities for two jets from Higgs decay $b_1 \cdot b_2 > 0.95$;
- b-tag probability for the b jet from spectator top decay $b_3 > 0.9$;
- sum of c-tag and b-tag probabilities for c quark from Higgs decay $c_4 + b_4 > 0.75$.

A final cut was then applied on the ratio of χ^2 values for the best signal and background hypothesis, $\chi^2_{\text{sig}}/\chi^2_{\text{bg}}$.

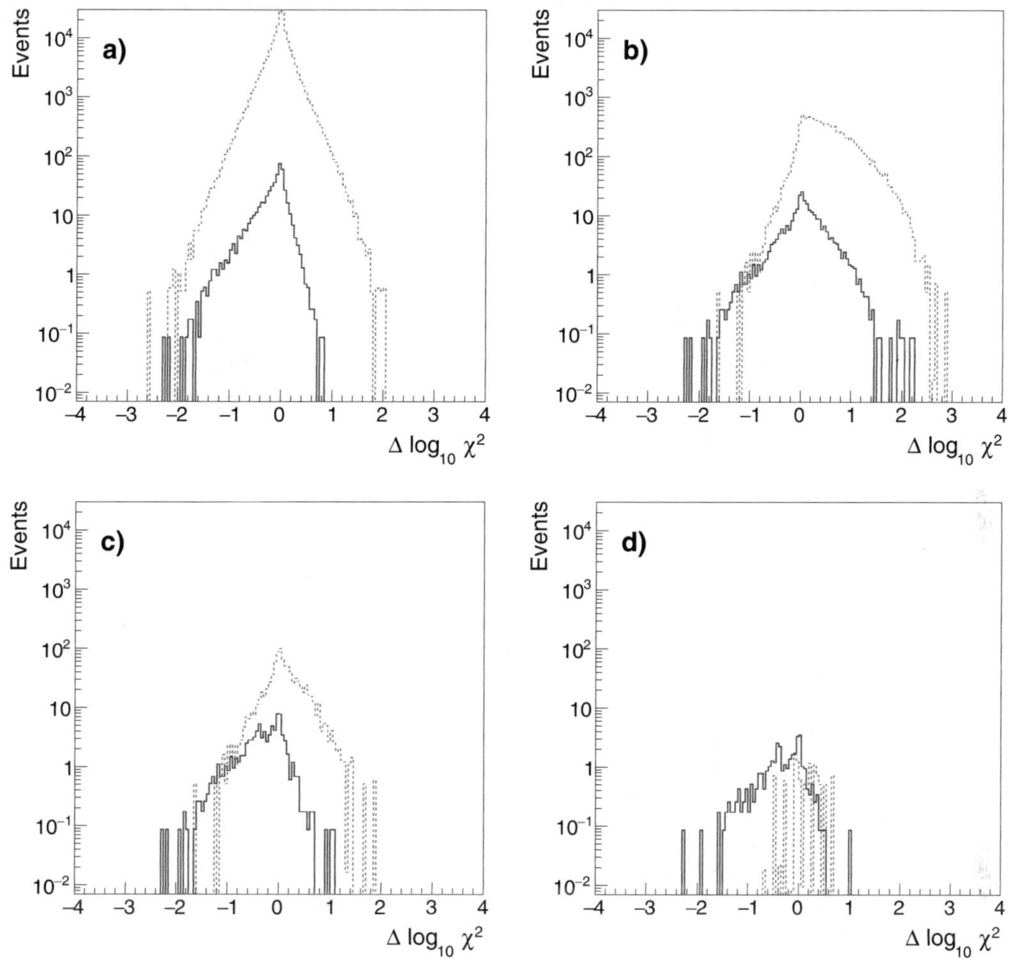

Figure 3. Distributions of $\Delta \log_{10} \chi^2 = \log_{10}(\chi^2_{\rm sig}/\chi^2_{\rm bg})$, for signal (solid blue histogram) and background (dashed red histogram) event samples, for all selected hadronic events (a), for events passing preselection cuts (b), for events with a good fit of the signal hypothesis (c) and the sample remaining after the final selection cuts (d).

6. Results

To discriminate between the background and the signal hypothesis the ratio of the χ^2 values of the two hypotheses was considered. Shown in figure 3 are the distributions of this ratio (in logarithmic scale; $\Delta \log_{10} \chi^2 = \log_{10}(\chi^2_{\rm sig}/\chi^2_{\rm bg})$), for signal and background samples corresponding to the integrated luminosity of 500 fb^{-1}. Before any selection cuts (figure 3a), the background level is three orders of magnitude higher than the signal one (which corresponds to the assumed branching ratio for the FCNC decays). Preselection cuts (figure 3b) and the kinematic fit (figure 3c) allow for background suppression by over two orders of magnitude, but it is still not sufficient to identify the signal. Also, the shape of the $\Delta \log_{10} \chi^2$ distribution is very similar for signal and background samples, showing that the kinematic fit itself is not sufficient for signal identification. The kinematic reconstruction is clearly not precise enough. Only after very tight cuts on the flavour tagging probabilities for the three b-quark jets and c jet, signal

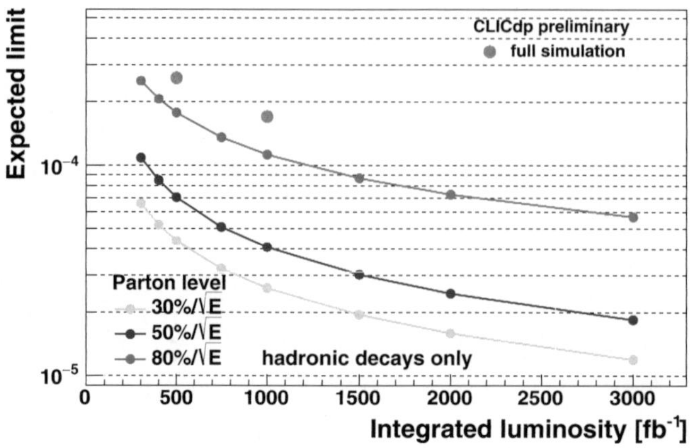

Figure 4. Expected 95% C.L. limits on $BR(t \to cH) \times BR(h \to b\bar{b})$ as a function of the integrated luminosity for e^+e^- collisions at $\sqrt{s} = 380$ GeV. Results based on the full detector simulation are compared to results from the parton level study restricted to the hadronic channel.

events can be selected with a reasonable purity (figure 3d). The final cut on $\Delta \log_{10} \chi^2$ was then optimised to obtain the best expected limit on the FCNC branching ratio.

For 500 fb^{-1} of integrated luminosity and the optimised cut value $\chi^2_{\text{sig}}/\chi^2_{\text{bg}} < 1.38$, the expected 95% C.L. limit is

$$BR(t \to cH) \times BR(h \to b\bar{b}) < 2.6 \cdot 10^{-4} .$$

The Standard Model background passing final selection cuts is estimated to be about 4.9 events while 31.8 events are expected for the signal (assuming $BR(t \to cH) \times BR(h \to b\bar{b}) = 10^{-3}$). The corresponding selection efficiencies are 3.9% for signal events and $1.2 \cdot 10^{-5}$ for the background sample.

Results from the detailed simulation study described in this contribution are compared to the earlier estimates from the parton level study in figures 4 and 5. For consistency, parton level results presented on these figures were obtained considering the hadronic decay channel only. The sensitivity of CLIC running at 380 GeV to FCNC top decays turns out to be significantly weaker than the one estimated from the parton level study, even assuming the worst (pessimistic) jet energy resolution. This is due to the much weaker discrimination power of the kinematic fit. Invariant mass resolutions for two- and three-jet configurations in the final state (W and top candiates), as obtained from the full simulation study, are not well described by the Gaussian distributions. Deterioration of the energy and mass resolution for the hadronic final state can be due to the large fraction of b jets in the considered sample as well as to the influence of the overlaid beam background. With the required background suppression factor of the order of 10^{-4}, tails of the mass distributions become very important and result in a significant fraction of background events fitting signal hypothesis better (two jets from W decay looking like H candidate). These events can only be suppressed by very tight flavour tagging cuts, but this reduces also the signal selection efficiency significantly. A comparison of parton level results obtained for different running energies, as shown in figure 5, indicates that the performance of the kinematic fit should improve when going to higher centre-of-mass energies. At $\sqrt{s} = 380$ GeV, top quarks are produced almost at rest and their decay products can easily mix. At higher energies, a larger boost should result in better separation of two top candidates and unambiguous result of the kinematic fit.

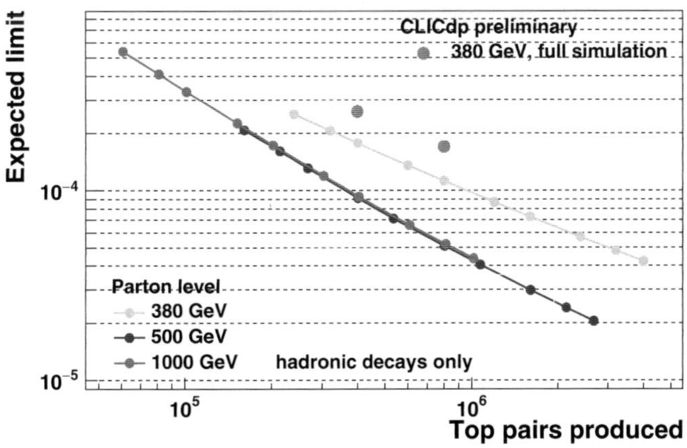

Figure 5. Expected 95% C.L. limits on $BR(t \to cH) \times BR(h \to b\bar{b})$ as a function of the collected sample of top pair production events. Results based on the full detector simulation for $\sqrt{s} = 380$ GeV are compared to parton level results for different running energies, assuming jet energy resolution parameter $S = 80\%$. Only the hadronic decay channel is considered.

7. Conclusions

Considered in this contribution is the feasibility of measuring the FCNC top decay $t \to cH$ at CLIC running at 380 GeV. Preliminary results based on the full detector simulation are presented. The main focus of the analysis was on the optimization of the kinematic event reconstruction in the hadronic channel. After optimisation of the event selection criteria, the expected 95% C.L. limit on the FCNC branching ratio, assuming an integrated luminosity of 500 fb^{-1}, is $2.6 \cdot 10^{-4}$. This limit is significantly worse than the initial estimates based on the parton level analysis, which is most likely due to the non-Gaussian tails of the mass resolution distributions. The analysis is ongoing and different options will be studied to improve signal to background discrimination and resulting limits. This includes optimization of the LCFI+ performance, analysis of the semi-leptonic decay channel and possible use of MultiVariate Analysis tools to optimise event selection. Finally, measurements at higher collision energies should be also considered.

Acknowledgments

This work benefited from services provided by the ILC Virtual Organisation, supported by the national resource providers of the EGI Federation.

References

[1] Aicheler M *et al.* 2012 A Multi-TeV Linear Collider Based on CLIC Technology CERN-2012-007, SLAC-R-985, KEK-Report-2012-1, PSI-12-01, JAI-2012-001
[2] Boland M J *et al.* (CLIC and CLICdp collaborations) 2016 Updated baseline for a staged Compact Linear Collider CERN-2016-004 (*Preprint* 1608.07537)
[3] Thomson M A 2009 *Nucl. Instrum. Meth.* **A611** 25–40 (*Preprint* 0907.3577)
[4] Linssen L, Miyamoto A, Stanitzki M and Weerts H 2012 Physics and Detectors at CLIC: CLIC Conceptual Design Report CERN-2012-003, ANL-HEP-TR-12-01, DESY-12-008, KEK-REPORT-2011-7 (*Preprint* 1202.5940)
[5] Glashow S L, Iliopoulos J and Maiani L 1970 *Phys. Rev.* **D2** 1285–1292
[6] Cabibbo N 1963 *Phys. Rev. Lett.* **10** 531–533
[7] Kobayashi M and Maskawa T 1973 *Prog. Theor. Phys.* **49** 652–657

[8] Agashe K *et al.* 2013 Working Group Report: Top Quark *Proc. Community Summer Study 2013: Snowmass on the Mississippi (CSS2013): Minneapolis, MN, USA, July 29-August 6, 2013* (*Preprint* 1311.2028)
[9] Bejar S, Guasch J and Sola J 2001 FCNC top quark decays beyond the standard model *Proc. 5th Int. Symp. on Radiative Corrections - RADCOR 2000* (*Preprint* hep-ph/0101294)
[10] Diaz-Cruz L and Pagliarone C 2006 Perspectives of detecting CKM-suppressed top quark decays at ILC *Proc. Summer School and Conference on New Trends in High-Energy Physics: Experiment, Phenomenology, Theory, Yalta, Crimea, Ukraine, September 16-23, 2006* (*Preprint* hep-ph/0612120)
[11] ATLAS Collaboration 2016 Expected sensitivity of ATLAS to FCNC top quark decays $t \to Zu$ and $t \to Hq$ at the High Luminosity LHC ATL-PHYS-PUB-2016-019
[12] Kilian W, Ohl T and Reuter J 2011 *Eur. Phys. J.* **C71** 1742 (*Preprint* 0708.4233)
[13] Moretti M, Ohl T and Reuter J 2001 O'Mega: An Optimizing matrix element generator (*Preprint* hep-ph/0102195)
[14] Staub F 2015 *Adv. High Energy Phys.* **2015** 840780 (*Preprint* 1503.04200)
[15] Suehara T and Tanabe T 2016 *Nucl. Instrum. Meth.* **A808** 109–116 (*Preprint* 1506.08371)
[16] Vos M *et al.* 2016 Top physics at high-energy lepton colliders (*Preprint* 1604.08122)
[17] Grefe C, Poss S, Sailer A and Tsaregorodtsev A (on behalf of the CLIC Detector and Physics Study, CLICdp) 2014 *J. Phys.: Conf. Ser.* **513** 032077
[18] Gaede F 2006 *Nucl. Instrum. Meth.* **A559** 177–180
[19] Boronat M, Fuster J, Garcia I, Roloff P, Simoniello R and Vos M 2016 Jet reconstruction at high-energy lepton colliders (*Preprint* 1607.05039)

IOP Conf. Series: Journal of Physics: Conf. Series **873** (2017) 012050 doi:10.1088/1742-6596/873/1/012050

Effective Field Theory in quest to parametrise Higgs properties: the transverse momentum spectrum case

Massimiliano Grazzini[a], **Agnieszka Ilnicka**[abc], **Michael Spira**[c], **Marius Wiesemann**[d]

[a] Physics Institute, University of Zürich, Winterthurerstrasse 190, CH-8057 Zürich
[b] Physics Department, ETH Zürich, Otto-Stern-Weg 5, CH-8093 Zürich
[c] Paul Scherrer Institute, CH-5232 Villigen PSI
[d] CERN Theory Division, CH-1211, Geneva 23

E-mail: grazzini@physik.uzh.ch, ailnicka@physik.uzh.ch, michael.spira@psi.ch, marius.wiesemann@cern.ch

Abstract. After the Higgs boson discovery, LHC turned into the precision machine to explore its properties. In case new resonances will not be found, the only access to New Physics would be via measuring small deviations from the SM predictions. A consistent approach which can be useful in both above cases is a bottom-up Effective Field Theory, with dimension six operators build of Standard Model fields. We present how this approach would work in case of the transverse momentum spectrum of the Higgs particle. In our calculation we augmented the Standard Model with three additional operators: modifications of the top and bottom Yukawa couplings, and a point-like Higgs coupling to gluons. We present resummed transverse-momentum spectra including the effect of these operators at NLL+NLO accuracy and study their impact on the shape of the distribution. We find that such modifications, while affecting the total rate within the current uncertainties, can lead to significant distortions of the spectrum. We also discuss the effect of the chromomagnetic operator on the Higgs production cross section at LO.
For more details see our paper [1].

1. Introduction

The scalar resonance discovered by ATLAS and CMS at the LHC in 2012 [2, 3] closely resembles the Higgs boson postulated in the Standard Model (SM). The SM, however, is not able to explain the existence of dark matter, the matter-antimatter asymmetry and the relatively low scale of electroweak symmetry breaking (hierarchy problem). Many theories beyond the SM (BSM) addressing the above issues have been developed, which manifest different patterns in the scalar sector and in the Higgs boson properties. As no strict argument exists for the discovery of new physics at the TeV scale, it is possible that new physics effects are accessible only by measuring small deviations from SM predictions. A consistent way to parametrise these deviations is offered by the Effective Field Theory (EFT), in which the unknown high-scale fields are integrated out leaving an infinite ladder of higher-dimensional operators (dim>4), with a well-defined hierarchy. The EFT can thus be used to build a bottom-up approach in which the usual dimension-four operators in the SM are augmented by leading (dimension-six) operators. Experimental data can be employed to fix the values of the so-called Wilson coefficients of these operators. A matching to the EFT allows then for the translation of the Wilson coefficients into bounds on the physical parameters of new physics models. The full set of dimension-six [4, 5] deformations of the SM was classified by 59 operators. The precision observables measured at LEP and the Tevatron put bounds on many of the Wilson coefficients, some even at the per-mille level [6, 7, 8]. However, several operators involving the Higgs field are still essentially unbounded. In the following, we therefore develop a strategy to determine such bounds by using the transverse-momentum (p_T) spectrum of the Higgs boson.

2. Transverse-momentum spectrum

Kinematical distributions provide an important handle on the determination of Higgs properties. Among the most important observables in this respect is the Higgs transverse-momentum distribution, that will be measured with high precision in Run II and High Luminosity stage of the LHC. First results from the LHC Run I were presented by the ATLAS collaboration in the 2γ and four-lepton final states [9, 10] and by the CMS collaboration in the 2γ final state [11], leaving still a significant amount of arbitrariness for the precise form of the distribution. The p_T spectrum provides more information than the total cross section, which is just one number; the shape, the position of the maximum and the normalisation allow us to disentangle effects that remain hidden in the total rates. For example, it is the simplest measurement to shed light on the nature of the Higgs coupling to gluons. The fact that the Higgs is a scalar, gives an additional simplification in the modeling of the Higgs p_T-spectrum, due to the factorization of production and decay in the narrow-width approximation.

The most important Higgs production channel at the LHC is gluon fusion, which, despite being a loop-induced process, is highly enhanced by the dominance of the gluon densities [12]. Therefore, we will concentrate on the spectra obtained in this production channel. In the past years a significant amount of work has been done to improve the precision of the calculations of the Higgs p_T spectrum. The first results at the lowest order ($\mathcal{O}(\alpha_S^3)$) were known since long time [13, 14]. It took nearly ten years until the $\mathcal{O}(\alpha_S^4)$ corrections were computed [15, 16, 17, 18]. These were carried out in the heavy-top limit (HTL, i.e. $m_t^2 \gg M_H^2, p_{TH}^2$).[1] Recently, results on Higgs+jet production at $\mathcal{O}(\alpha_S^5)$ were also obtained in HTL [21, 22, 23].

In the low-p_T region ($p_T \ll M_H$), the perturbative expansion is affected by large logarithmic terms of the form $\alpha_S^n \ln^m(m_H^2/p_T^2)$, with $1 \leq m \leq 2n$. This results in a singular behaviour of the distribution as $p_T \to 0$. To cure this problem one needs to resum these terms to all orders in α_S [24]. The resummation is carried out in impact parameter (b) space, and, in particular, we use the formalism of Ref. [25]. The resummed and fixed order results have to be properly matched at intermediate p_T to avoid double counting:

$$\left[\frac{\mathrm{d}\sigma}{\mathrm{d}p_T^2}\right]_{\text{f.o.+a.o.}} = \left[\frac{\mathrm{d}\sigma}{\mathrm{d}p_T^2}\right]_{\text{f.o.}} - \left[\frac{\mathrm{d}\sigma^{(\text{res})}}{\mathrm{d}p_T^2}\right]_{\text{f.o.}} + \left[\frac{\mathrm{d}\sigma^{(\text{res})}}{\mathrm{d}p_T^2}\right]_{\text{a.o.}} \tag{1}$$

where f.o. corresponds to fixed order, and a.o. to all orders calculations. In the formalism of Ref. [25], a unitarity constraint is enforced, such that the integral of the p_T-spectrum coincides with the corresponding total inclusive cross section computed at fixed order. More precisely, by performing the resummation at next-to-leading logarithmic accuracy (NLL) and including the fixed order result up to $O(\alpha_S^3)$ we obtain NLO+NLL accuracy, and the integral of the spectrum is fixed to the NLO total cross section. Top-and bottom-mass effects can be included in the resummed spectrum along the lines of Refs. [26, 27].

The inclusion of dimension-six and dimension-eight operators in the p_T-spectrum has been considered in Refs. [28, 29, 30, 31] and [32, 33], respectively. Strategies for extracting information

[1] Finite top-mass effects on the Higgs p_T distribution at $\mathcal{O}(\alpha_S^4)$ were estimated in Refs. [19, 20].

IOP Conf. Series: Journal of Physics: Conf. Series **873** (2017) 012050 doi:10.1088/1742-6596/873/1/012050

on the Higgs-gluon couplings from the measurements were studied in Ref. [30]. Most of the above studies, however, are limited to the high-p_T region of the spectrum, and do not include small-p_T resummation. Also our study is the first to include the analytic resummation rather that the parton shower. In this contribution we present the resummed p_T-spectrum at NLO+NLL accuracy, with the inclusion of a set of dimension-six parameters relevant for Higgs boson production.

3. Effective operators and their impact on the Higgs production cross section

We consider the effective Lagrangian

$$\mathcal{L} = \mathcal{L}_{SM} + \sum_i \frac{c_i}{\Lambda^2} \mathcal{O}_i \tag{2}$$

where the SM is supplemented by the inclusion of a set of dimension-six operators describing new physics effects at a scale Λ well above the electroweak scale. In our study we consider the following four operators

$$\mathcal{O}_1 = |H|^2 G^a_{\mu\nu} G^{a,\mu\nu}, \quad \mathcal{O}_2 = |H|^2 \bar{Q}_L H^c u_R + h.c., \tag{3}$$

$$\mathcal{O}_3 = |H|^2 \bar{Q}_L H d_R + h.c., \quad \mathcal{O}_4 = \bar{Q}_L H \sigma^{\mu\nu} T^a u_R G^a_{\mu\nu} + h.c. \tag{4}$$

These operators, in the case of single Higgs production, may be expanded as:

$$\frac{c_1}{\Lambda^2} \mathcal{O}_1 \to \frac{\alpha_S}{\pi v} c_g h G^a_{\mu\nu} G^{a,\mu\nu}, \tag{5}$$

$$\frac{c_2}{\Lambda^2} \mathcal{O}_2 \to \frac{m_t}{v} c_t h \bar{t} t, \tag{6}$$

$$\frac{c_3}{\Lambda^2} \mathcal{O}_3 \to \frac{m_b}{v} c_b h \bar{b} b, \tag{7}$$

$$\frac{c_4}{\Lambda^2} \mathcal{O}_4 \to c_{tg} \frac{g_S m_t}{2v^3} (v+h) G^a_{\mu\nu} (\bar{t}_L \sigma^{\mu\nu} T^a t_R + h.c). \tag{8}$$

The operator $\mathcal{O}_1$ corresponds to a contact interaction between the Higgs boson and gluons with the same structure as in the heavy-top limit (HTL) of the SM. The operators $\mathcal{O}_2$ and $\mathcal{O}_3$ describe modifications of the top and bottom Yukawa couplings. The operator $\mathcal{O}_4$ is the chromomagnetic dipole-moment operator, which modifies the interactions between the gluons and the top quark[2] (here $\sigma^{\mu\nu} = \frac{i}{2}[\gamma^\mu, \gamma^\nu]$). In our convention, based on the SILH basis [34, 35], we express the Wilson coefficients as factors in the canonically normalized Lagrangian.

The coefficients c_t, c_b and c_g can be probed in Higgs boson processes. In particular, c_t (and c_b) may be measured in the $t\bar{t}H$ (and $b\bar{b}H$) production modes.The coefficient c_b can also be accessed through the decay $H \to b\bar{b}$. The coefficient c_{tg}, instead, is constrained by top pair production [36].

We now consider the contribution of the effective operators $\mathcal{O}_1, \mathcal{O}_2$ and $\mathcal{O}_4$ on the production cross section, while omitting, for simplicity, the bottom contribution in $\mathcal{O}_3$. The relevant Feynman diagrams are displayed in Fig. 1. The corresponding amplitude can be cast into the form

$$\mathcal{M}(g(p_1) + g(p_2) \to H) = i \frac{\alpha_S}{3\pi v} \epsilon_{1\mu} \epsilon_{2\nu} \left[p_1^\nu p_2^\mu - (p_1 p_2) g^{\mu\nu} \right] F(\tau), \tag{9}$$

where $\tau = 4m_t^2/m_h^2$ and ϵ_1 and ϵ_2 are the polarization vectors of the incoming gluons. The contribution of the chromomagnetic operator to the function $F(\tau)$ has been addressed in the literature with contradicting results [37, 38] (see also Ref. [39]). In Ref. [37] it is found that the UV divergences in the bubble and triangle contributions cancel out. In the revised version of

IOP Conf. Series: Journal of Physics: Conf. Series **873** (2017) 012050 doi:10.1088/1742-6596/873/1/012050

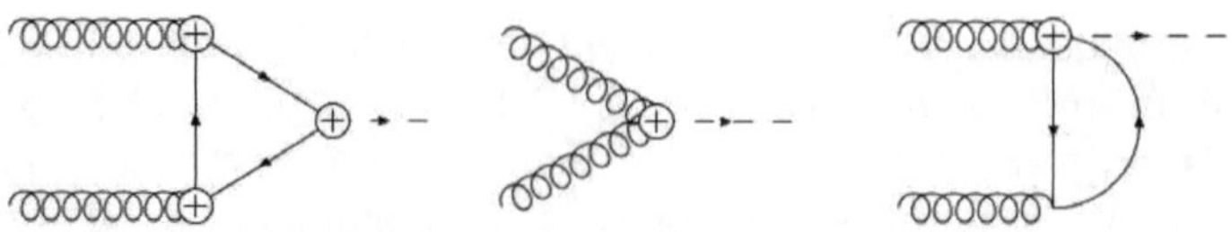

Figure 1. Feynman diagrams contributing to $gg \to H$ production at LO. The possible insertions of dimension-six operators are marked by a cross in a circle.

Ref. [38] it is instead stated that the UV divergence is present, and it has to be reabsorbed into the coefficient c_g.

Our results are consistent with the latter statement. We find

$$F(\tau) = \Gamma(1+\epsilon)\left(\frac{4\pi\mu^2}{m_t^2}\right)^{\epsilon}\left(c_t F_1(\tau) + c_{g0}F_2(\tau) + Re(c_{tg})\frac{m_t^2}{v^2}F_{30}(\tau)\right), \tag{10}$$

where

$$F_1(\tau) = \frac{3}{2}\tau\left[1 + (1-\tau)f(\tau)\right], \tag{11}$$

$$F_2(\tau) = 12, \tag{12}$$

$$F_{30}(\tau) = \frac{6}{\epsilon} + 3\left[1 - \tau f(\tau) - 2g(\tau)\right], \tag{13}$$

with the functions

$$f(\tau) = \begin{cases} \arcsin^2 \dfrac{1}{\sqrt{\tau}} & \tau \geq 1 \\ -\dfrac{1}{4}\left[\ln\dfrac{1+\sqrt{1-\tau}}{1-\sqrt{1-\tau}} - i\pi\right]^2 & \tau < 1 \end{cases}. \tag{14}$$

$$g(\tau) = \begin{cases} \sqrt{\tau - 1}\arcsin \dfrac{1}{\sqrt{\tau}} & \tau \geq 1 \\ \sqrt{1-\tau}\left[\ln\dfrac{1+\sqrt{1-\tau}}{1-\sqrt{1-\tau}} - i\pi\right] & \tau < 1 \end{cases}. \tag{15}$$

The $1/\epsilon$ divergence can be reabsorbed in the $\overline{\text{MS}}$ renormalization of the coefficient c_g:

$$c_{g0} = c_g(\mu_R) + \delta c_g \tag{16}$$

with

$$\delta c_g = \frac{m_t^2}{2v^2}Re(c_{tg})\Gamma(1+\epsilon)(4\pi)^{\epsilon}\left(-\frac{1}{\epsilon} - \ln\frac{\mu^2}{\mu_R^2}\right), \tag{17}$$

where μ_R denotes the renormalization scale of c_g. The final result reads

$$F(\tau) = c_t F_1(\tau) + c_g(\mu_R)F_2(\tau) + Re(c_{tg})\frac{m_t^2}{v^2}F_3(\tau), \tag{18}$$

where

$$F_3(\tau) = 3\left(1 - \tau f(\tau) - 2g(\tau) + 2\ln\frac{\mu_R^2}{m_t^2}\right). \tag{19}$$

[2] In this analysis we do not consider the contribution of the chromomagnetic dipole operator of the bottom quark.

In the HTL $m_t^2 \gg m_h^2$ we have

$$F_1(\tau) \to 1\,, \qquad F_2(\tau) \to 12\,, \qquad F_3(\tau) \to 6\left(\ln\frac{\mu_R^2}{m_t^2} - 1\right). \tag{20}$$

In the SM we have $c_t = 1$ and $c_g = c_{tg} = 0$, so that $F(\tau) \to F_1(\tau)$. In Ref. [36] data on top production are used to extract constraints on c_{tg}. The resulting region of allowed values of c_{tg} has been found to be

$$-0.04 \lesssim c_{tg} \lesssim 0.04\,. \tag{21}$$

The impact on the total cross section is less than 20%. We conclude that, although smaller than the impact of c_g, the effect of c_{tg} can still be important. We note, however, that the chromomagnetic operator provides a contribution which is formally $\mathcal{O}(\lambda_t^2)$ with respect to the others. In a strict expansion in α_S it can be neglected. This is what we will do in the next Section.

Focusing on the impact of c_t and c_g, we note that the total cross section alone does not allow us to disentangle the coefficients c_g and c_t:

$$\sigma \approx |12c_g + c_t|^2 \sigma_{SM} \quad (HTL)\,. \tag{22}$$

As already noted in the literature [29], the transverse momentum spectrum allows us to break this degeneracy.

4. Results

In this section we present the p_T spectra of Higgs particle, including the modification coming from the effective operators. As the reference the NLO+NLL SM predictions are also showed on the Figures, with the grey band on the lower panel showing the perturbative uncertainty. The uncertainties of the renormalization and factorization scales are estimated by performing the customary seven-point μ_R, μ_F variation, i. e. we consider independent variations within the range $\mu_0/2 \le \mu_F, \mu_R \le 2\mu_0$ with $1/2 < \mu_R/\mu_F < 2$, where $\mu_0 = \sqrt{p_T^2 + m_H^2}/2$. We then vary also three resummation scales by a factor of two. [3] It results in the uncertainty of about $\pm 20\%$ at the peak, to about $+50\%\ -30\%$ at $p_T = 400$ GeV. The values chosen for the effective coupling were guided by the (generalisation) of (22), to obtain the total cross section within 20% of SM value, what is the current experimental bound [4].

We start our analysis by considering the individual contribution of exactly one operator. Looking at the low-p_T interval (0 GeV$\le p_T \le$ 400 GeV) in Fig. 2 (a) we can directly deduce that modifications of the bottom Yukawa coupling through c_b dominantly affect the low-p_T shape of the distribution. In fact, at very low p_T we find effects that can even exceed the uncertainty of the SM prediction. As expected, $c_b < 1$ ($c_b > 1$) softens (hardens) the spectrum in that region.[5] The point-like Higgs-gluon coupling c_g, on the other hand, modifies the p_T-shape most notably at large transverse momenta (400 GeV$\le p_T \le$ 800 GeV), see Fig. 2 (b), where a positive (negative) c_g value hardens (softens) the spectrum. As expected, modifications of solely the top Yukawa through c_t have almost exclusively the effect of a rescaling of the total cross section. The deviations from the SM prediction through the dimension-six operators are within the scale uncertainty, although the differences in shape give some additional sensitivity to distinguish such effects. It is clear that in order to disentangle effects of this order it is necessary to start from a more accurate SM prediction.

[3] For the details of the calculation set up see our paper [1].

[4] Note that variations of the p_T distributions can be much larger due to the large experimental uncertainties.

[5] We point out, however, that this is true only when small deviations of c_b from its SM value $c_b = 1$ are considered. In this case the dominant effect of c_b is on the top-bottom interference. When c_b is significantly different from unity the squared bottom-loop contribution can change the picture.

IOP Conf. Series: Journal of Physics: Conf. Series **873** (2017) 012050 doi:10.1088/1742-6596/873/1/012050

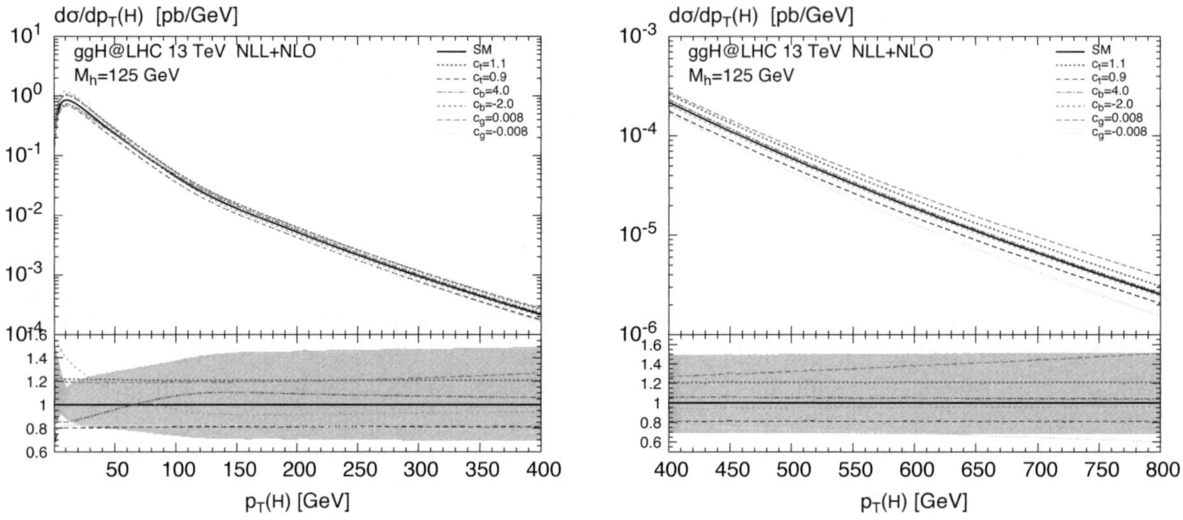

(a) (b)

Figure 2. Higgs transverse-momentum spectrum in the SM (black, solid) compared to separate variations of the dimension-six operators for (a) 0 GeV$\leq p_T \leq$ 400 GeV and (b) 400 GeV$\leq p_T \leq$ 800 GeV. The lower frame shows the ratio with respect to the SM prediction. The shaded band in the ratio indicates the uncertainty due to scale variations. See text for more details.

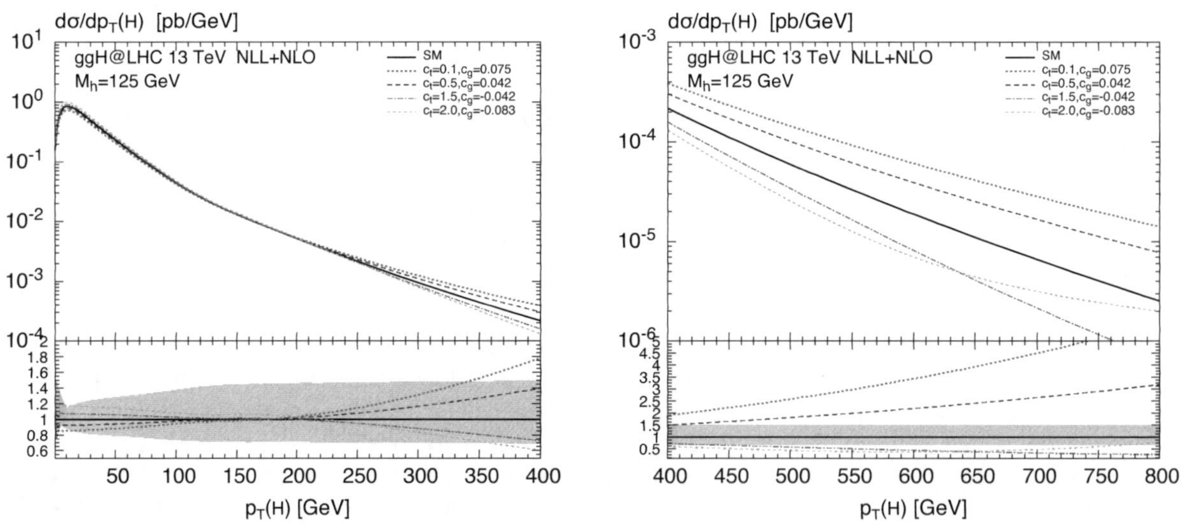

(a) (b)

Figure 3. Higgs transverse-momentum spectrum in the SM (black, solid) compared to simultaneous variations of c_t and c_g for (a) 0 GeV$\leq p_T \leq$ 400 GeV and (b) 400 GeV$\leq p_T \leq$ 800 GeV. The lower frame shows the ratio with respect to the SM prediction. The shaded band in the ratio indicates the uncertainty due to scale variations. See text for more details.

IOP Conf. Series: Journal of Physics: Conf. Series **873** (2017) 012050 doi:10.1088/1742-6596/873/1/012050

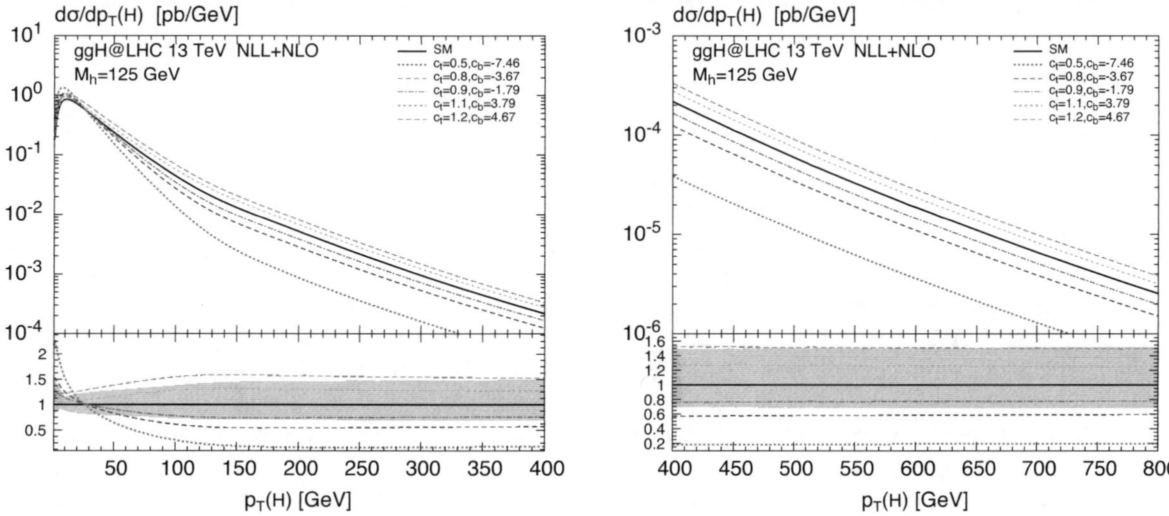

(a) (b)

Figure 4. Higgs transverse-momentum spectrum in the SM (black, solid) compared to simultaneous variations of c_t and c_b for (a) 0 GeV$\leq p_T \leq$ 400 GeV and (b) 400 GeV$\leq p_T \leq$ 800 GeV. The lower frame shows the ratio with respect to the SM prediction. The shaded band in the ratio indicates the uncertainty due to scale variations. See text for more details.

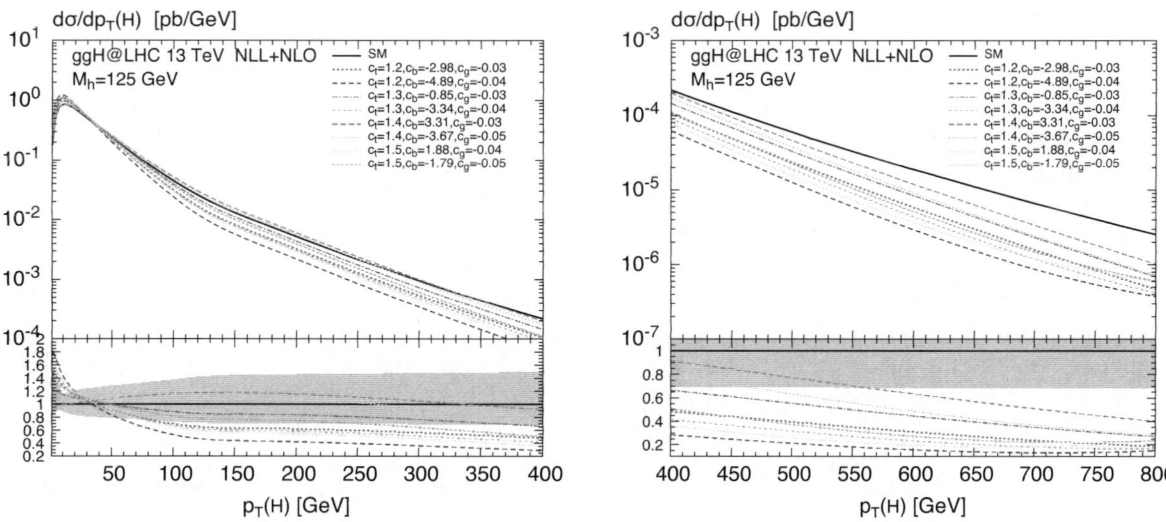

(a) (b)

Figure 5. Higgs transverse-momentum spectrum in the SM (black, solid) compared to simultaneous variations of c_t, c_g and c_b for (a) 0 GeV$\leq p_T \leq$ 400 GeV and (b) 400 GeV$\leq p_T \leq$ 800 GeV. The lower frame shows the ratio with respect to the SM prediction. The shaded band in the ratio indicates the uncertainty due to scale variations. See text for more details.

By contrast, the simultaneous variation of more than a single coefficient, as considered in Figs. 3-5, gives rise to more significant effects. The c_g, c_t and c_b parameters are chosen in the ballpark suggested by the studies of Refs. [6, 7, 8], while still keeping the inclusive cross section within about 20% of its SM value. Indeed, many combinations of c_g, c_t and c_b can be found which mildly affect the total cross section, while significantly changing the shape of the spectrum.

In Fig. 3 we present the simultaneous variation of c_t and c_g. In this case, both the small and high-p_T behaviour of the spectrum is altered by the different combinations of c_t and c_g coefficients. It is clear that in particular the large-p_T region offers a good discrimination between the different structures of c_t and c_g in terms of shape. Again, negative (positive) c_g values will soften (harden) the spectrum. The effects are well beyond the theoretical uncertainties already at NLL+NLO. We note that the yellow, short-dashed curve corresponding to $c_t = 2$, $c_g = -0.083$ develops a minimum in the ratio to the SM around $\sim 650\,\text{GeV}$. This is due to a compensation between the negative interference between the $\mathcal{O}_1$ and $\mathcal{O}_2$ operators, which is proportional to $c_g c_t$ and the contribution of $\mathcal{O}_1$ itself, which is proportional to c_g^2 and tends to produce a harder spectrum with respect to the SM prediction.

Fig. 4 shows spectra with modified top and bottom Yukawa couplings. In this case, the compensation of the BSM contributions is less straightforward, and it is difficult to compensate $c_t > 1$ without significantly affecting the inclusive cross section. For the magenta, long-dashed curve ($c_t = 1.2$, $c_b = 4.67$) we, thus, allow for a bigger change of the total cross section up to 30%. As pointed out before, the bottom-loop softens the spectrum and, since the variation of the bottom Yukawa coupling is rather large, the squared bottom-term is larger than the top-bottom interference term and the spectrum is softened also when negative c_b values are considered. The shape difference to the SM is very significant, but only in the small-p_T region, where the soft-gluon resummation is crucial to obtain a reliable prediction. Indeed, the contribution of the bottom loop decreases with growing p_T [40] and above 150 GeV the spectra have all very similar shapes with c_t driving their normalisation.

Finally, we discuss spectra obtained by switching on all three SMEFT operators, as shown in Fig. 5. Our focus here is on scenarios with increased top-quark Yukawa coupling (up to $c_t = 1.5$). These scenarios would be of particular interest in the case in which the excess on the $t\bar{t}H$ rate over the SM prediction [41, 42] should be confirmed. In order to compensate the increase in the cross section driven by $c_t > 1$ a negative c_g has been chosen. We observe a general tendency of the BSM spectra to fall below the SM prediction in the intermediate and high transverse-momentum regions, which is due to the negative c_g contribution. The total rate is compensated by the enhancement in the low p_T region, due to a combination of the negative c_g coefficient with both negative and positive c_b modifications. Overall, we find sizeable distortions of the p_T shapes due to the dimension-six operators far beyond the scale uncertainties of the NLL+NLO SM prediction, that exceed the previously considered scenarios with two simultaneous varied coefficients in both size and significance. Despite the similar overall behaviour, the predictions for the various scenarios may differ significantly, which enables their discrimination when compared to data.

We conclude this Section with a comment on the validity of the EFT approach. The computation we have performed is carried out under the assumption that we can consider the effects of higher-dimensional operators as a "small" perturbation with respect to the SM result. This implies in particular that the effect of dimension-eight operators can be neglected. This is not obvious, given that we are studying also the large transverse-momentum region. To check the above assumption we have repeated our calculations by dropping the $\mathcal{O}(1/\Lambda^4)$ suppressed terms originating from the square of the dimension-six contributions. We find that in most of the cases the differences with respect to the results shown in Figs. 2-5 are very small, even at high transverse momenta. Only in the scenarios considered in Fig. 3 ($c_t = 0.1, c_g = 0.075$ and $c_t = 2, c_g = -0.083$) the $\mathcal{O}(1/\Lambda^4)$ effects are important, and thus, the corresponding quantitative results should be interpreted with care.

5. Conclusions

New Physics might be not accessible at the LHC through direct searches, e.g., with the discovery of new resonances. In that case, it is crucial to fully exploit the data to study possible (small) deviations from the SM predictions. SMEFT offers a formalism for the parametrisation of high-scale BSM effects, which can be used for this purpose. In the SMEFT framework BSM effects are parametrised through appropriate higher-dimensional operators, and bounds on the corresponding Wilson coefficients can be set by comparing to the experimental data.

In this paper, we have presented a computation of the transverse-momentum spectrum of the Higgs boson in which the SM prediction is supplemented by possible BSM effects. Such effects are modeled by augmenting the SM Lagrangian with appropriate dimension-six operators. Our calculation consistently includes all the terms up to $\mathcal{O}(\alpha_S^3)$ accuracy and is supplemented by soft-gluon resummation at NLL accuracy, which is required to obtain reliable predictions at small transverse momenta. At the same level of accuracy we implement three dimension-six operators, related to the modifications of top and bottom Yukawa couplings and to the inclusion of a point-like ggH coupling. Additionally, we studied the impact of the chromomagnetic operator on the Higgs cross section at LO.

We have constructed a tool for reliable predictions of the Higgs p_T distribution including dimension-six operators and performed a comprehensive study of the possible effects due to the different dimension-six operators. We varied the c_t, c_b and c_g coefficients in the range suggested by recent global analyses and required the total cross section to meet the SM prediction at NLO within the current $\mathcal{O}(20\%)$ experimental uncertainty.

We found that variations of different SMEFT operators manifest themselves in different regions of the Higgs p_T spectrum:

- a modification of the Higgs-bottom Yukawa coupling ($\mathcal{O}_3$) induces effects almost exclusively at small transverse momenta
- a direct coupling of the Higgs boson to gluons ($\mathcal{O}_1$) changes the shape of the distribution in the high-p_T tail
- a change in the top-quark Yukawa coupling ($\mathcal{O}_2$) primarily affect the normalisation of the spectrum.

We can notice, from the presented spectra, that the shape of the transverse momentum distribution depends on the mass of the particle that mediates the Higgs-gluon coupling. The lower the mass of that particle, the softer is the resulting spectrum, and thus the enhancement of bottom loop leads to the softest spectra, while enhancement of point-like coupling (corresponding to infinite mass particle in the loop) to the hardest one.

To finalise our conclusions we mention the limitations of our study. Although with simultaneous contributions of the effective operators the spectra can easily exceed the uncertainties of NLL+NLO SM prediction, this is not the case for the separate operator's contribution. This calls for the increase of the precision of the calculations. The NNLL+NNLO predictions for the SM are known, however only on in the HTL approximation. Nonetheless, the relative BSM/SM effects we have obtained in this paper (i.e., the ratios plotted in the lower panels of Figs. 2-5) can be used to include BSM effects on top of NNLL+NNLO accurate SM predictions.

Another aspect which deserves some comments is the set of dimension-six operators we have considered. In the present calculation we have limited ourselves to consider the contributions of the operators related to modified top and bottom Yukawa couplings and of the additional direct Hgg interaction. As discussed in Sec. 3, although formally suppressed by two powers of the top Yukawa coupling, the chromomagnetic operator could still significantly contribute, within the current bounds on its Wilson coefficient, to the gluon fusion cross section. The extension of our calculation to include these effects is left for future work.

Acknowledgments

I would like to thank prof. Krawczyk for the invitation to the Discrete conference and acknowledge the financial support from the HARMONIA project under contract UMO-2015/18/M/ST2/00518 (2016-2019). This work is supported by the 7th Framework Programme of the European Commission through the Initial Training Network HiggsTools PITN-GA-2012-316704.

References

[1] Grazzini M, Ilnicka A, Spira M and Wiesemann M 2016 (*Preprint* 1612.00283)
[2] Aad G *et al.* (ATLAS) 2013 *Phys. Lett.* **B716** 1–29 (*Preprint* 1207.7214)
[3] Chatrchyan S *et al.* (CMS) 2013 *Phys. Lett.* **B716** 30–61 (*Preprint* 1207.7235)
[4] Buchmuller W and Wyler D 1986 *Nucl. Phys.* **B268** 621–653
[5] Grzadkowski B, Iskrzynski M, Misiak M and Rosiek J 2010 *JHEP* **10** 085 (*Preprint* 1008.4884)
[6] Dumont B, Fichet S and von Gersdorff G 2013 *JHEP* **07** 065 (*Preprint* 1304.3369)
[7] Falkowski A 2016 *Pramana* **87** 39 (*Preprint* 1505.00046)
[8] Butter A, boli O J P, Gonzalez-Fraile J, Gonzalez-Garcia M C, Plehn T and Rauch M 2016 *JHEP* **07** 152 (*Preprint* 1604.03105)
[9] Aad G *et al.* (ATLAS) 2014 *JHEP* **09** 112 (*Preprint* 1407.4222)
[10] Aad G *et al.* (ATLAS) 2014 *Phys. Lett.* **B738** 234–253 (*Preprint* 1408.3226)
[11] Chatrchyan S *et al.* (CMS) 2015 *arXiv:1508.07819* (*Preprint* 1508.07819)
[12] Georgi H M, Glashow S L, Machacek M E and Nanopoulos D V 1978 *Phys. Rev. Lett.* **40** 692
[13] Ellis R K, Hinchliffe I, Soldate M and van der Bij J J 1988 *Nucl. Phys.* **B297** 221
[14] Baur U and Glover E W N 1990 *Nucl. Phys.* **B339** 38–66
[15] Schmidt C R 1997 *Phys. Lett.* **B413** 391
[16] de Florian D, Grazzini M and Kunszt Z 1999 *Phys. Rev. Lett.* **82** 5209–5212 (*Preprint* hep-ph/9902483)
[17] Glosser C J and Schmidt C R 2002 *JHEP* **12** 016 (*Preprint* hep-ph/0209248)
[18] Ravindran V, Smith J and Van Neerven W L 2002 *Nucl. Phys.* **B634** 247–290 (*Preprint* hep-ph/0201114)
[19] Harlander R V, Neumann T, Ozeren K J and Wiesemann M 2012 *JHEP* **08** 139 (*Preprint* 1206.0157)
[20] Neumann T and Wiesemann M 2014 *JHEP* **11** 150 (*Preprint* 1408.6836)
[21] Boughezal R, Caola F, Melnikov K, Petriello F and Schulze M 2013 *JHEP* **06** 072 (*Preprint* 1302.6216)
[22] Boughezal R, Caola F, Melnikov K, Petriello F and Schulze M 2015 *Phys. Rev. Lett.* **115** 082003 (*Preprint* 1504.07922)
[23] Chen X, Gehrmann T, Glover E W N and Jaquier M 2015 *Phys. Lett.* **B740** 147–150 (*Preprint* 1408.5325)
[24] Collins J C, Soper D E and Sterman G F 1985 *Nucl. Phys.* **B261** 104
[25] Bozzi G, Catani S, de Florian D and Grazzini M 2006 *Nucl. Phys.* **B737** 73–120 (*Preprint* hep-ph/0508068)
[26] Mantler H and Wiesemann M 2013 *Eur. Phys. J.* **C73** 2467 (*Preprint* 1210.8263)
[27] Grazzini M and Sargsyan H 2013 *JHEP* **09** 129 (*Preprint* 1306.4581)
[28] Grojean C, Salvioni E, Schlaffer M and Weiler A 2014 *JHEP* **05** 022 (*Preprint* 1312.3317)
[29] Azatov A and Paul A 2014 *JHEP* **01** 014 (*Preprint* 1309.5273)
[30] Langenegger U, Spira M and Strebel I 2015 *arXiv:1507.01373* (*Preprint* 1507.01373)
[31] Maltoni F, Vryonidou E and Zhang C 2016 *JHEP* **10** 123 (*Preprint* 1607.05330)
[32] Harlander R V and Neumann T 2013 *Phys. Rev.* **D88** 074015 (*Preprint* 1308.2225)
[33] Dawson S, Lewis I M and Zeng M 2014 *Phys. Rev.* **D90** 093007 (*Preprint* 1409.6299)
[34] Giudice G F, Grojean C, Pomarol A and Rattazzi R 2007 *JHEP* **06** 045 (*Preprint* hep-ph/0703164)
[35] Contino R, Ghezzi M, Grojean C, Muhlleitner M and Spira M 2013 *JHEP* **07** 035 (*Preprint* 1303.3876)
[36] Buarque Franzosi D and Zhang C 2015 *Phys. Rev.* **D91** 114010 (*Preprint* 1503.08841)
[37] Choudhury D and Saha P 2012 *JHEP* **08** 144 (*Preprint* 1201.4130)
[38] Degrande C, Gerard J M, Grojean C, Maltoni F and Servant G 2012 *JHEP* **07** 036 [Erratum: JHEP03,032(2013)] (*Preprint* 1205.1065)
[39] Chien Y T, Cirigliano V, Dekens W, de Vries J and Mereghetti E 2015 (*Preprint* 1510.00725)
[40] Langenegger U, Spira M, Starodumov A and Trueb P 2006 *JHEP* **06** 035 (*Preprint* hep-ph/0604156)
[41] Aad G *et al.* (ATLAS) 2016 (*Preprint* ATLAS-CONF-2016-080)
[42] Chatrchyan S *et al.* (CMS) 2016 (*Preprint* CMS-PAS-HIG-16-022)

Institute of Physics
Dirac House, Temple Back
Bristol BS1 6BE UK

ISSN: 1742-6588
ISBN 978-1-5108-4559-6

CURRAN ASSOCIATES INC.
proceedings.com

Wake Conference 2017

Journal of Physics: Conference Series Volume 854

Visby, Sweden
30 May – 1 June 2017

IOP
Institute of Physics